ONE THOUSAND
AMERICAN FUNGI

ONE THOUSAND
AMERICAN FUNGI

TOADSTOOLS, MUSHROOMS, FUNGI: HOW TO
SELECT AND COOK THE EDIBLE; HOW TO DISTINGUISH
AND AVOID THE POISONOUS

by

Charles
McIlvaine
and
Robert K. Macadam

With a new essay on Nomenclatural Changes by

Robert L. Shaffer

University Herbarium
The University of Michigan
Ann Arbor, Michigan

DOVER PUBLICATIONS, INC.

NEW YORK

Published in Canada by General Publishing Company, Ltd., 30 Lesmill Road, Don Mills, Toronto,
Ontario.
Published in the United Kingdom by Constable
and Company, Ltd., 10 Orange Street, London WC 2.

This Dover edition, first published in 1973, is an
unabridged republication of the second revised
(1902) edition of the work originally published in
1900 by the Bowen-Merrill Company. This edition
also contains a new essay and table of nomenclatural changes by Robert L. Shaffer.

International Standard Book Number: 0-486-22782-0
Library of Congress Catalog Card Number: 72-87763

Manufactured in the United States of America
Dover Publications, Inc.
180 Varick Street
New York, N.Y. 10014

CONTENTS

LIST OF ILLUSTRATIONS

v

List of Illustrations

List of Illustrations

List of Illustrations

List of Illustrations

PREFACE

A SCORE of years ago (1880–1885) I was living in the mountains of West Virginia. While riding on horseback through the dense forests of that great unfenced state, I saw on every side luxuriant growths of fungi, so inviting in color, cleanliness and flesh that it occurred to me they ought to be eaten. I remembered having read a short time before this inspiration seized me a very interesting article in the Popular Science Monthly for May, 1877, written by Mr. Julius A. Palmer, Jr., entitled "Toadstool Eating." Hunting it up I studied it carefully, and soon found myself interested in a delightful study which was not without immediate reward. Up to this time I had been living, literally, on the fat of the land—bacon; but my studies enabled me to supplement this, the staple dish of the state, with a vegetable luxury that centuries ago graced the dinners of the Cæsars. So absorbing did the study become from gastronomic, culinary and scientific points of view, that I have continued it ever since, with thorough intellectual enjoyment and much gratification of appetite as my reward. I hope to interest students in the study as I am myself interested.

For twenty years my little friends—the toadstools—have been my constant companions. They have interested me, delighted me, fed me, and I have found much pleasure in making the public acquainted with their habits, structure, lusciousness and food value.

My researches have been confined to the species large enough to appease the appetite of a hungry naturalist if found in reasonable quantity; and my work has been devoted to segregating the edible and innocuous from the tough, undesirable and poisonous kinds. To accomplish this, because of the persistent inaccuracy of the books upon the subject, it was necessary to personally test the edible qualities of hundreds of species about which mycologists have either written nothing or have followed one another in giving erroneous information. While often wishing I had not undertaken the work because of the unpleasant results from personally testing fungi which proved to be poisonous, my reward has been generous in the discovery of many delicacies among the more than seven hundred edible varieties I have found.

For ten years I have planned to publish in book form what I know about toadstools; each effort to compile my information has shown me how much more I ought to know before going into print. Even now my work is still unfinished.

I am urged by my many toadstool friends (as I lovingly call those who, from all over the land, send me specimens for identification, and grow interested with me in the work), to publish what I already know upon the subject, that they, and others, may have a helpful book to guide them to a goodly portion of the edible species, and away from those that are inedible or poisonous.

In this book I comply with these requests. I have selected over seven hundred of the most plentiful and best varieties for the table, from my toadstool bill of fare; and I describe and caution against several species, some of which are deadly in their effects, if eaten; others of which induce ill-effects more or less serious. One thousand species and varieties are named and described.

Birds, flowers, insects, stones delight the observant. Why not toadstools? A tramp after them is absorbing, study of them interesting, and eating of them health-giving and supremely satisfying.

CHARLES McILVAINE.

INTRODUCTION

AMERICA is without a text-book of the American species of Fungi, among which the edible and poisonous varieties are found. Many excellent but expensive foreign volumes describe species common to both continents, and several special but widely scattered monographs have been published here. The need of the mycologist, mycophagist and amateur toadstool student is a book giving the genus, names and descriptions of the prominent American toadstools whose edibility has been tested, or whose poisonous qualities have been discovered. The absence of such a book, and the universal and rapidly-growing interest all over the United States in edible fungi, have led to the publication of the present work, which includes every species known to be esculent in North America. As a precautionary measure, full explications of all those known or suspected to be poisonous are included.

Many species found in this country only have been described and named by various authors, from the time of Schweinitz (1822) to the present day. These have been published in the botanical magazines and in the papers of scientific societies and colleges. The greater number have as author Professor Charles H. Peck, New York State Botanist, who has contributed an annual report each year from 1868. These appear in the reports of the State Museum of New York, and coming from the pen of our ablest mycologist are of great value to everyone interested in the study. The classifications and (in many instances) modified descriptions by such an eminent authority upon fungoid growth should therefore be the guides to American forms, that the confusion created by numerous descriptions of the same fungus by different observers may be avoided.

Professor N. L. Britton, editor of the Torrey Botanical Club, has courteously given permission to use the descriptions of new species given in its instructive Bulletins.

Professor A. P. Morgan and Laura V. Morgan, with equal courtesy, grant the use of text and illustrations contained in the most complete monograph published upon the Lycoperdaceæ (puff-balls, etc.) of America.

While the scientific classifications and descriptions have been strictly followed, the language has been simplified—with no sacrifice of scientific accuracy—that this volume may be fully adapted to popular use.

Professor Peck has given his valuable assistance in the identification of many species, all that were difficult or obscure having been submitted to him, and the writer is deeply indebted to him for many and long-continued courtesies, aiding in study and in the preparation of this work.

Several new species have been found by the writer, the greater part of excellent food value. He preferred that these should be named, described and placed in their proper genus and section by Professor Peck, believing it to be best for the discoverers of new species to defer to one whose vast experience enables him to name and classify in accordance with the demands of American species.

Where a species is vouched for as edible, it has been personally tested by the author and his willing undertasters up to eating full meals of it, or at least beyond all doubt as to its safety. Where others have eaten species which he has not had the opportunity to test, their names and opinions are given. When species heretofore under the ban of suspicion are in this volume, for the first time, announced to be edible (there are many of them), personal tests have not been considered sufficient, as idiosyncrasy might have affected the results. Others, at the writer's request, have eaten of the species until their innocence was fully established. In some cases, where the reputation of the fungi eaten was especially bad, scientists of note have made elaborate and exhaustive physiological tests of their substances, and in every instance confirmed the human testing.

While species which contain deadly poisons are few, their individuals are produced in great number. Nicety in distinguishing their botanic variance from edible species closely resembling them is necessary. No charm will detect the poison. Eating toadstools before their certain identification as belonging to edible species, is neither bravery nor common sense. The amateur should go slow.

The question often asked is: By what rule do you distinguish between edible and poisonous mushrooms? The answer usually surprises the questioner—there is no general rule. All such rules which have been given are false and unreliable. The quality of each was learned, one at a time. Sweet and sour apples alike grow on large and small trees, may be red or green, large or small, oblong or globular, and no visible appearance gives the least clue to the quality.

In a few genera certain rules may be applied, as in Clavaria—all not bitter or tough are edible. But such generalizations are each limited to its own genus.

The toadstools containing deadly poisons are thought to be confined to one genus of the gilled kind—Amanita, and to Helvella esculenta, now Gyromitra esculenta, to which are charged fatal results. The poisonous qualities of Gyromitra esculenta are not proven. Recent testings of this species prove it to be harmless and of good quality. By far the greater number of species contained in Amanita are notable for their tender substance and delicious flavor. By their stately beauty and unusual attractiveness both the poisonous and harmless kinds are seductive. *Any toadstool with white or lemon-yellow gills, casting white spores when laid—gills downward—upon a sheet of paper, having remnants of a fugitive skin in the shape of scabs or warts upon the upper surface of its cap, with a veil or ring, or remnants or stains of one, having at the base of its stem—in the ground—a loose, skin-like sheath surrounding it, or remnants of one, should never be eaten until the collector is thoroughly conversant with the technicalities of every such species, or has been taught by one whose authority is well known, that it is a harmless species.* This rule purposely includes the renowned Amanita Cæsaria, everywhere written as luscious. I regard it as the most dangerous of toadstools, because of its close resemblance to its sister plant—the Amanita muscaria—which is deadly. In the description of these species, other forcible reasons are given.

Another deadly species—the Amanita phalloides—is frequently mistaken by the inexperienced for the common mushroom. Safety lies in the strict observance of two rules: Never eat a toadstool found in the woods or shady places, believing it to be the common mushroom. Never eat a white- or yellow-gilled toadstool in the same belief. The common mushroom does not grow in the woods, and its gills are at first pink, then purplish-brown or black.

If through carelessness, or by accident, a poisonous Amanita has been eaten, and sickness results, take an emetic at once, and send for a physician with instructions to bring hypodermic syringe and atropine sulphate. The dose is $\frac{1}{180}$ of a grain, and doses should be continued heroically until the $\frac{1}{20}$ of a grain is administered, or until, in the physician's opinion, a proper quantity has been injected. Where the victim is critically ill the $\frac{1}{20}$ of a grain may be administered.

In every case of toadstool poisoning, the physician must be guided by the symptoms exhibited. Professor W. S. Carter, by numerous exhaustive trials upon animals, has proved that atropine, while valuable as against the *first*, is not an antidote for the *late* effects of the greater toadstool poisons. (See his chapter on toadstool poisons, especially prepared for this work.)

There are other species which contain minor poisons producing very undesirable effects. These are soon remedied by taking an emetic, then one or two doses of whisky and sweet oil; or vinegar may be substituted for the whisky. A few species of fungi are innocuous to the majority of persons and harmful to a few. So it is with many common foods—strawberries, apples, tomatoes, celery, even potatoes. The beginner at toadstool eating usually expects commendation for bravery, and fearfully watches for hours the coming of something dreadful. Indigestion from any other cause is always laid to the traditionary enemy, fright ensues, a physician is called, the scare spreads, and a pestilential story of "Severe Poisoning by Toadstools," gets into the newspapers. The writer has traced many such publications to imprudences in eating, with which toadstools had nothing to do.

The authoritative analysis of several common food species by Lafayette B. Mendel, of Sheffield Laboratory of Physiological Chemistry, Yale University, is given, and will correct the popular error about the great nutritive value of fungi, arising from previous erroneous analyses.

While species are reported as found in certain localities, it by no means follows that their growth is confined to these places. A species reported as found in the Adirondack mountains, unless belonging to the few peculiar to northern regions and high altitudes, is reasonably sure to be more plentiful in a like habitat south and west of them. South it will appear earlier and its season last longer.

Size is largely dependent upon latitude and may vary greatly in the same group. Temperature, moisture, favorable nourishment are important factors in growth.

Each species has its favorite habitat, and will thrive best upon it. There are few things under the sun upon which fungi do not grow. Their mission is particularly directed toward converting decaying matter, or matter which has accomplished its work in one direction, into usefulness in another. They are the wood-choppers, stewards, caterers of the forest, converters in the fields and chemists everywhere. They can not assimilate inorganic matter because of the absence of chlorophyl in their composition, but in organic matter they are omnivorous. When they feed on dead substances they are called saprophytes; when their support is derived from living tissues, parasites.

Scores of species of fungi were found in the forests, ravines and clearings of the West Virginia mountains from 1881 to 1885 inclusive, and eaten by the writer years before he had the opportunity to learn their names from books or obtain the friendly assistance of experts in identifying them. He knew the individuals without knowing their names, as one knows the bird song and plumage before formal introduction to the pretty creatures that charm him.

After he was able to get European publications upon the subject, and by their aid trace the species he had eaten to their names, descriptions and qualities, he was surprised to read that many of them were warned against as deadly. As informed by these books, he properly ought to have died several times. It soon became evident that authors had followed one another in condemning species, some because they bore brilliant hues, others because they were unpleasant when raw (just as is a potato), rather than investigate their qualities by testing them. Here was a realm of food-giving plants almost entirely unexplored. The writer determined to explore it. Instead of the one hundred and eleven species then recorded by the late Doctor Curtis as edible, my number of edible species now exceeds his by over six hundred.*

Let us clear away the rubbish and superstition that have so long obscured the straight path to a knowledge of edible toadstools. Let us bear in mind that a mushroom is a toadstool and a toadstool is a mushroom—the terms are interchangeable. If toads ever occupied the one-legged seat assigned them from time immemorial, they have learned in

---

* This book contains one hundred and fifty pages more than were originally estimated and promised to the subscribers. That all known edible and poisonous species might be fully described and published within one volume, the author was compelled to cut fifty thousand words from his manuscript. The localities from which species have been reported and the names of the reporters have been taken out, excepting where it was desirable to show that foreign species have been found in the United States, and where tested species have been found by the author. The principal cut has been from the notes of the author and of enlarged descriptions.

this enlightened age that the ground is much more reliable, and so squat upon it, except when exercising their constitutional right to hop. Snails, slugs, insects of many kinds, mice, squirrels and rabbits prey upon good and bad, each to its liking, notwithstanding oft-repeated assertion that snails and slugs infect noxious varieties only, or that animals select the innocuous only. We are warned against those which grow in the dark or damp; the mushroom of commerce is grown by the ton in the subterranean quarries of France, and everywhere in vaults and cellars for domestic use. The valued truffle never sees the light until it is taken from darkness to be eaten, and other varieties of the best prefer seclusion.

The wiseacres tell us that they must have equal gills, must not have thin tops, must not turn yellow when sprinkled with salt, must not blacken a silver spoon, that we must not eat of those changing color when cut or broken, of those exuding milk, or those which are acrid, hot, or bitter, and give many other specifics for determining the good from the bad. These tests are all worse than worthless, for if confidence is placed in them they will not only lead us away from esculent and excellent varieties but directly into eating venomous ones.

There are whole genera of fungi which are innocuous; but in the Family of Agaricaceæ, where the greatest variety of the edible and poisonous species are found, it is necessary to master one by one the details of their construction and learn to distinguish their differences as one does those of the many kinds of roses, or pinks, or hundreds of bright-faced pansies, and in the mastery of them lies the only charm that will safely guide.

Carefully remove the first toadstool found from whatever it is growing upon, and with it a portion of that from which it springs. If it is the earth a curious white network is discernible, fine as the delicate spinning of the spider, spreading its meshes throughout the mass. It will often remind of miniature vines climbing over miniature lattices. This is the mycelium from which the toadstool grew. In many instances it penetrates the earth to a considerable depth, and takes possession of large territory. It is often seen as the gardener turns up the soil or its fertilizer, and is perhaps taken for a mold. If the specimen is gathered from mat of wood leaves, the same white vine is observable slipping in between its layers. If taken from a tree, the decaying wood is traversed by it. From wherever a toadstool is plucked, it is removed from its mycelium.

This mycelium is but a thread-like mass of simple cells joined together at their ends and interlacing in a way a thousand-fold more intricate than a Chinese puzzle. Nothing in its structure indicates what its special product will be. The fungus which is plucked from it is in all its parts simply a mass of these threads—cells strung together, interlacing and ramifying.

When the season favors, the mycelium—which has, winter and summer and from year to year, lived its hidden life, or has sprung from a germinating spore—develops a number of its cells in a minute knob, small as a pin head. At this point the cells make special growth efforts to bring themselves within the favoring influences of heat and moisture; this tiny knob labors within itself, producing cell after cell, which takes shape and function for the future toadstool.

As it rapidly enlarges it pushes its way toward the surface of the ground, becomes more or less egg-shaped in this stage of its growth, and if cut in half longitudinally and examined, it will display what it is going to be when it grows up.

Suppose that it belongs to the first of the two great sections into which fungi are divided under the classification of Fries, who modified that of Persoon. The first has the spores—which represent the seeds in plants—naked, and it is called sporifera or spore-bearing. The second, which has the spores enclosed in cells or cysts, is called sporidifera or sporidia-bearing. If the cap of a gill-bearing toadstool be laid, gills downward, on a watch crystal or piece of white paper for a few hours, or, in some instances, a few minutes, a complete representation of the spaces between the gills will be found deposited as an impalpable powder. These are the spores.

The first section is divided into four cohorts. Two of these have hymeniums or spore-bearing surfaces more or less expanded. These are Hymenomycetes and Gastromycetes. In Hymenomycetes the hymenium is always exposed in matured plants, as with the common mushroom. When young, some plants are covered with a membrane. In Gastromycetes the hymenium is always concealed within a covering which bursts at maturity, as with the Lycoperdons or puff-balls. Cohort Coniomycetes includes rusts, smuts, etc., formed for the most part on living plants. There is no hymenium present. The spores are produced on the ends of inconspicuous threads, free or enclosed in a bottle-like receptacle called a perithecium. Cohort Hypomycetes is composed of those species of fungi commonly called molds. The spores are produced, naked, from the ends of inconspicuous threads.

In the Agaricaceæ—the first family in Hymenomycetes—the young plant is completely enveloped. (Plate III, fig. B, p. 2.) Its head is as yet undefined and its body may be classed as dumpy, but shut in and protected are a great quantity of knife-like plaits (Plate III, fig. C., p. 2), on the outer surface of which, when the plant matures, will be borne its spores. It therefore belongs to the Hymenomycetes, and to the Family Agaricaceæ—gill-bearing.

If the ground becomes moist or there comes a heavy dew or a rain, the young plant, closely compacted and very solid, which has been under the surface for many days waiting its chance to get forth to light and air, rapidly swells, breaks through the moistened earth, goes rapidly to cell-making, ruptures its outside covering, the head expands and in so doing spreads out its gills or hymenium. (Plate III, figs. C, D, E, p. 2.) The membrane which covered the gills either vanishes, or gathers round the stem in the form of a ring or circular apron, or it may partially adhere to the edges of the top, cap or pileus and hang as a fringe from it; the stem elongates; the whole plant assumes the colors of its species and in a few hours or days at most it stands forth, a marvel of beauty, structure and workmanship.

But little is known of how these spores reproduce themselves. The microscope fails to completely penetrate the mystery. A whole fungus is but a mass of cells, the spore is but one of them. That these simple cells do produce after their kind there is no doubt, but so minute is the germ and hidden its methods that science has failed to solve them.

The first Family of Hymenomycetes is Agaricaceæ. Its members always have gills or modifications of them. In some cases—notably in Cantharellus—the gills have the appearance of smooth, raised veins over which is the spore-bearing surface. The hymenium is but an extension of the fibers of the cap, folded up like the plaits and flutings of ruffles, and laundered with exquisite neatness. If it is carefully detached and spread out like a fan it will cover a large surface, many times the size of the cap from which it has been taken, and will show that what is a consumption of material in dress ornamentation is utilized by economical Dame Nature to increase the spore-bearing surface within a small space and for purely business purposes—spore-bearing. The color of these spores has much to do with the classification. The microscope with high light reveals the delicate shades of their coloring, but the main colors are readily distinguished by the naked eye when the spores are collected in a mass on glass or paper.

The Polyporaceæ have in place of gills closely packed tubes on the inside of which is the spore-bearing surface; each has a mouth from which to eject the spores.

The Hydnaceæ bear their spores from spines or spicules of various length protruding from the external surface of the cap. Sometimes the spines mock in miniature the stalactites of the Caverns of Luray, sometimes the shaggy mane of the lion, sometimes flowing locks of hair. These three Families belong to the Cohort Hymenomycetes, having their spore-bearing surface exposed early in life by the rupture of the universal veil.

The Lycoperdons or Puff-balls have the hymenium enclosed within an outer case, just as the apple with its seeds is enclosed for a dumpling. When the spores are matured the sack is ruptured and they escape as the dusty powder so well known to all. The Puff-ball belongs to the Cohort Gastromycetes, because its spores are protected within the hymenium until they are matured.

There are other Families which contain edible species. The Clavariaceæ—branched or club-shaped often found in as beautiful forms as delight us in coral, includes a few.

In Ascomycetes, of the covered spore division Sporidifera, there are several species which are excellent, and as they dry readily are much valued for flavoring purposes when winter forbids the growth of outdoor fungi. Of these the Morell has preference. The cap is covered with sinuosities and pits which bear the spores. There are several varieties of the Morell in the United States. They are known among the country people who cook and pickle them, as Honey-comb mushrooms.

The Tuberaceæ are subterranean fungi. The common truffle so much prized by epicures is a good representative. It is found a foot or more under the surface of the earth, and of such value is it that in some countries pigs are trained to hunt it from its hiding place. It is one of the few foreign growths apparently not taking kindly to our country. Efforts have been made to import and cultivate it, but without success.

It is possible, even probable, that it may yet be found in America by assiduous search.

I have said that there is but one way to distinguish the edible from the non-edible fungi; that is by mastering the characteristics of each species one by one. There are signs which point to the evil and those which point to the good, but they must be used as signals, not directors.

A nauseous, fetid odor should condemn a species as non-edible at once. Those having the flavor of flour or fresh meal are generally accepted as worthy of trial. Slimy, water-soaked, partially decomposed plants, or those impressing one as unpleasant in any way, should never find their place upon the table. Do not eat of any toadstool, unknown to the collector, beyond the careful and systematic testing required to determine whether it is edible or not.

A few species have a serious charge remaining against them; that of partiality. They unmistakably signify with whom they will agree and with whom they will not. These are notably Clitocybe illudens, Lepiota Morgani, Panæolus papilionaceus, all specialized in their places in the text.

Other species have hereditary taints upon their reputations. Most, if not all of them have stood present tests and relieved themselves of suspicion. But, alas that it should be so! The stigma must rest upon them for yet a while and until their defenders are so numerous that their purity, without a smirch, is popularly proclaimed.

Wherever wood grows and decays as it will, Polyporus, Panus, Lenzites, Schizophyllum and kindred genera stand prominently forth in countless numbers. The great majority of them are inedible because of their woody substance. A few are valued as food. Very many of them yield their soluble matter and flavor when boiled, and in this way make excellent soups and gravies, just as flax-seed and the bark of the slippery elm yield succulent matter. These, however, are not, with a few exceptions, mentioned in this book. Numbers of Clavarieæ and Hydneæ are in the same category. M. C. Cooke tersely says: "Fruits that are not peaches or apricots may be very good plums." In the introductions to genera their attributes are given; under "Instructions to Students" every guide to identification and selection will be found.

A Glossary, containing the botanic terms used in this book and, it is believed, all other terms used by mycologists in describing fungi, follows the descriptive text. It is strongly advised that it be carefully studied.

The roots and derivatives of the botanic terms are fully and carefully given by Dr. John W. Harshberger, professor of botany, University of Pennsylvania, to whom the author is specially indebted.

The excellent Glossary published by Dr. Edwin A. Daniels, Boston, has furnished many comprehensive definitions. It is the property of the Boston Mycological Club, and can be obtained from its secretary for twenty-five cents.

The determination of the proper accentuation of the generic and specific terms has been in many cases a difficult task, and, in some cases, owing to the dubious origin of the words in question, there is certainly room for difference of opinion. This task has been kindly and conscientiously performed by Prof. M. W. Easton, professor of Comparative and English Philology, University of Pennsylvania. Thanks are due to the Hon. Addison Brown, president of the Torrey Botanical Club, and Dr. Nathaniel L. Britton, professor of Botany in Columbia College, authors of "Illustrated Flora," for the determination of the accentuation of non-classical words ending in *inus*.

Three indexes are given: the first refers to the general contents, the second to the genera, the third to species and their genera, alphabetically arranged.

Mrs. Emma P. Ewing and Mrs. Sarah T. Rorer have kindly furnished some of their recipes for the preparation of several varieties of toadstools. The best results of the author's long experience in cooking toadstools are given in the chapter "Recipes for Cooking and Preparing for the Table," together with others selected from many sources. The personal taste of the server must be guide to the choice.

A child-friend of the writer, in telling him of her mother's cook, said: "She's a good cooker, but she has a bad temper." A good "cooker" will soon learn how to best display the individual flavor of each species. And be it known that each species of toadstool has a flavor of its own. These flavors vary as much as among meats and vegetables. No one species can be taken as standard of excellence.

The greatest care has been taken to secure illustrations correct in every botanic detail. With few exceptions the colored figures were drawn and painted by the writer. To obtain this important feature the requirements of art have frequently been sacrificed. An artist can make a picture of a toadstool; the mycologist must guide his brush or pencil in the making of a correct presentation. The happy combination of artist and mycologist occurs in Mr. Val. W. Starnes, Augusta, Ga., to whom this volume owes many of its illustrations. Mr. Frank D. Briscoe, widely known as an artist of rare ability, has arranged and painted in groups the studies made by the writer from typical plants, and added to the illustrations many excellent drawings of his own.

The unfailing reliability of the sun has been masterfully used by Dr. J. R. Weist, ex-Secretary of the American Society of Surgeons, Richmond, Ind.; H. I. Miller, Superintendent Terre Haute and Indianapolis Railroad, Terre Haute, Ind., and Mr. Luther G. Harpel, Lebanon, Pa., in making the unexcelled photographs generously contributed by them. The author is most thankful to them and to Mr. C. G. Lloyd, Cincinnati, Ohio—a scientific gentleman devoting lavishly of his time and money to the spread of mycological knowledge—for the privilege of selecting from his extensive collection of realistic photographs those adaptable to the species described herein.

The author's thanks are gratefully given to the many who have by help and encouragement furthered his efforts in producing this, the first American text-book upon fungi. Space precludes the naming of the many, but the few named do not outrank them in their interest, help and the author's appreciation:

Miss Lydia M. Patchen, President of the Westfield, N. Y., Toadstool Club (the first in America); Mrs. E. C. Anthony, Thomas J. Collins, E. B. Sterling, Berry Benson, Melvil Dewey, New York State Librarian; Dr. J. E. Schadle, Prof. J. P. Arnold, University of Pennsylvania; Prof. W. S. Carter, University of Texas; Boston School of Natural History; Massachusetts Horticultural Society; Prof. Wm. G. Farlow, University of Harvard.

Thus aided the author believes that his own conscientious, patient, loved labor in the study of edible and non-edible fungi and the production of this volume will be far-reaching in its one object—encouraging the study of toadstools.

The time for writing a complete flora of the United States has not yet come; a large part of the country remains as yet unexplored by mycologists; new species are being constantly discovered in the districts best known. Every book on the subject must be necessarily incomplete.

On the other hand, so far as concerns the known fungus-flora, there is imperative need of some guide to the student, which shall at least save him some part of the weary toil of hunting through the scattered literature in which alone, as things are at present, can be found the information he seeks. In this book I have tried to meet this need. It is not complete, but I have tried to so arrange the matter that the student can always decide whether the particular specimen in hand is or is not included, and, at least for all of our more conspicuous fungi, determine the family and genus. If the student can do so much, the task of finding the specific name, even when not included in this book, becomes very much simpler.

So much for the more scientific aspect of my book. But I have also kept in constant view the needs of the large and constantly growing number of persons who have no aim further than to learn to know the principal toadstools seen in their walks, just as they wish to know the principal trees and the more conspicuous birds. For such as these, the difficulty of deciding whether or no a particular individual fungus is described in the brief (sketching) manuals hitherto accessible is even more formidable than with the special student of botany.

Finally, I have kept in view throughout the work the needs of the mycophagists. They are not pot-hunters; they care much less for the physical pleasure of the appetite than for the close study of Nature that their inclination leads them into. Some day the delights of a mushroom hunt along lush pastures and rich woodlands will take the rank of the gentlest craft among those of hunting, and may perchance find its own Izaak Walton.

CHARLES MCILVAINE.

INSTRUCTIONS TO STUDENTS

To CATCH fish one must know more than the fish; to find toadstools one must know their season and habitats. They are propagated by their spores and from their mycelium—that web-like growth which is the result of spore germination.

The spores of ground-growing kinds, when shed upon the ground, are washed by rains along the natural drainage; therefore, when a specimen of one of these kinds is found, it is well to look up and down the natural water-shed, and follow it. Good reward will usually come of it. Few fungi are strictly solitary.

Careful observation of the habitats of the various genera and species will enable the student to know what may and may not be expected in a particular locality, and will save many a hunt.

When an unknown species is found, collect it carefully, examine it closely, note all its features. Determine to which division of fungi it belongs. If to the gilled family (Agaricaceæ) obtain the color of the spores (see directions). Look at the chart "Tabular View of Genera of Agaricaceæ," Plate I, p. 2 (after W. G. Smith, but enlarged, redrawn and emended). If the spores are white, it belongs to one of the genera in the first column—Leucosporæ; if pink, to one in the second column, and so on. It is often difficult to determine the spore color, because spores vary through many shades of the typical color. What are called white spores may be creamy, dirty, yellowish or brownish-white; pink spores will vary from almost white to reddish and salmon-color; brown spores from light-ochraceous through cinnamon to rusty; purple spores from dark-violet to purplish-black. Experience alone will enable the student to decide which color series is present. The Genera Charts, preceding the five different color series, show typical spore colors only. Again, authors describing the species frequently fail to see colors alike; if they do, their names for them frequently vary. For instance, few persons will agree upon a color expressed as "livid."

The color system principally used by botanists is Saccardo's ''Chromotaxia,'' costing fifty cents. It is decidedly inadequate. Ridgway's ''Nomenclature of Colors for Naturalists'' is far better, but it is out of print and obtainable only at the principal libraries. ''The Prang Standard of Color'' is the most complete ever issued, but it is inapplicable to existing descriptions of fungi.

To Make and Preserve Spore Prints. Take, to print upon, sheets of Bristol-board or any stiff, hard-surfaced white paper 6x9 inches or larger. Cut a round hole, four inches in diameter, in one of the sheets. Use this as a stencil. Lay it upon a print-sheet and where the opening occurs, paint with a weak solution of gum arabic—⅛ oz. (one teaspoonful) to one pint of water. Dry the print-sheets.

When a spore-print is to be taken, select a fully-grown specimen, remove the stem, place the spore-bearing surface upon the gummed paper, cover tightly with an inverted bowl or saucer, and allow to stand undisturbed for eight or ten hours. The moisture in the plant will soften the gummed surface; the spores will be shed and will adhere to it, making a perfect, permanent print. When the print is plain, remove the specimen carefully and dry the print. Number the print-cards to correspond with the number of the specimen in the ''Record of Fungi,'' and place them in a box or cover. Some genera shed their spores sooner and more freely than others. A surplus of spores is objectionable. In order to know when a print is plainly made, without disturbing the process, have either a specimen of the same age, or a piece of the one under the bowl, on another piece of gummed paper, covered in like manner. This can be examined and will give the desired information. A little experience will enable the student to obtain good and lasting prints.

The large black figures on some calendars, if cut with the white about them, are convenient as trial sheets for spore-printing. Lay the specimen partly on the white, partly on the black. If the spores are light, they show best on black ground, and if colored, they show best on the light.

Spore measurements, as given by different observers, vary to such a degree that they are of little value, excepting as determining a few species, but spore shapes and characteristics are of use as a last resort, in accurate determinations. A microscope of considerable power is needed.

xxx

A metrical scale and table of measures is here given, that the student may have a present guide to such measurements as are given in mycological publications.

Measures.

Decimetre.

1 Metre............39. 371....Inches
1 Decimetre.........5.9371   "
1 Centimetre (C M.) .39371   "
1 Millimetre (M. M.) .039371   "
1 Micron(μ) 1 Millionth of a Metre   25400 of an Inch.

1 Line (") ' 1⁄12 of an Inch

1 Gramme .....15.433   Troy Grains
1 Decigramme 1. 543   "   "
1 Centigramme....1543   "   "
1 Milligramme.....01543   "   "

The spore color being determined, turn to the Genera Chart, showing **Use of Charts of Genera.** spores of like color. Ascertain from the specimen whether or not its cap or hymenophore is distinct or easily separable from the stem and the gills free from the stem; if they are, it may belong to one of the genera in the upper row of figures; if the cap is not easily separable nor the gills free, look at the shape of the gills, and find on the chart a corresponding gill-shape. It is probable that the genus can thus be determined. Then turn to this genus in the text, read the heading, look over the "Analysis of Tribes," go to the tribe nearest in designating the properties of the specimen; comparing the specimen with the descriptions of species given thereunder, will probably enable the seeker to decide upon its name.

It should be remembered that the descriptions in the text are of the specimen or specimens which the author of the species saw. What the author says fixes the type of the species. Specimens of the species may, and very frequently do, vary greatly from the type. If the first attempt to fix the genus is not satisfactory, try again, and keep on trying until reasonably sure. The amateur will find, however good an opinion may exist in his mind of the stock of patience on hand, that the territory of patience has just been reached.

An excellent blank form for "Collectors' Notes" is published by the **Making and Preserving Notes.** Boston Mycological Club, at one cent. It is desirable that there should be uniformity in collectors' notes, and that they should be as full as possible. A form of this, or a similar kind, should be filled in and kept, and should also be used when specimens are sent to an expert for identification. Such specimens should be fresh, wrapped separately in tissue paper, numbered, and *a few* should be packed in a box that will *not crush in the mail*. The address of the sender should be upon the outside. The collector's notes should be sent in a letter, with a postage stamp for reply enclosed. If the specimens have to go a great distance, they should be partially dried in a slow, open oven, or they will be a rotten mass when they reach their destination.

There is but one way by which to determine the edibility of a species. **To Test Edibility of Species.** If it looks and smells inviting, and its species can not be determined, taste a very small piece. Do not swallow it. Note the effect on the tongue and mouth. But many species, delicious when cooked, are not inviting raw. Cook a small piece; do not season it. Taste again; if agreeable eat it (unless it is an Amanita). After several hours, no unpleasant effect arising, cook a larger piece, and increase the quantity until fully satisfied as to its qualities. Never vary from this system, no matter how much tempted. No possible danger can arise from adhering firmly to it. Recipes for preparing, cooking and serving are given in chapter on cooking.

It is better for the student to first become familiar with the common species, one at a time, than to attempt tracing the rare or many. Worry, fatigue and uncertainty are plentiful in an indiscriminate gathering of fungi. One species a day, properly traced and named, means learning three hundred and sixty-five species a year.

Unfamiliar terms will be encountered in the descriptive text. The **The Glossary.** Glossary defines them; and not only those in this book, but, it is believed, all those found in other books upon fungi. Where possible throughout the text, botanical terms have been anglicized. The meanings of those remaining unchanged should be memorized. It is quite as easy, and far better, to learn the botanical names of species and their characteristics, as to learn their common names; easier in fact, for the common names often vary with locality. The writer received a letter from an Alsatian living in St. Louis, telling him of favorite fungi he used to eat when in his own country. To all he gave local names, not one of which could be referred to the particular species meant.

Success and pleasure in the study of fungi will attend the student who observes carefully and who systematically records that which is observed.

ABBREVIATIONS OF THE NAMES OF AUTHORS OF SPECIES

A. and S.,	Albertini and Schweinitz
Arrh.,	Arrhenius
B. or Bull.,	Bulliard
Bad.,	Badham
Bagl.,	Baglietto
Bat. or Batsch,	Batsch
Batt.,	Battara
Berk. or M. J. B.,	Berkeley
Berk. and Br.,	Berkeley and Broome
Bolt.,	Bolton
Bon.,	Bonorden
Boud.,	Boudier
Boud. and Pat.,	Boudier and Patouillard
Bref.,	Brefeld
Bres.,	Bresadola
Brig.,	Briganti
Brond.,	Brondeau
Brot.,	Brotero
Cav. and Sech.,	Cavalier and Séchier
C. B. P.,	Plowright
Chev.,	Chevalier
Cke.,	Cooke
Cord.,	Corda
Crn.,	Crouan
Cum.,	Cumino
Curt.,	Curtis
D. and L.,	Durieu and Léveillé
D. C.,	De Candolle
De Guern.,	De Guernisac
Desm.,	Desmazieres
Dill.,	Dillenius
Dittm.,	Dittmar
Dun.,	Dunal
Ehrb.,	Ehrenberg
Ellis or J. B. E.,	J. B. Ellis
Eng.,	English Botany
Fayod,	Fayod
Fl. d.,	Flora danica

Forq.,	Forquignon
Fr.,	Elias Fries
Fckl. or Fuck.,	Fuckel
G. or Gill.,	Gillet
G. and R.,	Gillet and Rounreguére
God.,	Goddard
Grév.,	Gréville
H. and M.,	Harkness and Moore
Hazs.,	Hazslinsky
Hedw.,	Hedwig
Hoffm.,	Hoffmann
Holmsk.,	Holmskiold
Huds.,	Hudson
Huss.,	Mrs. T. J. Hussey
Jacq.,	Jacquin
Jungh.,	Junghuhn
Kalchb.,	Kalchbrenner
Karst.,	Karsten
Klotzsch,	Klotzsch
K.,	Krombholz
Lam.,	Lamark
Lang.,	Langlois
Lasch,	Lasch
Lenz,	Lenz
Let., Letell.,	Letellier
Lév.,	Léveillé
Leys.,	Leysser
Lib.,	Libert
Linn. or L.,	Linnæus
Mart.,	Martius
Mich.,	Micheli
M. J. B.,	Berkeley
Mont.,	Montagne
Morg.,	Morgan
Moug.,	Mougeot
Müll.,	Müller

Abbreviations of the Names of Authors of Species

Nees,	Nees	Schw.,	Schweinitz
		Scop.,	Scopoli
Osb.,	Osbeck	Sec.,	Secretan
		Somm.,	Sommerfelt
Pat.,	Patouillard	Sow.,	Sowerby
Paul.,	Paulet	Sw.,	Swartz
Pers.,	Persoon		
Pk.,	Peck	T. or Tul.,	Tulasne
Pol. or Poll.,	Pollini	Tod.,	Tode
		Tour.,	Tournefort
Q. or Quel.,	Quelet	Trat.,	Trattinik
Rab.,	Rabenhorst	U. and E.,	Underwood and Earle
Rav.,	Ravenel		
Relh.,	Relhan		
Retz.,	Retzius	Vent.,	Venturi
Riess,	Riess	Vill.,	Villars
Rost.,	Rostkovius	Vitt.,	Vittadini
Roz.,	Roze		
Roz. and Rich.,	Roze and Richon	Wahl.,	Wahlenberg
		Wall.,	Wallroth
Sacc.,	Saccardo	Weinm.,	Weinmann
Saund. and Sm.,	Saunders and Smith	Willd.,	Willdenow
Sch., Schaeff.,	Schaeffer	With.,	Withering
Schr. or Schrad.,	Schrader	W. P.,	Phillips
Schroet.,	Schröter	W. G. S., Sm. or Worth. Sm.,	
Schulz,	Schulz		Worthington Smith
Schum.,	Schumacher	Wulf.,	Wulfen

NAMES OF THE PRINCIPAL REPORTERS OF AMERICAN SPECIES

Alabama..Lucien M. Underwood, F. S. Earle
(U. and E.).
California..H. W. Harkness, Justin P. Moore
(H. and M.), Wm. Phillips.
Canada ....John Dearness.
Connecticut ....—— Wright.
Florida ....—— Calkins.
Georgia..Berry Benson, H. N. Starnes,
Val W. Starnes.
Illinois..Frederick J. Brændle.
Indiana..H. I. Miller, Dr. J. R. Weist.
Iowa ....Charles E. Bessey, T. H. Macbride.
Kansas..F. W. Cragin, Elam Bartholomew,
W. A. Kellerman.
Kentucky..C. G. Lloyd, A. P. Morgan.
Louisiana..Rev. A. B. Langlois.
Maryland ....Miss Mary E. Banning.
Massachusetts ....Charles C. Frost, W. G. Farlow,
James L. Bennett, Charles J. Sprague,
Robert K. Macadam,
Julius A. Palmer, Hollis Webster.
Minnesota ....Asa Emory Johnson.
Mississippi..U. S. Geological Survey.
Missouri..William Trelease.
Nebraska..Charles E. Bessey, F. E. Clements,
—— Webber.
New Brunswick ....A. C. Waghorne, James Fowler.
New England ....Boston Mycological Club.
New Jersey ....J. B. Ellis, Benjamin Everhart,
E. B. Sterling, Charles McIlvaine.
New York..Charles H. Peck, George F. Atkinson,
John Torrey.
North Carolina..Rev. M. A. Curtis,
Rev. Lewis de Schweinitz,
Charles McIlvaine.
Nova Scotia..Dr. John Somers.
Ohio..Charles G. Lloyd, A. P. Morgan,
W. S. Sullivant.

Names of the Principal Reporters of American Species

Oregon ....Dr. Harry Lane.
Pennsylvania.....................................Dr. William Herbst,
                                  Rev. Lewis de Schweinitz,
                                  Charles McIlvaine,
                                  Philadelphia Mycological Center.
Rhode Island ....James L. Bennett.
South Carolina ....Dr. H. W. Ravenel.
West Virginia...Charles McIlvaine, L. W. Nuttall.
Wisconsin ....W. F. Bundy, William Trelease.

NOMENCLATURAL CHANGES

by Robert L. Shaffer

UNIVERSITY HERBARIUM, THE UNIVERSITY OF MICHIGAN

ANN ARBOR, MICHIGAN

The seventy years that have elapsed since Charles McIlvaine's *One Thousand American Fungi* was first published have been a period of intense activity in the study of mushrooms and other fungi. The knowledge gained has resulted, among other things, in great changes in the system by which fungi are classified; and this, coupled with the concurrent development of an International Code of Botanical Nomenclature governing the application of scientific names of plants, has brought about changes in the names of many of the fungi treated in McIlvaine's book.[1] As an aid to correlating the wealth of information provided by McIlvaine with that in modern mycological literature I have compiled an alphabetical list of most of the scientific names of fungi used in the 1902 edition of *One Thousand American Fungi*.

In this list a specific or varietal name, with its author(s), appearing alone is a correct name used by McIlvaine for a species or variety still recognized as being a good one and as belonging to the genus indicated.[2] In some of these cases the author citation has been added or corrected, or the spelling of the name itself has been corrected. Authors' names are abbreviated according to the list on pages xxxiv and xxxv of the book; those not appearing there are spelled out completely.

If two specific names are connected by the symbol ≡, the first (*i.e.*, the one used by McIlvaine) is incorrect because it is contrary to the International Code of Botanical Nomenclature.

If two names connected by the symbol = have the same specific epithet (except perhaps for the endings indicating different genders, *e.g.*,

---

[1] See the nomenclatural appendix in the Dover edition (1967) of Louis C. C. Krieger's *The Mushroom Handbook* for a discussion of why scientific names change.

[2] The scientific name of a species (*e.g.*, *Morchella esculenta*) is binary, consisting of the name of the genus to which the species belongs (*Morchella*) and the specific epithet (*esculenta*).

abortivus and *abortivum*, *candida* and *candidum*, etc.), they are both correct according to the Code. However, the second is expressive of the more modern system of classification in which genera are usually more narrowly delimited.

If two specific names connected by the symbol = have different epithets, the second name is the correct one according to the Code, or the two species are no longer generally considered distinct, in which case the second name should be used. If, in addition, the generic names are different, the species is now often placed in the genus indicated by the second name.

Other situations are explained by a brief statement.

A name used by McIlvaine and missing from the following list could not be located in recent mycological literature. This is not necessarily to say either that the name is incorrect or that the species or variety should not be recognized, but merely that no recent evaluation of the name or of the species or variety it represents could be found.

Agaricus abruptus = *Agaricus silvicola* (Vitt.) Pk.

Agaricus arvensis Sch. ex Sec.

Agaricus californicus Pk.

Agaricus campester ≡ *Agaricus campestris* Linn. ex Fr.

Agaricus campestris Linn. ex Fr.

Agaricus comptulus Fr.

Agaricus diminutivus Pk.

Agaricus fabaceus Berk.

Agaricus haemorrhoidarius Schulzer in Kalchb.

Agaricus naucinus = *Lepiota leucothites* (Vitt.) Orton

Agaricus placomyces Pk.

Agaricus rodmani = *Agaricus bitorquis* (Q.) Sacc.

Agaricus silvaticus Sch. ex Sec.

Agaricus silvicola (Vitt.) Pk.

Agaricus subrufescens Pk.

Amanita abrupta Pk.

Amanita aspera (Fr.) S. F. Gray

Amanita caesarea (Scop. ex Fr.) Pers. ex Schw.

Amanita candida = *Amanita polypyramis* (Berk. & Curt.) Sacc.

Amanita chlorinosma (Pk. in Austin) Lloyd

Amanita citrina (Sch.) ex S. F. Gray

Amanita daucipes (Mont.) Lloyd

Amanita excelsa (Fr.) Kummer

Amanita frostiana (Pk.) Sacc.

Amanita lenticularis = *Limacella guttata* (Fr.) Konrad & Maublanc

Amanita magnivelaris Pk.

Amanita mappa = *Amanita citrina* (Sch.) ex S. F. Gray

Amanita monticulosa (Berk. & Curt.) Sacc.

Amanita muscaria (Linn. ex Fr.) Pers. ex Hooker

Amanita muscaria var. *regalis* (Fr.) Sacc.

Amanita nitida Fr.

Amanita pantherina (D.C. ex Fr.) Sec.

Amanita phalloides (Vaillant ex Fr.) Sec.

Amanita prairiicola Pk.

Amanita ravenelii (Berk. & Curt.) Sacc.

Amanita recutita = *Amanita porphyria* (A. & S. ex Fr.) Sec.

Amanita rubescens (Pers. ex Fr.) S. F. Gray

Amanita russuloides = *Amanita gemmata* (Fr.) G.

Amanita solitaria (B. ex Fr.) Sec.

Amanita spissa = *Amanita excelsa* (Fr.) Kummer

Amanita spreta (Pk.) Sacc.

Amanita strobiliformis (Paul. ex Vitt.) Bertillon

Amanita vaginata (B. ex Fr.) Vitt.

Amanita verna (B. ex Fr.) Pers. ex Vitt.

Amanita virosa Sec.

Amanita volvata (Pk.) Martin

Amanitopsis adnata = *Amanita gemmata* (Fr.) G.

Amanitopsis agglutinata = *Amanita agglutinata* (Berk. & Curt.) Lloyd

Amanitopsis farinosa = *Amanita farinosa* Schw.

Amanitopsis nivalis = *Amanita nivalis* Grev.

Amanitopsis pubescens = *Amanita pubescens* Schw.

Amanitopsis strangulata = *Amanita inaurata* Sec.

Amanitopsis vaginata = *Amanita vaginata* (B. ex Fr.) Vitt.

Amanitopsis vaginata var. *fulva* is recognized as a species, *Amanita fulva*
(Sch.) ex Sec.

Amanitopsis volvata = *Amanita volvata* (Pk.) Martin

Anellaria separata = *Panaeolus semiovatus* (Sow. ex Fr.) Lundell

Armillaria mellea (Vahl ex Fr.) Kummer

Armillaria mucida = *Oudemansiella mucida* (Schr. ex Fr.) Kummer

Armillaria ponderosa (Pk.) Sacc.

Armillaria robusta = *Tricholoma robustum* (A. & S. ex Fr.) Ricken

Bolbitius boltoni = *Bolbitius vitellinus* (Pers. ex Fr.) Fr.

Bolbitius fragilis = *Bolbitius vitellinus* (Pers. ex Fr.) Fr.

Boletinus appendiculatus Pk.

Boletinus cavipes (Opatowski) Kalchb.

Boletinus decipiens (Berk.) Pk.

Boletinus flavidus = *Suillus flavidus* (Fr.) Singer

Boletinus paluster (Pk.) Pk.

Boletinus pictus (Pk.) Pk.

Boletinus porosus = *Gyrodon merulioides* (Schw.) Singer

Boletus affinis Pk.

Boletus affinis var. *maculosus* Pk.

Boletus albellus = *Leccinum albellum* (Pk.) Singer

Boletus alboater = *Tylopilus alboater* (Schw.) Murrill

Boletus albus = *Suillus placidus* (Bon.) Singer

Boletus alutaceus Morg. in Pk.

Boletus alutarius = *Tylopilus felleus* (B. ex Fr.) Karst.

Boletus alveolatus = *Boletus frostii* Russell

Boletus americanus = *Suillus americanus* (Pk.) Snell

Boletus ananas = *Boletellus ananas* (Curt.) Murrill

Boletus auriflammeus = *Pulveroboletus auriflammeus* (Berk. & Curt.) Singer

Boletus auripes Pk.

Boletus auriporus = *Pulveroboletus auriporus* (Pk.) Singer

Boletus badiceps = *Tylopilus badiceps* (Pk.) Smith & Thiers

Boletus badius (Fr.) Fr.

Boletus barlae = *Boletus rubellus* K.

Boletus betula = *Boletellus betula* (Schw.) Gilbert

Boletus bicolor = *Boletus rubellus* K.

Boletus bovinus = *Suillus bovinus* (Linn. ex Fr.) O. Kuntze

Boletus braunii = *Phaeogyroporus braunii* (Bres.) Singer

Boletus brevipes = *Suillus brevipes* (Pk.) O. Kuntze

Boletus caespitosus = *Pulveroboletus caespitosus* (Pk.) Singer

Boletus calopus Fr.

Boletus castaneus = *Gyroporus castaneus* (B. ex Fr.) Q.

Boletus chromapes = *Leccinum chromapes* (Frost) Singer

Boletus chrysenteron (B. ex St. Amans) Q.

Boletus circinans = *Suillus granulatus* (Linn. ex Fr.) O. Kuntze

Boletus clintonianus is recognized as a variety, *Suillus grevillei* var. *clintonianus* (Pk.) Singer

Boletus conicus = *Tylopilus conicus* (Rav. in Berk. & Curt.) Beardslee

Boletus curtisii = *Pulveroboletus curtisii* (Berk.) Singer

Boletus cyanescens = *Gyroporus cyanescens* (B. ex Fr.) Q.

Boletus dichrous Ellis

Boletus duriusculus = *Leccinum duriusculum* (Schulzer ex Fr.) Singer

Boletus edulis B. ex Fr.

Boletus elbensis = *Suillus aeruginascens* (Sec.) Snell in Slipp & Snell

Boletus elegans = *Suillus elegans* (Schumacher ex Fr.) Snell

Boletus erythropus (Fr. ex Fr.) Sec.

Boletus eximius = *Tylopilus eximius* (Pk.) Singer

Boletus felleus = *Tylopilus felleus* (B. ex Fr.) Karst.

Boletus ferrugineus = *Tylopilus ferrugineus* (Frost) Singer

Boletus firmus Frost

Boletus fistulosus Pk.

Boletus flavidus = *Suillus flavidus* (Fr.) Singer

Boletus flavus = *Suillus nueschii* Singer

Boletus floccopus = *Strobilomyces floccopus* (Vahl ex Fr.) Karst.

Boletus fragrans Vitt.

Boletus fraternus = *Boletus rubellus* K.

Boletus frostii Russell

Boletus fulvus Pk.

Boletus fumosipes = *Tylopilus fumosipes* (Pk.) Smith & Thiers

Boletus glabellus Pk.

Boletus gracilis = *Porphyrellus gracilis* (Pk.) Singer

Boletus granulatus = *Suillus granulatus* (Linn. ex Fr.) O. Kuntze

Boletus griseus Frost

Boletus hemichrysus = *Pulveroboletus hemichrysus* (Berk. & Curt.) Singer

Boletus hirtellus = *Suillus hirtellus* (Pk.) O. Kuntze

Boletus holopus Rost.

Boletus illudens Pk.

Boletus impolitus Fr.

Boletus indecisus = *Tylopilus indecisus* (Pk.) Murrill

Boletus inflexus Pk.

Boletus isabellinus = *Boletellus ananas* (Curt.) Murrill

Boletus luridus Sch. ex Fr.

Boletus luridus var. *erythropus* is recognized as a species, *Boletus erythropus*
　(Fr. ex Fr.) Sec.

Boletus luteus = *Suillus luteus* (Linn. ex Fr.) S. F. Gray

Boletus magnisporus Frost

Boletus miniatoolivaceus Frost

Boletus mitis = *Suillus bovinus* (Linn. ex Fr.) O. Kuntze

Boletus modestus Pk.

Boletus mutabilis = *Boletus pulverulentus* Opatowski

Boletus nebulosus = *Tylopilus fumosipes* (Pk.) Smith & Thiers

Boletus nigrellus = *Tylopilus alboater* (Schw.) Murrill

Boletus ornatipes = *Pulveroboletus retipes* (Berk. & Curt.) Singer

Boletus pachypus = *Boletus calopus* Fr.

Boletus pallidus Frost

Boletus parasiticus B. ex Fr.

Boletus parvus Pk.

Boletus peckii Frost in Pk.

Boletus piperatus = *Suillus piperatus* (B. ex Fr.) O. Kuntze

Boletus porphyrosporus = *Porphyrellus pseudoscaber* (Sec.) Singer

Boletus punctipes = *Suillus punctipes* (Pk.) Singer

Boletus purpureus Pers.

Boletus radicans Pers. ex Fr.

Boletus ravenelii = *Pulveroboletus ravenelii* (Berk. & Curt.) Murrill

Boletus retipes = *Pulveroboletus retipes* (Berk. & Curt.) Singer

Boletus roxanae Frost

Boletus rubeus = *Boletus rubellus* K.

Boletus rubinellus = *Suillus rubinellus* (Pk.) Singer

Boletus rubropunctus = *Leccinum rubropunctum* (Pk.) Singer

Boletus russelli = *Boletellus russellii* (Frost) Gilbert

Boletus sanguineus = *Boletus rubellus* K.

Boletus satanas Lenz

Boletus scaber = *Leccinum scabrum* (B. ex Fr.) S. F. Gray

Boletus scaber var. *aurantiacus* is recognized as a species, *Leccinum aurantia-*
　cum (B. ex St. Amans) S. F. Gray

Boletus scaber var. *niveus* is considered a synonym of *Boletus holopus* Rost.

Boletus sensibilis Pk.

Boletus separans Pk.

Boletus serotinus = *Suillus aeruginascens* (Sec.) Snell in Slipp & Snell

Boletus sordidus Frost

Boletus spadiceus Fr.

Boletus speciosus Frost

Boletus spectabilis = *Boletinus spectabilis* Pk.

Boletus sphaerocephalus = *Boletus sulphureus* Fr.

Boletus sphaerosporus = *Paragyrodon sphaerosporus* (Pk.) *Singer*

Boletus spraguei Frost

Boletus striaepes = *Boletus subtomentosus* Linn. ex Fr.

Boletus subaureus = *Suillus subaureus* (Pk.) Snell

Boletus subglabripes = *Leccinum subglabripes* (Pk.) Singer

Boletus subluteus = *Suillus subluteus* (Pk.) Snell

Boletus subtomentosus Linn. ex Fr.

Boletus subvelutipes Pk.

Boletus sullivantii Berk. & Curt. in Mont.

Boletus tabacinus = *Tylopilus tabacinus* (Pk.) Singer

Boletus underwoodii Pk.

Boletus unicolor = *Suillus unicolor* (Frost in Pk.) O. Kuntze

Boletus variegatus = *Suillus variegatus* (Sow. ex Fr.) O. Kuntze

Boletus variipes Pk.

Boletus vermiculosus Pk.

Boletus versicolor = *Boletus rubellus* K.

Boletus versipellis = *Leccinum áurantiacum* (B. ex St. Amans) S. F. Gray

Bovista circumscissa = *Disciseda candida* (Schw.) Lloyd

Bovista minor Morg.

Bovista montana = *Bovista pila* Berk. & Curt.

Bovista nigrescens Pers. ex Pers.

Bovista pila Berk. & Curt.

Bovista plumbea Pers. ex Pers.

Bovistella ohiensis = *Bovistella radicata* (Mont.) Patouillard

Calvatia caelata = *Calvatia bovista* (Pers.) Kambly & Lee

Calvatia craniiformis (Schw.) Fr.

Calvatia cyathiformis (Bosc) Morg.

Calvatia elata (Massee) Morg.

Calvatia fragilis = *Calvatia cyathiformis* (Bosc) Morg.

Calvatia gigantea (Bat. ex Pers.) Lloyd

Calvatia pachyderma (Pk.) Morg.

Calvatia rubro-flava (Cragin) Morg.

Calvatia saccata = *Calvatia excipuliformis* (Scop. ex Pers.) Perdeck

Calvatia sigillata (Cragin) Morg.

xlv

Cantharellus albidus = *Hygrophoropsis albida* (Fr.) Maire

Cantharellus aurantiacus = *Hygrophoropsis aurantiaca* (Wulf. ex Fr.) Maire in Martin-Sans

Cantharellus aurantiacus var. *pallidus* = *Hygrophoropsis aurantiaca* var. *pallida* (Cke.) Kühner & Romagnesi

Cantharellus cibarius Fr.

Cantharellus cinereus (Pers.) ex Fr.

Cantharellus cinnabarinus Schw.

Cantharellus floccosus = *Gomphus floccosus* (Schw.) Singer

Cantharellus lutescens (Pers.) ex Fr.

Cantharellus minor Pk.

Cantharellus umbonatus = *Cantharellula umbonata* (Gmelin ex Fr.) Singer

Catastoma circumscissum = *Disciseda candida* (Schw.) Lloyd

Chitonia rubriceps = *Macrometrula rubriceps* (Massee) Donk & Singer

Claudopus byssisedus (Pers. ex Fr.) G.

Claudopus depluens (Bat. ex Fr.) G.

Claudopus nidulans = *Phyllotopsis nidulans* (Pers. ex Fr.) Singer

Claudopus variabilis = *Crepidotus variabilis* (Pers. ex Fr.) Kummer

Clavaria amethystina = *Clavulina amethystina* (Fr.) Donk

Clavaria aurantio-cinnabarino = *Clavulina aurantio-cinnabarina* (Schw.) Corner

Clavaria aurea = *Ramaria aurea* (Fr.) Q.

Clavaria botrytes = *Ramaria botrytis* (Fr.) Ricken

Clavaria cinerea = *Clavulina cinerea* (Fr.) Schroet.

Clavaria circinans = *Ramaria suecica* (Fr.) Donk

Clavaria clavata = *Clavulinopsis vernalis* (Schw.) Corner

Clavaria coralloides is recognized as a variety, *Clavulina cristata* var. *coralloides* Corner

Clavaria cristata = *Clavulina cristata* (Fr.) Schroet.

Clavaria densa = *Ramaria formosa* (Fr.) Q.

Clavaria dichotoma = *Clavulinopsis dichotoma* (Godey) Corner

Clavaria fastigiata = *Clavulinopsis corniculata* (Fr.) Corner

Clavaria flaccida = *Ramaria flaccida* (Fr.) Ricken

Clavaria flava = *Ramaria flava* (Fr.) Q.

Clavaria formosa = *Ramaria formosa* (Fr.) Q.

Clavaria fusiformis = *Clavulinopsis fusiformis* (Fr.) Corner

Clavaria herveyi = *Clavulina rugosa* (Fr.) Schroet.

Clavaria muscoides = *Clavulinopsis corniculata* (Fr.) Corner

Clavaria pistillaris = *Clavariadelphus pistillaris* (Fr.) Donk
Clavaria pyxidata = *Clavicorona pyxidata* (Fr.) Doty
Clavaria rugosa = *Clavulina rugosa* (Fr.) Schroet.
Clavaria spinulosa = *Ramaria spinulosa* (Fr.) Q.
Clavaria stricta = *Ramaria stricta* (Fr.) Q.
Clavaria subtilis = *Clavulinopsis subtilis* (Fr.) Corner
Clavaria tetragona = *Clavulinopsis tetragona* (Schw.) Corner
Clavaria vermicularis Fr.
Clitocybe adirondackensis (Pk.) Sacc.
Clitocybe aggregata = *Lyophyllum decastes* (Fr. ex Fr.) Singer
Clitocybe amethystina = *Laccaria amethystea* (B. ex Mérat) Murrill
Clitocybe brumalis (Fr. ex Fr.) Q.
Clitocybe candicans (Pers. ex Fr.) Kummer
Clitocybe candida = *Leucopaxillus candidus* (Bres.) Singer
Clitocybe catinus (Fr.) Q.
Clitocybe cerussata (Fr.) Kummer
Clitocybe clavipes (Pers. ex Fr.) Kummer
Clitocybe compressipes (Pk.) Sacc.
Clitocybe cyathiformis = *Cantharellula cyathiformis* (B. ex Fr.) Singer
Clitocybe dealbata (Sow. ex Fr.) Kummer
Clitocybe decastes = *Lyophyllum decastes* (Fr. ex Fr.) Singer
Clitocybe ditopus (Fr. ex Fr.) G.
Clitocybe flaccida (Sow. ex Fr.) Kummer
Clitocybe fragrans (Sow. ex Fr.) Kummer
Clitocybe fumosa = *Lyophyllum fumosum* (Pers. ex Fr.) Orton
Clitocybe gangraenosa = *Lyophyllum fumatofoetens* (Sec.) J. Schaeffer
Clitocybe geotropa (B. ex St. Amans) Q.
Clitocybe gigantea = *Leucopaxillus giganteus* (Sow. ex Fr.) Singer
Clitocybe gilva (Pers. ex Fr.) Kummer
Clitocybe illudens = *Omphalotus olearius* (D.C. ex Fr.) Singer
Clitocybe infundibuliformis = *Clitocybe gibba* (Pers. ex Fr.) Kummer
Clitocybe inversus = *Clitocybe flaccida* (Sow. ex Fr.) Kummer
Clitocybe laccata = *Laccaria laccata* (Scop. ex Fr.) Cke.
Clitocybe laccata var. *striatula* is recognized as a species, *Laccaria striatula*
  (Pk.) Pk.
Clitocybe maxima = *Clitocybe geotropa* (B. ex St. Amans) Q.
Clitocybe media = *Clitocybe clavipes* (Pers. ex Fr.) Kummer
Clitocybe metachroa (Fr.) Kummer

Clitocybe monadelpha = *Armillaria tabescens* (Scop. ex Fr.) Emel
Clitocybe morbifera = *Clitocybe dealbata* (Sow. ex Fr.) Kummer
Clitocybe multiceps = *Lyophyllum decastes* (Fr. ex Fr.) Singer
Clitocybe nebularis (Bat. ex Fr.) Kummer
Clitocybe ochropurpurea = *Laccaria ochropurpurea* (Berk.) Pk.
Clitocybe odora (B. ex Fr.) Kummer
Clitocybe phyllophila (Fr.) Kummer
Clitocybe pithyophila = *Clitocybe cerussata* (Fr.) Kummer
Clitocybe rivulosa (Pers. ex Fr.) Kummer
Clitocybe robusta Pk.
Clitocybe socialis (Fr.) G.
Clitocybe splendens = *Clitocybe gilva* (Pers. ex Fr.) Kummer
Clitocybe subzonalis = *Leucopaxillus subzonalis* (Pk.) Bigelow
Clitocybe tortilis = *Laccaria tortilis* (Bolt. ex S. F. Gray) Cke.
Clitocybe trullisata = *Laccaria trullisata* (Ellis) Pk.
Clitocybe truncicola (Pk.) Sacc.
Clitopilus abortivus = *Entoloma abortivum* (Berk. & Curt.) Donk
Clitopilus carneo-albus = *Leptonia sericella* (Fr. ex Fr.) Barbier
Clitopilus conissans = *Psathyrella subcernua* (Schulzer) Singer
Clitopilus micropus = *Entoloma micropus* (Pk.) Hesler
Clitopilus novaeboracensis = *Rhodocybe novaeboracensis* (Pk.) Singer
Clitopilus orcella = *Clitopilus prunulus* (Scop. ex Fr.) Kummer
Clitopilus popinalis = *Rhodocybe popinalis* (Fr.) Singer
Clitopilus prunulus (Scop. ex Fr.) Kummer
Clitopilus subvilis = *Entoloma subvile* (Pk.) Hesler
Collybia acervata (Fr.) Kummer
Collybia butyracea (B. ex Fr.) Kummer
Collybia confluens (Pers. ex Fr.) Kummer
Collybia dryophila (B. ex Fr.) Kummer
Collybia esculenta = *Pseudohiatula esculenta* (Wulf. ex Fr.) Singer
Collybia fusipes (B. ex Fr.) Q.
Collybia longipes = *Oudemansiella longipes* (B. ex St. Amans) Moser
Collybia maculata (A. & S. ex Fr.) Kummer
Collybia platyphylla = *Tricholomopsis platyphylla* (Pers. ex Fr.) Singer
Collybia radicata = *Oudemansiella radicata* (Relh. ex Fr.) Singer
Collybia spinulifera = *Marasmius cohaerens* (A. & S. ex Fr.) Cke. & Q.
Collybia velutipes = *Flammulina velutipes* (Curt. ex Fr.) Karst.
Coprinus atramentarius (B. ex Fr.) Fr.

Coprinus comatus (Müll. ex Fr.) S. F. Gray
Coprinus congregatus (B. ex St. Amans) Fr.
Coprinus domesticus (Bolt. ex Fr.) S. F. Gray
Coprinus ephemerus (B. ex Fr.) Fr.
Coprinus fimetarius = *Coprinus cinereus* (Sch. ex Fr.) S. F. Gray
Coprinus lagopus (r.) Fr. F
Coprinus micaceus (B. ex Fr.) Fr.
Coprinus niveus (Pers. ex Fr.) Fr.
Coprinus picaceus (B. ex Fr.) S. F. Gray
Coprinus plicatilis (Curt. ex Fr.) Fr.
Coprinus ovatus = *Coprinus comatus* (Müll. ex. Fr.) S. F. Gray
Coprinus silvaticus Pk.
Coprinus soboliferus = *Coprinus atramentarius* (B. ex Fr.) Fr.
Coprinus sterquilinus (Fr.) Fr.
Cortinarius alboviolaceus (Pers. ex Fr.) Fr.
Cortinarius armillatus (Fr. ex Fr.) Fr.
Cortinarius caerulescens (Sch. ex Sec.) Fr.
Cortinarius callisteus (Fr. ex Fr.) Fr.
Cortinarius castaneus (B. ex Fr.) Fr.
Cortinarius cinnabarinus Fr.
Cortinarius cinnamomeus (Linn. ex Fr.) Fr.
Cortinarius cinnamomeus var. *semisanguineus* is recognized as a species,
  Cortinarius semisanguineus (Fr.) G.
Cortinarius collinitus (Sow. ex Fr.) Fr.
Cortinarius corrugatus Pk.
Cortinarius distans Pk.
Cortinarius intrusus = *Conocybe intrusa* (Pk.) Singer
Cortinarius iodes Berk. & Curt.
Cortinarius multiformis (Fr. ex Sec.) Fr.
Cortinarius pholideus (Fr. ex Fr.) Fr.
Cortinarius purpurascens (Fr.) ex Fr.
Cortinarius sanguineus (Wulf. ex Fr.) Fr.
Cortinarius sebaceus Fr.
Cortinarius subpurpurascens (Bat. ex Fr.) Kickx
Cortinarius tofaceus Fr.
Cortinarius turbinatus (B. ex Fr.) Fr.
Cortinarius turmalis (Fr.) ex Fr.
Cortinarius varius (Sch. ex Fr.) Fr.

Cortinarius violaceus (Linn. ex Fr.) Fr.

Craterellus cantharellus = *Cantharellus odoratus* (Schw.) Fr.

Craterellus clavatus = *Gomphus clavatus* Pers. ex S. F. Gray

Craterellus cornucopioides (Linn. ex Fr.) Pers.

Craterellus dubius Pk.

Craterellus lutescens = *Cantharellus lutescens* (Pers.) ex Fr.

Craterellus sinuosus (Fr.) Fr.

Crepidotus fulvotomentosus = *Crepidotus calolepis* (Fr.) Karst.

Crepidotus mollis (Sch. ex Fr.) Kummer

Daedalea quercina Linn. ex Fr.

Entoloma clypeatum (Linn. ex Fr.) Kummer

Entoloma lividum = *Entoloma sinuatum* (B. ex Fr.) Kummer

Entoloma prunuloides (Fr.) Q.

Entoloma rhodopolium (Fr.) Kummer

Entoloma sinuatum (B. ex Fr.) Kummer

Fistulina firma = *Fistulina radicata* Schw.

Fistulina hepatica Sch. ex Fr.

Fistulina pallida = *Fistulina radicata* Schw.

Fistulina radicata Schw.

Fistulina spathulata = *Fistulina radicata* Schw.

Flammula alnicola = *Pholiota alnicola* (Fr.) Kummer

Flammula flavida = *Pholiota flavida* (Fr.) Singer

Flammula hybrida = *Gymnopilus hybridus* (Fr. ex Fr.) Singer

Fomes igniarius = *Phellinus igniarius* (Linn. ex Fr.) Q.

Galera hypnorum ≡ *Galerina hypnorum* (Schrank ex Fr.) Kühner

Galera lateritia ≡ *Conocybe lactea* (Lange) Métrod

Galera tenera ≡ *Conocybe tenera* (Sch. ex Fr.) Kühner

Galera vittaeformis ≡ *Galerina vittaeformis* (Fr.) Moser

Geaster hygrometricus ≡ *Astraeus hygrometricus* (Pers.) Morg.

Geaster minimus ≡ *Geastrum minimum* Schw.

Geoglossum glutinosum Pers. ex Fr.

Gomphidius glutinosus (Sch. ex Fr.) Fr.

Gomphidius oregonensis Pk.

Gomphidius rhodoxanthus = *Phylloporus rhodoxanthus* (Schw.) Bres.

Gomphidius viscidus = *Gomphidius rutilus* (Sch. ex Fr.) Lundell

Gyromitra brunnea Underwood

Gyromitra caroliniana (Bosc ex Fr.) Fr.

Gyromitra esculenta (Pers.) Fr.

1

Hebeloma crustuliniforme (B. ex St. Amans) Q.

Hebeloma fastibile (Pers. ex Fr.) Kummer

Hebeloma glutinosum = *Pholiota lenta* (Pers. ex Fr.) Singer

Helvella californica = *Gyromitra californica* (Phillips) Raitviir

Helvella crispa Fr.

Helvella elastica B. ex St. Amans

Helvella esculenta = *Gyromitra esculenta* (Pers.) Fr.

Helvella infula = *Gyromitra infula* (Sch. ex Fr.) Q.

Helvella lacunosa Afzelius ex Fr.

Helvella sulcata Afzelius ex Fr.

Hirneola auricula-judea = *Auricularia auricula* (Linn. ex Hooker) Underwood

Hydnum albidum Pk.

Hydnum albonigrum is recognized as a variety, *Phellodon niger* var. *alboniger*
(Pk.) Harrison

Hydnum caput-medusae = *Hericium erinaceus* (B. ex Fr.) Pers.

Hydnum caput-ursi = *Hericium ramosum* (B. ex Mérat) Let.

Hydnum coralloides = *Hericium coralloides* (Scop. ex Fr.) S. F. Gray

Hydnum erinaceum = *Hericium erinaceus* (B. ex Fr.) Pers.

Hydnum fennicum = *Sarcodon fennicus* (Karst.) Karst.

Hydnum ferrugineum = *Hydnellum ferrugineum* (Fr. ex Fr.) Karst.

Hydnum fragile = *Bankera fuligineo-alba* (Schmidt ex Fr.) Pouzar

Hydnum gelatinosum = *Pseudohydnum gelatinosum* (Fr.) Karst.

Hydnum imbricatum = *Sarcodon imbricatus* (Linn. ex Fr.) Karst.

Hydnum laevigatum = *Sarcodon laevigatus* (Swartz) Karst.

Hydnum repandum Linn. ex Fr.

Hydnum rufescens Fr.

Hydnum scabrosum = *Sarcodon scabrosus* (Fr.) Karst.

Hydnum septentrionale = *Climacodon septentrionalis* (Fr.) Karst.

Hydnum spongiosipes = *Hydnellum spongiosipes* (Pk.) Pouzar

Hydnum vellereum = *Phellodon confluens* (Pers.) Pouzar

Hydnum zonatum is recognized as a variety, *Hydnellum scrobiculatum* var.
zonatum (Bat. ex Fr.) Harrison

Hygrophorus borealis Pk.

Hygrophorus cantharellus (Schw.) Fr.

Hygrophorus caprinus = *Hygrophorus camarophyllus* (A. & S. ex Fr.) Dumée
et al.

Hygrophorus ceraceus (Wulf. ex Fr.) Fr.

li

Hygrophorus chlorophanus (Fr.) Fr.

Hygrophorus chrysodon (Bat. ex Fr.) Fr.

Hygrophorus coccineus (Sch. ex Fr.) Fr.

Hygrophorus conicus (Scop. ex Fr.) Fr.

Hygrophorus distans = *Hygrophorus fornicatus* Fr.

Hygrophorus eburneus (B. ex Fr.) Fr.

Hygrophorus erubescens (Fr.) Fr.

Hygrophorus flavodiscus Frost

Hygrophorus fuligineus Frost

Hygrophorus hypothejus (Fr. ex Fr.) Fr.

Hygrophorus laurae Morg.

Hygrophorus limacinus (Scop. ex Fr.) Fr.

Hygrophorus miniatus (Fr.) Fr.

Hygrophorus nitidus Berk. & Curt.

Hygrophorus niveus (Scop.) ex Fr.

Hygrophorus penarius Fr.

Hygrophorus pratensis (Pers. ex Fr.) Fr.

Hygrophorus puniceus (Fr.) Fr.

Hygrophorus speciosus Pk.

Hygrophorus sphaerosporus Pk.

Hygrophorus virgineus (Wulf. ex Fr.) Fr.

Hypholoma aggregatum = *Psathyrella velutina* (Pers. ex Fr.) Singer

Hypholoma appendiculatum = *Psathyrella candolleana* (Fr.) Maire

Hypholoma candolleanum = *Psathyrella candolleana* (Fr.) Maire

Hypholoma capnoides (Fr. ex Fr.) Kummer

Hypholoma dispersus = *Hypholoma marginatum* (Pers. ex Fr.) Schroet. in Cohn

Hypholoma elaeodes (Fr.) G.

Hypholoma epixanthum (Fr.) Q.

Hypholoma fasciculare (Huds. ex Fr.) Kummer

Hypholoma incertum = *Psathyrella candolleana* (Fr.) Maire

Hypholoma lachrymabundum = *Psathyrella velutina* (Pers. ex Fr.) Singer

Hypholoma perplexum = *Hypholoma sublateritium* (Fr.) Q.

Hypholoma sublateritium (Fr.) Q.

Hypholoma velutinus = *Psathyrella velutina* (Pers. ex Fr.) Singer

Hypomyces inaequalis = *Hypomyces hyalinus* (Schw. ex Fr.) T.

Hypomyces lactifluorum (Schw. ex Fr.) T.

Hypomyces purpureus = *Hypomyces lactifluorum* (Schw. ex Fr.) T.

Inocybe lanuginosa (B. ex Fr.) Kummer
Ithyphallus impudicus = *Phallus impudicus* Linn. ex Pers.
Lactarius aquifluus = *Lactarius helvus* (Fr.) Fr.
Lactarius blennius (Fr. ex Fr.) Fr.
Lactarius camphoratus (B. ex Fr.) Fr.
Lactarius chelidonium Pk.
Lactarius controversus (Fr. ex Fr.) Fr.
Lactarius corrugis Pk.
Lactarius deceptivus Pk.
Lactarius deliciosus (Linn. ex Fr.) S. F. Gray
Lactarius distans = *Lactarius hygrophoroides* Berk. & Curt.
Lactarius fuliginosus (Fr.) Fr.
Lactarius gerardii Pk.
Lactarius glyciosmus (Fr. ex Fr.) Fr.
Lactarius helvus (Fr.) Fr.
Lactarius hygrophoroides Berk. & Curt.
Lactarius hysginus (Fr. ex Fr.) Fr.
Lactarius indigo (Schw.) Fr.
Lactarius insulsus (Fr.) Fr.
Lactarius lignyotus Fr.
Lactarius luteolus Pk.
Lactarius mitissimus (Fr.) Fr.
Lactarius pallidus (Pers. ex Fr.) Fr.
Lactarius pergamenus = *Lactarius piperatus* (Scop. ex Fr.) S. F. Gray
Lactarius piperatus (Scop. ex Fr.) S. F. Gray
Lactarius plumbeus = *Lactarius turpis* (Weinm.) Fr.
Lactarius quietus (Fr.) Fr.
Lactarius rufus (Scop. ex Fr.) Fr.
Lactarius sanguifluus (Paul.) ex Fr.
Lactarius serifluus (D.C. ex Fr.) Fr.
Lactarius subdulcis (Pers. ex Fr.) S. F. Gray
Lactarius subpurpureus Pk.
Lactarius theiogalus (B. ex Fr.) S. F. Gray
Lactarius torminosus (Sch. ex Fr.) S. F. Gray
Lactarius turpis (Weinm.) Fr.
Lactarius vellereus (Fr.) Fr.
Lactarius volemus (Fr.) Fr.
Lentinus cochleatus = *Lentinellus cochleatus* (Pers. ex Fr.) Karst.

Lentinus lecomtei = *Panus rudis* Fr.

Lentinus lepideus (Fr. ex Fr.) Fr.

Lentinus magnus = *Lentinus lepideus* (Fr. ex Fr.) Fr.

Lentinus strigosus = *Panus strigosus* Berk. & Curt.

Lentinus tigrinus = *Panus tigrinus* (B. ex Fr.) Singer

Leotia chlorocephala = *Leotia atrovirens* Pers. ex Fr.

Leotia lubrica Pers.

Lepiota acutesquamosa (Weinm.) Kummer

Lepiota alluvina Pk.

Lepiota americana (Pk.) Sacc.

Lepiota cepaestipes (Sow. ex Fr.) Kummer

Lepiota clypeolaria (B. ex Fr.) Kummer

Lepiota cristata (Fr.) Kummer

Lepiota delicata = *Limacella delicata* (Fr.) Earle ex H. V. Smith

Lepiota excoriata (Sch. ex Fr.) Kummer

Lepiota felina (Pers. ex Fr.) Karst.

Lepiota friesii (Lasch) Q.

Lepiota granulosa = *Cystoderma granulosum* (Bat. ex Fr.) Fayod

Lepiota holosericea (Fr.) G.

Lepiota illinita = *Limacella illinita* (Fr. ex Fr.) Maire

Lepiota lenticularis = *Limacella guttata* (Pers. ex Fr.) Konrad & Maublanc

Lepiota mastoidea (Fr.) Kummer

Lepiota metulaespora (Berk. & Br.) Sacc.

Lepiota morgani = *Lepiota molybdites* (Meyer ex Fr.) Sacc.

Lepiota naucina = *Lepiota leucothites* (Vitt.) Orton

Lepiota naucinoides = *Lepiota leucothites* (Vitt.) Orton

Lepiota procera (Scop. ex Fr.) S. F. Gray

Lepiota rhacodes (Vitt.) Q.

Lepiota vittadini = *Amanita vittadinii* (Moretti) Vitt.

Leptonia incana (Fr.) G.

Lycoperdon acuminatum Bosc

Lycoperdon atropurpureum Vitt.

Lycoperdon bovista = *Calvatia bovista* (Pers.) Kambly & Lee

Lycoperdon caelatum = *Calvatia bovista* (Pers.) Kambly & Lee

Lycoperdon calyptriforme = *Lycoperdon acuminatum* Bosc

Lycoperdon coloratum Pk.

Lycoperdon curtisii Berk.

Lycoperdon cyathiforme = *Calvatia cyathiformis* (Bosc) Morg.

Lycoperdon echinatum Pers. ex Pers.

Lycoperdon elatum = *Calvatia elata* (Massee) Morg.

Lycoperdon excipuliforme = *Lycoperdon perlatum* Pers. ex Pers.

Lycoperdon frostii = *Lycoperdon pulcherrimum* Berk. & Curt.

Lycoperdon gemmatum = *Lycoperdon perlatum* Pers. ex Pers.

Lycoperdon giganteum = *Calvatia gigantea* (Bat. ex Pers.) Lloyd

Lycoperdon glabellum = *Lycoperdon umbrinum* Pers. ex Pers.

Lycoperdon molle = *Lycoperdon umbrinum* Pers. ex Pers.

Lycoperdon muscorum Morg.

Lycoperdon oblongisporum Berk. & Curt.

Lycoperdon peckii Morg.

Lycoperdon pedicellatum Pk.

Lycoperdon perlatum Pers. ex Pers.

Lycoperdon plumbeum = *Bovista plumbea* Pers. ex Pers.

Lycoperdon pulcherrimum Berk. & Curt.

Lycoperdon pusillum Pers.

Lycoperdon pyriforme Sch. ex Pers.

Lycoperdon rimulatum Pk.

Lycoperdon subincarnatum Pk.

Lycoperdon wrightii = *Lycoperdon curtisii* Berk.

Marasmius alliaceus (Jacq. ex Fr.) Fr.

Marasmius calopus (Pers. ex Fr.) Fr.

Marasmius oreades (Bolt. ex Fr.) Fr.

Marasmius peronatus = *Collybia peronata* (Bolt. ex Fr.) Kummer

Marasmius scorodonius (Fr.) Fr.

Marasmius urens (B. ex Fr.) Fr.

Marasmius wynnei Berk. & Br.

Merulius corium (Fr.) Fr.

Merulius rubellus = *Merulius incarnatus* Schw.

Merulius tremellosus Fr.

Mitrula vitellina var. *irregularis* is recognized as a species, *Spragueola irregularis* (Pk.) Nannfeldt

Morchella angusticeps Pk.

Morchella bispora = *Verpa bohemica* (K.) Schroet.

Morchella conica = *Morchella angusticeps* Pk.

Morchella crassipes = *Morchella esculenta* Pers. ex St. Amans

Morchella deliciosa = *Morchella esculenta* Pers. ex St. Amans

Morchella elata Fr.

Morchella esculenta Pers. ex St. Amans
Morchella semilibera (D.C. ex Fr.) Lév.
Mutinus bovinus = Mutinus curtisii (Berk.) Fischer
Mutinus caninus (Huds. ex Pers.) Fr.
Mycena alcalina (Fr. ex Fr.) Kummer
Mycena galericulata (Scop. ex Fr.) S. F. Gray
Mycena haematopus (Pers. ex Fr.) Kummer
Mycena latifolia (Pk.) Sacc.
Mycena parabolica (Fr.) Q.
Mycena rugosa = Mycena galericulata (Scop. ex Fr.) S. F. Gray
Mycenastrum spinulosum = Mycenastrum corium (Guersent ex Lam. & D.C.)
  Desvaux
Naucoria pediades = Agrocybe semiorbicularis (B. ex St. Amans) Fayod
Naucoria semi-orbicularis = Agrocybe semiorbicularis (B. ex St. Amans)
  Fayod
Naucoria striapes = Conocybe striaepes (Cke.) Lundell
Nolanea pascua (Linn. ex Fr.) Kummer
Nyctalis asterophora ≡ Asterophora lycoperdoides (B. ex Mérat) S. F. Gray
Nyctalis parasiticus ≡ Asterophora parasitica (B. ex Fr.) Singer
Omphalia campanella ≡ Xeromphalina campanella (Bat. ex Fr.) Maire
Omphalia oniscus ≡ Omphalina oniscus (Fr. ex Fr.) Q.
Omphalia umbellifera ≡ Omphalina ericetorum (Fr. ex Fr.) M. Lange
Pachyma cocos = Poria cocos (Schw.) Wolf
Panaeolus campanulatus (B. ex Fr.) Q.
Panaeolus fimicola (Fr.) Q.
Panaeolus papilionaceus (B. ex Fr.) Q.
Panaeolus retirugis (Fr.) G.
Panaeolus solidipes = Panaeolus sepulchralis (Berk.) Sacc.
Panus conchatus = Panus torulosus (Pers. ex Fr.) Fr.
Panus farinaceus = Panellus stipticus (B. ex Fr.) Karst.
Panus stipticus = Panellus stipticus (B. ex Fr.) Karst.
Panus strigosus Berk. & Curt.
Panus torulosus (Pers. ex Fr.) Fr.
Paxillus atrotomentosus (Bat. ex Fr.) Fr.
Paxillus involutus (Bat. ex Fr.) Fr.
Paxillus lepista = Lepista subaequalis (Britzelmayr) Singer
Paxillus panaeolus = Ripartites helomorpha (Fr.) Karst.
Paxillus porosus = Phylloporus rhodoxanthus (Schw.) Bres.

Peziza acetabulum = *Helvella acetabulum* (Linn. ex St. Amans) Q.

Peziza aurantia = *Aleuria aurantia* (Fr.) Fckl.

Peziza badia Pers. ex Mérat

Peziza coccinea = *Sarcoscypha coccinea* (Fr.) Lambotte

Peziza cochleata = *Otidea cochleata* (Linn. ex St. Amans) Fckl.

Peziza leporina = *Otidea leporina* (Bat. ex Fr.) Fckl.

Peziza macropus = *Helvella macropus* (Pers. ex Fr.) Karst.

Peziza odorata = *Peziza domiciliana* Cke.

Peziza onotica = *Otidea onotica* (Pers.) Fckl.

Peziza petersii Berk. & Curt.

Peziza repanda Pers.

Peziza saniosa Schr. ex Fr.

Peziza succosa Berk.

Peziza unicisa = *Otidea leporina* (Bat. ex Fr.) Fckl.

Peziza venosa = *Disciotis venosa* (Pers.) Boud.

Peziza vesiculosa B. ex St. Amans

Peziza vesiculosa var. *cerea* is recognized as a species, *Peziza cerea* Sow. ex Mérat

Phallus caninus = *Mutinus caninus* (Pers.) Fr.

Phallus duplicatus = *Dictyophora duplicata* (Bosc) Fischer

Phallus impudicus Linn. ex Pers.

Phallus indusiatus = *Dictyophora duplicata* (Bosc) Fischer

Phallus ravenelii Berk. & Curt.

Phallus rubicundus Bosc

Pholiota adiposa (Fr.) Kummer

Pholiota caperata = *Rozites caperatus* (Pers. ex Fr.) Karst.

Pholiota discolor (Pk.) Sacc.

Pholiota dura = *Agrocybe dura* (Bolt. ex Fr.) Singer

Pholiota flammans (Fr.) Kummer

Pholiota luteofolia = *Gymnopilus luteofolius* (Pk.) Singer

Pholiota marginata = *Galerina marginata* (Bat. ex Fr.) Kühner

Pholiota mutabilis = *Galerina mutabilis* (Sch. ex Fr.) Orton

Pholiota ornella = *Pholiota polychroa* (Berk.) Smith & Brodie

Pholiota praecox = *Agrocybe praecox* (Pers. ex Fr.) Fayod

Pholiota squarrosa (Müll. ex Fr.) Kummer

Pholiota squarrosoides (Pk.) Sacc.

Pholiota subsquarrosa (Fr.) Q.

Pholiota togularis = *Conocybe togularis* (B. ex Fr.) Kühner

Pilosace algeriensis Q.

Pleurotus circinatus = *Pleurotus lignatilis* (Pers. ex Fr.) Kummer

Pleurotus dryinus (Pers. ex Fr.) Kummer

Pleurotus lignatilis (Pers. ex Fr.) Kummer

Pleurotus mastrucatus = *Hohenbuehelia mastrucata* (Fr. ex Fr.) Singer

Pleurotus ostreatus (Jacq. ex Fr.) Kummer

Pleurotus petaloides = *Hohenbuehelia petaloides* (B. ex Fr.) Schulzer

Pleurotus pometi = *Pleurotus dryinus* (Pers. ex Fr.) Kummer

Pleurotus pulmonarius (Fr.) Q.

Pleurotus salignus = *Pleurotus ostreatus* (Jacq. ex Fr.) Kummer

Pleurotus sapidus = *Pleurotus cornucopiae* (Paul. ex Pers.) Rolland

Pleurotus serotinus = *Panellus serotinus* (Schr. ex Fr.) Kühner

Pleurotus subpalmatus = *Rhodotus palmatus* (B. ex Fr.) Maire

Pleurotus tessulatus (B. ex Fr.) Kummer

Pleurotus ulmarius (B. ex Fr.) Kummer

Pluteolus reticulatus = *Bolbitius reticulatus* (Pers. ex Fr.) Ricken

Pluteus admirabilis (Pk.) Pk.

Pluteus cervinus (Sch. ex Fr.) Kummer

Pluteus chrysophaeus (Sch. ex Fr.) Q.

Pluteus chrysophlebius (Berk. & Rav.) Sacc.

Pluteus granularis Pk.

Pluteus leoninus (Sch. ex Fr.) Kummer

Pluteus pellitus (Pers. ex Fr.) Kummer

Pluteus umbrosus (Pers. ex Fr.) Kummer

Polyporus abietinus = *Hirschioporus abietinus* (Dickson ex Fr.) Donk

Polyporus anax = *Bondarzewia berkeleyi* (Fr.) Bondartsev

Polyporus annosus = *Fomitopsis annosa* (Fr.) Karst.

Polyporus berkeleyi = *Bondarzewia berkeleyi* (Fr.) Bondartsev

Polyporus betulinus = *Piptoporus betulinus* (B. ex Fr.) Karst.

Polyporus chioneus = *Tyromyces semipileatus* (Pk.) Murrill

Polyporus circinatus = *Polystictus circinatus* (Fr.) Karst.

Polyporus confluens = *Albatrellus confluens* (A. & S. ex Fr.) Kotlaba & Pouzar

Polyporus cristatus = *Albatrellus cristatus* (Pers. ex Fr.) Kotlaba & Pouzar

Polyporus frondosus = *Grifola frondosa* (Dickson ex Fr.) S. F. Gray

Polyporus fumosus = *Bjerkandera fumosa* (Pers. ex Fr.) Karst.

Polyporus giganteus = *Grifola gigantea* (Pers. ex Fr.) Pilát

Polyporus hirsutus = *Coriolus hirsutus* (Wulf. ex Fr.) Q.

Polyporus immitis = *Tyromyces immitis* (Pk.) Bondartsev

Polyporus intybaceus = *Grifola frondosa* (Dickson ex Fr.) S. F. Gray

Polyporus leucomelas = *Boletopsis leucomelaena* (Pers. ex Pers.) Fayod

Polyporus maculatus = *Tyromyces albidus* (Sch. ex Sec.) Donk

Polyporus perennis = *Coltricia perennis* (Linn. ex Fr.) Murrill

Polyporus picipes Fr.

Polyporus sinuosus = *Coriolus sinuosus* (Fr.) Bondartsev & Singer

Polyporus squamosus Mich. ex Fr.

Polyporus sulphureus = *Laetiporus sulphureus* (B. ex Fr.) Bondartsev & Singer

Polyporus umbellatus = *Grifola umbellata* (Pers. ex Fr.) Pilát

Polysaccum pisocarpium = *Pisolithus arhizus* (Pers.) Rausch.

Polystictus versicolor = *Coriolus versicolor* (Linn. ex Fr.) Q.

Psathyrella atomata (Fr.) Q.

Psathyrella disseminata = *Coprinus disseminatus* (Pers. ex Fr.) S. F. Gray

Psathyrella gracilis (Fr.) Q.

Psathyrella graciloides = *Psathyrella subatrata* (Fr.) G.

Psilocybe semilanceata (Fr. ex Sec.) Kummer

Psilocybe spadicea = *Psathyrella spadicea* (Sch. ex Fr.) Singer

Russula abietina Pk.

Russula adusta (Pers. ex Fr.) Fr.

Russula albella Pk.

Russula alutacea (Pers. ex Fr.) Fr.

Russula atropurpurea = *Russula xerampelina* (Sch. ex Sec.) Fr.

Russula aurata (With.) ex Fr.

Russula basifurcata Pk.

Russula brevipes Pk.

Russula chamaeleontina Fr.

Russula citrina = *Russula ochroleuca* (Pers. ex Sec.) Fr.

Russula cyanoxantha (Sch. ex Sec.) Fr.

Russula decolorans (Fr.) Fr.

Russula delica Fr.

Russula drimeia = *Russula sardonia* Fr.

Russula elegans Bres.

Russula emetica (Sch. ex Fr.) Pers. ex S. F. Gray

Russula flavida Frost in Pk.

Russula foetens (Pers. ex Fr.) Pers. ex Fr.

Russula foetens var. *granulata* is recognized as a species, *Russula granulata* (Pk.) Pk.

Russula fragilis (Pers. ex Fr.) Fr.

Russula furcata = *Russula heterophylla* (Fr.) Fr.

Russula heterophylla (Fr.) Fr.

Russula integra (Linn.) ex Fr.

Russula lactea (Pers. ex Fr.) Fr.

Russula lepida Fr.

Russula lutea (Huds. ex Fr.) S. F. Gray

Russula mariae Pk.

Russula nauseosa (Pers. ex Sec.) Fr.

Russula nigricans (B. ex Mérat) Fr.

Russula nigricans var. *albonigra* is recognized as a species, *Russula albonigra* (K.) Fr.

Russula nitida (Pers. ex Fr.) Fr.

Russula ochroleuca (Pers. ex Sec.) Fr.

Russula ochrophylla Pk.

Russula olivacea (Sch. ex Sec.) Fr.

Russula pectinata (B. ex St. Amans) Fr.

Russula puellaris Fr.

Russula punctata = *Russula amoena* Q.

Russula purpurina = *Russula peckii* Singer

Russula pusilla Pk.

Russula roseipes (Sec.) Bres.

Russula rubra (Lam. ex Fr.) Fr.

Russula sanguinea (B. ex St. Amans) Fr.

Russula sordida = *Russula albonigra* (K.) Fr.

Russula subdepallens Pk.

Russula vesca Fr.

Russula virescens (Sch. ex Sec.) Fr.

Russula vitellina (Pers.) ex S. F. Gray

Schizophyllum commune Fr.

Scleroderma bovista Fr.

Scleroderma geaster = *Scleroderma polyrhizum* Pers.

Scleroderma verrucosum B. ex Pers.

Scleroderma vulgare = *Scleroderma citrinum* Pers.

Sparassis crispa Fr.

Sparassis herbstii = *Sparassis spathulata* (Schw.) Fr.

Sparassis spathulata (Schw.) Fr.

Spathularia clavata = *Spathularia flavida* Fr.

Spathularia flavida Fr.

Strobilomyces floccopus (Vahl ex Fr.) Karst.

Strobilomyces strobilaceus = *Strobilomyces floccopus* (Vahl ex Fr.) Karst.

Stropharia aeruginosa (Curt. ex Fr.) Q.

Stropharia semiglobata (Bat. ex Fr.) Q.

Stropharia spintriger = *Psathyrella spintrigera* (Fr.) Konrad & Maublanc

Stropharia stercoraria = *Stropharia semiglobata* (Bat. ex Fr.) Q.

Terfezia leonis (T.) T.

Terfezia spinosa Harkness

Trametes gibbosa (Pers.) ex Fr.

Tremella fimbriata = *Tremella foliacea* Fr.

Tremella intumescens = *Exidia glandulosa* Fr.

Tremella lutescens Fr.

Tremella mesenterica Fr.

Tremella mycetophila is a name applied to teratological formations on fruiting
  bodies of *Collybia dryophila* (B. ex Fr.) Kummer

Tremellodon gelatinosum = *Pseudohydnum gelatinosum* (Fr.) Karst.

Tricholoma album (Sch. ex Fr.) Kummer

Tricholoma albellum = *Tricholoma gambosum* (Fr.) Kummer

Tricholoma brevipes = *Melanoleuca brevipes* (B. ex Fr.) Patouillard

Tricholoma columbetta (Fr.) Kummer

Tricholoma equestre = *Tricholoma flavovirens* (Pers. ex Fr.) Lundell

Tricholoma flavescens = *Tricholomopsis flavescens* (Pk.) Singer

Tricholoma flavobrunneum (Fr.) Kummer

Tricholoma gambosum (Fr.) Kummer

Tricholoma grammopodium = *Melanoleuca grammopodia* (B. ex Fr.) Patouillard

Tricholoma humile = *Melanoleuca humilis* (Pers. ex Fr.) Patouillard

Tricholoma imbricatum (Fr. ex Fr.) Kummer

Tricholoma leucocephalum (Fr.) Q.

Tricholoma nudum = *Lepista nuda* (B. ex Fr.) Cke.

Tricholoma paedidum = *Melanoleuca paedida* (Fr.) Kühner & Maire

Tricholoma personatum = *Lepista personata* (Fr. ex Fr.) Cke.

Tricholoma pessundatum (Fr.) Q.

Tricholoma portentosum (Fr.) Q.

Tricholoma resplendens (Fr.) Karst.

Tricholoma russula = *Hygrophorus russula* (Sch. ex Fr.) Q.

Tricholoma rutilans = *Tricholomopsis rutilans* (Sch. ex Fr.) Singer

Tricholoma saponaceum (Fr.) Kummer

Tricholoma sejunctum (Sow. ex Fr.) Q.

Tricholoma subpulverulentum = *Melanoleuca subpulverulenta* (Pers.) Singer

Tricholoma sulphureum (B. ex Fr.) Kummer

Tricholoma terreum (Sch. ex Fr.) Kummer

Tricholoma transmutans = *Tricholoma flavobrunneum* (Fr.) Kummer

Tricholoma ustale (Fr. ex Fr.) Kummer

Tricholoma vaccinum (Pers. ex Fr.) Kummer

Trogia crispa = *Plicatura crispa* (Pers. ex Fr.) Rea

Tubaria furfuracea (Pers. ex Fr.) G.

Tuber aestivum Vitt.

Tuber californicum Harkness

Verpa digitaliformis = *Verpa conica* Swartz ex Pers.

Volvaria bombycina ≡ *Volvariella bombycina* (Sch. ex Fr.) Singer

Volvaria gloiocephala is recognized as a variety, *Volvariella speciosa* var. *gloiocephala* (D.C. ex Fr.) Singer

Volvaria loveiana ≡ *Volvariella surrecta* (Knapp) Singer

Volvaria speciosa ≡ *Volvariella speciosa* (Fr. ex Fr.) Singer

Volvaria taylori ≡ *Volvariella taylori* (Berk. & Br.) Singer

Volvaria volvacea ≡ *Volvariella volvacea* (B. ex Fr.) Singer

ONE THOUSAND
AMERICAN FUNGI

CLASS, FUNGI

SUB-CLASS BASIDIOMYCETES

COHORT *HYMENOMYCETES.* *Gr.*—a membrane, a fruit-bearing surface; *Gr.*—a mushroom. (So called from the hymenium or fruit-bearing surface.)

FUNGI composed of membranes, fleshy, woody or gelatinous, growing on wood or on the ground. The hymenium or spore-bearing surface exposed at an early stage. The spores are borne on basidia, spread over the surface. The common mushroom is typical of the family. All the members resemble it, more or less, in organization and reproductive organs. These latter, in the mushroom, are spread over lamellæ or gills. The spores, after ripening and dissemination, germinate and produce a mycelium or thread-like vine, which in turn develops the spore-producing part of the plant. Hymenomycetes is divided into the following six Families:—

a. HYMENIUM FIGURATE.

I. Spread over the surface of lamellæ or gills..................................AGARICACEÆ.
II. Lining the interior of tubes or pores.......................................POLYPORACEÆ.
III. Clothing the surface of spines or protuberances of various forms.....HYDNACEÆ.

b. HYMENIUM EVEN.

IV. Horizontal and mostly on the under surface.........................THELEPHORACEÆ.
V. Vertical and produced all over the surface................................CLAVARIACEÆ.
VI. Superior, gelatinous fungi..TREMELLACEÆ.

FAMILY I.—**AGARICACEÆ.**

In the Agaricaceæ the hymenium is spread over lamellæ or gills which radiate from a center or stem. The gills are composed of a double membrane, and are simple or branched.

The parts of an Agaric may all be present as in Amanitæ, or severally absent in other genera. When the young fungus is entirely enclosed in a wrapper or case, this case is called the *universal veil.* When this veil is ruptured by the growth of the stem, that part which remains

I

attached to the base is called the *volva*. The membrane reaching from the stem to the margin of the cap is the *partial veil;* when it ruptures by the expansion of the cap and all or a portion adheres to and about the stem it forms the *annulus or ring*. In some species one or both veils may be present, or one or both may be absent.

The stem is *central* when supporting the cap at its center; *excentric* when at one side of the center; *lateral* when it supports the cap from the side. If the stem is absent, the cap is said to be *sessile;* if the cap is horizontal and supported by a broad base it is *dimidiate;* if attached to its place of growth by its back it is *resupinate*.

Genera are largely distinguished by the manner in which the gills are attached to the stem. These distinguishing attachments are shown in the plates illustrating genera and in Plate IV. Gill-shapes.

For convenience Agaricaceæ is divided by the color of the spores into five series: white, pink, brown, purple, black. The last two, owing to the similarity of hue, are by some writers (preferably) included in the black-spored series. Spore color is a valuable assistant in determining species.

Series I. LEUCOSPORÆ. *Gr.*—white; *Gr.*—seed.

Spores white, rarely dingy or inclining to reddish. In the genus Russula the spores of some species are white, in some cream-color, and in several pale ochraceous. Variations from pure white are found in the spores of Tricholoma personatum and a few other species. Gill-color is not a guide to spore-color. Purple, yellow, brown, pinkish gills may produce white spores.

AMANITA.

(A name given to some esculent fungi by Galen, perhaps from Mount Amanus.)

Amanita. Universal veil (volva), which is at first continuous (completely enveloping the young plant), distinct from the skin of the cap. Hymenophore or cap, the part which bears the spore-bearing surface, distinct and easily separable from the stem, which leaves a socket in the flesh when it is removed. *All growing upon the ground. Fries.*

Pileus somewhat fleshy, convex then expanded. **Gills** free. Universal veil at first enclosing the entire plant, which as it grows bursts

2

PLATE I

TABULAR VIEW OF THE GENERA OF AGARICACEAE.

LEUCOSPORAE (WHITE)	RHODOSPORAE (PINK)	OCHROSPORAE (BROWN)	PORPHYROSPORAE (PURPLE)	MELANOSPORAE (BLACK)
AMANITA AMANITOPSIS	VOLVARIA	ACETABULARIA	CHITONIA	
LEPIOTA			AGARICUS (PSALLIOTA)	COPRINUS
	PLUTEUS	BOLBITIUS	PILOSACE	
ARMILLARIA		PHOLIOTA CORTINARIUS	STROPHARIA	GOMPHIDIUS
TRICHOLOMA LACTARIUS RUSSULA	ENTOLOMA	HEBELOMA INOCYBE	HYPHOLOMA	PANÆOLUS ANELLARIA
HYGROPHORUS CLITOCYBE XEROTUS NYCTALIS	CLITOPILUS	FLAMMULA PAXILLUS		
LENZITES LENTINUS PLEUROTUS PANUS TROGIA SCHIZOPHYLLUM	CLAUDOPUS	CREPIDOTUS		
COLLYBIA MARASMIUS HELIOMYCES	LEPTONIA	NAUCORIA	PSILOCYBE	
MYCENA HIATULA	NOLANEA	PLUTEOLUS GALERA	PSATHYRA	PSATHYRELLA
OMPHALIA	ECCILIA	TUBARIA	DECONICA	MONTAGNITES

PLATE II

LEUCOSPORAE.

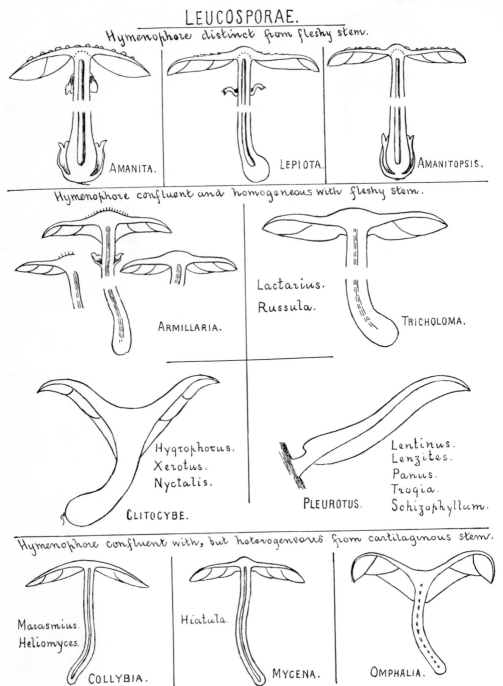

Hymenophore distinct from fleshy stem.

AMANITA.   LEPIOTA.   AMANITOPSIS.

Hymenophore confluent and homogeneous with fleshy stem.

ARMILLARIA.

Lactarius.
Russula.

TRICHOLOMA.

Hygrophorus.
Xerotus.
Nyctalis.

CLITOCYBE.

Lentinus.
Lenzites.
Panus.
Trogia.
Schizophyllum.

PLEUROTUS.

Hymenophore confluent with, but heterogeneous from cartilaginous stem.

Marasmius.
Heliomyces.

COLLYBIA.

Hiatula.

MYCENA.

OMPHALIA.

CHART OF GENERA IN WHITE SPORED SERIES — LEUCOSPORAE

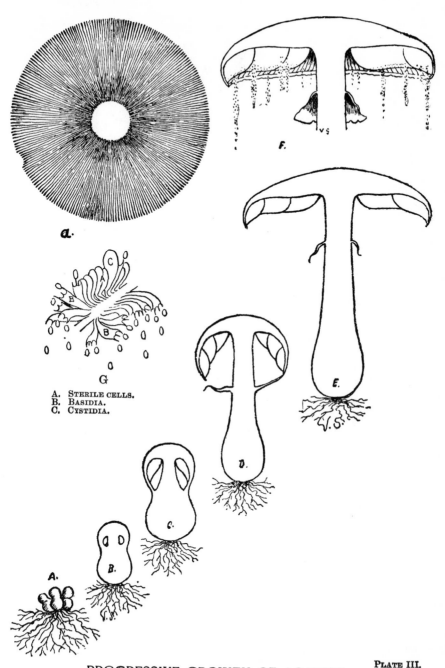

a.

G

A. STERILE CELLS.
B. BASIDIA.
C. CYSTIDIA.

F.

E.

D.

C.

B.

A.

PROGRESSIVE GROWTH OF AGARICS.

Figs.
A. B. C. D. E. STAGES OF DEVELOPMENT OF AN AGARIC.
F. GILLS SHEDDING SPORES.

Figs.
A. SPORE-PRINT.
G. SECTION OF GILL MAGNIFIED.

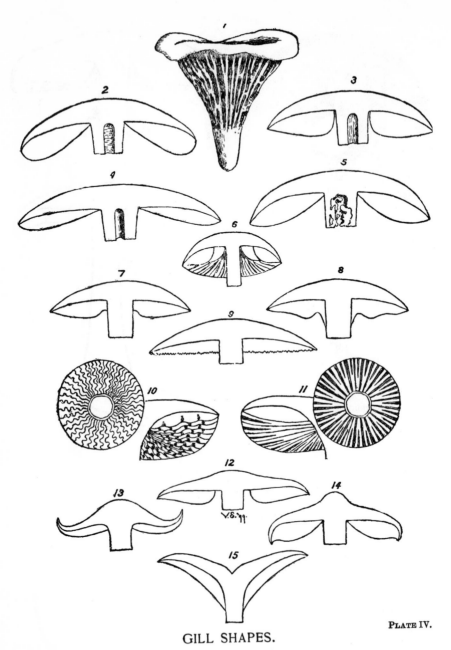

PLATE IV.

GILL SHAPES.

FIG. 1. GILLS AS VEINS; CAP INFUNDIBULIFORM.
2. GILLS ROUNDED IN FRONT (anteriorly.)
3. GILLS ROUNDED BEHIND (posteriorly.)
4. GILLS LANCEOLATE.
5. GILLS VENTRICOSE.
6. GILLS UNEQUAL; CAP CONVEX.
7. GILLS ADNEXED.
8. GILLS EMARGINATE, ALSO ADNATE AND HAVING DECURRENT TOOTH.

FIG. 9. GILLS SERRATE.
10. GILLS FLEXUOSE; WAVED.
11. GILLS DICHOTOMOUS.
12. GILLS FREE; CAP BROADLY UMBONATE.
13. GILLS NARROW; CAP MARGIN REFLEXED.
14. GILLS SLIGHTLY ADNEXED; CAP UMBONATE; MARGIN INVOLUTE.
15. GILLS DECURRENT; CAP UMBILICATE.

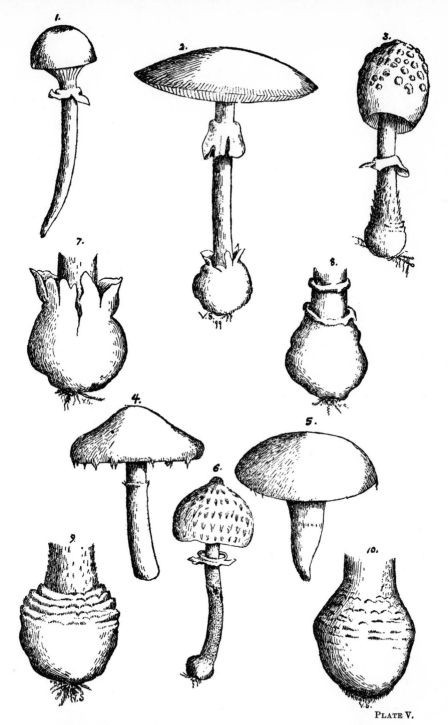

PLATE V.

RING SHAPES AND POSITIONS; VOLVA SHAPES.

FIG. 1. RING SUPERIOR, BROAD.
   2. RING MEDIAL, PENDULOUS.
   3. RING INFERIOR (low down).
   4. RING NARROW, FRAGMENTS APPENDICU-
      LATE.
   5. RING FIBRILLOSE.

FIG. 6. RING PERSISTENT, SOMETIMES MOVABLE.
   7. VOLVA FREE.
   8. VOLVA SEPARATING, CIRCUMSCISSILE.
   9. VOLVA IRREGULARLY, CIRCUMSCISSILE.

   10. VOLVA FRIABLE, DISAPPEARING.

through, generally carrying the upper part on the pileus, where it appears as patches or scales, the remainder enclosing the stem at the
base as a volva, either in a cup-like
form, closely adherent or friable
and evanescent. The partial veil in
youth extends from the stem to the
margin of the pileus, enclosing the
gills; when ruptured it depends from
the stem as a ring. **Stem** furnished
with a ring, and different in substance from that of the pileus.
Spores white.

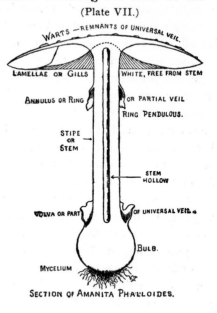

(Plate VII.)

SECTION OF AMANITA PHALLOIDES.

On the ground.

The nearest allied genus, Amanitopsis, is separated by the absence
of a ring, and Lepiota by its lack
of a volva; Volvaria, Acetabularia and Chitonia, possessing volvas,
are distinguished by the color of
their spores.

Amanitæ are the most beautiful and conspicuous of fungi. While
there are comparatively few species of them, the individual members
are plentiful in appearing from spring until the coming of frost. They
are solitary or gregarious in growth. Occasionally two or three are
found together. They frequent woods, groves, copse, margins of woods
and land recently cleared of trees. They are seldom found in open
fields. A careful study of all their botanic points should be the first
duty of the student of fungi. Familiarity with every characteristic of
the Amanitæ will insure against fatal toadstool poisoning, for it is the
well-grounded belief of those who have made thorough investigation
that, with the exception of Helvella esculenta, now Gyromitra esculenta, the Amanitæ, alone, contain deadly poisons.

*No Amanita, or piece of one, should be eaten before its identity is fully
established and its qualities ascertained by referring to the descriptions
herein given or to the opinion of an expert.*

They are the aristocrats of fungi. Their noble bearing, their beauty,
their power for good or evil, and above all their perfect structure, have
placed them first in their realm; and they proudly bear the three badges

Amanita. of their clan and rank—the volva or sheath from which they spring, the kid-like apron encircling their waists, and patch-marks of their high birth upon their caps. In their youth, when in or just appearing above the ground, they are completely invested with a membrane or universal veil, which is distinct and free from the skin of the cap. As the plant grows the membrane stretches and finally bursts. It sometimes ruptures in one place only and remains about the base of the stem as the volva. When such a rupture occurs the caps are smooth. In most species portions of the volva remain upon the cap as scruff or warts—pointed or rough—or as feathery adornment; any or all of which may in part or whole vanish with age or be washed away by rain.

Extending from the stem to the margin of the cap, and covering the gills, is the partial **veil**—a membranaceous, white texture of varying thickness. As the cap expands this veil tears from it. Portions frequently remain pendant from the edges, the rest contracts to the stem as a ring, or droops from it as a surrounding ruffle, or, if of slight consistency, may be fugacious and disappear, but marks, remains, or the veil itself will always be traceable upon the stem.

The Amanitæ are of all colors, from the brilliant orange of the A. Cæsarea, the rich scarlet or crimson of the A. muscaria, to the pure white of the A. phalloides in its white form.

Their stems are usually long, and taper from the base toward the top. In some forms the base is distinctly bulbous. The **volva** at the base is attached to the stem at its lower extremity. It may be visible as a cup or ruptured pouch with spreading mouth, or it may be of such friable texture as to appear like mealy scales. Often, when the plant is pulled from the ground, the volva remains, but the marks of its attachment will appear and should be carefully looked for. Their **gills** are commonly white, are of equal length and radiate from near the stem, which they do not reach, to the circumference of the cap. They are white, unless tinged with age, excepting upon A. Cæsarea and A. Frostiana where they are yellow.* Their caps are umbrella-shaped, flat or convex. Their flesh is white, does not change color when bruised. They are scentless and almost tasteless when fresh, when old they have a slightly offensive odor and taste.

The family is not a large one, not over thirty members complete its circle. Every feature, every part of its several members, should be thoroughly known before the intimacy of eating. While at least nine

---

* A. Frostiana is not always yellow gilled.

of the family are not only edible but delicate and sapid, far better will it be to leave all alone than to make a mistake. A piece of a poisonous variety the size of a dime will often cause serious disorders if eaten. Many persons have died from eating very small quantities.

Because of its ovate or button-like form when young, it is frequently mistaken for the common field mushroom; even experienced mycophagists have been deceived by it. No other poison has so puzzled scientists. Other varieties of fungi may interfere with digestion, but to the Amanitæ all deaths from toadstool-eating are traceable. Its subtle alkaloid is absorbed by the system, and in most cases lies unsuspected for from six to twelve hours, then its iron grip holds to the death. For centuries it has defied all remedies. The problem has been partially solved. At Shenandoah, Pa., August 31, 1885, a family of five were poisoned by toadstools; two died, three lived. Noting the sad account in the newspapers, I at once wrote to Shenandoah for specimens of the fungi eaten and a description of the treatment. I promptly received from Dr. J. E. Schadle (now Professor Schadle), the physician in charge of the cases, a box containing two harmless varieties and several fine specimens of the Amanita phalloides, all of which were gathered on the same spot and by the same person who gathered the toadstools doing the poisoning. They told the tale. A remarkably full and interesting account of the cases was sent to me by Dr. Schadle. After exhausting all other remedies, and after two of the five had died, he administered subcutaneously, by hypodermic injection, sulphate of atropine—a product of the deadly nightshade analagous to belladonna — $\frac{1}{180}$ to $\frac{1}{60}$ of a grain at a dose. It proved to be an antidote and saved the lives of the remaining three.

The action of atropine in arresting the deadly work of poisoning by amanitine had been foreshadowed by Schmidberg and Koppe, and dwelt upon in numerous published articles by Mr. Julius A. Palmer, to whom more than any other is due the branding of the murderous members of the Amanita family; but for the first time atropine was used upon the human system to ward their blows.

All of the species herein described are found in the United States. Of the twenty-seven, nine are edible, nine are either known to be deadly or are so closely allied to deadly species that it is unsafe to class them as other than poisonous until absolute proof is obtained of their harm-

Amanita. lessness.   The remaining nine I have not seen, neither is there any record of their qualities.

ANALYSIS OF SPECIES.

* Volva opening at the top or splitting all around, leaving a manifest, free border at the base of the stem.   Pileus naked or with broad membranaceous patches.

** Volva splitting regularly all round the lower portion, persistent, more or less closely embracing the base of the bulbous stem.   The upper portion being adnate to the pileus appears on it by expansion as scattered, thick warts.

*** Volva friable, entirely broken up into wart-like scales, therefore not persistent at the base of the stem, which is at first globose-bulbous, becoming less so as it lengthens.   Pileus bearing mealy patches, soon disappearing or with small, hard, pointed warts.

**** Volva rudimentary, flocculose, wholly disappearing.

* Volva bursting at top, etc.

A. viro'sa Fr.—*virus*, poison.

HINING white.   **Pileus** 3–4 in. broad, fleshy, *at first conical and acute*, afterwards bell-shaped, then expanded, naked, viscous in wet weather, shining when dry, *margin* always even, but most frequently *unequal*, turned backward and inflexed. **Flesh** white, unchangeable.   **Stem** 4–6 in. long, *wholly stuffed*, almost solid, split up into longitudinal fibrils, cylindrical from the bulbous base, often compressed at the apex, *torn into scales* on the surface, springing from a *lax, wide, thick volva*, which bursts open at the apex.   **Ring** close to the top, lax, silky, splitting up into floccose fragments.   **Gills** free, thin, narrow, narrowing at both ends, but a little broader in front, not decurrent on the stem (although the apex of the stem is often striate), crowded, somewhat floccose at the edge. *Fries.*

The pilei are most frequently oblique, extended and lobed on one side as in Hygrophorous conicus, scarcely ever depressed.   The pileus rarely becomes yellow.   The fragments of the veil often adhere to the edge of the gills.

6

In woods. Uncommon. August to October.
Fetid, poisonous. *Stevenson*.

Spores spheroid or subspheroid, 10–16µ *K.;* 8µ *W. P.;* sub-globose, 8–10µ *Massee*.

POISONOUS.

I think it a variety of A. phalloides.

A. phalloi´des Fr. *Gr.*—phallus-like. ([Color] Plate VI, figs. 2, 3.)
Pileus 3–4 in. broad, commonly shining white or lemon-yellow, fleshy, oval bell-shaped, then expanded, *obtuse*, covered over with a pellicle which is *viscid* (not glutinous) in wet weather, naked, rarely sprinkled with one or two fragments of the volva, *the regular margin even*. **Stem** 3–5 in. long, ½ in. and more thick, solid downward, bulbous, hollow and attenuated upward, *rather smooth*, white. **Ring** *superior*, reflexed, slightly striate, swollen, commonly entire, white. **Volva** more or less buried in the soil, bulbous, *semifree*, bursting open in a torn manner at the apex, with a lax border. **Gills** free, ventricose, 4 lines broad, shining white. *Fries*.

Pileus very variable in color, commonly white or yellow (A. citrina Pers.), becoming green (A. viridis Pers.), olivaceous and occasionally variegated with tiger spots; in late autumn with the disk almost black but whitish round the margin. Odor somewhat fetid, but little remarkable as compared with that of A. virosa.

In woods. Frequent. August to November.

A very POISONOUS and dangerous species. *Stevenson*.

Spores 8–9µ *W.G.S.;* 8–10µ *B.;* 7–9µ diam. *Massee;* globose, 7.6x6µ *Peck*.

Pileus at first ovate or subcampanulate, then expanded, slightly viscid when young and moist, smooth or rarely adorned by a few fragments of the volva, *even on the margin*, white, yellowish-brown or blackish-brown. **Lamellæ** rather broad, rounded behind, free, white. **Stem** equal or slightly tapering upward, stuffed or hollow, smooth or slightly floccose, ringed, *bulbous*, the ruptured volva either appressed loose or merely forming a narrow margin to the bulb.

Plant 4–8 in. high. **Pileus** 2–5 in. broad. **Stem** 3–6 lines thick.

This species is common and variable. It occurs everywhere in woods and assumes such different colors that the inexperienced mycologist is apt to mistake its different forms for distinct species. With us the pre-

7

Agaricaceæ

Amanita. vailing colors of the pileus are white, yellowish-white, grayish-brown and blackish-brown. It is remarkable that the form with a greenish pileus, which seems to be common enough in Europe, does not occur here. Fries also mentions a form having a white pileus with a black disk. A somewhat similar form occurs here, in which the pileus is grayish-brown with a black disk. Some of the variously colored forms were formerly taken to be distinct species, in consequence of which several synonyms have arisen, of which A. virescens Fl. Dan., Amanita viridis Pers., and Amanita citrina Pers., are examples. A. verna Bull. is a variety having a white pileus, a rather thick annulus and an appressed volva. It sometimes occurs early in the season; hence the specific name. It also occurs late in the season and runs into the typical form so that it is not easy to keep it distinct. The flesh and the lamellæ are white, the stem is white, pallid or brownish, and the annulus is either white or brownish. The bulb is generally very broad and abrupt or depressed, though it sometimes is small and approaches an ovate form. The large bulbs are sometimes split externally in two or three places and are, therefore, two- or three-lobed. In such cases the volva is less persistent than usual and its free portion then furnishes merely an acute edge or narrow margin to the bulb. Specimens sometimes occur in which the margin of the pileus is narrowly adorned with a slight woolly hairiness, but usually it is perfectly smooth and even. By this character, taken in connection with the membranous volva and bulbous base of the stem, the species is readily distinguished. Sometimes a strong odor is emitted by it, but usually the odor is slight. Authors generally pronounce this a poisonous and very dangerous species. Its appearance is attractive, but its use as food is to be avoided. *Peck*, 33d Rep. N. Y. State Bot.

Common in woods and recently cleared woodlands. Frequent over the United States. June to frost.

An exceedingly *poisonous*, *dangerous*, seductive species, responsible for most of the deaths from toadstool eating; because in its white form it is mistaken for the common mushroom—Agaricus campester. The real fault is with the collector, who should never eat any fungus found in the woods, believing it to be the mushroom. The mushroom does not grow in the woods. Neither has it *white gills*, nor *white spores*, nor a *volva* at the base of the stem as have Amanitæ.

The caps of A. phalloides vary in color—white, oyster-color, smoky

8

brown. The color of the commonest form is from white to a light hue Amanita.
of greenish yellow. The center of the cap, whatever may be the prevailing color, is usually several shades darker. In shape, the cap changes from a knob in youth, through the shapes of expansion, until it becomes fully spread, when it is umbrella-shaped, or almost flat. Some forms have a slightly raised portion or umbo in the center of the cap. The gills are white, of good width, rounded next to the stem and free from it.

The stem conforms in color to the cap, but in lighter shades. White-capped varieties have white stems. The stem has a sudden broad, distinct bulb at the base. On the upper side of the bulb there is usually a margin or rim. The stem tapers more or less toward the cap, from which it is easily separable. The cup, wrapper or volva is torn or split or irregular at the upper part, and is not pressed to the stem as in some forms.

Professor Peck, in his 48th Report, gives the following excellent synopsis of differences between the poisonous Amanita and edible fungi, for which it could only by great stupidity be mistaken:

Poison amanita. **Gills** persistently white. **Stem** equal to or longer than the diameter of the cap, with a broad, distinct bulb at the base.

Common mushroom. **Gills** pink, becoming blackish-brown. **Stem** shorter than the diameter of the cap, with no bulb at the base.

From all forms of the edible Sheathed amanitopsis the Poison amanita differs in its distinctly bulbous stem, in having a collar on the stem and in the absence of striations on the margin of the cap.

From the edible Reddish amanita, it is easily separated by the entire absence of any reddish hues or stains and of warts upon its cap.

From the Smooth lepiota its distinct, abrupt and marginal bulb at once distinguishes it.

A. ver′na Bull.—*vernus*, of spring. A variety of A. phalloides. POISONOUS. White. **Pileus** ovate then expanded, somewhat depressed, viscid, margin orbicular, even. **Stem** stuffed then hollow, equal, floccose, closely sheathed with the free border of the volva. **Ring** reflexed, swollen. **Gills** free. **Pileus** glabrous, even on the margin, white, viscid when moist. **Gills** white. **Stem** ringed, white, floccose, stuffed or hollow, closely sheathed at the base by the remains of the membranous volva, bulbous. **Spores** globose, 8μ broad.

Agaricaceæ

Amanita.

In woods. Spring and summer.

The Vernal Amanita scarcely differs from white forms of the A. phalloides except in the more persistent and more closely sheathing remains of the wrapper at the base of the stem. It is probably only a variety of that species, as most mycologists now regard it, and it should be considered quite as dangerous. I have not found it earlier than in July, although in Europe it is said to appear in spring, as its name implies. *Peck*, 48th Rep. N. Y. State Bot.

Common over the United States. West Virginia, New Jersey, Pennsylvania, May to November. It appeared at Mt. Gretna, Pa., on May 28, 1899. *McIlvaine.*

The absence of a ring separates white forms of A. volvata and A. vaginata.

The virulence of its poison is the same as that of A. phalloides.

A. magnivela′ris Pk.—*magnus*, large; *velum*, veil. **Pileus** convex or nearly plane, glabrous, slightly viscid when moist, even on the margin, white or yellowish-white. **Gills** close, free, white. **Stem** long, nearly equal, glabrous, white, furnished with a large membranous white annulus, sheathed at the base by the appressed remains of the membranous volva, the bulbous base tapering downward and radicating. **Spores** broadly elliptical, 10x6–8μ.

Pileus 3–5 in. broad. **Stem** 5–7 in. long, 4–6 lines thick.

Solitary in woods. Port Jefferson, Suffolk county. July.

The species resembles Amanita verna, from which it is separated by its large persistent annulus, the elongated downwardly tapering bulb of its stem, and especially by its elliptical spores. *Peck*, 50th Rep. N. Y. State Bot.

I have not seen this species. Its resemblance to A. verna is enough to place the ban upon it until it has been tested.

A. map′pa Fr.—*mappa*, a napkin. From the volva. **Pileus** 2–3 in. broad, commonly white or becoming yellow, slightly fleshy, convexo-plane, obtuse or depressed, orbicular, *dry*, margin for the most part even. **Stem** 2–3 in. long, 3–5 lines thick, stuffed then hollow, almost equal above the bulb, rather smooth, white. **Ring** superior, soft, lax, here and there torn. **Volva** regularly *circularly split*, somewhat obliterated; the globoso-bulbous base united with the stem, with an acute

and distant margin; the portion covering the pileus divided into broad,
irregular, somewhat separating scales. **Gills** annexed, crowded, narrow, shining, white. *Fries.*

Odor stinking. The color is that of A. phalloides, with which A. virosa exactly agrees, more rarely straw color, lemon-yellow, becoming green.

In mixed woods. Frequent. *Stevenson.*

Spores spheroid, 7–10μ *K.;* 8–9x6–8μ *B.;* subglobose, 7–9μ diameter *Massee.*

New York woods and fields, common, September to October, *Peck,* 22d Rep.; North Carolina, *Curtis;* New England, *Frost;* Minnesota, *Johnson;* Ohio, *Morgan;* District Columbia, *Miss Taylor.*
POISONOUS.
Probably but a variety of A. phalloides.

A. spre'ta Pk.—*spreta*, hated. ([Color] Plate VI, fig. 1.) **Pileus** subovate, then convex or expanded, smooth or adorned with a few fragments of the volva, substriate on the margin, whitish or pale-brown. **Gills** close, reaching the stem, white. **Stem** equal, smooth, annulate, stuffed or hollow, whitish, finely striate at the top from the decurrent lines of the lamellæ, not bulbous at the base, but the volva rather large, loose, subochreate. **Spores** elliptical, generally with a single large nucleus, 10–13x6–8μ.

Plant 4–6 in. high. **Pileus** 3–5 in. broad. **Stem** 4–6 lines thick.
Ground in open places. Sandlake and Gansevoort. August. *Peck,* 32d Rep. N. Y. State Bot.

This is a dangerous species, because containing a deadly poison and resembling the most common forms of Amanitopsis, therefore likely to be mistaken for them. Specimens sent by me to Professor Peck were identified as his species. I add my own description.

Pileus oval, broadly umbonate, date-brown toward and on umbo, soft, dry, smooth, more or less sulcate on edge. **Flesh** white, thin, except at center. **Stem** tapers rapidly above ring and at base, white reddish-brown toward middle, narrows toward volva from which it is almost free at the base, hollow, furfuraceous above ring. **Gills** white, crowded, free. **Ring** white, thin, persistent, but at times hard to distinguish because clinging to stem. **Volva** free, fitting close, upper

Amanita. margin thin, lower part quite thick, making stem appear bulbous, which it is not.   White forms occur.

Not as virulent as A. phalloides, but like it in its POISONOUS effects.   It differs from Amanitopsis in having a ring.

Grows in woods and on wood-margins.

Angora woods, West Philadelphia.   On ground in mixed woods, open and grassy places in wood and wood-margins.   August to September.   *McIlvaine.*

A. recuti'ta Fr.—having a fresh or new skin.   **Pileus** convex then plane, *dry*, smooth, frequently bearing fragments of the volva, margin nearly even.   **Stem** stuffed then hollow, attenuated, *silky*, volva circumscissile, becoming obliterated, margin closely pressed to stem; ring distant, white.   **Gills** striate-decurrent.

In pine woods.   Common.

No report upon quality.

A. Cæsa'rea Scop.—king-like.   (Called by the Greeks *Cibus Deorum*, food of the gods.)   CAUTION.   **Pileus** 3–8 in. across, hemispherical, then expanded, free from warts, distinctly striate on the margin, red or orange becoming yellow.   **Gills** free, yellow.   **Stem** 4–6 in. long, up to ¾ in. thick at base, slightly tapering upward, yellowish, flocculose, stuffed with white fibrils or hollow, with a conspicuous yellowish ring or veil.   **Volva** white, large, distinct and membranous.   **Spores** elliptical, 8–10µ *Peck.*

Open woods, under pines on lawns.   July to October.

Reported from North Carolina, South Carolina, Massachusetts, Maryland, New Jersey, Ohio, Alabama, Louisiana, Pennsylvania, New York. *Peck*, Rep. 23, 32, 33, 48.

This emperor of fungi is the most showy of its race.   It grows to 10 in. in height.   The cap reaches 8 in. in diameter and the stem over 1½ in. in thickness.   In very much smaller specimens about the same proportions occur.   The cap is at first ovate, then hemispherical, then expanded.   It has no warts or scales upon it.   The margin is distinctly striate.   The flesh is white, yellow or reddish under the skin; next to the gills it is usually yellow.

The stem tapers upward from the socket at its base.   It is yellowish and covered with loose fibrils of darker hue.   The ring is white, but frequently tinged with yellow. In taste and smell it is mild. Open Amanita woods is its favorite habitat, yet it is found growing luxuriantly under pines, maples, elms, on lawns. It is not often found, but when it is, it is solitary, or in groups or rings. In the latitude of Philadelphia it is found from July until October 1st. Further south its stay conforms to temperature, and it is more frequent. There is no doubt of its rare edibility abroad, and of its being eaten in America.

A specimen believed to be it should never be eaten until carefully distinguished from A. muscaria and A. Frostiana, which have warts or scales on the cap (which sometimes are not discernible after rain), white gills, and a volva which soon breaks up into fragments or scabs.

Appearing like a small form of A. muscaria, to which it was formerly referred, is A. Frostiana Pk. (Frost's Agaric). It closely resembles small A. Cæsarea, especially in the yellow tinge of stem, ring and gills. The volva and ring (persistent in A. Cæsarea) soon disappear, but are traceable by fluffy fragments, or yellow stains. It is extremely poisonous.

The differences, concisely, are these: A. Cæsarea (Orange Amanita). **Cap** smooth, though occasionally with a few fragments of the volva as patches upon it. **Gills** yellow. **Stem** yellow. **Volva** usually persistent, sometimes breaking up into soft, fluffy masses.

A. muscaria (Fly Amanita). Poisonous. **Cap** covered with remains of the volva as scales or wart-like patches. **Gills** white. **Stem** white or light-yellow. **Volva** not persistent, breaking up into fluffy fragments or scales.

A. Frostiana. Poisonous. Smaller and more delicate than the two preceding. **Cap** smooth or with yellow scales or wart-like patches. **Gills** yellow or tinged on edge with yellow. **Stem** white or yellow, the ring evanescent, but always leaving a yellow mark on stem. **Volva** yellow, breaking up into yellow fluffy fragments.

Far better for the amateur to let the A. Cæsarea, and anything resembling it, respectfully alone.

New York, *Gansvoort.* Circle forty feet in diameter. *Peck*, 32d Rep.; Maryland. There is not a doubt that this fungus can be eaten with impunity, *Banning;* Alabama, abundant. Edible. Alabama Bull. No. 80.

Roques and Cordier, French writers, regard it as the finest and most delicate of fungi, the perfume and taste being exquisite.

Agaricaceæ

Amanita. The writer has not had opportunity to eat A. Cæsarea. If such should occur he would go about it very cautiously. No suspicion attaches to it abroad, but evidence is accumulating in the hands of the writer (not yet convincing) that either locality may render it poisonous or that A. muscaria varies so much in appearance as to deceive even the expert into mistaking it for A. Cæsarea. It is possible that A. muscaria is, at times, in certain localities, harmless; but no such exception as this is noted in the entire fungoid realm. It is not so common that collectors should mourn its waste. It is better, far, to let it alone.

** Volva splitting regularly all around; pileus bearing thick warts, etc.

A. musca′ria Linn.—*musca*, a fly. ([Color] Plate VI, fig. 4. Plate IX.) POISONOUS. **Pileus** 4 in. and more broad, normally at first blood-red, soon orange and becoming pale, whitening when old, globose, then convex and at length flattened, covered with a *pellicle* which is *at first thick*, and in wet weather *glutinous*, but which gradually disappears, and sprinkled with thick, angular, separating fragments of the volva; *margin* when full-grown *slightly striate*. **Flesh** not compact, white, *yellow under the pellicle*. **Stem** as much as a span long, shining white, firm, torn into scales, at first stuffed with lax, spider-web fibrils, soon *hollow*; the *adnate base of the volva* forms an ovate bulb, which is *marginate with concentric scales*. **Ring** very soft, torn, even, inserted at the apex of the stem, which is often dilated. **Gills** free, but reaching the stem, decurrent in the form of lines, crowded, broader in front, white, rarely becoming yellow.

Var. *rega′lis*, twice as large. **Stem** stuffed, *solid when young*, as much as 1–2 in. thick, becoming light-yellow within; the volva terminates in 8–10 concentric squamoso-reflexed rows of scales. **Pileus** very glutinous, bay-brown or the color of cooked liver. **Gills** yellowish.

Var. *formo′sa*, soft, fragile. **Pileus** at first *lemon-yellow*, with mealy, lax, yellowish, easily-separating warts, often naked. **Gills** often becoming yellow. A. formosa, with the warts rubbed off.

Var. *umbri′na*, thinner and *more slender*. **Stem** hollow, often twisted, bulb narrowed. **Pileus** at first *umber*, then livid, with the exception of the disk, which is dingy-brown. **Gills** at length remote. *Stev.*

Pileus at first ovate or hemispherical, then broadly convex or nearly plane, slightly viscid when young and moist, *rough with numerous*

14

PLATE IX.

AMANITA MUSCARIA.

Photographed by Dr. J. R. Weist.

PLATE XIII.

LEPIOTA PROCERA.

whitish or yellowish warts, rarely smooth, narrowly and *slightly striate*
on the margin, white, yellow or orange-red. **Gills** white. **Stem** equal
or slightly tapering upward, stuffed with webby fibrils or hollow, bearing a white ring above, *ovate-bulbous* at the base, white or yellowish;
the volva usually breaking up into scales and adhering to the upper part
of the bulb and the base of the stem. **Spores** *elliptical*, 8–10x6–8μ.

Plant 5–8 in. high. **Pileus** 3–6 in. broad. *Peck,* 33d Rep. N. Y.
State Bot.

A white variety, with the pileus thickly studded with sharp warts,
occurs in Albany Rural Cemetery. July. *Peck,* 24th Rep.

Var. *al'ba* Pk. It also occurs on Long Island in two forms, the
normal one and a smaller one, in which the warts of the pileus are
evanescent or wanting. Not unfrequently it makes a close approach to
white forms of A. pantherina, in having the upper part of the bulb uniformly margined by the remains of the definitely circumscissile volva,
but this margin is more acute than in that species. *Peck,* 46th Rep.
N. Y. State Bot.

Spores spheroid-ellipsoid, 10–12x8–9μ *K.;* 6x9μ *W. G. S.;* elliptical, 8–10x6–8μ *Peck.*

"At Cincinnati, yellow A. muscaria are all we find." *Lloyd.*

Reported from most of the states. At Mt. Gretna I found it in great
quantity, and frequently three or four tightly crowded together. Many
pounds of it were sent to Professor Chittenden, Sheffield Laboratory,
Yale University. Near Haddonfield, N. J., large patches annually
grow under pines, gorgeous in their rich orange-red caps, usually scaly,
with at times lemon-yellow in the same clusters, smooth as A. Cæsarea.
It grows from July until after hard frosts.

It is undoubtedly *poisonous* to a high degree. Its juices in minute
quantity, carefully and scientifically injected into the circulation of ether-
ized cats, kill in less than a minute. A raw piece of the cap, the size
of a hazel nut, affects me sensibly if taken on an empty stomach. Dizziness, nausea, exaggeration of vision and pallor result from it. The
pulse quickens and is full, and a dreaded pressure affects the breathing.
I have not noticed change in the pupil of the eye. Nicotine from
smoking a pipe with me abates the symptoms, which entirely disappear in two hours, leaving as reminiscence a torturing, dull, skull-
pervading headache. If, as is asserted on good authority, the Siberians
use it as an intoxicant, they certainly suffer the accustomed penalty.

Amanita. It is possible that persons may, in a degree, become immune to its poison, as they do to arsenic, strychnia, opium, nicotine, or it may be that a portion of the poison is extracted by boiling. It is, however, extremely dangerous to rely upon extracting by any means the poison of the Amanita, and to eat the residue. Acetic acid or vinegar does *not* destroy the poison; it dissolves it to an extent and extracts it, and becomes as poisonous as the plant itself. There is no means of telling how much of the poison remains in the plant after such treatment. The safe plan is to eat, only, of toadstools which do not contain any poison to extract.

One redeeming virtue, alone, rests with A. muscaria—it kills flies.

A. Frost'iana Pk.—in honor of Charles C. Frost. POISONOUS. ([Color] Plate VI, fig. 5.)**Pileus** convex or expanded, bright-orange or yellow, warty, sometimes nearly or quite smooth, striate on the margin. **Gills** free, white or slightly tinged with yellow. **Stem** white or yellow, stuffed, bearing a slight, sometimes evanescent ring, bulbous at the base, the bulb *slightly margined* by the volva. **Spores** *globose*, 8–10μ in diameter.

Plant 2–3 in. high. **Pileus** 1–2 in. broad. **Stem** about 2 lines thick. June to October.

This appears like a very small form of the Fly Agaric, to which, as var. minor, it was formerly referred. The only decided characters for distinguishing it are its small size and globose spores. Our plant sometimes grows in company with A. muscaria, but it seems to prefer more dense woods, especially mixed or hemlock woods. It is generally very regular and beautiful and has the stem quite often of a yellow color, and the bulb margined above with a collar-like ring. *Peck*, 33d Rep. N. Y. State Bot.

West Virginia, New Jersey, North Carolina, *McIlvaine*.

A. Frostiana is found well over the land. It is frequent in shady woods and seems to favor ground under the prevailing tree—oak, chestnut, pine, hemlock, whichever it may be. From the many hundreds I have seen, I think it more likely to be mistaken by the novice for A. Cæsarea than A. muscaria, because of its often yellow gills and stem. It is much smaller and thinner than either. In the states I have found it, it is darker than described, being a rich reddish-orange or scarlet. The partial veil or ring is very evanescent but often found upon the stem as a yellow, floccose remnant. The stain of the ring is always Amanita. noticeable. The volva is seldom found entire. It, too, is evanescent, but, like the veil, is found yellow and fluffy, adhering to the fingers when touched.

It is probable that its highly colored cap has caused it to be gathered by the careless collector of bright-capped Russulæ, and that thus R. emetica got its bad name. Examine carefully any toadstool resembling it. The Russulæ have neither ring nor volva.

A. excel'sa Fr.—*excelsus*, tall. POISONOUS. **Pileus** 4–5 in. broad, *brownish-gray*, darker in the center, *fleshy*, soft, globose, then plane, *pellicle thin*, but viscous, and in reality separable in wet weather, then the surface is often *wrinkled-papillose*, or in a peculiar manner hollowed and pitted, sprinkled with angular, unequal, whitish-gray, easily separating warts, the remains of the friable volva; margin at first even, but when properly developed manifestly striate, even furrowed. **Flesh** soft, white throughout, unchangeable. **Stem** 4–6 in. long, 1 in. thick, at first stuffed, almost solid, but at length hollow, globose-depressed at the base, attenuated upward from the bulb, covered, sometimes as far as the ring, sometimes only on the lower part with *dense, squarrose, concentric scales* (from the epidermis of the stem being torn), striate at the apex. **Ring** superior, large, separating-free or at length torn. **Gills** quite *free, rounded* (not decurrent on the stem in the form of lines), very ventricose, ½ in. and more broad, shining white.

The *bulb when young* is *somewhat marginate*, but by no means separable, the margin proper, like that of A. muscaria, is marked with scales, buried in the soil, somewhat rooting, beneath the margin marked here and there with a concentric furrow. The shorter gills intermixed are more numerous than is usual among Amanitæ. There is a smaller variety, with the margin more frequently striate and the stem stuffed, then hollow. *Fries.*

Solitary, in woods, chiefly under beech. *Stevenson.*

Spores 6x9μ *W. G. S.;* 8–9x5–6μ *Massee.*

North Carolina, *Schweinitz, Curtis;* South Carolina, *Ravenel;* California, *Harkness and Moore;* Massachusetts, *Frost, Andrews;* Minnesota, *Johnson;* Rhode Island, *Olney.*

A. pantheri'na De C.—spotted like a panther. Doubtful. **Pileus**

17

Amanita. commonly olivaceous-umber when young, fleshy, convex then flattened or somewhat depressed, with a *sticky pellicle*, which is at first thick and olivaceous dingy-brown, then thinned out, almost disappearing and livid, the disk only becoming brownish; *margin evidently striate;* the fragments of the volva divided into small, equal, white, regularly arranged, moderately persistent warts. **Flesh** *wholly white*, never yellow beneath the pellicle. **Stem** 3–4 in. long, ½ in. thick, at first stuffed then hollow with spider-web fibrils within, equal or attenuated upward, slightly firm and sometimes scaly downward, *greaved* at the base by the separable *volva which has an entire and obtuse margin.* **Ring** more or less distant, adhering obliquely, white, rarely superior. **Gills** free, reaching the stem, broader in front, 3–4 lines broad, shining white.

It is readily distinguished from A. muscaria, var. umbrina, by the white flesh never becoming yellow beneath the pellicle. Variable in size and color, which, however, is never red or yellow, and in the position of the ring.

In woods and pastures. *Stevenson.*

Spores 7–8x4–5µ *K.;* 6–10µ *B.;* 8x4µ *W. G. S.;* 7.6x4.8µ *Morgan.* Not poisonous, *W. G. S.;* not edible, *Roze;* poisonous, *Leuba.*

North Carolina, Pennsylvania, Ohio, California, Wisconsin, Minnesota, Iowa, New York. *Peck.*

A. Ravenel'ii B. and C.—in honor of Henry W. Ravenel. **Pileus** 4 in. across, convex, broken up into distinct areas, each of which is raised into an acute, rigid, pyramidal wart. **Stem** 3 in. high, bulbous. **Volva** thick, warty, somewhat lobed. **Ring** deflexed.

South Carolina, June, *H. W. Ravenel;* a very fine species allied to A. strobiliformis, Vitt. Ann. and Mag. Nat. Hist., 1859; Alabama, *Atkinson* (Ll. Volvæ).

Properties not stated.

A. russuloi'des Pk.—resembling a Russula. **Pileus** at first ovate, then expanded or convex, rough with a few superficial warts, or entirely smooth, viscid when moist, widely striate-tuberculate on the margin, pale-yellow or straw color. **Gills** close, free, narrowed toward the stem, white. **Stem** firm, smooth, stuffed, annulate, equal or slightly tapering upward, bulbous; annulus thin, soon vanishing. **Volva** fragile, subappressed. **Spores** broadly elliptical, 10x8µ.

18

Plant 2–3 in. high. **Pileus** 1.5–2 in. broad. **Stem** 3–5 lines thick. Amanita.
Grassy ground in open woods. Greenbush. June.

This species is remarkable for the thin striate-tuberculate margin of the pileus, which causes it to resemble some species of Russula. *Peck*, 25th Rep. N. Y. State Bot.

Qualities not stated.

Massachusetts, *Francis*.

A. strobilifor′mis Vitt.—*strobilis*, a pine-cone, from the shape of the warts. ([Color] Plate VIII, fig. 3.) **Cap** 3–10 in. across, convex or nearly plane, white or cinereous, sometimes yellow on the disk, rough with angular, mostly persistent warts which sometimes fall away and leave the pileus nearly smooth; generally whitish, sometimes tinged with brown; the margin even and extending a little beyond the lamellæ. **Gills** free, rounded behind. **Veil** large and portions sometimes adhere to margin of cap. **Stem** 3–8 in. long, up to 1¼ in. thick, equal or slightly tapering upward, solid, floccose-scaly, white, bulbous, the bulb very large, sometimes weighing a pound, margined above and furnished with one or two concentric furrows, somewhat pointed below, firmly and deeply imbedded in the earth, floccose-mealy when young.

Spores elliptical, 13–15x8–10μ *Peck*.

Open woods and borders. June to October.

Edible. *W. G. Smith, Curtis, Peck*.

This is among the best of species. Its size, solidity, flavor are marked. I have found specimens weighing a pound and a half. It grows singly, but when one is found several are apt to be neighbors. When young, the cap is but a small knob upon a beet or top-shaped base, which is largely under ground. It cuts like a soft turnip, and has a strong, pungent, unmistakable odor, like chloride of lime, which entirely disappears in cooking. As the plant develops the bulb decreases in size. On all the many specimens the author has seen and eaten, the scabs are light brown and reddish-brown.

A. solita′ria Bull.—growing alone. **Pileus** convex or plane, warty, white or whitish, even on the margin. **Gills** reaching the stem, white or slightly tinged with cream color. **Stem** at first mealy or scaly, equal, solid, white, bulbous, the bulb scaly or mealy, narrowed below into a root-like prolongation. **Ring** lacerated, often adhering in fragments to the margin of the pileus and gills. **Spores** elliptical-oblong, 8–13x6.5μ.

19

Plant 4–8 in. high. **Pileus** 3–6 in. broad. **Stem** 4–6 lines thick. *Peck*, 33d Rep. N. Y. State Bot.

Solitary in woods and open places. July to October.

Georgia, *H. N. Starnes;* Indiana, *H. I. Miller;* West Virginia, New Jersey, Pennsylvania, *McIlvaine.*

Edible. *Curtis, H. N. Starnes*, Philadelphia Myc. Club.

In many localities I find it quite plentiful, and it is so reported from Georgia. Southern and middle New Jersey woods abound with it, and at Mt. Gretna, Pa., it is always present in its growing months.

The cap is sometimes tinged with brown as are the angular, erect warts which are generally numerous, but often falling off or few and scattered. The flesh is white and smells like chloride of lime, but not nearly so strong as A. strobiliformis. The volva is broken up into floccose scales which cling to bulb and lower part of stem. These scales may be white and mealy or brownish. The entire fungus has a fluffy exterior, which is easily removed by rubbing. The annulus is torn, a part often adhering to the margin of the pileus and the gills. This and the long, tapering, rooting bulb are marked characteristics. The bulb is brittle. It is difficult to get the fungus from the ground entire.

Stem and cap are juicy, tender, mild in flavor, wholesome. It is not equal in flavor to A. rubescens, but is more delicate.

By many its properties have been stated as poisonous, doubtful. Quantities of it have been eaten by myself and friends. Hypodermic injection of its juices into the blood circulation of live animals prove it perfectly harmless.

A. can'dida Pk.—shining white. **Pileus** thin, broadly convex or nearly plane, verrucose with numerous small, erect, angular or pyramidal, easily separable warts, often becoming smooth with age, white, even on the margin. **Flesh** white. **Gills** rather narrow, close, reaching to the stem, white. **Stem** solid, bulbous, floccose-squamose, white, the annulus attached to the top of the stem, becoming pendent and often disappearing with age, floccose-squamose on the lower surface, striate on the upper, the bulb rather large, ovate, squamose, not margined, tapering above into the stem and rounded or merely abruptly pointed below. **Spores** elliptical, 10–13x8μ.

Pileus 3–6 in. broad. **Stem** 2.5–5 in. long, 5–8 lines thick, the bulb 1–1.5 in. thick in the dried specimens.

COLOR PLATES

Grouped by F. D. Briscoe — Studies by C. McIlvaine.

PLATE VI.

FIG.   PAGE.   FIG.   PAGE.   FIG.   PAGE.

1. AMANITA SPRETA.   11   3. AMANITA PHALLOIDES (BROWN VAR.),   7   5. AMANITA FROSTIANA.   16
2. AMANITA PHALLOIDES (WHITE VAR.),   7   4. AMANITA MUSCARIA.   14   6. GYROMITRA ESCULENTA,   546

Grouped by F. D. Briscoe — Studies by C. McIlvaine.   PLATE VIII.

Grouped by F. D. Briscoe — Studies by C. McIlvaine.

PLATE X.

Grouped by F. D. Briscoe — Studies by C. McIlvaine.

PLATE XII.

FIG.		PAGE.
1.	LEPIOTA AMERICANA,	48
2.	LEPIOTA NAUCINOIDES,	45

FIG.		PAGE.
3.	LEPIOTA CEPAESTIPES,	46
4.	AMANITA RUBESCENS,	21

Grouped by F. D. Briscoe — Studies by C. McIlvaine and Val Starnes.

PLATE XVI

FIG.	PAGE.	FIG.	PAGE.
1. ARMILLARIA MELLEA,	55	3-4. LENTINUS LEPIDEUS,	230
2. ARMILLARIA MELLEA VAR. EXANNULATA,	56		

Grouped by F. D. Briscoe — Studies by C. McIlvaine.

PLATE XVIII.

Grouped by F. D. Briscoe — Studies by C. McIlvaine.

PLATE XXIV

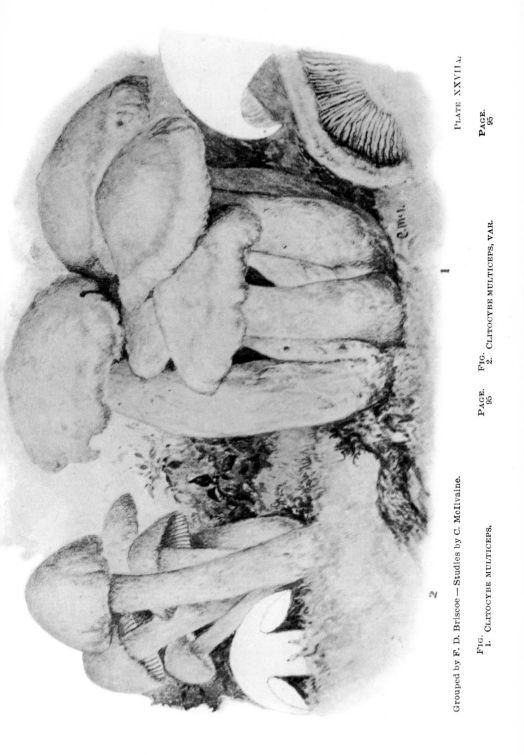

Grouped by F. D. Briscoe — Studies by C. McIlvaine.

FIG. 1. CLITOCYBE MULTICEPS.

PAGE. 95

FIG. 2. CLITOCYBE MULTICEPS, VAR.

PAGE. 95

PLATE XXVIIa.

Grouped by F. D. Briscoe — Studies by C. McIlvaine.

PLATE XXVIII

Grouped by F. D. Briscoe--Studies by C. McIlvaine.

PLATE XXXVII.

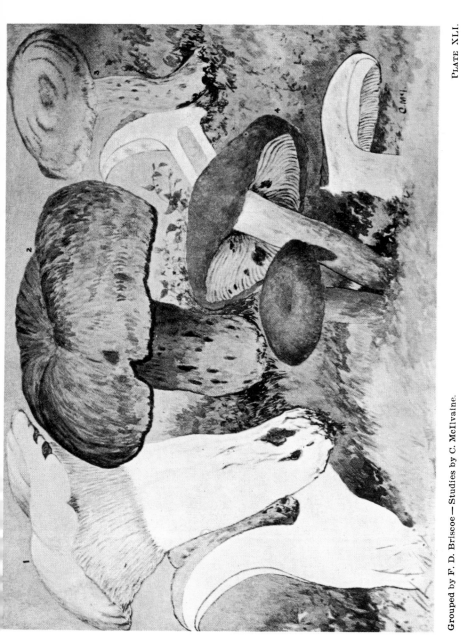

Grouped by F. D. Briscoe—Studies by C. McIlvaine.

PLATE XLI.

FIG.   PAGE.
1. LACTARIUS PIPERATUS,   168
2. LACTARIUS INDIGO,   171

FIG.   PAGE.
3. LACTARIUS DELICIOSUS,   170
4. LACTARIUS VOLEMUS,   180

Grouped by F. D. Briscoe — Studies by C. McIlvaine.

PLATE XLIV.

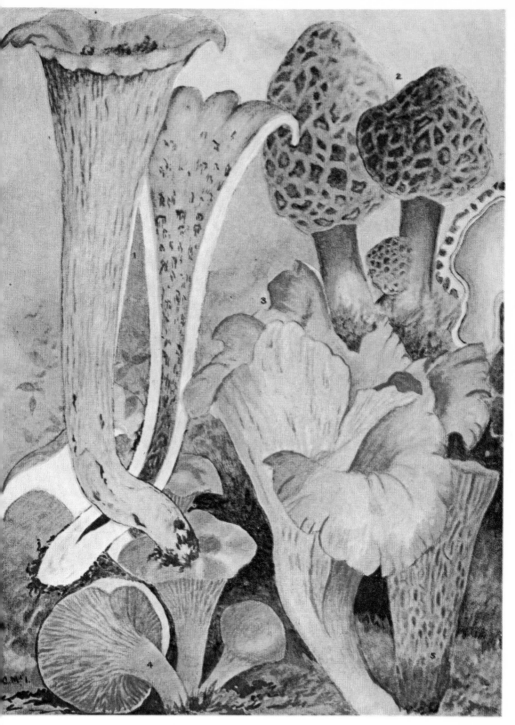

Grouped by F. D. Briscoe—Studies by C. McIlvaine.

Plate XLVI.

FIG.	PAGE.	FIG.	PAGE.
1. CANTHARELLUS FLOCCOSUS,	218	4. CANTHARELLUS CIBARIUS,	215
2. MORCHELLA ESCULENTA,	542	5. CANTHARELLUS BREVIPES,	219
3. CRATERELLUS CANTHARELLUS,	508		

Grouped by F. D. Briscoe — Studies by C. McIlvaine.

PLATE LXI.

FIG.
1. PLUTEUS CERVINUS,

PAGE.
243

FIG.
2. PLUTEUS CERVINUS, VAR.,

PAGE.
245

Grouped by F. D. Briscoe — Studies by C. McIlvaine.

PLATE LXIII.

Grouped by F. D. Briscoe — Studies by C. McIlvaine.

Grouped by F. D. Briscoe—Studies by C. McIlvaine.

PLATE LXXXII

FIG.
1. CORTINARIUS SQUAMULOSUS,
2. CORTINARIUS VIOLACEUS,

PAGE.
318
314

FIG.
3. CORTINARIUS OCHRACEUS,
4. CORTINARIUS TURMALIS,

PAGE.
319
309

FIG.
5. CORTINARIUS ARMILLATUS,

PAGE.
323

Grouped by F. D. Briscoe—Studies by C. McIlvaine

PLATE XCI.

FIG.
1. AGARICUS VARIABILIS,
2. AGARICUS SILVICOLA,

PAGE.
346
343

FIG.
3. AGARICUS PLACOMYCES,
4. AGARICUS CAMPESTER,

PAGE.
345
332

FIG.
5. AGARICUS CAMPESTER (SECTION),

PAGE.
332

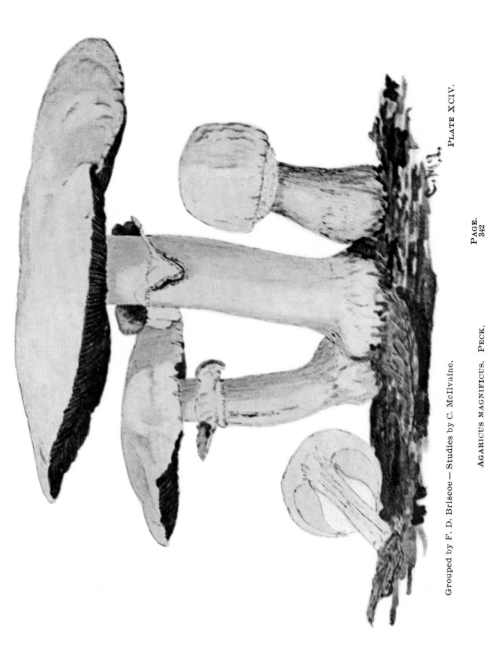

PLATE XCIV.

Grouped by F. D. Briscoe — Studies by C. McIlvaine.

AGARICUS MAGNIFICUS. PECK,

A new species of Agaricus.

PAGE.
342

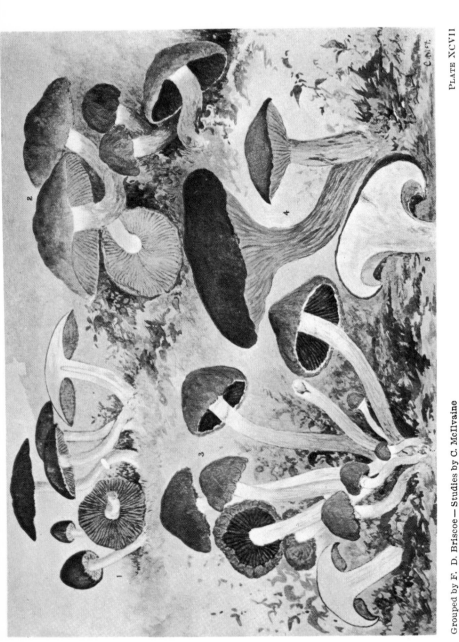

PLATE XCVII

Grouped by F. D. Briscoe — Studies by C. McIlvaine

FIG.
1. HYPHOLOMA APPENDICULATUM,
2. HYPHOLOMA PERPLEXUM,

PAGE.
363
354

FIG.
3. HYPHOLOMA SUBLATERITIUM,
4-5 GOMPHIDIUS RHODOXANTHUS,

PAGE.
359
394

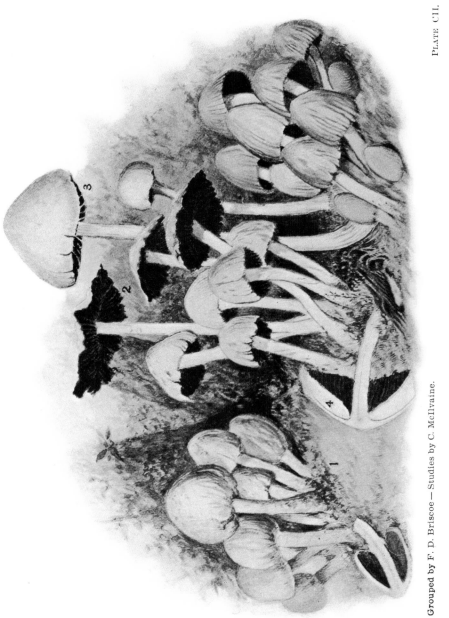

PLATE CII.

Grouped by F. D. Briscoe — Studies by C. McIlvaine.

FIG.

1. COPRINUS ATRAMENTARIUS,
2. COPRINUS MICACEUS,

PAGE
373
378

FIG.

3. PANAEOLUS SOLIDIPES.
4. PANAEOLUS SOLIDIPES (SECTION),

PAGE.
385
385

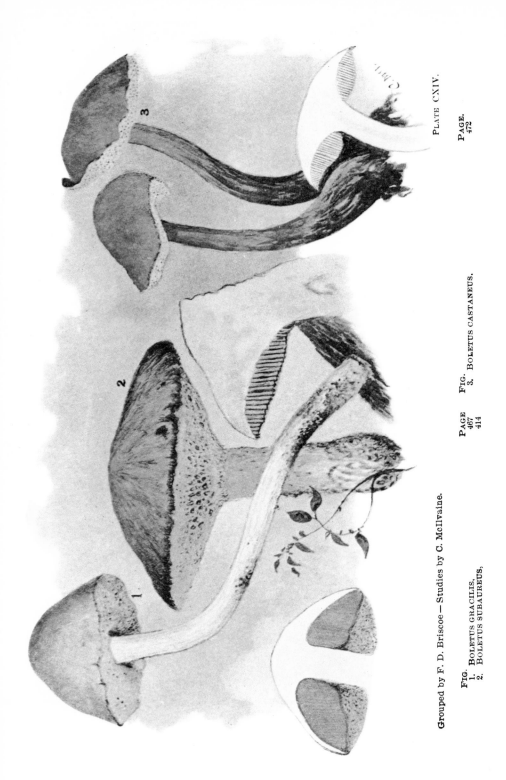

Grouped by F. D. Briscoe — Studies by C. McIlvaine.

Grouped by F. D. Briscoe — Studies by C. McIlvaine. New Species. PLATE CXVI.

PLATE CXVII.

PAGE.
429

Grouped by F. D. Briscoe — Studies by C. McIlvaine.

FIG.   PAGE.   FIG.
1-2. BOLETUS BICOLOR,   425   4. BOLETUS PALLIDUS,
3. BOLETUS RUBROPUNCTUS,   429

Grouped by F. D. Briscoe — Studies by C. McIlvaine.

PLATE CXVIII.

Grouped by F. D. Briscoe—Studies by C. McIlvaine.

FIG.
1. BOLETUS INDECISUS,

PAGE
468

FIG.
2-3-4. BOLETUS FELLEUS,

PAGE.
469

PLATE CXXII.

PLATE CXXV.

FIG.
1. FISTULINA HEPATICA.

PAGE.
477

FIG.
2. POLYPORUS SULPHUREUS.

PAGE.
485

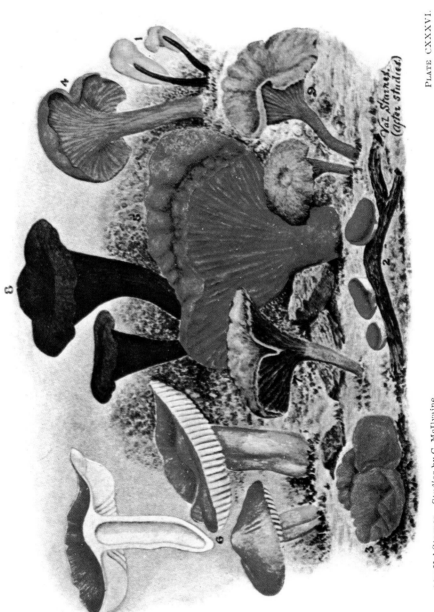

Grouped by Val Starnes — Studies by C. McIlvaine.

PLATE CXXXVI.

Grouped by F. D. Briscoe — Studies by C. McIlvaine.

PLATE CXXXVIII.

Grouped by F. D. Briscoe — Studies by C. McIlvaine.

FIG.
1. CLAVARIA AMETHYSTINA,
2. CLAVARIA AUREA,

PAGE.
516
520

FIG.
3. CLAVARIA FORMOSA,

PAGE.
520

PLATE CXXXIX.

This is a fine large species related to A. solitaria, but differing from it in the character of its bulb and of its annulus. The bulb is not marginate nor imbricately squamose. Its scales are small and numerous. Nor is it clearly radicating, though sometimes it has a slight abrupt point or myceloid-agglomerated mass of soil at its base. The veil or annulus is large and well developed, but it is apt to fall away and disappear with age. Its attachment at the very top of the stem brings it closely in contact with the lamellæ of the young plant and the striations of its upper surface appear to be due to the pressure of the edges of these upon it. It separates readily from the margin of the pileus and is not lacerated. In the mature plant the warts have generally disappeared from the pileus and sometimes its margin is curved upward *Peck*, Bull. Torr. Bot. Club, Vol. 24, No. 3.

Woods. Auburn, N. Y., Alabama, *U. and E.;* Pennsylvania, West Virginia, New Jersey, August to October, *McIlvaine.*

A dozen or more specimens were found in oak woods near Philadelphia, and carefully tested. Their edible qualities were found to be precisely the same as A. solitaria.

*** *Whole volva friable, etc.*

A. rubes′cens Pers.—*rubesco*, to become red. ([Color] Plate VIII, fig. 2. [Color] Plate XII, fig. 4. **Pileus** about 4 in. broad, dingy-reddish, becoming pale flesh-color, tan, scarcely pure, fleshy, convex, then plane, obtuse, moist but *not glutinous* in rainy weather and opaque when dry, covered with unequal, soft, mealy, whitish, easily-separating warts, which are smaller, harder and more closely adherent in dry weather; margin even and, when old, slightly striate only in wet weather. **Flesh** commonly soft, white when fresh, *reddening when broken*. **Stem** 4–5 in. long, as much as 1 in. thick, stuffed, somewhat solid, though soft within, conico-attenuated from the thickened base, reddish-*scaled*, becoming red-white, and without a trace of a distinct volva at the base. **Ring** superior, large, membranaceous, soft, striate and white within. **Gills** reaching the stem in an attenuated manner, forming decurrent lines upon it, thin, crowded, soft, as much as ½ in. broad, shining white.

Very changeable, but readily distinguished from all others of the same group *by the flesh being reddish when broken;* the stem and pileus are commonly spotted-red when wounded. In dry weather it is firmer, flesh reddening more slowly, warts minute. Odor scarcely any. There

Amanita. is a remarkable variety *circinata*, pileus becoming plane, umber-brown, warts adnate, crowded, roundish. A. circinatus Schum. *Stevenson.*

Spores spheroid-ellipsoid, 7–8x6µ *K.;* 8x6µ *W.G.S.;* 7–9x6–8µ *B.;* elliptical, 8–9µ long. *Peck.*

Not reported west of the Mississippi river.

Oak woods, borders and open places. July to September. Indiana, *H. I. Miller;* West Virginia, New Jersey, North Carolina, Pennsylvania, *McIlvaine.*

It is quite common, often growing in large patches. Recent authors agree upon the edibility and deliciousness of this species. The author knows it to be one of the most plentiful, useful and delicious, after several years of pleasant experience with it.

In July, 1899, at Mt. Gretna, I found, growing from the ground gregariously, a singular fungoid growth from 2–5 in. high; cap hemispherical, 1 in. in diameter, tightly fitting a solid stem of nearly the diameter of the cap. The whole was watery white, and evidently affected by a parasite. It was edible. September 1st Professor Peck wrote to me: "I think I have found the identity of the diseased Agaric, of which you sent me samples some time ago. I mean the one affected by *Hypomyces inæqualis* Pk. The host is Amanita rubescens, at least sometimes, and probably always."

The plant is very heavy for its size. The lack of a volva, the dingy color and reddish stains distinctly separate this from any poisonous Amanita.

A. spis'sa Fr.—compact, dense;—of the warts. **Pileus** umber, sooty or gray, fleshy, somewhat compact, convexo-plane, obtuse, smooth, even, but *marked with small, ash-colored, angular, adnate* warts; margin even, but often torn into fibers. **Flesh** *firm, white, quite unchangeable.* **Stem** 2–3 in. long, as much as 1 in. thick, *solid, turnip-shaped at the base*, somewhat rooting with a globoso-depressed not marginate bulb, curt, firm, shining white, at length *squamulose with concentric cracks.* **Ring** superior, large. **Gills** reaching the stem, *slightly striato-decurrent*, broad, crowded, shining white. *Fries.*

Spores 14µ *W.G.S.;* subglobose, 8–10µ *C.B.P.;* 6µ *W.P.;* rather pear-shaped, 9–10x6µ *Massee.*

Cap 2–3 in. across. **Stem** 2½–3 in. long, up to ¾ in. thick. New Jersey, oak woods, August and September. *McIlvaine.*

A. spissa has been reported from but few localities. It is rare in the latitude of Philadelphia. Half a dozen specimens have been found in neighboring New Jersey.

Taste and smell strong, but when cooked the dish is savory and not unlike one of A. rubescens.

A. as′pera Fr.—*asper*, rough. **Pileus** 2–3 in. across. **Flesh** rather thick at the disk, whitish, white or reddish with tints of livid or gray, *reddish or brownish under the cuticle;* convex then plane, margin thin and even, rough with firmly adnate, minute, closely crowded, angular warts, reddish-brown or livid-brownish, not pure white, unchangeable. **Gills** free and rounded behind, not striately decurrent, ventricose, white. **Stem** stuffed, striate above the ring, short at first, ovate, then elongating to 2–3 in., attenuated upward from a wrinkled bulb, squamulose, white without and within. **Ring** superior, entire.

Spores 8x6µ *Massee;* 8x6–7µ *W.G.S.*

The flesh of stem and bulb when eaten by insects is reddish, the bulb when old is a reddish-brown. The large ring and stem become red when touched. In these particulars it resembles A. rubescens. In smell it is somewhat strong, not unlike A. strobiliformis, but not nearly so pungent.

Cooked it is of excellent quality and flavor. I have eaten it since 1885.

A. abrup′ta Pk.—abrupt, of the bulb. **Pileus** thin, broadly convex or nearly plane, covered with small angular or pyramidal, erect, somewhat evanescent warts, white, slightly striate on the margin. **Flesh** white. **Gills** moderately close, reaching the stem and sometimes terminating in slightly decurrent lines upon it, white. **Stem** slender, glabrous, solid, bulbous, white, the bulb abrupt, subglobose, often coated below by the white persistent mycelium, the ring membranous, persistent. **Spores** broadly elliptical or subglobose, 8–10x6–8µ.

Pileus 2–4 in. broad. **Stem** 2.5–4 in. long, 3–4 lines thick.

The chief distinguishing mark of this species is the abrupt, nearly globose, bulbous base of the stem. This is somewhat flattened above and is sometimes longitudinally split on the sides. The small warts of the pileus are easily separable, and in mature specimens they have often wholly or partly disappeared. The remains of the volva are not present on the bulb in mature dried specimens, which indicates that the

23

Amanita. species should be placed in the same group with A. rubescens, A. spissa, etc. The latter species have the bulb of the stem similar to that of our plant, but the color of the pileus and other characters easily separate it. *Peck*, Bull. Torr. Bot. Club, Vol. 24, No. 3.

Alabama, *Underwood;* New Jersey, Pennsylvania, *McIlvaine*. July to September.

This species is edible and quite equal in quality to A. rubescens. Great care should be exercised in distinguishing it.

A. nit'ida Fr.—*niteo*, to shine. **Pileus** when flattened 4 in. broad, whitish, fleshy, *somewhat compact*, at first hemispherical, wrapped up, the *thick volva* forming a floccose crust, then *broken up into thick, remarkably angular, adhering warts, which become brownish*, dry, shining, without a glutinous pellicle, margin always even. **Flesh** *white, quite unchangeable*. **Stem** 3 in. long, 1 in. thick, solid, *firm*, conico-attenuated, *with a bulb-shaped base, squamulose*, white. **Ring** superior, thin, torn, slightly striate, white, villous beneath, at length disappearing. **Gills** *free*, crowded, *very broad*, as much as ½ in., ventricose, shining white. *Fries*.

Menands. Albany county. Our plant is more slender than the typical form, and has smaller but more numerous warts, but in other respects it exhibits the characters of this species. *Peck*, 43d Rep. N. Y. State Bot.

California, *H. and M.;* Maryland. Common in nearly every woods in Maryland. *Banning*.

From its likeness to poisonous species it should be suspected.

A. prairiic'ola Pk —*prairie, colo*, to inhabit. **Pileus** thin, convex, slightly verrucose, white, more or less tinged with yellow, even on the margin. **Flesh** white. **Gills** rather broad, subdistant, reaching the stem, white. **Stem** equal or slightly tapering upward, somewhat squamose toward the base, white or whitish, the annulus persistent. **Spores** large, broadly elliptical, 12–14μ long, 7–9μ broad.

Pileus 1.5–3 in. broad. **Stem** 2–2.5 in. long, 2–4 lines thick.

Bare ground on open prairies. Kansas. September. *E. Bartholomew*.

This species belongs to the same tribe as A. abrupta. The only evidence of the presence of a volva shown by the dried specimens is found in a few inconspicuous, but separable warts on the pileus. There is no

well marked bulb to the stem and no evidence remains of a volva at its Amanita.
base. *Peck*, Bull. Torr. Bot. Club, Vol. 24, No. 3.

Reported from Kansas only. Qualities unknown.

A. monticulo′sa Berk.—mountain, from the warts. **Pileus** 2.5–3 in.
across, convex, areolate, with a wart in the center of each areola; those
toward the margin consisting of soft threads meeting in a point, but
sometimes simply flocculent, the central warts angular, pyramidal, truncate, discolored. **Stem** bulbous, scaly, flocculent, white. **Veil** thick,
at length distant. **Gills** free, ventricose, remote, forming a well-defined
area around the top of the stem. The warts are not hard and rigid as
in A. nitida, and the free remote gills separate it from that and the
neighboring species. *Berk.*

North Carolina, sandy woods, common. *Curtis.*

Properties not known.

A. dau′cipes B. and M.—*daucum*, a carrot; *pes*, a foot. **Pileus** 2–5
in. broad, hemispherical, globose. **Flesh** white, soft, warts regular,
pyramidal, saffron color. **Gills** narrow, reaching the stem, broadest in
the middle. **Stem** 5–6 in. high, solid, base bulbous, with a restricted
cortina above, squamulose downward. **Veil** fibrillose, extending from
the margin of the pileus to the apex of the stem, fugacious.

In cultivated fields. Ohio. *Sullivant.* Properties not given.

A. lenticular′is Lasch.—resembling (the stem) a lentil.

Fries places this species in Amanita, in which Stevenson follows him.
Cooke and Massee place it in Lepiota, where it will be found.

**** *Volva rudimentary, wholly disappearing.*

A. chlorinos′ma Pk.—smelling like chlorine. ([Color] Plate VIII,
fig. 1.) **Pileus** convex or expanded, warty on the disk, covered on
the even margin with a light powdery, at length evanescent substance,
white. **Gills** white. **Stem** nearly cylindrical, stout, deeply penetrating
the earth. **Spores** broadly elliptical, 7–10μ long. Odor distinct, chlo-
rine-like.

Plant 6–7 in. high. **Pileus** 4–6 in. broad. **Stem** 1–2 in. thick.
Peck, Bot. Gaz., Vol. 4.

Amanita. Burnt ground in woods. August. Closter, N. J., *C. F. Austin;* Alabama, *U. and E.;* West Virginia, *Nuttall;* New Jersey, *Ellis;* Mt. Gretna, Pa., July, in a cluster of a dozen individuals, and afterward until frost, strong smelling, warts brownish-white. *McIlvaine.*

It is edible and equal to A. strobiliformis.

A. calyptra′ta Pk. **Pileus** fleshy, thick, convex or nearly plane, centrally covered by a large irregular persistent grayish-white fragment of the volva, glabrous elsewhere, striate on the margin, greenish-yellow or yellowish-brown tinged with green, the margin often a little paler or more yellow than the rest. **Lamellæ** close, nearly free, but reaching the stem and forming slight decurrent lines or striations on it, yellowish-white tinged with green. **Stem** stout, rather long, equal or slightly tapering upward, surrounded at the base by the remains of the ruptured volva, white or yellowish white with a faint greenish tint. **Spores** broadly elliptic, 10μ long, 6μ broad, usually containing a single large nucleus.

Pileus 10–20 cm. broad. **Stem** 10–15 cm. long, 12–20 mm. thick.

Rich ground in fir woods or their borders. Autumn. Oregon. *Dr. H. Lane.*

This is a large and interesting species, well marked and easily recognized by its large size, by the greenish tint that pervades the pileus, lamellæ, annulus and stem, and especially by the large persistent patch of grayish-white felty material that covers the center of the pileus and sometimes extends nearly to the margin. This is in fact the upper part of the ruptured volva that is carried up by the growing plant, and is very suggestive of the specific name. In the young state the plant is entirely enveloped in the volva, which then is similar to a goose egg in size and shape, and its walls are one-fourth to one-half inch thick. So thick and firm are they that the young plant appears sometimes to be unable to break through and it decays in its infancy.

Dr. Lane says that, having found that the Italians made use of this mushroom for food, he began eating it and introducing it to his friends, and he learned by personal trial that it is a thoroughly good and wholesome mushroom, which, when broiled with bacon, fried, baked or stewed, may be eaten with perfect safety and that it is a nutritious food. *Peck*, Bull. Torrey Bot. Club, Vol. 27, January, 1900.

A. crenula′ta Pk. **Pileus** thin, broadly ovate, becoming convex Amanita. or nearly plane and somewhat striate on the margin, adorned with a few thin whitish floccose warts or with whitish flocculent patches, whitish or grayish, sometimes tinged with yellow. **Lamellæ** close, reaching the stem, and sometimes forming decurrent lines upon it, floccose crenulate on the edge, the short ones truncate at the inner extremity, white. **Stem** equal, bulbous, floccose mealy above, stuffed or hollow, white, the annulus slight, evanescent. **Spores** broadly elliptic or subglobose, 7.5–10μ long, nearly as broad, usually containing a single large nucleus.

Pileus 2.5–5 cm. broad. **Stem** 2.5–5 cm. long, 6–8 mm. thick.

Low ground, under trees. Eastern Massachusetts. September. *Mrs. E. Blackford* and *George E. Morris.*

The volva in this species must be very slight, as its remains quickly disappear from the bulb of the stem. The remains carried up by the pileus form slight warts or thin whitish areolate patches. The annulus is present in very young plants, but is often wanting in mature ones, in which state the plant might be mistaken for a species of Amanitopsis. Its true affinity is with the tribe to which A. rubescens belongs. As in that species, the bulb soon becomes naked and exhibits no remains of the volva. It is similar to A. farinosa also in this respect, but quite unlike it in color, in the adornments of the pileus and in the character of its margin, which is even in the young plant and but slightly striate in the mature state. Its dimensions are said sometimes to exceed those here given, and it is reported to have been eaten without harm and to be of an excellent flavor. I have had no opportunity to try. *Peck,* Bull. Torrey Bot. Club, Vol. 27, January, 1900.

27

AMANITOP'SIS Roze.

Amanita; opsis, resembling.

Amanitopsis.

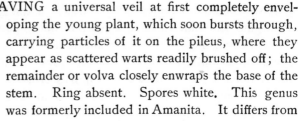

AVING a universal veil at first completely enveloping the young plant, which soon bursts through,
carrying particles of it on the pileus, where they
appear as scattered warts readily brushed off; the
remainder or volva closely enwraps the base of the
stem.   Ring absent.   Spores white.   This genus
was formerly included in Amanita.   It differs from
Amanita in the absence of a ring or collar upon the stem and in the
more sheathing volva.   It differs from Lepiota in having a volva.

Close observation is necessary in collecting Amanitopsis for the table.
It has no trace of ring or veil upon the stem.   So far as the species are
known no poisonous one exists.   But Amanita spreta Pk., which is
deadly, so closely resembles forms of Amanitopsis that those confident
of their knowledge will be deceived.   The veil or traces of veil, which
Amanita spreta always has, sometimes so adheres to and wraps the stem
that it is not noticeable without close examination, thus giving to it
every appearance of an Amanitopsis.

The volva of A. spreta is attached for a considerable distance to the
base of the tapering stem, and is not readily removed.   This is a guide
to detect it.   It is a wolf in sheep's clothing.

Amanitopsis corresponds to Volvaria in the pink-spored series, in
which, as far as known, there is no poisonous species.

All American species of Amanitopsis are given.   Several have not
been tested by the writer because of lack of opportunity.

A. vagina'ta Roze—*vagina*, a sheath. ([Color] Plate X, figs. 1, 2.)
Pileus thin, fragile, glossy, smooth except in rare instances where a
few fragments of the volva adhere to it for a time, deeply and distinctly
striate on the margin, sometimes umbonate.   **Flesh** white, in the dark
forms grayish under the skin.   **Stem** ringless, sometimes smooth, but
generally mealy or floccose, hollow or stuffed with a cottony pith, *not
bulbous.*   **Volva** long, thin, fragile, closely sheathing yet free from the
stem, except in the lower part, easily detachable and frequently remaining in the ground when the plant is pulled.   **Color** variable, generally
mouse-gray, sometimes livid, tawny-yellow or white, in one variety a

rich date-brown. **Spores** globose, 8–10μ broad *Peck;* elliptical 10x7–8μ Amanitopsis. *Massee.*

Var. *liv′ida* Pers.—livid. Leaden brown, gills dingy. ([Color] Plate X, fig. 2.)

Var. *ful′va* Schæff.—yellowish. Tawny-yellow or pale ochraceous.

This plant is widely dispersed, having been reported from many localities in the United States, also from Nova Scotia and Greenland.

On ground in woods and on margins of woods, under trees, in shaded grassy places. Sometimes in open stubble and pastures. June to frost. Mt. Gretna, September, 1899, found a cluster on decayed chestnut stump. Various colors abound—hazel, brown, gray, yellow, whitish. The caps and stems are tender as asparagus tips, but without much distinct flavor when cooked.

Great care must be taken to distinguish these forms from Amanita spreta Pk. which is poisonous. See heading of genus—Amanitopsis.

A. niva′lis Grev.—snowy. ([Color] Plate X, fig. 3.) **Pileus** at first ovate, then convex or plane, smooth, *striate on the thin margin, white,* sometimes tinged with yellow or ochraceous on the disk. **Flesh** white. **Gills** subdistant, white, free. **Stem** equal, rather tall, nearly smooth, *bulbous,* stuffed, white; the volva very fragile, *soon breaking up into fragments or sometimes persisting in the form of a collar-like ring at the upper part of the bulb.* **Spores** globose, 7.5–10μ in diameter.

Plant 4–6 in. high. **Pileus** 2–3 in. broad. **Stem** 2–4 lines thick. July to October.

It approaches in some respects A. Frostiana, but its larger size, smooth pileus, lighter color and the absence of an annulus will easily distinguish it from that species. *Peck,* 33d Rep. N. Y. State Bot.

Specimens have been repeatedly found by the writer in open oak woods near Philadelphia.

A strong, unpleasant bitter, which appears to develop while cooking, renders it unpalatable. It is harmless, but its use is not advised.

A. velo′sa Pk.—*velosus,* fleecy. **Pileus** at first subglobose, then bell-shaped or nearly plane, generally bearing patches of the remains of the whitish felty or tomentose volva, elsewhere glabrous, becoming sulcate-striate on the margin, buff or orange-buff. **Flesh** compact, white. **Gills** close, reaching the stem, subventricose, pale cream color. **Stem**

29

Amanitopsis. firm, at first attenuated and tomentose at the top, then nearly equal, stuffed, white or whitish, closely sheathed at the base by the thick volva. **Spores** globose, 10–13µ.

Pileus 2–4 in. broad. **Stem** 3–4 in. long, 3–4 lines thick.

Under oak trees. Pasadena, California. April. *A. J. McClatchie.*

This fungus is closely related to A. vaginata, from which it may be separated by the more adherent remains of the thicker volva which sometimes cover the whole surface of the pileus, and by the thicker gills which are somewhat adnate to the stem and terminate with a decurrent tooth. Bull. Torr. Bot. Club, Vol. 22, No. 12.

As it is probable this species will be found elsewhere than California, and from its close relation to A. vaginata likely to be edible, its description is here given.

A. strangula'ta (Fr.) Roze—choked, from the stuffed stem. ([Color] Plate X, fig. 4.) **Pileus** at first ovate or subelliptical, then bell-shaped, convex or *plane, warty*, slightly viscid when moist, *deeply and distinctly striate on the margin*, grayish-brown. **Gills** free, close, white. **Stem** equal or tapering upward, stuffed or hollow, nearly smooth, white or whitish, *the volva soon breaking up into scales or subannular fragments.* **Spores** globose, 10–13µ.

Plant 4–6 in. high. **Pileus** 2–4 in. broad. **Stem** 3–6 lines thick. *Peck*, 33d Rep. N. Y. State Bot.

A. Ceciliæ B. and Br. is a synonym.

Not distinct in color and general appearance from A. vaginata, but distinctly separated by its warty pileus and evanescent mouse-colored volva which does not sheath the stem. **Pileus** striate when young, then sulcate. **Stem** mealy, especially on the upper part.

Woods, open grassy places, wheat stubble, etc. June to September. Pennsylvania, New Jersey, West Virginia, *McIlvaine.*

In the latitude of Philadelphia the plant is found in great abundance. Its rather early appearance, staying quality, delicate consistency and flavor make it valuable as a food supply.

Pearl color, bluish-gray and gray are the prevailing **cap-coloring.**

A. adna'ta (W.G.S.) Roze—*adnatus*, adnate, of the gills. **Pileus** about 3 in. across. **Flesh** thick, whitish, firm, convex, then expanded, rather moist, pale yellowish-buff, often furnished with irregular, woolly

patches of volva; margin even, extending beyond the gills. **Stem** 2–4 Amanitopsis.
in. long, ½ in. thick, cylindrical, rough, fibrillose, pale buff, flesh distinct from that of the pileus, stuffed, then hollow; base slightly swollen.
Volva adnate, white, downy, margin free and lax, sometimes almost
obsolete. **Gills** truly adnate, crowded, with many intermediate shorter
ones, white. **Spores** subglobose, with an oblique point, 7–8μ *Massee.*

Tender, good flavor, yielding more substance when cooked than any
other Amanitopsis.

A. volva′ta Pk.—possessing a volva. **Pileus** convex, then nearly
plane, slightly striate on the margin, hairy or floccose-scaly, white or
whitish, the disk sometimes brownish. **Gills** close, free, white. **Stem**
equal or slightly tapering upward, stuffed, minutely floccose-scaly,
whitish, inserted at the base in a large, firm, cup-shaped, persistent
volva. **Spores** elliptical, 10x8μ.

Plant 2–3 in. high. **Pileus** 2–3 broad. **Stem** 3–4 lines thick.
Peck, 33d Rep. N. Y. State Bot.

The plant is easily recognized by its large, cup-shaped volva and cap,
which is not smooth, as is usual in a species with a persistent membranous volva, more or less scaly with minute tufts of fibrils or tomentose hairs. The gills are white in the fresh plant.

Professor Peck notes the species as quite rare. Numerous specimens
occur in the sandy oak woods of New Jersey, and in oak woods near
Angora, Philadelphia. July to October.

Care must be taken to determine the absence of an annulus or any
trace of one. Tender, delicate, without pronounced flavor. Equal to
Amanitopsis vaginata.

A. farino′sa Schw.—covered with *farina,* meal. **Pileus** nearly plane,
thin, *flocculent-pulverulent, widely and deeply striate on the margin,*
grayish-brown or livid-brown. **Gills** free, whitish. **Stem** whitish or
pallid, equal, stuffed or hollow, mealy, *sub-bulbous,* the volva *flocculent-
pulverulent,* evanescent. **Spores** variable, *elliptical ovate or subglobose,*
6–8μ long.

Plant about 2 in. high. **Pileus** 1 in. to 15 lines broad. **Stem** 1–3
lines thick. July to September.

This is our smallest Amanita (now Amanitopsis). It is neither very
common nor very abundant when it does occur. It is described by

Amanitopsis. Schweinitz as "solid," but I have always found it stuffed or hollow. *Peck*, 33d Rep. N. Y. State Bot.

A. pusil'la Pk.—small. **Pileus** thin, broadly convex or nearly plane, subglabrous, slightly umbonate, even on the margin, pale brown. **Gills** narrow, thin, close, free, becoming brownish. **Stem** short, hollow, bulbous, the bulb margined by the remains of the membranous volva. **Spores** broadly elliptical, 5–6x4μ.

Pileus about 1 in. broad. **Stem** 8–12 lines long, 1–2 lines thick.

Grassy ground. Gouverneur, St. Lawrence county. September. *Mrs. Anthony. Peck*, 50th Rep. N. Y. State Bot.

Edibility not tested.

A. pubes'cens Schw.—downy. **Pileus** yellow, covered with a thin pubescence, margin involute. **Stem** short, about 1 in. in length, at first white becoming yellowish, bulbous, bulb thick. **Volva** evanescent. **Gills** white.

In grassy grounds. Rare.

North Carolina, *Schweinitz, Curtis*.

A. agglutina'ta B. and C.—viscid. **Pileus** 1–2 in. broad, white, hemispheric then plane, viscid, areolate-scaly from the remains of the volva, margin thin, sulcate. **Stem** .5–1.5 in. long, 2 lines thick, short, solid, bulbous. **Volva** with a free margin. **Gills** broad, ventricose, rotundate-free. **Spores** elliptic.

In pine woods.

North Carolina, *Curtis*.

Resembling some of the dwarf forms of A. vaginata but at once distinguished by its solid stem and decidedly viscid, areolate-squamose pileus. Am. Jour. Sci. and Arts, 1848.

LEPIO'TA Fr.

Lepis, a scale.

Pileus *generally scaly* from the breaking up of the cuticle and the adherence of the concrete veil. **Gills** free, often very distant from the stem and attached to a *cartilaginous* collar. **Stem** hollow or stuffed, its flesh distinct from that of the pileus. **Ring** at first attached to the cuticle of the pileus, often movable, sometimes evanescent.

(Plate XI.)

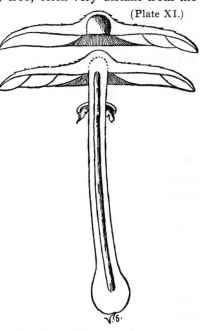

On the ground. Several are found in hot-houses and hot-beds, and are probably introduced species.

The universal veil, covering the entire plant when very young, is closely applied to the pileus, which from the breaking up of the cuticle is generally scaly. The **stem** in most species differs in substance from the pileus. This is readily seen by splitting the plant in half from cap to base. It is easily separated from the cap, leaving a cup-like depression therein. **Gills** usu-

SECTION OF LEPIOTA PROCERA.

ally white. In some species they are yellow-tinted. In others they become a dingy red when wounded or ageing.

The veil in this genus, being concrete with the cuticle of the pileus, never appears as loose warts or patches, neither is there a volva as in Amanita and Amanitopsis. These three genera are the only ones in the white-spored series having gills free from the stem. In a few species the gills are slightly attached to the stem, but are never decurrent upon it as in Armillaria. When the plant is young it is egg-shaped. It then gradually spreads, becomes convex, and opens until it is nearly flat, with a knob in the center.

The only species in this genus known to be poisonous to some persons is L. Morgani Pk., which is distinguished by its green spores and white

Lepiota. gills becoming green.   L. Vittadini has also been regarded with suspicion.

ANALYSIS OF TRIBES.

A. Pileus Dry.

Proceri (*L. procera*).   Page 35.

Ring movable.   The plant is at first entirely enclosed in a universal veil, which splits around at the base, the lower part disappearing on the bulb, the upper part attached to the pileus breaking up into scales. **Stem** encircled at the top with a cartilaginous collar to which the free, remote gills are attached.

Clypeolarii (*L. clypeolaria*).   Page 39.

Ring fixed, attached to the upper portion of the universal veil which *sheaths the stem* from the base upward, making it downy or scaly below the ring.   The remainder of the veil united with the pileus breaking up and becoming downy or scaly.   Collar at the apex of stem not so large as in Proceri, hence the gills are not usually so remote.   Taste and smell unpleasant, resembling that of radishes.

Annulosi (*annulus*, a ring).   Page 44.

Ring *fixed*, somewhat persistent, universal veil closely attached to the pileus.   Collar absent or similar in texture to the stem.   **Stem,** *not sheathed*.

Granulosi (*L. granulosa*).   Page 49.

Pileus *granular or warty*.   Universal veil sheathing the stem, at first continuous from the stem to the pileus, finally rupturing, forming a ring nearer the base.   Stem not so distinctly different from the pileus as in other sections.

Mesomorphi (*L. mesomorpha*).

Small, slender, stem hollow.   Pileus *smooth, dry*.

B. Pileus Viscid, Neither Scaly nor Warty.

Proce′ri.   Ring movable, etc.

L. proce′ra Scop.—*procerus*, tall.   (Plate XIII, p. 15)   Tall Lepiota.
Lepiota, Parasol Mushroom, in some localities Pasture Mushroom (a
misleading title).

HE **Flesh** not very thick, soft, permanently white.
Pileus at first ovate, finally expanded, cuticle soon
breaking up into brown scales, excepting upon the
umbo, umbo smooth, dark-brown, distinct.   The
caps vary in shades of brown, sometimes they
have a faint tinge of lavender.   **Gills** whitish,
crowded, narrowing toward the stem, and very remote from it.   **Stem** variable in length, often very
long, tubular, at first stuffed with light fibrils,
quite bulbous at base, generally spotted or scaly with peculiar snake-
like markings below the ring, which is thick, firm and readily movable.
When the stem is removed from pileus it leaves a deep cavity extending
nearly to the cuticle.

Pileus 3–6 in. broad.   **Stem** 5–12 in. high, about ½ in. thick.

White spores elliptical, 14–18x9–11µ *Peck;*   12–15x8–9µ *Massee;*
14x10µ *Lloyd*.

Readily known by its extremely tall stem, shaggy cap, distinct umbo
and the channel between the gills and stem.   Resembles no poisonous
species.

Before cooking the scurf should be rubbed from the caps, which alone
should be eaten, as the stem is tough.   Though the flesh is thin, the
gills are meaty and have a pleasant, nutty flavor.   Fried in butter it
has few equals.   It makes a superior catsup.

L. racho′des Vitt. *Gr.*—a ragged, tattered garment.   **Pileus** very
fleshy, but very soft when full grown, globose then flattened or depressed,
not umbonate, at first incrusted with a *thick, rigid*, even, very smooth,
bay-brown, wholly continuous *cuticle*, which remains entire at the disk
but otherwise *soon becomes elegantly reticulated with cracks;* these very
readily separate into *persistent*, polygonal, concentric *scales*, which are
revolute at the margin and attached to the surface with beautifully
radiating fibers, the surface remaining coarsely fibrillose-downy.   **Flesh**

Lepiota. white, *immediately becoming saffron-red* when broken, easily separating from the apex of the distinct stem, which is encircled with a prominent collar. **Stem** stout, at the first bulbous with a distinct margin upon the bulb, conical when young, then elongated, attenuated upward, as much as a span long, very robust, 1 in. thick, and more at the base, always even, and *without a trace of scales* or even of fibrils although the appearance is obsoletely silky, wholly whitish, hollow within, stuffed with spider-web threads, the walls remarkably and coarsely fibrous. **Ring** movable, adhering longer to the margin of the pileus than to the apex of the stem, hence rayed with fibers at the circumference, clothed beneath with one or two zones of scales. **Gills** *very remote*, tapering toward each end or broadest at the middle, crowded, whitish, sometimes reddening. *Stevenson.*

Veil remarkable in its development and thick margin.

Spores 6x8μ *W.G.S.*

Fort Edward, *Howe;* Westfield, N. Y., *Miss L. M. Patchen;* Pennsylvania, New Jersey, *McIlvaine.*

A heavier species than L. procera, of which by some writers it has been considered a variety, but it differs in the absence of umbo and flesh becoming tinged with red.

Stem is decidedly swollen downward. Veil heavy, apparently double, thickest at margin of cap to which it remains attached in heavy fragments. It tears from the stem, leaving no mark of ring.

Var. *puella'ris* Fr.—*puella*, a girl. Smaller than typical form, shining white, pileus with downy scales. Not yet reported in America.

Edible qualities similar to those of L. procera. It is sold indiscriminately with it in London markets.

L. excoria'ta Schaeff.—stripped of its skin. **Flesh** spongy, rather thick, white, unchangeable. **Pileus** at first globose, then flat, hardly umbonate, pale-fawn or whitish, disk dark; cuticle thin, silky or scaly, sometimes areolate, more or less peeled toward margin, hence its name. **Gills** ventricose, white, free, somewhat remote. **Stem** attenuated, hollow or stuffed, short, scarcely bulbous, smooth, white, not spotted, very distinct from flesh of pileus. **Ring** movable but not so freely as that of L. procera.

Stem 1½–2½ in. high, less than ¼ in. thick. **Pileus** 2–3 in. broad. **Spores** 14–15x8–9μ *Massee.*

PLATE XIV.

LEPIOTA MORGANI.

Photographed by Dr. J. R. Weist.

PLATE XV.

LEPIOTA NAUCINOIDES.

Photographed by Dr. J. R. Weist.

In pastures or grassy lawns.   May to September.

North Carolina, edible, *Curtis;* Massachusetts, *Frost;* California, *H. and M.;* Ohio, *Morgan;* Minnesota, *Johnson*.

Distinguished from the preceding by its smaller size and short stem which is scarcely bulbous.

Esculent qualities good.

L. mastoi'dea Fr. *Gr.*—breast-shaped.   **Pileus** rather thin, ovate, bell-shaped, then flattened, with a conspicuous acute umbo, cuticle thin, brownish, breaking up in minute scattered scales; the pileus appears whitish beneath.   **Stem** hollow, smooth, tough, flexible, attenuated from the bulbous base to the apex.   **Ring** entire, movable.   **Gills** very remote, crowded, broad, tapering at both ends, white.

Pileus 1–2 in. broad.   **Stem** 2–3 in. long, 3–4 lines thick at base, 1½–2 lines at apex.

North Carolina, edible, *Curtis*.   It is generally eaten in Europe.

In woods, especially about old stumps.   October.

The entire plant is whitish and is well marked by the prominent umbo, which generally has a depression around it.   It has the least substance of any in this section, and consequently not much value as food.

L. gracilen'ta Krombh.—*gracilis*, slender.   **Pileus** rather fleshy, thickest at the disk, ovate then bell-shaped, finally flattened, obscurely umbonate; at first brownish from the adnate cuticle, which, breaking up into broad adpressed scales, allows the whitish pileus to be seen beneath them.   **Gills** remote, very broad, crowded, pallid.   **Stem** whitish, obscurely scaly, hollow or containing slight fibrils, slightly bulbous. **Ring** thin, floccose, vanishing.

Stem 5–6 in. long, 3–5 lines thick.   In pastures, also in woods.

Spores 11x8μ *W.G.S.*

Almost as tall as L. procera, but slighter in stem and pileus; the ring, instead of being firm and persistent, is thin and fugacious, and the stem is hardly bulbous.

Edible, but not of the first quality.

L. Mor'gani Pk.—in honor of Professor Morgan.   (Plate XIV.) **Pileus** fleshy, soft, at first subglobose, then expanded or even depressed, white, the brownish or yellowish cuticle breaking up into scales except

37

Agaricaceæ

Lepiota. on the disk. **Gills** close, lanceolate, remote, white, then green. **Stem** firm, equal or tapering upward, subbulbous, smooth, webby-stuffed, whitish, tinged with brown. **Ring** rather large, movable. **Flesh** both of the pileus and stem white, changing to reddish and then to yellowish when cut or bruised. **Spores** ovate or subelliptical, mostly uninucleate, sordid green, 10–13x7–8μ.

Plant 6–8 in. high. **Pileus** 5–9 in. broad. **Stem** 6–12 lines thick. *Peck* in Bot. Gaz., March, 1879.

Open dry grassy places. Dayton, Ohio. *A. P. Morgan.*

This species is remarkable because of the peculiar color of the spores. No green-spored Agaric, so far as I am aware, has before been discovered, and no one of the five series, in which the very numerous species of the genus have been arranged, is characterized in such a way as to receive this species.

It seems a little hasty to found a series (Viridispori) on the strength of a single species. Until other species of such a supposed series shall be discovered it seems best to regard this as an aberrant member of the white-spored series. The same course has been taken with those Agarics which have sordid or yellowish or lilac-tinted spores.

It gives me great pleasure to dedicate this fine species to its discoverer Mr. Morgan. *Peck.*

Commonly 6–8 in. high, 5–9 in. diameter, though larger specimens are sometimes found. It is the most conspicuous Agaric in the meadows and pastures of the Miami valley; it appears to flourish from spring to autumn whenever there is abundance of rain.

It is heavier and stouter than L. procera and I am disposed to claim that it is the largest Agaric in the world. **Spores** 10–12x7–8μ. In immature specimens they are greenish-yellow. *Morgan.*

Kansas, *Bartholomew* (*Peck*, Rep. 50); Kansas, *Cragin;* Alabama, *U. and E.;* Georgia, *Benson;* Louisiana, *Rev. A. B. Langlois;* Michigan, *C. F. Wheeler* (*Lloyd*, Myc. Notes); Texas, *Prof. W. S. Carter;* Indiana, *H. I. Miller.*

L. Morgani is one of the largest, handsomest of the genus. It is very abundant in the western and southwestern states. Mr. H. I. Miller, Terre Haute, Ind., writes August 18, 1898: ''I have recently measured several which were more than twelve inches across. At the present time this mushroom is growing in more abundance throughout Indiana than any other. It grows luxuriantly in the pastures, generally

38

in grand fairy rings, five, ten, fifteen feet in diameter. We find it also Lepicta, in the woods. It is beautifully white and majestic, and these rings can be seen in meadows where the grass has been eaten close, for half a mile or more. The gills are white until the cap is almost opened, by which time the green spores begin to cause the gills to change to green. The meat is fine and is usually more free from worms than other mushrooms. Six families, here, have eaten heartily of them. The experience is that one or two members of each family are made sick, though in two families, who have several times eaten them, no one was made sick. I enjoy them immensely, and never feel any the worse for eating them. I doubt if we have a finer-flavored fungus. The meat is simply delicious. One fairy ring yields a bushel.''

Prof. W. S. Carter, University of Texas, Galveston, reported to me (and sent specimens of L. Morganii) the poisoning of three laboring men from eating this fungus. They were seriously sick, but recovered.

The conclusion is inevitable that this green-spored Lepiota contains a poison which violently attacks some persons, yet is harmless upon others.

I have not had opportunity to test it. It should be tested with great caution.

CLYPEOLA'RII. *Clypeus*, a shield. Ring fixed; stem sheathed, etc.

L. Frie'sii Lasch.—in honor of Fries. **Pileus** fleshy, soft, lacerated into appressed tomentose scales. **Stem** hollow, with a webby pith, subbulbous, scaly. **Ring** superior, pendulous, equal. **Gills** subremote, linear, crowded, branched. *Fries.*

Pileus fleshy but rather thin, convex or nearly plane, clothed with a soft, tawny or brownish-tawny down, which breaks up into appressed, often subconfluent scales, the disk rough with small acute, erect scales. **Flesh** soft, white. **Gills** narrow, crowded, free, white, some of them forked. **Stem** equal or slightly tapering upward, subbulbous, hollow, colored like the pileus below the ring, and there clothed with tomentose fibrils which sometimes form floccose or tomentose scales, white and powdered above. **Ring** well developed, flabby, white above, tawny and floccose-scaly below. **Spores** 7–8x3–4μ.

Plant 2–5 in. high. **Pileus** 1–4 in. broad. **Stem** 2–5 lines thick. Catskill mountains and East Worcester. July to September.

I have quoted the description of this species as it is found in Epicri-

39

Lepiota. sis, because the American plant which I have referred to it does not in all respects agree with this description, but comes so near it that it can scarcely be specifically distinct. In the American plant, so far as I have seen it, erect, acute scales are always present, especially on the disk, and the down of the pileus does not always break up into distinct areas or scales. Neither is the stem usually scaly, but rather clothed with soft tomentose or almost silky fibrils. The gills are crowded and some of them are forked. At the furcations there are slight depressions which interrupt the general level of the edges, and give them the appearance of having been eaten by insects. The plant has a slight odor, especially when cut or bruised. *Peck*, 35th Rep. N. Y. State Bot.

Remarks under L. acutesquamosa apply to L. Friesii, which Fries himself doubts being distinct from the first. The plants vary greatly in size, color and habitat. The name—acutesquamosa—carries a descriptive meaning with it that L. Friesii does not.

It does not appear to have been reported except by Professor Peck, but probably appears as L. acutesquamosa in other lists.

The edible qualities are excellent.

L. acutesquamo'sa Wein.—*acutus*, sharp; *squama*, a scale. **Pileus** fleshy, obtuse, at first hairy-floccose, then bristly with erect, acute, rough scales. **Stem** somewhat stuffed, stout, bulbous, powdered above the moderate-sized ring. **Gills** approximate, lanceolate, simple. *Fries*.

Pileus convex or nearly plane, obtuse or broadly subumbonate, clothed with a soft tawny or brownish-tawny tomentum, which usually breaks up into imperfect areas or squamæ, rough with erect, acute scales, which are generally larger and more numerous on the disk. **Gills** close, free, white or yellowish. **Stem** equal, hollow or stuffed with webby filaments, subbulbous. **Spores** about 7x3–4μ.

Woods and conservatories. Buffalo, *G. W. Clinton;* Albany, *A. F. Chatfield;* Adirondack mountains and Brewertown, *Peck.*

The form found in the hot-houses seems to have the tomentum of the pileus less dense and the erect scales more numerous than in the form growing in woods. The annulus is frequently lacerated. In the specimens of the woods the erect scales are sometimes blackish in color, and they then contrast quite conspicuously with the tawny or brownish-tawny tomentum beneath them. They vary in size and shape. Some resemble pointed papillæ, others, being more elongated, are almost spine-like. These are sometimes curved. They are generally larger Lepiota. and more numerous on the disk than elsewhere, and often they are wholly wanting on the margin. *Peck*, 35th Rep. N. Y. State Bot.

West Philadelphia, 1897, on lawn and growing from trunk of a maple tree; Mt. Gretna, Pa., mixed woods. *McIlvaine*.

I first saw specimens of L. acutesquamosa when sent to me by Miss Lydia M. Patchen, President Westfield Toadstool Club. It was later found by myself and tested. Specimens were sent to Professor Peck and identified as L. acutesquamosa.

Caps and stems brownish-purple. The pointed squamules or tufts have dark-brown points, shaded to a delicate purple at base. Gills light, faint flesh-color. Veil is silky, transparent, beautiful, quite tenacious—stretching until cap is well expanded, persistent, though at times fugacious. Smell like stewed mushrooms. The caps are of excellent substance and flavor.

L. his'pida Lasch.—rough. **Pileus** 2–3 in. across. **Flesh** thin, white, unchangeable; hemispherical then expanded, umbonate, tomentose or downy at first from the remains of the universal veil; during expansion the down becomes broken up into small, spreading, scaly points, which eventually disappear, umber-brown, sometimes with a tawny tinge. **Gills** free but near to the stem, the collar of the pileus prominent and sheathing the stem, crowded, ventricose, simple, white. **Stem** about 3–5 in. long, 3–5 lines thick, attenuated upward, densely squamosely-woolly up to the superior, membranaceous, reflexed ring, dingy-brown, stem tubular, but fibrillosely stuffed. **Spores** 6–7x4µ *Massee*.

In margins of and in open mixed woods, under pine trees, Haddonfield, N. J., July to September, 1892. Quite plentiful year after year in the same places. The American plant is taller than the English species, the stem reaching five inches, and the color of the cap a delicate tawnybrown. Smell slight, but pungent like radishes.

The whole fungus is tender and delicious. It is one of the few Lepiotæ that stews well.

L. feli'na Pers.—belonging to a cat. **Pileus** thin, bell-shaped or convex, subumbonate, adorned with numerous subtomentose or floccose blackish-brown scales. **Gills** close, free, white. **Stem** slender, rather

Lepiota. long, equal or slightly tapering upward, hollow, clothed with soft, loose, floccose filaments, brown. **Ring** slight, evanescent. **Spores** elliptical, 6–8x4–5µ.

Plant 2–3.5 in. high. **Pileus** .5–1.5 in. broad. **Stem** 1–2 lines thick.

Woods. Adirondack Mountains. August and September.

It is easily distinguished from A. rubrotincta by the darker color of the scales of the pileus, by the loose floccose filaments that clothe the brown stem, by the fugacious ring and the smaller spores. *Peck*, 35th Rep. N. Y. State Bot.

The caps compare favorably with other Lepiotæ in substance and flavor.

L. crista′ta A. and S.—*crista*, a tuft, crest. **Pileus** thin, bell-shaped or convex, then nearly plane, obtuse, at first with an even reddish or reddish-brown surface, then white adorned with reddish or reddish-brown scales formed by the breaking up of the cuticle, the central part or disk colored like the scales. **Gills** close, free, white. **Stem** slender, hollow, equal, smooth or silky-fibrillose below the ring, whitish. **Ring** small, white. **Spores** oblong or narrowly subelliptical, 5–7x3–4µ.

Plant 1–2 in. high. **Pileus** .5–1.5 in. broad. **Stem** 1–2 lines thick.

Grassy places and borders of woods. June to September.

This species is easily known by its small size and the crested appearance of the white pileus, an appearance produced by the orbicular unruptured portion of the cuticle that remains like a colored spot on the disk. The fragments or scales are more close near this central part and more distant from each other toward the margin, where they are often wholly wanting. The scales are sometimes very small and almost granular. In very wet weather the margin of the pileus in this and some other species becomes upturned or reflexed. *Peck*, 35th Rep. N. Y. State Bot.

Found in Woodland Cemetery, Philadelphia. June to September, 1897. *McIlvaine.*

Scales were appressed and slightly tinged with brown, often very small. Caps of same, upturned and bare near margin. Taste sweet, slightly like new meal. Odor strong.

Cooked it is of good consistency and pleasing to taste.

L. alluvi'na Pk.—*alluvies*, the over-flowing of a river. **Pileus** thin, Lepiota.
convex or plane, reflexed on the margin, white, adorned with minute
pale-yellow hairy or fibrillose scales. **Gills** thin, close, free, white or
yellowish. **Stem** slender, fibrillose, whitish or pallid, slightly thickened
at the base. **Ring** slight, subpersistent, often near the middle of the
stem. **Spores** elliptical, 6–7x4–5μ.

Plant 1–2 in. high. **Pileus** .5–1 in. broad. **Stem** 1–1.5 lines thick.
Alluvial soil, among weeds. Albany. July.

In the fresh plant the scales are of a pale yellow or lemon color, but
in drying they and the whole pileus take a deeper rich yellow hue. The
ring is generally remote from the pileus, sometimes even below the
middle of the stem. *Peck,* 35th Rep. N. Y. State Bot.

In 1897, I found it growing among weeds on lot near University of
Pennsylvania, Philadelphia. Seemingly it is a city resident.

The taste and smell are pleasant. Cooked it is tender and savory.
Both stems and caps are good.

L. metulæ'spora B. and Br.—*metula*, an obelisk. **Pileus** thin, bell-
shaped or convex, subumbonate, at first with a uniform pallid or brownish surface, which soon breaks up into small brownish scales, the margin
more or less striate, often appendiculate with fragments of the veil.
Gills close, free, white. **Stem** slender, equal or slightly tapering upward, hollow, adorned with soft floccose scales or filaments, pallid.
Ring slight, evanescent. **Spores** long, subfusiform.

Plant 2–3.5 in. high. **Pileus** .5–1.5 in. broad. **Stem** 1–2 lines
thick.

Woods. Adirondack mountains. August and September.

This species occurs with us in the same localities as L. felina, which
it very much resembles in size, shape and general characters, differing
only in color, the striate margin of the pileus and the character of the
spores.

The species has a wide range, having been found in Ceylon, England,
Alabama and Kentucky. *Peck,* 35th Rep. N. Y. State Bot.

This has not been elsewhere noted in the United States, probably
from neglect of the spore characters, being reported as L. clypeolaria.

New Jersey and Pennsylvania. *McIlvaine.*

43

ANNULO'SI.   Ring large, fixed; stem not sheathed.

Lepiota.   **L. holoseri'cea** Fr. *Gr.*—entire, silken.   **Pileus** 3 in. and more broad, whitish or clay-white, *fleshy*, soft, *convex then expanded*, rather plane, obtuse, *floccoso-silky*, somewhat fibrillose, *becoming even*, fragile, disk by no means gibbous; and wholly of the same color; margin involute when young.   **Flesh** soft, white.   **Stem** 2½–4 in. long, ½ in. and more thick, *solid*, bulbous and not rooted at the base, soft, fragile, silky-fibrillose, whitish.   **Ring** superior, membranaceous, large, soft, pendulous, the margin again ascending.   **Gills** wholly free, broad, ventricose, crowded, becoming pale-white.   *Fries.*

A species well marked from all others.   Inodorous.

On soil in flower beds.

Spores elliptical, 7–8x5µ *Massee;* 6x9µ *W.G.S.*

Wisconsin, *Bundy;* Minnesota, *Johnson.*

Considered esculent in Europe.

L. Vittadi'ni Fr.—in honor of the Italian mycologist.   **Pileus** 3–4 in. across.   **Flesh** 4–6 lines thick at the disk, becoming very thin at the margin, white;   convex then plane, obtuse or gibbous, densely covered with small, erect, wart-like scales, altogether whitish.   **Gills** free but rather close to the stem, 3–4 lines broad, rounded in front, thickish, ventricose, with a greenish tinge.   **Stem** 2½–3½ in. long, up to ⅔ in. thick, cylindrical, with numerous concentric rings of squarrose scales, up to the superior, large ring; whitish, or the edges of the scales often tipped with red, solid.   *Fries.*

In pastures, etc.

Intermediate between Lepiota and Amanita.

Noted by Fries as poisonous.   It may or may not be, but as a matter of precaution it is described.   A large species, pure white, extremely beautiful.

Massachusetts, *Farlow.*

L. nauci'na Fr.   No translation applicable.   **Pileus** 1–1½ in. broad, white, the disk of the same color, fleshy, soft, gibbous or obtusely umbonate when flattened, even, *the thin cuticle splitting up into granules.* **Stem** 1½–3 in. long, stuffed, at length *somewhat hollow*, but without a definite tube, *attenuated upward* from the thickened base, fibrillose, unspotted, white. **Ring** *superior*, tender, but persistent, *adhering to* Lepiota. *the stem*, at length reflexed. **Gills** free, approximate, crowded, ventricose, soft, white.

There is a prominent collar, as in the Clypeolarii, embracing the stem. Stature and appearance of L. excoriata, but commonly smaller, the superior ring adfixed, etc. *Fries.*

Spores subglobose, 6–7µ *Massee.*

L. naucina Fr. is the European species which has its American counterpart in L. naucinoides Pk. The variations in the American species are noted under L. naucinoides.

As Amanita phalloides—in its white form—the poisonous white Amanita, resembles L. naucina or L. naucinoides in some stages of its growth and may be confounded with it, careful note should be taken of their external differences. In L. naucinoides the bulb and stem are continuous, each passing into the other imperceptibly; in A. phalloides the junction of stem and bulb is abrupt and remains so, and the bulb is more or less enwrapped in the volva. The ring is also larger than in L. naucinoides and is pendulous, and the gills are permanently white. A certain means of distinguishing between them is by the application of heat as in cooking. On toasting both it will be found that the gills of the Amanita *remain white*, but those of the Lepiota *turn quickly brown*.

L. naucinoi'des Pk. No translation applicable.([Color] Plate XII, Plate XV, p.37.) **Pileus** soft, smooth, white or snowy-white. **Gills** free, white, slowly changing with age to a dirty pinkish-brown or smoky-brown color. **Stem** ringed, slightly thickened at the base, colored like the pileus. **Spores** subelliptical, uninucleate, white, 8–10 long x5–8µ broad. *Peck*, 48th Rep. N. Y. State Bot.

Kansas, *Cragin;* Wisconsin, *Bundy;* New Jersey, *Ellis;* Iowa, *Macbride;* New York, *Peck*, 23d, 29th, 35th Rep.,; Indiana, *H. I. Miller, Dr. J. R. Weist.*

L. naucinoides Pk. is the American counterpart of L. naucina Fr., a European species, excepting that the spores of the latter are described as globose. The caps are ovate when young and usually from 1½–3 in. across when expanded, but occasionally reach 4 in., smooth, but frequently rough or minutely cracked in the center, white or varying shades of white deepening in color at the summit. In a rare form var. squamo'sa, large, thick scales occur which are caused by the breaking

Lepiota. up of the cap surface. When young the gills are white or faintly yellow, becoming pinkish or dull brown in age. The pinkish hue is not always apparent. The outer edge of the veil or ring is thickest; usually it is firmly attached to the stem, but movable rings are frequently noticed. When the plant ages the ring is often missing, but traces of it are always discernible. Stem rarely equal, often it is distinctly bulbous, generally tapering upward from a more or less enlarged base, hollow when fully grown, until then containing cottony fibers within the cavity or appearing solid, 2–3 in. long, ¼–½ in. thick.

Its habitat is similar to that of the common mushroom—lawns, pastures, grassy places—though unlike the latter it is found in woods. Until thoroughly acquainted with it, specimens found in woods and supposed to be L. naucinoides should not be eaten. An Amanita might be mistaken for it. It is readily distinguishable from the common mushroom and its allies by the color of the gills and spores which are white, and differences in stem and veil.

It is found from July until after hard frosts. It was first reported edible by Professor Peck in 1875, under the name of Agaricus naucinus.

The L. naucinoides is rewarding the favor with which it has been received as an esculent, it being equal to the common mushroom and quite free from insects. Large crops of it are reported from all over the country, and from many sections it is told of as a stranger. During 1897–98 the author has found it in plenty upon ground familiar to him for years, upon which it had not previously shown itself. The common mushroom must look to its laurels.

Its cultivation as a marketable crop is possible and probable.

L. cepæsti′pes Sow.—*cepa*, an onion; *stipes*, stem. ([Color] Plate XII, fig. 3.) **Pileus** thin, at first ovate, then bell-shaped or expanded, umbonate, soon adorned with numerous *minute brownish scales*, which are often *granular or mealy, folded into lines* on the margin, white or yellow, the umbo darker. **Gills** thin, close, free, white, becoming dingy with age or in drying. **Stem** rather long, tapering toward the apex, generally *enlarged in the middle or near the base*, hollow. **Ring** thin, subpersistent. **Spores** subelliptical, with a single nucleus, 8–10x5–8μ.

Plant often cespitose, 2–4 in. high. **Pileus** 1–2 in. broad. **Stem** 2–3 lines thick.

46

Rich ground and decomposing vegetable matter. Also in graperies Lepiota. and conservatories. Buffalo, *G. W. Clinton;* Albany, *A. F. Chatfield. Peck,* 35th Rep. N. Y. State Bot.

Spores elliptical, 7–8x4µ *Massee;* 8x4µ *W.G.S.;* 8–10x5–8µ *Peck.* Haddonfield, N. J., Pennsylvania, *McIlvaine;* New York, *Mrs. E. C. Anthony;* Indiana, *H. I. Miller.* July to October.

Whoever has seen the seed-stalks of an onion knows the shape from which this fungus takes its name. The dense clusters are graceful, dainty, and contain many individuals of all ages—from the very young with egg-shaped heads, like pigmy C. comatus, to the fluff-capped eldest, willowy and fair to look upon. The out-door kind soon droops when matured; the young plants of a cluster will remain fresh for several days after taken from their habitat. Stems in these tufts are often quill-shaped, and the striations on the cap margins are shorter than those on their indoor cousins. These grow in hot-houses and stables. One of the two forms has a yellow cap, the other is white and fair.

These forms have often come to my table as a pleasant winter surprise. Children in the hot-houses of Haddonfield, N. J., watched for its appearance among the bedded plants, sure of a present when they brought me a meal of it. Both the white and yellow varieties were equally enjoyed.

The entire fungus is tender and delicious cooked in any way.

L. farino'sa Pk.—*farina,* meal. **Pileus** thin, rather tough, flexible, at first globose or ovate, then bell-shaped or convex, covered with a soft, dense, white veil of mealy down, which soon ruptures, forming irregular, easily-detersible scales, more persistent and sometimes brownish on the disk. **Flesh** white, unchangeable. **Gills** close, free, white, minutely downy on the edge. **Stem** equal or slightly tapering upward, somewhat thickened at the base, slightly mealy, often becoming glabrous, hollow or with a cottony pith above, solid at the base, white, pallid or straw-colored, the ring lacerated, somewhat appendiculate on the margin of the pileus, evanescent. **Spores** subovate, 10–13x8µ.

Pileus 1.5–2.5 in. broad. **Stem** 2–3 in. long, 2–4 lines thick.

Mushroom beds in a conservatory, Boston, Mass. March. Communicated by *E. J. Forster.*

This species is related to L. cepæstipes, from which it may be dis-

Lepiota. tinguished by its pileus, which is not folded on the margin, and by its larger spores. It is edible. It is very distinct from Amanita farinosa. *Peck*, 43d Rep. N. Y. State Bot.

Ohio, *Lloyd, Prof. William Miller* (Lloyd Myc. Notes).

L. America′na Pk.   ([Color] Plate XII, fig. 1.)   Pileus

at first ovate, then convex or expanded, umbonate, scaly, white, the umbo and scales reddish or reddish-brown. Gills close, free, white. Stem somewhat thickened at or a little above the base, hollow, bearing a ring, white. Spores subelliptical, uninucleate, 8–10x5–8μ.

The American lepiota belongs to the same genus as the parasol mushroom and the Smooth lepiota. It has one character in which it differs from all other species of Lepiota. The whole plant when fresh is white, except the umbo and the scales of the cap, but in drying it assumes a dull reddish or smoky-red color. By this character it is easily recognized.

In the very young plant the cap is somewhat egg-shaped and nearly covered by the thin reddish-brown cuticle, but as the plant enlarges the cuticle separates and forms the scales that adorn the cap. On the central prominence or umbo, however, it usually remains entire. The margin of the cap is thin and is generally marked with short radiating lines or striations. The gills do not quite reach the stem and are, therefore, free from it. Sometimes they are connected with each other at or near their inner extremity by transverse branches. They are a little broader near the margin of the cap than at their inner extremity. The stem affords a peculiar feature. It is often enlarged towards the base and then abruptly narrowed below the enlargement, as in the Onion-stemmed lepiota. In some instances, however, the enlargement is not contracted below and then the stem gradually tapers from the base upward. The stem is hollow and usually furnished with a collar, but sometimes this is thin and may disappear with advancing age. Wounds or bruises are apt to assume brownish-red hues.

The caps vary in width from 1–4 in.; the stems are from 3–5 in. long, and 2–5 lines thick. Sometimes plants attain even larger dimensions than these. The plants grow singly or in tufts in grassy ground or on old stumps. They may be found from July to October.

In flavor this species is not much inferior to the parasol mushroom, but when cooked in milk or cream it imparts its own reddish color to

48

Photographed by Dr. J. R. Weist.

LEPIOTA AMERICANA

PLATE XXVII.

CLITOCYBE MONADELPHA.

the material in which it is cooked.   It is, however, a fine addition to Lepiota. our list of esculent species.   *Peck*, 49th Rep. N. Y. State Bot.

I found several on a decaying willow trunk, and on the ground beside it, in Philadelphia.   In July, 1898, large quantities, often clustered, grew under the great, open auditorium of the Pennsylvania Chautauqua, at Mt. Gretna, Pa., from ground covered with crushed limestone.

The caps are meaty and excellent in flavor.   They should be broiled or fried.

GRANULOSI.   Pileus granular or warty.   Stem sheathed, etc.

L. granulo'sa Batsch.—*granosus*, full of grains.   **Pileus** thin, convex or nearly plane, sometimes almost umbonate, rough, with numerous granular or branny scales, often radiately wrinkled, rusty-yellow or reddish-yellow, often growing paler with age.   **Flesh** white or reddish-tinged.   **Gills** close, rounded behind and usually slightly adnexed, white.   **Stem** equal or slightly thickened at the base, stuffed or hollow, white above the ring, colored and adorned like the pileus below it.   **Ring** slight, evanescent.   **Spores** elliptical 4–5x3–4µ.

Plant 1–2.5 in. high.   **Pileus** 1–2.5 in. broad.   **Stem** 1–3 lines thick.   Woods, copses and waste places.   Common.   August to October.

This is a small species with a short stem and granular reddish-yellow pileus, and gills slightly attached to the stem, a character by which it differs from all the preceding.   The ring is very small and fugacious, being little more than the abrupt termination to the coating of the stem. *Peck*, 35th Rep. N. Y. State Bot.

Spores 5–6x3µ *B.;* 3x4µ *W.G.S.;* elliptical, 4–5x3–4µ *Peck.*

Var. *rufes'cens* B. and Br.   Pure white at first, then partially turning red and in drying acquiring everywhere a reddish tint.

Var. *al'bida* Pk.   Persistently white.

Though small many plants grow neighboring.   Being fleshy for their size, and of pleasing quality, they well repay gathering.   Remove stems.

Open woods, Angora, West Philadelphia; Haddonfield, New Jersey, *McIlvaine.*

A. CUTICLE VISCID. NEITHER SCALY NOR WARTY.

L. delica'ta Fr.—*delicatus*, delicate.   Up to 1 ½ in. across, reddish, becoming yellowish toward margin.   Flesh well proportioned to cap,

Lepiota. convex, obscurely umbonate, glabrous, slightly viscid. **Stem** 1 ½–2 in. long, very thin, but covered with dense downy scales, equal, lighter than cap. **Ring** usually entire, membranaceous, fluffy from scales. **Gills** free, crowded, ventricose, white.

Haddonfield, N. J., January, 1896–97, in hot-houses. *McIlvaine.*

A delicate, delicious Lepiota. Though small, it is meaty. Its appearance in hot-houses (it is found in woods) insures a crop at a time of year when other species are not plentiful, and when anything edible in the toadstool line is most welcome to their lovers.

L. lenticula'ris Lasch.—*lenticula*, a lentil. **Pileus** at first globose, then convex, even, naked, pinkish-tan color. **Flesh** thick, spongy, white. **Gills** close to stem, but free from it, ventricose, crowded, whitish. **Stem** 4–6 in. high, thick, equal or swollen at base, solid but spongy, more or less covered with scales; above the ring it is frequently covered with drops of water more or less green, which leave spots when they dry. **Veil** superior and very large.

Pileus 3–4 in. across. **Stem** 4–6 in. long, ½ in. and more thick. In damp woods.

Redman's Woods, Haddonfield, N. J. September, 1894. *McIlvaine.*

This species is included in Amanita by Fries and Stevenson. Massee places it in Lepiota. In the dozen or more specimens I have found, there was no trace of a volva, even when very young. I tested it carefully and at one time ate three good-sized caps without experiencing any indications of poison. I have seen it during but one season and not then (at one time) in sufficient quantity to make a meal off it. Cooked it has a slight cheesy flavor which is pleasant.

L. illi'nita Fr.—*illino*, to smear over. **Pileus** rather thin, soft, at first ovate, then campanulate or expanded, subumbonate, smooth, white, very viscid or glutinous, even or striate on the margin. **Gills** close, free, white. **Stem** equal or slightly tapering upward, stuffed or hollow, viscid, white. **Spores** broadly elliptical, 5x4µ broad.

Plant 2–4 in. high. **Pileus** 1–2.5 in. broad. **Stem** 2–3 lines thick. Thin or open woods. Adirondack mountains. July to September.

This is a smooth white species with the stem and pileus clothed with a clear viscid or glutinous veil. The margin of the pileus is often even, but the typical form of the species has it striate. The flesh is soft and white.   The species may be distinguished from the viscid white species Lepiota. of Hygrophorus by the free, not adnate nor decurrent lamellæ.   *Peck,* 35th Rep. N. Y. State Bot.

Springton and Mt. Gretna, Pa., 1887–1897.   *McIlvaine.*

Not yet found by me in quantity.   Several specimens eaten were of good flavor.

L. rugulo'sa Pk.   **Pileus** thin, submembranaceous, broadly convex or nearly plane, umbonate, rugulose, widely striate on the margin, whitish.   **Lamellæ** thin, narrow, close, free, whitish.   **Stem** short, equal, slightly silky, whitish, the annulus thin, persistent, white.   **Spores** elliptic, 7.5µ long, 4µ broad.

Pileus 12–20 mm. broad.   **Stem** about 2.5 cm. long, 2 mm. thick.

Moist grassy places under trees.   Washington, D. C.   July.   *Mrs. E. M. Williams.*   Perhaps in the fresh state the pileus is not as distinctly rugulose as when dry.   *Peck,* Bull. Torrey Bot. Club, Vol. 27, January, 1900.

ARMILLA'RIA Fr.

Armilla, a ring.

Armillaria. **Pileus** and **Stem** continuous. **Veil** partial, sometimes only indicated
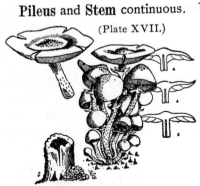
(Plate XVII.)
by the scales which clothe the stem terminating in the form of a ring. **Spores** white. On the ground or on stumps.

In the young plant the veil extends from the stem to the pileus, sometimes forming scaly patches upon it; below the ring it is attached to the stem often in scales.

But for the presence of the ring

ARMILLARIA MELLEA.

the species of this genus could be distributed in Tricholoma, Clitocybe and Collybia, with which they agree in all other characters.

In Amanita and Lepiota, the other ringed genera of the white-spored series, the flesh of the stem and pileus is not continuous; and their stems are therefore easily separated. Amanita is also distinguished by its volva.

ANALYSIS OF TRIBES.

TRICHOLOMATA. Page 52.

Gills sinuately adnexed, stem fleshy, ring often evanescent. (Like Tricholoma.)

CLITOCYBÆ. Page 55.

Gills not sinuate, more or less decurrent, narrowed behind; ring permanent. (Resembling Clitocybe.)

COLLYBIÆ. Page 58.

Gills adnate, equal behind; stem somewhat cartilaginous outside; ring permanent. (Resembling Collybia.)

I.—TRICHOLOMATA. Gills sinuately adnexed, etc.

A. robus'ta A. and S.—*robustus*, robust, sturdy. Substance of entire plant compact. **Pileus** 2–3 in. across, varying in shades of gray and

brown, scaly, fibrillose on margin, decreasing toward center or smooth, Armillaria.
convex or top-shaped and margin involute at first, expanding. **Flesh**
firm, very thick. **Gills** broad, emarginate, nearly free, crowded, whitish,
up to ½ in. broad. **Veil** large, membranaceous, sometimes floccose,
remaining adherent to the stem. **Stem** 1–2 in. long, obese, solid,
tapering at the base, brownish-white and fibrillose below veil, white and
flocculose above, flesh of stem continuous with that of the cap.

Stevenson gives var. *minor* with even cap with both gills and ring
very narrow.

Spores ovoid-spherical. 7µ. Q.

Edible, *Curtis;* District Columbia, *Mrs. M. Fuller.*

In mixed woods. Pennsylvania, West Virginia, New Jersey, *McIl-vaine.*

The substance of A. robusta differs from all other Armillaria in being
very compact. It is not acrid but has a marked flavor. Cut into small
pieces and well cooked it makes an acceptable dish. It is best in
croquettes and patties, or served with meats.

A. viscid'ipes Pk.—*viscidus*, sticky; *pes*, a foot. **Pileus** fleshy, compact, convex or nearly plane, glabrous, whitish with a slight yellowish
or reddish-yellow tint. **Flesh** white, odor peculiar, penetrating, sub-
alkaline. **Gills** narrow, crowded, sinuate or subdecurrent, whitish.
Stem equal, solid, viscid and slightly tinged with yellow below the
narrow membranous ring, whitish above. **Spores** elliptical, 8x5µ.

Pileus 3–6 in. broad. **Stem** 3–4 in. long, 6–12 lines thick.

In mixed woods. Rock City, Dutchess county. October.

It is a large fine fungus, easily known by its white and yellowish hues,
its crowded gills, viscid stem and peculiar penetrating almost alkaline
odor. The cuticle of the pileus is thin and soft to the touch, but it
sometimes cracks longitudinally and is sometimes slightly adorned with
innate fibrils. A. dehiscens is said to have a viscid stem, but it is also
squamose and the pileus is yellowish-ochraceous. *Peck*, 44th Rep
N. Y. State Bot.

Quite common in Pennsylvania and New Jersey. *McIlvaine.*

It loses its strong odor when cooked and is equal to other Armillaria
in edibility. Unless well cooked it has a slight saponaceous flavor.
This is easily overcome by a few drops of lemon juice or sherry.

A. appendicula'ta Pk.—bearing an appendicula or small appendage. **Pileus** broadly convex, glabrous, whitish, often tinged with rust color or brownish rust color on the disk. **Flesh** white or whitish. **Gills** close, rounded behind, whitish. **Stem** equal or slightly tapering upward, solid, bulbous, whitish, the veil either membranous or webby, white, commonly adhering in fragments to the margin of the pileus. **Spores** subelliptical, 8x5μ.

Pileus 2–4 in. broad. **Stem** 1.5–3.5 in. long; 5–10 lines thick. Auburn, Ala. October. *C. F. Baker.*

The general appearance of this species is suggestive of Tricholoma album, but the presence of a veil separates it from that fungus and places it in the genus Armillaria. The veil, however, is often slightly lacerated or webby and adherent to the margin of the pileus. *Peck*, Bull. Torrey Bot. Club, Vol. 24.

Mt. Gretna, Pa., Angora, Pa. On decaying roots in ground. August to November. Found plentifully in resorts of other Armillaria. Edibility the same. *McIlvaine.*

A. pondero'sa Pk.—*ponderosus*, weighty, ponderous. **Pileus** thick, compact, convex or subcampanulate, smooth, white or yellowish, the naked margin strongly involute beneath the slightly viscid, persistent veil. **Gills** crowded, narrow, slightly emarginate, white inclining to cream color. **Stem** stout, subequal, firm, solid, coated by the veil, colored like the pileus, white and furfuraceous above the ring. **Flesh** white. **Spores** nearly globose, 4μ in diameter.

Plant 4–6 in. high. **Pileus** 4–6 in. broad. **Stem** about 1 in. thick. Ground in woods. Copake, Columbia county. October.

The veil for a long time conceals the gills, and finally becomes lacerated and adheres in shreds or fragments to the stem and margin of the pileus. *Peck*, 26th Rep. N. Y. State Bot.

New England, *Frost;* New York, *Peck*, Repts. 26, 29, 41. West Virginia and Pennsylvania. Ground in woods. September to November. *McIlvaine.*

Professor Peck says in 26th Report: "This species has not been found since its discovery in 1872."

Where the Armillaria mellea frequents I have often found A. ponderosa. It was plentiful at Mt. Gretna, Pa., in September, 1898.

Young specimens are quite as edible as A. mellea, and rather more Armillaria. juicy.

II.—CLITOCYBÆ. Gills not sinuate, etc.

A. mel'lea Vahl.—*melleus*, of the color of honey. ([Color] Plate XVI, fig. 1.) **Pileus** adorned with minute tufts of brown or blackish hairs, sometimes glabrous, even or when old slightly striate on the margin. **Gills** adnate or slightly decurrent, white or whitish, beċoming sordid with age and sometimes variegated with reddish-brown spots. **Stem** ringed, at length brownish toward the base. **Spores** elliptical, white, 8–10μ long. *Peck*, 48th Rep. N. Y. State Bot.

Spores 9x5–6μ *W.G.S.;* 10x8μ *B.;* 8–10μ *Peck.*

The A. mellea is unusually prolific and is common over the United States and Europe. Specimens may be found in the spring-time, but in middle latitudes it is common from August until after light frosts. It is usually in tufts, some of which contain scores of plants and are showy over ground filled with roots, or on stumps or boles of decaying trees. It frequents dense woods and open clearings. I have seen acres of dense woodland at Mt. Gretna, Pa., so covered with it and its varieties that but few square yards were unoccupied.

A description of the typical A. mellea will rarely apply to any one plant. A combination of its variable features in one description would include something of nearly every white-spored Agaric under the sun. Yet there is something indescribable about it which once learned will unerringly betray it.

Its **Caps** vary from perfectly smooth, through tufts of scales and hairs, more or less dense, to matted woolliness. It may show any one of these conditions in youth and be bald in age. Some shade of yellow is the prevailing color, but this will vary from whitish to dark-purplish or reddish-brown. When water-soaked it is one color, when dry, another. Commonly the margins of the **Caps** are striated, sometimes they are smooth as a cymbal, and not unlike one, have a raised place or umbo in the center. **Flesh** white or whitish. **Gills** when young are white or creamy, usually running down the stem, sometimes slightly notched at attachment. They freckle in age and lose their fair complexion. The **Veil** or collar about the stem is as variable as fashion—thick and closely woven or flimsy as gossamer, or vanishing as the plant grows old. The

55

Armillaria. **Stems** may be even as a lead pencil, or swollen like a pen-holder, or bulbous toward the base, or distorted by pressure in the tufts. It is as variable in color as the cap, usually darkening downward in hues of brown. The outside is firm and fibrous, sometimes furrowed, inside soft or hollow.

Cap 1–6 in. across. **Stem** 1–6 in. long, ¼–¾ in. thick.

Var. *obscu'ra* has the cap covered with numerous small blackish scales.

Var. *fla'va* has the cap yellow or reddish-yellow, but in other respects it is like the type.

Var. *gla'bra* has the cap smooth, otherwise like the type.

Var. *radica'ta* has a tapering, root-like prolongation of the stem, which penetrates the earth deeply.

Var. *bulbo'sa* has a distinctly bulbous base to the stem, and in this respect is the reverse of var. radicata.

Professor Peck writes: "Var. *exannulata*([Color]Plate XVI, fig. 2.) has the cap smooth and even on the margin, and the stem tapering at the base. The annulus is very slight and evanescent or wholly wanting. The cap is usually about an inch broad, or a little more, and the plants grow in clusters, which sometimes contain forty or fifty individuals. It is more common farther south than it is in our state (N. Y.), and is reported to be the most common form in Maryland. This I call var. exannulata." From *Dr. Taylor*, Washington, D. C.; Indiana, *H. I. Miller*.

To these may be added also var. *al'bida* Pk. in which the pileus is white or whitish.

A variety, perhaps a variation of var. bulbosa was sent to me by E. B. Sterling, Trenton, N. J., and afterward found by myself at Mt. Gretna, Pa. The **Cap** purplish-brown, convex, striate and light on margin, edge irregular with parts of veil attached. **Flesh** white, very thin. **Gills** decurrent, arcuate, pinkish-gray. **Stem** stuffed, fibrous, white above, dense floccose veil, same color as cap below, swollen toward base which is pointed, sulcate, white inside, closely clustered and some of the stems distinctly bulbous. **Taste** decidedly unpleasant. An intense acridity develops and increases when the juices of raw pieces are swallowed, and the salivary glands are much excited. The acridity is not lost in cooking. It simply can not be eaten. Specimens were sent by me to Professor Peck who referred it to A. mellea.

I have never seen the abortive form of Clitopilus abortivus, though found in many places and in great quantity, showing any part or trace Armillaria. of the original plant. But that a similar monstrosity occurs upon A. mellea is shown by individuals and parts of individuals of a cluster being aborted. Without such positive proof, no one would suspect either of these odd formations to be abortive of either C. abortivus or A. mellea, or any other fungus. I consider the abortive form of A. mellea far superior in substance and flavor to it or any of its varieties.

The Armillaria can not be ranked among the tender or high-flavored toadstools, yet their abundance, meaty caps and nourishing qualities place them among our most valuable food species.

The caps when chopped into small pieces make good patties and croquettes. They have an impressive flavor of their own, and offer an esculent medium for seasoning and the gravies of various meats.

A. nardos'mia Ellis—*nardosmius*, of the odor of nardus. (A name applied by the ancients to several plants, especially *spica nardi*—spikenard.) **Pileus** fleshy, firm, thick and compact on the disk, thin toward the margin, whitish, variegated with brown spots, with a thick, tough and separable cuticle. **Flesh** white. **Gills** crowded, subventricose, slightly emarginate, whitish. **Stem** solid, fibrous, not bulbous, sheathed below by the brown velvety veil, the ring narrow, spreading, uneven on the edge. **Spores** subglobose, 6μ in diameter.

Pileus about 3 in. broad. **Stem** 1.5–3 in. long, 4–6 lines thick.

Ground in woods, Suffolk county. September. *Peck*, 43d Rep. N. Y. State Bot.

Several specimens from sandy grounds in pine woods, Haddonfield, N. J., were sent by me to Professor Peck and were identified by him. Plentiful at Mt. Gretna, Pa., September to frost, 1898. In mixed woods, on gravelly ground. Eaten in quantity by several persons. *McIlvaine*.

Cuticle of caps when dry breaking up into brownish, squamulose scales, margin involute. **Gills** subdecurrent. **Veil** thick, persistent. **Stem** short, subbulbous, solid. **Flesh** white. Very much resembles a short-stemmed Lepiota. Smell and taste strong, like almonds. Disappears in cooking.

III.—COLLYBIÆ. Gills adnate, stem somewhat cartilaginous.

Armillaria.   **A. mu'cida** Schrad.—*mucidus*, slimy.   **Pileus** commonly shining white, thin, almost transparent, hemispherical then expanded, obtuse, more or less radiato-wrinkled, smeared over with a thick tenacious gluten; margin striate when thinner.   **Stem** 1½–3 in. long, 1–2 lines thick at the apex, thickened at the base, stuffed, thin, rigid, curved ascending, smooth, white, but sooty scaly at the base when most perfectly developed.   **Ring** inserted at the apex of the stem, bent downward and glued close to the stem, furrowed, the white border again erect, with a swollen and entire margin, which sometimes becomes dingy brown. **Gills** rounded behind, obtuse, adhering to the stem and striato-decurrent, distant, broad, lax, mucid, always shining white.

Very variable in stature, from 1 in. (when of this size the stem is almost equal) to as much as 6 in. broad.   The color of the pileus varies gray, fuliginous, olivaceous.   The gills sometimes become yellow, but only from disease.   Sometimes solitary, sometimes a few are joined in a cespitose manner at the base.   *Stevenson.*

Spores elliptical, 15–16x8–9µ *Massee;* 17x14µ *W. G. S.*

North Carolina, *Schweinitz, Curtis;* Pennsylvania, *Schweinitz;* Maryland, *Miss Banning.*

West Virginia mountains, 1882, Haddonfield, N. J., 1891–94, on beech trees and roots.   *McIlvaine.*

Commonly considered esculent in Europe.

Dirt adheres so tenaciously to it that it is difficult to clean.   This, however, occurs only when the fungus grows from roots and pushes its way up through covering earth.   When growing from trees it is attractive and of good quality.

Should be chopped fine and well cooked.

TRICHOLO'MA Fr.

Gr.—a hair, a fringe.

Pileus symmetrical, generally fleshy, never truly umbilicate, seldom Tricholoma. umbonate. **Veil** absent or appearing only as fibrils or down on the margin of the pileus. **Gills** sinuate (the small sudden curve near the stem always apparent in the young plant), sometimes with a slightly decurrent tooth. **Stem** central, usually stout, fleshy-fibrous, without a bark-like skin. **Flesh** continuous with that of the pileus. **Ring** and **Volva** absent. **Spores** white or dingy.

(Plate XIX.)

SECTION OF TRICHOLOMA.

But one is known to be poisonous. Some are acrid or unpleasant in flavor. With one exception all grow on the ground in pastures and woods, appearing from May to late in the autumn.

Gills generally white or dingy, frequently spotted or stained. The pileus may be smooth or adorned with fibrous or downy scales, dry, moist, viscid or water-soaked.

The distinguishing feature of Tricholoma is the sinuate gills. In Collybia the stem bears a distinct bark-like skin; in Clitocybe the gills are never sinuate; species of Pleurotus are distinguished by growing on wood only, and Paxillus by their strongly-incurved margin and anastomosing gills.

In cooking Tricholoma consistency must be the guide to plan and time. The tougher varieties require to be cut into small pieces and to be well cooked, while the brittle and delicate varieties will cook quickly. Many of them make excellent soups.

ANALYSIS OF TRIBES.

A. PILEUS VISCID, FIBRILLOSE, SCALY OR DOWNY, NOT WATER-SOAKED.

Stem fibrillose from the remains of the adnate universal veil.

59

LIMACINA (*limas*, a slug or snail, slimy). Page 61.

Tricholoma. Cuticle of pileus viscid when moist, innately fibrillose or scaly, but not lacerated; flesh of pileus thick, firm; margin almost naked.
* Gills not discolored, nor becoming reddish.
** Gills discolored, usually spotted with reddish-brown.

GENUINA. Page 67.

Cuticle of the pileus never moist or viscid; torn into downy or floc-cose scales. Flesh soft, not water-soaked; margin involute and slightly downy at first.
* Gills not changing color, nor spotted with red or black.
** Gills becoming reddish or gray, the edge at last generally with reddish or black spots.

RIGIDA (*rigeo*, to be stiff). Page 74.

Pileus rigid, hard, somewhat cartilaginous when fleshy, very fragile when thin, cuticle rigid, granulated or broken up when dry into smooth scales, not torn into fibrils. Young specimens occur which are fibrillose from the veil, not from laceration of the cuticle.
* Gills white or pallid, not becoming spotted with red or gray.
** Gills becoming reddish, grayish or spotted.

SERICELLA (*sericeus*, silky). Page 74.

Pileus first slightly silky, soon becoming smooth, very dry, neither moist, viscid, water-soaked, nor distinctly scaly; rather thin, opaque, absorbing moisture, but is the same color as the gills. Stem fibrous, by which the smaller species resembling Collybia may be distinguished.
* Gills broad, rather thick, somewhat distant.
** Gills narrow, thin, crowded.

B. PILEUS EVEN, SMOOTH, NOT DOWNY NOR SCALY, NOT VISCID.

In rainy weather moist; when very young pruinose (but rarely con-spicuously) from the universal veil. Flesh soft and spongy or very thin when it is water-soaked.

GUTTATA (*gutta*, a drop). Page 76.

Pileus fleshy, soft, fragile, marked with drop-like spots or rivulose. Appearing in spring, rarely in autumn.

Cespitose, in troops or often in rings.
* Gills whitish.
** Gills becoming reddish or smoky-gray.

SPONGIOSA (*spongia*, a sponge). Page 78.

Pileus compact, then spongy, obtuse, even, smooth, moist but not hygrophanous; firm, growing in troops late in the autumn. Stem stout, base usually thickened, spongy fibrous. Gills at length decurrent but sinuate, by which character they are distinguished from Clitocybe.
* Gills not discolored.
** Gills discolored.

HYGROPHANA (*Gr.*, wet; to appear). Page 80.

Pileus thin, somewhat umbonate; flesh at length soft, watery. Stem rootless, containing a pith, entirely fibrous.

Flesh not exceeding in depth the width of the not broad, thin gills; thinnest toward the margin, hence somewhat umbonate. Color of the pileus either moist or dry, very variable in the same species. Pileus sometimes pulverulent from the persistence of the veil in dry weather.
* Gills whitish, not spotted.
** Gills more or less violet, gray or smoky. Not represented.

Series A.

PILEUS VISCID OR FIBRILLOSE, DOWNY OR SCALY.

I.—LIMA'CINA. Viscous when moist.

* Gills not becoming discolored, nor becoming reddish.

T. eques'tre Linn.—*equestre*, belonging to a horseman or knight, from distinguished appearance. **Pileus** fleshy, compact, convex becoming expanded, obtuse, pale-yellowish, more or less reddish tinged, the disk and central scales often darker, the margin naked, often wavy. **Flesh** white or tinged with yellow. **Gills** rounded behind, close, nearly free, *sulphur-yellow*. **Stem** stout, solid, pale-yellow or white, white within. **Spores** 6.5–8x4–5μ.

Pileus 3–5 in. broad. **Stem** 1–2 in. long, 6–10 lines thick.

Tricholoma. Pine woods, especially in sandy soil. Albany county. September to November.

This is a noble species but not plentiful in our state (N. Y.). The pileus is said to become greenish very late in the season. The stem, in the typical form, is described as sulphur-yellow in color, but with us it is more often white. The scales of the disk are sometimes wanting. In our plant the taste is slightly farinaceous at first, but it is soon unpleasant.

Var. *pinastreti* A. and S. is a slender form having a thin, even pileus, thinner and more narrow gills and a more slender stem. A. crassus Scop., A. aureus Schaeff., and A. flavovirens Pers. are recorded as synonyms of this species. *Peck*, 44th Rep. N. Y. State Bot.

Professor Peck later says in "Mushrooms and Their Use," p. 52: "I confidently add it to the list of edible species."

New Jersey, Pennsylvania and West Virginia. In pine forests and groves. September to frost. *McIlvaine.*

I have eaten it since 1883. All disagreeable odor about T. equestre (which I have seldom noticed) disappears upon cooking. The substance is rather tough, but good.

T. coryphæ'um Fr.—chief, leader. From its distinguished appearance. **Pileus** very fleshy but not compact, convex then plane, obtuse, viscid, yellowish, streaked with small brownish scales. **Stem** solid, attenuated upward. **Gills** emarginate, crowded, white, edge yellow.

Large and of striking appearance. In shady beech woods.

Pronounced a good edible by the Boston Myc. Club.

The color of the plants is given as greenish-yellow. Bull. Boston Myc. Club, 1896.

T. ustale Fr.—*uro*, to burn. **Pileus** fleshy, convex, then plane, obtuse, *even, smooth,* viscid, bay-brownish. **Stem** stuffed, equal, dry, rufo-fibrillose, apex naked, silky, nearly smooth. **Gills** emarginate, crowded, white, at length with reddish spots. *Cooke.*

Chiefly in pine woods.

Pileus 3 in. **Stem** 2–3 in. long, about ½ in. thick.

Spores 5x8μ *W. G. S.;* 7–8x5μ *Massee.*

North Carolina, *Curtis,* pine woods, *Schweinitz;* Kansas, *Cragin.* Massachusetts. Edible. Boston Myc. Club, Bull. No. 5.

T. resplen'dens Fr.—shining brightly. **Pileus** fleshy, convex then nearly plane, even, bare, *viscid, white,* sometimes hyaline-spotted or yellowish on the disk, shining when dry, the *margin straight.* **Flesh** white, taste mild, odor pleasant. **Gills** nearly free when young, then emarginate, somewhat crowded, rather thick, entire, white. **Stem** *solid,* bare, subbulbous, even, dry, white. **Spores** 8x4μ.

Pileus 2–4 in. broad. Stem 2–3 in. long, 4–8 lines thick.

Thin woods. Catskill mountains. September. *Peck,* 44th Rep. N. Y. State Bot.

Mt. Gretna, Pa., in mixed woods. October and November. *McIlvaine.* It is of excellent flavor, consistency and food value.

T. transmu'tans Pk.—changing. **Pileus** convex, *nearly bare,* viscid when moist, brownish, reddish-brown or tawny-red, usually paler on the margin. **Flesh** white, taste and odor farinaceous. **Gills** narrow, close, sometimes branched, whitish or pale yellowish, becoming dingy or reddish-spotted when old. **Stem** equal or slightly tapering upward, *bare* or slightly silky-fibrillose, stuffed or hollow, whitish, often marked with reddish stains or becoming reddish-brown toward the base, white within. **Spores** subglobose, 5μ.

Pileus 2–4 in. broad. Stem 3–4 in. long, 3–6 lines thick.

Woods. The plants are often cespitose.

I suspect that Agaricus frumentaceus of Curtis's catalogue belongs to this species. Both the pileus and stem, as well as the gills, are apt to assume darker hues with age or in drying, and this character suggested the specific name. The species is classed as edible. *Peck,* 44th Rep. N. Y. State Bot.

Curtis catalogues T. frumentaceum as edible.

T. transmutans is reported from many states. It has a mealy taste and odor. Wherever it is found it is a valuable food species.

T. sejunc'tum Sow.—separated; from the peculiar manner in which the gills separate from the stem. **Pileus** fleshy, convex then expanded, umbonate, slightly viscid, *streaked with innate brown or blackish fibrils,* whitish or yellowish, sometimes greenish-yellow. **Flesh** white, fragile. **Gills** *broad, subdistant,* rounded behind or emarginate, white. **Stem** solid, stout, often irregular, white. **Spores** subglobose, 6.5μ.

Pileus 1–3 in. broad. Stem 1–3 in. long, 4–8 lines thick.

63

Tricholoma. Mixed woods. Suffolk county, N. Y. September.

The plants referred to this species are not uncommon on Long Island, growing on sandy soil in woods of oak and pine. They are usually more or less irregular and the pileus becomes fragile. It is quite variable in color, sometimes approaching a smoky-brown hue, again being nearly white. The taste of the typical form is said to be bitter, but the flavor of our plant is scarcely bitter. In other repects, however, it agrees well with the description of the species. *Peck*, 44th Rep. N. Y. State Bot.

Spores 6µ *W.G.S.*

Flesh is tender. Cooked, of good body and peculiar but pleasant flavor. A valuable species, baked, scalloped, fried.

T. terri'ferum Pk.—*terra*, earth; *fero*, to bear. **Pileus** broadly convex or nearly plane, irregular, often wavy on the margin, glabrous, viscid, *pale-yellow*, generally soiled with adhering particles of earth carried up in its growth. **Flesh** white, with no decided odor. **Gills** thin, crowded, slightly adnexed, white, not spotted or changeable. **Stem** equal, short solid, white, *floccose-squamulose at the apex*. **Spores** minute, sub-globose, 3µ.

Pileus 3–4 in. broad. Stem 1–1.5 in. long, 6–8 lines thick.

Woods. Catskill mountains. September. *Peck*, 44th Rep. N. Y. State Bot.

Found in West Virginia, Pennsylvania, New Jersey. August to frost. *McIlvaine*.

Not inviting, hard to clean, nevertheless edible and good.

T. portento'sum Fr.—*portentosus*, strange, monstrous. **Pileus** 3–5 in. broad, *sooty*, livid, sometimes violaceous, fleshy, but thin in comparison with the stoutness of the stem, convexo-plane, somewhat umbonate, unequal and turned up, viscid, *streaked with black lines* (innate fibrils), but otherwise even and smooth, the very thin margin naked. **Flesh** not compact, white, fragile. **Stem** commonly 3 in. often 4–6 in. long, 1 in. thick, stout, *solid*, the whole remarkably fibrous-fleshy, somewhat equal, *naked*, but *fibrilloso-striate*, white; the base, which is occasionally attenuato-rooted, villous. **Gills** rounded, almost free, 3–4 lines to as much as 1 in. broad, *distant, white*, but varying, becoming pale-gray or yellow. *Fries*.

Spores 4–5x4µ *K.;* 5x4µ *W.G.S.*

West Virginia, 1882; New Jersey, Pennsylvania, in woods and open places. May to November. *McIlvaine.*

It is one of the first toadstools I experimented upon. I have been constant to it. Its caps fried in butter are unsurpassed.

** *Gills discolored, usually spotted with reddish-brown.*

T. fla′vo-brun′neum Fr.—*flavus,* yellow; *brunneus,* brown. **Pileus** fleshy, conical, then convex, at length expanded, subumbonate, viscid, *clothed with streak-like scales.* **Stem** *hollow, somewhat ventricose,* fibrillose, *at first viscid, yellowish within,* tip naked. **Gills** emarginate, *decurrent,* crowded, yellowish, then reddish. *Fries.*

Odor that of new meal. **Stem** 3–5 in. long, ½ in. thick, dull-reddish or brownish. **Pileus** 3–6 in. broad, disk darker, dingy dull-red or reddish-brown.

North Carolina, *Curtis;* damp woods, A. fulvus, *Schweinitz.*

Edible, *Cooke,* 1891.

T. rus′sula Schaeff.—reddish. ([Color] Plate XVIII, fig. 3.) **Pileus** fleshy, convex, becoming plane or centrally depressed, obtuse, viscid, even or dotted with granular squamules on the disk, *red or incarnate,* the margin usually paler, involute and minutely downy in the young plant. **Flesh** white, sometimes tinged with red, taste mild. **Gills** subdistant, rounded behind or subdecurrent, white, often becoming red-spotted with age. **Stem** solid, firm, whitish or rose-red, squamulose at the apex. **Spores** elliptical, 7x4µ.

Pileus 3–5 in. broad. **Stem** 1–2 in. long, 6–8 lines thick.

Mixed woods. Albany. Cattaraugus and Steuben counties. September and October.

According to the description the typical plant has the pileus incarnate and the stem rosy-red, but in the American plant the pileus is generally more clearly red and the stem white, though this is often varied by reddish stains. *Peck,* 44th Rep. N. Y. State Bot.

Mixed woods. August until after frost. At Mt. Gretna, Pa. 1897–1898 the patches were large, generous yielders.

Edible, *Cooke;* edible, *Cordier, Roques.*

T. russula is a dressy fungus and has a fashion of its own. The mot-

Tricholóma. lings upon its cap, gill and stem, in shades of red, subdued though they be, give it a handsome personality distinct from any other.

The species is a variable one in its minor markings. When moisture is prevalent the caps of all are viscid. Both young and old are often cracked. Stems frequently not squamulose at apex, frequently rosy when young, often flattened. The fibrous interior of the stem and its fibrous connection with the flesh of the cap are very marked. Gills emarginate in youth as well as in age. It is solitary, gregarious, occasionally bunched.

An excellent fungus, a free late grower, meaty, easily cooked, and of fine flavor.

T. frumenta'ceum Bull.—*frumentum*, made of corn. **Pileus** 2–3 in. broad, whitish or clay-color and variegated dull red, truly fleshy, convex then plane, obtuse, viscous, dry in fine weather, *even, smooth.* **Flesh** white. **Stem** 3 in. long, ½ in. thick, *solid, equal, fibrillose* when dry, whitish. **Gills** *rounded*, somewhat crowded, rather broad, white, at length spotted-red.

Wholly *becoming pale white*, but the stem and pileus are alike *marked-red*, and the gills are at length reddish, wherefore, as well as for the *strong smell of new meal*, it is undoubtedly nearest to A. pessundatus. When full grown it has all the appearance of Entoloma. On the ground. *Stevenson.*

Spores 6μ *W.G.S.*

North Carolina, *Curtis.* Edible. Porcher says Dr. Curtis was the first to declare it edible.

T. pessunda'tum Fr.—*pessum dare*, bent downward. **Pileus** fleshy, compact, convex, very obtuse, repand, viscid, *granulose or spotted.* **Stem** solid, firm, at first ovato-bulbous, *everywhere villose with whitish scales.* **Gills** emarginate, nearly free, crowded, white, at length spotted with red.

In pine woods. Odor and taste mealy.

Pileus bay, reddish, paler at the margin. Stature of Ag. equestris. *Fries.*

Spores 5x2.5μ *Massee;* very minute, globose, 2–3μ *C.B.P.*

Reckoned edible, but very rare. *Stevenson.*

California, *H. and M.*

II.—GENUI'NA.   Cuticle of pileus torn into downy or fibrillose scales. Tricholoma.

Gills not changing color nor becoming spotted.

T. decoro′sum Pk.—*decorus*, decorous.   **Pileus** firm, at first hemispherical, then convex or nearly plane, adorned with numerous *brownish subsquarrose tomentose scales*, dull ochraceous or tawny.   **Flesh** white.   **Gills** close, rounded and slightly emarginate behind, the edge slightly scalloped.   **Stem** solid, equal or slightly tapering upward, white and smooth at the top, elsewhere *tomentose-scaly* and colored like the pileus.   **Spores** broadly eiliptical, 5x4μ.

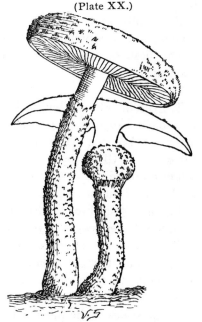

(Plate XX.)

TRICHOLOMA DECOROSUM.
Two-thirds natural size.

Pileus 1–2 in. broad.   **Stem** 2–4 in. long, 2–4 lines thick.

Decaying trunks of trees.   Catskill mountains and Alleghany county. September and October.

A rare but beautiful species.   It is often cespitose.   It departs from the character of the genus in growing on decaying wood.   *Peck*, 44th Rep. **N. Y.** State Bot.

Tricholoma decorosum is not rare in Pennsylvania.   I have found it at Angora, Philadelphia and in Chester county, Pa., growing in clusters and singly.   At first sight one might take it for one of the many forms of Armillaria, but even cursory examination shows the difference.

It is of good consistency and flavor, having a decided mushroom taste.

T. flaves′cens Pk.—pale yellow.   **Pileus** convex, firm, often irregular, dry, *slightly silky becoming bare*, sometimes cracking into minute scales on the disk, *whitish or pale yellow*.   **Flesh** whitish or yellowish. **Gills** close, white or pale-yellow, emarginate, floccose on the edge. **Stems** firm, solid, often unequal, central or sometimes eccentric, single or cespitose, colored like the pileus.   **Spores** subglobose, 5μ in diameter.

67

Tricholoma.   **Pileus** 2–3 in. broad.   **Stem** 1–2.5 in. long, 4–6 lines thick.

Pine stumps.   Albany and Rensselaer counties.   October.

The species seems to be related to T. rutilans but has not the red or purplish tomentum of that fungus.   It, like T. decorosum, is always lignicolous.   T. rutilans is sometimes so.   *Peck*, 44th Rep. N. Y. State Bot.

Frequently found in New Jersey, Pennsylvania and West Virginia. Pine stumps.   September to frost.   *McIlvaine*.

The flesh compares with that of T. rutilans, and makes an equally good dish.

T. gran'de Pk.   **Pileus** thick, firm, hemispherical, becoming convex, often irregular, dry, *scaly*, somewhat silky-fibrillose toward the margin, *white*, the margin at first involute.   **Flesh** grayish-white, taste *farinaceous*.   **Gills** close, rounded behind, adnexed, white.   **Stem** stout, solid, fibrillose, at first tapering upward, then equal or but slightly thickened at the base, pure white.   **Spores** elliptical, 9–11x6μ.

Pileus 4–5 in. broad.   **Stem** 2–4 in. long, 1–1.5 in. thick.

Among fallen leaves in woods.   Cattaraugus county.   September.

The plants are often cespitose, and then the pileus is more or less irregular and the gills somewhat lacerated.   The species is related to T. columbetta, from which its larger size, constantly scaly pileus, more cespitose mode of growth, larger spores and farinaceous taste separate it.   The scales of the pileus are brownish, and the pileus itself is sometimes slightly dingy on the disk.   The young margin is pure white like the stem, and both it and the upper part of the stem are sometimes studded with drops of moisture.

The plant was found on trial to be edible, but not of first quality. The flesh is not very tender, nor the flavor captivating even in young specimens.   *Peck*, 44th Rep. N. Y. State Bot.

Mt. Gretna, Pa.   Mixed woods.   August to frost.   *McIlvaine*.

Gross when old.   Young specimens of medium quality and flavor.

T. columbet'ta Fr.—*columba*, a pigeon. ([Color] Plate XVIII, fig. 5.) **Pileus** convex, then nearly plane, fleshy, obtuse, rigid, somewhat flexuous, dry, *at first bare, then silky-fibrillose*, becoming even or scaly, *white*, the margin at first *involute*, more or less tomentose.   **Flesh** white, taste *mild*. **Gills** close, emarginate, thin, white. **Stem** stout, solid, unequal nearly bare, white. **Spores** 7–8x4.5μ.

The species is very variable and the following varieties have been described:

Var. *A.* **Pileus** nearly always repand or lobed, at first bare, even, at length cracked-scaly, often reddish spotted, the margin when young inflexed, tomentose. **Stem** obese, even, unequal, swollen, an inch thick. The typical form.

Birch wood among mosses.

Var. *B.* **Pileus** subflexuous, silky-fibrillose, at length scaly, sometimes dingy-brown spotted, the margin scarcely tomentose. **Stem** longer, equal or slightly narrowed at the base.

Bushy places. Intermediate between *A* and *C*.

Var. *C.* **Pileus** regular, flattened, evidently fibrillose, sometimes spotted with blue, four inches broad. **Stem** equal, cylindrical, fibrillose-striate, four inches long.

Beech woods. A showy variety so diverse from variety *A* that it might be regarded as a distinct species, did not variety *B* connect them, and so much resemble both that it might with equal propriety be referred to either.

Pileus 2–4 in. broad. **Stem** 1–4 in. long, 3–12 lines thick.

Woods and pastures. Albany county, N. Y.

It may be distinguished from T. album by its mild taste. It is recorded as edible. *Peck*, 44th Rep. N. Y. State Bot.

Edible, *Curtis, Cooke, Stevenson.*

This much varied Tricholoma is as varied in its habitat. I have found it on vacant lots in Philadelphia, in mixed woods at Devon, Pa., and in the forests of the West Virginia mountains, and eaten it since 1881.

It cooks readily and is of mild, agreeable flavor.

T. ru′tilans Schaeff.—*rutilo*, to be reddish. **Pileus** fleshy, campanulate becoming plane, dry, at first *covered with a dark-red or purplish tomentum* then somewhat scaly, the margin thin, at first involute. **Flesh** yellow. **Gills** crowded, rounded, *yellow, thickened and downy on the edge.* **Stem** somewhat hollow, nearly equal or slightly thickened or bulbous at the base, soft, pale-yellow variegated with red or purplish floccose scales. **Spores** 6.5–8x6.5μ.

(Plate XXI.)

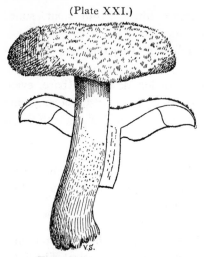

TRICHOLOMA RUTILANS.
About three-eighths natural size.

Pileus 2–4 in. broad. **Stem** 2–4 in. long, 5–8 lines thick.

On or about pine stumps, rarely on hemlock trunks. July to November. *Peck*, 44th Rep. N. Y. State Bot.

Spores subglobose, 5–6μ diameter *Massee;* 6–8x6μ *B.;* 6x9μ *W.G.S.*

West Virginia, Pennsylvania, New Jersey. May to November. *McIlvaine*.

Quite common in West Virginia mountains and in pine woods of New Jersey. The Boston Mycological Club reports it found in quantity in Massachusetts. The flesh when cooked is gummy, like the marshmallow confection. It is excellent.

** *Gills becoming reddish or gray, etc.*

T. vacci'num Pers.—*vacca*, a cow. **Pileus** fleshy, convex or campanulate, becoming nearly plane, umbonate, dry, floccose-scaly, reddish-brown, the margin *involute, tomentose*. **Flesh** white. **Gills** adnexed, subdistant, whitish, then reddish or reddish-spotted. **Stem** equal, *hollow*, covered with a fibrillose bark, naked at the apex, pale reddish.

Spores subglobose, 6μ.

Pileus 1–3 in. broad. **Stem** 2–3 in. long, 4–6 lines thick.

Under or near coniferous trees. Greene and Essex counties. September and October. *Peck*, 44th Rep. N. Y. State Bot.

Recorded as edible by Gillet.

Plentiful in New Jersey, Pennsylvania, West Virginia. Have eaten it since 1885. Fair.

T. fuligi'neum Pk.—*fuligineus*, resembling soot. **Pileus** convex or nearly plane, obtuse, often irregular, dry, minutely scaly, *sooty-brown*. **Flesh** grayish, odor and taste farinaceous. **Gills** subdistant, uneven on the edge, ash-colored *becoming blackish in drying*. **Stem** short, *solid*, equal, bare, ash-colored. **Spores** oblong-elliptical, 8x4μ.

Pileus 1–2.5 in. broad. **Stem** 1–1.5 in. long, 3–5 lines thick.   Tricholoma.
Among mosses in open places. Greene county. September. Rare.
Peck, 44th Rep. N. Y. State Bot.

Quite common in Pennsylvania and New Jersey on mossy wood
margins. It is of fair quality and flavor.

T. ter′reum Schaeff.—the earth. ([Color] Plate XVIII, fig. 4.) **Pi-**
leus fleshy, thin, soft, convex, cam-   (Plate XXII.)
panulate or nearly plane, obtuse or
umbonate, *innately fibrillose or floc-*
cose-scaly, ashy-brown, grayish-
brown or mouse color. **Flesh** white
or whitish. **Gills** adnexed, subdistant, more or less eroded on the edge,
white becoming ash-colored. **Stem**
equal, varying from solid to stuffed
or hollow, fibrillose, white or whitish.
Spores broadly elliptical, 6–7x4–5µ.

Pileus 1–3 in. broad. **Stem** 1–2
in. long, 2–4 lines thick.

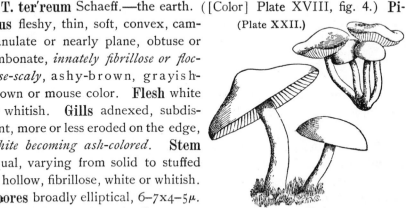

TRICHOLOMA TERREUM.
One-half natural size.

Woods. Albany, Rensselaer and Cattaraugus counties. September
to November. *Peck*, 44th Rep. N. Y. State Bot.

Spores 7x5.5µ *Morgan;* 5–6µ *Massee;* 6–7x4µ *K.;* 6µ *W.G.S.*
Eaten by Professor Peck. Eaten by McIlvaine. Quality fair.

T. ter′reum Schaeff.—var. *fra′grans* Pk. **Pileus** convex or nearly
plane, dry, innately-fibrillose or minutely floccose-scaly, grayish-brown
or blackish-brown. **Gills** rather broad, adnexed, whitish or ash-colored.
Stem equal, solid or stuffed, rarely hollow, whitish. **Spores** broadly
elliptical, 6–7x4–5µ.

The Fragrant tricholoma has a distinct farinaceous odor and flavor.
In other respects it closely resembles the Earth-colored tricholoma of
which it is considered a mere variety. The typical European plant is
said to be without odor or nearly so and has not been classed among
the edible species by European writers. But our variety, though not
high-flavored, is fairly good and entirely harmless. Its cap varies considerably in color but is some shade of gray or brown. Its center is
without any prominence or very bluntly prominent, and its surface is

Tricholoma. commonly very obscurely marked with innate fibrils or in small plants may have very small flocculose tufts or scales. The flesh is whitish as also are the gills, though these sometimes assume a more decided grayish hue. They are rather broad and loose and sometimes uneven on the edge or even split transversely. They are usually deeply excavated next the stem and attached to it by a narrow part. The stem is whitish or slightly shaded with the color of the cap. It often has a few longitudinal fibrils, but never any collar. It may be either solid, stuffed or spongy within, or in large specimens, hollow.

The plants grow gregariously or sometimes in tufts on the ground under or near trees or in thin woods, especially of pine, or in mixed woods. The caps vary from 1–4 in. broad, and the stems from 1–3 in. long and from 2–6 lines thick. The plants occur in autumn. In Europe there is a variety of this species which also has a farinaceous odor, but it differs from our plant in having reddish edges to the gills. It is called variety orirubens. *Peck*, 49th Rep. N. Y. State Bot.

Var. fragrans is plentiful and gregarious among New Jersey pines. October to frost. Other varieties are often found. Specimens found by me at Mt. Gretna, Pa., and sent to Professor Peck who identified them as var. fragrans Pk., were decidedly umbonate. Gills were easily separable from cap.

Var. fragrans is a favorite. It is pleasant to many, even raw. Plentiful salting while cooking develops a high and exquisite flavor.

T. fumes'cens Pk.—smoky. **Pileus** convex or expanded, dry, clothed with a very minute appressed tomentum, whitish. **Gills** narrow, crowded, rounded behind, whitish or pale cream color, *changing to smoky-blue or blackish* where bruised. **Stem** short, cylindrical, whitish. **Spores** oblong-elliptical, 5–6.5μ.

Pileus 1 in. broad. **Stem** 1–1.5 in. high, 2–3 lines thick.

Woods. Columbia county. October. Rare.

The species is remarkable for the smoky or blackish hue assumed by the gills when bruised and also in drying. It is apparently related to T. immundum Berk., but in that species the whole plant becomes blackish when bruised, and the gills are marked with transverse lines and tinged with pink. *Peck*, 44th Rep. N. Y. State Bot.

Mt. Gretna, Pa. September to November, 1898. *McIlvaine*.

The size of cap sometimes attains to 3 in and stem to ½ in. in thickness. Taste at first farinaceous then sweetish. The caps are of excellent Tricholoma. quality and flavor.

T. imbrica′tum Fr.—covered with tiles.   **Pileus** fleshy, *compact*, convex or nearly plane, obtuse, dry, innately scaly, fibrillose toward the margin, brown or reddish-brown, the margin thin, at first slightly *inflexed and pubescent then naked*. **Flesh** firm, thick, white.   **Gills** slightly emarginate, almost adnate, rather close, white when young, becoming reddish or spotted.   **Stem** *solid*, firm, nearly equal, fibrillose, white and mealy or pulverulent at the top, elsewhere colored like the pileus.   **Spores** 6.5x 4–5μ.

(Plate XXIII.)

TRICHOLOMA IMBRICATUM.
One-half natural size.

Pileus 2–4 in. broad.   **Stem** 2–3 in. long, 4–10 lines thick.   Under or near coniferous trees.   Greene and Essex counties.   September and October.

This is an edible species.   It has a farinaceous odor and taste when fresh.   *Peck*, 44th Rep. N. Y. State Bot.

Closely resembles T. transmutans in size, color and taste.   It is, however, easily separated by its dry cap and solid stem.   *Peck*.

Plentiful in pine woods of New Jersey, and among hemlocks in West Virginia.   Mt. Gretna, Pa., under pines.   October and November, 1898.   *McIlvaine*.

Specimens found at Mt. Gretna had caps dark umber when young, and margin incurved to stem.   Gills yellowish.   Stem up to 4 in. long, stout, solid, swollen at base, and having a short pointed ending, firm, fibrillose, white.   Flavor farinaceous.

Flesh of good texture and taste.

Tricholoma. III.—RIG'IDA. Pileus rigid, cuticle broken up into smooth scales, etc.

*Gills white or pallid, not becoming spotted with red or gray.
Not represented.*

**Gills becoming reddish or grayish, spotted, etc.*

T. sapona'ceum Fr.—*sapo*, soap. Strong, smelling of an undefinable soap. **Cap** 2–4 in. across, involute at first, convex then flattened, dry, glabrous, moist in wet weather, never viscid, brownish, more or less spotted or having the skin cracked into scales, occasionally covered with dark fibrils. **Flesh** firm, whitish becoming reddish when wounded. **Gills** emarginate, with a hooked tooth (uncinate) thin, distant, pale white. **Stem** 2–4 in. long, about ½ in. thick, often unequal, base sometimes long and rooting, usually smooth, at times reticulated with black fibrils, or is scaly. Distasteful.

The species is variable in size and color. Stevenson remarks: "Scarcely any species has been more confounded with others." It may always be safely distinguished by its odor, by its distant gills, by the smooth cuticle of the cap cracking into scales, and by the change of color to reddish when bruised.

West Virginia mountains. August to frost. 1881–85. New Jersey, Pennsylvania. *McIlvaine*.

This fungus is not extremely unpleasant when eaten—like T. sulphureum, but no one will care to eat it. There is nothing in the flavor to recommend it or to inspire a cultivation of taste for it.

IV.—SERICEL'LA. Pileus slightly silky, soon smooth, etc.

* *Gills broad, rather thick, somewhat distant.*

T. sulphu'reum Bull.—*sulphur*, brimstone. **Odor** strong, fetid or like gas tar. **Cap** 1–4 in. across, subglobose, then convex and plane, slightly umbonate, sometimes depressed, fleshy, margin at first involute. **Color** dingy or reddish sulphur-yellow, at first silky, becoming smooth or minutely tomentose. **Flesh** thick, yellow. **Gills** rather thick, narrowed behind, emarginate or acutely adnate, sometimes appearing arcuate from shape of cap. **Stem** 2–4 in. long, 3–5 lines thick, equal or slightly bulbous, often curved, smooth striate, sulphur-yellow, stuffed, Tricholoma. fibrous or hollow, yellow within, at times having yellow fibrous roots.

Spores 9–10x5μ *Massee.*

Very variable in size. Gregarious, common in mixed woods.

West Virginia, 1881. West Philadelphia, 1886. *McIlvaine.*

When quite young T. sulphureum is showy and inviting. Its smell is discouraging, its taste forbidding. No amount of cooking removes its unpleasant flavor. I have tried to eat enough of it to test its qualities, but was satisfied after strenuous efforts to mark it INEDIBLE.

T. chrysenteroi'des Pk.—like gold. **Pileus** fleshy, convex or plane, not at all umbonate, firm, dry, glabrous or slightly silky, *pale-yellow or buff*, becoming dingy with age, the margin sometimes reflexed, *flesh pale-yellow, taste and odor farinaceous.* **Gills** rather close, emarginate, yellowish, becoming dingy or pallid with age, *marked with transverse veinlets along the upper edge*, the interspaces veined. **Stem** equal, firm, *solid*, bare, fibrous-striate, yellowish without and within. **Spores** elliptical, 8–10x5–6μ.

Pileus 1–2 in. broad. **Stem** 2–3 in. long, 3–4 lines thick.

Woods. Lewis and Cattaraugus counties. September.

Nearly allied to T. chrysenterum, but separable by the gills, which are somewhat veiny and not free, by the entire absence of an umbo and by its farinaceous odor and taste. *Peck*, 44th Rep. N. Y. State Bot.

Frequently found at Angora, and in Woodland Cemetery, West Philadelphia.

Edible. Fair flavor and good quality.

T. o'picum Fr.—uncouth. **Pileus** 1–1 ½ in. across. **Flesh** rather thin, becoming grayish; convex, then expanded, obtusely-umbonate, at length usually upturned and split, very dry, even at first, then minutely scaly, gray. **Gills** broadly emarginate, ventricose, rather thick, scarcely distant, hoary. **Stem** 2–3 in. long, 2–3 lines thick, equal, fibrillose, becoming almost glabrous, pallid then grayish, stuffed. *Massee*

Among moss, in pine woods, etc.

Inodorous. Somewhat resembling T. saponaceum, but distinguished by the absence of smell.

Waretown, N. J. Under pines and open places in pine woods. August to September, 1889. *McIlvaine.*

When wet the caps become darker and have a mottled appearance. They are tender, but rather tasteless. The species serves to make quantity when cooked with others of higher flavor.

T. pipera'tum Pk.—*piper*, pepper. **Pileus** rather thin, firm, dry, convex, obtuse or subumbonate, virgate with innate brownish fibrils, varying in color from grayish-brown to blackish-brown, sometimes with greenish or yellowish tints. **Flesh** white or whitish, taste acrid. **Gills** broad, close, rounded behind, adnexed, whitish or yellowish. **Stem** generally short, equal, solid, silky, slightly mealy or pruinose at the top, white or slightly tinged with yellow. **Spores** elliptic, 6–7μ long, 5μ broad. **Pileus** 4–7 cm. broad. **Stem** 5–7 cm. long, 6–12 mm. thick.

The central part of the pileus is sometimes a little darker than the rest. The peppery or acrid taste is very distinct and remains in the mouth many minutes. This and the innately fibrillose character of the pileus are distinguishing characters of the species. The plants appear from September to November. *Peck*, Torr. Bull., Vol. 26.

Mt. Gretna, Pa. October to November, 1898, on damp ground among moss. *McIlvaine.*

Cap up to 3 in. across, bell-shaped, then convex, depressed in center and undulate, light-brown, darker toward center, dry, minutely fibrillose. **Flesh** thick, white, thin toward margin. **Gills** emarginate, unequal, not forked. **Stem** 1½–2 in. long, hard, equal or enlarging toward base, white, silky, striate.

Though peppery raw, this Tricholoma is of good substance and flavor when cooked.

B. Pileus Even, Smooth, Not Downy, Scaly, Nor Viscid, Etc.

V.—Gutta'ta. Pileus marked with drop-like spots or rivulose.

**Gills whitish.*

T. gambo'sum Fr.—*gambosus*, swelling near the hoof. **Pileus** 3–4 in. and more broad, *becoming pale-tan*, fleshy, *hemispherico-convex*, *then flattened*, obtuse, undulated and bent backward, even, smooth, but *spotted as with drops*, at length widely cracked (not, however, torn into squamules), the *margin* at *the first involute and tomentose*. **Flesh** thick, soft, fragile, white. **Stem** 2 in. and more long, ½–1 in. thick, *solid,* fleshy-firm, almost *equal*, often curved-ascending at the base, *white,* *downy at the apex.* **Gills** rounded or emarginato-adnexed, with a some- Tricholoma. what decurrent tooth and when old sinuato-decurrent, *crowded*, ventricose, 2–3 lines broad, *whitish.* *Fries.*

Odor pleasant, *of new meal.* Often forming large rings or clusters. A whitish form must not be confounded with T. albellus.

Spores 13x11µ *W.G.S.;* 13–14x8–9µ *Massee;* 13x10µ *Cooke.*

Angora, Philadelphia. Chester and Lebanon county, Pa. *McIlvaine.* Fair.

             ** *Gills becoming reddish or smoky-gray.*

T. tigri′num Schaeff.—spotted like a tiger. **Pileus** 2 in. broad, pallid-brown, variegated with crowded and *darker dingy-brown spots*, compactly fleshy, convex then expanded, obtuse, repand. **Flesh** thick, firm, white, unchangeable, but thin at the involute margin. **Stem** 1 in. long and thick, very compact, solid, pruinate, white. **Gills** rounded behind, at length decurrent with a tooth, crowded, narrow, white, at length darker.

Solitary or cespitose. Very distinguished, obese, and without any marked smell of new meal. In fir woods and open grassy ground. Rare. June to July. *Stevenson.*

Edible, *Cooke, Fries.*

T. albel′lum Fr.—*albus*, white. **Pileus** about 3 in. broad, becoming pale-white, passing into gray when dry, fleshy, thick at the disk, thinner at the sides, *conical then convex*, gibbous when expanded, when in vigor moist on the surface, *spotted (mottled) as with scales*, the thin margin naked. **Flesh** soft, floccose, white, unchangeable. **Stem** curt, 1½–2 in. long, 1 in. thick at the base, reaching ½ in. toward the apex, *solid*, fleshy-compact, ovato-bulbous (conical to the middle, cylindrical above the middle), fibrillose-striate, white. **Gills** very much *attenuated behind*, *not* emarginate, *becoming broad in front*, very crowded, quite entire, white. *Fries.*

Spores elliptical, 6–7x4µ *Massee;* ovoid, 3µ *W.G.S.;* ovoid, 3µ *Cooke.*

Pileus not becoming yellow. **Odor** weak when fresh, taste pleasant, almost that of cooked flesh. There are two forms: one larger, solitary, another smaller, connato-cespitose, quite as in A. albellus Sow. It is often confounded with smaller forms of A. gambosus. *Stevenson.*

North Carolina, *Curtis.* Damp woods. Edible.

VI.—Spongio′sa.   Pileus compact then spongy, smooth, moist.

* Gills not discolored.

T. vires′cens Pk.—*viresco*, to grow green.   **Pileus** convex or nearly plane, sometimes centrally depressed, moist, bare, *dingy-green*, the margin sometimes wavy or lobed.   **Gills** close, gradually narrowed toward the outer extremity, rounded or slightly emarginate at the inner, white. **Stem** subequal, *stuffed or hollow*, thick but brittle, whitish, sometimes tinged with green.   **Spores** broadly elliptical, 5x4μ.

Pileus 3–5 in. broad.   **Stem** 3–4 in. long, 6–12 lines thick.

Thin woods.   Essex county.   July.

The dull smoky-green hue of the pileus is the distinguishing feature of this species.   *Peck*, 44th Rep. N. Y. State Bot.

Quite common in West Virginia, New Jersey and Pennsylvania.   July to October.   *McIlvaine.*

Edible.   Tastes somewhat like many Russulæ, when cooked.   Flavor good.

T. fumidel′lum Pk.—smoky.   **Pileus** convex, then expanded, sub-umbonate, bare, moist, *dingy-white or clay-color clouded with brown*, the disk or umbo generally smoky-brown.   **Gills** crowded, subventricose, whitish.   **Stem** equal, bare, *solid*, whitish.   **Spores** minute, subglobose, 4.5x4μ.

Pileus 1–2 in. broad.   **Stem** 1.5–2.5 in. long, 2–3 lines thick.

Woods.   Albany county and Catskill mountains.   September and October.

The stem splits easily and the pileus becomes paler in drying.   It sometimes becomes cracked in areas.   *Peck*, 44th Rep. N. Y. State Bot.

On ground.   Mt. Gretna.   October and November.   1897.   *McIlvaine.*

The species was plentiful among the leaf mold, growing from the ground in mixed woods.

The caps are delicate in substance and flavor.

T. leucoceph′alum Fr.   *Gr.*—white; head.   **Pileus** 1 ½–2 in. across, convex then plane, even, moist, smooth, but when young covered with a satiny down; water-soaked after rain.   **Flesh** thin, tough, white.   **Gills** rounded behind and almost free, white.   **Stem** up to 2 in. long, ¼ in.

78

thick, exterior hard, shining, fibrous; interior hollow but solid at base Tricholoma.
which is attenuated and rooting, twisted. **Smell** strong of new meal.
Taste pleasant.

Spores 9–10x7–8μ.

Mt. Gretna, Pa. Grassy woods and borders. October to November,
1898. *McIlvaine.*

Quite common. The caps are excellent.

T. al'bum Schaeff.—*albus*, white. **Pileus** fleshy, tough, convex,
becoming plane or depressed, obtuse, very dry, even, *glabrous, white,*
sometimes yellowish on the disk, rarely wholly yellowish, the margin at
first involute. **Flesh** white, taste *acrid or bitter.* **Gills** emarginate, somewhat crowded, distinct, white. **Stem** solid, elastic, equal or tapering
upward, externally fibrous, obsoletely frosted at the apex, white. **Spores**
elliptical, 5–6μ long.

Pileus 2–4 in. broad. **Stem** 2–4 in. long, 4–6 lines thick.

Woods. Common. August to October. This species is variable in
color and in size, being sometimes robust, sometimes slender. It grows
singly, in troops or in tufts. It has no decided odor, but a bitter unpleasant taste. *Peck,* 44th Rep. N. Y. State Bot.

Cooked, tender and of fair flavor.

*** Gills becoming discolored.*

T. persona'tum Fr.—wearing a mask (from its many varieties of
colors). ([Color] Plate XVIII.) **Pileus** compact, becoming soft,
thick, convex or plane, obtuse, regular, moist, bare, variable in color,
generally pallid or ashy tinged with violet or lilac, the margin at first
involute and frosted with fine hairs. **Flesh** whitish. **Gills** broad,
crowded, rounded behind, free, *violaceous becoming sordid-whitish or
dingy-brown.* **Stem** generally thick, subbulbous, solid, fibrillose or
frosted with fine hairs, whitish or colored like the pileus. **Spores** dingy
white, subelliptical, 8–9x4–5μ. On white paper the spores have a
slight salmon tint, but they are regular in shape, not angular as in Entoloma.

Pileus 2–5 in. broad. **Stem** 1–3 in. long, 6–12 lines thick. *Peck,*
44th Rep. N. Y. State Bot.

Woods and open places, and growing from old, matted stable straw.
Common over the United States.

79

Tricholoma.   When T. personatum becomes known to the collector, either in the field or on the table, it is sure to become a favorite.   It is fleshy, rotund, stocky, moist and smooth, with a tendency in its cap to be wavy-rimmed and jauntily cocked in wet weather.   It grows singly or in troops, occasionally in tufts of from five to six individuals.   A patch of it is valuable and worth husbanding with covering of fine straw.   Cortinarius violaceus resembles it somewhat in color and shape, but it shows a spidery veil, and has brown spores.   It is edible.

The common name of T. personatum in England is Blewits, which translated into understandable English is believed to be "blue-hats." It is everywhere eaten, being of substantial substance, good flavor and cookable in any way.   It is especially fine in patties, stews and croquettes.

T. nu'dum Bull.—naked.   **Pileus** about 3 in. broad, becoming purple-violaceous then changing color, reddish, fleshy, comparatively *thin*, convexo-plane then *depressed*, obtuse, even, smooth, with a pellicle which is moist and manifest in rainy weather; margin inflexed, thin, naked. **Flesh** thin, pliant, colored.   **Stem** about 3 in. long, ½ in. thick, *stuffed*, *elastic*, equal, *almost naked*, mealy at the apex, *violaceous then becoming pale*.   **Gills** rounded then decurrent (on account of the depressed pileus), crowded, narrow, of the same color as the pileus or deeper *violaceous*, but soon changing color, *at length reddish* without the least tinge of violet.   *Stevenson*.

Spores 7x3.5µ *Massee;* 6–8x4µ *B.;* 6x3µ *W.G.S.*   On ground among leaves.   Esculent, very good and delicate.   *Cordier*.   Edible. *Roze*.   Edible, all American authorities.

VII.—HYGROPH'ANA.   Pileus thin, water-soaked, etc.

Gills whitish, not spotted.

T. grammopo'dium Bull.   *Gr.*—a line; *Gr.*—a foot.   **Pileus** 3–6 in. broad, *pallid-livid* or brownish-red when moist, whitish when dry, fleshy, very thin toward the margin, *campanulate then convex*, and at length flattened, obtusely umbonate, even, smooth, pellicle moist in rainy weather, not viscous, separating, flesh-colored when moist, white when dry, soft, fragile.   **Stem** *tall*, about 3–4 in. long, ½ in. and more

thick, *solid*, elastic, equal with exception of the *thickened base*, cyl- Tricholoma.
indrical, firm, smooth, *evidently longitudinally sulcate, whitish*. **Gills**
arcuato-adnate or broadly horizontally emarginate, acute at both ends,
very crowded, quite entire, very many shorter, somewhat branched
behind, white.

Odor moldy. Striking in appearance; the chief of this group.
There is a variety wholly white. In pastures and grassy woods. *Stevenson*.

Spores 5–6µ *Massee*.

Distinguished by the grooved stem and crowded gills, which are
adnate when the pileus is expanded. Often growing in rings.

North Carolina, *Curtis*. Not reported elsewhere. Esculent. *Cooke*.
Much eaten in Europe.

T. bre'vipes Bull.—*brevis*, short; *pes*, a foot. **Pileus** about 2 in.
broad, *umber then becoming pale*, fleshy, *soft, convex then becoming
plane*, even, smooth, moist (opaque when dry); flesh of the pileus *becoming brownish* when moist, becoming white when dry. **Stem** *solid*,
very *rigid*, at length fibrous, *pruinate at the apex, externally and internally fuscous;* otherwise very variable, sometimes *very short*, 2–3
lines only long and thick, attenuated downward; commonly 1 in.,
sometimes bulbous, sometimes equal, more slender. **Gills** emarginato-
free, *crowded*, ventricose, disappearing short of the margin, quite entire,
becoming fuscous then whitish. Solitary. *Inodorous*. The pileus is
often stained with soil. *Stevenson*.

Spores elliptical, 7.5x5µ *Peck;* 7–4µ *Massee*.

Esculent and very delicate. *Paulet*. Esculent. *Cooke*.

T. hu'mile Pers.—low, small. ([Color] Plate XVIII, fig. 6.) Very
variable in form and color. **Cap** 2–3 in. across, convex then expanded,
wavy, flattened, sometimes umbonate, sometimes depressed, glabrous,
occasionally powdered with thin white dust, fragments of veil, sometimes viscid. **Color** changes with moisture, blackish, grayish, and
having somewhat the appearance of an oyster. **Gills** rounded-adnexed,
with a slight tooth, arcuately decurrent, crowded, 2–3 lines broad,
whitish. **Flesh** soft, whitish or grayish. **Stem** 1–2 in. long, up to ½
in. thick, equal (misshapen by pressure when tufted), light gray, *cov-*

Tricholoma. *ered with fine down*, stuffed, becoming hollow, soft, fragile.   Gregarious, usually tufted.

Spores 7–8x5–6µ *K.*

Open woods, in gardens, among cinders, grass, etc., September to frost.

Woodland Cemetery, Philadelphia, 1897.   *McIlvaine.*

Its tufted habit and fair size, fleshy cap of good flavor, make it a desirable species.   It cooks readily and the caps are of fine flavor.

T. pæ′didum Fr.—*pædidus*, nasty.   **Pileus** about 1½ in. across. **Flesh** very thin, tough, becoming whitish; bell-shaped then convex, at length expanded, umbonate, at length depressed round the conical, prominent umbo, moist, virgate or streaked with innate fibrils radiating from the center, otherwise almost even, smoky-mouse color, opaque, margin naked.   **Gills** adnexed with a slight decurrent tooth, slightly sinuate, crowded, narrow, white then gray.   **Stem** about 1 in. long and 2 lines thick, base slightly bulbous, tough, slightly striate, naked, dingy-gray.   **Spores** elliptic-fusiform, 10–11x5–6µ.

In gardens, on dung-hills, etc.   Small, tough, color dingy, without a trace of violet tinge.   *Massee.*

Edible.   Cooks tender, and is of good flavor, notwithstanding its name, which in no way applies.

T. subpulverulen′tum Pers.—slightly dusty.   **Pileus** 1–2½ in. across, convex then plane or depressed in center, even, innately pruinose, hoary, white, whitish, grayish, margin extending as a slight rim incurved beyond gills.   **Flesh** white, thick, firm, hygrophanous.   **Gills** rounded without a tooth, close, narrow, white.   **Stem** 2–3 in. long, 3–5 lines thick, equal, solid, somewhat striate, whitish.

Spores 5x3µ *Massee;* 4x3µ *W.G.S.*

Biological grounds, University of Pennsylvania.   Philadelphia.   May to November, 1898.   *McIlvaine.*

A species one is glad to find.   It has a healthy substantial presence full of promise.   It is a solitary grower among grass on lawns and pastures, but its individuals are neighborly.   Caps and stems are excellent.

CLITO'CYBE Fr.

Gr.—sloping. (From the depression of the pileus.)

Pileus generally fleshy, becoming thin toward the margin, flexible or Clitocybe. tough, plane or depressed, margin involute. **Gills** adnate or decurrent, never sinuate. **Stem** confluent and homogeneous with flesh of pileus, somewhat elastic, with a spongy stuffing, frequently becoming hollow, externally fibrous. Universal veil when present conspicuous on the pileus like frost or silky dew, but commonly wanting.

Growing on the ground, frequently in groups. The thinner and hygrophanous species appear late in autumn. Some are quite fragrant. Collybia, Mycena and Omphalia are separated by their stems being cartilaginous, not externally fibrous as in Clitocybe. Tricholoma by its sinuate gills.

Variations in species of Clitocybe are great. A few are easily fixed in the genus, but many of them will puzzle the amateur and perplex the expert. The gills are always attached to the stem, and usually run down it. They are not notched next to the stem as in Tricholoma.

Like Tricholoma, Clitocybe has many species, most of which are common, and are probably edible. I therefore give Professor Peck's description of all Clitocybes thus far submitted to him.

I know of but one species which is injurious to some persons—Clitocybe illudens. Many eat and enjoy it. It does not agree with others. A few untried species are suspicious to a like extent. Clitocybe illudens possesses the property of phosphorescence.

Several species of Clitocybe have not been seen or tested by me, nor have I information that these have been tested.

ANALYSIS OF TRIBES.

A. Pileus Fleshy, Often Pallid When Dry, *not hygrophanous.*

Flesh firm, not watery, nor splitting into plates. Those which turn pale in drying differ from Series *B* by their silky luster.

Disciformes (disk-shaped). Page 85.

Pileus somewhat equally fleshy; convex then plane or depressed, obtuse, regular; gills at first adnate or regularly adnato-decurrent. Normally solitary.

Clitocybe.   * Pileus gray or brownish.
** Pileus violet or reddish.
*** Pileus becoming yellowish.
**** Pileus greenish, becoming pale.
***** Pileus white, becoming shining white.

Distinguished from white hygrophanous species and white species of Paxillus.

DIFFORMES (irregularly shaped). Page 94.

Pileus fleshy in the center, thin at the margin, at first umbonate, then expanded and depressed, irregular. Gills unequally decurrent, longer in some places than in others, sometimes rounded on one side of the stem or only reaching it as in Tricholoma. Stem somewhat cartilaginous externally, but fibrous.

Cespitose, often grown together at base, variable in form, sometimes solitary.

INFUNDIBULIFORMES (funnel-shaped). Page 98.

Pileus becoming thin from the fleshy center to the margin, at length funnel-shaped or deeply umbilicately depressed in the center. Stem spongy, externally fibrous. Gills deeply and equally decurrent from the first. Pileus often becoming discolored or pallid, not hygrophanous.

* Pileus colored or becoming pale, the surface (at least under a lens) innately flocculose or silky, bibulous, not moist.

** Pileus colored or pallid, smooth, moist in rainy weather.

*** Pileus shining whitish, with scattered superficial flocci or becoming smooth.

B. PILEUS FLESHY-MEMBRANACEOUS.

Flesh thin, soft, watery, hygrophanous.

CYATHIFORMES (cup-shaped). Page 104.

Flesh of pileus thin, consisting of two separable plates, disk not compact, hygrophanous, depressed then cup-shaped; gills at first adnate then decurrent, descending, straight. Color dingy when moist.

ORBIFORMES (round-shaped). Page 109.

Pileus somewhat fleshy, hygrophanous. convex then flattened or depressed, polished, not squamulose nor mealy; gills plane, horizontal, Clitocybe. thin, crowded, adnate or decurrent with a small tooth. Color dingy or becoming watery pale.

* Gills becoming ash-colored. Pileus at first dark.

** Gills whitish. Pileus becoming pale.

VERSIFORMES (variable in shape). Page 106.

Pileus thin, convex then deformed, tough, more or less squamulose or furfuraceous; gills adnate, broad, rather thick, generally distant. Color hygrophanous.

* Pileus squalid or brownish with dark squamules. None known to be edible.

** Pileus bright, of one color.

Series A.

I.—DISCIFOR'MES.

* *Pileus gray or brownish.*

C. nebula'ris Batsch.—*nebula*, a cloud. ([Color] Plate XXIV, fig. 7.) The Clouded clitocybe, Clitocybe nebularis, takes its name from the clouded-gray appearance of its thick cap, which is at first convex, but when mature, either flat or a little depressed. Its flesh is white, thickest in the middle, and in a vertical section is seen to taper rapidly downward into the stem. The gills are close together and rather narrow for the size of the plant. They are white or yellowish-white. The stout solid stem usually tapers upward from the base and is whitish.

The cap is two to four inches or more broad, the stem one to two inches long and about half an inch thick. The Clouded mushroom grows in woods, and sometimes forms large tufts or clusters among fallen leaves. It is found in autumn, but is not very common in this country. Authors differ in their estimate of the edible qualities of this mushroom, but the more recent ones generally agree in classing it as edible. "Mushrooms and Their Use," *C. H. Peck.*

Spores 4.5x3μ *Cooke;* elliptical 6x3.5μ *Massee;* 3x4μ *W.G.S.*

There has been great diversity of opinion as to the edibility of this species on the continent. Cordier and a friend suffered from it. Paulet counseled mistrust.

85

Clitocybe. This fungus is quite common in the West Virginia mountains and in some parts of Pennsylvania and New Jersey, where I have found it. It is, however, limited to localities. It is one of my favorites, being of marked flavor and agreeable consistency. I have not known it to harm anyone.

C. cla′vipes Pers.—*clava*, a club; *pes*, a foot. **Pileus** 1½–2½ in. (Plate XXV.)

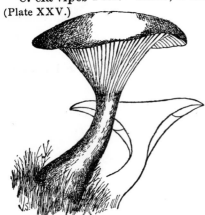

across, rather convex at first, soon plane, at length almost obconical, very obtuse, even, glabrous, dry, sometimes all one color, brown, sooty, livid-gray, etc., sometimes whitish towards the margin, very rarely entirely white. **Flesh** loose in texture, white, thin at the margin. **Gills** deeply decurrent, continued down the stem as straight lines, rather distant, flaccid, quite entire, broad, entirely and persistently white. **Stem** 2 in. long, base ½ in. and more thick, conically attenuated upward, rather fibrillose, livid, sooty, solid, spongy within. **Spores** elliptical, 6–7x4μ.

CLITOCYBE CLAVIPES.
About two-thirds natural size.

In woods, especially pine. Resembling C. nebularis in color, but quite distinct. Smell pleasant, entire substance soft and elastic. *Fries.*

Spores elliptical, 6–7x4μ *Massee;* sub-ellipsoid, 5–7x3–4μ *K.;* 6x8μ *W.G.S.*

Found in pine woods of New Jersey, and under spruce in West Virginia. Its substance is spongy, therefore does not stew well. Cooked in any other way it is delicate and of excellent flavor.

C. gangræno′sa Fr.—*gangræna*, gangrene. **Pileus** fleshy, convex then plane, obtuse, whitish, at first sprinkled with white powder, then naked, variegated, streaked. **Gills** slightly decurrent, arcuate, crowded, dingy-white. **Stem** somewhat bulbous, soft, striate, spongy, solid.

Stinking; large, flesh becoming blackish and variegated with black. Stem curved, sometimes excentric. Pileus whitish, here and there greenish, livid, etc. *Fries.*

Var. *nigres'cens* Lasch.   Whitish; pileus thin, soft, at first convex, Clitocybe.
obtuse then plane, somewhat umbonate, and somewhat depressed; gills
decurrent, very much crowded, narrow, stem solid, downy.

Pileus 2–3 in. broad.   **Stem** 1¼–1½ in. long, 2–3 lines thick.
Odor rather sweet, taste unpleasant.   *Cooke.*
New Jersey, Haddonfield, pine woods.   July to August.   *McIlvaine.*
This Clitocybe is in every way unattractive.   It is not poisonous, but
no one would care to eat it.

C. me'dia Pk.—*medius*, middle..   Because intermediate between C.
nebularis and C. clavipes.   **Pileus** (Plate XXVI.)
fleshy, convex, becoming plane or
slightly depressed, dry, dark grayish-
brown, the margin often wavy or irregular, flesh white, taste mild.   **Gills**
broad, subdistant, adnate or decurrent, whitish, the interspaces somewhat venose.   **Stem** equal or but
slightly thickened at the base, solid,
elastic, not polished, colored like or
a little paler than the pileus.   **Spores**
elliptical, 8x5μ.

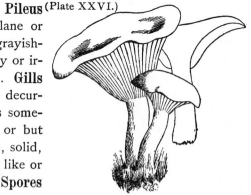

CLITOCYBE MEDIA.
One-half natural size.

Pileus 2–4 in. broad.   **Stem** 1–2
in. long, 4–8 lines thick.   Mossy ground in deep woods.   North Elba.
September.

This species is intermediate between C. nebularis and C. clavipes.
In its general appearance, and in the character of the pileus and stem,
it resembles C. nebularis, but in the character of the more distant gills
and in the size of the spores it is nearer C. clavipes, of which it might
perhaps be regarded as a variety.   Two forms are distinguishable.   In
one the gills are more distant, slightly rounded behind, and adnate or
abruptly terminated; in the other they are closer and more distinctly
decurrent.   The plant is edible.   *Peck*, 42d Rep. N. Y. State Bot.

I have known this fungus very favorably since 1883, and regard it as
one of the best.   I have seen it in the West Virginia mountains only,
but it will probably be found in cool, shaded, high localities all over the
country.   Both it and the C. nebularis are well worthy of search.

C. viles'cens Pk.—*vilesco*, of little value. **Pileus** convex, then plane or depressed, often irregular, glabrous, slightly pruinose on the involute margin, brown or grayish-brown, becoming paler with age, often concentrically rivulose. **Gills** close, adnate or decurrent, cinereous, sometimes tinged with dingy-yellow. **Stem** short, solid, sometimes compressed, grayish-brown, with a whitish tomentum at the base. **Spores** subglobose or broadly elliptical, 5–6.5μ; flesh whitish-gray, odor slight.

Plant gregarious, 1–2 in. high. **Pileus** 1–1.5 in. broad. **Stem** 1–2 lines thick. Grassy pastures. Jamesville, August. *Peck*, 33d Rep. N. Y. State Bot.

A pale form of this species grows on sandy soil, in which the pileus is smoky white, but it becomes grayish-brown in drying. The mycelium binds together a mass of sand, so that when the plant is taken up carefully a little ball of sandy soil adheres to the base of the stem. The stem is sometimes pruinose. The flavor is mild and agreeable. *Peck*, 50th Rep. N. Y. State Bot.

Sometimes plentiful about Philadelphia. Edible. Caps tender, slight flavor.

C. comitia'lis Fr.—belonging to an assembly. **Pileus** about 1½ in. across, fleshy, convex, then plane, obtuse, even, glabrous, rather moist but not hygrophanous, every part colored alike, sooty-umber, almost black. **Flesh** firm, white. **Gills** very slightly decurrent, horizontal, plane, thin, crowded, white. **Stem** 2–3 in. long, 3–4 lines thick, equally attenuated upward from the base, glabrous, sooty, elastic, stuffed. **Spores** elliptical, 7–8x4μ.

Damp places among mosses in pine woods, etc. Distinguished by the blackish color of the almost flat pileus, and the very slightly decurrent gills. Somewhat allied to C. clavipes, but firmer, smaller and inodorous. *Massee.*

Rather rare. Found in New Jersey among pines; in Pennsylvania in mixed woods.

Edible. Good texture and flavor.

** *Violet or reddish.*

C. cyanophæ'a Fr. *Gr.*—blue. **Pileus** 3–4 in. broad, becoming bluish-dusky-brown, compact, convex then plane, obtuse, smooth.

Stem 3 in. long, 1 in. thick at the base, attenuated upward, robust, Clitocybe. solid, smooth, *becoming azure-blue* when young, *abruptly white at the apex.* Gills deeply decurrent, crowded, violaceous, then becoming pale.

New York, Albion. In woods. October. Edible. *Dr. E. L. Cushing.*

Specimens sent to me by Dr. Cushing are the first and only ones of the species I have seen. The description is accurate. The spores were cream color.

C. monadel'pha Morg.—*monas*, single; *adelphos*, a brother. From its cespitose habit. (Plate XXVII.) Densely cespitose. Pileus fleshy, convex then depressed, at first glabrous, then scaly, honey color, varying to pallid-brownish or reddish. Stem elongated, solid, crooked, twisted, fibrous, tapering at the base, pallid-brownish or flesh color. Gills short, decurrent, not crowded, pallid flesh color. Spores white, a little irregular, 7.5x5.5μ.

On the ground in wet woods, spring to late autumn. Pileus 1–3 in. Stem 3–7 in. *Morgan.*

Grassy places. Menands. Albany county. September. Edible. Resembling Armillaria mellea, but distinguished from it by the absence of a collar from the stem, by the more decidedly decurrent lamellæ and by the solid stem. It is also more agreeable in flavor. It is related to C. illudens in habit and manner of growth. *Peck,* 51st Rep. N. Y. State Bot.

Spores 8x5μ *Peck.*

October 15, 1898. Identified by Professor Peck. September until frost.

Grows in great clusters about roots, etc., at Mt. Gretna. Frequently much water-soaked and uninviting. Taste variable, sometimes strong, woody.

It is edible, but care should be exercised in collecting to get young, fresh groups.

C. socia'lis Fr.—*socius*, a companion. Pileus about 1 in. broad, pale-yellowish with a reddish tinge, fleshy, convex then expanded, acutely umbonate especially when young, even, smooth, dry. Flesh moderately thin, white. Stem 1 in. long, 2 lines or a little more thick,

Clitocybe. solid, fibrous, commonly ascending, smooth, reddish, the rooting base hairy. **Gills** plano-decurrent, scarcely crowded, becoming yellow. *Fries.*

A very pretty species, densely gregarious, inodorous. The stem is sheathed-hairy at the base like Marasmius peronatus. Its greatest affinity is with A. vernicosus, of which it is perhaps a variety. *Stevenson.*

Quite common in pine woods of New Jersey. Though small, goodly messes of it may be gathered from its patches. The caps make a pleasing dish.

** *Pileus becoming yellow.*

None reported as tested for edibility.

**** *Pileus greenish or becoming pallid.*

C. odo'ra Bull.—*odorus*, fragrant. ([Color] Plate XXIV, fig. 9.) Fragrant. **Pileus** about 2 in. across, flesh rather thick, tough; soon plane and wavy, even, smooth, pale dingy green, silky when dry. **Gills** adnate, rather close, broad, greenish or pallid. **Stem** about 1–1 ½ in. long, 2 lines thick, base incrassated, elastic, stuffed. **Spores** elliptical, 6–8x4–5μ. In woods. *Massee.*

Readily distinguished by the strong, aniseed smell, dingy bluish-green pileus, and the pallid or greenish gills.

Sometimes somewhat cespitose. Tough; size variable, color varies between pale green and greenish-gray, usually all colored alike, but the gills are sometimes white; smell pleasant, spicy, especially when dry. *Fries.*

Spores 6x5μ *K.;* 8x4μ *B.*

A rather delicate, even exquisite dish. *Cooke.*

Edible. Exceedingly spicy. The flavor is pleasant, but rather strong. A few specimens mixed with others of like texture but less flavor make a tasty dish.

C. rivulo'sa Pers.—*rivus*, a stream. (Named from rivulet-like streaks on pileus.) **Pileus** 1–3 in. across, flesh thin, convex then plane and depressed, obtuse, often undulately lobed, dingy flesh-color or reddish, becoming pale, glabrous, then covered with a whitish down. **Gills** slightly decurrent, broad, rather crowded, pinkish-white. **Stem** about 2 in. long, 3–4 lines thick, rather fibrillose, tough, elastic, whitish, stuffed. **Spores** elliptical, 6x3.5μ. *Massee.*

Among grass by road-sides, etc.

Not common, but when found it is basket-filling. I have found it in Pennsylvania, New Jersey and West Virginia.

Edible. The caps are rather tough but become glutinous and tender when well cooked. Flavor fine.

***** *Pileus white, shining when dry.*

C. cerussa′ta Fr.—*cerussa*, white lead. **Pileus** 1½–3 in. across, flesh thick at the disk, becoming thin toward the margin; convex then almost plane, obtuse, even, minutely floccose then almost glabrous, white. **Gills** adnate, then decurrent, very much crowded, thin, permanently white. **Stem** about 2 in. long, 3–5 lines thick, smooth, tough, elastic, naked, spongy and solid, white. Among dead leaves, etc.

Taste mild, smell almost obsolete. Stem rather thickened at the base and often tomentose. Pileus said to be gibbous, but not umbonate nor becoming rufescent. Gills not changing to yellowish. *Fries.*

Spores 3µ *W.G.S.*

Edible. Good.

C. phylloph′ila Fr. *Gr.*—leaf-loving. Whitish-tan. **Pileus** 1–3 in. across, rather fleshy, convex then plane, becoming umbilicate and depressed, sometimes wavy, smooth and even. **Gills** thin, subdistant, white then tinged with ocher, rather broad, very slightly decurrent. **Stem** 2–3 in. long, equal, stuffed then hollow, whitish, tough, silky-fibrillose. **Spores** 6x4µ.

Among leaves in woods, etc.

Spores 6x4µ *Massee;* 6x3µ *W.G.S.;* 5.5x2.8µ *Morgan.*

Found at Devon, Pa., 1888; Angora, West Philadelphia, 1897. It is equal to the Pleurotus ostreatus (oyster mushroom) in texture, but not so high in flavor. Well cooked it is an agreeable and valuable food.

C. pithyoph′ila Secr. *Gr.*—pine-loving. **Pileus** 2–3 in. broad, dead-white when moist, shining whitish when dry, fleshy but *thin*, rather plane, *umbilicate*, at length irregularly shaped, repand and undulato-lobed, even, *smooth, flaccid*, the margin slightly striate when old. **Stem** *somewhat hollow*, rounded then compressed, equal, even, smooth, obsoletely or scarcely pruinose at the apex, white tomentose at the (not

91

Clitocybe. bulbous) base. **Gills** adnate, somewhat decurrent, *very crowded*, plane, 2–3 lines broad, distinct, quite entire, white.

Odor not remarkable, but pleasant. Gregarious, somewhat cespitose; *white* indeed, but when moist watery and *somewhat hygrophanous*, in which it evidently differs from A. phyllophila. A. tuba, which appears in the same places, is very like it. *Stevenson.*

Spores 6–7x4μ *B.*

Massachusetts, *Sprague ;* New York, *Peck*, Bull. 1887.

Albion, Orleans county, N. Y., October, 1898, *Dr. Cushing.*

Several specimens received were clearly referable to C. pithyophila, though varying in having caps deeply depressed but not umbilicate. The white tomentosity at base was present but indistinct.

Four specimens were eaten and found good. Eaten enjoyably by Dr. Cushing.

C. fus'cipes Pk.—*fuscus*, dirty; *pes*, a foot. **Pileus** thin, broadly convex or plane, umbilicate, glabrous, whitish and striatulate when moist, pure white when dry, odor and taste farinaceous. **Gills** nearly plane, subdistant, adnate or slightly decurrent, white. **Stem** equal, glabrous or slightly mealy at the top, hollow, dingy brown when moist, paler when dry. **Spores** globose, 5–6μ.

Pileus 4–8 lines broad. **Stem** about 1 in. long. Under pine trees. Carrollton. September. *Peck*, 44th Rep. N. Y. State Bot.

Edible. Its small size gives it minor importance, but a quantity of it makes an excellent meal.

C. can'dicans Pers.—*candico*, to be shining white. Entirely white. **Pileus** about 1 in. across, flesh thin, convex then plane or slightly depressed, umbilicate, regular or slightly excentric, even, with an adpressed silkiness, shining, shining white when dry. **Gills** adnate then slightly decurrent, crowded, very thin, narrow, straight. **Stem** 1–2 in. long, 1–2 lines thick, even, glabrous, cartilaginous, polished, equal, hollow, base incurved, rooting, downy. **Spores** broadly elliptical or subglobose, 5–6x4μ. *Massee.*

Among damp fallen leaves, etc.

Entirely white, small, rather tough; approaching Omphalia in the structure of the stem. The following form is described by Fries as occurring in pine woods: Stem thin, flexuous, base glabrous; pileus plane, not umbilicate, naked (without silky down). Gills scarcely Clitocybe. decurrent.

A remarkable form but scarcely to be separated as a species. *Fries.*

Quite common in West Virginia, Pennsylvania, New Jersey. The caps are excellent when well cooked.

C. dealba′ta Sow.—*dealbo*, to whitewash. **Pileus** about 1 in. or a little more broad, white, *slightly fleshy, tough*, convex then plane and at length revolute and undulated, always dry (not watery in rainy weather), even, smooth, *somewhat shining*, but as if innately pruinose under a lens. **Flesh** thin, arid, white. **Stem** 1 in. long, 2 lines thick, *stuffed, wholly fibrous*, at length also tubed, equal, but often ascending, whitish, mealy at the apex. **Gills** *adnate*, scarcely decurrent, thin, *crowded*, white.

Pileus sometimes orbicular, sometimes upturned and wavy. *Odor weak, pleasant*, but not very remarkable. Most distinct from A. candicans in the nature of the stem.

Edible. Its top is *exceedingly like ivory*. Its charming flavor is exceeded by very few other fungi. *Stevenson.*

Among leaves and grass. Woodland Cemetery, Philadelphia.

This charming fungus is common over the land. I have known it since 1881, and found it from North Carolina to West Virginia.

C. robus′ta Pk.—*robustus*, stout. **Pileus** thick, firm, at first convex, soon plane or slightly depressed in the center, glabrous, white, the margin at first involute or decurved, naked. **Flesh** white. **Gills** narrow, close, decurrent, whitish. **Stem** stout, rather short, solid, glabrous, equal or slightly tapering upward, often with a bulbous base, white. **Spores** elliptical, 8x4–5µ.

Pileus 3–4 in. broad. **Stem** 1–2 in. long, 8–12 lines thick.

Woods among fallen leaves. Catskill mountains. September to November.

This large and robust fungus is closely allied to C. candida Bres., from which it differs in the naked margin of the pileus, the absence of any marked odor and especially in the more elliptical shape of its spores. The same plant has been collected in Maryland by Mr. L. J. Atwater, who considers it edible, having eaten it with satisfaction and safety. *Peck*, 49th Rep. N. Y. State Bot.

Agaricaceæ

Clitocybe.   This fungus is quite plentiful in Pennsylvania and in open oak woods in New Jersey.   Its size and sometimes gregarious growth give it a permanent food value.   Its texture is coarse, but when cooked it is highly satisfactory.

C. gallina′cea Scop.—*gallina*, a hen.   Application not apparent. White; acrid.   **Pileus** 1–1 ½ in. across, rather fleshy at the disk, margin thin; convex then depressed, but not funnel-shaped, even, dry, opaque.   **Gills** slightly decurrent, narrow, crowded, thin.   **Stem** about 1 ½ in. long, 2 lines thick, equal, even, solid.   Among grass, moss, etc.

Resembling C. dealbata in form, but smaller, opaque, dingy-white, taste somewhat acrid.   Stem solid, but not cartilaginous, about 2 in. long, equal, ascending or flexuous, excentric, at first floccosely mealy, always opaque, white.   Pileus slightly fleshy, convex then plane, not depressed, obtuse, ½–1 in. broad, unequal, dry, pruinosely hoary; flesh white, compact, but thin.   Gills adnato-decurrent, thin, crowded, plane.   *Fries.*

It loses its acridity in cooking and is quite equal to C. dealbata.

C. trunci′cola Pk.—*truncus*, trunk of a tree.   **Pileus** thin, firm, expanded or slightly depressed in the center, smooth, dry, white.   **Gills** narrow, thin, crowded, adnate-decurrent.   **Stem** equal, stuffed, smooth, often excentric and curved, whitish.

Plant 1 in. high.   **Pileus** 1 in. broad.   **Stem** 1 line thick.

Trunks of frondose trees, especially maples.   *Croghan.*   September. *Peck*, 26th Rep. N. Y. State Bot.

Spores 5x3.5μ *Morgan.*

Found on maple trees in West Philadelphia, Pa.   Edible.   Good quality.

II.—Diffor′mes.

C. decas′tes Fr. *Gr.*—a decade; a number of ten.   From the stems being often joined in bundles of about ten.   Densely cespitose.   **Pileus** 5–12 in. across, soon almost plane, disk gibbous or obtuse; margin at first shortly incurved, then expanded, very much waved and often lobed, even, glabrous, dingy-brown or livid when moist, pale clay-color when dry.   **Flesh** exceedingly thin except at the disk, whitish.   **Stem** 4–7 in. long, ½–1 ½ in. thick, usually slightly thinner upward, rather soft,

94

entirely fibrous, solid, white, usually curved and ascending, coalescent Clitocybe. into a solid mass at the base. Gills adnato-decurrent, or often more or less adnexed, up to ½ in. broad, rather narrowed towards the margin, often wavy. Spores globose, smooth, 4µ diameter.

On the ground and on sawdust.

Albion, Orleans county, N. Y., *Dr. Cushing.* October, 1898.

On ground in grassy places (Woodland Cemetery, May 22, 1897). *McIlvaine.*

Particularly welcome to toadstool lovers are the early comers. The present species is among the first. It is rich in quantity, substance and flavor.

C. mul'ticeps Pk.—*multus*, many; *caput*, a head. ([Color] Plate XXVIIa.) Pileus fleshy, thin except on the disk, firm, convex, slightly moist in wet weather, whitish, grayish or yellowish-gray. Flesh white, taste mild. Gills close, adnate and slightly decurrent, whitish. Stems densely cespitose, equal or slightly thickened at the base, solid or stuffed, firm, elastic, slightly pruinose at the apex, whitish. Spores globose, 5–8µ.

Pileus 1–3 in. broad. Stem 2–4 in. long, 3–6 lines thick.

Open places, grassy ground, etc. Albany and Sandlake. June and October. This species forms dense tufts, often composed of many individuals. In this respect it is related to such species as C. tumulosa, C. aggregata and C. illudens. From the crowding together of many individuals the pileus is often irregular. Sometimes the disk is brownish and occasionally slightly silky. The gills are sometimes slightly sinuate, thus indicating a relationship to the species of Tricholoma. The taste, though mild, is somewhat oily and unpleasant. The plants appear in wet, rainy weather, either early in the season or in autumn. Specimens have been sent to me from Massachusetts by R. K. Macadam and Professor Farlow, and from Pennsylvania by Dr. W. Herbst. *Peck,* 43d Rep. N. Y. State Bot.

West Virginia, New Jersey, Mt. Gretna, Pa. In May, and in autumn months. Very variable in size, color, shape of gills, texture and taste. *McIlvaine.*

The early spring clusters are remarkable for their tenderness and excellence. Clusters of hundreds of individuals grew abundantly at Mt. Gretna in May, 1899. When the fungus was young the gills were

95

Clitocybe. sometimes adnate, almost free, often decurrent. The varying color of oysters is well seen in C. multiceps.

Edible. They should be well cooked. The addition of a little lemon juice or sherry conceals a slight raw taste sometimes present.

C. illu'dens Schw.—mocking, deceiving. (Plate XXIX*a*.) **Pileus** fleshy, convex or expanded, smooth, generally with a small umbo. **Gills** not crowded, unequally decurrent, some of them branched, narrowed toward each end, the edge, in dry specimens, discolored. **Stem** firm, solid, long, smooth, tapering at the base.

Height 5–8 in., breadth of pileus 4–6 in. **Stem** 6–8 lines thick. **Spores** 4–5μ *Peck.*

Grows in clumps or large masses about stumps or decaying trees from August to October. Its bright, deep yellow is attractive from a distance. As many as fifty plants may form a cluster. Cap from 2–6 in., fleshy, convex or expanded, often with a raised center directly over the stem; flesh juicy and yellow; gills yellow, widely separated, running down stem unequally; stem long, firm, solid, smooth, tapering toward base. When cooked the taste is rather saponaceous. Strong stomachs can retain a meal of them, but the fungus generally sickens the eater. Many testings show it to contain a minor poison. It is not deadly, but should not be eaten. Bull. No. 2, Phila. Myc. Center.

New York, *Peck*, Rep. 23–49. Well known in southern states. Indiana, *H. I. Miller.*

The mysterious property of phosphorescence is possessed by this fungus. As heat is known to develop in masses of the fungus it is of interest to know whether it is from the phosphorescence or a ferment. Its radiance by night surpasses its splendor by day. Mr. H. I. Miller, of Terre Haute, Ind., first drew the writer's attention to this quality. A large box of specimens sent by him retained their luminous quality after three days of travel to such an extent that the print of a newspaper could be read when held close to the mass.

Mr. Miller writes: "There is something about this fungus which generates heat. When I bring in a basketful of it, for the pleasure its phosphorescence affords my friends, I find that after having been in the basket for two or three hours, and while piled one bunch upon top of another, that to insert one's hand among the different clusters is like putting it close to a hot stove."

This fungus is so inviting in quantity and beauty that one turns from Clitocybe. it with a regret that lingers. Eaten in quantity it acts upon some persons as an emetic. I have several times eaten of it without other than pleasurable sensations, but persons partaking of the same cooking have been sickened.

C. fumo'sa Pers.—*fumus*, smoke. **Pileus** 1–3 in. across, fleshy, margin thin; convex, often gibbous when young, regular or wavy, even, pellicle not separable, glabrous, sooty-brown, soon livid or gray when dry. **Gills** adnate in regular forms, but often decurrent when the pileus is irregular, crowded, distinct, grayish-white from the first. **Stem** 2–3 in. long, 3–6 lines thick, almost equal, often twisted or curved, glabrous, dingy-white, apex mealy, solid, fibrous. **Spores** subglobose, 5–6µ diam.

In woods. Autumn.

Gregarious, somewhat cespitose, tough, rather cartilaginous. Pileus truly obtuse, never streaked, often regular. Smell none. *Fries.*

Var. *po'lius.* Densely and connately cespitose. **Pileus** convex, then plane, obtuse, smooth, gray. **Stem** flexuous, smooth. **Gills** crowded, whitish. Edible. *Cooke*, 1891.

Var. polius found growing in large quantities in Boston navy yard in stone barn. Determined by Professor Peck. A fair edible. *R. K. Macadam.*

This woods-growing Clitocybe has been many times found by me in a hot-house in Haddonfield, N. J. Professor Peck confirmed my identification. Either its spores or mycelium had evidently been carried thither in the wood-earth used by florists. The hot-house crops appeared in March, and continued until June.

Several of the plants showed an effort to comply with some condition unusual to them, by producing gills upon the upper side of the pileus. Those below were venose and crisped.

This wild species had thus been brought into cultivation. The cultivated plants were much more tender than the wild. Both are excellent.

C. connex'a Pk.—*connexus*, joined. From its relation to Tricholoma. **Pileus** thin, convex or expanded, subumbonate, clothed with a minute appressed silkiness, white, the margin sometimes faintly tinged with

Clitocybe. blue. **Gills** crowded, narrow, white inclining to yellowish. **Stem** equal or tapering downward, solid, whitish.

Plant 2–3 in. high. **Pileus** 2–3 in. broad. **Stem** 2 lines thick. Ground in woods. Croghan. September.

The gills sometimes terminate rather abruptly and are not strongly decurrent, hence it might easily be mistaken for a Tricholoma. The margin of the pileus is sometimes marked with slight ridges as in Ag. laterarius. The odor is weak but aromatic and agreeable. *Peck*, 26th Rep. N. Y. State Bot.

Found in plenty in oak woods near Philadelphia, and in West Virginia; a few specimens in southern New Jersey. Autumn.

Edible, and quite equal to most of the Clitocybes.

C. tumulo'sa Kalchbr.—*tumulus*, a mound. Cespitose. **Pileus** 1–2 in. across, disk fleshy, margin thin; conico-convex then expanded, obtusely umbonate or obtuse, even, glabrous, brownish-umber, becoming pale, margin drooping. **Gills** more or less decurrent or slightly emarginate, crowded narrow, white, then grayish. **Stem** 3–5 in. long, unequal, usually thicker below, minutely downy, pallid, solid.

On the ground in woods. Spring and autumnal months. Readily distinguished by the densely clustered habit, and the umber pileus. The gills are very variable, sometimes distinctly decurrent, at others rounded behind, and almost resembling a Tricholoma. **Spores** subglobose, 5–6μ. *Massee.*

California, *H. and M.;* New York, *Peck*, Rep. 42.

Sent to me by Mrs. Mary Fuller, Washington, D. C. The specimens eaten were of good consistency and flavor.

III.—INFUNDIBULFOR'MES.

* Pileus colored or becoming pale, etc., surface innately flocculose or silky ; not moist.

C. gigante'a Sow.—*giganteus*, of gigantic size. **Pileus** 6–10 in. across. **Flesh** rather thin in proportion to the size of the fungus, white, or tinged with tan, glabrous when moist, slightly flocculose when dry; margin involute then spreading, glabrous, rather coarsely grooved. **Gills** slightly decurrent, broad, very much crowded, branched and con-

nected by veins, whitish then pale tan-color, not separating spontane- Clitocybe.
ously from the hymenophore. **Stem** 1–2 in. long and nearly the same
in thickness, equal, pallid, solid. **Spores** white, 5x3μ.

In woods, etc.

A very distinct species, very showy, large, subcespitose, entirely
whitish tan-color; without close affinities. Stem solid, compact, and
firm inside and outside, 2½ in. long, ½ in. thick, equal, even, glabrous.
Pileus depressed from the first, then broadly, *i. e.*, plano-infundibuliform, thin but equally fleshy, soft, not flaccid, but easily splitting from
the margin toward the center (almost papery and involute when old),
upward of a foot broad, often excentric and generally sinuately lobed,
moist and adpressedly downy when growing, slightly flocculose and
cracked into scales when dry; margin at first very thin, involute,
pubescent, soon spreading, glabrous, at length revolute, coarsely furrowed or radiately wrinkled. Gills slightly decurrent, closely crowded,
almost 3 lines broad (2–3 times as broad as thickness of flesh of pileus),
connected by veins, thin, fragile, straight, but sometimes varying to
crisped and anastomosing, whitish then yellowish or tinged with rufous,
smell weak. *Fries.*

This species was placed in Clitocybe in Syst. Myc. and Epicrisis, but
in Hym. Europ. Fries removed it to Paxillus in which he is followed by
Stevenson. Cooke and Massee continue it in Clitocybe. Dr. Somers
found one measuring over 15 inches in diameter. *R. K. M.*

North Carolina, *Schweinitz.* Edible, *Curtis;* Wisconsin, *Bundy;* California, *H. and M.;* Nova Scotia, *Dr. Somers.*

Large quantities of Clitocybe gigantea grow in the West Virginia
mountains, and in woods around Philadelphia. July to November.

Its substance is coarse, but of good flavor. It should be chopped fine.

C. max′ima Gärtn and Meyer. (Fl. Wett.)—greatest. ([Color]
Plate XXIV, figs. 5, 6.) **Pileus** as much as 1 foot broad, becoming
pale-tan or whitish, *fleshy*, compact at the disk, otherwise thin, *somewhat flaccid* (not capable of being split), broadly funnel-shaped, gibbous *with a central umbo*, always very dry, the surface *becoming silky-
even* or squamulose; margin involute, pubescent, always *even*. **Flesh**
white, at length soft. **Stem** as much as 4 in. long, 1 in. thick, solid,
compact, but internally spongy, *elastic*, attenuated upward, fibrillose-

99

Clitocybe. striate, whitish. **Gills** *deeply decurrent*, pointed at both ends, some-what crowded, soft, simple, *whitish*, not changeable.

The pileus is always very dry because the surface absorbs moisture. Odor weak, pleasant, almost that of A. infundibuliformis. On account of its gigantic stature and color, it has often been interchanged with A. gigantea Sow.; it is in no wise, however, allied to that species, but is so closely allied to A. infundibuliformis that it might be taken for a very luxuriant form of it. *Stevenson.*

Spores 6x4µ *Massee;* 5x3µ *W.G.S.*

New England, *Frost;* California, *H. and M.*

Common in the West Virginia mountains, mixed woods in New Jersey and Pennsylvania. June to November. *McIlvaine.*

It is coarse, dry, hard, but chopped fine and cooked in various ways, either by itself or with meats, it is a good food.

C. infundibulifor'mis Schaeff.—*infundibulum*, a funnel; *forma*, form. ([Color] Plate XXIV, fig. 11.) The Funnel-form clitocybe, Clitocybe infundibuliformis, is a neat and pretty species easily recognized by the funnel shape of its mature cap and by its pale red color. When very young the cap is slightly convex and often adorned with a slight umbo in its center. As it matures the margin becomes elevated so that the cap assumes a shape somewhat resembling that of a wine glass. The margin is sometimes wavy. The flesh is thin and white. The gills are close, thin, white or whitish and decurrent. The stem is smooth, colored like or a little paler than the cap and mostly tapering from the base upward.

The cap is 2–3 in. broad, the stem 1½–3 in. long and ¼–½ in. thick.

The funnel-shaped mushroom grows in woods or copses in summer and autumn, especially in wet seasons. It is somewhat variable in color, but is usually a pale-red, tinged with buff, and sometimes becoming more pale with age. It delights to grow among fallen leaves, and often there is an abundant white cottony mycelium at the base of the stem. When it grows in clusters the caps are apt to be irregular because of mutual pressure. "Mushrooms and Their Use." *Peck.*

Spores 5–6x3–4 *B.*

Very common and in plenty after rains, when large patches of it may be found. I have usually found the light pinkish-buff color to abound, and the stem thinner than described by Prof. Peck. Size of cap from Clitocybe. 1–3 in.

It is a good, reliable food species. The stem should be removed, and the caps well cooked.

** *Pileus colored or pallid, smooth, moist in wet weather.*

C. subzonal'is Pk.—*sub*, under; *zonalis*, pertaining to a zone. **Pileus** thin, centrally depressed or subinfundibuliform, marked with two or three obscure zones, with a slight appressed silkiness, pale yellow. **Gills** close, narrow, equally decurrent, some of them forked, pallid or yellowish. **Stem** equal, slightly fibrillose, stuffed, pale yellow.

Plant 2 in. high. **Pileus** 2–3 in. broad. **Stem** 2–3 lines thick.

Ground in woods. *Croghan.* September. *Peck*, 26th Rep. N. Y. State Bot.

Found in oak woods, Angora, West Philadelphia, growing singly. Specimens few. Edible; pleasant.

C. gil'va Pers.—*gilvus*, pale brownish-yellow. **Pileus** 2–4 in. broad, *pale yellowish, fleshy, compact, convex then depressed,* very obtuse, even, smooth, *dampish when fresh,* polished and *shining* when dry, here and there spotted as with drops, the margin remaining long involute. **Flesh** compact, not laxly floccose, but at length fragile, *somewhat of the same color as the pileus.* **Stem** 1–2 in. and more long, ½ in. and more thick, *solid, fleshy,* stout, not elastic, somewhat equal, smooth, paler than the pileus, villous at the base. **Gills** decurrent, thin, *very much crowded,* often *branched,* arcuate, narrow, *pallid then ochraceous.*

Odor not remarkable. The stem has been noticed at length also hollow, perhaps eroded by larvæ. It corresponds with the Paxilli. The primary form, which is very different from all the rest, is curt, obese, robust, scarcely ever infundibuliform. *Stevenson.*

Spores 4–5x5µ *K.;* 4–5µ *Massee.*

North Carolina, *Schweinitz, Curtis;* Pennsylvania, *Schweinitz;* New York, *Peck,* R. 51, under pines. July to September.

Mt. Gretna, Pa. July, 1898, ground, mixed woods. *McIlvaine.*

Pileus 1–2½ in. across, depressed, almost infundibuliform, smooth. **Color** varied lemon to bright orange. **Flesh** lemon color throughout. **Gills** varying in color, usually same color as pileus. **Stem** all of one

Clitocybe. color, same as pileus, stuffed, sometimes short, and pointed, sometimes thickened at base. Taste and smell pleasant. Edible; good.

C. subinvolu'ta Batsch.—turned under at the margin. **Pileus** brick color, convex, depressed, smooth, margin closely involute. **Flesh** pallid. **Stem** paler, stout, straight, somewhat equal, veined on the lower part with oblique coalescing slightly elevated wrinkles, tomentose and inclining to flesh color above toward the gills, base obtuse. **Gills** decurrent, rather broad, of the same color as the pileus.

The stem is rough on the surface and destitute of luster. It resembles Paxillus involutus in size and habit, in the crenate and involute margin of the pileus, and in the stem being obsoletely veined at the base and tomentose toward the gills. *Stevenson.*

New England, *Frost;* New York, *Peck*, Rep. 22.

Edible, *Cooke.*

C. geo'tropa Bull. *Gr.*—the earth; *Gr.*—to turn. From the turned down margin. **Pileus** 2–5 in. across. **Flesh** thick, white, convex, then plane and finally more or less depressed, obtusely umbonate, the prominence remaining after the pileus becomes depressed, very smooth, even, margin thin, incurved, downy, pale pinkish-tan or buff. **Gills** decurrent, crowded, narrow, simple, white, then colored like the pileus. **Stem** 3–5 in. long, 1 in. or more thick at the base, slightly attenuated upward, compact, fibrillose, colored like the pileus or paler, solid. **Spores** elliptical, 6–7x4–5µ. *Massee.*

In woods and on their borders. Often in rings or troops.

Differs from C. maxima in being firmer, glabrous, and color much more variable; from C. gilva in the thinner pileus, less crowded gills, and white flesh.

Spores 5–7µ *W.G.S.*

In England and on the continent it is considered excellent and superior to most edible fungi.

Found in West Virginia, 1881; Haddonfield, N. J., 1891. Spring and autumn. *McIlvaine.*

Edible, coarse, dry. In stews and mixed to form croquettes or patties, it is a desirable species, owing to its plentifulness.

C. splen'dens Pers.—*splendens*, shining. Solitary. **Pileus** 2–3 in. across, flesh rather thick, white, plane then depressed or funnel-

shaped, glabrous, shining, yellowish. **Gills** deeply decurrent, narrow, crowded, simple, white. **Stem** about 1 in. long, 3 lines thick, glabrous, colored like the pileus, solid, slightly thickened at the base or
equal. *Massee.*

In woods, among pine leaves, etc.

Intermediate between C. gilva and C. flaccida. The typical form of
C. gilva differs in the compact pileus, often with drop-like markings,
the very much crowded, somewhat branched, pale ochraceous gills and
flesh. *Fries.*

Sent to me from Trenton, N. J., by E. B. Sterling.

Edible; quality good, deficient in flavor.

C. inver'sus Scop.—*inverto*, inverted. **Pileus** 2–3 in. across. **Flesh**
thin, fragile; convex, soon funnel-shaped, margin involute, glabrous,
even, reddish or dull brownish-orange. **Gills** decurrent, simple, pallid
then reddish. **Stem** about 1 ½ in. long, 2 lines thick, glabrous, rather
rigid, paler than the pileus, stuffed, soon hollow. **Spores** subglobose,
4µ diameter. *Massee.*

Among leaves, etc.

Gregarious, subcespitose, forming very large tufts, especially late in
the autumn, deformed. Smell peculiar, slightly acid. Stem sometimes
stuffed, usually hollow, hence compressed, rather rigid and corticated
outside, not elastic, without a bulb, glabrous, whitish; the somewhat
rooting base with white down, and often growing together in tufts,
variously deformed, curved, ascending, etc. *Fries.*

Spores subglobose, 4µ *Massee;* 3µ *W.G.S.*

Closely resembles C. infundibuliformis, but differs from it in the color
of gills and flesh. The entire plant is dark in color. Solitary; in troops;
cespitose.

Found in mixed woods. Haddonfield, N. J. Summer and autumn.

That part of the plant which readily breaks away from the stem is
tender and of good flavor. The remainder is tough.

C. flac'cida Sow.—*flaccidus*, limp. **Pileus** 2–3 in. across, flaccid,
orbicular, umbilicate, umbo persistently absent, margin spreading,
arched, glabrous, even, rarely cracking into minute squamules, tawny-
rust colored, shining, not becoming pale. **Flesh** thin, pallid, rather
fragile when fresh, but quite flaccid when dry. **Gills** deeply decurrent,

Clitocybe. arcuate, crowded, narrow, about 1 line broad, white, then tinged yellowish.   **Stem** imperfectly hollow, elastic, tough, 1–2 in. long, 2–3 lines thick somewhat equal, polished, naked, reddish-rust color, base thickened, downy.   **Spores** subglobose, 4–5x3–4μ.

Among leaves, etc.   Gregarious, stems often grown together at the base.   Sometimes solitary and regular.   Summer and autumn.   *Massee.*   **Spores** subglobose, 4–5x3–4μ.

Found in 1886 in West Philadelphia—oak woods.   Since in New Jersey, North Carolina, and interior of Pennsylvania.

Edible.   Well cooked it compares favorably with C. infundibuliformis and others of like texture.

*** *Pileus shining white.*

C. cati'na Fr.—*catinus*, a bowl.   **Pileus** 2 in. broad, at first *white,* *in no wise hygrophanous*, then passing into pale flesh-color during rain, and into tan-color in dry weather, *fleshy*, moderately thin, plane then funnel-shaped, always obtuse, even, *smooth*.   **Flesh** thin, *flaccid*, white.   **Stem** 3 in. long, 1 ½ in. thick, *stuffed*, internally spongy, *elastic*, tough, thickened and tomentose at the base.   **Gills** decurrent, *straight, descending*, not horizontal, broad, not much *crowded*, persistently white.   *Fries.*

Ray Brook, Adirondack mountains.   August.   The pileus is at first white, but in wet weather it becomes pallid or discolored with age.   The plants were found growing among pieces of bark of arbor vitæ lying on the ground.   *Peck*, 43d Rep. N. Y. State Bot.

Quite common in West Virginia, Pennsylvania, New Jersey.   Woods among dead leaves.   August until frost.

Edible.   Excellent in flavor and quality.

Series B.

IV.—CYATHIFOR'MES.

C. cyathifor'mis Bull.—*cyathus*, a cup; *formis*, form.   **Pileus** 1 ½–3 in. across, flesh thin, plano-depressed when young, then infundibuliform, even, glabrous, hygrophanous, rather slimy and usually dark brown when moist, becoming pale and opaque when dry, undulate in large specimens, the margin remains involute for a long time. **Flesh** Clitocybe. watery, similar in color to the pileus, splitting. **Gills** adnate, becoming decurrent with the depression of the pileus, joined behind, distant, grayish-brown, sometimes branched. **Stem** spongy and stuffed inside, elastic, at length often hollow, 2–4 in. long, 3–4 lines thick, attenuated upward, brownish-fibrillose, fibrils forming an imperfect reticulation, colored like the pileus or a little paler, apex naked (not mealy), base villous. *Massee.*

On the ground in pastures and woods, rarely on rotten wood.

Usually blackish-umber, but varies to paler grayish-brown, pinkytan, pale cinnamon or brownish; then dingy-ochraceous or tan-color. Margin expanded when old, and also indistinctly striate. *Fries.*

Var. *cineras'cens* Fr. **Pileus** up to 1 in. across, thin, infundibuliform, pale smoky-brown. **Gills** decurrent, yellowish-white. **Stem** 1–2 in. long, 1½ line thick, grayish, reticulately fibrillose, hollow.

Spores 8x5µ *W.G.S.;* 10–12x5–6µ *B.;* 9x6µ *Morgan.*

Mt. Gretna, Pa. Among leaves in woods. September to October. Gregarious. *McIlvaine.*

Fair in quality.

C. bruma'lis Fr.—*bruma*, winter. From its late appearance. **Pileus** about 1 in. across. **Flesh** thin, expanded, umbilicate then infundibuliform and usually variously waved and lobed, glabrous, flaccid, hygrophanous, livid, whitish or yellowish when dry, disk often darker. **Gills** decurrent, about 1 line broad, crowded, pallid. **Stem** up to 2 in. long and about 2 lines thick, nearly equal, slightly curved, glabrous, whitish, often compressed, imperfectly hollow. **Spores** 4–5x3–4µ.

In woods, etc.

Truly autumnal, being most abundant in November. There are two forms: (*a*) on pine leaves in pine woods; (*b*) among heather. (*a*) Stem rather firm, hollow, about 2 in. long, 2 lines thick, equal or slightly thickened at the apex, at length compressed, somewhat incurved, glabrous, naked, becoming livid, white when dry, base white and downy. Flesh of pileus membranaceous, at first convex, umbilicate, margin reflexed, about 1 in. across, then funnel-shaped, often irregular and undulate, up to 2 in. broad, glabrous, even, livid when moist, whitish then becoming yellowish when dry, disk at first usually darker. Gills decurrent, at first arcuate, then descending, 1 line broad, crowded,

105

Clitocybe. distinct, livid then yellowish-white, smell weak, not unpleasant. (*b*) Entirely watery white; stem hollow, somewhat striate, base glabrous; pileus infundibuliform, margin deflexed, milky-white when dry. Gills less crowded, but rather broader, whitish. *Fries.*

Spores 3µ *W.G.S.;* 4–5x3–4µ *Massee.*
Edible. *Cooke.*

C. morbi′fera Pk.—*morbus*, disease; *fero*, to bear. **Pileus** thin, fragile, glabrous, convex, becoming plane or centrally depressed, slightly hygrophanous, grayish-brown when moist, whitish or cinereous when dry, sometimes slightly umbonate. **Gills** narrow, close, adnate or slightly decurrent, whitish or pallid. **Stem** short, equal, hollow, colored like the pileus or a little paler. **Spores** minute, broadly elliptical, 4µ long, almost as broad.

Pileus .5–1.5 in. broad. **Stem** about 1 in. long, $\frac{1}{8}$–$\frac{1}{4}$ in. thick.

Grassy ground and lawns. November. Washington, D. C. *F. J. Braendle.*

The species seems related to C. expallens, but the margin of the pileus is not striate as in that fungus. The taste is very disagreeable and remains in the mouth a long time. Two persons were made ill by eating it, but their sickness lasted only about three hours. *Peck.*

I have not seen this species. Its reputation is bad. Caution should be observed.

V.—VERSIFOR′MES.

** *Pileus bright, of one color.*

C. trullisa′ta Ellis. **Pileus** fleshy, plano-convex, at length depressed in the center, innate fibrous-scaly, becoming smoother on the disk, margin thin. **Gills** unequal, not crowded, coarse and thick, adnate with a decurrent tooth, at length white pulverulent, purple-violet at first, becoming dark brick-red. **Stem** stuffed, fibrillose, with a long club-shaped base penetrating deeply into the sand. **Spores** large, cylindric-oblong, 15–20µ.

In old sandy fields. September to October.

The interior of the stem in the young plant is like the gills, violet-purple, and the club-shaped base is covered with a tomentose coat, to which the sand adheres tenaciously.

Related to A. laccatus and A. ochropurpureus B.

Resembles the larger forms of A. laccatus, but it has a stouter habit, the pileus is more squamulose, the stem is bulbous or thickened at the base, the mycelium is violet-colored and the spores are oblong.   Bull. Torrey Bot. Club, November, 1874.

New Jersey, *Ellis;* New York, *Peck*, Rep. 33.

Haddonfield, Watertown, N. J.   Sandy soil in pine woods.   *McIlvaine.*

Densely cespitose.   Caps and stems brown, glutinous and so incrusted with sand that it is almost impossible to clean them.   Edible, but not desirable.

C. lacca'ta Scop.—made of lac.   ([Color] Plate XXIV, fig. 10.) **Pileus** thin, fleshy, convex, sometimes expanded, even or slightly umbilicate, smooth or minutely tomentose-scaly, hygrophanous when moist, dull reddish-yellow or reddish flesh-colored, sometimes striatulate when dry, pallid or pale dull ochraceous.   **Gills** broad, rather thick and distant, attached, not decurrent, flesh-colored. **Stem** slender, firm, fibrous, stuffed, equal, concolorous.

Height 1–6 in., breadth of pileus 6 lines to 2 in.   Common.   June to October.

An extremely variable and abundant species occurring almost everywhere throughout the season.   *Peck*, 23d Rep. N. Y. State Bot.

Spores 8–9μ *Massee;* 8–10μ *B.*

Var. *pallidifo'lia* Pk.—*pallidus*, pale; *folium*, leaf. Gills whitish or pallid, decurrent.

Var. *stria'tula* Pk.—*stria*, a furrow.   Pileus moist, smooth, thin, showing shading radiating lines, extending from near the center to the margin.   In wet or damp places.

A form occurs with a decidedly bulbous base. Gills appearing emarginate with a decurrent tooth.

Clitocybe laccata is made the type of a new genus by Berkeley and Broome.   Massee accepts the genus but it is not generally accepted by the standard authors.   It is a well defined genus, and a fitting place for C. laccata, C. amethystina, C. ochropurpurea, C. tortilis, which it puzzles anyone to identify as Clitocybe.

C. amethys'tina Bolt.—*amethystinus*, color of an amethyst. ([Color]

Agaricaceæ

Clitocybe. Plate XXIV, fig. 8.)   **Pileus** 1–2½ in. across, dark-purple, umbilicate, smooth, minutely tomentose, involute.   **Gills** dark-purple, decurrent, broad.   **Stem** 2–3 in. high, fibrillose, purple, streaked with white fibrils, equal, densely covered with white tomentum at base.

Also written *Clitocybe laccata amethystina* Sacc.

"In my opinion it is a good species and should be kept distinct as Bolton gave it, and not be tacked on to C. laccata as a variety.   I should write it Clitocybe amethystina Bolt."   *Peck*, letter September 17, 1897.

New York, *Peck*, Rep. 41; New Jersey, *Sterling;* Mt. Gretna, Pa., on wood soil, June to frost, 1897–1898, *McIlvaine*.

Generally included in C. laccata as a variety, and has therefore been reported under that name.

Great quantities of C. amethystina grew in troops on beds made up of wood earth about the cottages at Mt. Gretna, Pa.   The woods over them is dense.

The caps are tough, but they cook readily and make a pleasing dish.

C. tor'tilis Bolt.—*tortilis*, twisted.   **Pileus** membranaceous, convexo-plane then depressed, obscurely marked with radiating striæ.   **Stem** hollow, twisted, fragile.   **Gills** adnate, thick, distant, fleshy-rose, cespitose, small, irregular, pileus and stem rusty in color.

Hard ground in an old road.   Sandlake.   August.   A species closely allied to C. laccata and appearing like an irregular dwarf form of that species.   Sometimes cespitose.   *Peck*, 41st Rep. N. Y. State Bot.

Excepting that this fungus is frequently found with C. laccata, and might be taken for a new species if not here described, it would not be separated from C. laccata.

Its edible qualities are similar.

C. ochropurpu'rea Berk.—*ochra*, ocher; *purpureus*, purple.([Color] Plate XXIV, figs. 1, 2, 3, 4.)   **Pileus** subhemispherical, at length depressed, fleshy, compact, tough, pale yellow, slightly changing to purplish, cuticle easily separable; margin inflexed, at first tomentose. **Stem** paler, here and there becoming purplish, solid, swollen in the middle, occasionally equal.   **Gills** thick, purple, broader behind, decurrent.   **Spores** white or pale yellow.

Pileus 2 in. broad.   **Stem** 2½ in. high, ¾ in. thick in the center.

August.   On clayey soil in woodlands.

108

Its spores darken when shed in quantity, have a granulated and light-
lilac appearance. It is a solitary grower, sometimes reaching the height
of six inches. The upturned, wavy pileus, showing the purple gills in
contrast with the pale Naples-yellow of the cap is markedly attractive.
The stem is often rough with fibers, hard and tough. The caps are
tough. It grows in grassy woods and open places. The novice, even
the expert, will be puzzled to place it in its genus.

Specimens were sent to me by Miss Lydia M. Patchen, Westfield,
N. Y., and E. B. Sterling, Trenton, N. J. I afterward found many at
Mt. Gretna, Pa. I reported their edible qualities to Prof. Peck who
wrote, September 3, 1897: "I have often wished it was edible, but it
has such a disagreeable flavor when fresh that I have never ventured to
eat it. I have known it to be mistaken for the common mushroom,
but not eaten."

Though tough it cooks tender and is excellent. Stew and put in
patties or croquettes.

VI.—Orbifor′mes.

* Gills becoming ash-colored.

C. di′topa Fr. *Gr.*—twofold; *Gr.*—a foot. Probably from stems
growing two together. Pileus thin, submembranaceous, convex, rarely
with a small umbo, smooth, hygrophanous, brown when young and
moist, grayish-white when dry. Gills grayish, close, thin, attached,
not decurrent. Stem slender, equal, smooth, hollow.

Height 1–2 in., breadth of pileus 6–18 lines. Stem 1–2 lines thick.
Pine woods. West Albany. October.

The plant has the odor and taste of new meal. I have seen no specimens with the pileus depressed. *Peck*, 23d Rep. N. Y. State Bot.

C. meta′chroa Fr. *Gr.*—changing color. Separated from C. ditopa
by its thicker, depressed pileus, its thicker, less close gills, and the absence of odor.

Pine woods. West Albany. October. *Peck*, 23d Rep. N. Y.
State Bot.

Moderately plentiful in New Jersey pines. September to October.
Edible, tough; when well stewed of good flavor.

***Gills whitish.**

Clitocybe. **C. compres'sipes** Pk.—*compressus*, pressed together; *pes*, a foot. **Pileus** thin, convex or expanded, umbilicate, glabrous, hygrophanous, brownish when moist, whitish or pale yellow when dry, margin thin. **Gills** close, subarcuate or horizontal, adnate or subdecurrent, whitish. **Stem** firm, hollow, generally compressed, slightly pruinose. **Spores** elliptical, 5–6.5x4–4.5μ. **Flesh** white when dry, odor slight, farinaceous.

Plant gregarious, 1–1.5 in. high. **Pileus** 6–16 lines broad. **Stem** 1–2 lines thick.

Grassy places. Albany. July.

The moist pileus is sometimes obscurely zonate. The odor is not always perceptible unless the pileus is moist or broken. The stem is sometimes compressed at the top only, sometimes at the base only, and rarely it is wholly top-shaped. *Peck*, 33d Rep. N. Y. State Bot.

Found on open lots in West Philadelphia. Though small it usually grows in troops which yield fair quantity. The caps are tender and of good flavor.

C. fra'grans Sow.—*fragrans*, fragrant. Smell strong, spicy. **Pileus** about 1 in. across. **Flesh** rather thick; convex, soon expanded and slightly depressed or umbilicate, even, glabrous, hygrophanous, uniform watery-white, disk not darker, whitish when dry. **Gills** slightly decurrent, rather crowded, 1 line broad, distinct, whitish. **Stem** about 2 in. long, equal, slightly curved, elastic, glabrous, whitish, stuffed then hollow.

In woods among moss, etc.

Distinguished from other species resembling it in color and size, by the fragrant smell resembling aniseed. *Massee.*

Spores 6x4μ *W.G.S.*

Found in West Virginia, New Jersey, Pennsylvania. July to severe frosts. *McIlvaine.*

Edible. The strong taste of anise is not lost in cooking.

C. pino'phila—pine loving. **Pileus** thin, convex, umbilicate or centrally depressed, glabrous, moist, pale tan-color, paler or alutaceous when dry. **Gills** moderately close, subarcuate, adnate or slightly decurrent, whitish. **Stem** equal, stuffed or hollow, glabrous or subpruinose, colored like the pileus. **Spores** nearly elliptical, 4–6μ long; odor Clitocybe. and taste resembling that of fresh meal.

   Plant 1–2 in. high.   **Pileus** about 1 in. broad.   **Stem** 1–2 lines thick.

   Ground under pine trees.   Albany and   Ticonderoga.   July and August.   *Peck*, 31st Rep. N. Y. State Bot.

   Quite plentiful in pine woods of New Jersey.   Edible; pleasant.

COLLY'BIA Fr.

Gr.—a small coin.

Collybia. **Pileus** fleshy, usually thin, *margin incurved* at first, not corrugated. **Stem** different in substance from the pileus, but confluent with it; hollow, with a cartilaginous bark, internally cartilaginous or soft, often rooting. **Gills** free or obtusely adnexed, membranaceous, soft.

Growing on the ground, wood, leaves and decaying fungi.

In Clitocybe and Tricholoma the substance of the stem and pileus is alike; they differ in the character of the stem. Tricholoma has no distinct bark-like coat, and in Clitocybe the stem is covered with minute fibers. In Mycena as in Collybia the stem is different in substance from the pileus, but is distinguished by the margin of the pileus being straight. It is most closely allied to Marasmius, which is characterized by its tough coriaceous substance, which when dried fully revives and expands on being moistened. The line between them can not always be closely drawn, and there are numerous species which it is difficult to place with certainty in either genus. This does not apply to the fleshy edible species of this genus as they are quite distinct from Marasmius.

Peck's 49th Report contains a monograph of the New York species of Collybia, supplemented by one of those found in other states.

Several common, prolific, long-season, delicious fungi occur in this genus. They vary in size from "a small coin" to five inches across. They grow in woods, on wood, on ground, on leaves, on lawns and among moss and grass in shaded places. The writer has tested many species raw, and eaten small quantities cooked, which are not herein described for the reason that not enough of a species was found to test to full extent. So far as is reported and as his experience goes, there is not a poisonous species in Collybia. Many of them are strong in odor.

ANALYSIS OF TRIBES.

Series A. GILLS WHITE OR BRIGHTLY COLORED, NOT GRAY. FLESH WHITE.

STRIÆPEDES (striate-stemmed). Page 113.

Stem stout, hollow or imperfectly filled with a spongy pith; grooved or striate with fibers.

PLATE XXIX.

Photographed by Dr. J. R. Weist.

COLLYBIA RADICATA.

Photographed by Dr. J. R. Weist.

PLATE XXXV.

PLEUROTUS OSTREATUS.

* Gills broad, rather distant.
** Gills narrow, crowded.

VESTIPEDES (clothed-stemmed). Page 118.

Stem thin, equal, hollow or with a pith, even, velvety, downy or covered with a bloom.
* Gills broad, rather distant.
** Gills very narrow, closely crowded.

LÆVIPEDES (even-stemmed). Page 120.

Stem thin, equal, hollow, naked, smooth—except the base—apparently not striate, but some species are minutely striate under a lens.
* Gills broad, lax, usually more or less distant.
** Gills narrow, crowded.

Series B. GILLS BECOMING GRAY. HYGROPHANOUS.

TEPHROPHANÆ. Page ——.

Color brownish becoming gray. Allied to the last section of Tricholoma and Clitocybe, but distinguished from them by the cartilaginous stem.
Some are strong scented. None known to be edible.

STRIÆ'PEDES.

*Gills broad, rather distant.

C. radica'ta Relh.—*radix*, a root. (Plate XXIX, p. 112.) **Pileus** 1½–4 in. across, from convex to nearly plane, broadly umbonate, frequently wrinkled toward and at the umbo, glutinous when moist. Color variable, usually brown in grayish shades, from dark to almost white. **Flesh** thin, white, elastic. **Gills** white, thick, tough, distant, ventricose, adnexed, rounded or notched behind like Tricholoma, sometimes with a decurrent tooth. **Stem** 4–8 in. long, 3–5 lines thick, smooth, firm, same color as pileus, tapering upward, becoming vertically striate or grooved, often twisted, ending in a long, tapering, pointed root deeply planted in the earth.
Spores elliptical, 14–15x8–9μ *Massee;* 11x17μ *W.G.S.;* 11x9μ *W. P.;* 16–17x10–11μ *B.*

113

Collybia. Often sombre, but erect, neat and handsome. Growing solitary and in troops in woods, usually near stumps, if much decayed, sometimes on them, or on shaded lawns and grassy places. June to October.

Var. *furfu'racea* Pk. **Stem** furfuraceous, less distinctly striate.

Var. *pusil'la* Pk. Plant small. **Pileus** about 1 in. broad, passing gradually into the typical form. **Stem** slender.

Professor Peck says: "The variety furfuracea is common and connects this species with C. longipes, which has a villose stem and dry velvety pileus." 49th Rep.

Common to the United States. Edible. *Curtis*, according to Dr. F. Peyre Porcher of Charleston, S. C., was the first to declare this edible.

A very attractive species. The purity of its gills is especially noticeable. I began eating it in 1881, and it has continued to be a favorite. The caps should be broiled or fried. They are sweet, pleasing in texture, and delicately flavored.

C. platyphyl'la Fr. *Gr.*—broad; a leaf. ([Color] Plate XXVIII, fig. 1.) **Pileus** 3–4 in. broad, dusky and gray then whitish, fleshy-membranaceous, *thin, fragile*, soon flattened, obtuse, watery when moist, *streaked with fibrils.* **Stem** 3–4 in. long, ½ in. thick, stuffed, soft, equal, fibrilloso-striate, otherwise smooth, naked or obsoletely powdered at the apex, whitish, shortly and bluntly rooted at the base. **Gills** obliquely cut off behind, *slightly* adnexed, ½ in. and more *broad, distant*, soft, white.

Odor not remarkable. It inclines toward the Tricholomata in the *somewhat membranaceous cuticle of the soft stem. Fries.*

Spores 13x19µ *W.G.S.*

Solitary, gregarious, rarely clustered. On rotten wood, roots, ground near stumps, among leaves, etc. June to October.

Distinguished by the very broad and deeply emarginate gills, which frequently slope up behind to near the cap then with a short turn downward connect with the stem which is either stuffed or hollow, and by the abundant, cord-like rooting mycelium. The gills are very broad. Professor Peck says: "The species is quite variable. The pileus is sometimes irregular and even eccentric, the thin margin may be slightly striate, is often split and in wet weather may be upturned or revolute. The lamellæ are sometimes ½ in. broad or more and transversely split. They may be obscurely striated transversely and even veiny above with

venose interspaces. Occasionally a slight anise-like odor is perceptible, Collybia.
but in decay the plants have a very disagreeable odor and disgusting
appearance.'' 49th Rep. N. Y. State Bot.

West Virginia, 1880–1885; Haddonfield, N. J., 1896. Gregarious, and in large bunches. Mt. Gretna and Eagle's Mere, Pa., 1897,
McIlvaine.

When fresh, in good condition, the caps are good, but they are not
nearly equal in substance or flavor to C. radicata and C. longipes. They
are best broiled or fried.

Var. *re'pens* Fr. **Pileus** more fleshy, depressed. **Stem** hollow,
compressed, pruinate at the apex, with a *creeping, string-like mycelium.*

It is best distinguished by its white, villous, anastomosing, very
much branched mycelium which creeps a long distance in a rooting
string-like manner. The so-called roots are quite different from the
stem, not a prolongation of the stem itself. *Fries.*

Clearly a variety of C. platyphylla. C. platyphylla is quite variable,
even puzzling. Edible qualities the same.

C. long'ipes Bull.—*longus*, long; *pes*, a foot. **Pileus** 1–2 in. across,
conical then expanded, umbonate, dry, minutely, beautifully velvety.
Color from pale to date-brown, sometimes umber. **Flesh** white, thin,
elastic. **Gills** white, broad, tough, thick, adnexed, distant, ventricose,
rounded behind, emarginate. **Stem** 4–6 in. long, 2–4 lines thick, tapering upward, usually densely and minutely velvety like the cap, nearly
same color, with a long, tapering root.

On much decayed stumps and logs. July to October. Closely resembles C. radicata. It is readily distinguished by its velvety cap and
stem. It is more glutinous.

Spores spheroid, 12μ Q.
California. Edible. *H. and M.*

West Virginia mountains, 1880–1885; Cheltenham, Pa., 1889. *Mc-
Ilvaine.*

Excepting from California, C. longipes has not previously been reported as found in the United States. It is not plentiful in the forests
of West Virginia, yet I often found it upon rotting stumps and logs,
solitary, but up to a dozen in the same vicinity. It is unmistakable. Its
rich yet dull velvety cap and stem and the purity of its gills hold the
finder's admiration.

Collybia. The caps fried or broiled are delicious, resembling in every way those of C. radicata.

C. fu'sipes Bull.—*fusus*, a spindle; *pes*, a foot. ([Color] Plate XXVIII, fig. 4.) **Pileus** 1–3 in. broad, *reddish-brown*, becoming pale and also dingy-tan, fleshy, convex then flattened, umbonate (the umbo at length vanishing), even, smooth, dry, here and there broken up in cracks when dry. **Stem** 3 in. and more long, commonly ½ in., but here and there as much as 1 in. broad, *fibrous-stuffed then hollow*, remarkably cartilaginous externally, *swollen, ventricose in the middle, attenuated at both ends*, often twisted, longitudinally *furrowed*, red or reddish-brown, *rooted in a spindle-shaped manner at the base*. **Gills** *annulato-adnexed* (joined into a ring), soon separating, free, broad, distant, firm, connected by veins, crisped, white then becoming somewhat of the same color as the pileus, often spotted. *Stevenson.*

Spores 6x3µ *W.G.S.;* 4–5x2–4µ *B.*

Solitary, gregarious, usually densely clustered on decaying wood, roots, etc. August until after heavy frosts.

West Virginia, 1882, *McIlvaine.*

In the West Virginia mountains C. fusipes is frequent. Caps in the clusters rarely exceed 1½ in. across. They show an auburn or burgundy shade of brown in their color. When young they are smooth and appear to remain so unless rained upon or moistened, when they crack more or less finely in drying. At first the connection of the gills with the stem is peculiar—they join in a collar-like ring at the top of the stem. As the cap expands the gills part more or less and separate from the stem. The stem is markedly spindle-shaped, though variously flattened by compression in dense clusters; the outside often splitting, breaking and turning out from the stem.

The caps, alone, are good, the stem being hard and refractory. The caps are very fine, cooked in any way.

The caps dry well, and are a pleasant addition to gravies, soups and other dishes. They make a choice pickle.

***Gills narrow, crowded.*

C. macula'ta A. and S.—*macula*, a spot. **Pileus** fleshy, firm, convex or nearly plane, even, glabrous, white or whitish, sometimes varied

with reddish spots or stains. **Flesh** white. **Gills** narrow, crowded, Collybia.
adnexed, sometimes nearly or quite free, white or whitish. **Stem** generally stout, firm, equal or slightly swollen in the middle, striate, white, stuffed or sometimes hollow, commonly narrowed at the base, rooting, often curved at the base, rarely slightly thickened and blunt. **Spores** subglobose, 4–6μ broad, sometimes showing a slight point at one end.

Pileus 2–4 in. broad. **Stem** 2–4 in. long, 3–6 lines thick.

Var. *immacula'ta* Cke. This differs from the type in having no reddish spots or stains.

This species is easily recognized by its large size, firm or compact substance and white color. It grows in soil filled with decaying vegetable matter or on much decayed wood. *Peck*, 49th Rep. N. Y. State Bot.

West Philadelphia, Pa. Weed grown lot near University of Pennsylvania. September to frost. Grew gregariously over a large lot. The plants varied greatly in size and appearance. The gills of most were crenulate (scalloped). Assorted specimens were sent Professor Peck who wrote: "They are all forms of C. maculata."

The caps were stewed and eaten in abundance by many, and pronounced "Fine."

C. butyra'cea Bull.—*butyrum*, butter; buttery to the touch. **Pileus** 2–3 in. broad, normally *reddish-* (Plate XXX.)
brown, but becoming pale, fleshy, convex then expanded, more or less *umbonate*, dry, *even*, smooth. **Flesh** buttery, soft, somewhat hygrophanous, flesh-color then white. **Stem** 2–3 in. long, *attenuated* upward from the thickened white downy base, hence much thinner at the apex, 2–3 lines only, but at the base ½–1 in. thick, externally covered over with a *rigid cartilaginous cuticle*, internally stuffed with soft *spongy pith*, or hollow only when old, *striate, reddish*, commonly smooth, but varying with

COLLYBIA BUTYRACEA.

white deciduous scales, and occasionally wholly downy with soft hairs.

Collybia. **Gills** slightly adnexed, *somewhat free*, thin, *crowded*, notched at the edge, white, *never spotted-reddish.* *Stevenson.*

Spores 6–10x3–5μ *B.;* elliptical, 7–9x4–5μ.

Cap greasy looking. **Umbo** dark.

The color of the cap is variable. The species differs from C. dryophila in having an umbonate pileus, slightly uneven gill-edges and stem which tapers upward.

Solitary and in troops under coniferous trees. Spring, autumn.

West Virginia, Chester county and Eagle's Mere, Pa., *McIlvaine.*

The caps cook quickly, are tender and have a good flavor.

VESTI'PEDES.

* Gills broad, rather distant.

C. velu'tipes Curt.—*velutum*, velvet; *pes*, a foot. **Pileus** 1–4 in. broad

(Plate XXXI.)

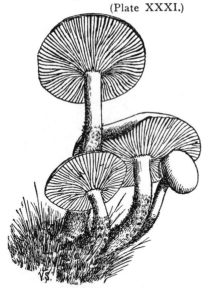

COLLYBIA VELUTIPES.
Natural size.

in the same cluster, *tawny*, sometimes paler at the margin, moderately fleshy at the disk, but thin at the circumference, convex then soon becoming plane, often eccentric, irregular and bent backward, smooth, *viscous;* margin spreading and at length slightly striate. **Flesh** watery, soft, slightly tawny-hyaline. **Stem** 1–3 in. long, 1–4 lines thick, tough, externally cartilaginous, *umber then becoming black, densely, minutely velvety,* commonly ascending or twisted, commonly equal, even, internally fibrous-stuffed and hollow. **Gills** broader and rounded behind, slightly adnexed, so as at first sight to appear free, *somewhat distant*, very unequal, *becoming pallid-yellow or tawny. Fr.*

Spores ellipsoid, 7μ *W.G.S.;* 6x4μ *B.*; elliptical, 7x3–3.5μ *Massee.*

Our American plant, common to the states, is rarely found attaining such dimensions. Its usual size is from 1–2 in. across, more frequently

at 1–1¼ . It is generally found in clusters more or less dense. The
color varies from yellowish to a dark yellowish-brown. The center is
darker than the margin. The cap viscid when moist, often irregular
from crowding. Gills may be rounded or notched at their attachment
to the stem, whitish or yellowish. Stem usually hollow, 1–4 in. long,
1–3 lines thick, whitish when young becoming colored with the dense
brownish velvety hairs.

It grows on stumps, roots in the ground, trunks and earth heavily
charged with wood matter. I have found it in every month of the year.
The heavier crop appears in September, October and November, and
lasts until long after heavy frosts. Then sporadic clusters spring up
wherever the winter sun gives them encouragement.

It sometimes does considerable damage to the tree so unfortunate as
to be its host. It begins its growth upon some injured or decayed spot
and by continually insinuating itself under the surrounding bark it, by
its mycelium and growth, pries the bark away from the wood until the
tree is entirely denuded.

It is a valuable species, not only on account of its continuous growth,
but because of its plentifulness and excellent substance.

** *Gills very narrow, closely crowded.*

C. con′fluens Pers.—**Pileus** ¾–1½ in. broad, thin, tough, flaccid,
convex or nearly plane, obtuse, rarely somewhat umbonate, glabrous,
hygrophanous, reddish grayish-red or reddish-brown and often striatulate on the margin when moist, pallid, whitish or grayish when dry.
Lamellæ narrow, crowded, free, whitish or yellowish-gray. **Stem** 2–5
in. long, 1–2 lines thick, equal, cartilaginous, hollow, clothed with a short
dense somewhat pulverulent whitish pubescence or down. **Spores**
minute ovate or subelliptical, slightly pointed at one end, 5–6x3–4μ.

Among fallen leaves in woods. Common. July to October.

The plants commonly grow in tufts, but sometimes in lines or arcs of
circles or scattered. They revive under the influence of moisture and
thereby indicate an intimate relationship to the genus Marasmius. The
pileus varies much in color, but commonly has a dull reddish or russety
tinge when moist, sometimes approaching bay-red. It fades in drying
and becomes almost white or grayish-white, but sometimes the center
remains more deeply colored than the margin. The stem is commonly

119

Collybia. rather long in proportion to the width of the pileus. Occasionally it is somewhat flattened either at the top or throughout its entire length. Sometimes the stems become united at the base which union is suggestive of the specific name. *Peck*, 49th Rep.

West Virginia, New Jersey, Pennsylvania, *McIlvaine*. July to frost.

The caps of C. confluens are of excellent substance and flavor. Their quantity makes up for their small size. I have gathered them 2 in. across, but their average size is about 1 in. They dry well.

LÆVI'PEDES.

** Gills broad, more or less distant.*

C. esculen'ta Wulf.—*esculent.* **Pileus** ½ in. and more broad, *ochraceous-clay*, often becoming dusky, *slightly fleshy*, convex then plane, orbicular, *obtuse*, smooth, even or when old slightly striate. **Flesh** tough, white, savory. **Stem** 1 in. and more long, scarcely 1 line thick, or thread-like and wholly equal, *obsoletely tubed*, tough, *stiff and straight*, even, smooth, slightly shining, *clay-yellow*, with a *long perpendicular*, commonly *smooth*, tail-like *root*. **Gills** adnexed, even decurrent with a very thin small tooth, then separating, *very broad*, limber, *somewhat distant, whitish*, sometimes clay-color.

Gregarious but never cespitose. The tube of the stem is very narrow. *Stevenson.*

The smallest edible Collybia. *Cooke.* Edible. In dense woods. *Curtis.* It is dried and preserved. *Cordier.*

In pastures and grassy places. Spring and early summer.

Edible, but rather bitter flavor. In Austria, where it is in great plenty in April, large baskets are brought to market under the name of Nagelschwämme—nail mushrooms.

Professor Peck describes C. esculentoides Pk., 49th Rep. N. Y. State Bot., which he states: "Differs from the type in its paler and more ochraceous color and in its farinaceous flavor, and is related to the European C. esculenta from which it differs essentially in the umbilicate pileus and in the absence of any radicating base to the stem."

*** Gills narrow, crowded.*

C. dryophil'a Bull. *Gr.*—oak-loving. ([Color] Plate XXVIII, fig. 3.) **Pileus** 1–3 in. across, bay-brown-rufous, etc., *becoming pale*, but not hygrophanous, slightly fleshy, tough, convexo-plane, *obtuse, commonly depressed in the center,* even, smooth; margin at first inflexed then flattened. **Flesh** thin, white. **Stem** 1–3 in. long, 1–3 lines thick, cartilaginous, *remarkably tubed,* thin, even, smooth, somewhat rooting, commonly *becoming yellow or reddish.* **Gills** *somewhat free,* with a small decurrent tooth, but appearing adnexed when the pileus is depressed, *crowded, narrow,* distinct, plane, *white* or becoming pale.

There are numerous monstrous forms which are very deceiving: *a.* **Stem** elongated, waved, decumbent, inflated at the base; **pileus** broader, *lobed;* **gills** white. *b. Funicularis,* larger, cespitose, the lax and decumbent **stem** equal and hairy at the base, **gills** sulphur-yellow. These forms, analogous with A. repens Bull., occur on heaps of leaves. *c.* Countless specimens growing together in a large cluster; **stems** thick, inflated, irregularly shaped, *sulcate,* brown, the mycelium collecting the soil in the form of a ball; **pileus** very irregularly shaped, full of angles, undulated, blackish then bay-brown. In gardens. *Stevenson.*

Spores elliptic-fusiform, 7–8x4μ; 6μ *W.G.S.*

Professor Peck, 49th Rep. N. Y. State Bot., gives the following: **Pileus** thin, convex or nearly plane, sometimes with the margin elevated, irregular, obtuse, glabrous, varying in color, commonly some shade of bay-red or tan-color. **Flesh** white. **Lamellæ** narrow, crowded, adnexed or almost free, white or whitish, rarely yellowish. **Stem** equal or sometimes thickened at the base, cartilaginous, glabrous, hollow, yellowish or rufescent, commonly similar in color to the pileus. **Spores,** 6–8x3–4μ.

Pileus 1–2 in. broad. **Stem** 1–2 in. long, 1–2 lines thick.

Woods, groves and open places. Common. June to October.

West Virginia, North Carolina, New Jersey, Pennsylvania. *McIlvaine.*

C. dryophila is so common and variable that descriptions would fail to cover it in its eccentricities. The writer has eaten it in all the forms obtained since 1881. A very pretty form grew in large quantities among pine needles at Eagle's Mere, Pa., in August, 1897. It was cooked and served at the hotel table. Many ate it and were delighted.

Dr. Badham refers to a case in which illness was caused by eating it. In my eighteen years' experience with it, knowing it to have been enjoyably eaten by scores of persons, I have not heard of the slightest discomfort from it.

Collybia.   **C. spinulif′era** Pk.—*spinula*, a little thorn. **Pileus** fleshy, thin, convex or nearly plane, glabrous, hygrophanous reddish tan-color tinged with pink and slightly striatulate on the margin when moist, paler when dry, adorned with minute colored spinules or setæ. **Gills** narrow, close, rounded behind and free, pale cinnamon-color, becoming somewhat darker with age, spinuliferous. **Stem** slender, tough, glabrous, shining, hollow, reddish-brown, often paler or whitish at the top, especially in young plants, with a whitish myceloid tomentum at the base. **Spores** elliptical or nearly so, 4μ.

Plant cespitose. **Pileus** 8–16 lines broad. **Stem** 2–3 in. long, about 1 line thick.

Prostrate trunks and ground among leaves in woods. Lewis county. September.

In this species the lamellæ, under a lens, appear to be minutely pubescent or velvety. This is due to the colored spinules or setæ which clothe them. *Peck*, 49th Rep. N. Y. State Bot.

Angora, Pa. September, 1897. Among moss in mixed woods. September to frost. *McIlvaine.*

Specimens identified by Professor Peck. Stems of some tapered at base.

Excepting the extreme base of stems the whole plant is tender and of good flavor.

(Plate XXXIa.)

COLLYBIA ACERVATA (young).

C. acerva′ta Fr.—*acervus*, a heap. **Pileus** fleshy but thin, convex or nearly plane, obtuse, glabrous, hygrophanous, pale tan-color or dingy pinkish-red and commonly striatulate on the margin when moist, paler or whitish when dry. **Gills** narrow, close, adnexed or free, whitish or tinged with flesh-color. **Stem** slender, rigid, hollow, glabrous, reddish, reddish-brown or brown, often whitish at the top, especially when young, commonly with a white matted down at the base. **Spores** elliptical, 6x3–4μ.

Plant cespitose. **Pileus** 1–2 in. broad. **Stem** 2–3 in. long, about 1 line thick.

Decaying wood and ground among fallen leaves in woods. Adirondack Collybia mountains. August and September. *Peck*, 49th Rep. N. Y. State Bot.

This very pretty plant resembles forms of C. dryophila. The coloring of the stems is often extremely delicate, like paintings upon rice paper.

West Virginia mountains; Eagle's Mere, Pa. August to frost. *McIlvaine*.

The entire plant is tender, delicate and of fine flavor. In these qualities it is not distinguishable when cooked from the smaller forms of C. dryophila.

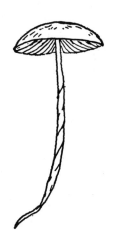

MYCE'NA Fr.

Gr.—a fungus.

Mycena.

Pileus regular, rarely depressed in the center, thin, usually streaked with longitudinal lines, at first conico-cylindrical, *margin at the first straight*, closely embracing the stem which is attenuated upward. **Stem** hollow, slender, cartilaginous. **Gills** adnate or adnexed, sometimes with a small tooth, never decurrent. **Spores** white.

(Plate XXXII.)

MYCENA GALERICULATA.

Generally small and slender, growing on branches, twigs, heaps of leaves, sometimes on the ground, some minute species on single dead leaves. Long, rooting stems are not uncommon. Clitocybe and Omphalia are separated by their decurrent gills and in Collybia the margin is at first incurved.

In this genus the species of the various sections are not always distinguished by single sharply defined characteristics, so that it will sometimes be necessary to pay attention to all the features. Species with a thread-like stem are found in other sections than Filipedes and some of the Lactipedes are slippery when moist, but not truly viscous.

ANALYSIS OF TRIBES.

CALODONTES (*kalos*, beautiful; *odontes*, teeth). Page 126.

Stem juiceless, not dilated into a disk at the base. Edges of gills darker, minutely toothed.

ADONIDEÆ (*Adonis*, referring to beauty). Page 126.

Stem juiceless, not dilated at the base. Gills of one color, not changing color. Color pure-colored, bright, not becoming brownish or gray. On the ground.

RIGIPEDES (rigid-stemmed). Page 126. Mycena.

Stem firm, rigid, rather tough, juiceless, more or less rooting. Gills changing color, white, then gray or reddish, generally at length connected by veins.

Tough, persistent, inodorous, usually on wood, very cespitose, but individuals of the same species sometimes grow singly on the ground.

FRAGILIPEDES (fragile-stemmed). Page 130.

Stem fragile, juiceless, fibrillose at the base, scarcely rooting. Pileus hygrophanous. Gills becoming discolored, at length somewhat connected by veins.

Thin, fragile, often soft, normally growing singly on the ground. A few strong smelling, cespitose on wood.

FILIPEDES (thread-stemmed). Page 130.

Stem thread-like, flaccid, somewhat tough, rooting, juiceless, generally extremely long in proportion to the pileus. Gills becoming discolored, paler at the edge.

Straight, growing singly on the ground; inodorous. Pileus dingy-brown, becoming paler.

LACTIPEDES (milky-stemmed). Page 130.

Gills and rooting stem milky when broken.

GLUTINIPEDES (glutinous-stemmed). Page 131.

Stem juiceless but externally sticky with gluten. Gills at length decurrent with a tooth.

BASIPEDES (base-stemmed). Page 131.

Stem dry, rootless, the base naked and dilated into a disk or small hairy bulb. Growing singly, slender, soon becoming flaccid.

INSITITIÆ (*insero*, to insert or graft). Page 131.

Stem very thin, dry, growing as if inserted in the supporting surface, not downy, not disk-like at the base.

Gills adnate with a small decurrent tooth. Small, very tender, becoming flaccid with the first touch of the sun.

Mycena is a large genus composed of small species. About sixty members have been found in America. They are from ½ to 1 in. across the cap, with thin stems and altogether delicate appearance. Yet the flesh of most of them has a gummy consistency in the mouth, and they shrink but little in stewing. Heretofore not any appear to have been reported as edible, probably because the size of the species has not attracted experimenters. While some have a strong odor and taste of radishes, and one species is bitter, it is probable that all are edible. The writer has eaten, raw and cooked, small quantities (all he has found) of many species not here reported as edible, which will, when further tested, be reported upon.

The substance and flavor of those here given is remarkably pleasant. Their late coming, hardiness and abundance are commendable qualities.

I.—CALODON'TES. Stem juiceless. Gills minutely toothed. None tested.

II.—ADONI'DEÆ. Stem juiceless. Gills of one color, etc. None tested.

III.—RIGIDI'PEDES. Stem rigid. Gills at first white, changing color, etc.

M. prolif'era Sow.—*proles*, offspring; *fero*, to bear. ([Color] Plate X, figs. 6, 7.) **Pileus** ⅔–1¼ in. across, slightly fleshy, expanded bell-shape, dry, the broad umbo darker (dingy-brown), slightly striate, and at length furrowed or rimosely split at the margin (pale yellowish or becoming brownish-tan). **Stem** 2½–3 in. long, firm, rigid, *smooth, shining, slightly striate*, rooted. **Gills** adnexed, somewhat distinct, becoming pale white.

Inodorous, only at length nauseous. Very closely allied to M. galericulata, in habit approaching nearest to M. cohærens. The stems are pallid at the apex, but slightly tawny-bay-brown below, and glued together by hairy down at the base. There is a *white* form with transparent stem—on trunks. *Fries.*

Mt. Gretna, Pa. On ground in grass. Mycelium spreading on leaves. *McIlvaine.*

Found in great plenty. Base of stems is sometimes white when in Mycena. dense tufts.

The whole plant is tender, cooking in fifteen minutes, and is of fine flavor. No one will want a better fungus.

M. rugo'sa Fr.—*ruga*, a wrinkle. **Pileus** ash-color but becoming pale, very tough, slightly fleshy at the disk, otherwise membranaceous, bell-shaped then expanded, at length rather plane, somewhat obtuse, more or less corrugated (unequal with elevated wrinkles), always dry, not moist even in rainy weather, striate at the circumference. **Stem** commonly short, remarkably cartilaginous, tubed, rigid, tough, straight, at length compressed, even, smooth, pallid, with a short oblique hairy root. **Gills** *arcuato-adnate*, with a decurrent tooth, united behind in a collar, somewhat distant, connected by veins, broad, ventricose, white then gray, edge sometimes quite entire, sometimes with saw-like teeth.

Always inodorous. Formerly connected with M. galericulata. M. rugosa is arid, very tough, more rarely cespitose, the pileus firm, somewhat obtuse, wrinkled but without striæ, the gills arcuato-adnate with a hooked tooth, *white then ash-color*. The genuine M. galericulata is fasciculato-cespitose, somewhat fragile, the pileus thinner, at first conical and umbonate, striate without wrinkles, the gills adnate, with a decurrent tooth, white then *flesh-color*. Between these there is a long series of intermediate forms. *Fries.*

California, *H. and M.;* Kansas, *Cragin;* Wisconsin, *Bundy;* New York, September, *Peck,* 46th Rep.; West Virginia, New Jersey, Pennsylvania. On decaying wood and ground near stumps. August to November. *McIlvaine.*

The tenacity frequently occurring in Mycena is well shown in this species. The caps and stem cook tender, but it is better to discard the stems, as the two do not become tender at the same time.

M. galericula'ta Scop.—*galericulum*, a small peaked cap. ([Color] Plate X, fig. 5.) **Pileus** somewhat membranaceous, conical bell-shaped then expanded, striate to the umbo, dry, smooth, becoming brownish-livid or changeable in color. **Stem** rigid, *polished, even, smooth*, with a spindle-shaped root at the base. **Gills** *adnate, decurrent with a tooth*, connected by veins, whitish and flesh-colored.

Very protean. Normally growing in bunches, the numerous stems

127

Mycena. (never sticky) glued together with soft hairy down at the base. But it occurs also solitary, larger, pileus as much as 2 in. broad, wrinkled-striate. The essential marks by which it is distinguished from A. rugosa are these: **Stem** in general thinner, less tense and straight, often curved, more fragile. **Pileus** membranaceous, conico bell-shaped, umbonate, striate but not corrugated, moist in rainy weather. **Gills** adnate, with a decurrent tooth, more crowded, *whitish then flesh-colored*. The color both of the pileus (normally dingy-brownish then livid) and of the stem (normally becoming livid-brownish) is much more changeable than that of A. rugosa, becoming yellow, rust colored, etc. It is not so tough and pliant as A. rugosa. Forms departing from the type are very numerous; the most beautiful is var. *calopus* (*Gr.*, beautiful; *Gr.*, a foot) with chestnut-colored stems, united in a spindle-shaped tail. *Fries.*

Spores spheroid or subspheroid, 9–10x6–8μ *K.;* 8–11x4–6μ *B.;* 6–7x4μ *Massee*

Common. Autumnal. Very variable. On trunks, fallen leaves.

Two well-marked varieties of this very variable species were observed the past season. One grows on the ground among fallen leaves. It has a dark brown pileus, close lamellæ and a very long stem, generally of a delicate pink color toward the top. It might be called var. *longipes*. The other grows under pine trees, has a broadly convex or expanded grayish-brown pileus and a short stem. It might be called var. *expansus*. *Peck*, 26th Rep. N. Y. State Bot.

"*M. alcalina* is closely allied to it (M. galericulata), but has a stronger alkaline odor and a rather more fragile stem. In one of your specimens I detect a slight incarnate tint to the gills, and this is pretty conclusive evidence that it belongs to M. galericulata. Species of Mycena are not generally reckoned among edible fungi or even promising fungi; I suppose on account of the thin flesh of the cap, but of course it is possible to make up in numbers what is lacking in size. I am glad to know you have found this to be an esculent one." Letter Professor Peck to C. McIlvaine, October 5, 1893.

The caps and stems when young make as good a dish as one cares to eat. The substance is pleasant, and the flavor delicate. They are best stewed slowly in their own fluids, after washing, for ten minutes and seasoned with pepper, salt and butter.

M. parabo'lica Fr.—shaped like a parabola. **Pileus** becoming black at the disk, inclining to violaceous, otherwise becoming pale, whitish, Mycena. somewhat membranaceous, at first erect and oval, then parabolic, obtuse, never expanded, moist, somewhat shining when dry, smooth, even, striate toward the entire margin. **Stem** 2–3 in. long, 1 line thick, tubed, tense and straight but not very rigid, thickened and bearded-rooted at the base, pale below, dark violaceous above, when young white-mealy, otherwise even, smooth, dry. **Gills** simply adnate, ascending, somewhat distant, rarely connected by veins, quite entire, white, somewhat gray at the base.

Stem less rigid than that of A. galericulatus. Truly gregarious or cespitose. *Fries.*

Spores 12x6µ *B.;* elliptical, 11–12x6µ *Massee.*

Trenton, N..J. June. *E. B. Sterling;* West Virginia, New Jersey, Pennsylvania, on decaying stumps, trunks of oak, chestnut, poplar, pine. June until far into the winter. *McIlvaine.*

Plant up to 2½ in. high. **Caps** usually about ½ in., but reaching ¾ in.

A neat, attractive plant, whether single or in dense tufts. Its smell is strong of fresh meal, and taste of that delicate flavor one finds in the succulent base of the round, swamp rush, when pulled from its sheath —one that every country school boy and girl knows. It is pleasant raw, and delicious when cooked.

M. latifo'lia Pk.—*latus,* broad; *folium,* a leaf. **Pileus** convex, rarely somewhat umbonate, striatulate, grayish-brown. **Gills** white, broad, hooked, decurrent-toothed. **Stem** slender, smooth, hollow, subconcolorous, white-villous at the base.

Height 1–1.5 in., breadth of pileus 4–6 lines. **Stem** .5 lines thick. Under pine trees. Center. October.

A small species with quite broad gills, growing among the fallen leaves of pine trees. Gregarious. *Peck,* 23d Rep. N. Y. State Bot.

Mt. Gretna, Pa. Among pine needles, scattered, sometimes four or five in a cluster. September to October. *McIlvaine.*

Autumnal. Not rare. The caps though small are tenacious in the mouth and lose little in cooking. The substance is agreeable and flavor fine.

IV.—Fragili′pedes. Stem fragile, juiceless, etc. None tested.

V.—Fili′pedes. Stem thread-like, etc.

Mycena. **M. collaria′ta** Fr.—*collare*, a collar. **Pileus** ½ in. and more broad, typically dingy-brown, but becoming pale, commonly gray-whitish, becoming brownish only at the disk, membranaceous, bell-shaped then *convex*, somewhat umbonate, striate, when dry rigid, smooth, *not soft nor slightly silky*. **Stem** about 2 in. long, tubed, *thread-like* but almost 1 line thick, *tough*, dry, smooth, even or slightly striate under a lens, becoming pale. **Gills** adnate, *joined in a collar* behind, thin, crowded, *hoary-whitish or obsoletely flesh-colored*.

The gills are somewhat distant when the pileus is expanded. There is not a separate collar as in Marasmius rotula; the gills are only joined in the form of a collar, and remain *cohering* when they separate from the stem. *Fries.*

Spores 8–10x4–6µ B.

New York. Old stumps and rotten logs. June. *Peck*, 23d Rep. Mt. Gretna, Pa. Cespitose on decaying wood. July, September and October. *McIlvaine.*

Very much like M. galericulata, but gills not connected by veins. The caps usually have a pinkish hue, often brownish. The stems are not as tender as the caps. The flavor is excellent.

VI.—Lacti′pedes. Stem and gills milky, etc.

M. hæma′topa Pers. *Gr.*—blood; *Gr.*—a foot. **Pileus** about 1 in. broad, white flesh-color, fleshy-membranaceous, *slightly fleshy* chiefly *at the disk*, conical then bell-shaped, *obtuse*, nay convex and spuriously umbonate, naked, even or slightly striate at the margin, which is *at the first elegantly toothed*. **Stem** 2–4 in. long, 1 line and more thick, remarkably tubed, rigid, normally everywhere *powdered with whitish, delicate, soft hairy down*, sometimes, however, denuded of it. **Gills** adnate, often with a small decurrent tooth, the alternate ones shorter, in front disappearing short of the slight margin of the pileus, whitish and wholly of the same color at the edge.

Cespitose (very many of the stems conjoined and hairy at the base), firm, stature almost that of M. galericulata, wholly abounding with Mycena. dark blood-colored juice.

On stumps. Frequent. September. *Stevenson.*

Spores spheroid-ellipsoid, 10x6–7μ *K.*

I find a non-cespitose form of this species with **red-margined gills.** Its red juice, however, will serve to distinguish it and show its true relations. *Peck*, 31st Rep.

Common in tufts like M. galericulata and of about the same size, but is readily distinguished by its red juice. This pretty plant can often be gathered in considerable quantity, and well repays the collector.

VII.—GLUTINI'PEDES. Stems gelatinous, etc.
None tested.

VIII.—BASI'PEDES. Stem dilated at base, etc.
None tested.

IX.—INSITI'TIÆ. Stem inserted.
None tested.

HIA'TULA Fr.

Hio, to gape.

Hiatula.

(Plate XXXIII.)

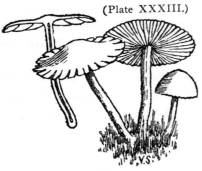

HIATULA WYNNIÆ.

Pileus symmetrical, very thin, without a distinct pellicle, formed by the union of the backs of the gills, splitting when expanded. **Gills** almost or quite free, white. **Stem** central. **Spores** white.

Allied to Lepiota in the thin pileus and free gills, but differing in the entire absence of a ring. Not at all deliquescent as in the genus Coprinus, near to which it was at one time placed by Fries. *Massee.* Reported from North Carolina.

OMPHA'LIA Fr.

Gr.—belonging to an umbilicus.

Omphalia.

Pileus generally *thin*, usually umbilicate at first, then funnel-shaped,

(Plate XXXIV.)

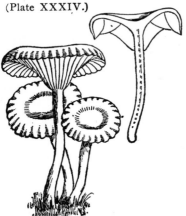

OMPHALIA UMBELLIFERA.
Enlarged about two sizes.

often hygrophanous, margin incurved or straight. **Gills** *truly decurrent* from the first, sometimes branched. **Stem** distinctly cartilaginous, polished, tubular, often stuffed when young. **Flesh** continuous with that of the pileus but differing in character. **Spores** white, somewhat elliptical, smooth.

Generally on wood, preferring hilly woods and a damp climate.

Resembling Collybia and Mycena in the flesh of stem and pileus being different in texture and in the externally cartilaginous stem. It is perfectly separated by the gills being markedly decurrent from the first.

The American species of Omphalia number between thirty-five and Omphalia. forty. Many of them are common. Few woods are free from them. Several of them are beautiful. They are usually small and lacking in substance. Raw, the writer has not found one that is objectionable in any way; a few have a woody taste. But two species have been found by him in sufficient quantity to make a dish. It is probable that all are edible. At best the species of Omphalia are valuable in emergency only.

ANALYSIS OF TRIBES.

COLLYBARII.

Pileus dilated from the first, margin incurved.

MYCENARII.

Pileus campanulate at first, margin straight and pressed to the stem.

COLLYBA'RII.

* Pileus dilated from the first; margin incurved.

O. onis'cus Fr. *Gr.*—a wood-louse. From the ashy color. **Pileus** scarcely 1 in. broad, dark *ashy* becoming pale, gray-hoary when dry, somewhat membranaceous, or slightly fleshy, *flaccid*, fragile when old. *convexo-umbilicate* or funnel-shaped, often irregular, undulato-flexuous, even-lobed, *smooth*, *even*, margin striate. **Stem** 1 in. long, 1 line and more thick, stuffed then tubed, *slightly firm*, moderately tough, sometimes round, curved, sometimes unequal, compressed, ascending, undulated, *gray.* **Gills** shortly *decurrent*, somewhat distant, quaternate, *ash-color.* Not cespitose. *Fries.*

Spores 12x7–8μ *B.*

Massachusetts, *Sprague;* California, *H. and M.*, who record it as edible.

O. umbellif'era—*umbella*, a little shade; *fero*, to bear. From its umbrella-like shape. (Plate XXXIV, p. 132.) **Pileus** about ½ in. broad, commonly whitish, *slightly fleshy-membranaceous*, convex then plane, *broadly obconic* with the decurrent gills, not at all or only slightly umbilicate, hygrophanous, when moist watery, *rayed with darker striæ*,

133

Omphalia. when dry even, changeable in appearance, silky, flocculose, rarely squamulose, *the margin, which is at first inflexed, crenate* (scalloped). **Stem** *short,* not exceeding 1 in. long, almost 1 line thick, stuffed then soon tubed, slightly firm, equal or dilated toward the apex into the pileus, of the same color as the pileus, commonly *smooth,* but varying pubescent, white villous at the base. **Gills** *very broad behind, triangular,* decurrent, *very distant,* edge of the gills straight.

Cosmopolitan. The common form is to be found everywhere from the sea level to 4,000 feet. *Stevenson.*

Spores 3x4µ *W.G.S.;* 10x4µ *W. P ;* green variety 10x6µ *W. P.;* broadly elliptical, 8–10x5–6µ *Peck.*

O. umbellifera is known the world over. It is very variable in size and color. With us it is seldom over ¾ in. broad. **Stem** ½–1 line thick. It grows on decaying wood and ground full of decaying material. There are several varieties. All are edible, but not worth describing. This description is given that the student may recognize one of our common plants, and eat it, if very hungry.

Mycena'rii.

O. campanel'la Batsch.—*campana,* a bell. **Pileus** thin, rather tough, hemispherical or convex, glabrous, umbilicate, hygrophanous, rusty yellow-color and striatulate when moist, paler when dry. **Gills** moderately close, arcuate, decurrent, yellowish, the interspaces venose. **Stem** firm, rigid, hollow, *brown,* often paler at the top, *tawny-strigose at the base.* **Spores** elliptical, 6–7x3–4µ.

Pileus 4–8 lines broad. **Stem** about 1 in. long, scarcely 1 line thick. Much decayed wood of coniferous trees. Very common. May to November. *Peck,* 45th Rep. N. Y. State Bot.

Spores ellipsoid, 6–8x3–4µ *C.B.P.;* 7x3µ *W.P.;* 6–9x3–4µ *B.*

The quantity alone, in which this small species can be found, makes it worth mentioning as an edible species. It is common over the United States where coniferous trees abound. Its favorite habitat is upon the rotting debris of these trees. Occasionally it grows from the ground, but only from that which is heavily charged with woody material. It is social in troops, or affectionate in clusters, or maintains a single existence.

It is edible, of good substance when stewed, tender and of fair flavor.

134

PLEURO'TUS.

Gr.—a side; *Gr.*—an ear.

Stem excentric, lateral or none. *Epiphytal (very rarely growing on* Pleurotus. *the ground)*, irregular, fleshy or membranaceous. *Fries.*

The excentric, generally lateral stem, absent in some of the species, separates this from other genera of the white-spored series.

Pileus varying from fleshy in the larger to membranaceous in the smaller forms, but never becoming woody. **Veil** generally wanting, when present its remains sometimes appear on the margin of the pileus, or as an evanescent ring on the stem. **Gills**, edge acute, generally decurrent, in some species with a well-marked tooth, rarely simply adnate. **Stem** fleshy, confluent and homogeneous with the pileus.

Wood, dead or alive; a few species appear on the ground.

P. ulmarius and others of the larger forms, when growing in an upright position, may have the stem central and the pileus horizontal. The stems of some species of Clitocybe and Omphalia if growing laterally are sometimes excentric and oblique.

This genus is analogous to Claudopus, pink-spored, and Crepidotus, brown-spored.

Spores white, but those of P. sapidus are faintly tinged with lilac, and of P. ostreatus, var. euosmus, with purple.

ANALYSIS OF TRIBES.

EXCENTRICI.   Page 137.

Pileus entire, laterally extended, excentric, not truly lateral.
* Veil fugacious, fragments adhering to stem or margin of pileus.
** Veil none; gills sinuate or obtusely adnate.
*** Veil none, gills very decurrent, stem distinct, almost verticai.
**** Veil none, gills very decurrent, stem proper absent, pileus lateral, extended behind into a short, stem-like oblique base.

DIMIDIATI.   Page 144.

Pileus not at first resupinate, lateral, prolonged without a definite margin behind, into a very short lateral, stem-like base.

RESUPINATI. Page 146.

Pileus resupinate from the first, then reflexed.

Pleurotus. If any odium attaches to the word toadstool, it should be forgotten and forever banished in presence of this cleanly, neat, handsome genus, choice in its growing places from lichen-covered stumps, or bark-clad boles, or highly perched limbs, or the scented surfaces of decaying wood. Several of its species perfume themselves throughout with pleasant spicy odors. Many are most accommodating in their constant coming.

Mr. H. I. Miller, superintendent Terre Haute and Indianapolis Railroad, writes: ''Most of the mushroom books give greatest space to the A. campester. For some parts of the country this may be desirable, but for Indiana and Ohio, considering the food value, the P. ostreatus is the best fungus we have in these states, from the fact that anybody wanting a mess can nearly always obtain a basketful of this variety, whereas the others depend upon a good many weather conditions. Having located a few logs and stumps in the spring, where the P. ostreatus grows, these same stumps and logs can be used all season. The crops are successive, and while some of the spots seem to be barren for a few days at a time, the others will be bearing. It does not make much difference what the kind of log or stump, whether it be beech, oak or elm, or what the species of tree. I think I have found them on all our forest trees, and it is not necessary for the tree to be dead. If there is a decaying portion, the spores seem to be carried by the little black beetle that infests the ostreatus, from one place to another, and wherever a small spot of dead wood is found we are likely to find the P. ostreatus. This being the only edible mushroom that we can find in large quantities all through the season in this neck of the woods, it seems to me that a general knowledge of it will serve the economic purpose more than any other fungi.''

The presence of the P. ostreatus and its esculent companions is noted from our northern boundary to the gulf. Poplar, maple, birch, hickory, ash, apple, laburnum and oak trees are its favored residences. Deer feed upon it, and kine are attracted by its scent even when deep under snow. When properly selected and *slowly* cooked, the Pleuroti are toothsome.

From the fact that the spores of this fleshy and valuable genus find fostering lodgment in many trees when in decay, it is more than prob- Pleurotus. able that the several species can be propagated by planting their spores upon such decaying woods, or by transplanting the mycelium.

Growths of P. ostreatus, P. sapidus, P. salignus, and probably other species of Pleurotus, can be forced, by watering the spots upon which they are known to grow. Dr. Kalchbrenner mentions that the P. sapidus is in this way cultivated in Hungary. Acting upon this mention the writer had good success with P. ostreatus. Experiments in this direction are likely to be interesting and rewarding.

No species is suspected of being noxious.

An analysis of P. ostreatus is given by Lafayette B. Mendel, Sheffield Laboratory of Physiological Chemistry, Yale University, as follows:

```
Water ...............................................................73.70%
Total solids..............................................26.30
    The dry substance contained:
Total nitrogen............................................. 2.40
Extractive nitrogen...................................... 1.27
Protein nitrogen.......................................... 1.13
Ether extract............................................... 1.6
Crude fiber................................................. 7.5
Ash .......................................................... 6.1
Material soluble in 85% alcohol .....................31.5
```

American Journal of Physiology, Vol. 1, No. 11, March 1, 1898.

<h3 style="text-align:center">I.—EXCEN'TRICI.</h3>

*Veil fugacious, etc.

P. dry'inus Pers. *Gr.*—oak. **Pileus** 2 in. broad, whitish, variegated with spot-like scales which become dingy-brown, lateral, oblique, rather plane. **Flesh** thick. **Stem** very curt and obese, commonly 1 in. long and thick, somewhat lateral, somewhat woody, squamulose, white, with a short, blunt root. **Veil** scarcely conspicuous on the stem, but appendiculate round the margin of the pileus when young. **Gills** not very decurrent, somewhat simple, not anastomosing behind, narrow, white, becoming yellow when old.

On trunks, oak, ash, willow, etc. *Stevenson.*

Spores 10x4μ *Massee.*

Edible. *Cordier, Cooke.*

When young the caps are tender; of the consistency, when cooked, of

Pleurotus. Polyporus sulphureus. In taste and smell the species varies from other Pleuroti, in having a distinct musk-like flavor. This is agreeable, reminding one of the common mushroom—A. campester.

** *Veil none, gills sinuate, etc.*

P. ulma'rius Bull.—*ulmus*, an elm. **Pileus** 3–5 in. and more broad, *becoming pale-livid*, often marbled with round spots, fleshy, *compact*, horizontal, moderately regular although more or less excentric, convex then plane, disk-shaped, even, smooth. **Flesh** white, tough. **Stem** 2–3 in. long, 1 in. thick, solid, firm, *elastic*, somewhat excentric, curved-ascending, *thickened* and tomentose *at the base*, not rarely villous throughout, white. **Gills** horizontal, *emarginate* or rounded *behind*, slightly adnexed, broad (broader in the middle), somewhat crowded, whitish.

The pileus is sometimes cracked in a tessellated manner. *Stevenson.*

Spores nearly globose, 5μ long *Morgan;* 5–6.5μ broad *Peck;* 6μ *W.G.S.*

Var. *aceri'cola—acer*, maple; *colo*, to inhabit. Plant smaller, cespitose.

Trunks and roots of maple trees. Adirondack mountains. September.

Var. *populi'cola—populus*, poplar; *colo*, to inhabit. Plant subcespitose, stem wholly tomentose. West Albany. *Peck*, Monograph, N. Y. Species of Pleurotus, Rep. 39.

The gills are sometimes torn across like those of Lentinus.

The historic elms of Boston Common have borne copious crops of this well-known and easily distinguished species from time immemorial. Every fall, about the first of September, if the season is favorable, later if not, copious crops appear decorating the trunks, and branches, sometimes at a height of thirty or forty feet. Growth takes place where branches have broken off or the trees have been wounded from other causes. They occur very generally on elms in the outlying districts of the city, but are rare in the country, seeming to be distinctly urban in their tastes. No damage is apparent from their growth.

Immediately in the rear of Independence Hall, Philadelphia, a fine cluster appears with equal autumnal regularity.

Though the elm tree is the chosen habitat of this fungus, it is little less select in its choice than other members of its genus.

When young and small P. ulmarius is tender and of acceptable flavor. Pleurotus. The stems and centers of older specimens should be cut away, and the tender parts of the caps, only, used.

P. tessula′tus Bull.—*tessela*, a small cube for pavement. **Pileus** *becoming pale-tawny*, horizontal, compactly fleshy, convex then plane, and in a form which is somewhat lateral depressed behind, irregular, even, smooth, *variegated* with round and hexagonal paler *spots*. **Flesh** thick, white. **Stem** short, 1 in. or little more long, solid, *compact*, *equal* or attenuated at the base, very excentric, curved-ascending, even, *smooth*, white. **Gills** *sinuate behind*, uncinato-adnate, thin, *crowded*, white or becoming yellow.

Solitary; according to some cespitose. The pileus is not cracked in a tessellated manner, as one might easily imagine from the name, but variegated with spots. Smaller than A. ulmarius (to which it is too closely allied), but almost more compact, with a smell of new meal.

On trunks. *Stevenson.*

North Carolina, *Schweinitz.* Edible. *Curtis.* Edible. *Cordier.*

On specimens growing cespitose and singly, found at Haddonfield, N. J. September, 1895, on trunk of apple tree, and at Eagle's Mere, Pa., singly on sugar maple, August, 1898, the margin of caps were beautifully marked, but not cracked.

In quality it is better than P. ulmarius.

P. subpalma′tus Fr.—*sub* and *palma*, a palm. **Pileus** 3–5 in. across. **Flesh** thick, soft, variegated; convex then more or less flattened, irregularly circular, obtuse, wrinkled, smooth, with a gelatinous cuticle, rufescent. **Stem** excentric or almost lateral, but the pileus is always marginate behind, fibrillose, short, equal, flesh fibrous, soft. **Gills** adnate, 3–4 lines broad, crowded, joined behind, dingy. *Massee.*

On old trunks, squared timber, etc.

Very remarkable for having the flesh variegated as in Fistulina hepatica. Pileus, especially when young, covered with a viscid pellicle. *Fr.*

Spores minutely echinulate, nearly globose, 5.6x7μ *Morgan.*

Ohio, *Morgan;* Wisconsin, *Bundy.*

I frequently found this species in North Carolina, growing from oak ties and standing oak timber. I did not notice distillation of rufescent drops from the cap. The soft flesh had good flavor. The gelatinous

Pleurotus. cuticle imparts its character to the dish. Mixed with Lentinus lepideus, a much tougher plant, which grows in great abundance in the same localities, it makes toothsome food.

P. lignati'lis Fr.—*lignum*, wood.   Dingy whitish.   **Pileus** 1–4 in. broad, rarely central, commonly more or less excentric, occasionally wholly lateral, often kidney-shaped, fleshy, thin, but compact and tough, fissile, convex then plane, obtuse and often umbilicate, *flocculoso-pruinate*, at length denuded with rain, repand, margin at first involute then expanded, undulato-lobed when luxuriant.   **Stem** sometimes 2–3 in., sometimes 3–4 lines long (even obliterated), *stuffed then hollow*, always *thin*, unequal, curved, curved or flexuous, tough and flexile, whitish, everywhere pruinato-villous, rooting and somewhat tomentose at the base.   **Gills** *adnate*, very *crowded* and narrow, unequal, divergent in the lobes, shining white.   *Fries.*

Exceedingly variable, wholly inconstant in form; substance thin and pliant; commonly densely cespitose, but also single.   Odor strong of new meal.

On wood, beech, etc.   *Stevenson.*

Parasitic on a rotten plant of Polyporus annosus on elm.   *W.G.S.*

White and grayish-white, margin faintly striate; white-spotted, odor distinctly farinaceous.   *C.M.*

Spores 3–4µ long, *Morgan, Cooke, W.G.S.;* 4–5µ *K.*

Var. *abscon'dens* Pk.—obscure.   New York, *Peck*, Rep. 31, 39.

On trunks, scattered, sometimes loosely clustered.   Griffins, Delaware county, N. Y.   September.   New York, *Peck*, Rep. 31, 39.

Kingsessing, near Philadelphia; Mt. Gretna, Pa.   *McIlvaine.*

This is a good species in every way.   I have not found it in extended quantity, but it is probable that it will be found in plenty when closer observed and better known.

P. circina'tus Fr.—to make round.   *Wholly white*, not hygrophanous. **Pileus** about 3 in. broad, *orbicular,* horizontal, fleshy, tough, convex then plano-disk-shaped, obtuse, even, but *covered over with a shining whitish slightly silky luster*.   **Stem** 1–2 in. long, 3–4 lines thick, *stuffed, elastic*, equal, *central* or slightly excentric, commonly *straight, smooth*, bluntly rooted at the base.   **Gills** adnate, slightly decurrent, crowded, broad (as much as 3 lines), white.   *Fries.*

An exceedingly distinct species. Regular, solitary, with a weak, Pleurotus.
pleasant, not mealy odor. The pileus is a little thicker than that of
A. lignatilis, but less compact; the gills are twice as broad. As A.
lignatilis is changeable, this is always constant in form.

On rotting birch stump. *Stevenson.*

California, *H. and M.*

Found at Eagle's Mere, Pa., August, 1898, on birch trees. Generally
solitary; sometimes six or eight on one tree, beautifully shining white,
at a distance resembling young Polyporus betulinus. Large quantities
of it grow in the extensive birch forests at Eagle's Mere, yielding a
ready food supply. Its flavor is pleasant, and texture, when cooked,
quite tender.

P. pubes'cens Pk.—*pubes*, down or soft hair. **Pileus** fleshy, convex, suborbicular, pubescent, yellowish. **Gills** broad, subdistant,
rounded behind, sinuate, pallid tinged with red. **Stem** short, firm,
curved, eccentric, colored like the pileus. **Spores** globose, 8μ broad.

Pileus about 2 in. broad. Stem scarcely 1 in. long.

Trunks of trees. Lyndonville. *C. E. Fairman.* *Peck*, 44th Rep.
N. Y. State Bot.

West Virginia, on oak trunks. *McIlvaine.*

High, agreeable flavor; texture about as in P. ostreatus.

***Gills decurrent; stem distinct, etc.*

P. sa'pidus Kalchb.—savory. Cespitose, or several pilei appearing
to spring from a common branched (Plate XXXVI.)
stem. **Pileus** 1–3 in. across. **Flesh**
thick, excentric, regular, convex or
obtusely gibbous then depressed,
glabrous, white or brownish. **Stem**
stout, solid, several usually springing from a thickened knob, whitish,
1–2 in. long, expanding upward into
the pileus. **Gills** decurrent, rather
distant, narrow, whitish. **Spores** elliptical, 10–11x4–5μ.

On elm trunks.

A very variable species; accord-

SECTION OF PLEUROTUS SAPIDUS.
One-half natural size.

141

Pleurotus. ing to Kalchbrenner, the spores have a faint tinge of lilac, and the pileus is white, tawny, brownish, or umber on the same trunk. The white form only has been met with in this country. *Massee.*

Spores with a lilac tinge, oblong or a little curved and pointed, 8.3x3.7μ *Morgan;* oblong, 9–11.5x4–5μ *Peck;* 10–11x4–5μ *Massee.*

Not observed in England until 1887.

Quite common throughout the United States, growing upon decaying wood, whether above or under ground. It has few distinct features. The only positive one distinguishing it from P. ostreatus is its lilac-tinted spores. The tint is faint but noticeable upon white background. Excepting for purposes of the student, its separation, as a species, from P. ostreatus is not necessary. When old it has more body than the latter, but is equally superior as a food fungus.

Professor Peck remarks of it: "A stew made of it is a very good substitute for an oyster stew."

It can be cultivated by watering the places upon which it is known to appear.

P. pome'ti Fr.—*pometum*, an orchard. **Pileus** white, fleshy, soft, sub-flaccid, irregular, involute, convex, even, smooth, disk depressed. **Gills** decurrent, crowded, separate behind. **Stem** 2–3 in. high, 3–4 lines thick, excentric, solid, tough, ascending, rooting.

On trunks of pear and apple trees.

Especially distinguished by the rooting stem.

North Carolina, edible, *Curtis;* California, *H. and M.*

**** *Gills decurrent. Stem lateral, etc.*

P. ostrea'tus Jacq.—*ostrea*, an oyster. (Plate XXXV, p. (113) **Pileus** 3 – 5 in. broad, when young almost becoming black, *soon becoming pale*, brownish-ash color, passing into yellow when old, fleshy, *soft, shell-shaped*, somewhat dimidiate, *ascending*, smooth, moist, even, but sometimes with the cuticle torn into squamules. **Stem** shortened or obliterated, firm, elastic, ascending obliquely, *thickening upward*, white, strigoso-villous at the base. **Gills** *decurrent, anastomosing behind, somewhat distant*, broad, white, sometimes turning light yellow, *and without glandules*.

For the most part cespitose, imbricated, very variable, sometimes almost central. The pileus is at first convex and horizontal, then Pleurotus. expanded and ascending. *Stevenson.*

Spores 10–12x4–5µ *Massee;* 7.5–10x4µ *Peck.*

General over the United States.

Var. *glandulo'sus* Ag. g. Bull.—With the habit of the typical form, but larger. Pileus dark brown, becoming pale. Gills white, with scattered small wart-like or glandular bodies.

On trunks. A very constant but somewhat rare variety; easily known by the dark-brown pileus. The gland-like bodies on the gills are due to the outward growth of the hyphæ of the trama in minute patches here and there. *Massee.*

Var. *euos'mus* Berk.—strong-smelling. Strong scented, imbricate. Pileus fleshy, depressed, shining, silky when dry, at first white with a tinge of blue, then brownish. Stem short or obsolete. Gills decurrent, ventricose, dingy, white. **Spores** 12–14x5µ, pale pinkish-lilac.

On elm trunks. Pilei very much crowded, 2 in. or more across, deeply depressed, unequal, at first white, invested with a slight blue varnish, at length of a pale brown. Stems distinct above, connate below. Gills rather broad; running down to the bottom of the free portion of the stem. Spores oblong, narrow, oblique, white, tinged with purple. The whole plant smells, when first gathered, strongly of tarragon. *B. and Br.*

Found at Richmond, Ind., *Dr. J. R. Weist.* On hickory stump at Mt. Gretna, Pa., *McIlvaine;* Haddonfield, N. J., *T. J. Collins.*

This esculent fungus closely allied to P. ostreatus, and differing only in having lilac spores, has been followed from book to book by a bad reputation, probably because of its "rosy" or lilac spores—all fungi having pink spores having been, until recently, ignorantly branded by authors as poisonous. The writer has eaten meals of it many times, as have his friends. It is in every way equal to P. ostreatus.

The rare qualities of this species are stated in the descriptive heading of the genus. Its very name implies excellence. The camel is gratefully called the ship of the desert; the oyster mushroom is the shellfish of the forest. When the tender parts are dipped in egg, rolled in bread crumbs, and fried as an oyster they are not excelled by any vegetable, and are worthy of place in the daintiest menu.

P. salig'nus Schwam.—*salix*, willow. **Pileus** 2–3 in. broad, sooty

Pleurotus. ash-color or ochraceous, fleshy, compact, *spongy*, somewhat dimidiate, *horizontal*, at first pulvinate, even, at length depressed behind and here and there strigose, the incurved margin entire. **Stem** always short, firm, more or less tomentose. **Gills** horizontal, hence less manifestly decurrent, separate behind, but *branched in the middle*, crowded, dingy, often eroded at the edge, not glandular.

Among the larger and firmer species. Solitary, scarcely ever cespitose. It is commonly confounded with A. ostreatus, but is certainly a different species. Although the stature is in general the same, it is easily distinguished by the pileus being more compact, and more pulverulent when young, then depressed, by the gills being thinner, more crowded, somewhat branched, but not anastomosing behind, and dingy soot-color; the spores also are dingy. *Stevenson.*

Spores oblong or cylindrical-oblong, 8x4µ *W.G.S.;* 8–10x3–4µ *B.*

Dr. Curtis wrote of this: ''Indeed I have found several persons who class this among the most palatable species. To such persons a dish of fresh mushrooms need seldom be wanting, as this one can be had every month of the year in this latitude.''

In New Jersey, in the vicinity of Philadelphia, Pa., I have found P. salignus in quantity. It has been sent to me by Dr. J. R. Weist, of Richmond, Ind., who writes, ''I have eaten it with great enjoyment.''

In 1881 I found it frequently on water beeches and willows, and thoroughly tested its edible qualities. *R. K. Macadam*, Boston.

When young or fresh, it is quite equal to any Pleurotus. When old, as with others of the Pleuroti, it is tough. Nevertheless their margins are always edible unless decaying.

II.—DIMIDIA'TI.

P. petaloi'des Bull.—petal of a flower. **Pileus** 1–2 in. long, *dingy-brown*, becoming pale, dimidiate, fleshy, but in no wise compact, rather plane, *somewhat spathulate*, continuous with the stem and *depressed behind*, hence the villous down of the stem ascends to this point (the disk) of the pileus, otherwise smooth, even, margin at first involute then expanded. **Stem** about ½ in. long, sometimes however very short, solid, firm, *compressed, channeled* when larger, more or less villous, whitish. **Gills** *decurrent, very crowded*, very narrow (scarcely beyond 2 mm. broad), linear, very unequal, white then ash-color.

Taste bitter. The form on wood is somewhat horizontal, gregarious Pleurotus. here and there imbricated. *Stevenson.*

Spores 9–10x4µ *Massee;* 8x4µ *W.G.S.;* minutely globose, 3–4µ *Peck.*

Edible. *Cooke, Cordier.*

P. spathula′tus Pers.—shaped like a spathula. **Pileus** rather thin, 1–2 in. broad, ascending, spathulate, tapering behind into the stem, glabrous, convex or depressed on the disk and there sometimes pubescent, alutaceous or brownish tinged with gray, red or yellow. **Gills** crowded, linear, decurrent, whitish or yellowish. **Stem** compressed, sometimes channeled above, grayish-tomentose. **Spores** elliptical, 7.6x 4–5µ broad; odor and taste farinaceous.

Ground. Sandlake. June. Edible.

It grows singly or in tufts and is an inch or more in height. The margin is thin and sometimes striatulate and reflexed. Toward the base the flesh is thicker than the breadth of the gills. The cuticle is tough and separable. The flesh is said by Gillet to be tender and delicate. Persoon describes the disk as spongy-squamulose, but in our specimens it is merely pubescent or tomentose. *Peck*, 39th Rep. N. Y. State Bot.

Recorded as edible by Professor Peck. At Eagle's Mere, Pa., I found many specimens agreeing with this description. They grew from decaying wood under ground, yet had the appearance of growing from the earth. It is probable that others have been deceived. In quality I found this to be one of the best.

P. sero′tinus Fr.—late. **Pileus** fleshy, 1–3 in. broad, compact, convex or nearly plane, viscid when young and moist, dimidiate kidney-shaped or suborbicular, solitary or cespitose and imbricated, variously colored, dingy-yellow, reddish-brown, greenish-brown or olivaceous, the margin at first involute. **Gills** close, determinate, whitish or yellowish. **Stem** very short, lateral, thick, yellowish beneath and minutely tomentose or squamulose with blackish points. **Spores** minute, elliptical, 5µ long, 2.5µ broad.

Dead trunks of deciduous trees. *Peck*, 39th Rep. N. Y. State Bot.

Mt. Gretna, Pa., 1887, and at Mt. Moriah, near Philadelphia, from August until November, 1898. Upon these findings the pileus was tomentose at base, as was the short stem.

Pleurotus. The species is not noticeably viscid after its youth. The viscidity can be detected in old specimens by moistening the pileus. Its flavor is not marked, nor is its texture as pleasing as most others of its genus, but being a late species it satisfies the longing of the mycophagist for his accustomed food.

P. pulmona′rius Fr.—*pulmo*, lung, from texture. **Pileus** 2–3 in. broad, *ash-colored*, continuous with the stem, fleshy, soft, but tough, *flaccid, obovate* or kidney-shaped, plane or reflexo-conchate at the margin, even, *smooth*. **Flesh** thin, soft, white. **Stem** very short, solid, exactly lateral, *horizontal* or ascending, *round, villous,* expanded into the pileus. **Gills** decurrent but *ending determinately, moderately broad,* distinct, not branched or anastomosing at the base, livid or *ash-color*.

The primary form is solitary. The pileus is ashy-tan when dried. It differs from A. salignus alike in the definitely lateral stem and in the thin flaccid pileus. *Fries.*

Not previously reported.

Found by Miss Madeleine Le Moyne, Washington, Pa., September, 1898, and sent to writer. Gills 3 lines broad, not narrow in proportion to flesh.

Taste and smell similar to P. ostreatus. Cooked it is tender, and more succulent than P. ostreatus.

III.—Resupina′ti.

P. mastruca′tus Fr.—*mastruca*, a sheepskin. **Pileus** up to 2 in. long and 1 in. broad, sessile, at first resupinate then expanded and horizontal, often lobed, upper stratum of pileus gelatinous, brown, bristling with squarrose or erect squamules. **Flesh** thickish. **Gills** radiating from the point of attachment, broad, rather distant, grayish-white.

On old trunks. Imbricated. Readily distinguished by the brown, squarrosely scaly pileus. *Massee.*

Spores oblong, oblique, 8x5μ *Morgan.*

In June, 1886, the writer found this species in oak woods near Philadelphia. It grew on fallen trunks and on decaying spots of living timber.

It is edible, and of good flavor, but is rough in the mouth. If found in quantity, the extract of it would make a delicate soup.

HYGROPH'ORUS Fr.

Gr.—moist; *Gr.*—to bear.

Pileus regular or undulated and wavy, often viscid or moist. **Flesh** <small>Hygrophorus.</small>
of the pileus continuous with that of
the stem and descending as a trama
into the gills. **Gills** adnate or adnexed, more or less decurrent, *waxy*,
often thick and forked, *edge always
thin and sharp*, often branched.

(Plate XXXVIII.)

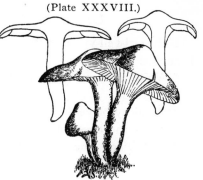

HYGROPHORUS PRATENSIS.

On the ground. Many species are
brightly colored. **Spores** white.

This genus differs from the preceding genera in the manifest trama,
the substance of which is similar to
that of the pileus; from Lactarius
and Russula by the trama not being vesicular, but somewhat floccose
with granules intermixed; from Cantharellus, its nearest ally, by the
sharp edge of the gills. The Cortinarii, Paxilli and Gomphidii are at
once distinguished from it by their colored spores and the changing color
of their gills, as well as by other marks. From all the other genera of
Agaricini it is distinguished by a mark peculiar to itself, viz., by the
hymeneal stratum of the gills changing into a waxy mass, which is at
length removable from the trama. This altogether singular character is
specially remarkable in H. caprinus, coccineus, murinaceus, etc. Hence
the gills seem full of watery juice, but they do not become milky like
those of the Lactarii. *Fries.*

From the description by Fries, the author of the genus, it is manifest
that one has to wait the ripening of the fungus before the peculiar characteristic mark of the genus, *i. e.*—gills turning into a waxy mass,
easily removable from the cap—can be observed. Many of the species
are difficult to determine when fresh. Nevertheless, there is an indescribable, watery, waxy, translucent appearance about the gills which
catches the eye of the expert, and is soon learned by the novice. The
white spores readily separate the genus from kindred shapes in the col-
ored-spored genera.

So far as tested none of the species is poisonous. One English spe-

Hygrophorus. cies is fetid. It is probable that they are all edible, varying in quality only. Fries well, and is superior in croquettes and patties.

ANALYSIS OF TRIBES.

LIMACIUM (*limax*, a slug). Page 148.

Universal veil viscid, with occasionally a floccose partial one, which is annular or marginal.
* White or becoming yellowish.
** Reddish.
*** Tawny or yellow.
**** Olivaceous-umber.
***** Dingy cinereous or livid.

None known to be edible.

CAMAROPHYLLUS (*Gr.*—a vault; a leaf). Page 152.
(From the arched shape of the gills.)

Veil none. Stem even, smooth or fibrillose, not rough with points. Pileus firm, opaque, moist after rain, not viscid. Gills distant, arcuate.
* Gills deeply and at length obconically decurrent.
** Gills ventricose, sinuately arcuate or plano-adnate.

HYGROCYBE (*Gr.*—moist; *Gr.*—the head). Page 155.

Veil none. Whole fungus thin, watery, succulent, fragile. Pileus when moist viscid, shining when dry, rarely floccoso-scaly. Stem hollow, soft, without dots. Gills soft. Most of the species are brightly colored and shining. This tribe is the type of the genus.
* Gills decurrent.
** Gills adnexed, somewhat separating.

LIMA'CIUM.

* *White or yellowish-white.*

H. chry'sodon Fr. *Gr.*—gold; a tooth. From tooth-like squamules. **Pileus** 2–3 in. broad, *white*, shining when dry, but commonly yellowish with minute adpressed squamules at the disk, light yellow-*flocculose at the involute margin*, fleshy, convex then plane, obtuse, viscid. **Flesh**

148

white, sometimes reddish. **Stem** 2–3 in. long, about ½ in. thick, Hygrophorus. stuffed, soft, somewhat equal (sometimes, however, irregularly shaped or thickened at the base), white, with minute *light yellow squamules*, which are more crowded and arranged in the form of a ring *toward the apex*. **Gills** decurrent, distant, 3 lines broad, thin, white, somewhat yellowish at the edge, sometimes crisped.

Odor not unpleasant. There is a manifest *veil*, not woven into a continuous ring, but *collected in the form of floccose squamules at the apex of the stem and the margin of the pileus*. Var. leucodon with white squamules. *Fries.*

In woods.

The lamellæ are said to be crisped, and when young, to have the edge yellow-floccose; but I have seen no such specimens. *Peck*, 23d Rep. N. Y. State Bot.

Spores 8x4µ *Cooke.*

West Virginia, New Jersey, Pennsylvania. *McIlvaine.*

A pleasant, excellent species, whose rarity is regrettable.

H. ebur′neus Bull. Fr.—*ebur*, ivory. Wholly *shining white*. **Pileus** fleshy, sometimes thin, sometimes somewhat compact, convexo-plane, somewhat repand, even, *very glutinous* in rainy weather, *margin soon naked*. **Stem** sometimes short, sometimes elongated, stuffed then hollow, unequal, *glutinous* like the pileus, *rough at the apex with dots in the form of squamules*. **Gills** decurrent, distant, veined at the base, 3–4 lines broad, tense and straight, quite entire. *Fries.*

Odor mild, not unpleasant. Very changeable. The veil is absent, unless the *very plentiful gluten* which envelops the stem be regarded as a universal veil; *margin of the young pileus* involute, only at the first *pubescent, soon naked*. The stem is soft internally, at length hollow, attenuated toward the base.

In woods and pastures. Frequent. September to October. *Stevenson.*

The whole plant is pure white when fresh, but in drying the gills assume a cinnamon-brown hue. *Peck*, Rep. 26.

Spores 6x5µ *Cooke;* 4x5µ *W.G.S.;* 5–6µ *K.;* 6x4µ *C.B.P.*

A common and wide-spread species frequenting woods and pastures. Edible. *Curtis.*

The author ate it in West Virginia, in 1882; at Devon, Pa., 1887;

_{Hygrophorus.} Haddonfield, N. J., 1890. It is well flavored but in texture is not of first quality.

H. pena'rius Fr.—*penus*, food. **Pileus** *tan-color, opaque*, fleshy, especially when young, at first umbonate, then very obtuse, hemispherical then flattened, even, smooth, *commonly dry*, margin at first involute, exceeding the gills, undulated when flattened. **Flesh** thick, hard, whitish, unchangeable. **Stem** curt, 1 ½ in. or more long, about ½ in. thick at the apex, *solid, compact*, hard, *attenuated at the base into a spindle-shaped root*, ventricose to the neck, again attenuated upward or wholly fusiform-attenuated, pale-white, smeared with tenacious, easily dried slime, *warty*. **Flesh** firm, but *externally more rigid*, cuticle somewhat fragile. **Veil** not conspicuous. **Gills** adnato-decurrent, acute behind, *distant, thick*, 3–4 lines broad, veined, tan inclining to pale. *Fries.*

Odor pleasant, taste sweet. The fusiform root is as long as the stem. In mixed woods. *Stevenson.*

Spores 7–8x4–5μ.

Edible. *Cooke.*

Large specimens occurred in mixed woods, in November, 1898, at Mt. Gretna. The caps varied from 1 ½–5 in. across. The color was white, tinged with yellow, much lighter than described. The caps look coarse and the stems are not inviting; but the caps have a pleasant odor. When stewed for twenty minutes they are meaty and tasty.

** *Reddish.*

H. erubes'cens Fr.—*erubesco*, to become red. **Pileus** 2–4 in. and more broad, white becoming everywhere red, fleshy, gibbous then convexo-plane, viscid, *adpressedly dotted with squamules or becoming smooth*, sometimes wholly compact, sometimes thin towards the *margin which is at the first naked*. **Flesh** firm, white. **Stem** sometimes short, robust, 2 in. long, 1 in. thick and attenuated upward, sometimes elongated, 4 in. long, equal or attenuated at the base, *solid*, flexuous, *with red fibrils, dotted with red upward*. **Gills** decurrent, distant, *soft, white, with red spots*. *Fries.*

Veil none. The ground color is white, as it is also internally, but it

everywhere becomes red and the pileus often rosy blood-color.   Hand-Hygrophorus.
some, growing in troops, commonly forming large lax circles.

In pine woods.   *Stevenson.*

Spores ellipsoid, very obtuse at both ends, 8–10x4–5μ *K.;* 8x4μ
Cooke.

Edible.   *Cooke.*

*** *Tawny or yellow.*

H. ni'tidus B. and Rav.—*shining.*   **Pileus** thin, fleshy, convex,
broadly umbilicate, smooth, shining, viscid, pale yellow with the margin
striatulate when moist, nearly white when dry.   **Gills** arcuate, decurrent,
yellow.   **Stem** slender, brittle, smooth, viscid, hollow, yellow.   **Flesh**
yellow.

Height 2–4 in., breadth of **Pileus** 8–12 lines.   **Stem** 1–2 lines thick.
Swamps.   Sandlake.   August.

The cavity of the stem is very small.   *Peck*, 23d Rep. N. Y. State Bot.

Found in many states and places, usually on moist ground beside
streams, or spring heads.   It sometimes parades itself in irregular pro-
cessions, at others in sparse patches.   It is delicate in flavor, and tender
cooked.

**** *Olivaceous-umber.*

H. limaci'nus Fr.—*limax,* a slug.   **Pileus** 1½–2½ in. broad, *disk
umber then sooty,* paler round the margin, fleshy, convex then flattened,
obtuse, smooth, viscid.   **Flesh** rather firm, white.   **Stem** 2–3 in. long,
½ in. thick, *solid,* firm, ventricose, *sticky,* flocculose, fibrilloso-striate,
roughened with squamules at the apex.   **Gills** adnate, then decurrent,
somewhat distant, thin, *white inclining to ash-color.*   *Fries.*

Veil entirely viscous, not floccose.

In woods among damp leaves.   *Stevenson.*

Spores 12x4μ *Cooke.*

New York, *Peck*, Rep. 34.   Thin woods and open places.

Reported edible Bulletin No. 5, 1897, Boston Mycological Club.

H. hypoth'ejus Fr. *Gr.*—under;   *Gr.*—sulphur (under gluten).
Pileus 1–2 in. broad, *at first smeared with olivaceous gluten,* ash-colored, when the gluten disappears, becoming pale and yellowish, orange

151

Hygrophorus. or rarely (when rotting) rufescent, fleshy, *thin*, convex then depressed, *obtuse*, even, somewhat streaked. **Flesh** thin, white then becoming light yellow. **Stem** 2–4 in. long, 2–3 lines and more thick, *stuffed*, equal, *even*, *viscous*, but rarely spotted with the veil, at length hollow. Partial *veil* floccose, at the first *cortinate and annular, soon fugacious*. **Gills** decurrent, *distant*, distinct, at first pallid (even whitish) soon *yellow*, sometimes flesh-color. *Fries.*

Very protean, changeable in color and variable in size. Stem not scabrous. There is no trace of the veil when the plant is full grown. Appearing after the first cold autumn nights, and lasting even till snow.

In pine woods. Frequent. *Stevenson.*

Spores 10x6µ *Cooke;* 12x4µ *W.G.S.*

Hollis Webster, in Bulletin No. 5, 1897, Boston Mycological Club, writes: "H. hypothejus Fr., when dried, is crisp and nutty, and very good to carry in the pocket for occasional nibble."

II.—CAMAROPHYL'LUS.

* Gills deeply decurrent, etc.

H. praten'sis Fr.—*pratum*, a meadow. ([Color] Plate XXXVII, figs. 1, 2, 3.) Plate XXXVIII, p. 147.) **Pileus** 1–2 in. and more broad, somewhat pale yellowish, *compactly fleshy especially at the disk, thin toward the margin*, convex then flattened, *almost top-shaped* from the stem being thickened upward, even, smooth, moist (but not viscous) in rainy weather, when dry often rimosely incised, here and there split regularly round. **Flesh** firm, white. **Stem** 1½–2 in. long, ½ in. and more thick, *stuffed*, internally spongy, externally polished-evened and firmer, *attenuated downward*, even, smooth, naked. **Gills** *remarkably decurrent*, at *first arcuate, then extended in the form of an inverted cone*, very distant, thick, firm, brittle, connected by veins at the base, very broad in the middle, of the same color as the pileus. *Fries.*

Very protean. Veil none. The flesh of the pileus is formed as it were of the stem dilated upward. The typical form resembles the Cantharelli. *Everywhere becoming light yellow-tawny*, but varying with the stem and gills pale-white.

In pastures. Common. *Stevenson.*

Spores 6x4µ *Cooke;* 6–10x4–6µ *K.*

Common over the United States.   West Virginia, 1881, North Caro-_{Hygrophorus.}

lina, 1890, Pennsylvania, 1887, Mt. Gretna, 1897–1898. *McIlvaine.*
Gregarious, and often in tufts, sometimes in partial rings.

An exceedingly variable species.   White, buff, smoky, pinkish colors
are common.   The cap shapes are also diverse.   The margins of some
are incurved; of others repand.   The weather seems to have much to
do with their shapes.

M. C. Cooke says: "It requires careful cooking, as it is liable to be
condemned as tough, unless treated slowly, but it is a great favorite
abroad."   He calls them "Buff Caps."

All fungi are the better for slow cooking.   The H. pratensis in all its
forms is excellent, but particularly so in croquettes and patés.

H. virgin'eus Fr.—*virgo*, a virgin. ([Color] Plate XXXVII, fig. 6.)

Wholly white. **Pileus** fleshy, convex then plane, *obtuse*, moist, *at length
depressed*, cracked into patches, floccose when dry. **Stem** *curt, stuffed,
firm*, attenuated at the base, externally becoming even and naked. **Gills**
decurrent, distant, rather thick. *Fries.*

Flesh sometimes equal, sometimes abruptly thin.   Commonly confounded with H. niveus, but it is more difficult to distinguish it from
white forms of H. pratensis.   It is distinguished chiefly by its smaller
stature, by the color being constantly white, sometimes becoming pale,
by the *obtuse pileus* being scarcely turbinate, *at length cracked into patches
and floccose when dry*, and by the gills being thinner, etc.

In pastures.   Common.   *Stevenson.*

Spores 12x5–6µ *Cooke.*

Tastes like M. oreades.   *M. J. B.*   Delicious broiled or stewed. *Cooke.*

"Mony littles make muckle," says the Scotch proverb.   It applies
well to the brave little toadstool looking through the first grass of lawns
for the coming of spring, and coming again in the autumn, defiant of
early frosts.   Small though it be, its numbers soon fill the basket.

The "Ivory Caps" are plentiful, and extend their haunts to the
woods, where thick mold or grassy places abound.

H. ni'veus Fr.—*niveus*, snow-white. ([Color] Plate XXXVII, fig. 7.)

Wholly white. **Pileus** scarcely reaching 1 in. broad, *somewhat membranaceous*, and without a more compact disk, hence truly *umbilicate*,

Hygrophorus. bell-shaped then convex, smooth, striate and viscid when moist, not cracked when dry. **Flesh** thin, everywhere equal, white, hygrophanous. **Stem** 2 in. or a little more long, 1–2 lines thick, *tubed, equal,* even, smooth, tense and straight. **Gills** decurrent, *distant, thin,* scarcely connected by veins, arcuate, quite entire.

Thinner, *tougher,* and later than H. virgineus, etc. Being hygrophanous the pileus is shining white when dry. Very tender forms occur.

In pastures. *Stevenson.*

Spores 7x4µ *Cooke.*

The H. niveus, H. virgineus, "Ivory Caps" as M. C. Cooke calls them, are pretty and plentiful in some sections. In the West Virginia mountains, along grass-grown road-sides, their purity and exquisite perfume attracted me in 1881. I have them and a few others to thank for seducing me into becoming a mycophagist. I think of them affectionately. I have seldom met with them since. They are found on lawns and in pastures and on grassy edges of woods, early in spring and late in autumn.

H. boreal'is Pk.—northern. **Pileus** thin, convex or expanded, smooth, moist, white, sometimes striatulate. **Gills** arcuate-decurrent, distant, white. **Stem** smooth, equal or tapering downward, stuffed, white.

Plant 2 in. high. **Pileus** 8-12 lines broad. **Stem** 1 line thick.

Ground in woods. Croghan and Copake. September and October.

The species is related to H. niveus but the pileus is not viscid. *Peck,* 26th Rep. N. Y. State Bot.

Found at Mt. Gretna, Pa., October 20, 1898, ground in mixed woods. The cap is white, silky, smooth, *not* viscid. Stem likewise.

A neat species pleasant in every way.

*** Gills ventricose, adnate, etc.*

H. dis'tans Berk.—distant (of the gills). **Pileus** about 2 in. broad, white, with a silky luster, here and there stained with brown, somewhat fleshy, plane or depressed, viscid. **Stem** white above, *gray* below, and attenuated, not spotted. **Gills** decurrent, *few, very distant,* somewhat ventricose, pure white then tinged with ash-color, interstices obscurely wrinkled.

Often umbilicate. Remarkable for the few and distant gills. *Stevenson.*
Spores 10x8μ *Cooke.*

Caps white, shaded to light pinkish-brown toward **center.** **Gills** very distant. Leaves adhere to cap.

Specimens tested were of mild, pleasant flavor.

H. sphæro′sporus Pk. **Pileus** fleshy and thick in the center, sub-obconic, convex, obtuse or slightly umbonate, whitish, inclining to reddish-brown, the margin incurved. **Flesh** firm, white. **Gills** rather broad, subdistant, adnate or slightly decurrent, white. **Stems** tufted, flexuous, solid, glabrous, often slightly thickened at the base, colored like the pileus. **Spores** globose, 6–8μ broad.

Pileus 6–12 lines broad. **Stem** 1–2 in. long, 2–3 lines thick.

Iowa. October. Communicated by C. McIlvaine.

The fresh plant is said to have no decided odor, but when partly dried it emits a slight but rather unpleasant odor. It belongs apparently to the section Camarophyllus, and is related to Hygrophorus Peckii. *Peck*, Torr. Bull., Vol. 22, No. 12.

Received by the writer from Hon. Thomas Updegraff, MacGregor, Iowa, and forwarded to Professor Peck as a new species.

The fungus has but slight taste and is without odor when fresh.

It is probably edible. Not received in sufficient quantity to test.

III.—HYGRO′CYBE.

*Gills decurrent.

H. cera′ceus Fr.—*cera*, wax. **Pileus** about 1 in. broad, *waxy-yellow, shining,* slightly fleshy, thin, but slightly firm, convexo-plane, obtuse, slightly pellucid-striate, viscid. **Stem** 1–2 in. and more long, about 2 lines thick, *hollow,* often *unequal,* flexuous and at length compressed, even, smooth, of the same color as the pileus, never darker at the apex. **Gills** *adnato-decurrent, broad, almost triangular,* distinct, yellow. *Fries.*

Fragile; easily distinguished from others by its waxy (not changeable) color. *Stevenson.*

Spores 8x6μ *Cooke.*

Eaten in Germany.

Agaricaceæ

Hygrophorus. Found at Angora and Kingsessing, Philadelphia, 1887. August to October. Open grassy places in woods, and in pastures. Scattered and in troops. Excellent. Stew slowly.

H. cantharel'lus Schw. *Gr.*—a small vase. ([Color]PlateXXXVII, fig. 5.) **Pileus** thin, convex, at length umbilicate or centrally depressed, minutely squamulose, moist, bright red, becoming orange or yellow. **Gills** distant, subarcuate, decurrent, yellow, sometimes tinged with vermilion. **Stem** smooth, equal, subsolid, sometimes becoming hollow, concolorous, whitish within.

Height 2–4 in., breadth of pileus 6–12 lines. **Stem** 1–2 lines thick.

Swamps and damp shaded places in fields or woods. July to September. Common. *Peck*, 23d Rep. N. Y. State Bot.

Var. *fla'va.* Pileus and stem pale yellow. Gills arcuate, strongly decurrent.

Var. *fla'vipes.* Pileus red or reddish. Stem yellow.

Var. *fla'viceps.* Pileus yellow. Stem red or reddish.

Var. *Ro'sea.* Has the pileus expanded and the margin wavy scalloped. Swamps. Sandlake. *Peck*, 23d Rep.

Common in the Adirondack region, and throughout Pennsylvania and New Jersey, in all its varieties.

The resemblance to H. miniatus in color is great, but there is a marked difference in the gills, which extend further down the thinner stem. It is tougher, and takes longer to cook. It has a flavor of its own which is enjoyed by some and condemned by others.

H. cocci'neus Schaeff.—of a scarlet color. ([Color] Plate CXXXVI, fig. 6.) **Pileus** 1–2 in. and more broad, *at first bright scarlet, then soon changing color and becoming pale*, slightly fleshy, convex, then plane and often unequal, *obtuse*, at first viscid and even, *smooth*, not floccose-scaly. **Flesh** of the same color as the pileus. **Stem** 2 in. long, 3–4 lines thick, *hollow*, then *compressed* and rather even, not slippery, *scarlet upward, always yellow at the base.* **Gills** wholly adnate, *decurrent with a tooth*, plane, distant, connected by veins, watery-soft as if fatty, when full grown *purplish at the base, light yellow in the middle, glaucous at the edge. Fries.*

Flesh of the pileus descending into the gills and forming a trama of the same color. Fragile. Varying in stature, easily mistaken for some

156

of the following species which are of the same color.   Pileus at length Hygrophorus.
becoming yellow.   *Stevenson.*

Spores 10–12x6μ *Cooke;* 7x4μ *Morgan.*

Edible.   *Cooke, Peck.*

In woods and pastures.   In troops.   Common in West Virginia, Pennsylvania, New Jersey.   *McIlvaine.*

Excellent when stewed for twenty minutes.

H. fla'vo-dis'cus Frost—*flavus,* yellow ; *discus,* disk.   **Pileus** convex or plane, smooth, glutinous, white (Plate XXXIX.) with a pale-yellow or reddish-yellow disk.   **Flesh** white.   **Gills** adnate or decurrent, subdistant, white, sometimes with a slight flesh-colored tint, the inter-spaces sometimes veiny. **Stem** subequal, solid, glutinous, white, sometimes slightly stained with yellow.   **Spores** elliptical, 6–8x4μ.

Plant 2–3 in. high.   **Pileus** 1–3 in. broad.   **Stem** 2–8 lines thick.

Pine woods.   West Albany.   November.

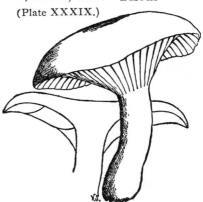

HYGROPHORUS FLAVO-DISCUS.
About two-thirds natural size.

This, like H. fuligineus, has a short white space at the top of the stem, free from the viscidity that exists elsewhere.   It resembles in many respects Hygrophorus speciosus, which has the pileus red, fading to yellow with advancing age. Perhaps the three may yet prove to be forms of one very variable species, for the most conspicuous differences between them consist in the colors of the pileus.   The constancy with which the three styles of coloration has thus far been maintained indicates a specific difference, but color alone is not generally regarded as having any specific value. *Peck,* 35th Rep. N. Y. State Bot.

Spores 6.4–7.6x4μ *Peck.*

I find this very good but its dirty pellicle should be peeled before using.   *Peck,* in letter, 1896.

Mr. Hollis Webster writes of H. flavo-discus (Yellow Sweet Bread) in Bull. No. 45, of the Boston Mycological Club, 1897 : ''This is a mushroom worth going a long way to get.   It is abundant in rich woods

Hygrophorus. under pines in certain localities, and is a great favorite with those who know it. It is easily prepared and requires little cooking.''

I have eaten enjoyably of it since 1881.

Plentiful in the Jersey pines, in West Virginia and Pennsylvania, and equal to any toadstool of its size.

H. fuligi'neus Frost—resembling soot.

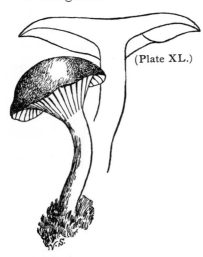

HYGROPHORUS FULIGINEUS.
About one-half natural size.

(Plate XL.)

Pileus convex or nearly plane, glabrous, very viscid or glutinous, grayish-brown or soot-color, the disk often darker or almost black. **Gills** subdistant, adnate or decurrent, white. **Stem** solid, viscid or glutinous, white or whitish. **Spores** elliptical, 7–9x5μ.

The Sooty hygrophorous resembles the Club-stemmed clitocybe in the color of its cap, but in nearly every other respect it is different. When moist the cap is covered with an abundant gluten which when dry gives it a shining appearance as if varnished. The color varies from grayish-brown to a very dark or sooty-brown with the central part usually still darker or almost black, but never with an umbo. The flesh and the gills are white. The stem also is white or but slightly shaded toward the base with the color of the cap. It is variable in length and shape, being long or short, straight or crooked, everywhere equal in thickness or tapering toward the base. It is glutinous and unpleasant to handle.

The cap is 1–4 in. broad, the stem 2–4 in. long, and 4–8 lines thick. The plants grow either singly or in tufts. In the latter case the caps are often irregular from mutual pressure.

The plants occur early in October and November, in pine woods or woods of pine and hemlock intermixed.

This mushroom is tender and of excellent flavor, but its sticky and often dirty covering should be peeled before cooking. *Peck*, 49th Rep. N. Y. State Bot.

Found at Angora, near Philadelphia, August 1, 1897. Densely ces- Hygrophorus.
pitose.

Raw it tastes like dead leaves. Tender and of fine flavor when cooked.

H. minia′tus Fr.—*minium*, red lead.   ([Color] Plate XXXVII,
fig. 4.) **Pileus** thin, fragile, at first convex, becoming nearly plane, glabrous or minutely squamulose, often umbilicate, generally red. **Gills**
distant, adnate, yellow, often tinged with red. **Stem** slender, glabrous,
colored like the pileus. **Spores** elliptical, white, 8μ long.

Cap ½–2 in. broad. **Stem** 1–2 in. long, 1–2 lines thick. *Peck*, 48th
Rep. N. Y. State Bot.

Var. *lutes′cens*. Pileus yellow or reddish-yellow. Stem and gills yellow. Plant often cespitose. *Peck*, 41st Rep. N. Y. State Bot.

Spores 10x6μ *Cooke;* elliptical, white.

Grows where it pleases and abundantly throughout the land. In wet
weather I have found it in July and late in autumn.

Professor Peck says: It is scarcely surpassed by any mushroom in
tenderness of substance and agreeableness of flavor.

The gunner for partridges will not shoot rabbits; the knowing toadstool seeker will pass all others where H. miniatus abounds.

** *Gills adnexed, etc.*

H. puni′ceus Fr.—blood-red. **Pileus** 2–4 in. broad, glittering blood-
scarlet, in dry weather and when old becoming pale especially at the
disk, slightly fleshy for its breadth, at first bell-shaped, obtuse, commonly
repand or lobed, very irregular, even, smooth, viscid. **Flesh** of the same
color, fragile. **Stem** 3 in. long, ½–1 in. thick, solid when young, at
length hollow, very stout (not compressed), ventricose (attenuated at
both ends), striate, and for the most part squamulose at the apex, when
dry light yellowish or of the same color as the pileus, always white and
often incurved at the base. **Gills** ascending, ventricose, 2–4 lines
broad, thick, distant, white-light yellow or yellow and often reddish at
the base. *Fries.*

The largest of the group and very handsome. It certainly differs
from H. coccineus, for which it is commonly mistaken, in stature, in
the adnexed gills, and in the white base of the striate stem. The attachment of the gills varies, but from the form of the pileus they ascend to
the base of the cone and appear free.

Hygrophorus.
In pastures. *Stevenson.*

Spores 8x5μ *Cooke.*

Edible. *Cooke.* No harm would come of confusing it with the vermilion mushroom—H. miniatus Pk.

H. con'icus Fr.—conical. **Pileus** thin, submembranaceous, fragile, smooth, conical, generally acute, sometimes obtuse, the margin often lobed. **Gills** rather close and broad, subventricose, narrower toward the stem, free, terminating in an abrupt tooth at the outer extremity, scarcely reaching the margin, yellow. **Stem** equal, fibrous-striate, yellow, hollow.

Height 3–6 in., breadth of pileus 6–12 lines. **Stem** 1–2 lines thick.

Ground in woods and open places. North Elba and Center. August to October.

The color of the pileus is variable. I have taken specimens with it pale sulphur-yellow and others with it bright red or scarlet. The plant turns black in drying. *Peck*, Rep. 23, New York State Bot.

Spores 10x7μ *Cooke;* 10x6μ *Morgan.*

An old-time cure-all had medicinal virtues proportionate to its offensiveness. Old-time writers, contrariwise, gave every toadstool a bad name which changed color or displeased their noses. The pretty little Hygrophorus conicus, for these reasons, has, until now, been under the ban of suspicion. M. C. Cooke, in his handy book, Edible and Poisonous Mushrooms, was the first to lighten its sentence and make it a sort of ticket-of-leave culprit.

The writer has frequently eaten it, and is glad to vouch for its harmlessness and testify to its eminent respectability.

H. chloroph'anus Fr. *Gr.*—greenish-yellow. **Pileus** 1 in. broad, commonly bright sulphur-yellow, sometimes, however, scarlet, not changing color, somewhat membranaceous, very fragile, at first convex, then plane, obtuse, orbicular and lobed, and at length cracked, smooth, viscid, striate. **Stem** 2–3 in. long, 2–3 lines thick, hollow, equal, round, rarely compressed, wholly even, smooth, viscid when moist, shining when dry, wholly unicolorous, rich light yellow. **Gills** emarginato-adnexed, very ventricose, with a thin decurrent tooth, thin, distant, distinct. *Fries.*

Very much allied to H. conicus, but never becoming black, and other-

160

wise certainly distinguished by its convex, obtuse, striate pileus, by its Hygrophorus. even and viscous stem, and by its emarginato-free, thin, somewhat distant, whiter gills. Like H. ceraceus in appearance.

In grassy and mossy places. **Common.** August to October. *Stevenson.*

Spores 8x5μ *Cooke;* 8μ *Q.*
Received from E. B. Sterling, Trenton, N. J., August, 1897.
Open grassy woods.
But three specimens were tested. They were in every way agreeable.

LACTA′RIUS Fr.

Giving *lac* (milk).

HE hymenophore continuous with the stem. **Pileus** Lactarius. somewhat rigid, fleshy, becoming more or less depressed, often marked with concentric zones. **Gills** unequal, membranaceous-waxy, slightly rigid, milky, edge acute, decurrent or adnate and often branched. **Stem** stout, central, rarely excentric except in those growing on trunks. **Spores** globose, minutely echinulate, white, rarely yellowish.

Nearly all grow on the ground.

Distinguished from all other fungi by the presence of a granular **milk** which pervades every part of the plant and especially the gills; it is commonly white, sometimes changing color and in section Dapetes highly colored from the first. The nature of the milk, especially its taste, whether acrid, subacrid or mild, must be carefully noted in distinguishing species, as it is the most useful characteristic.

In Russula, the only allied genus, the milk-bearing cells are present, but their contents do not appear as milk.

Many of the species are peppery, acrid, astringent; some mildly so, others will be long remembered if tasted raw. Yet not a species is hotter than some radishes, onions, and others of our favorite vegetables. Who would condemn them because they are peppery? There is not a single species of Lactarius which retains its pepperiness after cooking. This quality has to be and is supplied by one of our favorite condiments

161

Lactarius. —pepper itself.   Simply because they are *toadstools* and *hot*, they have been condemned without trial.   It is remarkable that not one of the fungi known to be deadly gives any warning by appearance or flavor of the presence of a poison.   The day will probably come when it can be said that if toadstool eaters will confine themselves to *hot* species, otherwise attractive, they will run no risk.   Panus stypticus is astringent, not hot.

ANALYSIS OF TRIBES.

PIPERITES (peppery, after *piperitis*, pepperwort).   Page 163.

Stem central.   Gills unchangeable, not pruinose nor becoming discolored.   Milk white at first, usually acrid.

   * TRICHOLOMOIDEI—inclining to Tricholoma.   Pileus moist, viscid, margin incurved and downy at first.

   ** LIMACINI—*limax*, a slug.   Pileus viscid when moist, with a pellicle, margin naked.

   *** PIPERATI.   Pileus without a pellicle, hence absolutely dry, often more or less downy or unpolished.

DAPETES (*daps*, a feast).   Page 170.

Stem central.   Gills naked.   Milk highly colored from the first.

RUSSULARIA (inclining to Russula).   Page 173.

Stem central.   Gills pallid then discolored, at length dark and powdered with the white spores.   Milk at first white, mild, or from mild becoming acrid.

   * VISCIDI—*viscidus*, viscid, sticky.   Pileus viscid at first.

   ** IMPOLITI—*impolitus*, unpolished.   Pileus squamulose, downy or pruinose.

   *** GLABRATI—*glaber*, smooth.   Pileus polished, smooth.

PLEUROPUS (*pleura*, side; *pous*, a foot).

Stem excentric or lateral.   Growing on trunks.   None known to be edible.

I.—Piperi'tes.

* Tricholomoi'dei. *Pileus viscid, margin incurved, etc.*

L. tormino'sus Fr.—*tormina*, gripes. **Pileus** 2–4 in. broad, convex, Lactarius.
then depressed, viscid when young or moist, yellowish-red or pale-
ochraceous tinged with red or flesh color, often varied with zones or
spots, the at first involute *margin persistently tomentose-hairy*. **Gills**
thin, close, narrow, whitish, often tinged with yellow or flesh color.
Stem 1.5–3 in. long, 4–8 lines thick, equal or slightly tapering downward, hollow, sometimes spotted, whitish. **Spores** subglobose or
broadly elliptical, 9–10μ. **Milk** white, taste acrid.

Woods. Adirondack mountains and Sandlake. August. *Peck*,
38th Rep. N. Y. State Bot.

Poisonous, and Gillet declares it to be deleterious and even dangerous, and that in the raw state it is a very strong drastic purgative. On
the other hand, Cordier states that almost all authors agree in stating
that it is eaten with impunity, and that Letellier has eaten it more than
once without inconvenience.

Cooke states: "Whether it is poison is rather uncertain, and probably assumed from its acridity."

Bulliard says: "It is very acrid and this is changed by heat into an
astringent of such power that a very little suffices to produce the most
terrible accidents." On the other hand, Boudier says that the presence of an acrid milk is an indication of no importance, that in certain parts of the country they eat such Lactaria as even L. piperatus
and do not experience any trouble. Certain Russulæ as acrid as any
Lactaria are known to be inoffensive.

The Russians preserve it in salt and eat it seasoned with oil and vinegar.

L. tur'pis Fr.—*turpis*, base, from its ugly appearance. **Pileus** large,
as much as 3–12 in. broad, *olivaceous inclining to umber*, fleshy, rigid,
convex becoming plane, disk-shaped or umbilicate, at length depressed,
innately hairy at the circumference or wholly covered over with tenacious gluten, *zoneless*, sometimes tawny toward the margin, *at length*
entirely *inclining to umber; margin for a long time involute, at the first*
villous, olivaceous-light-yellow, then more or less flattened, at length

Lactarius. often densely furrowed. **Flesh** compact, white, then slightly reddish. **Stem** 1½–3 in. long, ½–1 in. and more thick, *solid*, hard, equal or *attenuated downward*, even or pitted and uneven, but not spotted, viscid or dry, *pallid or dark olivaceous*, ochraceous-whitish at the apex. **Gills** adnato-decurrent, thin, 1–2 lines broad, much crowded, forked, *white straw-color*, spotted brownish when broken or bruised. **Milk** acrid, white, unchangeable. *Fries.*

Gregarious, *rigidly and compactly fleshy;* habit almost that of Paxillus involutus. It varies with the stem hollow, and the pileus somewhat zoned.

Spores spheroid or subspheroid, uniguttate, echinulate, 6–8μ *K.;* minutely spinulose, 6–8μ *Massee.*

New Jersey, Trenton, *E. B. Sterling;* North Carolina, *Curtis, Schweinitz;* Mt. Gretna, Pa. September, 1898. Along road in woods, moist places. *McIlvaine.*

The species is attractive by its very homeliness and odd individuality. It is not inviting. Cooked it is coarse and resembles L. piperatus. An emergency species.

L. controver′sus Fr.—*contra*, against; *verto*, to turn. **Pileus** 3 in. and more broad, fleshy, compact, rigid, at the first convex, broadly umbilicate, when fuller grown *somewhat funnel-shaped, oblique*, on emerging from the ground dry, flocculose, *whitish*, then with rain smooth, viscid, *reddish, with blood-colored spots and zones* (especially toward the margin), margin acute when young, closely involute, more or less villous. **Flesh** *very firm.* **Stem** commonly 1 in. long and thick, sometimes, however, 2 in. long and then manifestly attenuated toward the base and often excentric, *solid, obese*, even but pruinate and as if striate at the apex from the obsoletely decurrent tooth of the gills, wholly *white*, never pitted. **Gills** decurrent, thin, very crowded, 1–2 lines broad, with many shorter ones intermixed, but rarely branched, pallid-white-flesh-color. **Milk** white, unchangeable, plentiful. *Fries.*

Odor weak but pleasant, taste very acrid. Allied to L. piperatus.
In woods. Uncommon. August to October. *Stevenson.*
Spores echinulate, 8x6μ *W.G.S.;* globose, rough, 6–8μ *Massee.*
California, *H. and M.*
Edible, rather deficient in aroma and flavor. *Cooke.*

L. blen'nius Fr.   *Gr.*—slimy.   **Pileus** 3–5 in. across.   **Flesh** thick, Lactarius.
firm; soon expanded and more or
less depressed, glutinous, dingy
greenish-gray, often more or less
zoned with drop-like markings; margin at first incurved and downy.
Gills slightly decurrent, crowded,
narrow, whitish or with an ochraceous tinge.   **Stem** 1–2 in. long, up to
1 in. thick at the apex, where it expands into the thick flesh of the pi-

(Plate XL*a*.)

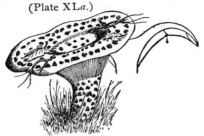

LACTARIUS BLENNIUS.
About one-fourth natural size.

leus, often attenuated at the base, viscid, colored like the stem or paler,
soon hollow.   **Milk** persistently white, very acrid.   **Spores** subglobose,
7–8x6μ.

In woods, on the ground, very rarely on trunks.

L. turpis somewhat resembles the present species but differs in the
darker olive-brown pileus and the yellow down on the incurved margin,
especially when young.   *Massee.*

Pileus 2–4 in. broad, fleshy, rarely subzonate, convex, the margin
generally involute and adpresso-tomentose (quite smooth, *Fries*); at
length more or less depressed, dull cinereous-green, at first viscid, more
or less pitted.   **Milk** white, not changeable.   **Gills** rather narrow, pale
ochraceous, scarcely forked, not connected by veins.   **Stem** 1 in. long,
¼–½ in. thick, paler than the pileus, attenuated downward, obtuse,
smooth, at length hollow, sometimes pitted, very acrid.   *Berk.*

Edible.   Coarse.

** LIMACI'NI.   *Pileus viscid, etc.*

L. insul'sus Fr.—tasteless.   **Pileus** 2–4 in. broad, convex and umbilicate, then funnel-shaped, glabrous, viscid, *more or less zonate, yellowish*, the margin naked.   **Gills** thin, close, adnate or decurrent, some of
them forked at the base, whitish or pallid.   **Stem** 1–2 in. long, 4–6
lines thick, equal or slightly tapering downward, stuffed or hollow,
whitish or yellowish, generally spotted.   **Spores** 7.6–9μ.   **Milk** white,
taste acrid.

Thin woods and open, grassy places.   Greenbush and Sandlake, N.Y.
July and August.

165

Lactarius.   Our plant has the pileus pale yellow or straw color, and sometimes nearly white, but European forms have been described as having it orange-yellow and brick-red.   It is generally, though often obscurely, zonate.   The zones are ordinarily more distinct near the margin, where they are occasionally very narrow and close.   The milk in the Greenbush specimens had a thin, somewhat watery appearance.   *Peck*, 38th Rep. N. Y. State Bot.

West Virginia, Pennsylvania, New Jersey.   July to September.   Common in mixed woods and grassy places.   *McIlvaine*.

Edible.   *Cordier, Curtis*.

L. insulsus is another peppery member of Lactarius which has suffered unjustly.   I have eaten it since 1881, and think it the best of the hot milk species.   Its flesh is not as coarse as others, and is of better flavor.   There is little difference in quality between it and L. deliciosus.

L. hys′ginus Fr.   *Gr.*—a crimson dye.   **Pileus** 2–3 in. broad, rigid, at first convex, then nearly plane, umbilicate or slightly depressed, even, viscid, zoneless or rarely obscurely zonate, *reddish-incarnate, tan-color or brownish-red*, becoming paler with age, the thin margin inflexed. **Gills** close, adnate or subdecurrent, whitish, becoming yellowish or cream-colored.   **Stem** 1–2 in. long, 4–8 lines thick, equal, glabrous, stuffed or hollow, colored like the pileus, or a little paler, sometimes spotted.   **Milk** white, taste acrid.

Woods.   Sandlake and Canoga, N. Y.   July and August.   Not common.

The reddish hue of the pileus distinguishes this species from its allies. The gluten or viscidity of the pileus in our specimens was rather tenacious and persistent.   *Peck*, 38th Rep. N. Y. State Bot.

Spores subglobose, whitish on black paper, yellowish on white paper, 9–10µ *Peck;* 10x7–8µ *Massee*.

Mt. Gretna, Pa., 1897.   Mixed woods.   August, September.

Not very acrid.   The entire acridity disappears in cooking   Several specimens were found and eaten, enough to prove it esculent and of good quality.

*** PIPERATI. *Pileus dry, etc.*

L. plum'beus Fr.—like *plumbum*, lead. **Pileus** 2–5 in. broad, com- Lactarius.
pact, convex, then infundibuliform, dry, unpolished *sooty or brownish-
black*. **Gills** crowded, white, or yellowish. **Stem** 1.5–3 in. long, 3–6
lines thick, solid, equal, thick. **Milk** white, acrid, *unchangeable*.
Spores 6.3–7.6μ.

The specimens which I have referred to this species were found in the
Catskill mountains several years ago, growing in hemlock woods, under
spruce and balsam trees. I have not met with the species since. The
pileus in the larger specimens had a minutely tomentose appearance,
but in the dried specimens this has disappeared. They also varied in
color from blackish-brown to pinkish-brown and grayish-brown, but
they can scarcely be more than a mere form or variety of the species
the description of which, as given by Fries, I have quoted. In the
Handbook the pileus is described as dark fuliginous-gray or brown, and
Gillet describes it as black-brown, dark fuliginous or lead color, and
adds that the plant is poisonous and the milk very acrid and burning.
Cordier says that the flesh is white and the taste bitter and disagreeable.
Peck, 38th Rep. N. Y. State Bot.

Poisonous. *Gillet.*

L. pergame'nus Fr.—parchment. *White*. **Pileus** fleshy, pliant,
convex then plano-depressed, spread, zoneless, slightly wrinkled,
smooth. **Stem** stuffed, smooth, changing color. **Gills** *adnate*, very
narrow, *horizontal*, very crowded, branched, white, then straw-color.
Milk white, acrid.

Very much allied to L. piperatus, but differing in the *stem* being
stuffed, at length softer internally, elongated, 3 in., unequal, attenuated downward and here and there ascending, *quite smooth;* in the *pileus*
being *thinner, pliant*, elastic, most frequently irregular and excentric,
for the most part flexuous, at first convex (not umbilicate), then *rather
plane, the surface very smooth*, but unpolished and *wrinkled* in a peculiar manner; and in the *gills* being adnate, not decurrent, *very crowded,
very narrow* (scarcely 1 line broad), always *straight and horizontal*, not
arcuate or extended upward, *soon straw-color*. The flesh is very milky,
but the gills are sparingly so. *Fries.*

In woods. October.

Agaricaceæ

Lactarius. **Spores** subglobose, rather irregular, 6–8μ *C.B.P.;* broadly elliptical, echinulate, 7x5–6μ *Massee.*

Eaten on the continent and Nova Scotia. Edible. *Cooke.*

North Carolina, *Curtis;* New England, *Frost;* Ohio, *Morgan.*

L. pipera′tus Fr.—*piper,* pepper. ([Color] Plate XLI, fig. 1.) **Pileus** 4–9 in. broad, *white,* fleshy, rigid, umbilicate when young, reflexed (margin at first involute) at the circumference, when full grown wholly *funnel-shaped,* for the most part regular, even, smooth, zoneless. **Flesh** white. **Stem** 1–2 in. long, 1–2 in. thick, solid, obese, equal or obconical, even, obsoletely pruinose, white. **Gills** *decurrent, crowded, narrow,* scarcely broader than 1 line, obtuse at the edge, *dividing by pairs,* arcuate then all *extended upward* in a straight line, white, here and there with yellow spots. **Milk** white, unchangeable, plentiful and very acrid.

Compact, firm, dry, inodorous. The pileus becomes obsoletely yellow when old. Although the gills are spotted with yellow, they do not change to straw color like those of L. pergamenus. *Fries.*

Spores white, nearly smooth, 6.3–7.6μ *Peck;* subglobose, 8–9μ diameter *Massee;* 5x6μ *W.G.S.*

Pennsylvania, West Virginia, 1881–1885. New Jersey, Pennsylvania in woods and on grassy places. July to October. *McIlvaine.*

Edible. *Curtis.*

L. piperatus is a readily distinguished species. It is very common. In 1881, after an extensive forest fire in the West Virginia forests, I saw miles of the blackened district made white by a growth of this fungus. It was the phenomenal growth which first attracted my attention to toadstools. I collected it then in quantity and used it, with good results, as a fertilizer on impoverished ground.

It has been eaten for many years in most countries, yet a few writers continue to warn against it. It is the representative fungus of its class— meaty, coarse, fair flavor. It is edible and is good food when one is hungry and can not get better. It is best used as an absorbent of gravies.

L. decepti′vus Pk.—deceiving. **Pileus** 3–5 in. broad, compact, at first convex and umbilicate, then expanded and centrally depressed or subinfundibuliform, *obsoletely tomentose or glabrous* except on the margin, white or whitish, often varied with yellowish or sordid stains, the

168

margin at first involute and *clothed with a dense, soft or cottony tomentum*, Lactarius. then spreading or elevated and more or less fibrillose. **Gills** rather broad, distant or subdistant, adnate or decurrent, some of them forked, whitish, becoming cream-colored. **Stem** 1–3 in. long, 8–18 lines thick, equal or narrowed downward, solid, pruinose-pubescent, white. **Spores** white, 9–12.7µ. **Milk** white, taste acrid.

Woods and open places, especially under hemlock trees. **Common.** July to September.

Trial of its edible qualities was made without any evil consequences. The acridity was destroyed by cooking. *Peck*, 38th Rep. N. Y. State Bot.

Alabama, *U. and E.;* New York, *Peck*, 38th Rep.; West Virginia, 1881–1885, Pennsylvania, New Jersey. Woods and open places. July to October. *McIlvaine.*

In common with all peppery Lactarii the present species loses the quality in cooking. The edible qualities then depend upon texture, substance, flavor. The species is coarse but meaty and of fair flavor.

L. velle'reus Fr.—*vellus*, fleece. **Pileus** 2–5 in. broad, compact, at first convex and umbilicate, then expanded and centrally depressed or subinfundibuliform, the *whole surface minutely velvety-tomentose, soft to the touch*, white or whitish, the margin at first involute, then reflexed. **Gills** distant or subdistant, adnate or decurrent, sometimes forked, whitish becoming yellowish or cream-colored. **Stem** .5–2 in. long, 6–16 lines thick, firm, solid, equal or tapering downward, pruinose-pubescent, white. **Milk** white, taste acrid. **Spores** white.

Woods and open places. Common. July to September. *Peck*, 38th Rep. N. Y. State Bot.

Spores white, nearly smooth, 7–9µ *Peck;* 4x8µ *W.G.S.*

West Virginia, Pennsylvania, New Jersey. Woods and open places. July to October. *McIlvaine.*

Poisonous according to some authors. *Cordier.* Edible. *Leveille.* Eaten it for eighteen years. *McIlvaine.*

This common, very acrid species is characterized by the downy covering of its cap.

It is a coarse species, but **meaty.** Its acridity is lost in cooking, when it makes a fair dish.

Lactarius. **L. involu'tus** Soppitt.—involved.   Every part white or with a very slight ochraceous tinge.   **Pileus** 1–2 in. across, flesh about 1½ lines thick, equal up to the margin, compact, rigid, convex, soon becoming plane or slightly depressed, margin strongly and persistently involute, extreme edge minutely silky, remainder even and glabrous.   **Gills** very slightly decurrent, densely crowded, not ½ line broad, sometimes forked. **Stem** ⅔–1 in. long, 2–3 lines thick, equal, or slightly thickened at the base, glabrous, even, solid, very firm.   **Milk** white, unchangeable, not scanty, very hot.   **Spores** obliquely elliptical, smooth, 5x3µ.

Very firm and rigid, resembling in habit L. vellereus in miniature. Most nearly allied to L. scoticus, but known at once by the exceedingly narrow, densely-crowded gills and the smooth, elliptical spores. *Massee.*

West Virginia, 1881–1885, plentiful.   Angora, West Philadelphia. August, September, 1897.   In mixed woods. *McIlvaine.*

Much smaller than L. piperatus.   **Pileus** convex, then plane with depressions in center, margin involute.   **Gills** slightly decurrent, densely crowded, very narrow.   **Stem** short, firm, solid.   **Milk** white, very hot.

L. involutus is readily mistaken for small forms of L. vellereus and L. piperatus.   The extremely narrow gills, so close and firm that it takes sharp eyes to follow them, are a distinguishing **mark.**

Its flesh is of same consistency as L. piperatus—hard and coarse.   It loses its pepperiness in cooking and is a good emergency plant, or solvent.

II.—DAPETES — *daps*, food.   Milk highly colored, etc.

America is rich in this section.   Fries records but two species, L. deliciosus and L. sanguifluus, while America has four.   The edible properties of three are known to be good; L. subpurpureus has not come under observation, but is added to complete the series as it is probably edible and is well marked by its dark-red milk. *McIlvaine.*

L. delicio'sus Fr.—delicious. ([Color] Plate XLI, fig. 3.)   **Pileus** 2–6 in. broad, *orange-brick-color, yellowish or grayish-orange,* becoming pale, fleshy, when quite young *depressed in the center,* margin naked, involute, then plano-depressed or broadly funnel-shaped with the margin unfolded, smooth, slightly viscid, *zoned* (zones sometimes obsolete).   **Flesh** soft, not compact, pallid, colored at the circumfer-

ence only by the juice. **Stem** 1–2 in. and more long, 1 in. thick, Lactarius.
stuffed then hollow, at length fragile, equal or attenuated at the base,
spotted in a pitted manner, of the same color as the pileus or paler.
Gills somewhat decurrent, crowded, narrow, arcuate, often branched,
typically *saffron-yellow*, but *becoming pale and always becoming green
when wounded.* **Milk** *aromatic, from the first red-brick-saffron. Fries.*

Spores white, spheroid, echinulate 7–8μ *K.;* 6μ *W.G.S.;* echinulate,
9–10x7–8μ *Massee;* subglobose, 7.6–10μ *Peck.*

In woods, under firs, etc.

Pileus dingy orange-red becoming pale, often greenish. **Every** part
turns to a homely green when bruised. It is from 3 to 5 in. across,
thick, convex, then depressed in center, margin at first curved in. **Gills**
decurrent, narrow, saffron-color. **Milk** saffron-red or orange changing
to green; sweet scented but slightly acrid. I have never seen but one
specimen with milk distinctly orange, and changing to green. The
milk in this species varies in color, much depending upon moisture. It
grows in patches, sometimes in clusters.

Edible. *Curtis.*

There is no question of its edibility. Old and modern writers applaud
it. Each cooks to his liking and thinks his own way best. It requires
forty minutes' stewing or baking; less time if roasted or fried. It can
be cooked in any way, but, like all Lactarii, it must be well cooked.

L. in'digo Schw.— ([Color] Plate XLI, fig. 2.) **Pileus** 2–5 in.
broad, at first umbilicate with the margin involute, then depressed or
infundibuliform, *indigo-blue with a silvery-gray luster*, zonate, especially
on the margin, sometimes spotted, becoming paler and less distinctly
zonate with age or in drying. **Gills** close, *indigo-blue*, becoming yellowish and sometimes greenish with age. **Stem** 1–2 in. long, 6–10
lines thick, short nearly equal, hollow, often spotted with blue, colored
like the pileus. **Milk** *dark-blue.*

Dry places, especially under or near pine trees. Not rare but seldom
abundant. July to September. *Peck*, 38th Rep. N. Y. State Bot.

Spores subglobose, 7.6–9μ long *Peck.*

West Virginia, North Carolina, New Jersey, Pennsylvania. Solitary
and in groups, in pine and mixed woods. July to September. *McIlvaine.*

The exceptional color of L. indigo will halt anyone with ordinary
observing power. It ˙is unnecessary to describe it further. Being a

Lactarius. large, stout plant it frequently lifts the leaf mat as it pushes upward, making leaf-mounds under which it is hidden, as do many of the Cortinarii. But even in such instances there are usually a few solitary plants standing prominently forth as sentinels.

It is edible, but coarse. Good flavor.

L. chelido'nium Pk. **Pileus** 2–3 in. broad, at first convex, then nearly plane and umbilicate or centrally depressed, *grayish-yellow or tawny*, at length varied with bluish and greenish stains, often with a few narrow zones on the margin. **Gills** *narrow*, close, sometimes forked, anastomosing or wavy at the base, *grayish-yellow*. **Stem** 1–1.5 in. long, 4–6 lines thick, short, subequal, hollow, colored like the pileus. **Spores** globose, 7.5μ. **Milk** sparse, *saffron-yellow;* taste mild.

Sandy soil, under or near pine trees. Saratoga and Bethlehem.

The milk of this species resembles in color the juice of celandine, Chelidonium majus. It is paler than that of L. deliciosus. By this character and by the dull color of the pileus, the narrow lamellæ, short stem and its fondness for dry situations, it may be separated from the other species. Wounds of the flesh are at first stained with the color of the milk, then with blue, finally with green. A saffron-color is sometimes attributed to the milk of L. deliciosus, which may indicate that this species has been confused with that, or that the relationship of the two plants is a closer one than we have assigned to them. *Peck*, 38th Rep. N. Y. State Bot.

Mt. Gretna, Pa. In mixed woods, gravelly low ground. September, October. *McIlvaine.*

A score or more solitary specimens were found and eaten. The substance and flavor are not distinguishable from L. deliciosus, which is lauded to the summit of good toadstools.

L. subpurpu'reus Pk.—*sub*, under; *purpureus*, purple. **Pileus** at first convex, then nearly plane or subinfundibuliform, more or less spotted and zonate when young, and moist *dark-red with a grayish luster*. **Gills** close, *dark-red*, becoming less clear and sometimes greenish-stained with age. **Stem** equal or slightly tapering upward, soon hollow, often spotted with red, colored like the pileus, sometimes hairy at the base. **Spores** subglobose, 9–10μ. **Milk** *dark-red*.

Pileus 2–3 in. broad. **Stem** 1.5–3 in. long, 3–5 lines thick.

Damp or mossy ground in woods and swamps. July and August.

At once known by the peculiar dark-red or purplish hue of the milk, which color also appears in the spots of the stem and in a more subdued tone in the whole plant. The color of the pileus, gills and stem is modified by grayish and yellowish hues. In age and dryness the zones are less clear, and dried specimens can scarcely be distinguished from L. deliciosus. *Peck*, 38th Rep. N. Y. State Bot.

I have not seen this species.

III.—RUSSULARIA.

* VISCIDI. *Pileus viscid.*

L. pal'lidus Fr.—*pale.* **Pileus** 3–6 in. broad, flesh-color or clay-color to *pallid, somewhat tan*, fleshy, umbilicato-convex, depressed, obtuse, margin broadly and for a long time involute, smooth, gluey, *zoneless.* **Flesh** pallid. **Stem** 2 in. and more long, about ¾ in. thick, somewhat equal, stuffed then *hollow*, even, smooth, of the same color as the pileus. **Gills** somewhat decurrent, arcuate, rather broad, 1½–2 lines and more; somewhat thin, crowded, somewhat branched, whitish at length of the same color as the pileus. **Milk** white, unchangeable. *Fries.*

Taste *somewhat mild.* Stature that of L. deliciosus, *but more lax in texture and always pallid.* There is a variety with the pileus inclining to dingy-brown. *Stevenson.*

Mixed woods. September to October.

Spores echinulate, almost round, 8µ *W.G.S.;* 7–11µ *Cooke;* 9–10x7–8µ *Massee.*

North Carolina, *Schweinitz, Curtis;* Massachusetts, *Frost;* Minnesota, *Johnson;* Rhode Island, *Bennett.*

Edible. *Cooke.*

L. quie'tus Fr.—calm, mild. **Pileus** 3 in. broad, fleshy, depressed, obtuse, margin deflexed, smooth, at first viscid, *somewhat cinnamon*, flesh-color, disk darker, *somewhat zoned*, soon dry, *somewhat silky*, opaque, *becoming pale.* **Flesh** white then reddish. **Stem** 2–3 in. long, ½ in. and more thick, stuffed, *spongy*, smooth, reddish, *at length beautifully rust-color.* **Gills** adnato-decurrent, somewhat forked at the

173

Lactarius. base, 1 ½ –2 lines broad, *white then soon brick-red.* **Milk** white, unchangeable, *sweet. Fries.*

> In woods. August to November. *Stevenson.*
>
> **Spores** echinulate, 8–10x6–7μ *Massee;* 10–12μ *Cooke.*
>
> Nova Scotia, *Somers;* New York, *Peck*, Rep. 42.
>
> Edible. *Cooke.* Eaten in France and held in estimation.

L. theio′galus Fr. *Gr.*—brimstone; milk. **Pileus** 2–5 in. broad, fleshy, thin, convex, then depressed, even, *glabrous*, viscid, *tawny-reddish.* **Lamellæ** adnate or decurrent, close, pallid or reddish. **Stem** 1–3 in. long, 4–10 lines thick, stuffed or hollow, even, colored like the pileus. **Spores** *yellowish, inclining to pale flesh-color*, subglobose, 7.5–9μ. **Milk** white, *changing to sulphur-yellow*, taste tardily acrid, bitterish.

> Woods and groves. Common. July to October.
>
> Our plant does not fully accord with the description of the species as given by Fries. *Peck*, 38th Rep. N. Y. State Bot.
>
> **Spores** subglobose, 7–8μ diameter *Massee;* subglobose, 7.5–9μ *Peck.*
>
> West Virginia, 1881–1885; Mt. Gretna, Pa. July, 1897; New Jersey, common in mixed woods. July to frost. *McIlvaine.*
>
> L. theiogalus possesses all the good qualities of the hot milk species. While I ate it whenever I chose in West Virginia, I did not again eat it until 1897 at Mt. Gretna. There several partook of it and thought it rather coarse, but of good flavor. It requires long cooking.

L. fuligino′sus Fr.—*fuligo*, soot. **Pileus** 1–2.5 in. broad, firm, becoming soft, convex plane or slightly depressed, even, *dry*, zoneless, *dingy ash-color or buff-gray*, appearing as if covered with a dingy pruinosity, the margin sometimes wavy or lobed. **Gills** adnate or subdecurrent, subdistant, whitish then yellowish, becoming *stained with pink-red or salmon-color where wounded.* **Stem** 1–2 in. long, 3–5 lines thick, equal or slightly tapering downward, firm, stuffed, colored like the pileus. **Spores** globose, *yellowish*, 7.5–10μ. **Milk** white, taste tardily and sometimes slightly acrid.

> Thin woods and open grassy places. Greenbush and Sandlake, N. Y. July and August. *Peck*, 38th Rep. N. Y. State Bot.
>
> A form with the pileus colored like that of L. lignyotus, but with the gills much closer than in that species, was found in a swamp near Sevey. July. *Peck*, 43d Rep.
>
> POISONOUS. *Barla and Reveil, Cordier.*

L. fumo′sus Pk. **Pileus** 1.5–2.5 in. broad, firm, convex, then expanded and slightly depressed in the center, smooth, dry, smoky-brown
or sordid-white. **Gills** close, adnate or slightly rounded behind, white,
then yellowish. **Stem** 3–5 lines thick, firm, short, smooth, stuffed,
generally tapering downward. **Spores** distinctly echinulate, yellow, 6μ
in diameter. **Flesh** and **Milk** white; taste at first mild, then acrid.
Plant 1.5–2 in. high.

Grassy ground in open woods. Greenbush. July.

The peculiar smoky hue of the pileus and yellow spores enable this
species to be easily recognized. The flesh when wounded slowly
changes to a dull pinkish-color. Related to L. fuliginosus. *Peck*, 24th
Rep. N. Y. State Bot.

**IMPOLITI. *Pileus downy, etc.*

L. ru′fus Fr.—red. **Pileus** 2–4 in. broad, convex and centrally
depressed, then funnel-shaped, generally with a small umbo, glabrous,
sometimes slightly floccose or pubescent when young, especially on the
margin, zoneless, *bay-red or brownish-red*, shining. **Gills** narrow or
moderately broad, sometimes forked, close, subdecurrent, yellowish or
reddish. **Stem** 2–4 in. long, 3–5 lines thick, nearly equal, firm, stuffed,
paler than or colored like the pileus. **Spores** white, 7.6–10μ. **Milk**
white, taste very acrid.

Low woods and swamps. North Elba. August. Rare.

The red Lactarius is known by its rather large size, dark-red pileus
and intensely acrid taste. It has been found but once in our state. The
flesh is pinkish and the stem sometimes pruinose. It is designated by
authors as very poisonous and extremely poisonous. Cordier even says
that worms never attack it. *Peck*, 38th Rep. N. Y. State Bot.

Massachusetts, *Frost;* New York, *Peck*, Rep. 23, Rep. 38.

I have not recognized this species. It is given as markedly
POISONOUS.

L. glycios′mus Fr. *Gr.*—sweet; *Gr.*—scent. **Pileus** ½–1½ in.
broad, thin, convex nearly plane or depressed, often with a small umbo
or papilla, *minutely squamulose*, ash-colored, grayish-brown or smoky-
brown, sometimes tinged with pink, the margin even or slightly and
distinctly striate. **Gills** narrow, close, adnate or decurrent, whitish or

Lactarius. yellowish. **Stem** ½–1½ in. long, 1–3 lines thick, equal, glabrous or obsoletely pubescent, stuffed, rarely hollow, whitish or colored like the pileus. **Milk** white, taste acrid and unpleasant, sometimes bitterish, odor *aromatic*. *Peck*, 38th Rep. N. Y. State Bot.

Smell agreeable, of melilot, as that of L. camphoratus.

Spores spheroid, echinulate, 6–8μ *K.;* subglobose, size variable, 6–10μ *Massee.*

The American plant, so far as observed, does not have the red hues ascribed to the European.

Haddonfield, N. J., *T. J. Collins;* Scranton, Pa., *Dr. J. M. Phillips;* Chester county, Pa., September, 1887, on ground in woods, *McIlvaine.*

This small Lactarius was found on several occasions. Its odor is attractive, but its taste is not. Cooked it is of high flavor, but will not be liked by many.

L. aqui'fluus Pk.—watery. **Pileus** fragile, fleshy, convex or expanded, at length centrally depressed, dry, smooth, or sometimes appearing as if clothed with a minute appressed tomentum, reddish tan-colored, the decurved margin often flexuous. **Gills** rather narrow, close, whitish, becoming dull reddish yellow. **Stem** more or less elongated, equal or slightly tapering upward, colored like the pileus, smooth, hollow, the cavity irregular as if eroded. **Spores** subglobose, rough, 7.6μ. **Flesh** colored like the pileus. **Milk** sparse, watery.

Plant 3–8 in. high. **Pileus** 3–6 in. broad. **Stem** 5–10 lines thick.

Swamps and wet mossy places in woods. Sandlake and North Elba. August and September.

The relationship of this plant is with L. serifluus, to which it was formerly referred, but from which I am now satisfied it is distinct. The hollow stem is a constant character in our plant, and affords a ready mark of distinction. The plant, though large, is very fragile, and breaks easily. The taste is mild or but slightly acrid. Sometimes there is an obscure zonation on the pileus, which, in large specimens, is apt to be irregular and much worm-eaten. The milk looks like little drops of water when first issuing from a wound, but it becomes a little less clear on exposure to the atmosphere. The decided but agreeable odor of the dried specimens persists a long time. *Peck*, 28th Rep.

This plant is sometimes cespitose. The pileus when dry is tawny-gray and scaly or cracked scaly. The margin may be even or coarsely sulcate-striate. The flesh is grayish or reddish-gray. The color of the Lactarius. lamellæ varies from creamy-white to tawny-yellow. The stem often has a conspicuous white myceloid tomentum at its base. I have never found this plant with a white or milky juice, and therefore I am disposed to regard it not as a variety of L. helvus, but as a distinct species. Its mild taste and agreeable odor suggested a trial of its edible qualities. It is harmless, but the lack of flavor induces me to omit it from the list of edible species. *Peck*, 50th Rep. N. Y. State Bot.

Var. *brevis'simus* Pk. **Pileus** 1–1.5 in. broad, **grayish-buff. Gills** crowded, adnate, yellowish or cream-color. **Stem** very short, 6–8 lines long.

Black mucky soil in roads in woods. Township 24, Franklin county. September.

Plant fragrant; sometimes cespitose. *Peck*, 51st Rep. N. Y. State Bot.

Angora, West Philadelphia, in moist oak woods. August, 1897, Philadelphia Myc. Center.

Flesh rather hard when cooked, and insipid. Good as an absorbent or in emergency.

L. lignyo'tus Fr.—*lignum*, wood. **Pileus** 1–4 in. broad, broadly convex plane or slightly depressed, dry, with or without a small umbo, generally rugose-wrinkled, *dark-brown, appearing subpulverulent or as if suffused with a dingy pruinosity*, the margin sometimes crenately lobed and distinctly plicate. **Gills** moderately close or subdistant, adnate, white or yellowish, *slowly changing to pinkish-red or salmon color where wounded*. **Stem** 1–3 in. long, 2–6 lines thick, equal or abruptly narrowed at the apex, even, glabrous, stuffed, colored like the pileus, sometimes plicate at the top. **Milk** white, taste mild or tardily and slightly acrid.

Var. *tenu'ipes*. **Pileus** about 1 in. broad. **Stem** slender, 2–3 in. long and about 2 lines thick.

Wet or mossy ground in woods and swamps. Adirondack mountains and Sandlake. July and August. Not rare in hilly and mountainous districts. *Peck*, 38th Rep. N. Y. State Bot.

Spores globose, yellowish, 9–11.3µ *Peck;* pale ochraceous, subglobose, minutely echinulate, 9–10µ diameter *Massee.*

West Virginia mountains, 1881–1885; Eagle's Mere; Mt. Gretna,

Lactarius. Pa.   Solitary and gregarious, moist woods and wooded places.   July to September.   *McIlvaine.*

In my long experience with the plant I have not seen any change of color, save that, like the white milk of other species, it darkens slightly to a cream color.   I have found it distinctly umbilicate and quite umbonate in the same patch.

L. lignyotus is one of the best of Lactarii and quite equal to L. volemus.

L. corru'gis Pk.—having wrinkles or folds.   **Pileus** 3–5 in. broad, firm, convex, then nearly plane or centrally depressed, *rugose reticulated*, covered with a *velvety pruinosity or pubescence, dark reddish-brown or chestnut-color*, fading with age to tawny-brown.   **Gills** close, dark cream-color or subcinnamon, *becoming paler* when old, sordid or brownish where bruised or wounded.   **Stem** 3–5 in. long, 6–12 lines thick, equal, solid, glabrous or merely pruinose, paler than but similar in color to the pileus.   **Spores** subglobose, 10–13μ.   **Milk** copious, white, taste mild.

Thin woods.   Sandlake, Gansevoort and Brewerton, N. Y.   August and September.

This curious Lactarius is related to L. volemus, from which it may be separated by its darker colors and its corrugated pileus.   The flexuous reticulated rugæ present an appearance similar to that of the hymenium of a Merulius.   The pileus is everywhere pruinose-pubescent and the gills bear numerous spine-like or acicular cystidia or spicules, 4–5μ long.   These are so numerous on and near the edges of the gills that they give them a pubescent appearance.   *Peck*, 38th Rep. N. Y. State Bot.

I found many at Mt. Gretna, Pa., up to 6½ in. in diameter.   Flesh not so firm as L. volemus.   Stem equal, rugulose, flattened in old specimens.   Milk very slightly acrid.

Better in taste and quality than L. volemus.

L. lute'olus Pk.—yellowish.   **Pileus** 2–3 in. broad, fleshy, rather thin, convex or nearly plane, commonly umbilicately depressed in the center and somewhat rugulose, pruinose or subglabrous, buff-color. **Flesh** white, taste mild.   **Milk** copious, flowing easily, white or whitish. **Gills** close, nearly plane, adnate or slightly rounded behind, whitish, becoming brownish where wounded. **Stem** 1–1.5 in. long, 3–5 lines Lactarius. thick, short, equal or tapering downward, solid, but somewhat spongy within, colored like the pileus. **Spores** globose, 7.6µ broad.

Dry woods. East Milton, Mass. August. *H. Webster.*

This species is related to Lactarius volemus and L. hygrophoroides, but its smaller size and short stem will distinguish it from the former and its close gills from the latter. Its paler buff-color will separate it from both. Some specimens have a narrow encircling furrow or depressed zone near the margin and a slightly darker shade of color on the margin. The milk constitutes a remarkable feature of the species. According to the notes of the collector it is exceedingly copious, rather sticky, serous in character with white particles in suspension. It flows from many points as soon as the plant is disturbed and it stains the gills. It is impossible to collect an unstained specimen, so free is the flow of the milk. He, Mr. Webster, says: "I have never succeeded in picking a specimen so quietly as to prevent an instant and copious flow of its milk." Torrey Bull., Vol. 23, No. 10, 1896.

Angora, West Philadelphia, August, 1897. In oak woods. August, September. *McIlvaine.*

Quite frequent there. My attention was directed to it by the "narrow encircling furrow or depressed zone near the margin."

It is of like quality to L. volemus.

L. Gerar'dii Pk. **Pileus** 1.5–4 in. broad, broadly convex plane or slightly depressed, dry, generally rugose-wrinkled, with or without a small umbo or papilla, *dingy-brown*, the thin spreading margin sometimes flexuous lobed or irregular. **Gills** *distant*, adnate or decurrent, *white or whitish*, the interspaces generally uneven. **Stem** 1–2 in. long, 3–6 lines thick, subequal, stuffed or hollow, colored like the pileus. **Spores** globose, *white*, 9–11.3µ. **Milk** white, *unchangeable*, taste *mild*.

Woods and open places. Poughkeepsie, *W. R. Gerard.* Greenbush, Sandlake and Croghan, N. Y. July to September.

This Lactarius closely resembles the Sooty lactarius in color, but differs from it in its more distant gills, white spores and constantly mild taste. Wounds of the flesh and gills do not become pinkish-red as in that plant. From L. hygrophoroides its darker color, hollow stem and more globose rougher spores separate it. *Peck*, 38th Rep. N. Y. State Bot.

Lactarius. In the color of the pileus and stem this species is like the larger forms of L. fuliginosus. *Peck*, 26th Rep.

Edible. Boston Myc. Club Bull.

*** GLABRA'TI. *Pileus smooth.*

L. vole'mus Fr.—*volema pira*, a kind of large pear.([Color] Plate XLI, fig. 4.) **Pileus** 2–5 in. broad, firm, convex, nearly plane or centrally depressed, rarely funnel-shaped, sometimes with a small umbo, generally even, *glabrous*, dry, *golden-tawny or brownish-orange*, sometimes darker in the center, often becoming rimose-areolate. **Gills** *close*, adnate or subdecurrent, white or yellowish, becoming sordid or brownish where bruised or wounded. **Stem** 1–4 in. long, 4–10 lines thick, subequal, variable in length, firm, solid, glabrous or merely pruinose, colored like the pileus, sometimes a little paler. **Milk** *copious*, white, taste mild, flat.

Var. *subrugo'sus*. **Pileus** rugose-reticulated on the margin. *Peck*, 38th Rep. N. Y. State Bot.

Spores globose, white, 9–11.3μ *Peck;* 5–6μ diameter *Massee*.

Very delicious raw and celebrated from early times. *Fries*.

Common over the United States, well known everywhere and distinguished for its edible qualities. It is crisp and unless carefully cooked is hard and granular. It should have long, slow cooking, though it may be roasted or fried.

L. hygrophoroi'des B. and C.—resembling Hygrophorus. **Pileus** (Plate XLII.) 1–4 in. broad, firm, convex or nearly plane, umbilicate or slightly depressed, rarely funnel-shaped, glabrous or sometimes with a minute velvety pubescence or tomentum, dry, sometimes rugose-wrinkled and often becoming cracked in areas, *yellowish-tawny or brownish-orange*. **Gills** *distant*, adnate or subdecur-

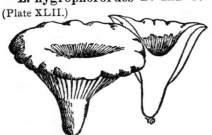

LACTARIUS HYGROPHOROIDES.

rent, white or cream-color, the interspaces uneven or venose. **Stem** .5–1 in. long, 4–8 lines thick, short, equal or tapering downward, *solid*, glabrous or merely pruinose, colored like the pileus. **Spores** subglobose or broadly elliptical, *nearly smooth*, 9–11.3μ. **Milk** white, taste Lactarius. mild.

Grassy ground and borders of woods. Albany, Greenbush and Sand-lake. July and August.

This plant has almost exactly the color of L. volemus, but differs from it in its distant gills, short stem, less copious milk and less globose spores. Its flesh is white, with a thickness about equal to the breadth of the gills. It is probably edible, but has not yet been tested. The typical L. hygrophoroides is described as having the pileus yellowish-red and pulverulent, and the gills luteous. It is also represented as a small plant; but our specimens, while not fully agreeing with this description, approach so closely to it in some of their forms that they doubtless belong to the same species. We have therefore extended the description so that it may include our plant. In wet weather the pileus sometimes becomes funnel-form by the elevation of the margin. *Peck,* 38th Rep. N. Y. State Bot.

Mt. Gretna, Pa., 1897, grassy grounds and borders of woods. Mixed, moist woods and grassy borders. July to September. *McIlvaine.*

Pileus up to 4 in. across. **Stem** 1–2½ in., tapering, equal or tapering downward. When growing in woods the stem is longer than when growing on borders.

Its edible qualities are excellent.

L. mitis'simus Fr.—*mitis*, mild. **Pileus** 1–3 in. broad, *golden-tawny,* zoneless, fleshy, thin, somewhat rigid, convex, *papillate*, depressed, papilla vanishing, even, smooth, somewhat slippery when moist. **Flesh** pallid. **Stem** elongated, 1–3 in. long, ⅓–½ in. thick, stuffed, then hollow, even, smooth, of the same color as the pileus. **Gills** adnato-decurrent, somewhat arcuate, then tense and straight, 1–1½ lines and more broad, thin, crowded, a little paler than the pileus, most frequently stained with minute red spots. **Milk** white, *mild*, plentiful.

Thin; very much allied to L. subdulcis, but distinguished by the *taste* being *mild, then somewhat bitterish*, and especially by the *bright, golden-tawny, resplendent* color of the pileus and stem. *Fries.*

In mixed and pine woods. August to November. *Stevenson.*

Spores 6–8x5–6μ *Massee;* 10μ *Cooke;* spheroid, echinulate, 6–7μ *C.B.P.*

Lactarius.   California, *H. and M.*
Edible.   *Cooke.*   Eaten on the continent.

L. subdul'cis Fr.—*sub; dulcis,* sweet.   **Pileus** .5–2 in. broad, thin,
(Plate XLIII.)

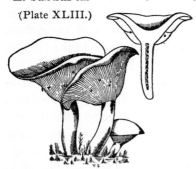

LACTARIUS SUBDULCIS.

convex, then plane or slightly funnel-shaped, with or without a small umbo or papilla, glabrous, even, zoneless, moist or dry, tawny-red, cinnamon-red or brownish-red, the margin sometimes wavy or flexuous.   **Gills** rather narrow, thin, close, whitish, sometimes tinged with red.   **Stem** 1–2.5 in. long, 1–3 lines thick, equal or slightly tapering upward, slender, glabrous, sometimes villous at the base, stuffed or hollow, paler than or colored like the pileus.   **Spores** 7.6–9µ.   **Milk** white, taste mild or tardily and slightly acrid, sometimes woody or bitterish and unpleasant.   **Flesh** whitish, pinkish or reddish gray, odor *none.*

Fields, copses, woods, swamps and wet places.   July to October. Very common.

This species grows in almost every variety of soil and locality.   It may be found in showery weather on dry, rocky soil, on bare ground or among mosses or fallen leaves.   In drier weather it is still plentiful in swamps and wet, shaded places, and in sphagnous marshes.   It sometimes grows on decaying wood.   It is also as variable as it is common. Gillet has described the following varieties:

Var. *cinnamo'meus.*   **Pileus** cinnamon-red, sub-shining.   **Stem** stuffed, then hollow; taste mild, becoming slightly acrid or bitter.

Var. *ru'fus.*   **Pileus** dull chestnut-red; becoming more concave. **Stem** spongy; taste mild.

Var. *ba'dius.*   **Pileus** bay-red, shining as if varnished, with an obtuse disk and an inflexed, elegantly crenulate margin.   **Stem** very glabrous, hollow.

The first and second varieties have occurred within our limits.   The first also has the stem elastic and furnished with a whitish or grayish tomentum or strigose villosity at the base, when growing among moss in swamps.   A form occurred in Sandlake, in which some of the speci-

mens were proliferous.   The umbo had developed into a minute pileus. Lactarius.
With us the prevailing color of the pileus is yellowish-red or cinnamon-
red.   Sometimes the color is almost the same as that of L. volemus
and L. hygrophoroides, and again it is a tan-color or a bay-red, as in
L. camphoratus, from which such specimens are scarcely separable, except by their lack of odor.   In young plants the pileus usually has a
moist appearance, which is sometimes retained in maturity.   Cordier
pronounces the species edible, and says that he has tested it several
times without inconvenience.   *Peck*, 38th Rep. N. Y. State Bot.

Spores 10μ *Cooke;* 7μ *W.G.S.*

West Virginia mountains, 1881–1885; Pennsylvania, New Jersey,
everywhere on moist ground.   July to October.   *McIlvaine.*

Edible.   *Curtis.*

The description of Fries as enlarged and modified by Professor Peck,
together with that of the varieties placed to the credit of the species by
Gillet, are given above in full.   The species with its ascribed varieties
is common and well known.   Var. *ba'dius* occurs in West Virginia and
Pennsylvania.   They are all edible and vary but little in quality.
L. subdulcis requires long cooking.

L. muta'bilis Pk.—changeable.   **Pileus** 2–4 in. broad, thin, convex
or nearly plane, zonate when moist, reddish-brown, the disk and zones
darker, zoneless when dry, flesh colored like the pileus.   **Milk** sparse,
white, taste mild.   **Gills** narrow, close, adnate, whitish, with a yellowish or cream-colored tint when old.   **Stem** 1–2 in. long, 3–5 lines thick,
equal or tapering upward, stuffed or spongy within, glabrous, colored
like the pileus.   **Spores** subglobose, rough, 7.6μ broad.

Low, damp places.   Selkirk and Yaphank, N. Y.   June and September.

The species is allied to L. subdulcis, from which the larger size and
zonate pileus separate it.   The zones disappear in the dry plant, and
this change in the marking of the pileus suggests the specific name.
They appear to be formed by concentric series of more or less confluent
spots and are suggestive of such species as L. deliciosus and L. subpurpureus.   *Peck*, 43d Rep. N. Y. State Bot.

West Virginia, Pennsylvania.   Solitary but frequent.   In moist woods
and margins of woods.   June to October.   *McIlvaine.*

Lactarius. I have been familiar with and eaten this plant since 1882, but thought it might be a variety of L. deliciosus, with light-colored milk.

L. mutabilis is an excellent species, equal to any Lactarius.

L. camphora′tus Fr.—*camphor*. **Pileus** 1–2 in. across, *brown*-brick-red, *somewhat zoned, sometimes zoneless*, fleshy, thin, depressed, dry, smooth. **Stem** short, 1–2 in., stuffed, somewhat undulated, of the same color as the pileus. **Gills** adnate, crowded, *yellowish-brick-color*. **Milk** mild, white, odor agreeable, spicy. *Fries.*

Strong smelling. So like L. subdulcis that it can be distinguished safely only by its odor of melilot when dried. *Stevenson.*

Pileus .5–1.5 in. broad. **Stem** 1–2 in. long, 2–3 lines. *Peck*, 38th Rep. N. Y. State Bot.

Spores spherical, echinulate, 6–7μ *Q.*; subglobose, 8–9μ *Massee;* 7.6–9μ *Peck.*

Taste and smell not of camphor, but of melilot.

North Carolina, *Curtis;* South Carolina, *Ravenel;* Wisconsin, *Bundy;* New York, *Peck*, Rep. 23, Mon. 38th Rep.

West Virginia, Pennsylvania, July to October, in moist places. Mixed woods, etc. *McIlvaine.*

Edible. *Gillet.*

Its mild taste distinguishes it at once from L. rufus.

It has high but pleasant flavor. If the flavor is too evident to suit some tastes, it is well to mix milder species with it.

RUS'SULA Pers.

Reddish.

Pileus regular, rigid, usually becoming more or less depressed. Russula.
Flesh of the pileus descending into
the gills forming a cellular trama.
Veil and consequently the ring absent.   Stem smooth, stout, rigid,
brittle, spongy within. **Gills** rigid,
fragile, edge thin and acute. **Spores**
rounded, often echinulate, white or
yellowish.   On the ground.

(Plate XLV.)

RUSSULA.

Closely allied to Lactarius but
separated by the absence of milk.
The gills of some species exude watery drops in moist weather. Owing
to the similarity of form and the variable coloring many species are diffi-
cult to determine; all the characters should be carefully noted, not
omitting that of the taste.

Russulæ are readily distinguished by the stout, short, brittle stem and
the fragility of the pileus and gills.   They especially love open woods
and appear during the summer and fall months, some being found until
sharp frosts occur.

It has been claimed by mushroom growers, until within a few years,
that the spores of the mushroom have to pass through the digestive
apparatus of the horse before they will germinate.   It has been conclu-
sively demonstrated that such a transmission is not a necessity.   It was
for a long time my opinion—following the opinion of others—that such
assistance was necessary.   In my many efforts to propagate valuable
food species of the wild toadstools I endeavored to find the method by
which the spores were disseminated, and through what digestive medium
they passed—either of insect or animal—before germination. Noticing
that the Russulæ were fed upon by a small black beetle, I planted in
suitable places, not the toadstools, but the beetles found upon them.
The result was that in several instances I grew the Russulæ.   My
experiments, while interesting, are not conclusive, because I later found
that the same results could be obtained from the toadstool itself when

185

Russula. planted under its own natural life conditions. It is certain that beetles can not be raised by planting Russulæ.

The beetles known as tumble-bugs—canthon lævis—deposit eggs in the center of balls made of animal droppings; dig a hole in the ground and drop them into it. These droppings frequently contain the spores of the meadow mushroom. Thus planted with the proper surrounding of manure, and at the proper depth, the spores germinate, spread mycelium, and a crop of mushrooms is the result. The beetle becomes a horticulturist. No wonder the Egyptians, thousands of years ago, made it—the scarabeus—their sacred emblem, and that, today, the *fleur-de-lis* of France, so the Rosicrucians say, perpetuates its glorious worth and calling.

Most Russulæ are sweet and nutty to the taste; some are as hot as the fiercest of cayenne, but this they lose upon cooking. To this genus authors have done especial injustice; there is not a single species among them known to be poisonous, and, where they are not too strong of cherry bark and other highly flavored substances, they are all edible; most of them are favorites. Where they present no objectionable appearance or taste, their caps make most palatable dishes when stewed, baked, roasted or escalloped. The time of cooking should be determined by the consistency of the variety; some will cook in five minutes, others not under thirty. Salt, butter and pepper are the only necessaries as seasoning.

ANALYSIS OF TRIBES.

I.—COMPACTÆ (*compingo*, to put together; compact). Page 187.

Pileus fleshy throughout, hence the margin is at first bent inward and always without striæ, without a distinct gluey pellicle (in consequence of which the color is not variable, but only changes with age and the state of the atmosphere). Flesh compact, firm. Stem solid, fleshy. Gills unequal.

II.—FURCATÆ (*furca*, a fork. With *forked* gills). Page 191.

Pileus compact, firm, covered with a thin, closely adnate pellicle, which at length disappears, margin abruptly thin, at first inflexed, then spreading, *acute, even*. Stem at first compact, at length spongy-soft within. Gills *somewhat forked*, with a few shorter ones intermixed, commonly attenuated at both ends, thin and normally narrow.

III.—RIGIDÆ (*rigidus*, rigid). Page 194.

Pileus without a viscid pellicle, *absolutely dry, rigid, the cuticle commonly breaking up into flocci or granules.* Flesh thick, compact, firm, vanishing away short of the *margin which is straight* (never involute), soon spreading, and always *without striæ.* Stem solid, at first hard, then softer and spongy. Gills, a few dimidiate, others divided, rigid, *dilated in front and running out with a very broad, rounded apex,* whence the margin of the pileus becomes obtuse and is not inflexed. *Exceedingly handsome,* but rather rare.

IV.—HETEROPHYLLÆ (*R. heterophylla,* the typical species of the section). Page 198.

Pileus fleshy, firm, with a thin margin which is at first inflexed, then expanded and striate, covered with a thin adnate pellicle. The gills consist of many shorter ones mixed with longer ones, along with others which are forked. Stem solid, stout, spongy within

V.—FRAGILES (*fragilis,* fragile or brittle). Page 201.

Pileus more or less fleshy, rigid-fragile, covered with a pellicle which is always continuous, and in wet weather viscid and somewhat separable; margin membranaceous, at first convergent and not involute, in full-grown plants commonly sulcate and tubercular. Flesh commonly floccose, lax, friable. Stem spongy, at length wholly soft and hollow. Gills almost all equal, simple, broadening in front, free in the pileus when closed. Several doubtful forms occur. R. integra is specially fallacious from the variety of its colors.

* Gills and spores white.
** Gills and spores white, then light-yellowish or bright lemon-yellowish.
*** Gills and spores ochraceous.

COMPAC'TÆ.

R. ni'gricans Bull.—*nigrico,* to be blackish. **Pileus** 2–4 in. and more broad, olivaceous-fuliginous, *at length black,* fleshy to the margin which is at first bent inwards, convex then flattened, umbilicato-depressed, when young and moist slightly viscid and even (without a separable pellicle), at length cracked in scales. **Flesh** firm, white,

Russula. when broken becoming red on exposure to the air. **Stem** 1 in. thick, persistently solid, equal, pallid when young, *at length black.* **Gills** *rounded* behind, slightly adnexed, *thick, distant,* unequal, paler, reddening when touched. *Fries.*

Compact, obese, inodorous, within and without *at length wholly black,* in which it differs from all others. The flesh becomes red when broken because it is saturated with red juice, although it does not exude milk. Sometimes a very few of the gills are dimidiate.

In woods. Common. June to November. *Stevenson.*

Var. *albo'nigra* Krombh.—*albo,* white; *negro,* to be black. **Pileus** fleshy, convexo-plane, depressed in the middle, at length funnel-shaped, viscid, *whitish, smoky about the margin.* **Flesh** white, turning black when broken. **Stem** solid, stout, dusky, becoming blackened. **Gills** decurrent, crowded, unequal, dusky-whitish. In grassy places.

Spores papillose, 8μ *W.G.S.;* subglobose, rough, 8–9μ *Massee.*

New York. Our specimens agree with the description in every respect, except that the gills are not distant. *Peck,* 32d Rep.

Mild when raw, but with a heavy woody taste.

Cooked it makes a good dish, but does not equal most Russulæ.

R. purpuri'na Quel. and Schulz.—purple. (Plate XLV*a.*) **Pileus** fleshy, margin acute, subglobose, then plane, at length depressed in the center, slightly viscid in very wet weather, not striate, often split, pellicle separable, rosy-pink, paling even to light yellow. **Gills** crowded in youth, afterward subdistant, white, in age yellowish, reaching the stem, 2–4 lines broad in front, not greatly narrowed behind, almost equal, not forked. **Stem** spongy, stuffed, very variable, cylindrical, attenuated above and below the middle, rosy-pink becoming paler (rarely white) toward the base, color obscure in age. **Flesh** fragile, white, reddish under the skin; odor slight, taste mild. **Spores** white, globose, sometimes sub-elliptical, 4–8μ long, minutely warted.

Pileus 1.5–2.5 in. across. **Stem** up to .4 in. thick, 1.2 in. long.

"This is a beautiful and very distinct species easily known by its red stem, mild taste and white spores." *Peck,* 42d Rep. N. Y. State Bot.

R. adus'ta Fr.—*aduro,* to scorch. **Pileus** pallid or whitish, *grayish-sooty,* equally fleshy, compact, depressed then somewhat infundibuliform, margin at first inflexed, smooth, then erect, without striæ. **Flesh**

PLATE XLV*a*.

RUSSULA PURPURINA.

Photographed by Dr. J. R. Weist.

unchangeable. **Stem** solid, obese, of the same color as the pileus. Russula.
Gills adnate then decurrent, *thin, crowded*, unequal, white then dingy, not reddening when touched. *Fries.*

Spores subglobose, almost smooth, 8–9μ *Massee.*

In pine and mixed woods.

West Virginia, Pennsylvania, New Jersey, in pine woods and in mixed woods. August to frost. *McIlvaine.*

R. adusta is solitary but often in small troops. It is easily recognized by the brownish blotches upon its cap, and the crowding of its thin gills.

The solid flesh must be well cooked. It is then of good flavor.

R. bre'vipes · Pk.—*brevis*, short; *pes*, a foot. **Pileus** 3–5 in. broad, at first convex and umbilicate, then infundibuliform, dry, glabrous or slightly villose on the margin, white, sometimes varied with reddish-brown stains. **Flesh** whitish, taste mild, slowly becoming slightly acrid. **Lamellæ** thin, close, adnate or slightly rounded behind; white. **Stem** solid, white.

(Plate XLV*b*.)

RUSSULA BREVIPES.
After Prof. Peck.

Spores globose, verruculose, 10–13μ.

Stem 6–10 lines long, 6–10 lines thick.

Sandy soil in pine woods. Quogue. September.

This species is related to Russula delica, but is easily distinguished by its short stem and crowded gills. The pileus also is not shining and the taste is tardily somewhat acrid. From Lactarius exsuccus it is separated by the character of the gills and the very short stem which is about as broad as it is long. The spores also are larger than in that species. The gills in the young plant are sometimes studded with drops of water. They are not clearly decurrent. Some of them are forked at the base. The pileus is but slightly raised above the surface of the ground and is generally soiled by adhering dirt and often marked by rusty or brownish stains. The plants grew in old roads in the woods where the soil had been trodden and compacted. *Peck*, 43d Rep. N. Y. State Bot.

Russula. West Virginia, 1882; Pennsylvania, 1887–1894; New Jersey, 1892. Solitary in pine and hemlock woods, generally on bare, compact ground. August to October. *McIlvaine.*

This species is a sparse grower, but its good size and respectable numbers soon fill the basket. When fresh it is of good substance and flavor.

R. del'ica Fr.— *delicus*, weaned. (Milkless, juiceless in gills.) **White. Pileus** 3–5 in. broad, fleshy throughout, firm, umbilicate then infundibuliform, regular, everywhere even, smooth with a *whitish luster*, the involute margin without striæ. **Flesh** firm, juiceless, not very thick, white. **Stem** curt, 1–2 in. long, ½ in. and more thick, solid, even, smooth, white. **Gills** *decurrent, thin, distant,* very unequal, white, exuding small watery drops in wet weather. *Fries.*

Spores minutely echinulate, white, broadly elliptical, 8–10x6–7μ *Massee.*

In appearance it resembles Lactarius vellereus and L. piperatus, but its gills do not distill milk or juice. It differs, too, in its mild taste. It is related to R. brevipes Pk.

A large, coarse species, cup-shaped at maturity. I have found it in several localities in Massachusetts in July and August. It is of fair quality cooked, but much inferior to R. virescens, etc. *Macadam.*

West Virginia, Pennsylvania, New Jersey, in mixed woods, August to October. *McIlvaine.*

Edible. Taste mild. From the juiceless variety of L. vellereus its mild taste alone furnishes a separate character. *Peck.*

I have eaten it since 1882, but it is not a favorite. Its quality is fair.

R. sor'dida Pk.—dirty. ([Color] Plate XLIV, fig. 4.) **Pileus** firm, convex, centrally depressed, dry, sordid-white, sometimes clouded with brown. **Gills** close, white, some of them forked. **Stem** equal, solid, concolorous. **Spores** globose, 7.5μ. Taste acrid. **Flesh** changing color when wounded, becoming black or bluish-black.

Plant 4–5 in. high. **Pileus** 3–5 in. broad. **Stem** 6–12 lines thick. Ground under hemlock trees. Worcester. July.

It resembles L. piperatus in general appearance. The whole plant turns black in drying. *Peck,* 26th Rep. N. Y. State Bot.

Ohio, *Morgan;* Pennsylvania, *Herbst;* West Virginia, 1881–1885,

Pennsylvania, New Jersey, pine, hemlock and mixed woods, July to
September. *McIlvaine.*

It is of better quality than most coarse-grained Russulæ.

FURCA'TÆ.

R. furca'ta Fr.—*furca*, a fork. **Pileus** 3 in. broad, sometimes greenish, sometimes umber-greenish, fleshy, compact, gibbous then plano-depressed or infundibuliform, *even*, smooth, but often *sprinkled with slightly silky luster*, pellicle here and there separable, margin thin, at first inflexed, then spreading, always *even*. **Flesh** firm, somewhat cheesy, white. **Stem** 2 in. or a little more long, solid, firm, equal or attenuated downward, even, white. **Gills** *adnato-decurrent, rather thick,* somewhat distant but broad, attenuated at both ends, frequently forked, shining white. *Fries.*

Spores globose, echinulate, 6–7µ *C.B.P.; 7–8x9µ Massee.*

In woods, and grass under trees.

The frequently forked gills, from which the species takes its name, their being thick and slightly decurrent, help to distinguish it. It is quite common in its several varieties.

Taste mild at first. A slight bitter develops which disappears in cooking. It is then of good quality, not equal to R. virescens. Older writers marked it poisonous, doubtless for no other cause than its slight bitter. I have eaten it freely for fifteen years.

R. sangui'nea Fr.—*sanguis*, blood. **Pileus** 2–3 in. broad, blood-red or becoming pale round the *even*, spreading, *acute margin*, fleshy, firm, at first convex, obtuse, then depressed and infundibuliform and commonly gibbous in the center, polished, even, *moist* in damp weather. **Flesh** firm, cheesy, white. **Stem** stout, spongy-stuffed, at first contracted at the apex, then equal, slightly striate, white or reddish. **Gills** at first adnate, then truly decurrent, very crowded, very narrow, connected by veins, fragile, somewhat forked, shining white. *Fries.*

Spores 9–10µ diameter *Massee.*

In pine and mixed woods. July to October.

Color same as R. rubra but differs in its hard cheesy flesh, rigid, slightly yellowish gills in age. The gills of R. sanguinea are truly decurrent, and pointed in front.

Russula.   Poisonous. *Stevenson.* Krapp says he has experienced grave inconveniences from eating it.

Myself and very many friends eat all fresh inviting Russulæ. We do not discriminate against a single peppery or acrid species, not even the R. emetica which has been severely maligned. In fact the peppery Russulæ are usually substantial in flesh and choice in substance.

The opinion of many is that R. sanguinea is one of the best. I have eaten it for years.

R. depal'lens Pers.—*palleo,* to be pale.   **Pileus** 3–4 in. across, pallid-reddish or inclining to dingy-brown, etc., fleshy, firm, convex, then plane, more rarely depressed, but commonly *irregularly shaped and undulated,* even, the thin, adnate pellicle presently changing color, especially at the disk, the spreading margin even, but slightly striate when old.   **Flesh** white.   **Stem** about 1½ in. long, solid, firm, commonly attenuated downward, *white, becoming cinereous* when old.   **Gills** adnexed, broad, crowded, distinct, but commonly forked at the base, often with shorter ones intermixed.   Inodorous, taste mild.   The color of the pileus is at first pallid-reddish, or inclining to brownish, then whitish or yellowish, opaque in every stage of growth.   It approaches nearest to the Heterophyllæ.   *Fries.*

In beech woods, pastures, etc.   August to September.

Spores subglobose, echinulate, 7–8μ *Massee.*

R. depallens somewhat resembles R. heterophylla.   Both are edible. It is a solitary grower and not common, but when found it occurs in good quantity.   It belongs to the best class of Russulæ.

R. subdepal'lens Pk.—*sub, de* and *palleo,* to be pale.   **Pileus** fleshy, at first convex and striate on the margin, then expanded or centrally depressed and tuberculate-striate on the margin, viscid, blood-red or purplish red, mottled with yellowish spots, becoming paler or almost white with age, often irregular.   **Flesh** fragile, white, becoming cinereous with age, reddish under the cuticle, taste mild.   **Lamellæ** broad, subdistant, adnate, white or whitish, the interspaces venose.   **Stem** stout, solid but spongy within, persistently white.

Spores white, globose, rough, 8μ broad.

Pileus 3–6 in. broad.   **Stem** 1.5–3 in. long, 6–12 lines thick.

Under a hickory tree.   Trexlertown, Pa.   June.   *W. Herbst.*

Closely related to Russula depallens, from which it differs in having Russula. the margin of the pileus striate at first and more strongly so when mature, also in the pileus being spotted at first, the gills more distant, the stem persistently white and the spores white. Bull. Torrey Bot. Club. Vol. 23, No. 10. October, 1896.

I do not doubt its edibility. See R. depallens.

R. ochrophyl la Pk.—*ochra*, a yellow earth; *phyllon*, a leaf. **Pileus** 2–4 in. broad, firm, convex becoming nearly plane or slightly depressed in the center, even or rarely very slightly striate on the margin when old, purple or dark purplish red. **Flesh** white, purplish under the adnate cuticle, taste mild. **Gills** entire, a few of them forked at the base, subdistant, adnate, at first yellowish, becoming bright ochraceous buff when mature, dusted by the spores, the interspaces somewhat venose. **Stem** equal or nearly so, solid or spongy within, reddish or rosy tinted, paler than the pileus. **Spores** bright ochraceous buff, globose-verruculose, 10μ broad.

The ochery-gilled Russula is a large fine species, but not a common one. It differs but little in color and size from the European pungent Russula, Russula drimeia, but it is easily distinguished from it by its mild taste.

The cap is dry, convex or a little depressed in the center, purple or purplish red, the white flesh purplish under the cuticle, which, however, is not easily separable.

The gills are nearly all entire, extending from the stem to the margin of the cap. They are therefore much closer together near the stem than at the margin. They are at first yellowish, but a bright ochraceous buff when mature. They are then dusted by the similarly colored spores.

The stem is stout, nearly cylindric, firm but spongy in the center and colored like the cap, but generally a little paler. There is a variety in which the stem is white and the cap deep red. In other respects it is like the typical form. Its name is Russula ochrophylla albipes.

The ochery-gilled Russula grows in groups under trees, especially oak trees, and should be sought in July and August. *Peck*, 51st Rep. N. Y. State Bot.

West Virginia, Pennsylvania, July to September, *McIlvaine*.

Edible. *Peck*, 50th Rep. N. Y. State Bot.

RI'GIDÆ.

Russula.   **R. lac'tea** Fr.—*lac*, milk. **Pileus** 2 in. broad, at the first *milk-white*, *then tan-white*, *throughout compactly fleshy*, bell-shaped, then convex, often excentric, without a pellicle, always dry, at the first even, then slightly cracked when dry, margin straight, thin, obtuse, even. **Flesh** compact, white. **Stem** 1 ½–2 in. long, 1 ½ in. thick, solid, very compact, but at length spongy-soft within, equal, even, always white. **Gills** *free*, very broad, *thick*, *distant*, rigid, forked, white. *Fries.*

Spores subglobose, echinulate, 7–9µ *Massee.*

Closely allied to R. albella Pk. from which it differs in its shorter stem, and pileus cracking into areolæ, and gills not being entire.

In mixed woods, in patches, not common.

Botanic creek, West Philadelphia, Pa., patches, *McIlvaine*, 1887.

Edible and of good flavor. *Macadam.*

Raw, it has a raw, rather unpleasant taste and odor, a little like some acorns. But its firm, thick flesh, meaty gills and stem, and good flavor when well cooked, rank it equal to any.

R. albel'la Pk.—whitish. **Pileus** 2–3 in. broad, thin, fragile, dry, plane or slightly depressed in the center, even or obscurely striate on the margin, commonly white, sometimes tinged with pink or rosy-red, especially on the margin. **Flesh** white, taste mild. **Lamellæ** entire, white, becoming dusted by the spores. **Stem** 1–2 in. long, 3–4 lines thick, equal, solid or spongy within, white.

Spores white, globose, 7.6µ broad.

Dry soil of frondose woods. Port Jefferson. July.

Closely allied to R. lactea, but differing in its fragile texture, entire lamellæ, more slender stem, and in the pileus not cracking into areas. *Peck*, 50th Rep. N. Y. State Bot.

R. vires'cens Fr.—*viresco*, to be green. ([Color] Plate XLIV, fig. 6.) **Pileus** green, compactly fleshy, globose then expanded, at length depressed, often unequal, always dry, not furnished with a pellicle, wherefore the *flocculose cuticle is broken up into patches or warts*, margin straight, obtuse, *even.* **Flesh** white, not very compact. **Stem** solid, internally spongy, firm, *somewhat rivulose*, white. **Gills** free, somewhat crowded, sometimes equal, **sometimes forked, with a** few shorter Russula. ones intermixed, *white. Fries.*

Taste mild; good, raw.

Spores scarcely echinulate, almost globular, 6µ *W.G.S.* Spores 8–10µ *Massee;* 6–7.6µ *Peck.*

Cap round when young, very hard, then convex or becoming dished, sometimes repand. It is without a separable skin, covered with various sized areas of mouldy looking patches which are at times distinctly cracked. The color varies from a bright bluish-green to grayish-green, such shades remind one of mouldy cheese or the shades of Roquefort; again the color may vary in shades of light leather brown, occasionally the caps are almost white, opaque in each shade of color. Flesh crisp, brittle, thick, white, mild, good raw. Gills and stem as described.

R. virescens is common in the United States but not generally plentiful. It is a solitary grower, usually but few are found in a patch. Striking in appearance when its green colors are present, and always clean looking and inviting. It sometimes attains the size of 5 in. across. It is a hot weather Russula and rarely appears before the latter part of June, then after rains.

To eat, it should be in a healthy, fresh condition. All Russulæ impart a stale flavor if any part of gills or cap is wilting, drying or decaying. It requires forty minutes' slow stewing, or it can be dressed raw as a salad. Roasted or fried crisp in a hot buttered pan it is at its best. It should be well salted.

R. lep'ida Fr.—*lepidus*, neat, elegant. Pileus 3 in. broad, *blood-red-rose*, becoming pale, whitish especially at the disk, somewhat equally fleshy, convex then expanded, scarcely depressed, obtuse, opaque, unpolished, *with a silky appearance, at length often cracked scaly*, margin spreading, obtuse, without striæ. Stem as much as 3 in. long, often 1 in. thick, *even, white or rose-color.* Gills rounded behind, rather thick, somewhat crowded, often forked, connected by veins, white, often red at the edge.

Taste mild; wholly compact and firm, but the flesh is cheesy, not somewhat clotted. The gills are often red at the edge, chiefly toward the margin, on account of the margin of the pileus being continuous with the gills. *Fries.*

Spores 8–10x6–8µ *Syll.*

195

Agaricaceæ

Russula. Frequent. July to October, in mixed woods.

A common and variable species in size and color, but the cap is always some shade of rose-red or lake. The flesh is compact and cheesy. The gills sometimes edged with pink as they near the margin. Taste mild.

The crisp flesh of R. lepida requires forty minutes' slow stewing, if stewed. It yields a delicate pink shade to the dish. Roasted or cooked in a hot buttered pan it is excellent.

R. ru'bra Fr.—*ruber*, red. **Pileus** unicolorous, a cinnabar-vermilion, but becoming pale (tan) when old, disk commonly darker, compact, hard but fragile, convex, then flattened, here and there depressed, absolutely dry, *without a pellicle, but becoming polished-even*, often sinuously cracked when old, margin spreading, obtuse, even, always persistent. **Flesh** white, *reddish under the cuticle.* **Stem** 2–3 in. long, about 1 in. thick, solid, even, varying white and red. **Gills** obtusely adnate, somewhat crowded, whitish, then yellowish, with dimidiate and forked ones intermixed.

Very *acrid*, very hard and rigid, most distinct from all the others of this group in the *pileus becoming polished-even*, although without a pellicle, in the *flesh being somewhat clotted*, and in the *very acrid taste.* **Gills** often red at the edge. *Fries.*

Spores whitish, *Fries;* spheroid, 8–10µ *K.*

Krapp says he has experienced grave inconveniences from eating it. European authorities mark "poisonous."

I do not hesitate to cook it either by itself or with other Russulæ and serve it at my table. It is easier cooked than R. virescens and others of the crisp species, and has equal flavor.

R. Linnæ'i Fr.—in honor of Linnæus. **Pileus** 3–4 in. broad, unicolorous, dark purple, blood-red or bright rose, opaque, not becoming pale, everywhere fleshy, rigid, plano-depressed, sometimes spread upward, even, smooth, *dry, without a separable pellicle*, margin spreading, obtuse, without striæ. **Flesh** thick, *spongy-compact, white.* **Stem** 1 ½ in. and more long, 1 in. and more thick, stout, firm, but spongy-soft within, somewhat ventricose, *obsoletely reticulated* with fibers, intensely blood-red. **Gills** *adnate, somewhat decurrent*, rather thick, not crowded, *broad* (more than ½ in.), fragile, sparingly connected by veins, white,

196

becoming yellow when dry, with a few dimidiate ones intermixed, somewhat anastomosing behind. *Fries*.

Spores wholly white, *Fries;* ellipsoid, spheroid, echinulate, 11μ *Q.;* 9–11x8–9μ *Massee*.

West Virginia, 1881–1885. West Philadelphia, Pa., on Bartram's Botanic creek. *McIlvaine*.

R. Linnæi is one of our handsomest and best Russulæ. European authors state its habit to be exactly that of R. emetica, but though I have known it intimately for many years I have not been struck with this in the American plant. Its large size, its more or less red stem never entirely white, at times hollow, cavernous, its less solid flesh, habit of growing in troops, sometimes parts of rings, flourishing best where the leaf mat is heaviest, loving the leaf drift in fence-corners, are well marked distinctions.

When young there is no better Russula. As it ages the stem becomes soft, spongy and should be thrown away. The caps, only, eaten.

R. oliva′cea Fr.—*oliva*, an olive; *olivaceus*, the color of an olive. **Pileus** 2–4 in. across, dingy-purple then olivaceous or wholly brownish-olivaceous, fleshy, convexo-flattened and depressed, *slightly silky and squamulose*, margin spreading, even. **Flesh** *white, becoming somewhat yellow*. **Stem** firm, ventricose, rose-color to pallid, spongy-stuffed within. **Gills** adnexed, wide, *yellow*, with shorter and forked ones intermixed.

Mild. Near to R. rubra, but certainly distinct in the stem being definitely spongy, in the pileus being unpolished, and in the gills being soft and brightly colored; corresponding with R. alutacea. *Fries*.

Spores light yellow, *Fries;* spheroid, punctate, 10μ *Q.;* globose, minutely granulate, yellow, 9–10μ diameter *Massee*.

Mt. Gretna, Pa., 1897–1898.

Pileus 2–4 in. across, 2–3 in. long, ½–⅓ in. thick.

The caps are equally good with R. alutacea. They **must** be fresh, and similarly cooked.

R. fla′vida Frost—yellow. ([Color] Plate XLIV, fig. 3.) **Pileus** fleshy, convex, slightly depressed, unpolished, bright yellow. **Gills** white, adnate, turning cinereous. **Stem** yellow, solid, white at the extreme apex. *Frost* Ms.

Russula. **Pileus** fleshy, convex, slightly depressed in the center, not polished, yellow, the margin at first even, then slightly striate-tuberculate. **Gills** nearly entire, venose-connected, white, then cinereous or yellowish. **Stem** firm, solid, yellow, sometimes white at the top.

Spores yellow, subglobose, $6.5-7.6\mu$ in diameter. **Flesh** white, taste mild.

Plant 2–3 in. high. **Pileus** 2–3 in. broad. **Stem** 4–6 lines thick. *Frost* Mss.

Ground in **woods**. Sandlake. August. *Peck*, 32d Rep. N. Y. State Bot.

R. flavida is showy, solitary and in patches. The stem when young and solid is equally good with the cap. Cooks in twenty-five minutes and is of good flavor.

HETEROPHYL'LÆ.

R. ves'ca Fr.—*vesco*, to feed. **Pileus** *red-flesh-color, disk darker*, fleshy, slightly firm, plano-depressed, *slightly wrinkled with veins*, with a viscid pellicle, margin at length spreading. **Flesh** cheesy, firm, shining white. **Stem** *solid*, compact, externally rigid, *reticulated and wrinkled* in a peculiar manner, often attenuated at the base, shining white. **Gills** adnate, crowded, thin, shining white, with many unequal and forked ones intermixed, but scarcely connected by veins.

Of middle stature. *Taste mild*, pleasant. *Fries.*

Spores globose, echinulate, white, $9-10\mu$ diameter *Massee.*

In mixed woods. Common. August to frost.

R. vesca is frequent in woods or margins, and under trees in the open. It is especially fond of growing in the grass under lone chestnut trees. The caps seldom exceed 2½ in. across.

It is one of the best.

R. cyanoxan'tha (Schaeff.) Fr. *Gr.*—blue; *Gr.*—yellow. (From the colors.) ([Color] Plate XLIV, fig. 1.) **Pileus** 2–3 in. and more broad, *lilac or purplish then olivaceous-green*, disk commonly becoming pale often yellowish, *margin* commonly becoming *azure-blue or livid purple*, compact, convex then plane, then depressed or infundibuliform, sometimes even, sometimes wrinkled or streaked, viscous, margin deflexed then expanded, remotely and slightly striate. **Flesh** firm,

cheesy, white, commonly reddish beneath the separable pellicle. **Stem** Russula.
2–3 in. long, as much as 1 in. thick, *spongy-stuffed*, but firm, often
cavernous within when old, equal, smooth, *even*, shining white. **Gills**
rounded behind, connected by veins, not much crowded, broad, forked
with shorter ones intermixed, shining white.

Allied to R. vesca in its *mild*, pleasant *taste* and in other respects,
but constantly different in the color of the pileus, which is very variable,
whereas in R. vesca it is unchangeable. The peculiar combination of
colors in the pileus, though very variable, always readily distinguishes
it. *Fries.*

Spores 8–9μ, cystidia numerous, pointed, *Massee;* 8–10x6–8μ *Sacc.*
In mixed woods. Common. August to October.
Pronounced one of the best esculent species by all authorities.

R. heterophyl'la Fr. *Gr.*—differing; *Gr.*—a leaf. (Gills differing
in length.) **Pileus** very variable in color, but *never becoming reddish
or purple*, fleshy, firm, convexo-plane then depressed, *even, polished*,
the very thin pellicle disappearing, margin thin, even or densely but
slightly striate. **Flesh** white. **Stem** solid, firm, somewhat equal, *even*,
shining white. **Gills** *reaching the stem in an attenuated form, very narrow, very crowded*, forked and dimidiate, shining white.

Taste *always mild*, as in R. cyanoxantha, from which it differs in its
smaller stature, in the pileus being thinner, even, *never reddish* or purplish, with a thin closely adnate pellicle, in the *stem being firm and solid*,
and in the *gills* being *thin, very narrow, very crowded*, etc. The apex
of the stem is occasionally dilated in the form of a cup, so that the gills
appear remote. *Fries.*

Spores echinulate, 5x7μ *W.G.S.;* 7–8μ diameter *Massee.*
Common. Woods. July to November.
Edible, of a sweet nutty flavor. *Stevenson.*
R. heterophylla is very common. Its smooth, even pileus, colored in
some dingy shade of green, distinguishes it. It is much infested by
grubs. Specimens for the table should be young and fresh. Wilted
specimens are unpleasant.

R. fœ'tens Fr.—*fœtens*, stinking. **Pileus** 4–5 in. and more broad,
dingy yellow, often becoming pale, thinly fleshy, at first bullate, then
expanded and depressed, covered with a pellicle which is adnate, not

Russula. separable, and viscid in wet weather, margin broadly membranaceous, at the first bent inward *with ribs which are at length tubercular.* **Flesh** thin, *rigid*-fragile, pallid. **Stem** 2 in. and more long, ½–1 in. thick, stout, stuffed then hollow, whitish. **Gills** adnexed, crowded, connected by veins, with very many *dimidiate and forked* ones intermixed, whitish, at the first *exuding watery drops.*

Fetid. Taste acrid. Very *rigid*, most distinct from all others in *its very heavy empyreumatic* odor. In very dry weather the odor is often obsolete. The margin is more broadly membranaceous and hence marked with *longer furrows* than in any other species. It differs from all the preceding ones in the gills at the first exuding watery drops. The gills become obsoletely light yellow, and dingy when bruised. *Fries.*

Pileus fleshy, with a wide thin margin, hemispherical or convex, then expanded or depressed, viscid when moist, widely striate-tuberculate on the margin, dull pale yellow or straw color. **Lamellæ** rather broad, close, venose-connected, some of them forked, whitish. **Stipe** nearly cylindrical, whitish, hollow. **Spores** white. **Plant** sometimes cespitose.

Height 2–4 in.; breadth of pileus 2–3 in. **Stipe** 4–6 lines thick.

Pine woods. West Albany. October.

Taste mild at first, then slightly disagreeable. *Peck*, 23d Rep. N. Y. State Bot.

Spores minute, echinulate, almost globular, 8μ *W.G.S.;* 8–10μ *Massee.*

In woods. Common. July to October.

Var. *granula'ta* has the pileus rough with small granular scales. *Peck*, Rep. 39.

A very coarse and easily recognized species. Reckoned poisonous, though eaten by slugs. *W.G.S.*

The verdict is against it. Both smell and taste are usually unpleasant. Cooked it retains its flavor, more closely resembling wild cherry bark than anything else. On two occasions I ate enough to convince me that it was not poisonous.

R. el'egans Bresad.—*elegans*, pretty. Mild at first, becoming acrid with age. **Pileus** 2–3 in. across. **Flesh** rather thick; convex then depressed; margin tuberculose and striate when old, viscid, bright rosy flesh-color, soon ochraceous at the circumference, everywhere densely granulated. **Gills** adnexed or slightly rounded, narrow behind, very **Russula.** much crowded, equal, rarely forked, whitish, becoming either entirely or here and there ochraceous-orange. **Stem** 1½–2 in. long, 5–7 lines thick, a little thickened at the base, rather rugulose, white, base ochraceous. **Flesh** white, turning ochraceous and acrid when old.

Spores 8–10μ diameter *Massee.*

Allied to R. vesca. Known by the bright rose-colored, densely granular pileus and tuberculose margin. When old the pileus is almost entirely ochraceous. *Massee.*

Frequent in the West Virginia forests, 1881–1885. Chester county, Pa., 1887–1890. In mixed woods. July to September. *McIlvaine.*

It differs from R. vesca in its cap being minutely granulated instead of streaked, and in becoming acrid with age.

The caps are of good quality, needing to be well cooked.

FRA'GILES.

Gills and spores white.

R. eme'tica Fr.—an emetic. ([Color] Plate XLIV, fig. 2.) **Pileus** 3–4 in. broad, at first rosy then *blood-color*, tawny when old, sometimes becoming yellow and at length (in moist places) white, at first bell-shaped then flattened or depressed, polished, *margin* at length *furrowed and tubercular.* **Flesh** *white, reddish under the separable pellicle.* **Stem** spongy-stuffed, stout, elastic when young, fragile when older, even, white or reddish. **Gills** somewhat *free*, broad, somewhat distant, shining white.

Handsome, regular, moderately firm, but fragile when full grown, *taste* very *acrid. Fries.*

Spores shining white, *Fries;* spheroid, echinulate, 8–10μ *K.;* 7μ *W.G.S.*

Maryland, *Miss Banning;* New York, *Peck*, Rep. 22; Indiana, Illinois, *H. I. Miller.*

Said to act as its name implies as an emetic. Certainly poisonous. *Stevenson.*

Krapp says he has himself experienced rare inconveniences from eating it. Preferred to others in Indiana and Illinois. *H. I. Miller*, 1898.

The varying reports upon R. emetica are quoted above. In 1881, in

Russula. the West Virginia mountains, I began testing this Russula and soon found that it was harmless. At least twenty persons ate it in quantity, during its season, for four years. Yet, in my many published articles, I continued, out of regard for the opinions of others and in excess of caution, to warn against all bitter and peppery fungi. But from that time until the present I have eaten it, and I have made special effort to establish its innocence by getting numbers of my friendly helpers to eat it.

It was suggested by one of its prosecutors that perhaps I was mistaking another fungus for it. In October, 1898, I sent to Professor Peck a lot of the Russula I was eating. He wrote: "It seems to be R. emetica as you state. It certainly is hot enough for it."

R. pectina′ta Fr.—*pecten*, a comb. **Pileus** 3 in. broad, at first gluey, *toast-brown*, then dry, becoming pale, tan, with the *disk* always *darker*, fleshy, *rigid*, convex then flattened and depressed or concavo-infundibuliform (basin-shaped); margin thin, *pectinato-sulcate* (deeply ribbed), here and there irregularly shaped. **Flesh** *white, light yellowish under the pellicle*, which is not easily separable. **Stem** curt, 2 in. long, ¾–1 in. thick, *rigid*, spongy-stuffed, longitudinally *slightly striate, shining white*, often attenuated at the base. **Gills** *attenuato-free* behind, broader toward the margin, somewhat crowded, *equal*, simple, white.

Odor weak, but nauseous, approaching that of R. fœtens. *Fries.*

Spores 8–9µ diameter *Massee*.

New York, *Peck*, 43d Rep. West Virginia, Pennsylvania, New Jersey. Common in woods, grassy, mossy places. July to frost. *McIlvaine.*

Named from the furrows of the margin being like the teeth of a comb.

Both the appearance and smell of this Russula will detect it. The peculiar comb-like furrows of its margin, viscid or varnished-looking cap, and strong but more spicy smell than cherry-bark are noticeable.

It is edible, but so strong in flavor that a piece of one will spoil a dish if cooked with other kinds.

R. ochroleu′ca Fr. *Gr.*—pale yellow; *Gr.*—white. **Pileus** *yellow, becoming pale*, fleshy, flattened or depressed, polished, with an adnate pellicle, the spreading margin *becoming even*. **Stem** spongy, stuffed, firm, *slightly reticulato-wrinkled, white, becoming cinereous*. **Gills** *rounded behind*, united, broad, *somewhat equal*, white becoming pale.

Odor obsolete, but pleasant. The pileus is never reddish. It agrees

wholly with R. emetica in structure and stature, as well as in the *acrid* Russula.
taste; it differs however in the stem being slightly recticulato-wrinkled,
white becoming cinereous, in the adnate pellicle of the pileus, in the
margin remaining for a long time *even* (remotely striate, but not tubercular, only when old), and in the gills being rounded behind and becoming pale. The color of the pileus is constant. The gills remain
free and do not exude drops. *Fries.*

Cap 2–4 in. across. **Stem** 2–3 in. long, up to ¾ in. thick.
Spores papillose, 7µ *W.G.S.*, 8x9µ *Massee.*
Frequent in woods. July to October.

Not as common as R. emetica, yet frequently found, usually solitary,
at times gregarious. It is quite peppery, but loses pepperiness in cooking. Myself and others have frequently eaten it.

R. ci'trina Gillet—*citrina*, citron colored. **Mild. Pileus** 2–3 in.
across, slightly fleshy at the disk, margin thin; convex then more or
less expanded and slightly depressed, rather viscid when moist, smooth,
slightly wrinkled at the margin when old, bright lemon-yellow, color
usually uniform, sometimes paler at the margin, occasionally with a
greenish tint, center of pileus at length becoming pale-ochraceous; pellicle separable. **Gills** slightly decurrent, broadest a short distance from
the margin, and gradually becoming narrower towards the base, forked
at the base and also sometimes near the middle, white, 1½ lines deep
at broadest part. **Stem** 2–3 in. long, about 4 lines thick, equal or
slightly narrowed at the base, slightly wrinkled, straight or very slightly
waved, solid.
Spores subglobose, echinulate, 8µ diameter.
In woods.

Known by the clear lemon-yellow or citron-colored pileus and the
persistently white gills and stem. The taste is mild at first, but becomes slightly acrid if kept in the mouth for a short time. *Massee.*

R. citrina can hardly be classed among the acrid species. The taste
is slightly of cherry-bark and disappears in cooking. It is usually found
in patches which contain ten to twenty individuals. It is a species of
fair quality.

R. fra'gilis Fr.—fragile. **Pileus** 1–1½ in. broad, rarely more, flesh-
color, changing color, very thin, fleshy only at the disk, at the first con-

Russula. vex and often umbonate, then plane and depressed, pellicle thin, becoming pale, slightly viscid in wet weather; *margin* very thin, *tuberculoso-striate*. **Stem** 1 ½ –2 in. long, spongy within, soon hollow, often slightly striate, white. **Gills** slightly adnexed, very *thin, crowded,* broad, *ventricose,* all equal, shining white. *Fries.*

Very acrid. Smaller and more fragile than the rest of the group, directly changing color. The color is variable, often opaque, typically flesh-color, when changed in color white externally and internally, often with reddish spots. Among varieties of color is to be noted a livid flesh-colored form, with the disk becoming fuscous.

It is not easy to define it from fragile forms of R. emetica, but the gills are much more crowded, thinner, and often slightly eroded at the edge, ventricose; the pileus thinner and more lax, etc. *Stevenson.*

Var. *nivea* Fr.—*nivea,* snowy. Whole plant white.

Spores minutely echinulate 8–10x8μ *Massee.*

Though one of the peppery kind, I have not, after fifteen years of eating it, had reason to question its edibility. The caps are not meaty, but what there is of them is good.

R. puncta′ta Gillet—*punctata,* dotted. **Mild. Pileus** 1 ½–2 ½ in. across. **Flesh** thin, white, reddish under the cuticle; convex then flattened, viscid, rosy, disk darkest, punctate with dark reddish point-like warts, pale when old; margin striate. **Gills** slightly adnexed, 2 lines broad, white then yellowish, edge often reddish. **Stem** about 1 in. long, 4–5 lines thick, attenuated and whitish at the base, remainder colored like the pileus, stuffed.

Spores 8–9μ diameter *Massee.*

Among grass.

Edible. Boston Myc. Club Bull. 1896.

***Gills and spores white then yellowish or bright lemon.*

R. in′tegra Fr.—*integer,* entire, whole. **Pileus** 4–5 in. across, typically red, changing color, fleshy, campanulato-convex then expanded and depressed, fragile when full-grown, with a gluey pellicle, at length *furrowed and somewhat tubercular* at the margin. **Flesh** *white,* sometimes yellowish above. **Stem** at first short, conical, then club-shaped or *ventricose*, as much as 3 in. long, up to 1 in. thick, spongy-stuffed, Russula. commonly stout, *even*, shining *white*. **Gills** somewhat free, very broad, up to ¾ in., equal or bifid at the stem, somewhat distant, connected by veins, pallid-white, at length light yellow, *somewhat powdered yellow with the spores.*

Taste mild, often astringent. The most changeable of all species, especially in the color of the pileus which is typically red, but at the same time inclining to azure-blue, bay-brown, olivaceous, etc. Sometimes the gills are sterile and remain white. *Fries.*

Spores ellipsoid-spheroid or spheroid echinulate, globose, rough, 8–9μ *C.B.P.;* 9–10μ diameter, pale ochraceous. *Massee.*

It is difficult to separate R. integra from R. alutacea. The spores usually show upon the gills as pale dull yellow powder. It is of equal excellence.

R. decolo'rans Fr.—*de* and *coloro*, to color. **Pileus** 3–5 in. broad, color various, at first orange-red, then light yellow and becoming pale, fleshy, spherical then expanded and depressed, remarkably regular, viscid when moist, thin and at length striate at the margin. **Flesh** *white, but becoming somewhat cinereous* when broken, and more or less *variegated with black spots* when old. **Stem** *elongated*, 3–5 in., cylindrical, solid, but spongy within, often *wrinkled-striate, white then becoming cinereous* especially within. **Gills** adnexed, often in pairs, thin, crowded, fragile, white then yellowish.

Taste mild. Colors changeable according to a fixed rule, but not variable. The gills are not ochraceous-pulverulent as in R. integra, nor shining and pure yellow as in R. aurata, etc. *Fries.*

Spores yellow, 8.3μ *Morgan.*

New York, *Peck*, 23d Rep. Angora, West Philadelphia, Pa., 1897, in mixed woods. August to October. *McIlvaine.*

Esculent and of good quality. *Morgan.*

Meals of it make one regret its scarcity.

R. basifurca'ta Pk.—forked near stem. **Pileus** 2–3 in. broad, firm, convex, umbilicate, becoming somewhat funnel form, glabrous, slightly viscid when moist, the thin pellicle scarcely separable except on the margin, dingy-white, sometimes tinged with yellow or reddish-yellow, the margin nearly even. **Lamellæ** rather close, narrowed toward the

Russula. base, adnate or slightly emarginate, many of them forked near the base, a few short ones intermingled, white becoming yellowish. **Stem** 8–12 lines long, 5–6 lines thick, firm, solid, becoming spongy within, white.

Spores elliptical, pale yellow, uninucleate or shining, 9x6.5μ. **Flesh** white, taste mild, then bitterish.

Dry hard ground in paths and wood roads. Canoga, N. Y. July.

This species closely resembles pale forms of R. furcata, from which it is separated by the absence of any silky micor and by the yellowish color and elliptical shape of the spores and by the yellowish hue of the lamellæ. *Peck*, 38th Rep. N. Y. State Bot.

Mt. Gretna, Pa., September, 1898, to frost. Gravelly ground. Solitary. Gills adnate. Identified as his species by Professor Peck.

The slight bitterish taste disappears in cooking. It is edible and of fair quality.

R. aura′ta Fr.—*aurum*, gold. **Pileus** 2–3 in. broad, varying *lemon-yellow, orange and red*, disk darker, fleshy, *rigid*, brittle however, hemispherical then plane, disk not depressed, pellicle thin, adnate, viscid in wet weather, *margin even*, and slightly striate only when old, but sometimes wrinkled. **Flesh** *lemon-yellow* under the pellicle, white below. **Stem** 2–3 in. long, solid, *firm*, but spongy within, cylindrical, obsoletely striate, white or lemon-yellow. **Gills** rounded free, connected by veins, broad, equal, shining, never pulverulent, whitish inclining to light yellow, but vivid *lemon-yellow at the edge*. *Fries*.

West Virginia, 1881–1885; Pennsylvania, 1887–1898. In woods under pines. July to October. *McIlvaine*.

Pileus sometimes depressed in center, very viscid when wet.

A troop of this Russula upon brown wood mat is a pretty sight. Its rich and brightly-colored cap attracts the eye from a distance. The yellow edge of its gills is the distinctive mark of the species.

The smell is pleasant, the taste slightly of cherry bark.

Cooked it is one of the best Russulæ.

R. atropurpu′rea Pk.—*atre*, black; *purpureus*, purple. Dark purple Russula. **Pileus** 3–4 in. broad, at first convex, then centrally depressed, glabrous, dark purple, blackish in the center, the margin even or slightly striate. **Flesh** white, grayish or grayish-purple under the separable pellicle, taste mild, odor of the drying plant fetid, very un-

pleasant. **Lamellæ** nearly equal, subdistant, sometimes forked near the Russula.
stem, at first white, then yellowish, becoming brownish where bruised.
Stem 2–3 in. long, 5–8 lines thick, equal, glabrous, spongy within,
white, brownish where bruised. **Spores** subglobose, minutely rough,
pale ochraceous with a salmon tint, 8–10μ.

Open woods. Gansevoort. July.

In color this species resembles R. variata, but in other respects it is
very different. It is very distinct in the peculiar color of its spores,
and in the brownish hue assumed by wounds. *Peck*, 41st Rep. N. Y.
State Bot.

West Philadelphia, Pa. July, 1897. Open woods. Solitary. Philadelphia Myc. Center.

Many were eaten and enjoyed. Only fresh plants are acceptable, and
they should be cooked as soon as gathered. Even in wilting they become unpleasant.

***Gills and spores ochraceous.*

R. aluta′cea Fr.—*aluta*, tanned leather. **Pileus** 2–4 in. broad,
commonly bright blood-color or *red*, even black-purple, but becoming
pale, especially at the disk, fleshy, bell-shaped then convex, flattened
and somewhat umbilicate, even, *with a remarkably sticky pellicle,
margin thin, at length striate, tubercular.* **Flesh** *snow-white.* **Stem**
2 in. long, solid, stout, equal, even, white, most frequently *variegated-
reddish*, even purple. **Gills** at first free, *thick, very broad*, connected
by veins, all equal, somewhat distant, at first pallid light yellow, then
bright ochraceous, not pulverulent.

It is distinguished from R. integra by its gills not being pulverulent.
Fries.

Spores yellow 7–9μ *Massee;* 11–14x8–10μ *Sacc., Syll.*
July to frost. *McIlvaine.*

R. alutacea is easily recognized among Russulæ by its mild taste and
broad yellow gills. In young specimens one sometimes has to look at
the gills at an angle to detect the yellow. It is quite common but a
solitary grower. It is everywhere eaten as a favorite. Only fresh
plants yield a good flavor. When the stem is soft, it should be thrown
away.

R. puella'ris Fr. ([Color] Plate XLIV, fig. 7.) **Mild. Pileus** 1–1 ½ in. across, flesh almost membranaceous except the disk; conico-convex then expanded, at first rather gibbous, then slightly depressed, scarcely viscid, color peculiar, purplish-livid then yellowish, disk always darker and brownish; tuberculosely striate, often to the middle. **Gills** adnate but very much narrowed behind, thin, crowded, white then pale-yellow, not shining nor powdered with the spores. **Stem** 1–1 ½ in. long, 2–4 lines thick, equal, soft, fragile, wrinkled under a lens, white or yellowish; stuffed, soon hollow; taste mild.

Spores subglobose, pale-yellow, echinulate, 10x8–9μ *Massee.*

In woods.

Among the most frequent and readily recognized of species, occurring in troops. Always small, thin, taste mild. Allied to R. nitida, but more slender; color paler, and not shining. *Fries.*

Distinguished from R. nitida and R. nauseosa by the absence of smell. *Massee.*

Var. *inten'sior* Cke. Nearly the same size as the typical form; pileus deep purple, nearly black at the disk.

The stem has a tendency to become thickened at the base, and turns yellowish when touched.

Var. *rose'ipes* Sec., given by Massee, has been retained as a distinct species by Professor Peck, Rep. 51, and is described in place. R. pusilla Pk., 50th Rep., is closely allied to it.

West Virginia, Pennsylvania, New Jersey, North Carolina. Common in woods and under trees in short grass. July to September. *McIlvaine.*

This little Russula is ubiquitous. It does not amount to much when other fungi are plenty, because of its very thin cap, but it thrives in all sorts of summer weather. When its companions are scarce or parched R. puellaris is gladly gathered by the mycophagist, its numbers making up for its lightness and lack of flavor.

R. pusil'la Pk.—little. **Pileus** very thin, nearly plane or slightly and umbilicately depressed in the center, glabrous, slightly striate on the margin, red, sometimes a little darker in the center, the thin pellicle separable. **Flesh** white, taste mild. **Lamellæ** broad for the size of the plant, subventricose, subdistant, adnate or slightly rounded behind, white, becoming yellowish-ochraceous in drying. **Stem** short, soft, solid or spongy within, white.

Spores faintly tinged with yellow, 7.6µ broad.
Pileus scarcely 1 in. broad. Stem 6–12 lines long, 2–3 lines thick. Bare ground in thin woods. Port Jefferson. July.

The coloring matter of the pileus may be rubbed upon paper and produce on it red stains if the surface is previously moistened with water or dilute alcohol. This is one of the smallest Russulas known to me. The pileus was less than an inch broad and the stem less than an inch long in all the specimens seen by me. The species is closely allied to R. puellaris, and especially resembles the variety intensior in color. It differs in its smaller size, even or but slightly striate margin, broad lamellæ and in the stem or flesh not becoming yellowish spotted where touched. *Peck*, 50th Rep. N. Y. State Bot.

West Virginia, 1881–1885. Pennsylvania, 1896–1897. July to September. *McIlvaine.*

It makes up in quality what it lacks in quantity.

R. rose'ipes (Secr.) Bres.—*rosa*, a rose; *pes*, a foot. ([Color] Plate XLIV, fig. 5.) **Pileus** 1–2 in. broad, convex becoming nearly plane or slightly depressed, at first viscid, soon dry, becoming slightly striate on the thin margin, rosy-red variously modified by pink orange or ochraceous hues, sometimes becoming paler with age, taste mild. **Gills** moderately close, nearly entire, rounded behind and slightly adnexed, ventricose, whitish becoming yellow. **Stem** 1 ½–3 in. long, 3–4 lines thick, slightly tapering upward, stuffed or somewhat cavernous, white tinged with red.

Spores yellow, globose or subglobose.

The plants grow in woods of pine and hemlock and have been collected in July and August. The flesh is tender and agreeable in flavor. *Peck*, 51st Rep. N. Y. State Bot.

Spores globose, minutely echinulate, pale ochraceous, 8–10µ diameter *Massee*.

R. roseipes is common in West Virginia under hemlocks and spruces. At Mt. Gretna, Pa., it grew sparingly under pines. It is excellent.

R. Ma'riæ Pk. **Pileus** fleshy, convex, subumbilicate, at length expanded and centrally depressed, minutely pulverulent, bright pink-red (crimson lake), the disk a little darker, margin even. **Lamellæ** rather

Russula. close, reaching the stem, some of them forked, venose-connected, white, then yellowish. **Stem** equal, solid, colored like the pileus except the extremities which are usually white. **Spores** globose, nearly smooth, 7.6µ in diameter; flesh of the pileus white, red under the cuticle, taste mild.

Plant 2 in. high. **Pileus** 1.5–2 in. broad. **Stem** 3–6 lines thick. Dry ground in woods. Catskill mountains. July.

The minute colored granules, which give the pileus a soft pruinose appearance, are easily rubbed off on paper, and water put upon the fresh specimens is colored by them. *Peck*, 24th Rep. N. Y. State Bot.

New York, *Peck*, 24th and 50th Rep.; West Virginia, 1882–1885; Mt. Gretna, Pa., solitary in mixed woods. July to September. 1897–1898. *McIlvaine*.

It is on a par with most Russulæ.

R. ochra′cea Fr.—*ochra*, a yellow earth. **Mild. Pileus** about 3 in. across. **Flesh** rather thick at the center, becoming thin toward the margin, pale ochraceous, soft; convex then expanded and depressed, margin coarsely striate, pellicle thin, viscid, ochraceous with a tinge of yellow, disk usually becoming darker. **Gills** slightly adnexed, broad, scarcely crowded, ochraceous. **Stem** about 1½ in. long, 5–7 lines thick, slightly wrinkled longitudinally, ochraceous, stuffed, soft.

Spores globose, echinulate, ochraceous, 10–12µ diameter.

In pine and mixed woods.

The mild taste and ochraceous color of every part, including the flesh, separate the present from every other species.

Commonly confounded with Russula fellea, but known at once by its mild taste. Agreeing most nearly with R. lutea in color, but differing in the softer flesh, which becomes ochraceous upward; sulcate margin of the pileus, and broader, less crowded gills. **Pileus** persistently ochraceous, disk usually darker. **Stem** sometimes yellow, sometimes white. *Fries*.

North Carolina, borders of woods, *Curtis;* California, *Harkness and Moore*.

Fries says that the flavor is mild, but Roze places it in the list of suspected species, although he notes it as not acrid; it may be inferred that he considers the flavor unpleasant. *Macadam*.

"Like chicken," not common. Boston Myc. Club Bull. 1896.

R. lu′tea (Huds.) Fr.—*luteus*, yellow.   **Pileus** 1–2 in. broad, *yel-* Russula.
low, at length becoming pale, and occasionally wholly white, thinly
fleshy, soon convexo-plane or plano-depressed, sticky when moist, *even*
or when old obsoletely striate *at the margin*.   **Flesh** white.   **Stem** ½
in. long, 3–4 lines thick, stuffed then *hollow*, soft, fragile, equal, even,
white, never reddish.   **Gills** somewhat free, connected by veins, *crowded*,
narrow, all equal, ochraceous-egg-yellow.

Always small, very regular, taste mild.   *When young the pileus is*
always of a beautiful yellow.   Fries.

Spores yellow, echinulate, 8μ *W.G.S.;* globose, rough, 6–7μ *C.B.P.;*
8–10x7–8μ *Massee.*

Allied to R. vitellina, but differs in having the margin of the cap
even, and but little odor.

The plant I have so referred has the gills at first white and the stem
yellow like the pileus; it may be a new species.   In beech woods,
Morgan; West Virginia, Pennsylvania, New Jersey, in mixed woods,
often under beeches, August to November, *McIlvaine.*

The plants I have found have white gills when young (few species
have not), but rapidly become yellow.   The stem is usually white when
young, and sometimes remains so, but often becomes more or less
yellow.

It is a pretty species.   The flavor is not as strong as in some species,
but is delicate.

R. nauseo′sa Fr.   **Pileus** variable in color, typically *purplish at the*
disk, then livid, but becoming pale and often whitish, laxly fleshy, thin,
at first plano-gibbous, then depressed, viscid in wet weather, *furrowed*
and somewhat tubercular at the somewhat membranaceous *margin*.
Flesh soft, white.   **Stem** short, about 1 in. long, 4 lines thick, spongy-
stuffed, slightly striate, white.   **Gills** adnexed, ventricose, *somewhat*
distant, here and there with a few shorter ones intermixed, light yellow
then dingy ochraceous.

The taste is mild, but also nauseous, as the odor often is.   The habit
is that of R. nitida, of the same color of pileus, but differing in the color
of the gills.   *Fries.*

Cap about 2 in. across.   **Stem** 1–2 in. long, ¼–½ in. thick.
Spores dingy yellow, 8–9μ diameter.   *Massee.*
North Carolina and Pennsylvania, *Schweinitz;* West Virginia, Penn-

Russula. sylvania, New Jersey, in pine and mixed woods.   August to October. *McIlvaine.*

The odor and taste of R. nauseosa are misnamed, therefore the plant. They are heavy at times, when the plant is wet or old, as is the case with R. fœtens, but they are always of cherry bark.   Both odor and taste disappear in cooking.   The species is as good as any Russula of its texture.

R. vitelli'na Fr.—*vitellus*, yolk of egg.   Pileus 1 in. broad, *unicolorous*, light yellow then wholly pallid, somewhat membranaceous, at length *tuberculoso-striate*, somewhat dry, disk very small, slightly fleshy. Stem thin, scarcely exceeding 1 in. long, 2 lines thick, equal.   Gills separating-free, equal, *distant*, rather thick, connected by veins, saffron-yellow.

Pretty, very fragile, strong-smelling, mild.   *Fries.*

Spores 7–8µ diameter *Massee.*

West Virginia, New Jersey, Pennsylvania, August to October.   In pine and mixed woods, July to October.   Not common in number.

This pretty species has a cherry-bark taste and smell like R. fœtens, though not so offensively heavy.   It is not poisonous.   A small piece of it will affect a whole dish of other Russulæ.

R. chamæleonti'na Fr.—changing color like a chamæleon.   Pileus 1–2 in. broad, thinly fleshy, soon flattened, sometimes oblique with a thin, separable, viscid pellicle, which is at first flesh-color, then presently changing color, becoming yellow at the disk and at length wholly yellow, margin even, then slightly striate.   Stem as much as 3 in. long, but thin, somewhat hollow, slightly striate, white.   Gills more or less adnexed, *thin, crowded*, equal, narrow, somewhat forked, light-yellow-ochraceous.

Mild, inodorous, very fragile.   Pileus *rosy blood-red, purplish lilac*, etc.   Sometimes even at the first yellowish at the disk.   *Fries.*

Spores globose, ochraceous, 7–8µ diameter *Massee.*

In pine and in mixed woods.   August to October.   *McIlvaine.*

The change in color of the cap which gives name to this species is not remarkable.   Most species of Russulæ are sensitive to light.   An otherwise highly colored cap will be almost white when a leaf adheres to it.   If in youth it grows under dense shade it will be very much

lighter than if where light is generous, and will remain so.   If in grow- Russula.
ing it thrusts itself out of shadow, its color will change and it will
deepen.   The apparent rarity of R. chamæleontina I think due to the
close observation necessary to detect its changes in color, which, as I
have found it, are by no means constant.   It is quite plentiful in the
pines of southern New Jersey, and at Mt. Gretna, Pa., it is frequently
found.

It is a good esculent species.

CANTHAREL'LUS Adans.

Gr.—a vase, a cup.

Cantharellus. Hymenophore continuous with the stem, descending unchanged into the trama. **Gills** thick, fleshy, waxy, *fold-like*, somewhat branched, *obtuse at the edge.* **Spores** white. Fleshy, putrescent fungi, without a veil. *Fries.*

CANTHARELLUS CIBARIUS.

In Cantharellus the gills—vein-like and generally thick with an obtuse edge—are entirely different from those of all the preceding genera. In those they are thin, and distinct from the pileus and from each other. In Hygrophorus the gills are frequently thick, but the edge is always sharp. The species of Craterellus are funnel-shaped, resembling some of those in Cantharellus, but are distinguished by their lack of evident gills.

Monograph New York Species of Cantharellus, *Peck*, Bull. 1887.

The members of this genus are few, but they are choice. Of them is the Cantharellus cibarius, of which Trattinik quaintly says: "Not only this same fungus never did any one harm, but might even restore the dead."

The writer first made its acquaintance when among the West Virginia mountains in 1881. The golden patches of single and clustered cibarius, fragrant as ripened apricots, tufting the short grass or mossy ground under beeches, oaks and like-growing trees, through which the sunlight filtered generously, were so tempting, that he determined there must be luxury, even in death, from such toadstools.

Experiments made by the writer in West Virginia where the species grows luxuriantly and is of much higher flavor than any he has found elsewhere, prove that it is easy to transplant within congenial habitats, either by the mycelium or spores. Nature, there, resorts to washing masses of leaves containing the propagating parts of the fungus along the depressions of the water-sheds, and it is found growing plentifully where the wind has drifted forest leaves against trees, brush, and fence-corners.

Other species of the genus do not, as a rule, grow so plentifully, neither are they of equal excellence, but several of them are equal to

any other species.   Suspicion has been thrown upon C. aurantiacus. Cantharellus.
There is such a marked difference between the excellence of the genus
in West Virginia and other localities, that it is possible C. aurantiacus
may be noxious elsewhere, but the writer has not found it so; and it
would be an astonishing contradiction of Nature's ways if it was.

Stevenson says: "It (C. cibarius) must have four hours slow cooking." The writer has found thirty minutes to be sufficient; and it will
fry in butter as quickly as any other fungus.

ANALYSIS OF TRIBES.

MESOPUS (*mesos*, middle; *pous*, a foot).   Page 215.
Stem central.

*Stem solid.
**Stem tubular.

PLEUROPUS (*pleura*, the side; *pous*, a foot).
Stem lateral.

RESUPINATUS (*resupinatus*, lying on the back).
Stem absent.

All the species known to be edible belong to Mesopus.

ME'SOPUS.

* Stem solid.

C. ciba′rius Fr.—*cibaria*, food.   ([Color] Plate XLVI, fig. 4.
Plate XLVII.)   **Pileus** fleshy, obconic, smooth, egg-yellow, slightly
depressed.   **Gills** thick, distant, more or less branching and anastomosing, concolorous.   **Stem** firm, solid, often tapering downward, concolorous.   **Flesh** white.

Height 2–4 in., breadth of **pileus** 2–3 in.   **Stem** 3–6 lines thick.
In open woods and grassy places.   Common.   July and August.

Edible.   The smell of apricots is not always clearly perceptible in
American specimens.   *Peck*, Monograph New York Species of Cantharellus, Rep. 23.

Cantharellus. **Spores** 6x8μ *W.G.S.;* 7.6x5μ *Morgan;* spheroid-ellipsoid, 8–9x5–6μ *K.;* 11μ *Q.*

(Plate **XLVII.**)

CANTHARELLUS CIBARIUS.

Reported from the Atlantic to the Pacific and from Columbia river to Louisiana. June to September.

Wherever grown C. cibarius is one of the best. In European countries it is highly rated, and is expensive. Its mode of growth varies with its plentifulness. In the West Virginia mountains large patches of it closely cover the ground. Clusters weighing ½ pound are frequent.

When shredded, or cut across the fibers, slow cooking for half an hour is sufficient, if the plants are fresh. If gathered for some hours, they should be soaked for a time.

C. mi′nor Pk. **Pileus** fleshy, thin, convex then expanded and depressed, egg-yellow. **Gills** very narrow, distant, sparingly branched, yellowish. **Stem** slender, subflexuous, equal, smooth, hollow or stuffed, concolorous.

Height 1–2 in., breadth of pileus 6–12 lines.
In open woods. July. *Peck*, 23d Rep. N. Y. State Bot.
Spores 6.4–7.6x4–5μ *Peck.*
West Virginia, New York, Pennsylvania. *McIlvaine.*
Grows in the West Virginia mountains, along with C. cibarius, and separate from it. It is more tender than C. cibarius, and not equal in flavor to those found there. I usually cooked them together and thus got quantity well flavored.

C. auranti′acus Fr.—orange-yellow. ([Color] Plate CXXXVI, fig. 4.) **Pileus** fleshy, obconic, nearly plane above, smooth or minutely tomentose, dull orange with the disk usually brownish, the margin decurved

and sometimes yellowish. **Gills** narrow, close, repeatedly forked, Cantharellus
orange, sometimes yellowish. **Stem** inequal, generally tapering upward,
colored like the pileus. **Flesh** yellowish, taste mild.

Height 2–3 in., breadth of **Pileus** 1–3 in. **Stem** 2–4 lines thick.

Ground and very rotten logs in woods or in fields. Common. *Peck,*
23d Rep. N. Y. State Bot.

Spores 6.4–7.6x4–5µ *Peck,* 10x5µ *Massee.*

Var. *pallidus* Pk. **Pileus** and gills pale yellow or whitish yellow.

Stevenson says of the English species, "Unpleasant, reckoned poisonous." The writer's acquaintance with C. aurantiacus has been principally confined to West Virginia. There its taste is mild, scent but little, flavor not distinguishable from eastern C. Cibarius. There it is perfectly safe and wholesome; neither have the writer and his friends any
reason for condemning it.

C. umbona'tus Fr.—having an umbo. **Pileus** 1 in. and more broad,
ashy-blackish, slightly fleshy, convex when young, *umbonate, at length
depressed,* even, dry, *flocculoso-*silky on the surface, shining brightly
especially under a lens. **Flesh** soft, white, often becoming red when
wounded. **Stem** 3 in. long, about 4 lines thick, *stuffed,* equal, elastic,
villous at the base, *ash-colored,* but paler than the pileus. **Gills** decurrent, thin, tense and straight, *crowded,* repeatedly divided by pairs,
shining-white.

Odor and taste scarcely notable. Gregarious. Among the taller
mosses the stem is longer. Often overlooked from its habit being that
of an agaric. It varies with the pileus squamulose and blackish.

In woods. April to August. *Fries.*

The rather prominent gills of this small species are likely to confuse
those not familiar with its variance from the genuine type. Reddish
tinge to flesh not noticed in the American species. The writer has
gathered it in several states and enjoyed it for many years.

C. rosel'lus Pk.—rosy. **Pileus** thin, funnel-shaped, regular, glabrous,
pale pinkish-red. **Flesh** white. **Gills** narrow, close, dichotomous, deeply
decurrent, whitish, tinged with pink. **Stem** equal, slender, solid,
subglabrous, often flexuous, colored like the pileus. **Spores** minute,
broadly elliptical, 3.5x2.5µ.

Pileus 4–8 lines broad. **Stem** about 1 in. long, scarcely 1 line thick.

Cantharellus.

(Plate XLVIII.)

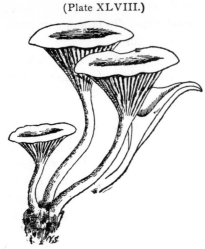

CANTHARELLUS ROSELLUS.
Natural size.

Mossy ground in groves of balsam. North Elba. September. This small species belongs to the section Agaricoides, and is apparently closely allied to C. albidus, from which its smaller size and different color distinguish it. The pileus is sometimes deeply umbilicate. *Peck*, 42d Rep. N. Y. State Bot.

Frequent in pine woods of New Jersey, near Haddonfield, where the plant is sturdier than described. Though small it grows gregarious and in troops from which appetizing quantities can be gathered.

It makes a pretty dish of pinkish hue and one of rare excellence.

C. lutes′cens Bull.—yellowish. ([Color] Plate CXXXVI, fig. 9.) **Pileus** thin, fleshy, convex, umbilicate, brownish-floccose, yellowish. **Gills** very distant, sparingly branched, arcuate-decurrent, pale ochraceous. **Stem** slender, slightly tapering downward, smooth, shining, bright orange-tinted yellow, stuffed or hollow.

Height 2–3 in., breadth of **Pileus** 8–15 lines.

Mossy ground in woods. Catskill and Adirondack mountains, also Sandlake. August to October.

This is regarded by some as a variety of A. tubæformis. *Peck*, 23d Rep. N. Y. State Bot.

In mixed and scrub-pine woods near Haddonfield, N. J.; mixed woods Angora and Kingsessing, Philadelphia.

Perhaps constancy to C. cibarius has influenced the writer in favor of members of its family, and accounts for the gusto in ''Fine'' set opposite his notes to the present species. Nevertheless such is his opinion.

** *Stem tubular.*

C. flocco′sus Schw.—woolly. ([Color] Plate XLVI, fig. 1.) **Pileus** fleshy, elongated funnel-form or trumpet shape, floccose-squamose,

218

ochraceous-yellow. Gills vein-like, close, much anastomosing above, Cantharellus. long decurrent and subparallel below, concolorous. Stem very short, thick, rarely deeply rooting.

Height 2–4 in., breadth of Pileus at the top 1–3 in.

Woods and their borders. Not rare. Utica, *Johnson.* Albany and Sandlake. July and August. *Peck*, 23d Rep. N. Y. State Bot.

Spores 12.5–15x7.6μ *Peck.*

New York, *Peck*, Rep. 23; Maine, *Mrs. Stella F. Fairbanks;* West Virginia, *McIlvaine.*

A beautiful species of good quality.

C. bre'vipes Pk.—*brevis*, short; *pes*, a foot.   ([Color] Plate XLVI, fig. 5.) Pileus fleshy, obconic, gla-brous, alutaceous or dingy cream-color, the thin margin erect, often irregular and lobed, tinged with lilac in the young plant; folds nu-merous, nearly straight on the mar-gin, abundantly anastomosing be-low, pale umber tinged with lilac. Stem short, tomentose-pubescent, ash-colored, solid, often tapering downward. Spores yellowish, oblong-elliptical, uninucleate, 10–12μx5μ.

(Plate XLIX.)

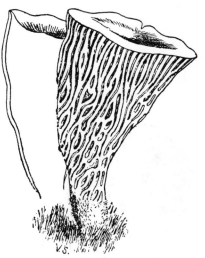

CANTHARELLUS BREVIPES.
Small plant, two-thirds natural size.

Plant 3–4 in. high. Pileus 2–3 in. broad. Stem 4–6 lines thick.

Woods. Ballston, Saratoga coun-ty. July.

This interesting species is related to the C. floccosus, both by its short stem and its abundantly anastomosing folds. The two species should be separated from the others and constitute a distinct section. The flesh in C. brevipes is soft and whitish, and the folds are generally thin-ner than in C. floccosus. *Peck*, 33d Rep. N. Y. State Bot.

Plentiful in West Virginia mountains in 1884, growing in patches. Found in mixed woods near Cheltenham, Pa., and at Springton, Pa., 1887.

Cantharellus.   In West Virginia it is prolific and rivals the C. cibarius in excellence. The flesh is softer, not so fibrous, and cooks more readily.

In that locality there is a marked difference between C. brevipes and C. floccosus. The latter is much longer, and markedly resembles the large end of a gold lined cornet. Like the C. cibarius it is not of as good quality in eastern states.

Nyctalis.

NYCTALIS Fr.

Gr.—night.   From inhabiting dark places.

(Plate L.)

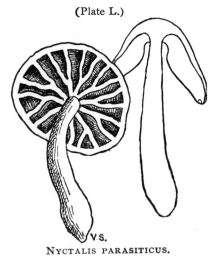

NYCTALIS PARASITICUS.

Hymenophore continuous with the stem. Gills fleshy, thick, juicy, obtuse at the edge, not decurrent on the stem nor fold-like. Veil (in species which have been fully observed) floccoso-pruinose.

Fleshy fungi, not reviving, of uncertain and irregular occurrence, differing in many respects from one another and from the rest of the Agaricini. Fries.

The typical species are saprophytic on decaying fungi. But one species, Nyctalis asterophora, reported in America. See *Peck*, 26th Rep. N. Y. State Bot.

MARAS'MIUS Fr.

Gr.—to wither or shrivel.

Pileus regular, thin, tough and pliant. **Gills** pliant, rather tough, Marasmius. somewhat distant, variously attached or free, with an acute entire edge. **Stem** cartilaginous or horny, continuous with the pileus but of different texture. Not putrescent but drying up with lack of moisture, reviving and assuming the original form with the advent of rain. This character distinguishes Marasmius from all other genera of Agaricaceæ.

(Plate LI.)

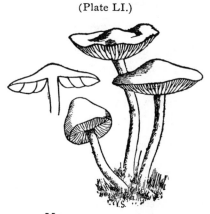

MARASMIUS OREADES.
About one-half natural size.

Its nearer relations are Collybia and Mycena.

Fries says that all Agaricaceæ having the smell of garlic are found in this genus. On the ground, but generally on wood or leaves.

Professor Peck reports over forty species of this genus found in New York state. Several not found in New York are reported from other states. The writer has found a few such species in Pennsylvania and West Virginia. Many untried species will probably prove to be edible; the majority are too small to be of food value. M. urens, reported poisonous, and M. peronatus, heretofore considered poisonous, have been found by the writer to be edible. Several species not described herein have been tested for edibility to a limited extent only.

In this genus occurs the famed M. oreades, the Mousseron of France, the Champignon and Scotch bonnet of England, the Fairy-ring mushroom of America.

ANALYSIS OF TRIBES.

COLLYBIA (inclining to Collybia). Page 223.

Flesh of pileus pliant, at length rather leathery, grooved or wrinkled, margin incurved at first. **Stem** somewhat cartilaginous; mycelium woolly, absent in some species.

221

Marasmius.

A. Scortei (*scorteus*, leathery). Page 223.

Stem solid or stuffed, then hollow, fibrous within, outside covered with down. Gills separating from the stem, free.

* Base of stem woolly or strigose.

** Stem naked at the base, often interwoven with twisted fibers

B. Tergini (*terginus*, leathery). Page 225.

Stem rooting, distinctly *tubular*, not *fibrous*, distinctly *cartilaginous*. Gills receding then free. Pileus thinner than in the preceding group, hygrophanous, even or with the margin striate.

* Stem woolly below, smooth above.

** Stem when dry covered with velvety down.

C. Calopodes (*Gr.*—beautiful; *Gr.*—a foot). Page 226.

Stem short, not rooting, often with a floccose or downy, tubercular base. Pileus convex, involute, then plane and more or less depressed, in which state the gills typically adnate are subdecurrent. On twigs, branches, etc. Gregarious.

* Stem quite smooth above, shining, base not swollen.

** Stem covered with velvety down, rather swollen at the base.

Mycena (inclining to Mycena). Page 227.

Stem horny, hollow, often filled with pith, tough, dry. Mycelium rooting, not floccose. Pileus somewhat membranaceous, bell-shaped, then expaned, margin at first straight and pressed to the stem.

A. Chordales (*chorda*, a gut). Page 227.

Stem rigid, rooting or dilated at the base. Pileus bell-shaped or convex. Type manifestly that of Mycena.

B. Rotulæ (*rotula*, a little wheel).

Stem thread-like, flaccid, base not dilated or floccose but appearing to enter the matrix abruptly. Pileus soon becoming plane or umbilicate. On leaves.

* Stem quite smooth, shining.

** Stem minutely velvety or hairy.

Apus (*a*, without; *pous*, a foot).

Pileus sessile, resupinate.

I.—COLLY'BIA.

A. SCORTEI.

* Stem woolly or strigose at base.

M. u'rens Fr.—*uro*, to burn. **Pileus** 2–3 in. broad, unicolorous, Marasmius. pale yellowish, *becoming pale*, slightly fleshy, *moderately compact at the disk, even*, but here and there scaly or cracked in wavy lines when dry, smooth, the thin margin involute. **Stem** 2–3 in. long, 3 lines thick, *solid*, composed of crisp tough fibers, rigid, equal, sometimes however ventricose, ½ in. thick, *everywhere clothed with white flocci, pale*, white-downy at the base. **Gills** free, united behind, *at length remote* from the stem, *distant*, tough, at first pale-wood-color, *then brown*.

Gregarious, somewhat cespitose. *Taste very stinging*. The stem is not strigosely sheathed at the base. *Fries*.

In mixed woods. Frequent. June to September.

A curious form occurred with the pileus turning very dark when full-grown. *B. and Br.* POISONOUS. Worthington Smith has tested it by accident. It produced headache, swimming of brain, burning in throat and stomach, followed by severe purging and vomiting. *Stevenson*.

Gregarious or cespitose. Taste very pungent, a feature which separates the present from M. oreades. Not coarsely tomentose at the base, as in M. peronatus, but only downy. *Massee*.

Spores 3x4µ *W. G. S.;* elliptical, 8x4µ *Massee*.

Pennsylvania, New Jersey, West Virginia. *McIlvaine*.

I have not known it to disagree with myself or friends. That it may not agree with some persons is unquestioned. Collectors should carefully test it upon themselves.

M. perona'tus Fr.—*pero*, a kind of boot. **Pileus** 1–2 in. and more broad, light yellowish or pallid brick-red, then becoming pale, *wood-color* or tan, at first fleshy-pliant, then *coriaceo-membranaceous*, convex then plane, obtuse, flaccid, slightly wrinkled, even at the disk, *at length pitted, striate at the margin*. **Flesh** white. **Stem** 2–3 in. long, 1–2 lines thick, stuffed, fibrous, tough, attenuated upward, *at length hollow* and compressed, *furnished with a bark*, light yellow then pallid, *cuticle villous* but separating and reddish when rubbed, somewhat incurved at the base, where it is *clothed with dense, somewhat strigose*, yellowish or

Marasmius. white *villous down*. **Gills** *adnexed, then separating*, free, moderately *thin*, and *crowded*, when young whitish, *pallid wood-color*, at length somewhat remote, reddish.

B. Woolly sheathed at the base. Taste acrid like that of M. urens, odor none. *Fries*.

In woods. Common. *Stevenson*.

Spores pip-shaped, 7x4μ *W.G.S.;* 10x6–7μ *Massee*.

New York. Thin woods. North Elba. August. September. *Peck*, 42d Rep.; West Virginia, June to December, West Philadelphia and Mt. Gretna, Chester county, Pa. *McIlvaine*.

M. peronatus is the wood-cousin of M. oreades. It is still reputed poisonous by all writers upon the subject, though M. C. Cooke gives it the benefit of a doubt. The name is given because of the base of the stem being densely covered with short hairs or a woolly down, and is thus easily recognized. It is common in woods, among decaying leaves, especially of the oak, from May until after frosts. It is usually solitary, but a few individuals are sometimes clustered. It is quite peppery to the taste, but pleasantly so. I have repeatedly eaten it, as have my friends. It loses its acridity in cooking, and though the caps are tougher than M. oreades, they make a highly flavored and delicious dish. Collectors should carefully test it for themselves.

*** *Stem naked at the base, etc.*

M. ore'ades Fr. *Gr*.—mountain-nymphs. Scotch bonnet. Champignon. Mousseron. (Plate LI, p. 221.) **Pileus** 1–2 in. broad, *reddish then becoming pale, absorbing moisture, whitish when dry, fleshy*, pliant, convex then plane, somewhat umbonate, even, smooth, slightly striate at the margin when moist. **Stem** 2–3 in. long, 1 ½ lines thick, *solid*, very tough, *equal*, tense and straight, *everywhere clothed with a villous-woven cuticle* which can be rubbed off, pallid; bluntly rooted at the base, naked, not villous or tomentose. **Gills** free, broad, *distant*, the alternate ones shorter, *at first soft*, then firmer, pallid-white.

Odor weak, but *pleasant*, stronger when dried, *taste mild*. Commonly growing in circles or rows. *Fries*.

Spores 6–7x5–6μ *K.;* elliptical, 8x5μ *Massee;* nearly elliptical, white, 7.6–9μ long *Peck*.

Common throughout the states during the summer **months after rains,** and in rings, but can be found from May until after frost. If one knows Marasmius. where the rings are to be found M. oreades can be gathered when shriveled, and are quite as good, after soaking, as when fresh.

M. oreades must be sought for where the grass is luxuriant. It hides among it. It is well worthy of the search. Raw, fresh or shriveled, it is sweet, nutty, succulent when eaten; stewed well it is delicious. Though tough its consistency is agreeable. The most delicate stomachs can digest it. The writer saved the life of a lovely woman by feeding her upon it when nothing else could be retained; and of another, by feeding Coprinus micaceus, after a dangerous operation. He introduced these species, together with a few others, into a large hospital in Philadelphia, where they were used with marked beneficial effect, and such use is now widespread.

When dried, by exposure to the air or sun, it can be kept indefinitely, neither losing its aroma or flavor, which it graciously imparts to soups or any other dish.

Collybia dryophila, Stropharia semi-globata, and Naucoria semi-orbicularis are sometimes found growing with it. These species are delicious and harmless.

Lafayette B. Mendel in the Am. Jour. of Physiology, March, 1898, gives the following analysis:

Twenty freshly gathered specimens (from New Haven) weighed 9 grams, an average weight of 0.45 grams each. The analysis gave:

Water 74.96%
Total solids...25.04
Total nitrogen of dry substance...................................... 5.97
Ash of dry substance...................................... 7.23

B. Tergini.

** *Stem downy when dry, etc.*

M. Wyn'nei B. and Br. **Pileus** 1–1½ in. broad, *lilac*-brown, tardily changing color, fleshy, convexo-plane, somewhat umbonate. **Stem** 2 in. long, 1½ line thick, tubed, *furfuraceous*, somewhat of the same color as the pileus. **Gills** adnexed, thick, distant, bright-colored, beautifully tinged with *lilac;* interstices even.

Inodorous. Gregarious or cespitose. The stem springs from a white mycelium, but is by no means strigose or tawny at the base. Quite distinct from M. fusco-purpureus. *Fries.*

225

Marasmius. Among leaves, twigs, etc. *Stevenson.*

Spores elliptical, 7–8x4µ *Massee.*

Kingsessing, West Philadelphia. Gregarious and cespitose, among leaves, etc., in oak woods. September to October, 1885.

This very pretty fungus very much resembles at first sight the small purplish Clitocybes, but is readily distinguished on examination. I ate the caps and enjoyed them during the seasons of 1885 and 1887, but have not seen the plant since.

The caps are equal to M. oreades.

C. CALOPODES.

* *Stem smooth, etc.*

M. scorodo'nius Fr. *Gr.*—a plant that smells like garlic. **Pileus** 2 in. and more broad, rufous when young, but soon becoming pale, whitish (not hygrophanous), slightly fleshy, pliant, convex then soon plane, obtuse, always arid; even when young, at length wrinkled and crisped. **Stem** 1 in. long, scarcely 1 line thick, *horny,* tough, tubed, equal, *very smooth throughout, shining, reddish, inserted and naked* at the base. **Gills** *adnate,* often separating, connected by veins, at length crisped in drying, whitish.

Commonly gregarious. *Readily distinguished from neighboring species by its strong odor of garlic. Fries.*

Heaths and dry pastures on twigs, etc. Rare.

Edible. Esteemed for flavoring. *Stevenson.*

Spores elliptical, 6x4µ *Massee.*

North Carolina, *Schweinitz, Curtis;* New England, *Frost;* New Jersey, *Ellis;* New York, August, *Peck,* 23d Rep.

M. ca'lopus Fr. *Gr.*—beautiful; a foot. **Pileus** about 4 lines broad, *whitish,* slightly fleshy, tough, convex then flattened, obtuse, rarely depressed, even, smooth, slightly wrinkled when dried. **Stem** 1 in. long, 1 line thick, tubed, slightly attenuated upward, even, *smooth,* tough, dull-red or *bay-brown-red,* shining, *somewhat rooted.* **Gills** slightly emarginate, in groups of 2–4, thin, white.

Inodorous. Almost smaller than M. scorodonius, but the stem is longer, otherwise very like it. *Fries.*

Spores elliptical, 7x4µ *Massee.*

226

Twigs and stems among fallen leaves in woods.   Ticonderoga.   Au- Marasmius. gust.

This might easily be mistaken for M. scorodonius, but it is without odor, and has a different insertion of the lamellæ.   It is sometimes cespitose.   The pileus in our specimens is whitish.   *Peck*, 31st Rep. N. Y. State Bot.

Because of its similarity to M. scorodonius, which is edible, it is given here.

II.—MYCENA.

A. CHORDALES.

M. allia′ceus Fr.—*allium*, garlic.   **Pileus** 1–1 ½ in. broad, whitish inclining to fuscous, often milk-white when young, somewhat membranaceous, campanulate then expanded, somewhat umbonate, even, at length striate and sulcate, smooth, dry.   **Stem** as much as 8 in. long, *horny*, rigid, fistulose, attenuated upward, *pruinato-velvety*, *blackish*, rooted at the base where it is somewhat incurved and naked.   **Gills** adnexed in the form of a ring, then *free*, slightly ventricose, arid, slightly distant, fuscous-whitish, crisped when dry.

Odor strong, of garlic, persistent.   *There is nothing of a reddish tinge in the whole plant.*   The stem is not tomentose at the base as in the Tergini.   *Fries.*

Among leaves and on rotten wood.   Frequent.   August to October. *Stevenson.*

Spores 14–16x8µ *Massee.*

North Carolina, *Schweinitz*, *Curtis;* Pennsylvania, *Schweinitz;* Minnesota, *Johnson;* Novia Scotia, *Somers.*

Edible.   Bull. Boston Myc. Club.

227

HELIOMYCES Lev.

Helios, the sun; *myces*, a fungus.

Heliomyces. **Pileus** membranaceous, between leathery and gelatinous, radiately sulcate. **Gills** equal, edge acute. **Stem** somewhat woody, cylindrical, central.

Allied to Marasmius, but differing in its sub-gelatinous substance. None reported edible.

LENTI'NUS Fr.

Lentus, tough or pliant.

Lentinus. **Pileus** fleshy-coriaceous, pliant, tough and hard when old, persistent.

(Plate LII.)

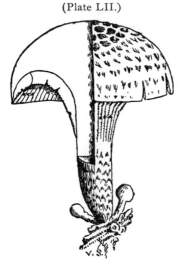

SECTION OF LENTINUS.

Gills becoming dry, tough, simple, unequal, thin, margin acute, *toothed*, more or less decurrent. **Stem** when present central, excentric or lateral, hard and firm, continuous with the flesh of the pileus.

Growing on wood.

Spores somewhat round, even, white.

Distinguished from other coriaceous genera by its serrated and torn gills.

"The genera Lentinus and Lenzites are found in every region of the world; their principal center, however, is in hot countries, where they attain a splendid development. On the contrary, toward the north they rapidly decrease in number." Fungi. *Cooke and Berkeley.*

In habitat and mode of growth Lentinus closely resembles Pleurotus, and parallel genera with colored spores. When young the species are inviting, and when well cooked are meal-giving. They are not delicacies, but substantials. They dry well. Grated they make soups, and give their pleasant flavor to any dish.

ANALYSIS OF TRIBES.

MESOPODES (*mesos*, middle; *pous*, a foot). Page 229.   Lentinus.
Stem distinct.

PLEUROTI (*pleura*, a side; *ous*, an ear).
Stem lateral or absent.   None known to be edible.

I.—MESO'PODES (center-stemmed).

L. Lecom'tei Fr.   **Pileus** coriaceous, funnel-shaped, regularly reflexed, hairy, tawny. **Gills** crowded, pallid. **Stem** short, hairy, tawny.
Common to the states.

Professor Peck writes to me: "This plant, by reason of its rather
tough substance, has commonly been referred to Lentinus, under the
name L. Lecomtei Schw., but this reference is scarcely satisfactory to
me, since the edge of the lamellæ is scarcely at all serrate as required
by that genus. It seems to me it would go better under the genus
Panus. It is variable—sometimes eccentric or even lateral. It is sometimes called Lentinus strigosus, but I do not think the two are distinct
species, however distinct they may be in form."   February 26, 1894.

Like all Lentinus the present species is rather tough, yet chopped
into small pieces, well cooked and seasoned, it is quite equal to P.
ostreatus and many others of high renown.

L. tigri'nus Fr.—*tigris*, a tiger. From the markings. **Pileus** commonly 2 in. broad, white, *variegated*
with somewhat adpressed, *blackish,*
hairy squamules, fleshy-coriaceous,
thin, commonly orbicular and central, at first convexo-plane, *umbili-*
cate, at length funnel-shaped, often
split at the margin when dry. **Stem**
about 2 in. long, *thin*, solid, very
hard, commonly attenuated downward, minutely *squamulose*, whitish,
often ascending and becoming dingy-
brown at the base, at first furnished

(Plate LIII.)

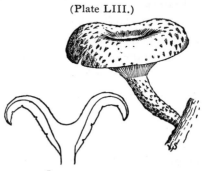

LENTINUS TIGRINUS.
About one-half natural size.

at the apex with an entire reflexed ring, which soon falls off. **Gills** de-

Lentinus. current (*by no means sinuate*), narrow, crowded, unequal, toothed like a saw, white.

Somewhat gregarious, even cespitose, thinner and more coriaceous and regular than L. lepideus B., wholly blackish with squamules. *Fries.*

On old stumps. Rare. *Stevenson.*

When fresh very tender and easily torn, when dry coriaceous. *Sow.* Smell strong, acrid, like that of some Lactarii. *M. J. B.*

Spores 6.6x3.3μ *Morgan;* elliptical, smooth, 7x3.5μ *Massee.*

Agreeable taste and odor, eaten in Europe. *Roques.*

Edible, tough when old and never very delicate or digestible.—*M. C. Cooke.*

Not found in sufficient quantity to test.

L. lepi′deus Fr. *Gr.*—scaly. ([Color] Plate XVI, figs. 3, 4.)   **Pileus** 2–4 in. broad, pallid-ochraceous, variegated with adpressed, *darker, spot-like scales, fleshy*, very *compact* and firm, *irregular*, commonly excentric, convex then depressed, but not truly umbilicate, sometimes broken up into cracks. **Flesh** pliant, white.   **Stem** short, commonly 1 in. long, solid, *stout, very irregularly formed*, almost woody, tomentose-scaly, whitish, rooted at the base, *at the first furnished with a veil toward the apex*.   **Gills** decurrent, but *sinuate behind*, crowded broad, transversely striate, whitish, edge torn into teeth.

Odor pleasant. *Fries.*

Spores 11x5μ *W.G.S.*, 7x3μ *Massee.*

Lentinus lepideus is a sort of commercial traveler. It is common wherever railroads are. It is partial to oak ties and its mycelium is injurious to them. It is found upon pine and other timbers. The writer has collected large clusters of it from oak sawdust. The European plant is noted as "almost always solitary." In the United States it is seldom so. It is noted as growing in damp, dark places, but it loves the sun.

As a food it is about on a par with P. ulmarius, not as tough, but harder when old. It is a reliable species from spring until late autumn, is persistent and dries well. It is neat, handsome, prolific. When young it makes a good dish, and when old can be used to advantage in soups.

L. cochlea′tus Fr.—*cochlea*, a snail. **Pileus** 2–3 in. broad, flesh- Lentinus.
color, but becoming pale, somewhat tan, fleshy-pliant, thin, commonly excentric, imbricated, very unequal, somewhat lobed or contorted, sometimes plane, sometimes funnel-shaped-umbilicate, but not pervious, *smooth*. **Stem** solid, firm, sometimes central, most frequently excentric, sometimes wholly lateral, *always sulcate, smooth*, flesh-colored upward, reddish-brown downward. **Gills** decurrent, crowded, serrated, white-flesh-color. *Fries*.

Pliant, tough, flaccid, very changeable in form, sometimes solitary, sometimes cespitose, imbricated, growing into each other. From very small forms which are commonly solitary, with the stem and pileus scarcely 1 in. it ranges to 3 in.

On stumps. Frequent. August to October.

According to Fries the odor is weak, of anise; but it is generally strong and very pleasant. *Stevenson*.

Spores nearly globular, 4µ diameter *Morgan;* spheroid or ellipsoid-spheroid, uniguttate, 4–6µ *K.;* almost globular, 4µ *W.G.S.*

The dense clusters of all sized members are usually plenty in favored localities. It is inviting in appearance, taste and spicy odor. It retains a suspicion of the latter when cooked which gives the dish a flavor pleasant to many. It must be young to be tender. When dry—like others of its kind—it can be grated and used in many ways.

L. Un′derwoodii Pk. **Pileus** fleshy, tough, convex or nearly plane, the glabrous surface cracking into areola-like scales which are indistinct or wanting toward the margin, whitish or slightly tinged with buff or pale ochraceous. **Flesh** white. **Gills** moderately close, decurrent, slightly connecting or anastomosing at the base, somewhat notched on the edge, whitish, becoming discolored in drying. **Stem** stout, hard, solid, eccentric, squamose, colored like the pileus. **Spores** oblong, 13–15x5–6.5µ.

Plant cespitose. **Pileus** 3–6 in. broad. **Stem** 1.5–3 in. long, about 1 in. thick.

This differs from L. magnus in its cespitose habit, eccentric stem, longer spores, less distinctly areolate-squamose pileus and in its habitat. The gills are connected at the base very much like those of Pleurotus ostreatus. *Peck*, Torr. Bull. Vol. 23, No. 10.

North Carolina, Pennsylvania, *McIlvaine*.

Lentinus. The writer first met with it in North Carolina, near Washington, on oaks and railroad timbers, and in Fairmount Park, Philadelphia. It attains quite a size, grows singly and in clusters. Its clean, cake-like appearance is attractive. Cooked it ranks with P. ulmarius, L. lepideus, and Panus strigosus.

PA'NUS Fr.

A name given to a tree-growing fungus by Pliny.

Panus. Whole fungus between fleshy and leathery, tough, not woody, texture fibrous. **Gills** unequal, tough, becoming leathery, edge acute and unbroken. **Stem** present or absent.

(Plate LIV.)

Growing on wood. Various in form, lasting long. Allied to Lentinus but differing in the tough and very entire gills.

Spores even, white.

ANALYSIS OF SPECIES.

PANUS TORULOSUS.
About one-fourth natural size.

* Stem excentric.

** Stem lateral.

*** Stem absent. Pileus resupinate or dimidiate.

Species of this genus are among our most observable fungi. Their settlements are frequent on decaying trees, stumps, branches, on fences, cut timber, etc. Most of them are small, but their coriaceous build prevents their shrinking in cooking. Most species have a pleasant farinaceous taste and odor, which they yield, together with a gummy substance, to soups and gravies.

Tasting a small piece will immediately tell, if the species is not known, whether it is edible or of the styptic kind.

* Stem excentric.

P. concha'tus Fr.—Formed like *concha*, a shell-fish. **Pileus** about

                    PLATE LV.

PANUS STRIGOSUS.

3 in. across, tough and flexible, unequal, excentric or dimidiate, margin Panus. often lobed, cinnamon-color becoming pale, at length more or less scaly. **Flesh** thin. **Gills** narrow, forming decurrent lines on the stem, somewhat branched; pinkish-white then pale-ochraceous. **Stem** about ⅔ in. long, 3–4 lines thick, solid, unequal, pale, base downy. *Massee*.

On trunks of beech, poplar, etc.

Often imbricated and more or less grown together. Allied to Panus torulosus, but distinguished by the much thinner pileus, more expanded and excentric, also dimidiate, flaccid, cinnamon becoming pale, but the form not constant. **Stem** about ½ in. long, 4 lines thick, often compressed, downy at the base. **Pileus** 2–4 in. broad, scaly when old. **Gills** decurrent in long, parallel lines, not at all resembling those of Pleurotus ostreatus, which anastomose behind, but frequently unequally branched, at first whitish or pale flesh-color, then wood-color, crisped when dry. *Fries*.

Always known by its shell-like form and its tough substance.

Sent to the writer by Mr. E. B. Sterling, Trenton, N. J. September, 1898.

The appearance of scales upon the pileus was scarcely noticeable. Taste pleasant. The fungus is tough when old, but yields an excellent gravy.

P. torulo′sus Fr.—a tuft of hair. (Plate LIV, p. 232.) **Pileus** 2–3 in. broad, somewhat flesh-color, but varying reddish-livid and becoming violet, *entire*, but very excentric, fleshy, somewhat compact when young, *plano-infundibuliform, even*, smooth. **Flesh** pallid. **Stem** short, commonly 1 in., solid, oblique, tough, firm, commonly with gray, but often violaceous *down*. **Gills** decurrent, somewhat distant, simple, separate behind, reddish then tan-color.

Very changeable in form, at first fleshy-pliant, at length coriaceous. In the covering of the stem it approaches Paxillus atro-tomentosus, but there is no affinity between them. *Fries*.

On old stumps.

Spores 6x3μ *W. G. S.*

North Carolina, *Curtis;* Massachusetts, *Frost;* Minnesota, *Johnson;* Kansas, *Cragin;* New York, *Peck*, Rep. 30.

Much esteemed in France, *W.D.H.* Edible, but tough. *M.C.C.*

233

Panus. **P. lævis** B. and C.—light. **Pileus** 3 in. broad, orbicular, slightly depressed, white, clothed in the center with long, intricate, rather delicate hairs, which are shorter and more matted toward the inflected margin; substance rather thin. **Stem** 3 in. high, ½ in. thick, attenuated upward, generally excentric, sometimes lateral, not rooting, solid, hairy below like the margin of the pileus. **Gills** rather broad, entire, decurrent, but not to a great degree, the interstices even above, behind clothed with the same coat as the top of the stem. **Spores** white.

On oak and hickory trunks.

A most distinct species, remarkable for its great lightness when dry and the long villous but not compressed or compound flocci of the pileus. Sometimes the center of the pileus becomes quite smooth when old.

One of the prettiest of fungi. The markings upon the white margin are more precise than those of the finest bee comb. One does not tire looking at the work of Nature's geometrician. It is not plentiful, but is of useful size. It has good flavor and cooks quite tender.

P. strigo'sus B. and C.—covered with stiff hairs. **Pileus** white, excentric, clothed with coarse strigose pubescence, margin thin. **Stem** strigose like the pileus. **Gills** broad, distant, decurrent. Allied to P. lævis.

(Plate LV*a*.)

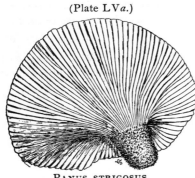

PANUS STRIGOSUS.
One-third natural size.

Pileus 8 in. broad. **Stem** 2–3 in. long, 1 in. or more thick.

On oak stumps.

Decaying wood of deciduous trees. September.

It is remarkable for its large size and the dense hairy covering of the pileus and stem. *Peck*, 26th Rep. N. Y. State Bot.

A remarkably handsome fungus. A specimen taken from a cluster growing upon an apple tree measured 10 in. across. Its creamy whiteness, and short hairy stem make it unmistakable among other tree-fungi.

When very young it is edible, but soon becomes woody. Even when aged it yields a well flavored gravy.

** *Stem lateral.*

P. farina'ceus Schum.—*farina*, meal. From the scurf on the pileus.
Pileus cinnamon-umber, somewhat coriaceous, flexuous, cuticle separating into whitish-bluish-gray scurf. **Stem** short, lateral, of the same color as the pileus. **Gills** determinately free, distinct, paler.
The habit is that of P. stipticus. *Stevenson.*
Pennsylvania, A. pleurotus f., *Schweinitz;* Ohio, *Morgan.*
Var. albido-tomentosus. See Panus albido-tomentosus.

P. al'bido-tomento'sus CKE. MASS.—*albidus*, white; *tomentum*, down. **Pileus** about ⅔ in. long, ½ in. broad, horizontal, sometimes imbricated, semi-circular, subcoriaceous, flexuous or regular, pale umber, densely clothed with a short, whitish, velvety down, which seems to be persistent, but thinner and shorter toward the shortly incurved margin. **Stem** lateral, very short, or entirely absent, and attached by a downy base. **Gills** radiating from the point of attachment; narrowed behind, lanceolate, honey-colored, margin entire, rigid, scarcely crowded, shorter ones intermixed. **Spores** subglobose, smooth, 5μ diameter.
On trunks and branches.
Pileus about 1 in. broad, often in imbricated tufts. It is doubtful whether this is not a distinct species from the type described by Fries. *Cooke and Massee.*
Panus albido-tomentosus is given by Cooke and Massee as a variety of Panus farinaceus. The writer decides to give it place as a species.
It has been sent to me by Mr. H. I. Miller, from Terre Haute, Ind., by Dr. E. L. Cushing, Albion, N. Y., Miss Madeleine Le Moyne, Washington, Pa. I have found it in West Virginia, New Jersey and many parts of Pennsylvania. It is plentiful in patches upon branches and boles of deciduous trees. Long, slow cooking makes it tender. It makes a luscious gravy after thirty minutes' stewing.

*** *Stem absent, pileus resupinate or dimidiate.*

P. betuli'nus Pk.—*betula*, birch. **Pileus** thin, suborbicular or dimidiate, nearly plane, glabrous, prolonged behind into a short stem, grayish-brown, darker or blackish toward the stem. **Gills** narrow, close, decur-

Panus. rent, whitish. Stem adorned with a slight tawny hairiness which is more fully developed toward the base. Spores minute, 4–5x1.5–2µ.

Decaying wood of birch. Newfoundland. October, *Rev. A. C. Waghorne. Peck*, Bull. Torrey Bot. Club, Vol. 23, No. 10.

Common in West Virginia mountains on birches, 1882; found at Eagle's Mere, Pa., August, 1898. Quite plentiful on decaying birch trees, which abound there. Size from ½–1½ in. across.

Eaten raw it has a gummy quality and very pleasant nutty flavor. I did not have opportunity to cook it, but regard it as a species well worth trying.

P. stip'ticus Fr.—*stypticus*, astringent. Pileus ½–1 in. broad, cinnamon becoming pale, arid, thin, but not membranaceous, kidney-shaped, pruinose, the *cuticle separating into furfuraceous scales*. Stem not reaching 1 in. long, solid, *definitely lateral*, compressed, *dilated upward*, ascending, pruinose, paler than the gills. Gills ending determinately (not decurrent), thin, very narrow, crowded, *elegantly connected by veins*, cinnamon. *Fries*.

Gregarious, cespitose, remarkable for *its astringent taste*. The pileus sometimes has a funnel-shaped appearance with lobes all around.

On stumps, etc. Common. August to February.

Reckoned poisonous. *Stevenson*.

Spores obovoid-spheroid, 2–3x1–2µ *K.;* 3x4µ *W.G.S.*

Plentiful and general. The markings upon the cap in moist weather are sometimes exquisitely regular.

The immediate and lasting unpleasantness of this fungus to mouth and throat, whether cooked or raw, will cancel all desire to eat of it forevermore. A nibble will detect it. It is reckoned poisonous, and may be. No one but a determined suicide would resort to it. Dr. Lambotte asserts that it is a violent purgative.

XER'OTUS Fr.

Gr.—dry; **Gr.**—an ear.

(Plate LVI.)

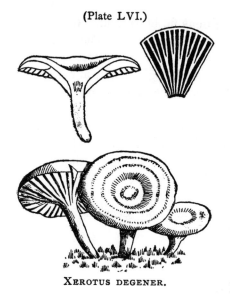

Hymenophore continuous with the stem, descending into the trama which is homogeneous with the *coriaceous pileus.* **Gills** coriaceous, broadly plicæform, dichotomous, edge quite entire, obtuse. *Rigid, persistent, analogous with the Cantharelli, but differing in the whole structure. Fries.*

The gills are more distant than in any species of Agaricaceæ.

None edible.

XEROTUS DEGENER.

TRO'GIA Fr.

After *Trog*, a Swiss botanist.

Gills fold-like, edge longitudinally channelled (in the single Eu- ropean species only crisped). In other respects agreeing with Xerotus. *Soft, flaccid, but arid and persistent, texture fibrillose. Fries.*

Reviving when wet. **Spores** white. *Stevenson.*

Spores elongated or cylindrical.

American representative, Trogia crispa, var. variegata.

(Plate LVIII.)

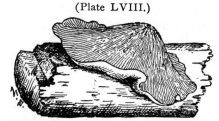

TROGIA CRISPA.
Natural size.

Pileus and gills variegated with bluish or greenish-blue stains. Sandlake. September. *Peck*, 38th Rep. N. Y. State Bot.

Not edible.

237

SCHIZOPHYL'LUM Fr.

Gr.—to split; *Gr.*—a leaf.

Schizophyllum.   (Plate LVIII *a.*)

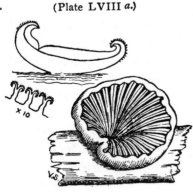

SCHIZOPHYLLUM COMMUNE.

Pileus fleshless, arid. **Gills** coriaceous, fan-wise branched, united above by the tomentose pellicle, bifid, split longitudinally at the edge. **Spores** somewhat round, white. *Fries.*

The two lips of the split edge of the gills are commonly revolute. The farthest removed of all the Agaricini from the type.

Growing on wood. *Stevenson.*

Common on decaying wood. Tough.

LENZITES Fr.

After *Lenz*, a German botanist.

Lenzites. **Pileus** corky or coriaceous, texture arid and floccose. **Gills** coriaceous, firm, sometimes simple and unequal, sometimes anastomosing and forming pores behind, trama floccose and similar to the pileus, edge somewhat acute. The European species are dimidiate, sessile, persistent, growing on wood, quite resembling Dædalea. *Fries.*

Allied most nearly to Trametes and Dædalea and forming as it were the transition from Agaricaceæ to Polyporaceæ. In tropical countries they are more woody in texture. *Stevenson.*

Very common. None edible.

(Plate LVII.)

RHODOSPORAE.

Hymenophore distinct from fleshy stem.

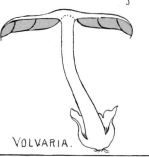

VOLVARIA.

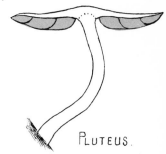

PLUTEUS.

Hymenophore confluent and homogeneous with fleshy stem.

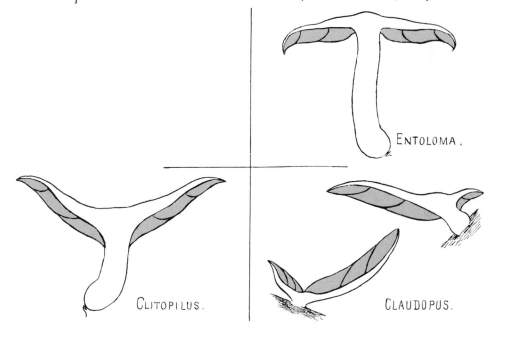

ENTOLOMA.

CLITOPILUS.

CLAUDOPUS.

Hymenophore confluent with, but heterogeneous from cartilaginous stem.

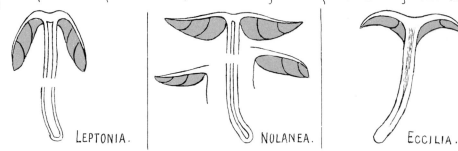

LEPTONIA.

NOLANEA.

ECCILIA.

CHART OF GENERA IN PINK-SPORED SERIES—RHODOSPORAE.

Series II. **RHODOSPORÆ.** *Gr.*—rose; *Gr.*—seed. Or **HYPORHO'DII—** *hypo*, under; *rhodon*, rose.

Spores pink or salmon-color.

In Volvaria, Pluteus and most of Clitopilus, the spores are regular in shape, as in the white-spored series, in the rest of the subgenera they are generally angular and irregular.

Though European writers, generally, condemn the rosy-spored series as inedible, a few of our best American edibles are found in it—notably Pluteus cervinus.

VOLVA'RIA   Fr.

Volva, a wrapper.

Spores regular, oval, pink, or salmon.   **Veil** universal, forming a Volvaria. perfect *volva*, distinct from the cuticle of the pileus.   **Stem** separating easily from the pileus.   **Gills** *free*, rounded behind, at the very first white then pinkish, soft.   Analogous with Amanita.

Growing in woods and on rich mold, rotten wood and damp ground, hence often found in hot-houses and gardens.   **V.** Loveiana Berk. is parasitic on Clitocybe nebularis.

There are thirteen species reported from different parts of the United States.   Most of them grow upon wood.   Two species have previously been reported as edible, to which I have added V. Taylori, tested by myself.

One species, V. gloiocephala, is upon the authority of Letellier, given as poisonous.   It is found in several parts of the United States, but no comment has been made upon its edibility.   I have not seen it. A careful study of its botanic characters is urged.   It should be regarded as poisonous until its reputation is cleared up, as it probably will be.

ANALYSIS OF SPECIES.

* Pileus dry, silky or fibrillose.

** Pileus more or less viscid, smooth.

239

*Pileus dry, silky or fibrillose.

Volvaria.

V. bombyci'na Schaeff.—*bombyx*, silk. Pileus 3–8 in. broad, *wholly white*, fleshy, soft, at first globose, soon bell-shaped, at length convex, somewhat umbonate, *everywhere silky* or, *when older, hairy-scaled*, more rarely becoming smooth at the vertex. Flesh not thick, white. Stem 3–6 in. long, ½ in. thick or more at the base, solid, equally attenuated from the base to the apex, even, smooth, white. Volva soon torn asunder, ample, 2–3 in. broad, membranaceous, lax, slashed, somewhat viscid, persistent. Gills free, very crowded when young, almost cohering, ventricose, in groups of 2–4, then toothed, flesh-colored.

(Plate LIX*a*.)

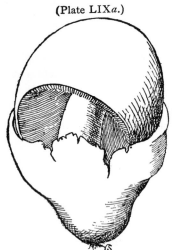

VOLVARIA BOMBYCINA.
Natural size.

Ovate when young. According to some becoming brownish. The stem is curved-ascending on vertical trunks and straight on prostrate ones. Commonly solitary, sometimes however cespitose. *Stevenson*.

Spores elliptic, smooth, 6–7x4µ *Massee;* 6–8µ *Lloyd*.
Considered edible. *Stevenson.* Edible. *Curtis.*

Very general but not common over the United States. It is a large plant, from 3 in. upward across cap. Growing from wood, oaks, maples, beech, etc.

The writer has not been successful in finding it. Drawing, spore-print and description received from *H. I. Miller*, Terre Haute, Ind.

Upon such an authority as the late Dr. Curtis there is no doubt of its edibility.

V. volva'cea Bull.—*volva*, a wrapper. Pileus 2–3 in. across. Flesh white, thick at the disk, very thin elsewhere, soft, bell-shaped then expanded, obtuse, grayish-yellow, virgate or streaked with adpressed blackish fibrils. Gills free, about 2 lines broad, pale flesh-color. Stem 2–4 in. long, about 4 lines thick, almost equal, white, solid.

240

Plate LIX.

VOLVARIA BOMBYCINA.

Photographed by Dr. J. R. Weist.

Volva large, loose, whitish. **Spores** smooth, elliptical, 6–8x3.5–4μ; no cystidia. *Massee.*

(Plate LX.)

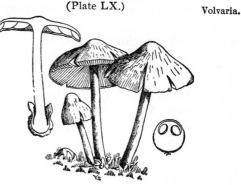

On the ground by roadsides, etc., also in stoves.

Allied to V. bombycina, but constantly different in the less ample and less persistent, brownish volva. **Pileus** 3 in. across, rarely more, gray, elegantly virgate with blackish fibrils; flesh-color of the gills not so pure. *Fries.*

VOLVARIA VOLVACEA.
Two-fifths natural size.

Once found in woods at roots of a tree. It occurs every year in the cellar of our drug store. *Lloyd* "Volvæ."

North Carolina, *Schweinitz;* Minnesota, *Johnson;* Ohio, *Morgan.*
Probably edible, should be carefully tested.

V. Tay'lori Berk. **Pileus** 1¾ in. high and broad, livid, conico-campanulate, obtuse, striately cracked from the apex, thin, margin lobed and sinuated. **Stem** 2½ in. long, ¼ in. thick, pallid, solid, nearly equal, slightly bulbous at the base. **Volva** date-brown, lobed, somewhat lax, small. **Gills** uneven, broad in front, very much attenuated behind, rose-color.

Pileus beautifully penciled and cracked. The dark volva, bell-shaped pileus, and uneven, attenuated gills are marked characters. The habit is rather that of some Entoloma than of its more immediate allies. *Fries.*

Spores 6x9μ *W.G.S.;* broadly elliptical, smooth, 5x3.5–4μ *Massee.*
Indiana, *Mrs. L. H. Cox;* West Philadelphia, in much decayed stump of maple. *McIlvaine.*

Caps 1½–2 in. across and beautifully penciled and cracked. **Stem** 1½–3 in. long. **Gills** up to ⅛ in. wide. The spores when shed in body are a beautiful maroon. Resembling V. volvacea, but lighter in color, and having a brown volva. Specimens sent me by J. J. Newbaker, Steelton, Pa., had snow-white caps and when young were velvety to the touch. Gills tinged with pink; volva dark brown.

The few specimens eaten were of good flavor, somewhat resembling Pluteus cervinus.

Agaricaceæ

** *Pileus more or less viscid, smooth.*

Volvaria. **V. specio′sa** Fr.—*speciosus*, handsome. **Pileus** 3–5 in. broad, whitish, *gray* or umber *at the disk*, fleshy, globose when young, then bell-shaped, at length plane and somewhat umbonate, even, *smooth, gluey.* **Flesh** soft, floccose, white. **Stem** 4–8 in. long, as much as 1 in. thick, solid, firm, slightly attenuated from the base as far as the apex, when young, *white-villous* and tomentose at the base, then becoming smooth, white. **Volva** bulbous rather than lax, free however, variously torn into loops, membranaceous, ½–1 in. broad, externally tomentose, white. **Gills** free, flesh-colored.

The gills are wholly the same as those of A. bombycinus. It occurs also thinner, with the pileus wholly gray. *Fries.*

Spores 12–18x8–10µ *K.;* elliptical or subglobose, smooth, 14–16x 8µ *Massee.*

Distinguished by the whitish, viscid pileus, and the downy volva and stem. *Massee.*

"Common in cultivated soil, especially grain fields and along roads. A fine edible agaric and our most abundant one in California." *McClatchie.* Volvæ, U. S., Lloyd.

V. gloioceph′ala Dec. Fl. *Gr.*—sticking; head. **Pileus** dark opaque brown, fleshy, bell-shaped then expanded, umbonate, smooth, *glutinous*, striate at the margin. **Stem** solid, *smooth*, becoming brownish or tawny; the *volva*, which is *circularly split*, pressed close. **Gills** free, reddish.

Fragments of the volva are sometimes seen on the pileus. The stem is commonly more slender than that of A: speciosus. *Fries.*

On the ground. Uncommon. June to October. *Stevenson.*

Pileus about 3 in. across, with a strong regular, obtuse umbo in the center, of a delicate mouse-gray, viscid when moist, but when dry shining, quite smooth, margin striate in consequence of the thinness of the flesh. **Stem** 6 in. or more high, about ½ in. thick in the center, attenuated upward, bulbous at the base, clothed with a few slight fibers, easily splitting, solid, rather dingy, ringless. **Volva** loose, villous like the base of the stem, splitting into several unequal lobes; the gills are broad, especially in front, narrower behind and quite free, so as to leave a space round the top of the stem, white, tinged with grayish-pink;

242

margin slightly toothed. Smell strong and unpleasant, and taste disa- Volvaria. greeable. *M.J.B.* VERY POISONOUS according to Letellier. *Stevenson.*

Spores 19x9μ *W.G.S.;* elliptical, smooth, 10–12x6–7μ *Massee.*

Distinguished by the smoky, glutinous pileus. The measurement of the spores as given by Saccardo (19x9μ) is certainly too large, and is probably an uncorrected error. *Massee.*

North Carolina, *Curtis;* South Carolina, *Ravenel;* Ontario, *Dearness;* California, *Harkness and Moore;* Ohio, *Morgan;* Mississippi, Minnesota, *Johnson.*

PLUTEUS Fr.

(*Pluteus*, a shed. From the conical shape of the pileus.)

Stem fleshy, distinct from the pileus. **Gills** free, rounded behind Pluteus. (never emarginate), at first cohering, white, then colored by the spores.

Generally growing on or near trunks of trees.

Resembling Volvaria in all respects but the volva. **Spores** rosy.

Several of the genus are edible. Pluteus cervinus is one of our earliest, persistent, plentiful, delicious food species. The caps of those tested are tender, easily cooked and best fried.

ANALYSIS OF SPECIES.

*Cuticle of the pileus separating into fibrils or down, which at length disappear.

**Pileus frosted with atoms, somewhat powdery.

***Pileus naked, smooth.

*Cuticle of pileus fibrillose, etc.

P. cervi′nus Schaeff.—*cervus*, a deer. ([Color] Plate LXI, fig. 1.) **Pileus** fleshy, at first campanulate, then convex or expanded, *even, glabrous, generally becoming fibrillose or slightly floccose-villose* on the disk, occasionally cracked, variable in color. **Lamellæ** broad, somewhat ventricose, at first whitish, then flesh-colored. **Stem** equal or slightly tapering upward, firm, solid, fibrillose or subglabrous, variable in color. **Spores** broadly elliptical, 6.5–8x5–6.5μ.

Pluteus. **Plant** 2–6 in. high. **Pileus** 2–4 in. broad. **Stem** 3–6 lines thick.

The typical form has the pileus and stem of a dingy or brown color and adorned with blackish fibrils, but specimens occur with the pileus white, yellowish, cinereous, grayish-brown or blackish-brown. I have never seen it of a true cervine color. It is sometimes quite glabrous and smooth to the touch and in wet weather it is even slightly viscid. It also occurs somewhat floccose-villose on the disk, and the disk, though usually plane or obtuse, is occasionally slightly prominent or subumbonate. The form with the surface of the pileus longitudinally rimose or chinky is probably due to meteorological conditions. The gills, though at first crowded, become more lax with the expansion of the pileus. They are generally a little broader toward the marginal than toward the inner extremity. Their tendency to deliquesce is often shown by their wetting the paper on which the pileus has been placed for the purpose of catching the spores. The stem is usually somewhat fibrous and striated but forms occur in which it is even and glabrous. When growing from the sides of stumps and prostrate trunks it is apt to be curved. Two forms deserve varietal distinction.

Var. *al'bus*. Pileus and stem white or whitish.

Var. *al'bipes*. Pileus cinereous yellowish or brown. Stem white or whitish, destitute of blackish fibrils.

In Europe there are three or four forms which have been designated as species under the names of A. rigens, A. patricius, A. eximius and A. petasatus, but Fries gives them as varieties or subspecies of A. cervinus, though admitting that they are easily distinguished. None of these have occurred in our state. *Peck*, 38th Rep. N. Y. State Bot.

Var. *visco'sus*. The normal character of the cuticle of the species is slightly viscid in wet weather, but the specimens we collected and photographed were exceedingly viscid. They also differed from the normal form in their lighter color, flesh much thicker at the disk and thin at the margins, and cuticle not appearing fibrillose. It is close to petasatus, but differs, however, in its narrower gills and in having no striæ. It is a good variety if it is not a good species. *Lloyd*, Myc. Notes.

Spores 7–8x5–6μ *K.;* 6–8x4–5μ *B.;* 4x5μ *W.G.S.;* 5.8x4.6μ *Morgan*.

Frequent on decaying stumps, roots and wood, May to frost. *McIlvaine*.

Its free gills should distinguish it from any Entoloma, though both have pink spores and eventually pink gills. Among the earliest of

large species.   The sight of it is stimulating to the mycophagist.   He **Pluteus.**
then knows the toadstool season to be truly opened.

Caps only are tender.   The stems are edible, but they are not of the
same consistency as the caps, therefore will not cook with them. Fried
in a buttered pan or broiled, they are exceedingly toothsome.

In October, 1898, a beautiful variety   ([Color] Plate LXI, fig. 2.)
occurred which I had not previously seen.   It was sent by me to Professor Peck.   The plants grew in large clusters from rotting, refuse
straw in the ruin of a stable; the white, cottony mycelium running
upon and through the straw.   The solid stems of some were straight,
others curved, ranging from 2–6 in. long, the taller ones tapering from
base to spindling apex, the shorter ones decidedly bulbous and ending
abruptly.   They were twisted and delicately marked.   These markings
break up into dark thread-like fibrils, leaving the stem striate and satin-
glossy.   **Pileus** from 2–4 in. across, dark Vandyke-brown when young,
lighter in age, streaked, glossy.   **Gills** at first white, tardily changing
to light salmon color, broad, ventricose, free.

Taste and smell pleasant of almonds.   Good, delicious.

Professor Peck wrote of it: ''It has the general appearance of
Pluteus cervinus, but these specimens seem to depart from the usual
form of growing in clusters from the ground, and in having an almond
flavor.   Without knowing more about it I would scarcely feel justified
in separating it from such a variable species.   As Fries sometimes remarks concerning variable species: Perhaps several species are concealed under the one name, but a pretty full and accurate knowledge of
them is desirable if one is to split them up.''

This is excellent judgment.   While I believe the above to be a distinct species, the disposition to make new species of varieties is regrettable in many botanists.

Var. *Bul'lii* Berk., MS.   **Pileus** 4–6 in. across, flesh thick, convex
then expanded, smooth, even, pallid, the disk darker.   **Gills** free,
rounded behind, rather distant from the stem, crowded, ½ in. broad,
pale salmon-color.   **Stem** 3–4 in. long, 1 in. and more thick, slightly
swollen at the base, fibrillose, pale brown, darkest at the base, solid.
Massee.

Pileus 6 in. across, expanded from bell-shape, ashy-white (oyster
color), glossy, like floss silk, silky fibrillose, irregularly corrugated.
Skin separable.   **Flesh** spongy, pure white, like shreds of cotton, sep-

Pluteus. arable into plates, very brittle, ½ in. thick at stem, immediately thinning to ⅛ in., very thin toward margin. **Gills** thin, elastic, rounded behind, close to stem, free, ½ in. wide, close, alternate short and long, white, then tinged and spotted pink with spores which when cast in mass are a pinkish-brown with slight lavender shade. **Stem** 5 in. long, ½–¾ in. thick, subequal, spreading at top, white, silky-fibrillose, changing to very light yellowish brown from center to base, exterior hard, skin thin, tough, interior filled with continuous, cottony fibers, snow-white, brittle, watery, slightly swollen at base. Taste pleasant.

Mt. Gretna, Pa., July, 1898, on chestnut stump and in woods on ground among leaves. Leaves adhere to base of stem which is powdery-white. *McIlvaine.*

Cooked, it is as good as P. cervinus.

Var. *petasa'tus* Fr. **Pileus** 3–4 in. across, flesh rather thick, campanulate then expanded, umbonate, grayish-white, very smooth, with a viscid cuticle, at length striat to the middle. **Gills** free, ½ in. and more broad, crowded, becoming dry, white then reddish. **Stem** 4–5 in. long, ½–⅔ in. thick, rigid, very slightly and equally attenuated from the base, whitish, fibrillosely striate, solid.

On heaps of straw and dung, sawdust, etc.

Color verging on bay when old. Stem and margin of gills at length with a tawny tinge. *Fries.*

Haddonfield, New Jersey, Bell's Mill, sawdust, 1890; Mt. Gretna, Pa., August, 1898, among sawdust from ice-house. **Caps** 6 in. across. **Stem** easily split, exterior hard, fibrillose, streaked, whitish, shining, stuffed with cottony fibers. **Spores** dark pink. *McIlvaine.*

Equal to P. cervinus.

P. umbro'sus Pers.—shady, from its dark color. **Pileus** fleshy, at first bell-shaped, then convex or expanded, *roughly wrinkled* and more or less villose on the disk, fimbriate on the margin, *blackish-brown.* **Gills** broad, somewhat ventricose, at first whitish, then flesh-colored, *blackish-brown and fringed or toothed on the edge.* **Stem** solid, colored like or paler than the pileus, fibrillose or villose-squamose. **Spores** elliptical, 8x5μ.

Decaying woods and swamps, especially of pine, both in shaded and open places. Not rare. *Peck,* 38th Rep. N. Y. State Bot.

Spores broadly elliptical, smooth, 6–7x5µ; cystidia ventricose, 65– Pluteus. 75x18–20µ *Massee.*

New York, *Peck*, Rep. 32, 38; West Virginia, Pennsylvania, North Carolina, New Jersey, frequent on decaying logs, stumps, pine and other woods. *McIlvaine.*

At times the caps are a deep sepia-brown. It is readily distinguished from P. cervinus by the wrinkled, downy disk of the cap and the gills having dark-brown edges. Smell rather strong. Professor Peck says he has not seen it with the margin fimbriate. Neither have I, though this is prominent in the European species.

P. umbrosus is a fine species, equal in every way to P. cervinus, which is seldom excelled. Caps only are tender.

P. pelli'tus Fr. **Pileus** 1–2 in. across. **Flesh** thin, soft, white, convex then plane, somewhat umbonate, regular, silky-fibrous, dry, white. **Gills** free, rounded behind, crowded, 1½ line broad, ventricose, white then flesh-color, margin slightly toothed. **Stem** about 2 in. long, 2–3 lines thick, slightly thickened at the base, even, glabrous, shining, white, stuffed. **Spores** elliptical, smooth, 10x6µ.

Among grass at the roots of trees, etc.

Our only Pluteus with a pure white, even pileus and stem. Superficially resembling Entoloma prunuloides, which differs in the broadly emarginate—not free—gills, and in the strong smell of new meal. *Massee.*

Mt. Gretna, Pa., October, 1898. *McIlvaine.*

Pileus up to 3 in. across. **Gills** ¼ in. broad, free, moist, imbricated. **Stem** up to 5 in. long, easily detachable from cap, solid, juicy, solitary and cespitose. On very old sawdust, upon which grass was growing.

Tender, excellent.

**Pileus frosted, etc.

P. granula'ris Pk.—sprinkled with grains. **Pileus** convex or nearly plane, subumbonate, *rugose-wrinkled, granulose or granulose-villose,* varying in color from yellow to brown. **Lamellæ** rather broad, crowded, ventricose, whitish, then flesh colored. **Stem** equal, solid, colored like the pileus, often paler at the top, *velvety-pubescent,* rarely scaly. **Spores** subglobose or broadly elliptical, 6.5–8x5–6.5µ.

Pluteus. **Plant** 1.5–3 in. high. **Pileus** 1–2 in. broad. **Stem** 1–2 lines thick. Decaying wood and prostrate trunks in woods. Hilly and mountainous districts. June to September.

The species is closely related to P. cervinus and P. umbrosus, but is readily distinguished from them by the peculiar vesture of the pileus and stem. The granules are so minute and so close that they form a sort of plush on the pileus, more dense on the disk and radiating wrinkles than elsewhere. The clothing of the stem is finer, and has a velvety-pubescent appearance, but in some instances it breaks up into small scales or squamules. The color of the pileus and stem is usually some shade of yellow or brown, but occasionally a grayish hue predominates. The darker color of the granules imparts a dingy or smoky tinge to the general color. The disk is often darker than the rest of the pileus. *Peck*, 38th Rep. N. Y. State Bot.

West Virginia mountains. Eagle's Mere and Springton Hills, Pa. Frequent. July to October, on decaying wood. *McIlvaine.*

P. granularis is a much smaller species than P. cervinus and its allies. At Eagle's Mere, Pa., August, 1898, it was quite plentiful in mixed woods. Its caps are excellent.

*** *Pileus naked.*

P. admira'bilis Pk.—admirable. **Pileus** thin, convex or expanded, generally broadly umbonate, glabrous, *rugose-reticulated*, moist or hygrophanous, striatulate on the margin when moist, often obscurely striate when dry, yellow or brown. **Lamellæ** close, broad, rounded behind, ventricose, whitish or yellowish, then flesh-colored. **Stem** slender, glabrous, *hollow*, equal or slightly thickened at the base, yellow or yellowish white, with a white mycelium. **Spores** subglobose or broadly elliptical, 6.5–8x6.5μ.

Var. *fus'cus.* **Pileus** brown or yellowish-brown.

Plant 1–2 in. high. **Pileus** 6–10 lines broad. **Stem** .5–1 line thick. Decaying wood and prostrate trunks in forests. Common in hilly and mountainous districts. July to September.

This beautiful Pluteus is closely related to P. chrysophlebius B. and R., a southern species, which, according to the description, has the veins of the pileus darker colored than the rest of the surface and the stem enlarged above and hairy at the base, characters not shown by our Pluteus. plant.

In our plant small young specimens sometimes have the stem solid, but when fully developed it is hollow, though the cavity is small. This character, with its small size, distinguishes it from P. leoninus. *Peck*, 38th Rep. N. Y. State Bot.

Springton Hills, Chester county, Pa., Mt. Gretna, Pa. Frequent. June to frost. *McIlvaine.*

Possesses the same rare edible qualities as P. cervinus, P. umbrosus. The caps, only, are tender.

P. chrysophæ'us Schaeff. *Gr.*—gold. **Pileus** 1–2½ in. across. **Flesh** very thin except at the disk, bell-shaped then expanded, glabrous, naked, slightly wrinkled, margin striate, cinnamon-color. **Gills** free, 2–3 lines broad, whitish then pale salmon-color. **Stem** 2–3 in. long, 2–3 lines thick, whitish, glabrous, equal, more or less hollow.

On beech trunks, etc.

Resembling P. leoninus in size, but differing in the cinnamon color of the pileus, which is often obtusely umbonate. *Massee.*

Spores 5µ *W.P.*

Haddonfield, N. J. June to October, beech roots and trunks. *Mc-Ilvaine.*

Excellent.

ENTOLO'MA Fr.

Gr.—within; *Gr.*—a fringe.

(Probably referring to the innate character of the pseudo veil.)

Pileus rather fleshy, margin incurved, without a distinct veil. **Stem** fleshy or fibrous, soft, sometimes waxy, continuous with the flesh of the pileus. **Gills** *sinuate*, adnexed, often separating from the stem. **Spores** rosy, elliptical, smooth or subglobose and coarsely warted.

Corresponding in structure with Tricholoma, Hebeloma and Hypholoma; separated from other rosy-spored genera by the sinuate gills.

About twenty species of Entoloma are given in the states; of them seventeen are described by Professor Peck, as found in New York. I have not found a single species in sufficient quantity to test its edibility.

Two of the European species, E. sinuata Fr. and E. livida Bull., are reputed to be very poisonous, producing headache, dizziness, vomiting, etc. Worthington Smith ate ¼ oz., which nearly proved fatal.

Professor Peck reports a species, E. grande Pk.,which he considers suspicious.

Even the reported poisonous species have a pleasant odor corresponding to those of the esculent species. This makes them the more deceptive and dangerous. The pinkish or flesh-colored spores and gills distinguish Entoloma from Hebeloma, which has brown spores, and Tricholoma, which has white. Pluteus, which has pink spores and gills, is readily separated from it.

Great caution should be observed. Entolomas should be thrown away or carefully tested.

ANALYSIS OF TRIBES.

GENUI'NI (genuine, typical species). Page 251.

Pileus smooth, moist or viscid; not hygrophanous.

LEPTONI'DEI (inclining to Leptonia).

Pileus flocculose or squamulose; absolutely dry.

NOLANI'DEI (inclining to Nolanea). Page 252.

Pileus thin, hygrophanous, somewhat silky when dry.

I.—Genui'ni.

E. gran'de Pk.—**Pileus** fleshy, thin toward the margin, glabrous, Entoloma. nearly plane when mature, commonly broadly umbonate and rugosely wrinkled about the umbo, moist in wet weather, dingy yellowish-white verging to brownish or grayish-brown. **Flesh** white, odor and flavor farinaceous. **Lamellæ** broad, subdistant, slightly adnexed, becoming free or nearly so, often wavy or uneven on the edge, whitish becoming flesh-colored with maturity. **Stem** equal or nearly so, solid, somewhat fibrous externally, mealy at the top, white. **Spores** angular, 8–10μ.

Pileus 4–6 in. broad. **Stem** 4–6 in. long, 8–12 lines thick.

Thin mixed woods. Menands. August.

The flavor of this mushroom is not at first disagreeable, but an unpleasant burning sensation is left in the mouth for a considerable time after tasting. It is therefore to be regarded with suspicion. *Peck*, 50th Rep. N. Y. State Bot.

SUSPICIOUS. I have not seen this species. It is given that it may be guarded against until tested for edibility.

E. sinua'tum Fr.—waved. **Pileus** 6 in. broad, *becoming yellow-white*, very fleshy, *convex then expanded*, at first gibbous, at length depressed, repand and sinuate at the margin. **Stem** 3–6 in. long, 1 in. thick, *solid*, firm, stout, equal, compact, *at first fibrillose*, then smooth, naked, shining white. **Gills** *emarginate*, slightly adnexed, ½–¾ in. broad, crowded, distinct, pale yellowish-red. *Fries*.

(Plate LXII.)

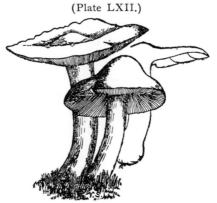

ENTOLOMA SINUATUM.
About one-fourth natural size.

Gregarious, compact, handsome.

Odor *strong, pleasant, almost like that of burnt sugar*, not of new meal. The pileus becomes broken into squamules when dry. There is a variety with a shorter stem.

In mixed woods. Uncommon. July to October.

The gills are often irregular in their attachment. Very poisonous; producing headache, swimming of the brain, stomach pains, vomiting,

Entoloma. etc.   Worthington Smith, who first experimented with it, ate about ¼ oz., which very nearly proved fatal.   *Stevenson.*

Spores 9µ *W.G.S.*

Rhode Island, *Olney* (Curtis Am. Jour.); Massachusetts, *Sprague;* Connecticut, *Wright;* Minnesota, *Johnson;* New York, *Peck*, Rep. 35.

"This and E. fertilis, which are closely allied, are deserving of more than suspicion, for they are veritably dangerous."   *Cooke.*

"Wholesome and very good to eat."   *Cordier.*

In the presence of such opposite opinions it is better to choose the safer.   Do not eat it.

E. prunulo'ides Fr.—*prunus*, a plum.   **Pileus** 2 in. and more broad, whitish, becoming yellow or livid, fleshy, *bell-shaped then convex*, at length flattened, somewhat umbonate, unequal (but not repand), even, *viscid*, smooth, at length longitudinally cracked, at length slightly striate at margin.   **Stem** 3 in. long, 3–4 lines thick, fibrous-fleshy, solid, equal, even or slightly striate, smooth, naked, white.   **Gills** somewhat free, emarginate, rarely rounded, at first only slightly adnexed, 3–4 lines broad, crowded, ventricose, white then flesh-color.   *Fries.*

Odor strong of new meal, wholly that of A. prunulus.   Very scattered in growth.   Like A. lividus, but very different, thrice as small. It differs entirely from A. cervinus.

On the ground in woods.   Autumn.   **Spores** subglobose, coarsely warted, 10µ *Massee;* regularly six-angled or one angle more marked, 8µ *B.;* 9µ *W.P.*

North Carolina, dry swamps, *Curtis;* Minnesota, *Johnson.*

POISONOUS.   *Roze.*

I have not seen this species.   Do not eat it before carefully testing.

III—NOLANI'DEI.

Pileus thin, hygrophanous, repand, etc.

E. clypea'tum *Linn.*—resembling a shield.   **Pileus** as much as 3 in. broad, *lurid* when moist, when dry gray and *variegated or streaked with darker spots or lines*, fleshy, *bell-shaped then flattened*, umbonate, smooth, fragile.   **Flesh** thin, white when dry.   **Stem** almost 3 in. long, 3–4 lines and more thick, stuffed, at length hollow, *wholly fibrous*, equal, round, fragile, *longitudinally fibrillose*, becoming ash-colored, pulverulent at the very apex. **Gills** *rounded-adnexed*, separating-free, 3–4 Entoloma. lines broad, ventricose, somewhat distant, dingy, then red-pulverulent with the spores, serrulated at the edge chiefly behind.

It has occurred in May cespitose; better developed and solitary in the end of August.

In woods, gardens and waste places. Frequent. Spring, autumn. *Stevenson.*

North Carolina, *Schweinitz, Curtis;* Ohio, *Morgan;* New England, *Frost;* California, *H. and M.;* Rhode Island, *Bennett;* New York, *Peck,* Rep. 23.

POISONOUS. *Leuba.*

I have not seen this species. It should not be eaten before careful testing.

E. rhodopo'lium Fr. *Gr.*—rose; *Gr.*—gray. **Pileus** 2–5 in. broad, hygrophanous, when moist dingy-brown (young) or livid, becoming pale (when full grown), *when dry isabelline-livid, silky-shining,* slightly-fleshy, bell-shaped when young, then expanded and somewhat umbonate or gibbous, at length rather plane and sometimes depressed, *fibrillose* when young, *smooth when full grown,* margin at the first bent inwards and when larger undulated. **Flesh** white. **Stem** 2–4 in. long, 3–5 lines thick, *hollow,* equal when smaller, when larger attenuated upwards and *white-pruinate at the apex,* otherwise *smooth,* slightly striate, *white.* **Gills** adnate then separating, somewhat sinuate, slightly distant, 2–4 lines broad, *white then rose-color. Fries.*

Fragile, commonly large and often handsome, almost **inodorous.**

In mixed woods. Frequent. August to October.

Spores pretty regular, 8–10x6–8µ *B.;* 7µ *W.G.S.*

New England, *Frost;* Minnesota, *Johnson;* Iowa, *Brændle;* Rhode Island, *Bennett;* Ohio, *Morgan;* New York, *Peck,* Rep. 23d, 38th, A. rhodopolius, var. umbilicatus Pk., the same as Clitopilus subvilis Pk., Rep. 40.

Edible. *Paulet.* Edible. *Cooke.*

CLITOPI′LUS Fr.

Gr.—a declivity; *Gr.*—a cap.

Clitopilus. **Pileus** more or less excentric or regular, margin at first involute.

(Plate LXIV.)

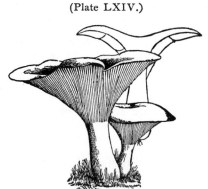

Gills more or less decurrent, never sinuate nor seceding from the stem, salmon-color. **Stem** fleshy or fibrous, not polished and cartilaginous externally, central, expanded upward into the flesh of the pileus. **Spores** smooth or warted.

Closely resembling Eccilia, differing mostly in the stem not being cartilaginous at the surface. Distinguished from Entoloma by the gills not being sinuate.

Agrees in structure with Clitocybe in the Leucosporæ. *Massee.*

CLITOPILUS PRUNULUS.
One-third natural size.

Growing on the ground, often strong smelling. Caps usually depressed or umbilicate and waved on margin.

Some of the best of edible kinds are within this genus; a few are unpleasant raw, none poisonous.

Most authors follow Fries in the arrangement of the species, dividing them into two groups, the Orcelli, distinguished by deeply decurrent gills and an irregular, scarcely hygrophanous pileus, with the margin at first flocculose; and Sericelli, distinguished by adnate or slightly decurrent gills and a regular silky or hygrophanous-silky pileus with a naked margin. This arrangement is not strictly applicable to some of our species. C. abortivus, C. erythrosporus and C. Noveaboracensis have the gills deeply decurrent in some individuals, adnate or slightly decurrent in others, and therefore the same species might be sought in both groups. For this reason the primary grouping of our species has been made to depend upon the variation in the spore colors. By far the greater number of our species appear to be peculiar to this country, only two of them occurring also in Europe.

ANALYSIS OF SPECIES.

Spores and mature gills flesh-colored.......................... I Clitopilus.
Spores and mature gills rosy-red............................. 9
Spores very pale flesh-colored...............................10

1. Pileus hygrophanous..................................... 8
1. Pileus not hygrophanous................................. 2
  2. Pileus gray or grayish-brown......................... 5
  2. Pileus some other color............................. 3
3. Pileus white or whitish................................. 4
3. Pileus pale tan-color...........................C. pascuensis
  4. Pileus firm, dry, pruinate....................C. prunulus
  4. Pileus soft, slightly viscid when moist........... C. Orcella
5. Pileus large, more than 1.5 in. broad.............C. abortivus
5. Pileus small, less than 1.5 in. broad..................... 6
  6. Spores even .... C. unitinctus
  6. Spores angular...................................... 7
7. Stem longer than the width of the zoneless pileus...C. albogriseus
7. Stem shorter than the width of the commonly zonate
      pileus....................................C. micropus
  8. Pileus brown or grayish-brown..................C. subvilis
  8. Pileus white or yellowish-white..............C. Woodianus
9. Stem colored like the pileus................C. erythrosporus
9. Stem white, paler than the pileus...............C. conissans
  10. Pileus even...11
  10. Pileus rivulose....................... C. Noveboracensis
11. Stems cespitose, solid ....C. cæspitosus
11. Stems not cespitose, hollow.................C. Seymourianus
Peck, 42d Rep. N. Y. State Bot.

SPORES FLESH-COLOR.

A. SPORES EVEN.

C. pru'nulus Scop.—*prunus*, plum. ([Color] Plate LXIII, figs. 4, 5.)
Pileus fleshy, *compact*, at first convex and regular, then repand, *dry*,
pruinate, white or ashy-white. **Flesh** white, unchangeable, with a
pleasant farinaceous odor. **Gills** deeply decurrent, subdistant, flesh-

255

Clitopilus. colored. Stem solid, naked, striate, white. Spores subelliptical, pointed at each end, 10–11x5–6μ.

Pileus 1.5–3 in. broad. Stem 1–2 in. long, 3–4 lines thick. Woods.

Not abundant, but edible, and said to be delicious and one of the best of the esculent species. *Peck*, 42d Rep. N. Y. State Bot.

June to October. Most plentiful in August and September.

Very plentiful in oak woods at Angora, West Philadelphia, moderate crops at Mt. Gretna, Pa.

An abortive form ([Color] Plate LXIII, figs. 2, 3.) occurs not distinguishable from that of Armillaria mellea. It grows singly and in tufts, very variable in shape, white, tinged with brown on ruptured surfaces. This form equals its original.

C. prunulus has a strong smell of fresh meal. It is a delicious species. Stew. It is one of the very best in patties, croquettes, etc.

C. Orcel'la Bull.—Pileus fleshy, *soft*, plane or slightly depressed,

(Plate LXV.)

CLITOPILUS ORCELLA.
Two-thirds natural size.

often irregular, even when young, *slightly silky, somewhat viscid when moist*, white or yellowish-white. Flesh white, taste and odor farinaceous. Gills deeply decurrent, *close*, whitish then flesh-colored. Stem short, solid, flocculose, often eccentric, thickened above, white. Spores elliptical, 9–10x5μ.

Generally a little smaller than the preceding species, softer and more irregular, but so closely allied that by some it is considered a mere variety of it. It is said to be edible and of delicate flavor. It occurs in wet weather in pastures and open places. *Peck*, 42d Rep. N. Y. State Bot.

Grows in oak woods, Angora, West Philadelphia; Mt. Gretna, Pa.

Qualities same as C. prunulus. Delicious.

C. pascuen'sis Pk.—pasture. Pileus fleshy, compact, centrally depressed, *glabrous, reddish or pale-yellowish*, the cuticle of the disk cracking into minute areas. Gills rather narrow, close, decurrent,

whitish, becoming flesh-colored. **Stem** short, equal or tapering down- Clitopilus. ward, solid, glabrous, colored like the pileus. **Spores** subelliptical, pale incarnate, 7.5–10x5–6μ.

Pileus 2–3 in. broad. **Stem** 8–18 lines long, 4–6 lines thick.

Pastures. Saratoga county.

The species is related to C. prunulus from which it is distinct by its shorter, paler spores, its glabrous pileus cracked in areas on the disk and tinged with red or yellowish and by its paler gills. From C. pseudo-orcella it differs in its glabrous pileus with no silky luster and in its closer gills. Its odor is obsolete but it has a farinaceous flavor. It is probably esculent, but has not been found in sufficient quantity to afford a test of qualities. *Peck*, 42d Rep. N. Y. State Bot.

C. unitinct'us Pk.—one-colored. **Pileus** thin, *submembranaceous*, flexible, convex or nearly plane, centrally depressed or umbilicate, glabrous, subshining, often concentrically rivulose, grayish or grayish-brown. **Flesh** whitish or grayish-white, odor obsolete, taste mild. **Gills** narrow, moderately close, *adnate or slightly decurrent*, colored like the pileus. **Stem** slender, straight or flexuous, subtenacious, equal, slightly pruinose, grayish-brown, with a close white myceloid tomentum at the base and white root-like fibers of mycelium permeating the soil. **Spores** elliptical, 7.5x5μ.

Var. *al'bidus*. Whitish or grayish-white, not rivulose. Gills broader. Spores brownish flesh-color.

Pileus 6–16 lines broad. Stem about 1 in. long, 1 line thick.

Woods of pine or balsam. Albany and Essex counties. Autumn.

The variety is a little paler than the typical form, with gills a little broader, but is probably not specifically distinct. *Peck*, 42d Rep. N. Y. State Bot.

I have not seen this species. Edibility not reported.

B. SPORES ANGULAR OR IRREGULAR.

1. *Pileus not hygrophanous.*

C. aborti'vus B. and C.—abortive. ([Color] Plate LXIII, figs. 1, 2, 3.) **Pileus** fleshy, firm, convex or nearly plane, regular or irregular, dry, *clothed with a minute silky tomentum*, becoming smooth with age, gray or grayish-brown. **Flesh** *white*, taste and odor subfarinace-

257

Clitopilus. ous. **Gills** thin, close, slightly or deeply decurrent, at first whitish or pale gray, then flesh-colored. **Stem** nearly equal, solid, minutely flocculose, sometimes fibrous-striated, colored like or paler than the pileus. **Spores** irregular, 7.5–10x6.5μ.

Pileus 2–4 in. broad. **Stem** 1.5–3 in. long, 3–6 lines thick.

Ground and old prostrate trunks of trees in woods and open places. August and September.

Our species has been found to be edible, but its flavor is scarcely as agreeable as that of some other species. *Peck*, 42d Rep. N. Y. State Bot.

It requires longer cooking than C. prunulus, and is then quite equal in excellence.

The fungus is so named because of the abortive form of it frequently found associated with it. This is faithfully portrayed on Plate LXIII. This is in every way similar to the aborted forms of C. prunulus and Armillaria mellea.

Both forms plentiful near Philadelphia. The undeveloped masses are also similar to those of C. prunulus.

The abortive form is a superior edible to the original.

C. popina'lis Fr.—*popina*, a cook-shop. **Pileus** 1–2 in. across, flesh thin, flaccid, convex then depressed, somewhat wavy, glabrous, opaque, gray, spotted and marbled. **Flesh** grayish-white, unchangeable. **Gills** very decurrent, broader than the thickness of the flesh of the pileus, lanceolate, crowded, dark-gray, at length reddish from the spores. **Stem** stuffed, 1–2 in. long, 2 lines thick, equal, often flexuous, naked, paler than the pileus. **Spores** subglobose, slightly angular, 4–5μ *Massee*.

Solitary or gregarious, smell pleasant like new meal, entirely gray. *Fries.*

Woods. Gansevoort. July. The whole plant is of a grayish color except the mature gills, which have a flesh-colored hue, and the base of the stem, which is clothed with a white tomentum. It has a farinaceous odor. *Peck*, 51st Rep. N. Y. State Bot.

Scattered. Mt. Gretna, Pa. September to November. *McIlvaine.*
Edible, pleasant.

C. carneo-al'bus Wither.—light flesh color. **Pileus** up to 1 **in.**

across, convex then expanded, center becoming depressed and the mar- Clitopilus.
gin drooping, even, polished, white, the disk becoming usually tinged
with red. **Flesh** thin. **Gills** slightly decurrent, 1 line broad, crowded,
salmon color. **Stem** 1–1½ in. long, 1 line thick, about equal, solid,
white. **Spores** globose, nodulose, 7–8μ diameter.

Inodorous; gregarious.

In the section given in Cke. Illustr., the stem is represented as being
distinctly hollow. *Massee.*

New York, shaded ground. June. *Peck*, 45th Rep.

C. al'bogri'seus Pk.—pale-gray. **Pileus** firm, convex or slightly depressed, *glabrous*, pale-gray, odor farinaceous. **Gills** moderately close,
adnate or slightly decurrent, grayish then flesh-colored. **Stem** solid,
colored like the pileus. **Spores** angular or irregular, 10–11x7.5μ.

Pileus 6–12 lines broad. **Stem** 1.5–2.5 in. long, 1–2 lines thick.

Woods. Adirondack mountains. August. *Peck*, 42d Rep. N. Y.
State Bot.

Scattered. Mt. Gretna, Pa., woods. August to October. *McIlvaine.*

Edible, pleasant.

C. mi'cropus Pk.—short-stemmed. **Pileus** thin, fragile, convex or
centrally depressed, *umbilicate, silky*, gray, usually with one or two narrow zones on the margin, odor farinaceous. **Gills** narrow, close, adnate or slightly decurrent, gray, becoming flesh-colored. **Stem** *short*,
solid, slightly thickened at the top, pruinose, gray with a white mycelium at the base. **Spores** angular or irregular, 10x6μ.

Pileus 6–12 lines broad. **Stem** 8–10 lines long, 1 line thick.

Thin woods. Essex and Rensselaer counties. August.

This species is closely allied to the preceding one, but may be sepa-
rated from it by its short stem and silky umbilicate subzonate pileus.
Both species are rare and have been observed only in wet, rainy weather.
Peck, 42d Rep. N. Y. State Bot.

Scattered; markedly umbilicate. Mt. Gretna, Pa., woods. August,
September. *McIlvaine.*

Edible, pleasant.

2. *Pileus hygrophanous.*

Clitopilus. **C. subvi'lis** Pk.—small value. **Pileus** thin, centrally depressed or umbilicate, with the margin decurved, hygrophanous, *dark-brown* and striatulate on the margin when moist, grayish-brown and silky shining when dry, taste farinaceous. **Gills** *subdistant*, adnate or slightly decurrent, whitish when young, then flesh-colored. **Stem** slender, brittle, rather long, *stuffed or hollow*, glabrous, colored like the pileus or a little paler. **Spores** angular, 7.5-10μ.

Pileus 8–15 lines broad. **Stem** 1.5–3 in. long, 1–2 lines thick.

Damp soil in thin woods. Albany county. October.

The species is allied to C. vilis, from which it is separated by its silky-shining pileus, subdistant gills and farinaceous taste. *Peck*, 42d Rep. N. Y. State Bot.

Scattered. Mt. Gretna, Pa. September to November. *McIlvaine.* Edible, pleasant.

C. Wood'ianus Pk. **Pileus** thin, convex or nearly plane, umbilicate or centrally depressed, hygrophanous, striatulate on the margin when moist, *whitish or yellowish-white* and shining when dry, the margin often wavy or flexuous. **Gills** close, adnate or slightly decurrent, whitish, then flesh-colored. **Stem** equal, flexuous, shining, *solid*, colored like the pileus. **Spores** subglobose, angular, 6-7.5μ.

Pileus 1–2 in. broad. **Stem** 2–3 in. long, 2 lines thick.

Ground and decayed prostrate trunks in woods. Lewis county. September.

This species is perhaps too closely allied to the preceding, but it may easily be separated by its paler color, closer gills and solid stem, though this is sometimes hollow from the erosion of insects. *Peck*, 42d Rep. N. Y. State Bot.

C. Un'derwoodii Pk.—in honor of L. M. Underwood. **Pileus** rather thin but fleshy, nearly plane or slightly depressed in the center, even, whitish. **Gills** narrow, close, slightly decurrent, pale flesh-colored. **Stem** rather short, equal or slightly tapering upward, solid, whitish. **Spores** subglobose, 4–5μ long.

Pileus 6–18 lines broad. **Stem** about 1 in. long and 2 lines thick.

Syracuse and Jamesville.   September and October.   *L. M. Under-* Clitopilus.
wood.   Peck, 49th Rep. N. Y. State Bot.

SPORES ROSY-RED.

C. erythro'sporus Pk.   *Gr.*—red-spored.   **Pileus** thin, hemispherical or strongly convex, glabrous or merely pruinose, pinkish-gray.
Flesh whitish tinged with pink, taste farinaceous.   **Gills** narrow,
crowded, arcuate, *deeply decurrent,* colored like the pileus.   **Stem**
equal or slightly tapering upward, hollow, slightly pruinose at the top,
colored like the pileus.   **Spores** elliptical, 5x3–4μ.

Pileus 1–2 in. broad.   **Stem** 1–1.5 in. long, 2–3 lines thick.

Decayed wood and among fallen leaves in woods.   Albany and
Ulster counties.   September and October.

The species is easily recognized by its peculiar uniform color, its narrow, crowded and generally very decurrent gills and by its bright rosy-
red spores.   Sometimes individuals occur in which the gills are less
decurrent.   *Peck,* 42d Rep. N. Y. State Bot.

Mt. Gretna, Pa., among fallen leaves.   Sparsely gregarious.   September to November.   *McIlvaine.*

Edible, good.

C. conis'sans Pk.—dusted.   **Pileus** thin, convex, glabrous, pale
alutaceous, often *dusted by the copious spores.*   **Gills** close, *adnate,* reddish-brown.   **Stem** slender, brittle, hollow, cespitose, *white.*   **Spores**
narrowly elliptical, 7.5x4μ.

Pileus 1–1.5 in. broad.   **Stem** 1–2 in. long, 1–2 lines thick.

Base of an apple tree.   Catskill mountains.   September.

Remarkable for the bright rosy-red spores which are sometimes so
thickly dusted over the lower pilei of a tuft as to conceal their real color.
The species is very rare.   *Peck,* 42d Rep. N. Y. State Bot.

SPORES VERY PALE FLESH-COLORED, MERELY TINTED.

C. cæspito'sus Pk.—tufted.   **Pileus** at first convex, firm, nearly **reg-**
ular, shining, white, then nearly plane, fragile, often irregular or eccentric, glabrous but with a slight silky luster, *even,* whitish.   **Flesh** white,
taste mild.   **Gills** narrow, thin, crowded, often forked, adnate or slightly

261

Agaricaceæ

Clitopilus. decurrent, whitish, becoming dingy or brownish-pink. **Stems** *cespitose,* solid, silky-fibrillose, slightly mealy at the top, white. **Spores** 5x4μ.

> **Pileus** 2–4 in. broad. **Stem** 1.5–3 in. long, 2–4 lines thick.

> Thin woods and pastures. Ulster county. September.

This is a large, fine species, very distinct by its cespitose habit, white color and very pale sordid-tinted spores. But for the color of these the plant might easily be taken for a species of Clitocybe. The tufts sometimes form long rows. *Peck,* 42d Rep. N. Y. State Bot.

> Mt. Gretna, Pa. October. *McIlvaine.*

Tender, not much flavor.

C. Noveboracen′sis Pk.—New York Clitopilus. **Pileus** thin, convex, then expanded or slightly depressed, dingy white, *cracked in areas or concentrically rivulose,* sometimes obscurely zonate, odor farinaceous, *taste bitter.* **Gills** narrow, close, deeply decurrent, some of them forked, white, becoming dingy, tinged with yellow or flesh-color. **Stem** equal, solid, colored like the pileus, the mycelium white, often forming white branching root-like fibers. **Spores** globose, 4–5μ broad.

> Var. *brevis.* Margin of the pileus, in the moist plant, pure white. **Gills** adnate or slightly decurrent. **Stem** short.

> **Pileus** 1–2 in. broad. **Stem** 1–2 in. long, 1–3 lines thick.

> Woods and pastures. Adirondack mountains, Albany and Rensselaer counties. August to October.

The plant is gregarious or cespitose. Sometimes, especially in the variety, it grows in lines or arcs of circles. The margin is often undulated, and in the variety it is, when fresh and moist, clothed with a film of interwoven webby white fibrils which give it a peculiar appearance, and if the spore characters are neglected it might be mistaken for Clitocybe phyllophila. The disk is often tinged with reddish-yellow or rusty hues when moist, and its rivulose character is then more distinct. A farinaceous odor is generally present, especially in the broken or bruised plant, but its taste is bitter and unpleasant. Sometimes bruises of the fresh plant manifest a tendency to assume a smoky-brown or blackish color. The base of the stem is sometimes clothed with a white myceli-oid tomentum. *Peck,* 42d Rep. N. Y. State Bot.

C. Sey′mourianus Pk.—**Pileus** fleshy, thin, broadly convex or slightly depressed, even, *pruinose, whitish with a dark lilac tinge,* sometimes lobed and eccentric. **Gills** narrow, crowded, decurrent, some of them Clitopilus. forked at the base, whitish with a pale flesh-colored tint. **Stem** equal, silky-fibrillose, *hollow*. **Spores** minute, globose or nearly so, 3.5–4μ long.

Pileus 1–2.5 in. broad. **Stem** 1.5–2.5 in. long, 3–4 lines thick. Woods. Lewis county. September. *Peck*, 42d Rep. N. Y. State Bot.

———

LEPTO'NIA Fr.

Gr.—slender.

Rosy-spored. **Stem** *cartilaginous*, tubular (the tube stuffed or hollow), polished, somewhat shining. **Pileus** *thin*, umbilicate or with a darker disk, cuticle fibrillose or separating into darker scales, *margin at first incurved*. **Gills** at first adnexed or adnate but readily separating. *Fries.*

The Leptoniæ are related to the Clitopili as the Collybiæ are to the Clitocybæ. The species are small, elegant, brightly colored, inodorous (except A. incanus), and abound *in rainy weather*. Gregarious or growing in troops; on the ground, commonly on dry mossy pastures, but also in marshy places. *Stevenson.*

Six American species reported. I have not seen any.

(Plate LXVI.)   Leptonia.

LEPTONIA.

NOLA'NEA Fr.

Nola, a little bell.

Nolanea.

(Plate LXVII.)

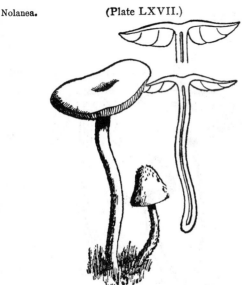

NOLANEA PASCUA.
About natural size.

Rosy-spored. **Stem** *tubed*, the tube more rarely stuffed with a pith, *cartilaginous*. **Pileus** somewhat membranaceous, *bell-shaped*, somewhat papillate, striate and sometimes even, sometimes also clothed with flocci, *margin straight and at the first pressed to the stem*, and not involute. **Gills** free or adfixed, and not decurrent. *Fries*.

Nolanea agrees with Leptonia and Eccilia among the pink-spored species. It corresponds with Mycena, Galera and Psathyra. Several Entolomata are nearly allied. The species are thin and slender, commonly inodorous and fragile, though some of them are tough. Growing on the ground in summer and autumn. *Stevenson*.

Seven American species reported. None seen by writer. *Peck*, Rep. 24, 26, 35, 39, 50.

ECCI'LIA Fr.

Gr.—I hollow out.

(Plate LXVIII.)

ECCILIA ATROPUNCTA.
Two-thirds natural size.

Stem *cartilaginous*, tubular (the tube hollow or stuffed), expanded upward into the *pileus*, which is somewhat membranaceous and at the first turned inward at the margin. **Gills** attenuated behind, truly decurrent, becoming more so when the pileus is depressed, and not separating as those of Nolanea.

Corresponding in structure with Omphalia of the white-spored and Tubaria of the brown-spored series. Allied to Clitopilus in the decurrent

gills, but separated by the cartilaginous, smooth stem.

E. car'neo-gri'sea B. and Br.—*caro*, flesh; *griseus*, gray. **Pileus** about 1 in. broad, gray flesh-color, umbilicate, striate, delicately dotted, margin slightly glittering with dark particles. **Stem** about 1 ½ in. long, slender, fibrous-hollow upward, wavy, of the same color as the pileus, shining, smooth, white-downy at the base. **Gills** adnato-decurrent, somewhat undulated, distant, rosy, the irregular margin darker. *Stevenson.*

Spores irregularly oblong, rough, 7x5μ *Massee.*

Nova Scotia, *Dr. Somers.*

New Jersey, *E. B. Sterling*, August, 1897; Eagle's Mere, Pa., common under pines, *McIlvaine.*

(Plate LXIX.)

ECCILIA CARNEO-GRISEA.
Natural size.

Eccilia. This neat little species is sweet and pleasant raw, and when cooked makes an agreeable dish. European authorities give the taste as unpleasant, but there is nothing of the sort about the American representative.

CLAU'DOPUS Smith.

Claudus—lame; *pous*—a foot.

Claudopus. **Pileus** eccentric, lateral or resupinate. **Spores** pinkish.
The species of this genus were formerly distributed among the Pleuroti and Crepidoti, which they resemble in all respects except the color of the spores. The genus at first was made to include species with lilac-colored as well as pink spores, but Professor Fries limited it to species with pink spores. In this sense we have taken it. The spores in some species are even, in others rough or angulated. The stem is either entirely wanting or is very short and inconspicuous, a character indicated by the generic name. The pileus often rests upon its back and is attached by a point when young, but it becomes turned backward with age. The species are few and infrequent. All inhabit decaying wood.

(Plate LXX.)

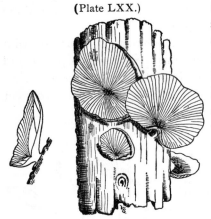

CLAUDOPUS VARIABILIS.
Natural size.

SYNOPSIS OF THE SPECIES.

Pileus yellow..................................C. nidulans
Pileus white or whitish.................................. I
  I. Spores even.................................C. variabilis
  I. Spores angulated.............................C. depluens
Pileus gray or brown.. 2
  2. Pileus striatulate when moist.................C. Greigensis
  2. Pileus not striatulate.......................C. byssisedus
Peck, 39th Rep. N. Y. State Bot.

C. ni'dulans Pers.—*nidus*, a nest. **Pileus** 1–3 in. broad, stemless, Claudopus. attached by the pileus or rarely narrowed behind into a short stem-like base, caps often overlapping one another, suborbicular or kidney-shaped, *downy*, somewhat pointed-hairy or scaly-hairy toward the margin, *yellow or buff color*, the margin at first turned inward. **Lamellæ** rather broad, moderately close or subdistant, *orange-yellow*. **Spores** even, slightly curved, 6–8μ long, about half as broad, delicate pink.

Decaying wood. Sandlake. Catskill and Adirondack mountains. Autumn.

This fungus was placed by Fries among the Pleuroti, and in this he has been followed by most authors. But the spores have a delicate pink color closely resembling that of the young lamellæ of the common mushroom, Agaricus campestris. We have, therefore, placed it among the Claudopodes, where Fries himself has suggested it should be placed if removed at all from Pleurotus. Our plant has sometimes been referred to Panus dorsalis Bosc., but with the description of that species it does not well agree. The tawny-color, spoon-shaped pileus, pale floccose scales, short lateral stem and decurrent lamellæ ascribed to that species are not well shown by our plant. The substance of the pileus, though rather tenacious and persistent, can scarcely be called leathery. The flesh is white or pale yellow. The hairy down of the pileus is often matted in small tufts and intermingled with coarse hairs, especially toward the margin. This gives a scaly or pointed-hairy appearance. The color of the pileus is often paler toward the base than it is on the margin. *Peck*, 39th Rep. N. Y. State Bot.

Mt. Gretna, Pa., November, 1898, decaying stumps. *McIlvaine*.

An autumnal species growing upon wood. Not common.

The light yellow tomentosity of the cap arranges itself into shapes as fascinating as crystals of snow.

Taste pleasant, mild. Texture more solid than P. ostreatus, consequently tougher. It is edible but not desirable. Must be chopped fine and cooked well.

Series III. **OCHRO'SPORÆ** (Dermini). Spores brown.

Ochrosporæ, third in color series, ranges in spore color from dull ochraceous, through bright ocher, to rusty orange and ferruginous or iron-rust. The various shades will tax even a color expert.

There are no species in the series corresponding to Amanitæ. In Acetabularia there is a cup-like volva; in Pholiota there is a distinct interwoven ring on the stem; in Cortinarius the secondary veil is like a cobweb, and may form an imperfect zone around the stem, or hang as fibers from the margin of the cap; Pluteolus exactly resembles Pluteus.

There are many edible species of good quality in the series. None are known to be poisonous. The substance, as a rule, is tougher than in most of the preceding genera, and in many instances has a strong woody flavor. Several species are late growers, and are among the best of fungi. Notably in Pholiota.

ACETABULA'RIA Berk.

Acetabulum, a vinegar-cup. From the cup-like volva.

Acetabularia. Universal veil distinct from the pileus; hymenophore distinct; gills free; spores pallid, tawny or brown.

Analogous to Volvaria and Chitonia.

No American species reported.

Ochrosporae.

Hymenophore distinct from fleshy stem.

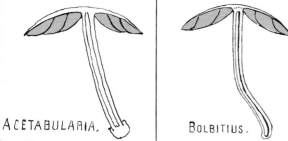

Acetabularia. Bolbitius.

Hymenophore confluent and homogeneous with fleshy stem.

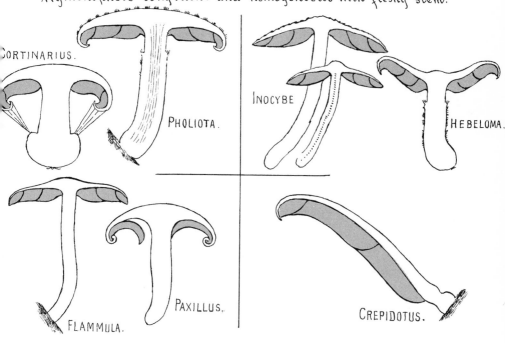

Cortinarius. Pholiota. Inocybe Hebeloma. Flammula. Paxillus. Crepidotus.

Hymenophore confluent with, but heterogeneous from cartilaginous stem.

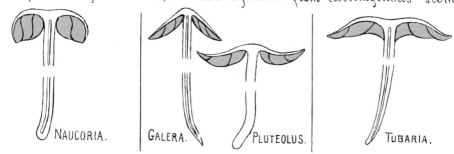

Naucoria. Galera. Pluteolus. Tubaria.

Chart of genera in brown-spored series—Ochrosporae.

PHOLIO'TA Fr.

Gr.—a scale.

Pileus more or less fleshy. **Gills** adnate, with or without a decur- Pholiota. rent tooth, tawny or rust colored at maturity from the spores. **Flesh** of stem continuous with that of the pileus. **Ring** distinct, interwoven. **Spores** sepia-brown, bright yellowish-brown or light red.

Generally on wood, sometimes on the ground in damp moss, frequently densely cespitose. Some of the species are large and bright colored. Distinguished from all other genera of the brown-spored series by the possession of a distinct ring. In Cortinarius the veil and ring are web-like.

Stevenson notes in his description of the genus: "None are to be commended as edible." My investigation shows that there are several delicious species, notably P. squarrosa and subsquarrosa. Their lateness and plentifulness make them valuable food fungi. I have nothing but praise for the entire genus.

ANALYSIS OF TRIBES.

A. HUMIGENI (*humus*, ground; *gigno*, to bear). Page 270.

On the ground, rarely cespitose.
* Eudermini. *Gr.*—well; *dermini*, the brown-spored series.
Spores ferruginous.
** Phæoti. *Gr.*—dusky.
Spores dusky rust-colored.

B. TRUNCIGENI (*truncus*, a trunk; *gigno*, to bear). Page 273.

On wood; subcespitose.
* Ægeritini. *P. ægerita*, the type of the section.
Pileus naked, not scaly, sometimes cracked. Gills pallid, then reddish or dusky. None known to be edible.
** Squamosi—*squama*, a scale.
Pileus scaly, not hygrophanous. Gills becoming discolored.
* Gills not becoming purely rust-colored.
** Gills yellow, then rust-color or tawny.
*** Hygrophani. *Gr.*—moist; to appear.
Gills cinnamon, not at first yellow.

C. MUSCIGENI (*muscus*, moss; *gigno*, to bear).

Pholiota. Hygrophanous. Like Galera with a ring.

A. HUMIGENI. On ground.

* Eudermini. *Spores ferruginous.*

P. capera′ta Pers.—*capero*, to wrinkle. ([Color] Plate LXXI, fig. 2.) **Pileus** 3–5 in. broad, more or less intensely yellow, fleshy, but thin in proportion to its size and robust stem, ovate then expanded, obtuse, viscid only when moist and not truly so, even at the disk, wrinkled in pits at the sides, *incrusted with white superficial flocci.* **Stem** 4–6 in. long, more than 1 in. thick, solid, stout, cylindrical with exception of the base which is often tuberous, shining white, *scaly above the ring, which is membranaceous, reflexo-pendulous, and broken into squamules at the apex.* **Gills** adnate, crowded, thin, somewhat serrated, *clay*-cinnamon.

When young the pileus is incrusted with the veil or with white mealy-floccose soft, hairy down, which is crowded on the even disk and scaly towards the thin pitted-furrowed margin; and as this separates the pileus is naked. Veil universal, floccoso-mealy, at the first cohering in the form of a volva but not continuous; in rainy weather remaining in the form of a volva at the base. **Spores** dark ferruginous on a white ground, paler on a black ground. There is a smaller form (A. macropus Pers.) in pine woods, pileus even and paler. **Stem** 3 in. long, and without a tuberous base. **Ring** oblique and often incomplete. *Stev.*

Spores 10µ *B. and Br.;* 12x4µ *W. P.;* spheroid-ellipsoid, uniguttate, 11–12x8–9µ *K.;* 12x4.5µ *Massee.*

Not previously reported.

This fungus occurs sparingly in rich woods near Boston. It is much esteemed in Germany, and eagerly sought by the common people, who call it familiarly the "Zigeuner" (Gypsy). Boston Myc. Club Bull. 1896.

I have found this species in but one place—on the south hill of the great Chester valley, Pa., where it grows plentifully in woods. The taste raw was slightly acrid, but when cooked this disappeared. Many ate of the species and enjoyed it.

P. togula′ris Bull.—*togula*, a little cloak.   From the ample ring.   Pholiota.
Pileus 1 ½ in. broad, *pallid ochraceous*, fleshy, soft, bell-shaped then expanded, obtuse, orbicular, *without striæ*, smooth.   **Flesh** thin, soft, becoming yellow.   **Stem** 3–4 in. long, 2 lines thick, tubed, rigid, equal, cylindrical, rough with stiff fibers, naked and becoming yellow at the apex, becoming dingy brown downward.   **Ring** medial, more than 1 in. distant, entire, spreading-reflexed.   **Gills** adnato-separating, ventricose, crowded, narrowed in front, becoming yellow, at length pale rust-color, never becoming dingy brown.

Protean, slender, very variable in stature, growing in troops.   *b.*   More slender, but densely gregarious, with the wholly pallid smooth stem thinner, often flexuous.   This form is exactly A. mesodactylus Berk. *c.* Very small.   Pileus 1 in.   **Stem** 1 in. or a little more, scarcely 1 line thick, very flexuous, becoming rust-color.   *Stevenson.*
Spores elliptical, 8x3.5μ *Massee.*
New Jersey, on decayed chips mixed with dirt.   **May, 1898.**   *E. B. Sterling.*
Not previously reported.
The specimens sent were tested **and found to be of good quality.**

** Phæ′oti.   *Spores fuscous—ferruginous (dingy rust-color).*

P. du′ra Bolt.—*durus*, hard.   **Pileus** 3 in. and more broad, tawny, tan-color, becoming dingy brown, fleshy, *somewhat compact*, convexo-plane, obtuse, smooth, *then cracked into patches*, margin even.   **Stem** commonly curt, 2 in. long, about ½ in. thick, *stuffed*, even solid, hard, becoming silky-even, then longitudinally cracked when dry, thickened at the *apex, mealy* and more than usually widened into the pileus, vary-ing ventricose and irregularly-shaped.   **Ring** *torn*.   **Gills** *adnate*, striato-decurrent with a tooth, ventricose, ½ in. broad, *livid then dingy rust-color.*

The stem is abundantly furnished with fibrillose rootlets at the base. Although very closely allied to A. præcox, it is readily distinguished by its rust-color or brown-rust spores.   *Stevenson.*
Spores 9x5μ *W.G.S.;* 8–9x5–6μ *Massee.*
Haddonfield, N. J.   June to October.   Florist's garden, *McIlvaine.*
After rains P. dura appears, solitary, from spring to autumn.   The

Pholiota. cracked cap, in mature specimens, distinguishes it from other species found on its habitat. It varies in size from 1½ in. up to 4 in. across. The caps are excellent.

P. præ'cox Pers.—*præcox*, early.
(Plate LXXII.)

PHOLIOTA PRÆCOX.
After Peck.

Pileus 1–2 in. broad, convex or nearly plane, soft, nearly or quite glabrous, whitish, more or less tinged with yellow or tan-color. **Gills** close, adnexed, at first whitish, then brownish or rusty-brownish. **Stem** 1.5–3 in. long, 2–2.5 lines thick, rather slender, mealy or glabrous, stuffed or hollow, whitish. **Spores** elliptical, rusty-brown, 10–13x 6–8μ.

The Early Pholiota is a small but variable species. From other similarly colored species that appear in grassy ground early in the season, the collar on the stem will generally distinguish it. Its cap is usually convex when young but nearly flat in the mature plant. It is rather pale in color but not a clear white, being tinted with yellow or pale tan-colored hues. The gills are whitish when the cap first opens, but they soon change to a rusty-brown hue in consequence of the ripening of the spores. They are excavated at the inner extremity and slightly attached to the stem. They are ventricose when the cap is fully expanded. The stem is rather slender, nearly or quite straight and soon smooth and hollow. It is pale or whitish, and usually furnished with a small collar. Sometimes the collar is slight and disappears with age and sometimes the fragments of the veil remain attached to the margin of the cap leaving nothing for a collar.

The plants usually grow in grassy ground, lawns and gardens, and appear from May to July.

Var. *minor* Batt. is a small form having the cap only about 1 in. broad and the remnants of the veil adherent to the margin of the cap. It is represented by figures 6 to 12.

Var. *sylvestris* Pk. has the center of the cap brownish or rusty-brown, and grows in thin woods. *Peck*, 49th Rep. N. Y. State Bot.

Spores inclining to fuscous, spheroid-ellipsoid, 8–13x5–7µ *K.;* 8x6µ Pholiota. *W.G.S.;* 8–13x6–7µ *Massee.*

West Virginia, New Jersey, North Carolina, Pennsylvania, May to August. On rich ground, lawns, gardens, etc. *McIlvaine.*

Coming as it does in early spring, it is a prized species wherever found.

The caps only are good.

B. TRUNCIGENI. On wood.

** Squamosi. *Scaly*.

P. squarro'sa Mull.—*squarrosus,* scurfy. ([Color] Plate LXXI, fig. 3.) Pileus 3–5 in. broad, saffron-rust-color, scaly with *innate, crowded, revolute, darker* (becoming dingy brown), persistent *scales,* fleshy, convex bell-shaped then flattened, commonly obtusely umbonate or gibbous, dry. Flesh light-yellow, compact when young, sometimes thin. Stems curt when young, as much as 8 in. long when full-grown, as much as 1 in. thick at the apex, remarkably attenuated downwards, stuffed, scaly as far as the ring with crowded, revolute, darker scales. Ring only slightly distant from the apex, rarely membranaceous, entire or often slashed,

(Plate LXXIII.)

PHOLIOTA SQUARROSA.
One-half natural size.

generally floccoso-radiate, of the same color as the scales. Gills adnate with a decurrent tooth, crowded, narrow, *pallid-olivaceous* then rust-color.

Spores ferruginous. Very cespitose, forming large heaps. Stems commonly cohering at the base, varying very much in stature in the

273

Pholiota. same cluster; varying also much thinner, scarcely ever curved-ascending. Odor heavy, stinking; sometimes, however, obsolete. *Stevenson.*

Spores ellipsoid, 7–8x4–5µ *K.;* 4x5µ *W.G.S.;* 8x4µ *Massee.*

On trunks of trees, on and near stumps, etc. Common. August to December.

West Virginia, 1881–1885, New Jersey, Pennsylvania. On rotten wood and stumps. August to long after frost. *McIlvaine.*

Edible. *Curtis.*

The American species, as I have repeatedly found it, is not so large as given in the European description, and the habitat is more closely confined to the trunks of standing trees and stumps not much decayed. It is a showy species, to be seen from afar off, especially after the leaves fall. Taste when young, raw, is sweet, mealy; when mature, like stale lard.

Cooked, the caps are of good substance and flavor. One of the very best.

P. squarrosoi'des Pk.—*squarrosus,* scurfy; *eidos,* form. Pileus firm, convex, viscid when moist, at first densely covered by erect papillose or subspinose tawny scales, which soon separate from each other, revealing the whitish color and viscid character of the pileus. Lamellæ close, emarginate, at first whitish, then pallid or dull cinnamon color. Stem equal, firm, stuffed, rough with thick squarrose scales, white above the thick floccose ring, pallid or tawny below. Spores minute, elliptical, 5x4µ.

Densely cespitose, 3–6 in. high. Pileus 2–4 in. broad. Stem 3–5 lines thick.

Dead trunks and old stumps of maple. Adirondack and Catskill mountains. Autumn.

This is evidently closely related to A. squarrosus, with which it has, perhaps, been confused, but its different colors and viscid pileus appear to warrant its separation. *Peck,* 31st Rep. N. Y. State Bot.

Occurred in large clusters on sugar maples at Eagle's Mere in October, and on stumps at Mt. Gretna. It very closely resembles P. squarrosa. Its caps are of the very best.

P. subsquarro′sa Fr.—*sub*, under; *squarrosus*, scurfy. ([Color] Plate Pholiota. LXXI, fig. 4.)　**Pileus** 2 in. and more broad, *brown rust-color*, with darker, *adpressed*, floccose *scales*, fleshy, convex, obtuse or gibbous, viscid.　**Stem** 3 in. long, 4–5 lines thick, stuffed (often hollow when old), equal, yellow-rust-color, clothed with darker scales which are adpressed, or spreading only at the apex, not rough, furnished with an annular zone at the apex, becoming yellow-rust-color within.　**Gills** deeply sinuate, emarginate, *almost free*, arcuate, crowded, at first pale then dingy yellow.

Spores rust-color.　The pileus is viscid, but not glutinous like that of A. adiposus.　It holds a doubtful place between A. aurivellus and A. squarrosus, departing from both, however, in the gills being at the first yellow; and from A. squarrosus, to which it is more like, in the gills being emarginato-free, not decurrent. Somewhat cespitose. Almost inodorous.　*Fries.*

Spores ferruginous, size not stated.

West Philadelphia, Mt. Gretna, Pa., Haddonfield, N. J.　September until after frosts.　*McIlvaine.*

Not previously reported.

The maple trees in West Philadelphia frequently show large clusters of it up to twenty feet from ground; to be seen from afar after the leaves have fallen.　Our American species differs somewhat from the European. American species:

Pileus 1–3 in. across, fleshy, convex, *very viscid*, rich brownish-yellow, covered with darker adpressed floccose scales.　**Flesh** slightly yellow.　**Gills** white when very young slightly emarginate, adnexed, crowded, ¼ in. broad, brown.　**Stem** 2–3 in. long, ½ in. thick, equal or tapering toward base, stuffed, then hollow, covered with squamose scales as far up as the slight ring, smooth above ring.　**Ring** membranaceous, slight.

Spores rust-color.

The species is variable and differs greatly in youth and maturity.

The caps, fried in hot buttered pan, are unexcelled.

Equally fine in croquettes and patties.

** *Gills yellow, then rust-color.*

Pholiota. **P. adipo′sa** Fr.—*adeps*, fat.

(Plate LXXIV.)

PHOLIOTA ADIPOSA.
About natural size.

Pileus fleshy, firm, at first hemispherical or subconical, then convex, very viscid or glutinous when moist, scaly, yellow. **Flesh** whitish. **Gills** close, adnate, yellowish becoming rust-color with age. **Stem** equal or slightly thickened at the base, scaly below the slight radiating floccose ring, solid or stuffed, yellow, generally rust-color at the base. **Spores** elliptical, $7.6\times5\mu$.

The Fat pholiota is a showy species. Its tufted mode of growth, rather large size, yellow color and rusty-brown scales make it a noticeable object. The stem is somewhat and the cap very viscid when moist, and this viscidity when dry gives it a shining appearance. The scales of the cap become erect or reflexed and sometimes appear blackish at the tips. They sometimes disappear with age. The flesh is firm and white or whitish. The gills when young are yellow or pale-yellow, but when mature they assume a ferruginous or rusty color like that of the spores. The stem is similar in color to the cap, but paler or nearly white at the top and usually reddish-brown or rusty-brown at the base. The collar is slight and often scarcely noticeable in mature specimens.

The **Cap** is 2–4 in. broad, the **Stem** 2–4 in. long and 4–6 lines thick. The plants commonly grow in tufts on stumps or dead trunks of deciduous trees in or near woods. They may be found from September to November. It is well to peel the caps before cooking. This species is not classed as edible by European authors, but I find its flavor agreeable and its substance digestible and harmless. *Peck*, 49th Rep. N. Y. State Bot.

Spores $8\times5\mu$ *W.G.S.;* elliptical, ferruginous, $7\times3\mu$ *Massee.*

Mt. Gretna, Pa. October until after frost. About trees and stumps and on logs. *McIlvaine.*

P. adiposa yields a substantial substance of good flavor.

P. flam'mans Fr.—*flamma*, flame. **Pileus** 2-4 in. broad, yellow- Pholiota. tawny, fleshy, convex then plane, somewhat umbonate, *absolutely dry*, sprinkled with *superficial, pilose,* somewhat concentric, *paler* or *sulphur-yellow, rough* or curly *scales;* margin at first inflexed, then spread when larger. **Flesh** thin, *light yellow.* **Stem** 3 in. long, 2-3 lines thick, stuffed then *hollow, equal,* most frequently flexuous, *very light yellow as are also the crowded rough scales.* **Ring** membranaceous, entire, not far removed from the pileus, of the same color. **Gills** *adnate* and without a tooth, somewhat thin, crowded, at the first *bright sulphur-yellow,* at length rust-color, edge quite entire.

Pileus by no means hygrophanous. It is distinguished from all others by the *sulphur-yellow scales on the tawny pileus.* Forming small clusters. Inodorous. The ring is sometimes only indicated by an annular zone. *Fries.*

Spores ellipsoid, 4x2µ *K.;* ellipsoid, 3-4x2-2.5µ *C.B.P.;* 4x2µ *W.P.;* 8x4µ *Massee.*

Quite plentiful in the New Jersey pines, from October until after heavy frosts. Caps seldom over 3 in. across. Solitary, and in clusters of not over half a dozen.

The caps fried are delicious.

P. luteofo'lia Pk.—*luteus*, yellow; *folium*, a leaf. **Pileus** firm, convex, dry, scaly, fibrillose on the margin, pale-red or yellowish. **Lamellæ** broad, subdistant, emarginate, serrate on the edge, yellow, becoming bright rust-color. **Stem** firm, fibrillose, solid, colored like the pileus, often curved from the place of growth. **Ring** obsolete. **Spores** bright rust-color, 7x4µ.

Plant subcespitose, 2-3 in. high. **Pileus** 1-2 in. broad. **Stem** 3-5 lines thick.

Trunks of birch trees. Forestburgh. September.

The general appearance of this plant is like A. variegatus or reddish forms of A. multipunctus. The reddish color appears sometimes to fade with age. *Peck,* 27th Rep. N. Y. State Bot.

Eagle's Mere, Pa. In clusters, on birch trees. August, 1898. *McIlvaine.*

Grows in quantity in the birch forests. The caps are delicious.

Pholiota. **P. ornel'la** Pk. (Agaricus ornellus Pk., 34 Rep., p. 42.) **Pileus** convex or nearly plane, slightly squamose, reddish-brown tinged with purple, the margin paler, floccose-appendiculate. **Gills** moderately close, yellowish or pallid, becoming brown. **Stem** equal or slightly thickened upward, solid, squamulose, pale-yellow, sometimes expanded at the base into a brownish disk margined with yellowish filaments. **Spores** brown, elliptical, 6–7.5x4–5μ.

Plant 1–2 in. high. **Pileus** about 1 in. broad. **Stem** 1 line to 1.5 lines thick.

Decaying wood. South Ballston, Saratoga county. October.

The scales of the pileus are sometimes arranged in concentric circles. The purplish tint is not always uniform, but in some instances forms spots or patches. *Peck*, 34th Rep. N. Y. State Bot.

Specimens, clustered, found by me on railroad ties at Haddonfield, N. J., September, 1897, had caps 1–1½ in. broad, of a dull green without tinge of purple; skin minutely cracked, showing the white flesh in the interstices; stem 1–2 in. long, 3–4 lines thick, slightly thickened upward, pale orange, solid, squamulose; ring floccose; taste when raw, slightly bitter. These were sent to Professor Peck who wrote: ''Appears to be a form of P. ornella Pk., but it differs some in color, being more of a green hue than of purple or olivaceous. It is pretty and I would like to know more about it before deciding on it fully.''

I have not since found it. Very palatable when cooked.

*** Hygrophani. *Gills cinnamon, etc.*

P. muta'bilis Schaeff.—changeable. **Pileus** about 2 in. broad, cinnamon when moist, becoming pale when dry, hygrophanous, slightly fleshy, convex then flattened, commonly obtusely umbonate, sometimes depressed, even and *smooth*, but when young occasionally scaly throughout. **Stem** about 2–3 in. long, 2 lines and more thick, *rigid*, stuffed then hollow, equal or attenuated downward, *scaly-rough as far as the ring*, *rust-color*, *blackish* or umber *downward*, often ascending or twisted. **Ring** membranaceous, externally scaly. **Gills** *adnato-decurrent*, crowded, rather broad, pallid then cinnamon. *Stevenson.*

Densely cespitose, variable in stature.

Spores ellipsoid-obovate, 6x11μ *W.G.S.*; 7x4μ *W.P.*; 9–11x5–6μ *Massee*; 11x7μ *Morgan*.

Edible. *Curtis.* Considered excellent in Europe.

P. margina′ta Batsch.—*marginatus*, margined. **Pileus** 1 in. and Pholiota.
more broad, honey-colored when moist, tan when dry, hygrophanous,
slightly fleshy, convex then expanded, obtuse, even, *smooth*, margin
striate. **Stem** about 2 in. long, 1–2 lines thick, *tubed*, equal, *fibrillose*
or slightly striate, *not scaly, of the same color as the pileus*, but becoming dingy-brown, and *commonly white velvety at the base*. **Ring** 1–2
lines distant from the apex, often in the form of a cortina and fugacious.
Gills *adnate*, crowded, thin, *narrow*, at first pallid, then darker cinnamon.

It varies much, and is deceptive on account of the vanishing veil. In
hedges there is a very small cespitose form with the pileus only ½ in.
broad, and the stem tough and smooth, with exception of the remains
of the fugacious cortina. There also occur on the ground among
mosses smaller and paler forms, which must be carefully distinguished
from A. unicolor, etc. *Stevenson.*

Spores 7–8x4µ *Massee.*

Haddonfield, N. J., November, December, 1896. In pine woods.
McIlvaine.

The caps of this small Pholiota, seldom over 1½ in. across, can be
gathered in goodly numbers where it frequents. They are of excellent
quality.

P. dis′color Pk.—changing color. **Pileus** thin, convex, then expanded or slightly depressed, smooth, viscid, hygrophanous, watery-
cinnamon and striatulate on the margin when moist; bright ochraceous-
yellow when dry. **Lamellæ** close, narrow, pallid then pale rust-color.
Stem equal, hollow, fibrillose-striate, pallid. **Ring** distinct, persistent.
Spores elliptical, 7x5µ.

Plant subcespitose, 2–3 in. high. **Pileus** 8–16 lines broad. **Stem**
1 line thick.

Old logs in woods. Greig. September.

The change of color from the moist to the dry state is very marked.
This species resembles Agaricus autumnalis, in which the annulus is
fugacious and the spores are longer. The edge of the gills in both is
white-flocculose. *Peck*, 25th Rep. N. Y. State Bot.

Two forms of this species are found. One has a scattered form of
growth, the other found on decaying wood of birch is cespitose. The

Pholiota. species is allied to P. marginata, from which it is readily distinguished by its viscid pileus. *Peck*, Rep. 44.

Var. *discolor minor* Pk. Small. Pileus 6–10 lines broad, chestnut color when young or moist. Stem about 1 line thick, at first clothed with whitish fibrils.

Among mosses about or on the base of stumps. September. *Peck*, Rep. 46.

West Virginia. Eagle's Mere, Mt. Gretna, Pa. August to frost. On decaying wood. *McIlvaine*.

This little Pholiota is abundant where it does grow. In the West Virginia forests I have seen logs with many tufts of it upon each. The caps are fairly good.

INO'CYBE Fr.

Gr.—fiber; *Gr.*—head.

Universal veil somewhat fibrillose, concrete with the cuticle of the Inocybe. pileus, often free at the margin, in the form of a cortina. **Gills** somewhat sinuate (but they occur also adnate and in two species decurrent), changing color, but not powdered with cinnamon. **Spores** often rough, but in others even, more or less brownish-rust color.

Inocybe (with Hebeloma) corresponds with Tricholoma. Inocybe and Hebeloma have some common features, but they are really very distinct. Inocybe is readily distinguished by the fibrillose covering of the pileus, which never has a distinct pellicle, by the

(Plate LXXV.)

INOCYBE LANUGINOSA.
One-fourth natural size.

veil which is continuous and homogeneous with the fibrils of the pileus, and by the rusty-brown spores. All grow on the ground. They are (mostly) strong-smelling (commonly nauseous). None are edible. *Stevenson.*

None reported as either edible or poisonous. Those I have tested are not pleasant.

PLUTE'OLUS Fr.

Dim. of *pluteus*, a shed.

Pluteolus. **Pileus** conical or bell-shaped, then expanded, rather fleshy, viscid, margin at first straight and pressed to the stem. **Gills** *free, rounded behind.* **Stem** somewhat cartilaginous, its substance different from that of the pileus.

Growing on wood.

Spores rust or saffron color. Pluteus, the only genus having the same structure, is separated by its salmon-colored spores.

P. reticula'tus Pers.—*rete*, a net. From the net-work of veins on the pileus. **Pileus** slightly fleshy, bell-shaped, then expanded, sticky, reticulate with anastomosing veins, pale violaceous, striate on the margin. **Lamellæ** free, ventricose, crowded, rusty-saffron. **Stem** hollow, fragile, fibrillose, mealy at the top, white. **Spores** elliptical, ferruginous, 10–13x5–6.5μ.

(Plate LXXVI.)

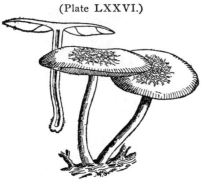

PLUTEOLUS RETICULATUS.
About natural size.

Pileus 1–2 in. broad. **Stem** 1–2 in. long, 1–2 lines thick.

Decaying wood. Cattaraugus county. September.

The specimens which I have referred to this species appear to be a small form with the pileus scarcely more than an inch broad and merely wrinkled on the disk, not distinctly reticulate as in the type. In the dried specimens the pileus has assumed a dark violaceous color. The dimensions of the spores have been taken from the American plant. I do not find them given by any European author. *Peck*, 46th Rep. N. Y. State Bot.

In October, 1897, P. reticulatus grew in large quantities on a fallow lot close by the University of Pennsylvania, Philadelphia. The lot was thickly covered with tall heavy-stemmed weeds, a mat of which, from the year before was present. The reticulations upon the cap are intricate and distinct. I have not seen it since.

The whole plant is tender and of fine flavor.

HEBELO'MA Fr.

Hebe, youth; *loma*, fringe.

Partial veil fibrillose or absent. **Pileus** smooth, continuous, some- Hebeloma. what viscid, margin at first incurved. Flesh of stem continuous with that of the pileus; fleshy, fibrous, clothed, top rather mealy. **Gills** attached, notched at the stem, edge inclined to be pale. **Spores** clay-colored.

On the ground.

Closely allied to Inocybe, formerly included in Hebeloma, but differing in the character of the cuticle of the pileus which in Inocybe is scaly or fibrillose. Many of the species are strong in smell and taste. None have hitherto been considered edible and some have been regarded as poisonous

ANALYSIS OF TRIBES.

INDUSIATI (*indusium*, a garment). Page 283.

Furnished with a ring from the manifest veil, which often makes the margin of the pileus superficially silky.

DENUDATI (*denudo*, to lay bare). Page 286.

Pileus smooth. Veil absent. None known to be edible.

PUSILLUS (*pusus*, a little boy).

Pileus scarcely an inch broad. None known to be edible.

The writer has not as yet investigated the edible qualities of this genus to his satisfaction. Much work remains to be done. But two species of Hebeloma are given as edible. They are good, but do not rank above second-class. Several others have been tested, but not in sufficient quantity to report upon their quality with perfect safety. So far as tested the species have been harmless.

INDUSIA'TI. With a ring, etc.

H. mus'sivum Fr.—*mussivus*, undecided. (Uncertain in generic place.) **Pileus** 2–4 in. broad, either of one color, yellow or darker at the disk which is like a smooth sugar-cake, fleshy, *compact*, firm, con-

Hebeloma. vex then plane, unequal, very obtuse, viscid, at first *smooth* and even, margin bent inward, even, then commonly turning upward and broken up into scales. **Flesh** thick, becoming yellow. **Stem** 4 in. long, commonly 1 in. thick, *very fleshy*, sometimes stuffed, sometimes hollow at the top, equal or broad in the middle, *wholly fibrillose and powdered at the top, light yellow.* **Veil** fibrillose, very evanescent. **Gills** emarginate, somewhat crowded, 3 lines broad, dry (not distilling drops), *at first light yellow*, then together with the spores somewhat rust-colored.

Odor weak, not unpleasant. Very distinct. It departs widely from all the following species in its habit and bright colors. The habit is that of a Flammula or Cortinarius, but the gills are emarginate and not powdered; from the turned up pileus and from the stem being powdered at the top, and from other marks it is to be referred to Hebeloma. *Fries.*

Spores elliptical, 12x6μ *Massee.*

New Jersey, Haddonfield. Under pine trees. Solitary. Frequent. September, 1896. *McIlvaine.*

Not previously reported.

Taste, even raw, is pleasant. It is meaty and the meat is good. It requires slow cooking and is best chopped fine and served in patties or croquettes.

H. fasti'bile Fr.—*fastidibilis*, loathsome. From the smell. **Pileus** 2 in.
(Plate LXXVIa.)

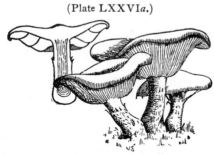

HEBELOMA FASTIBILE.
One-fourth natural size.

and more broad, pale yellowish, tan or becoming pale, compactly fleshy, convexo-plane, obtuse, somewhat wavy, even, smooth, the turned-in margin downy. **Stem** 2–3 in. long, ½ in. thick, *solid*, wholly fleshy-fibrous, stout, somewhat bulbous, often twisted, everywhere *white-silky and fibrillose*, white, but varying pallid, white-scaly upward. **Cortina** remarkable, white, occasionally in the form of a ring. **Gills** remarkably *emarginate, somewhat distant*, rather broad, at first becoming pale-white, then dingy clay-color, edge whitish, *distilling drops* in rainy weather.

Somewhat cespitose   Odor and taste of radish, bitterish. Like A.

crustiliniformis; the odor is the same except that it is stronger, but it Hebeloma. differs conspicuously *in the manifest veil and somewhat distant gills.*

Var. *al'ba*, stem longer, equal, somewhat hollow, fibrous-scaly at the apex, gills distant. A. spiloleucus Krombh., A. sulcatus Lindgr. is an elegant form with the margin of the pileus sulcate or rugoso-plicate. In mixed woods. Common. July to October. *Stevenson.*

Spores 11x8μ *W.G.S.;* elliptical, pointed, 10x8μ *Morgan.*

Var. *elegans.* Pileus purple-brown.

This sometimes appears on disused mushroom beds in large quantities, but the method by which the spores gain access is involved in darkness.

"A very suspicious species and has the reputation of being noxious." *Cooke.*

"There is considerable external resemblance between this and A. campestris. No fungus is so often mistaken for A. campestris as this dangerous plant. *W. G. Smith.*

This species is considered noxious abroad. No test is reported of its qualities here.

I have not seen it.

H. glutino'sum Lind.—*gluten*, glue. ([Color] Plate LXXI, fig. 1.) **Pileus** about 3 in. broad, yellow-white, the disk darker, fleshy, convex then plane, *regular*, obtuse, with a tenacious *viscous* gluten, and slimy in wet weather, *sprinkled with white superficial scales.* **Flesh** whitish, becoming light-yellow. **Stem** 3 in. long, *stuffed*, firm, *somewhat bulbous*, *white-scaly* and fibrillose, and white-mealy at the top, often rough with bundles of hairs at the base, at length rust-color within. Partial thread-like veil manifest, in the form of a cortina. **Gills** sinuato-adnate, somewhat decurrent, crowded, broad, *pallid then light-yellowish*, at length clay-cinnamon. Odor peculiar, mild.

On branches and among leaves, oak and beech. Frequent. September to December. *Stevenson.*

Spores 5x4μ *W. P.;* plum-shaped, 7μ *Q.;* elliptical, 10–12x5μ *Massee;* ellipsoid, 6–7x3–4μ *K.*

New York. Among fallen leaves and half-buried decaying wood, in thin woods. Conklingville. September. In wet weather the gluten is sufficiently copious to drop from the pileus. *Peck,* Rep. 40.

Haddonfield, N. J., among leaves in mixed woods. Frequent. 1896.

Hebeloma. Mt. Gretna, Pa., among leaves under oaks. Frequent. September to November. *McIlvaine.*

Caps 1 ½–3 in. across. Remarkably glutinous, shining as if varnished when wet. Partial veil not always noticeable.

The odor and taste are pleasant. The caps when well cooked are meaty, good, but of second quality.

DENUDA'TI. Pileus smooth, etc.

H. crustulinifor'me Bull.—*crustulum*, a small pie; *forma*, form. **Pileus** pale-whitish tan, most frequently pale-yellowish or brick-color at the disk, fleshy, convexo-plane, obtuse or slightly gibbous with an obtuse umbo, somewhat spreading with an uneven margin, even, smooth, at first slightly viscid, not zoned. **Flesh** transparent when moist. **Stem** *stuffed then hollow*, stout, somewhat bulbous, white, naked, white-scaly at the top. **Gills** *rounded-adnexed*, crowded, *narrowed*, 1 line broad and linear, thin, whitish then clay-color, at length date-brown, *the unequal edge distilling watery drops in wet weather, spotted when dry.*

Veil quite *wanting*. Odor strong, fetid, of radish. Very variable in stature; the stem, however, is never elongated as in A. elatus, etc.; in smaller specimens equal, pileus regular, gills almost adnate.

In mixed woods. Common. August to November. *Stevenson.*

Spores ellipsoid, 10–12x5–7µ *K.;* 9x5µ *W.G.S.*

Var. *mi'nor* Cke. Smaller than the type.

Minnesota, common in woods, *Johnson;* California, *H. and M.;* Wisconsin, *Bundy;* New Jersey, *Ellis;* Vermont, *Burt* (Lloyd); New York, *Peck,* 41st Rep.; Mt. Gretna, Pa., November, 1898. In woods. *McIlvaine.*

But one specimen found and that was sent to Professor Peck. Taste bitter.

Regarded as poisonous by European writers. It is not reported as tested in America.

FLAM'MULA Fr.

Flamma, a flame.

(In reference to the bright colors of many of the species.)

Pileus fleshy, margin *at first turned inward*. **Veil** fibrillose or none. _{Flammula.} Stem fleshy-fibrous, not mealy at the top. **Gills** decurrent or attached without a tooth. **Spores** mostly pure rust color; some brownish-rust, others tawny-ochraceous.

A few species grow on the ground, the majority on wood.

ANALYSIS OF TRIBES.

GYMNOTI (naked). Page 288.

Pileus dry, generally scaly. Spores not yellowish.

LUBRICI (*lubricus*, slimy). Page 289.

Pileus covered with a continuous, *viscid*, smooth, partly separable cuticle. Veil fibrillose. Spores not yellowish. Gregarious, on the ground, rarely on wood. Distinguished from Hebeloma by the gills not being sinuate and the top of the stem not mealy.

UDI (*udus*, moist). Page 290.

Veil slight, generally hanging in fragments. Cuticle of the pileus continuous, not separable, smooth, in places superficially downy, moist or slightly viscid in rainy weather. Spores not yellowish. Cespitose, growing on wood.

SAPINEI (*sapinus*, pine). Page 291.

Veil silky, very slight, adpressed to the stem or forming a silky ring on it. Cuticle of pileus thin, the flesh splitting at the surface into scales, not viscid. Distinguished by the gills and spores being light yellow or tawny. Somewhat cespitose; always on pine or on the ground among pine branches.

SERICELLI (*sericeus*, silky).

Cuticle of the pileus slightly silky, dry or at the first viscid None known to be edible.

Flammula. The genus **Flammula** is not represented in our territory by a large number of species. It is, nevertheless, not very sharply distinct from the allied genera, Pholiota, Hebeloma and Naucoria. From Pholiota it is especially separated by the slight development of the veil which is merely fibrillose or entirely wanting. It never forms a persistent membranous collar on the stem. From Hebeloma it may be distinguished by the absence of a sinus at or near the inner extremity of the gills, by the absence of white particles or mealiness from the upper part of the stem and by the brighter or more distinctly rusty or ochraceous color of the spores. From Naucoria the fleshy or fibrously fleshy stem affords the most available distinguishing character. The genus belongs to the Ochrosporæ or ochraceous-spored series, but the spores of its species vary in color from ochraceous or tawny-ochraceous to rust-color or brownish-rust color. The three things to be especially kept in mind in order to recognize the species are the color of the spores, the adnate or decurrent but not clearly sinuate gills and the fleshy or fibrously fleshy stem without a membranous ring.

Our species are mostly of medium size, none being very small and one only meriting the appellation large. They appear chiefly in late summer or in autumn and grow in woods or in wooded regions either on the ground or more often on decaying wood. Many are gregarious or cespitose in their mode of growth. Some have a bitterish or unpleasant flavor and none of our species has yet been classed as edible. Fries arranged the species in five groups, of which the names and more prominent characters are here given. *Peck*, 50th Rep. N. Y. State Bot.

The few species which the writer has found to be edible, and the two new species found by him, were tested after the publication of the above. Several of the species found are not mentioned herein for the reason that a sufficient quantity was not obtained to make certain their quality as a food. The bitterness, as far as observed, with which most of the species are tainted disappears in cooking.

GYMNO'TI. Veil absent, pileus dry, etc.

F. alie'na Pk. **Pileus** thin, flexible, broadly convex, umbilicate, dry, bare, slightly striate on the margin when old, grayish or pale grayish-brown. **Flesh** white, fibrous. **Gills** thin, subdistant, bow-shaped, decurrent, ochraceous-brown. **Stem** firm, fibrous-striate, solid, slightly tapering upward, colored like the pileus, covered at the base Flammula. with a dense white tomentum. **Spores** rusty-brown, globose, 5µ broad.

Pileus 3–5 cm. broad. **Stem** 5 cm. long, 4–6 mm. thick.

Gregarious on partly burned anthracite coal, Mt. Gretna, Pa. September. *C. McIlvaine.*

The species is peculiar in its color and habitat. In the dried specimen the gills have assumed a brown color with no ochraceous tint. Mr. McIlvaine remarks that it is an edible species, dries well, and is excellent when cooked. Its relationship is with F. anomala Pk., but it is a larger plant with darker color and a different habitat. *Peck*, Bull. Torr. Bot. Club, Vol. 26, F. 1899.

It grows on partly *burned* anthracite coal, not *buried*, as printed in the Torrey Bulletin. The mycelium completely involves the pieces of coal, holding them tightly in its meshes. Patches of it were strictly limited to the size of the ash-pile containing the partly burned coal. Quite fifty were found.

As stated, it is edible, and it is of remarkably fine substance for a Flammula.

LU'BRICI. Pileus viscid, etc.

F. edu'lis Pk.—eatable. **Pileus** fleshy, convex, obtuse, glabrous, moist, brown, grayish-brown or yellowish-brown, sometimes rimose. **Flesh** whitish. **Lamellæ** rather broad, close, decurrent, bright tan color, becoming brownish-rusty. **Stems** cespitose, equal, stuffed or hollow, brown. **Spores** subelliptical, 13×5–6μ.

Pileus 2–3 in. broad. **Stem** 2–3 in. long, 3–6 lines thick.

Grassy ground, along pavements, in gutters and by the side of wooden frames of hotbeds. Haddonfield, N. J. October. *C. McIlvaine.*

The collector of this species informs me that the flavor of the fresh plant is slightly bitter, but that this disappears in cooking and the fungus furnishes a very good and tender article of food. Successive crops continued to appear for a month. In the dried specimens the stem is striate. *Peck*, Bull. Torr. Bot. Club, Vol. 24, No. 3.

This new species appears annually in the same place. I have not found it elsewhere. It is meaty and excellent.

UDI.   Pileus smooth, not viscid; veil fragmentary, etc.

Flammula.   **F. alni cola** Fr.—*alnus*, alder; *colo*, to inhabit.   **Pileus** 2–3 in.

(Plate LXXVI*b*.)

FLAMMULA ALNICOLA.
Two-thirds natural size.

broad, *yellow*, at length becoming rust-color and sometimes green, fleshy, convex then flattened, obtuse, slimy when moist, but not truly viscous, at the first superficially fibrillose toward the margin.   **Flesh** not very compact, of the same color as the pileus.   **Stem** 2–3 in. and more long, ½ in. thick, *stuffed then hollow*, attenuato-rooted, commonly curved-flexuous, *fibrillose*, at first yellow, then becoming rust-color.   **Veil** *manifest*, sometimes fibrillose, sometimes woven into a spider-web veil.   **Gills** somewhat adnate, broad, plane, at first *dingy-pallid* or yellowish-pallid, at length together with the plentiful spores rust-colored.

The gills vary decurrent and rounded according to situation.   Odor and taste bitter.   There are two forms: *a*. Pileus irregular, fibrillose round the margin; gills at first dingy-pallid.   *b*. Salicicola, pileus somewhat convex, smooth, rarely at the first downy-scaly; gills at first yellowish-pallid.   *Fries.*

Spores subelliptical, 8x5µ *K.;* 8–10x5–6µ *Peck.*

New York, swampy woods about base of alders, October, *Peck*, Rep. 35; at base of alders, with adnate gills, and on birch stumps, with the gills rounded behind, Rep. 39.   Mt. Gretna, Pa., New Jersey, mixed woods, August to November, 1898, *McIlvaine*.

Gregarious and in loose tufts, not plentiful.   It is a pretty plant, usually of a bright yellow, sometimes darker at the center of cap. Traces of an evanescent fibrillose ring are occasionally found or the fibrils adorn the margin of the cap.   The gills next to the stem are either rounded, attached or slightly decurrent.

Raw the taste is slightly bitter.   This disappears in long cooking.

290

F. fla'vida Schaeff. (Pers.)—*flavidus*, light yellow. **Pileus** fleshy, Flammula. thin, broadly convex or nearly plane, glabrous, moist, pale yellow. **Flesh** whitish or pale yellow, taste bitter. **Lamellæ** moderately close, adnate, pale or yellowish becoming rust-color. **Stem** equal, often more or less curved, hollow, fibrillose, whitish or pale yellow, with a white mycelium at the base. **Spores** 8x5μ.

Pileus 1–2 in. broad. **Stem** 1–3 in. long, 1–3 lines thick.

Decaying wood of various trees. Commonly in wooded or mountainous districts. Summer and autumn.

Our specimens were found on wood of both coniferous and deciduous trees. The plants are sometimes cespitose. The pileus becomes more highly colored in drying. The spores are pale rust-colored approaching ochraceous. In Sylloge the spores of this species are described as pale yellowish. *Peck*, 50th Rep. N. Y. State Bot.

Spores broadly elliptical, 6–8x5μ *Massee.*

New York, decaying wood, *Peck*, Rep. 32, 50; *Mrs. E. C. Anthony*, August. West Virginia, 1881–1885; Mt. Gretna, Pa. August to October. *McIlvaine.*

F. flavida is a frequent species, gregarious and tufted on decaying vood, either standing, fallen, or as roots in the ground. The texture and substance are good. The slight bitter when raw disappears in cooking. The caps, only, are tender.

SAPIN'EI. Gills and spores yellowish, etc.

F. hy'brida Fr.—*hybrida*, a hybrid. **Pileus** about 2 in. broad, at first tawny-cinnamon, then tawny-orange, fleshy, hemispherical with the margin involute, then expanded, obtuse, regular and well formed, even, *smooth, moist.* **Flesh** moderately compact, pallid. **Stem** 2–3 in. long, 4–5 lines thick, at first *stuffed* with a soft pith, then hollow, *attenuated* (almost conico-attenuated) *upward*, whitish with adpressed silky-hairy down (becoming tawny when the down is rubbed off) slightly striate, with white hairs at the base, and somewhat mealy at the apex. **Veil** *manifest in the form of an annular zone at the apex of the stem*, white or at length colored with the spores. **Gills** adnate, somewhat crowded, *light yellow then tawny, not spotted. Fries.*

Spores elliptical, tawny-ochraceous, 7–8x4–5μ *Massee;* 6x4μ *W. P.*

291

Flammula. Mt. Gretna, Pa., August, September, 1898. On ground under pine trees. Gregarious. *W. H. Rorer.* Not elsewhere reported.

This is a handsome plant, quite prolific in the large pine groves at Mt. Gretna, Pa. The caps are of good flavor.

F. mag'na Pk.—*magnus*, large. **Pileus** fleshy, broadly convex, soft, dry, fibrillose and somewhat streaked, pale yellow or buff, the margin commonly becoming revolute with age. **Flesh** whitish or yellowish. **Gills** close, adnate or slightly decurrent, often crisped or wavy toward the stem, about three lines wide, ochraceous. **Stem** equal or thickened toward the base, fleshy-fibrous, solid, elastic, fibrillose, colored like the pileus, brighter yellow within. **Spores** subelliptical, ochraceous, 10x6μ.

Cespitose. **Pileus** 4–6 in. broad. **Stem** 3–4 in. long, 8–12 lines thick.

About the base of trees. Westchester county. October.

This is a large and showy species. The stems are sometimes united at the base into a solid mass. The young gills are probably yellow, but I have seen only mature specimens. *Peck,* 50th Rep. N. Y. State Bot.

New Jersey, Trenton, ground in clearing, in pairs and singly. November, *E. B. Sterling;* Mt. Gretna, Pa. Mixed thin woods. October to November. Near trees. Cespitose, *McIlvaine.*

Individuals of all ages were found and eaten. The young gills are very light yellow, darkening to a deep, rich yellow.

The caps are of good substance and flavor. When very young the stems are edible.

TUBA'RIA W.G.S.

Tuba, a trumpet.

(Plate LXXVII.)   Tubaria.

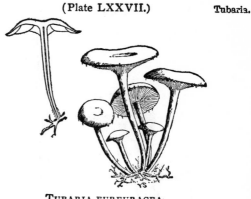

Stem *somewhat cartilaginous, fistulose.* **Pileus** somewhat membranaceous, often clothed with the universal floccose veil. **Gills** *somewhat decurrent.* **Spores** rust-color or (in Phæoti) brownish-rust color.

The species referred to this subgenus were taken from Naucoria and Galera because they correspond with Omphalia and Eccilia. The pileus is, however, distinctly umbilicate or depressed in only a few of them; the others are placed here on account of their somewhat decurrent gills, which are broadest behind and triangular. *Fries.*

Small and unimportant.

TUBARIA FURFURACEA.
Natural size.

NAUCO'RIA Fr.

Naucum, a nut-shell.

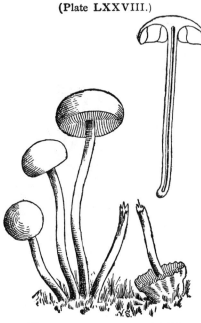

(Plate LXXVIII.)

Naucoria.

Pileus more or less fleshy, conical or convex, then expanded, *margin at first incurved*. **Gills** free or adnate, not decurrent. **Veil** fugacious or absent, sometimes attached in minute flakes to the edge of the young pileus. **Stem** cartilaginous, hollow or with a spongy stuffing. Growing on wood or on the ground, sometimes rooted. **Spores** various shades of brown, dull or bright.

Naucoria corresponds with Collybia, Leptonia and Psilocybe; from the latter it is distinguished by the spore colors and from Galera in the brown-spored series by the margin of the pileus being at first incurved.

"The spores are rust-color, or brownish rust-color. The color of the pileus is some shade of yellow. The stem is not distinctly ringed, but sometimes a slight spore-stained band marks the place of the obsolete ring." *Peck*, 23d Rep. N. Y. State Bot.

NAUCORIA SEMI-ORBICULARIS.
Natural size.

The members of this genus are with two or three exceptions very common, and common over the land. The greater number grow on the ground among grass; a few grow upon decaying wood. The stems are not of the same texture as the cap and frequently will not cook tender. The caps, however, are, of all species tested, tender and of good flavor. Species of the genus are among the first to appear in spring, and well reward the enterprising mycophagist for his early tramps.

ANALYSIS OF TRIBES.

GYMNOTI (*Gr.*—naked). Page 295.

Pileus smooth. Veil absent. Spores rust-color, not becoming dusky-rust-color.

PHÆOTI (*Gr.*—dusky).   Page 296.

Pileus smooth. Gills and spores dusky rust-color. Veil rarely mani- Naucoria.
fest.

LEPIDOTI (*lepis*, a scale).

Pileus flocculose or squamulose. Veil manifest.
None known to be edible.

I.—GYMNO'TI.

N. hama'dryas Fr.—*Gr.*, a nymph attached to her tree. **Pileus**
1½–2 in. broad, *bay-brown-ferruginous* when young and moist, pale
yellowish when old and becoming pale, slightly fleshy, convex then expanded, gibbous, even, smooth. **Stem** 2–3 in. long, 3 lines thick,
somewhat fragile, hollow, equal, naked, smooth, *pallid*. **Gills** *attenuato-adnexed*, somewhat free, slightly ventricose, almost 2 lines broad,
crowded, rust-color, opaque. **Veil** none. Widely removed from neighboring species. Pileus *somewhat separate* as in Plutei. *Fries.*

Spores elliptical, rust-color, 13–14x7μ *Massee.*

Haddonfield, N. J. Frequent. Solitary. On ground along pavements, under trees, in woods. Spring to autumn. *McIlvaine.*

Massee gives it as hygrophanous. I have not found it so. It is
moist after rain and dew.

The caps and upper part of the stem are tender, easily cooked and of
good flavor.

N. cero'des Fr. *Gr.*—wax. **Pileus** ½–1 in. broad, watery cinnamon when moist, tan-color *when dry*, somewhat membranaceous, *convex bell-shape* and flattened, at length depressed, *obtuse*, when moist
smooth, pellucid-striate at the circumference, *when dry* even, *slightly
silky-atomate*. **Stem** 2–3 in. long, 1–2 lines thick, slightly firm, tubed,
equal, somewhat flexuous, fibrilloso-striate under a lens, *becoming dingy
bay-brown* sometimes for the most part, sometimes only at the base,
pallid upward, mealy at the apex. **Gills** adnate, separating, *very broad
behind*, hence almost triangular, *somewhat distant*, broad, plane, soft,
distinct, pallid then cinnamon very finely fimbriated at the edge under
a lens. *Fries.*

The typical form, growing among damp mosses, is quite early, gregarious, with the colors almost those of Galera hypnorum, but other-

Naucoria. wise very different. *b.* Another form occurs on naked, commonly burnt soil, in late autumn, with almost the habit of N. pediades, but with a different color of gills and spores; this form is firmer. **Stem** 1 in. long, tense and straight, and color more ochraceous. *Stevenson.*

Spores 9µ *B. and Br.;* smooth, 6x3µ *Massee.*

West Virginia, New Jersey, Pennsylvania, in grass and moss, along damp wood margins. August to October. *McIlvaine.*

N. cerodes is not plentiful where I have found it. Enough has been collected at a time to prove it esculent. It is tender, but has not much flavor.

(Plate LXXVIIIa.)

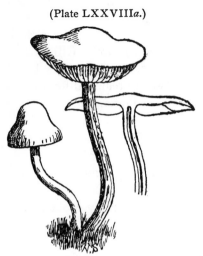

NAUCORIA STRIAPES.

N. stri'apes Cke.—*stria,* a line; *pes,* a foot. **Pileus** 1–1½ in. broad, ochraceous, bell-shaped, obtuse, then expanded, smooth, even. **Stem** 2–3 in. long, 2 lines thick, hollow, equal, erect or flexuous, white, *longitudinally striate*. **Gills** slightly adnate behind, rather distant, tawny rust-color.

Cespitose or gregarious. Among grass on lawn. *Stevenson.*

Spores narrowly elliptical, 10–12 x4µ *Massee.*

New Jersey, Trenton. Growing among leaves near dump. May to November. *E. B. Sterling.*

The few specimens tested were delicate and of slight flavor.

II.—Phæ'oti.

N. pedi'ades Fr.—*Gr.,* a plain. **Pileus** 1–2 in. broad, *yellow* or pale yellowish-ochraceous then becoming pale, slightly fleshy, convex then plane, obtuse, even, dry, smooth, at length crookedly cracked, but always without striæ. **Flesh** white. **Stem** 2–3 in. long, 1–2 lines thick, *stuffed with a pith, somewhat flexuous*, tough, equal, but with a small bulb at the base, *slightly silky becoming even, yellowish*. **Gills** adnexed, 2 lines broad, at first crowded, at length somewhat distant, *somewhat dingy-brown, then dingy cinnamon.*

Spores brownish-rust-color. The small bulb at the base is formed by Naucoria. the mycelium being rolled together. Stature variable. *Fries.*

Spores dingy rust-color, elliptical, 10–12x4–5µ *Massee.*

West Virginia, New Jersey, Pennsylvania, in grassy places, pastures and along pavements. Common. May to November. *McIlvaine.*

In 1897 Fairmount Park, Philadelphia, abounded with N. pediades, which were collected and eaten by many. The caps are tender and of a mushroom flavor.

N. semi-orbicula′ris Bull.—*semi*, half; *orbicularis*, round. (Plate LXXVIII, p. 294.) **Pileus** 1–2 in. broad, *tawny rust-color* then ochraceous, slightly fleshy, convexo-expanded, obtuse, dry, even, smooth, corrugated when dry. **Stem** 3–4 in. long, scarcely beyond 1 line thick, cartilaginous, tough, slender, tense and straight, equal, even, smooth, becoming pallid *rust-color*, shining, often darker at the base, *internally containing a separate narrow tube* which is easily broken up into fibrils. **Gills** adnate, rarely sinuate behind, almost 3 lines *broad*, and many times broader than the flesh of the pileus, crowded, *pallid then rust-color*.

The pileus is slightly viscid when fresh and moist. Easily distinguished from S. semi-globatus, with which it has been confounded, by the stem. *Stevenson.*

Spores 14x8µ *W.G.S.;* 10x5–6µ *Massee.*

Allied to N. pediades, distinguished by its viscid cap when moist, and dark stem.

Common over the states. Washington, D. C., *Mrs. Mary Fuller.*

West Virginia, Pennsylvania, North Carolina, New Jersey. Solitary, sometimes cespitose, very common on lawns, rich pastures, etc. April until frost. *McIlvaiue.*

This is one of our first appearing toadstools, coming up when the grass shows its full spring hue. It is found after rains until the coming of frost. Its hemispherical caps, precise, neat, dark gills and brown spores readily distinguish it. While usually small, patience and pick= ing will soon gather quarts. The caps cook easily and are of excellent flavor.

N. platysper′ma Pk.—*platys*, broad; *sperma*, seed. **Pileus** convex, becoming nearly plane, glabrous, slightly tinged with ochraceous or reddish-yellow when young, soon whitish, the margin at first adorned with

Naucoria. vestiges of a white flocculent veil. **Flesh** white. **Lamellæ** moderately close, slightly rounded behind, pallid, becoming brownish. **Stem** equal, stuffed with a white pith, slightly flocculent or furfuraceous above when young, whitish, the mycelium sometimes forming white thread-like strands. **Spores** broadly elliptical, 15μ long, 12.5μ broad.

Pileus 1–1.5 in. broad. **Stem** 3–5.5 in. long, 1.5–2 in. thick.

On the ground. Compton, Cal. *Prof. A. J. McClatchie.*

'This species differs from N. pediades and N. semi-orbicularis, to which it is related, by its larger, broader spores and paler color. *Peck*, Bull. Torr. Bot. Club, Vol. 25, No. 6.

This new species reported from California is so closely allied to N. semi-orbicularis and N. pediades, both of which are edible, that it is here given, that it may be recognized by students on the Pacific coast or wherever it occurs.

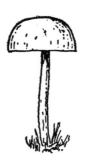

GALE'RA Fr.

Galerus, a cap.

Pileus more or less membranaceous, conical or oval, then expanded, Galera. striate, margin at the first straight, then adpressed to the stem. **Gills** not decurrent. **Stem** somewhat cartilaginous, continuous with the pileus, but differing in texture, tubular. **Veil** none or fibrillose. **Spores** tawny-ochraceous.

Slender, fragile, generally growing on the ground.

Galera corresponds with Mycena, Nolanea, Psathyra and Psathyrella, which are distinguished by their spore colors. In the brown-spored series Naucoria is separated by the margin of the pileus being at first incurved, and Tubaria by the decurrent gills.

The genus is composed of small species, but many grow in clusters, and are of a consistency which decreases but little in quick cooking. Those tested are delicate in texture and flavor.

G. lateri'tia Fr.—*later*, a brick. **Pileus** 1 in. high, *pale yellowish when moist*, ochraceous when dry, hygrophanous, membranaceous, *acorn-shaped then bell-shaped*, obtuse, even, smooth, slightly and densely striate at the margin when moist. **Stem** 3 in. and more long, 1 line thick, tubular, attenuated upward, tense and straight, even, but *white-pruinose, whitish*. **Gills** *adnexed* in the top of the cone, hence appearing as if free, ascending, very narrow, *crowded, cinnamon*.

Gills almost adpressed to the stem, almost pendulous. Remarkably analogous with A. ovalis, but easily distinguished by the *linear gills* and the absence of a veil; very fragile. *Fries*.

Spores 11x5µ *W.P.;* 11–12x5–6µ *Massee.*

West Virginia, New Jersey, North Carolina, Pennsylvania. On dung and rich pastures. June to frost. *McIlvaine.*

The narrow conical cap, distinctly striate, distinguishes this species from G. tenera. In quality there is no difference. It is a well-flavored, delicate species.

Agaricaceæ

Galera.

G. te'nera Schaeff.—*tener*, tender.

(Plate LXXIX.)

GALERA TENERA.
Two-thirds natural size.

Pileus ½ in. and more high, *of one color, pallid rust-color when damp*, becoming pale when dry, hygrophanous, somewhat membranaceous, *conico-bell-shaped*, commonly smooth, slightly striate when moist, wholly even when dry, opaque, somewhat atomate. **Stem** commonly 3–4 in. long, 1 line thick, tubular, fragile, equal or when larger thickened downward, *tense and straight, somewhat shining*, striate upward, of the same color as the pileus when moist, and like it becoming pale when dry. **Gills** *adnate in the top of the cone*, appearing *as if free*, ascending, somewhat crowded, *linear, cinnamon*.

Pastures and grassy places in woods. Common. May to November. *Stevenson.*

Spores ellipsoid, 14–21x8–12μ *K.;* 14–8μ *W.G.S.;* 14x7μ *W.P.;* 12–13x7μ *Massee;* elliptical, dark rust-color, almost rubiginous, 13–16.5x8–10μ *Peck.*

Var. *pilosella* (Agaricus pilosellus Pers.), has both pileus and stem clothed with a minute erect pubescence when moist. A form is sometimes found in which the center of the pileus is brown or blackish-brown. *Peck*, 46th Rep. N. Y. State Bot.

Var. *obscu'rior* Pk. A notable form of this species was found growing in an old stable of an abandoned lumber camp. The plants were large, the pileus in some being more than an inch broad, the stems were 3–6 in. long and the color was rust-colored as in G. ovalis, to which the plants might be referred but for the large spores. Essex county. July. I have labeled the specimens variety *obscurior*. *Peck*, 50th Rep.

Haddonfield, N. J.; Chester county; West Philadelphia, Pa.; West Virginia. In rich pastures, on lawns, dung in woods. Common. June to October. *McIlvaine.*

Very variable in size and in color when wet and dry. The color of gills and spores readily distinguishes it in its habitats. From spring to

300

frost it can usually be gathered in quantity. It is small, tender, shrivels Galera. in cooking, but makes a savory, excellent dish.

Var. obscurior found cespitose on very old manure at a ruined stable, Mt. Gretna, Pa., August. *McIlvaine.*

G. fla′va Pk.—*flavus*, yellow. **Pileus** membranous, ovate or bell shaped, moist or subhygrophanous, obtuse, plicate striate on the margin, yellow. **Lamellæ** thin, narrow, crowded, adnate, at first whitish, then yellowish-cinnamon. **Stem** equal or slightly tapering upward, hollow, slightly striate at the top, sprinkled with white mealy particles, white or yellowish. **Spores** ovate or subelliptical, brownish-rust-color, 13x8μ.

Pileus 6–12 lines broad. **Stem** 2–3 in. long, 1–1.5 lines thick.

Damp vegetable mold in woods. Tompkins county. July.

This species is well marked by the pale-yellow color of the pileus and its plicate striations which are very distinct even in the dried specimens. They extend half way to the disk or more. When dry the pileus is seen to be sprinkled with shining atoms as in some other species of the same genus. Occasionally the yellow cuticle cracks into squamules or small scales. *Peck*, 46th Rep.

Trenton, N. J., *Sterling;* Haddonfield, N. J.; Pennsylvania. Among chips in woods and on woods ground. *McIlvaine.*

This species is frequent, and when plentiful well worth gathering. It has a more woody flavor than other Galera, but is tasty.

G. vittæfor′mis Fr.—*vitta*, a chaplet; *forma*, form. **Pileus** ½–1 in. broad, *date-brown* when moist, membranaceous, conical then hemispherical, obtuse, *even at the disk*, striate toward the margin, smooth. **Stem** 1½–3 in. long, ½–1 line thick, tubular, equal, *somewhat straight*, but not tense and straight, smooth or sometimes pubescent, slightly striate under a lens, *opaque, rust-color.* **Veil** scarcely conspicuous. **Gills** adnate, broader at the middle, in the form of a segment when larger, somewhat ascending, somewhat distant, at first *watery-cinnamon*, at length rust-color. *Fries.*

Spores elliptical, 12x6μ *Massee.*

Haddonfield, N. J.; Mt. Gretna, Pa. On pastures, lawns, etc. June to September. *McIlvaine.*

Not previously reported.

Galera. Though small it makes up in quantity when found. The stems are not as tender as the caps. Quality good.

BOLBI'TIUS Fr.

Gr.—cow's dung.

Bolbitius. **Pileus** membranaceous. **Gills** adnexed or free, membranaceous, soft, salmon-color or rusty, dissolving (not dripping as in Coprinus), powdered with the rusty spores. **Stem** central; universal veil absent, partial veil often obsolete.

Very delicate and fragile, remarkable among the Ochrosporæ for the gills dissolving into mucus, and in this respect analogous with Coprinus among the Melanosporæ, and Hiatula amongst the Leucosporæ. Growing on dung or amongst grass where dung abounds.

A small but very natural genus, with the vegetative portion like Coprinus and the fructification resembling Cortinarius, hence occupying an intermediate position between these two genera. *Fries.*

B. Bol'toni Fr.—after Bolton. **Pileus** rather fleshy, viscid, at first even, then with the membranaceous margin sulcate, disk darker, subdepressed. **Stem** attenuated, yellowish, at first floccose from the remains of the fugacious veil. **Gills** subadnate, yellow then livid-brown. *Fries.*

Haddonfield, N. J., cespitose among manure on sawdust.

Of small substance but good consistency and flavor.

B. fra'gilis Fr. **Pileus** 2 in. broad, *light-yellow*, then becoming Bolbitius.
pale, somewhat membranaceous,
(Plate **LXXX.**)

BOLBITIUS FRAGILIS.
Two-thirds natural size.

almost pellucid, conical then expanded, somewhat u m b o n a t e, smooth, viscous, *striate round the margin* (which is often crenulated). **Stem** 3 in. long, 1 line or little more thick, fistulose, attenuated upward, *naked, smooth* (and without a manifest veil), yellow. **Gills** attenuato-adnexed, almost free, ventricose, *yellow then pale cinnamon.* **Spores** rust-colored. *Fries.*

(Plate LXXX*a*.)

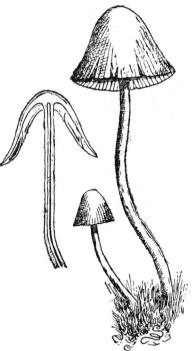

BOLBITIUS NOBILIS.
About two-thirds natural size.

Thinner than B. Boltoni, etc., **very** fragile, rapidly withering.

On dung. Common. June to October. *Stevenson.*

Spores subspheroid-ellipsoid, elliptical, 7x3–5µ *Massee.*

West Virginia; Pennsylvania. June to frost. On rich grass and dung.

Pileus usually not over 1.5 in. across. Often in plenty. Its substance does not cook away as with C. micaceus. It amply repays gathering, being highly flavored.

B. no'bilis Pk.—noble. **Pileus** thin, fleshy on the disk, ovate then bell-shaped, smooth, plicate-striate, pale-yellow, the disk tinged with red, the margin at length recurved and splitting. **Gills** subdistant, tapering outwardly, attached, the alternate ones

303

Bolbitius. more narrow, pale-yellow with a darker edge. **Stem** long, equal, smooth, striate at the top, hollow, white.

Plant cespitose, 3–5 in. high. **Pileus** 1 in. broad. **Stem** 1 line thick. Ground in woods. Greig. September.

A fine large species, but probably rare. *Peck*, 24th Rep. N. Y. State Bot.

I have not seen this species. Figure after Professor Peck.

CREPIDO'TUS Fr.

Gr.—a slipper.

Crepidotus. **Veil** wanting or not manifest. **Pileus** eccentric, lateral or resupinate.
(Plate LXXXI.) **Spores** rust-color.

CREPIDOTUS MOLLIS.
Natural size.

The Crepidoti correspond in shape and habit to the smaller Pleuroti and the Claudopodes, but they are distinguished from both by the rust-color of their spores. These are globose in several species, in others they are elliptical. In some there is a depression on one side which gives them a naviculoid character and causes the spore to appear slightly curved when viewed in a certain position. In consequence of the similarity of several of our species, the character of the spores is of much importance in their identification, and it is unfortunate that European mycologists have so generally neglected to give the spore characters in their descriptions of these fungi. In most of the species the pileus is at first resupinate, but it generally becomes reflexed as it enlarges. It is generally sessile or attached by a mass of white fibrils or tomentum. For this reason it is usually somewhat tomentose or villose about the point of attachment, even in species that are otherwise glabrous. In several species the pileus is moist or hygrophanous and then the thin margin is commonly striatulate. This character is attributed to but one of the

304

dozen or more European species. Their mode of growth is usually Crepidotus. gregarious or somewhat loosely imbricated, in consequence of which the pileus, which in most species is white or yellowish, is often stained by the spores, and then it has a rusty, stained or squalid appearance. The species occur especially on old stumps, prostrate trunks and soft much decayed wood in damp, shaded places. *Peck*, 39th Rep. N. Y. State Bot.

C. ful′vo-tomento′sus Pk.—tawny-tomentose. **Pileus** ¾–2 in. broad, scattered or gregarious, suborbicular, kidney-shaped or dimidiate, sessile or attached by a short, white-villose tubercle or rudimentary stem, hygrophanous, watery-brown and sometimes striatulate on the margin when moist, whitish, yellowish or pale ochraceous when dry, *adorned with small, tawny, hairy or tomentose scales.* **Lamellæ** broad, subventricose, moderately close, rounded behind, radiating from a lateral or eccentric white villose spot, whitish becoming brownish-ferruginous. **Spores** *elliptical* often uninucleate, 8–10x5–6µ.

Decaying wood of poplar, maple, etc. Common. June to October.

A pretty species, corresponding in some respects to the European C. calolepis, but much larger and with tawny, instead of reddish scales. The cuticle is separable and is tenacious, though it has a hyaline gelatinous appearance. The pileus is subpersistent, and specimens dried in their place of growth are not rare. *Peck*, 39th Rep. N. Y. State Bot.

Haddonfield, N. J.; Angora, West Philadelphia. On decaying hickory. *McIlvaine.*

Substance fair. Taste strong but pleasant.

CORTINA'RIUS Fr.

Cortina, a veil or curtain.

Cortinarius. **Veil** resembling the consistency of a cob-web, superficial, distinct from the cuticle of the pileus. **Flesh** of pileus and stem continuous. **Gills** persistent, dry, changing color, powdered with the spores. **Trama** fibrillose. **Spores** globose or oblong, somewhat ochraceous on white paper. *Fries.*

This genus is not easily confounded with any other, the cob-webby veil stretched from stem to pileus in the young plant not being found in other fungi. This must be looked for only in youth, as from its tender character it soon breaks and often appears only as a very indistinct collar on the stem, colored from catching the falling spores. The colors are generally pronounced and often extremely bright, there being very few prettier toadstools than those inclined to the blue or purple shades, which are not uncommon in the immature form. The color of the spores is also a marked feature, being rusty or brownish-ochraceous, turning the gills to the same color at maturity. On account of this change it is generally necessary to have specimens at both stages of growth to accurately determine the species. The gills are thin, attached to the stem in various manners, rarely slightly decurrent.

Cortinarius is distinguished from Flammula by growing on the ground and by the bright ferruginous color of its spores.

Cortinarius is a sturdy, hardy genus preferring northern latitudes and autumnal months, though several of its species grow as far south as Alabama, and one, a new species described by Professor Peck, is found on the Helderberg mountains in May. The genus contains many species, most of which produce in great numbers, yet being woods-growing, and coming as they do when leaves are falling, they are often missed because of their similarity to their surroundings.

Heretofore, less than a dozen species have been reported as eaten. This number is now doubled. While several species are bitter and others equally unpleasant, not one has been accused of harm. It is highly probable that other varieties than those herein given will prove equally acceptable as food. I have tested all I have found in sufficient quantity to warrant passing judgment upon them.

The genus does not contain as many species of superior excellence as other fleshy genera of like numbers. The flesh is frequently dry and of a strong woody or musky flavor, which it does not lose in cooking. The Cortinarius. stems are seldom cookable. All can be fried in butter, but cut in small pieces and well stewed, or stewed and served in patties, or made into croquettes are certain ways of keeping them in palate memory.

ANALYSIS OF TRIBES.

PHLEGMACIUM (*Gr.*—shiny or clammy moisture). Page 308.
Pileus viscid. Stem firm, dry. Veil partial, cobweb-like.

A. CLIDUCHII (*Gr.*—holding the keys—the typical subdivision). Page 308.

Partial veil as a ring on the upper part of the stem which is equal or slightly expanded above. Not distinctly bulbous.
* Gills pallid then clay-colored.
** Gills purplish then clay-colored.

B. SCAURI (*Gr.*—club-footed). Page 310.

Bulbous. Bulb depressed or top-shaped, with a distinct margin caused by the pressure of the pileus before expansion. Veil generally ascending from the margin of the bulb. Gills somewhat sinuate.
* Gills whitish then cinnamon.
** Gills blue then cinnamon.
*** Gills brownish-white then cinnamon.

MYXACIUM (*Gr.*—mucus). Page 313.
Universal veil glutinous. Pileus and stem viscid. Stem slightly bulbous. Gills adnate.

INOLOMA (*Gr.*—a fibrous fringe). Page 314.
Pileus dry, not hygrophanous or viscid, covered at first with innate silky scales or fibrils, becoming smooth. Veil simple. Pileus and stem fleshy, rather bulbous.
* Gills violaceous, then cinnamon.
** Gills pinkish-brown, then cinnamon.
*** Gills yellow, then cinnamon.

DERMOCYBE. Page 320.

Cortinarius.　Pileus thin, equally fleshy, at first silky with a fine down, becoming smooth when adult.   Not hygrophanous, but flesh watery when moist or colored.   Stem equal or larger above, externally rigid, elastic or brittle, internally stuffed or hollow.   Veil single, thread-like.

TELAMONIA. Page 323.

Pileus moist, hygrophanous, at first smooth or sprinkled with the whitish superficial evanescent fibrils of the veil.   Flesh thin, or when thick it becomes abruptly thin toward the margin, scissile.   Stem ringed below or coated from the universal veil, slightly veiled at the apex, hence with almost a double veil.

HYGROCYBE. Page 325.

Pileus hygrophanous, smooth or covered with superficial white fibrils, not viscid, moist when fresh, becoming discolored when dry.   Flesh very thin or scissile, rarely more compact at the center.   Stem rather rigid, bare.   Veil thin, rarely collapsing and forming an irregular ring on the stem.

PHLEGMA'CIUM.   (*Gr.*—clammy moisture.)

A. CLIDUCHII.

**Gills pallid, then clay-colored.*

C. seba'ceus Fr.—*sebum*, tallow.   **Pileus** 2½–5 in. broad, unicolorous, *pale*, of the color of tallow, equally fleshy, convex then rather plane, commonly very repand, viscid, smooth, but at the first *covered over with a whitish pruinose luster*.   **Flesh** white. **Stem** 3–4 in. long, ½–1 in. thick, solid, stout, compact, never bulbous, often twisted and compressed, slightly fibrillose, pale white. **Cortina** delicate, fugacious, adhering only to the margin of the pileus. **Gills** emarginate, *not crowded*, connected by veins, 4 lines broad, clay-color or pallid-cinnamon, paler at the sides.   *Fries.*

The flesh of the pileus is not compact at the disk and abruptly thin at the circumference, but equally attenuated toward the margin.   The flesh of the stem is white.   The gills never turn bluish-gray.   Taste mild. *Stevenson.*

Spores pip-shaped, 9x7µ *Cooke.*

A very common and prolific species in West Virginia, New Jersey, Pennsylvania, North Carolina. *McIlvaine.*

Pushing from the earth in great clusters it raises the mat of leaves above it into hut-like mounds through which it seldom bursts. Yet side openings to its huts show its coziness, and reveal the ground thickly dusted with its spores. Detecting these mounds is part of the wood-craft of a toad-stool hunter.

Where clusters are not dense, or the fungus is solitary, the stem is frequently swollen at the base, even bulbous.

Both caps and stems are edible, but the stems are not equal to the caps. It is a valuable food species, because of its lateness and quantity. It is not of best quality.

C. tur'malis Fr.—*turma*, a troop. ([Color] Plate LXXXII, fig. 4.) **Pileus** yellow-tan, most frequently darker at the disk, not changeable, compact, convex then plane, very obtuse, even, smooth (sometimes obsoletely piloso-virgate), when young veiled with pruinate but very fugacious villous down, soon naked, viscid. **Flesh** white. **Stem** sometimes 3 in., sometimes 6 in. long, 1 in. thick, solid, very hard, rigid, *cylindrical*, here and there attenuated at the base, shining white when dry, *when young sheathed with a white woolly veil*, naked when full grown. Cortina entirely fibrillose, superior and persistent in the form of a ring, at length ferruginous with the spores. **Gills** variously adnexed, rounded or emarginate, even decurrent with a tooth, crowded, *serrated*, white then clay-color. *Fries.*

I find it edible and of great value, being plentiful in pine woods, Maryland. I have collected a bushel in less than an hour in October. Under pine needles forming mounds. *Taylor.*

The localities and the habit of C. turmalis are very like that of C. sebaceus. The leaf mat broods the clusters.

C. turmalis is on a par with C. sebaceus. Personally I prefer the latter.

**Gills purplish, then clay-colored.

C. va'rius (Schaeff.) Fr.—*varius*, changeable. **Pileus** 2 in. and more broad, bright *ferruginous-tawny*, compact, hemispherico-flattened, very

Cortinarius. obtuse, regular, slightly viscid, even, smooth, the thin margin at first incurved, appendiculate with the cortina. **Flesh** firm, white. **Stem** curt, 1½–2½ in. long, 1 in. and more thick, *bulbous*, absolutely immarginate, compact, *shining white*, adpressedly flocculose, the superior veil pendulous. **Gills** emarginate, thin, somewhat crowded, *at first* narrow, *violaceous-purplish*, then broader and ochraceous-cinnamon, always quite entire.

Variable in stature, but the habit and colors are always unchangeable. It varies with the stem taller and somewhat equal, the pileus yellow-tawny, and the gills dark blue. *Fries*.

In woods. Uncommon. September to November. *Stevenson.*
Minnesota; Ohio.
Edible. *Cooke*, 1891.

B. SCAU'RI.

* Gills whitish then cinnamon.

C. intru'sus Pk. **Pileus** fleshy, rather thin, convex, then expanded, glabrous, somewhat viscid when moist, even or radiately wrinkled on the margin, yellowish or buff, sometimes with a reddish tint. **Flesh** white. **Lamellæ** thin, close, rounded behind, at first whitish or creamy-white, then cinnamon, often uneven on the edge. **Stem** equal or slightly tapering either upward or downward, stuffed or hollow, sometimes beautifully striate at the top only or nearly to the base, minutely floccose when young, soon glabrous, white. **Spores** broadly elliptical, brownish-cinnamon, 6–8x4–5μ.

Pileus 1–2.5 in. broad. **Stem** 1–3 in. long, 3–6 lines thick.

Mushroom beds, manured soil in conservatories or in plant pots. Boston, Mass. *R. K. Macadam.* Haddonfield, N. J. *C. McIlvaine.*

This interesting species is closely allied to Cortinarius multiformis and belongs to the Section Phlegmacium. It has a slight odor of radishes and is pronounced edible by Mr. McIlvaine. Its habitat is peculiar, but it possibly finds its way into conservatories and mushroom beds through the introduction of manure or soil, or leaf mold from the woods. It seems strange, however, that it has not yet been detected growing in the woods or fields. Hebeloma fastibile is said sometimes to invade mushroom beds, and our plant resembles it in so many particulars that it is with some hesitation I separate it. The chief differences are in the stem and spores. The former, in Hebeloma fastibile, is described

310

as solid and fibrous-squamose and the latter as 10x6 micromillimeters Cortinarius. in size. The brighter color of the smaller spores and the stuffed or hollow smooth stem of our plant will separate it from this species. *Peck*, Bull. of the Torrey Bot. Club, October, 1896.

Cortinarius intrusus was a happy find. Several pints of it were collected by the author in February—usually a famine month for the mycophagist. They grew on the ground, in beds among plants, and with potted plants in a hot-house in Haddonfield, N. J. The crop continued well into the spring. The species is delicate, savory, and a most accommodating renegade from its kind. I have never found it elsewhere.

**Gills blue, then cinnamon.*

C. cærules cens Fr. Pileus 2–3 in. across, equally fleshy, convex then plane, obtuse, regular, even, almost glabrous, but often fibrilloso-streaked; viscid, when dry shining or opaque, dingy yellow, almost tan-colored, varying to yellowish-brown, etc. **Gills** slightly rounded behind, adnexed, thin, closely crowded, 2 lines broad, at first clear intense blue then becoming purplish, at length dingy cinnamon. **Stem** about 2 in. long, ½ in. thick (bulb more than an inch), firm, equally attenuated upward, at first fibrillose, bright violet, then becoming pale and whitish, naked, bulb often disappearing with age; veil fibrillose, fugacious. **Spores** elliptical, 9–10x5µ.

Amongst moss in woods, etc.

Neither the gills nor the flesh change color when broken, a point which distinguishes the present from C. purpurascens. When young every part is generally blue. Smell scarcely any. *Fries.*

Spores 10–12x5µ *Cooke.*

Haddonfield; West Virginia; Mt. Gretna, Pa. In woods September to frost. *McIlvaine.*

The American species seldom entirely loses the bluish-purple color of its cap. The beautiful color fades somewhat or becomes splotched with yellow. Neither does the bulb ordinarily disappear with age. It is common. Taste of cap is mild, somewhat woody. They require long, slow stewing, and are better made into patties and croquettes.

C. purpuras'cens Fr. — gills becoming purple when bruised. **Pileus** 4–5 in. across, fleshy, disk compact, obtuse, wavy, variable,

Agaricaceæ

Cortinarius. covered with a dense layer of gluten, but opaque when dry, bay or reddish then tawny-olivaceous, spotted; often depressed round the margin, which is at first incurved then wavy, marked with a raised brown line. **Flesh** entirely clear blue. Gills broadly emarginate, 3 lines and more broad, crowded, bluish-tan, then cinnamon, violet-purple when bruised. **Stem** about 3 in. long, ⅔ in. and more thick, solid, bulbous, everywhere fibrillose, intensely pallid clear blue, very compact, juicy, becoming purplish-blue when touched, bulb submarginate. **Spores** elliptical, 10–12x5–6µ. *Fries.*

Var. *subpurpuras'cens*. Massachusetts. *Frost.*

Plentiful in West Virginia mountains in mixed woods, 1882. On South Valley Hill, near Downington, Pa., October, 1887. Haddonfield, N. J., 1892. In woods. September to frost. *McIlvaine.*

Both stems and caps are juicy when young and of agreeable flavor. It is among the best edible species of Cortinarius.

*** *Gills brownish-white, then ferruginous.*

C. turbina'tus Fr.—*turbo*, a top. **Pileus** *unicolorous*, dingy-yellow or green, *becoming pale*, hygrophanous, opaque when dry, fleshy, convex then flattened, obtuse, at length depressed, orbicular, even, *smooth*, viscid. **Flesh** soft, *white*. **Stem** commonly curt, 2 in., but varying elongated, yellowish, springing from a globoso-depressed distinctly marginate bulb, otherwise equal, cylindrical, *stuffed then hollow*. **Gills** attenuato-adnate, thin, crowded, broad, *quite entire*, at first pallid light-yellowish, at length somewhat ferruginous.

The typical form is *regular*, distinct from its allies in the *hygrophanous pileus, in the gills being isabelline-ferruginous and quite entire, and in being without any dark-purple or purple color*. Easily distinguished by its turbinate bulb. *Fries.*

In woods. Uncommon. *Stevenson.*

Spores rough, 14–16x7µ; rough, *Cooke.*

Cap 2–4 in. across. **Stem** commonly about 2 in. long, sometimes longer. *Massee.*

North Carolina, *Schweinitz;* Pennsylvania, *Schweinitz;* Massachusetts, *Frost;* Minnesota; Nova Scotia.

Edible. *Cooke.*

312

MYXA'CIUM. (*Gr.*—mucus.)

C. collin'itus Fr.—*collino*, to besmear.
glabrous, glutinous when moist, shining when dry. **Gills** rather broad,
dingy-white or grayish when young.
Stem cylindrical, solid, viscid or glutinous when moist, transversely cracking when dry, whitish or paler than
the pileus. **Spores** subelliptical, 13–
15μ.

Pileus convex, obtuse, Cortinarius.

(Plate LXXXIII.)

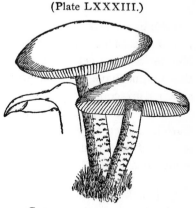

CORTINARIUS COLLINITUS.
About natural size.

The Smeared cortinarius is much
more common than the Violet cortinarius and has a much wider range.
Both the cap and stem are covered
with a viscid substance or gluten
which makes it unpleasant to handle.
The cap varies in color from yellow
to golden or tawny-yellow and when the gluten on it has dried it is
very smooth and shining. The flesh is white or whitish. The young
gills have a peculiar bluish-white or dingy-white color which might be
called grayish or clay color, but when mature they assume the color of
the spores. They are sometimes minutely uneven on the edge.

The stem is straight, solid, cylindrical and usually paler than the cap.
When the gluten on it dries it cracks transversely, giving to the stem a
peculiar scaly appearance.

The cap is 1½–3 in. broad, and the stem 2–4 in. long, and ¼–½
in. thick.

The plant grows in thin woods, copses and partly cleared lands and
may be found from August to September.

It is well to peel the caps before cooking, since the gluten causes dirt
and rubbish to adhere tenaciously to them. *Peck*, 48th Rep. N. Y.
State Bot.

In 41st Rep. N. Y. State Mus. Nat. Hist., p. 71, Professor Peck describes a closely allied species, C. muscigenus, n. sp., "separated by its
more highly-colored pileus, striate margin and even, not diffracted-
squamose stem."

Cortinarius. Prof. L. B. Mendel gives the following analysis: "Young specimens gathered in New Haven early in November, 1897, gave:

Water 91.13%
Total solids.. 8.87
Total nitrogen of dry substance.................................... 3.63

Edible. *Cooke.*

In appearance the Smeared cortinarius does not appeal to be eaten. Neither does an eel. But peeled both are inviting. Raw, the caps of this fungus have a strong woody smell and taste. This is somewhat subdued by cooking.

I have found the plant in West Virginia, Pennsylvania and North Carolina, often among the leaves in mixed woods, but it prefers a goodly supply of light and the freedom of open places. It is often gregarious, sometimes tufted.

C. io'des B. and C. **Pileus** 1½–2 in., convex, at length plane, viscid, firm, violet-purple. **Flesh** white, thick. **Veil** fugacious, spider-web. **Stem** 2–3 in. long, 1½ in. thick, solid, thickened below. **Gills** violet, at length cinnamon, ventricose, adnate, sub-emarginate, irregular, sometimes forked. *B. and C.*

This is a small but beautiful species, the pileus, lamellæ and stem being of a bright-violet or purplish-violet hue. The spores are sub-elliptical, generally uninucleate, 10x6μ. *Peck*, 32d Rep. N. Y. State Bot.

The pileus in this species is sometimes spotted with white. The bulbous white stem is adorned with lilac-colored fibrils. *Peck*, 35th Rep. N. Y. State Bot.

Sparingly found among roots at Mt. Gretna, Pa., September, 1897–1898.

The caps are fairly good.

INOLO'MA. (*Gr.*—fiber; *Gr.*—a fringe.)

* *Gills violaceous then cinnamon.*

C. viola'ceus Fr. ([Color] Plate LXXXII, fig. 2.) One of our most plentiful and beautiful autumnal fungi. As the American plant differs somewhat from the European, Professor Peck's description is given.

Pileus convex, becoming nearly plane, dry, adorned with numerous

314

persistent hairy tufts or scales, dark violet.   **Lamellæ** rather thick, dis- Cortinarius.
tant, rounded or deeply notched at the inner extremity, colored like the
pileus in the young plant, brownish-cinnamon in the mature plant.
Stem solid, fibrillose, bulbous, colored like the pileus.   **Spores** subelliptical, 12.5µ long.

The Violet cortinarius is a very beautiful mushroom and one easy of
recognition.   At first the whole plant is uniformly colored, but with age
the gills assume a dingy ochraceous or brownish-cinnamon hue.   The
cap is generally well formed and regular and is beautifully adorned with
little hairy scales or tufts.   These are rarely shown in figures of the
European plant, but they are quite noticeable in the American plant and
should not be overlooked.   The flesh is more or less tinged with violet.

The gills when young are colored like the cap.   They are rather
broad, notched at the inner extremity and narrowed toward the margin
of the cap.   When mature they become dusted with the spores whose
color they take.

The stem also is colored like the cap.   It is swollen into a bulb at
the base and sometimes a faint ochraceous band may be seen near the
top.   This is due to the falling spores which lodge on the webby filaments of the veil remaining attached to the stem.

Cap 2–4 in. broad.   **Stem** 3–5 in. long, about ½ in. thick.   *Peck*,
48th Rep. N. Y. State Bot.

Minerva, Essex county.   A form of this species occurs here, having
the pileus merely downy or punctate-hairy under a lens, no squamules
being distinguishable by the naked eye.   July.   *Peck*, 50th Rep. N. Y.
State Bot.

Spores 12–14x10µ *Cooke*.

The spider web veil is exquisitely displayed in this species.   This,
with its strongly bulbous base and violet tinge throughout, easily mark
it.   Though usually solitary great numbers of it are found in its settlements.   The mixed woods of central New Jersey abound with it in July,
August and September.   Throughout Pennsylvania and West Virginia
it is common, and is reported from several other states.   In Redman's
woods, near Haddonfield, N. J., a densely clustered form of singular
beauty occurs.   A dozen individuals of various forms and sizes with
swollen stems form a compact mass, rich in color, and cutting crisp and
juicy as an apple.   They are far better than other Cortinarii I have
eaten.   I have not seen it elsewhere.

Agaricaceæ

Cortinarius. C. violaceus is everywhere eaten, and is in my opinion the best of its genus. The American plant is not inodorous, but has a decided mushroom smell and taste.

C. albo-viola'ceus Pers.

(Plate LXXXIV.)

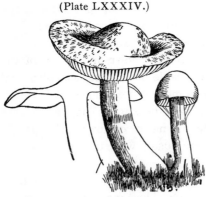

CORTINARIUS ALBO-VIOLACEUS.
One-half natural size.

Pileus fleshy, rather thin, convex, then expanded, sometimes broadly subumbonate, smooth, silky, whitish, tinged with lilac or pale violet. Lamellæ generally serrulate, whitish-violet, then cinnamon-color. Stem equal or a little tapering upward, solid, silky, white, stained with violet, especially at the top, slightly bulbous, the bulb gradually tapering into the stipe.

Height, 3–4 in. ; breadth of pileus, 2–3 in. ; stipe, 3–6 lines thick.

Ground in thin woods, more frequently under poplars. Center. October.

The stem is sometimes subannulate, and being violet above and white below the obscure ring, it appears as if sheathed with a silky-white covering. Inodorous. Sometimes the stem gradually tapers from the base to the top, so that it can scarcely be called bulbous. *Peck*, 23d Rep. N. Y. State Bot.

Spores 12x5–6μ *Cooke;* 6–9x4–5μ *K.;* pruniform, 10μ *Q.*

An allied species C. (Inoloma) lilacinus, *Peck*, with the stem and bulbous part much broader than the cap, is not as common, but of far better flavor.

Common in West Virginia, Pennsylvania, New Jersey, in mixed woods. September to frost. *McIlvaine.*

A mushroom flavor develops in cooking. The consistency of the flesh is good. It is of medium grade.

C. lilaci'nus Pk.

Pileus firm, hemispherical, then convex, minutely silky, lilac-color. Lamellæ close, lilac, then cinnamon. Stem stout, bulbous, silky-fibrillose, solid, whitish, tinged with lilac. Spores nucleate, 10x6μ.

316

Plant 4–5 in. high. **Pileus** 3 in. broad. **Stem** 4–6 lines thick. Cortinarius. Low mossy ground in woods. Croghan. September. This is a rare but beautiful plant, allied to C. alboviolaceus, from which it may be distinguished by its stouter habit, deeper color and bulbous stem. In the young plant the bulb is much broader than the undeveloped pileus that surmounts it. *Peck*, 26th Rep. N. Y. State Bot.

Massachusetts, *Frost;* Minnesota, Nova Scotia.

I have found a few specimens in several places: West Virginia, Redman's woods, Haddonfield, N. J., in which place it is more plentiful than in any locality I have noted. Near lake at Eagle's Mere, Pa., August, and at Springton, Pa. Excellent.

C. as′per Pk.—rough. **Pileus** fleshy, firm, hemispherical, then convex, rough with minute, erect, brown scales, ochraceous. **Gills** close, rounded behind and slightly emarginate, dull violaceous, then pale cinnamon. **Stem** equal, bulbous, solid, fibrillose-scaly, colored like the pileus but smooth and violaceous at the top, the bulb white with an abundant mycelium. **Spores** broadly elliptical, with a pellucid nucleus, 8μ long.

Plant 3–4 in. high. **Pileus** 2–3 in. broad. **Stem** 3–5 lines thick. Ground in cleared places. Greig. September.

A fine species. The flesh of the stem is violaceous. *Peck*, 24th Rep. N. Y. State Bot.

This plant sometimes grows in tufts or clusters and bears a very close resemblance to Armillaria mellea, both in color and in the character of the scales of the pileus. *Peck*, 27th Rep.

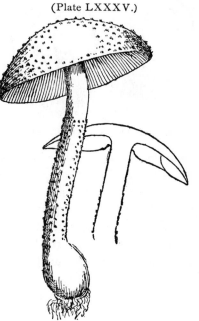

(Plate LXXXV.)

CORTINARIUS ASPER.
About two-thirds natural size.

In thin woods and clearings, West Virginia, New Jersey, Pennsylvania. The whole fungus is edible when young, and ranks high in Cortinarii. When full grown the stem is hard. Cut in thin, transverse slices it

Cortinarius. cooks tender, but does not equal the cap. Like most of the Cortinarii it is found in the autumn until frost kills it.

** *Gills pinkish-brown then cinnamon.*

C. squamulo′sus Pk.   ([Color] Plate LXXXII, fig. 1.)   **Pileus** thick, fleshy, convex, densely fibrillose-squamulose, cinnamon-brown, the scales darker. **Lamellæ** not crowded, deeply emarginate, pale pinkish-brown, then cinnamon-colored. **Stipe** thick, solid, shreddy, subsquamulose, concolorous, swollen at the base into a very large tapering or subventricose bulb.

(Plate LXXXVI.)

CORTINARIUS SQUAMULOSUS.

Height 4–6 in., breadth of pileus 2–4 in., stipe 6–9 lines thick at the top, 12–18 lines at the bottom.

Borders of swamps in woods. Sandlake. August.

Related to C. pholideus and C. arenatus, but distinct by the deep emargination of the lamellæ. It gives out a strong odor while drying. The color of the flesh is pinkish-white. *Peck*, 23d Rep. N. Y.

This species was discovered in 1869, and had not since been observed by the writer until the past season. It is manifestly a species of rare occurrence. *Peck*, 28th Rep.

Massachusetts, *Frost;* Wisconsin, Minnesota. Ranges from New England to Kentucky unchanged. *Morgan.*

Specimens from E. B. Sterling, Trenton, N. J., September, 1897. Asylum grounds. Several found at Mt. Gretna, August and September, 1897. Solitary in oak woods, gravelly soil. *McIlvaine.* Sent to Professor Peck and identified. Specimens were much darker than Professor Peck's plates.

C. squamulosus is not attractive in appearance. The caps, only, are edible. Their consistency is very pleasant and flavor fairly good.

318

C. autumna′lis Pk. **Pileus** fleshy, convex or expanded, dull rusty-yellow, variegated or streaked with innate rust-colored fibrils. **Gills** rather broad, with a wide shallow emargination. **Stem** equal, solid, firm, bulbous, a little paler than the pileus.

Height 3–4 in., breadth of pileus 2–4 in. **Stem** 6 lines thick.

Pine woods. Bethlehem. November. The plant is sometimes cespitose. The flesh is white. *Peck*, 23d Rep. N. Y. State Bot.

Mt. Gretna, Pa.,1899. *McIlvaine.* Quality fair. Caps meaty.

(Plate LXXXVI*a*.)   Cortinarius.

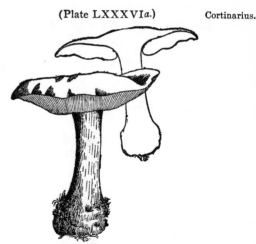

CORTINARIUS AUTUMNALIS.

C. ochra′ceus Pk. ([Color] Plate LXXXII, fig. 3.) **Pileus** fleshy, convex, at length broadly subumbonate or gibbous, smooth, even or obscurely wrinkled, pale ochraceous. **Stem** solid, fibrillose, ochraceous at the top, white below, gradually enlarged into a thick bulbous base.

Height 2–4 in., breadth of pileus 2–3 in. **Stem** 4–6 lines thick at the top, 12–18 lines at the base.

Under balsam trees in open places. Catskill mountains. October.

The stem appears as if sheathed. In some specimens the stem is short and rapidly tapers from the base to the top. *Peck*, 23d Rep. N. Y. State Bot.

Many of the species were found by the writer in mixed woods among leaves at Mt. Gretna, Pa., September, 1898. Specimens were identified by Professor Peck.

The gills are bright yellow when young. Cap smooth, innately fibrillose, not viscid. **Spores** light brown.

Tasteless; smell faint. Good consistency. A fair flavor develops in cooking.

*** *Gills yellow.*

C. (Inoloma) annula′tus Pk. **Pileus** broadly convex, dry, villose-squamulose, yellow. **Flesh** yellowish. **Lamellæ** rather broad, subdistant, adnexed, yellow. **Stem** solid, bulbous, somewhat peronate by the

319

Cortinarius.

(Plate LXXXVI*b*.)

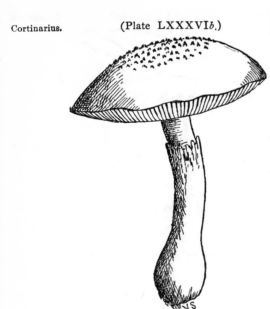

CORTINARIUS ANNULATUS.
Natural size.

yellow fibrillose annular-terminated veil. **Spores** broadly elliptical or subglobose, 8μ long.

Pileus 1–3 in. broad. **Stem** 1.5–3 in. long, 3–6 lines thick.

Thin woods. Whitehall. August. The whole plant is yellow inclining to ochraceous. It has the odor of radishes. The squamules of the pileus are pointed and erect on the disk, and often darker-colored there. The species is allied to C. tophaceus and C. callisteus, from which it is separated by its persistently annulate stem and more yellow color. *Peck*, 43d Rep.

Specimens received from E. B. Sterling, Trenton, N. J., September 5, 1897. Identified by Professor Peck. Mixed woods Kingsessing, near Bartram's Garden, Philadelphia, September, 1897.

Solitary among grass and leaves. The permanent marking of the veil is conspicuous. Eight specimens were found and eaten. The caps cook tender, and have a decided but not unpleasant flavor.

DERMO′CYBE. (*Gr.*—skin; *Gr.*—a head.)

C. cinnabari′nus Fr.—*cinnabaris*, dragon's blood. **Pileus** 2–3 in. broad, *scarlet-red*, truly fleshy, campanulate, then flattened, obtuse or very obtusely umbonate, silky, then becoming smooth and shining, or obsoletely scaly; the firm flesh paler. **Stem** 1½–2 in. long, 3–4 lines and more thick, solid, equal, sometimes however bulbous, fibrillose or striate, scarlet-red, reddish brick-color internally. Cortina fibrillose, lax, cinnabar. **Gills** wholly adnate, somewhat decurrent, 3 lines broad, somewhat distant, connected by veins, unequal and darker at the edge, dark blood-color when bruised.

Odor of radish. Readily distinguished from all others by its *splendid scarlet color*, and from C. sanguineus by its short solid and firm

stem, its broad pileus and *somewhat distant gills.* Stem never becom- Cortinarius.
ing yellow. *Fries.*

Spores 7–8x4µ *Cooke.*

It is a variable species with us.

Cap 1 ½ in. across, convex, broadly umbonate, margin involute, yellowish-brown, silky, innately fibrillose, shining, when young the cap is round, margin involute. **Veil** white, fibrillose, fugacious, leaving no trace on stem. **Flesh** thick in center, solid, close-grained, white, tinged with brown. Tastes strongly as radishes. Skin partially detachable.

Gills exceedingly beautiful in their deep claret-color, which is permanent, decurrent.

Stem 3 in. long, shining, smooth, white near top, brownish below, equal, fibrous, stuffed, skin removable.

On ground among pines, near station, **Mt. Gretna, Pa.** August to frost. Solitary, gregarious and cespitose.

Taste and smell like radishes. The caps cook well and are of fair flavor. Makes good patties and croquettes.

C. cinnabarinus, Var. 1. Mt. Gretna, Pa., August to frost. On decaying chestnut stumps.

Cap 1 in. across, shining, convex, orange-brown, white on margin and under minute appressed squamules, but few on margin; apparent remnant of a veil on cap, as a viscid skin.

Gills rounded behind, slightly emarginate, like Tricholoma, grayish-brown when young, becoming a brilliant scarlet, unequal.

Stem 2 in. high, over ¼ in. thick, white, covered with brownish-orange appressed squamules, often with stained marking of veil or fragments of veil as ring. Cespitose, connate.

Taste and smell strong like radishes. Flavor in dish is decided but pleasant. Makes good patties and croquettes.

Specimens were identified by Professor Peck as C. cinnabarinus, as were those of the preceding. The variations are so great that I give this place as a variety.

C. sanguin′eus Fr.—*sanguis,* blood. **Pileus** 1–1 ½ in. broad, *blood-color,* becoming slightly pale when dry, fleshy, thin, convex then plane, obtuse, occasionally depressed, silky or squamulose. **Flesh** reddish, paler. **Stem** 2–3 in. long, 2–3 lines thick, stuffed then hollow, equal (rather attenuated than thickened at the base), here and there

Cortinarius. flexuous, with fibrils of the same color, almost darker than the pileus. Cortina arachnoid, fugacious, red blood-color. **Gills** adnate, crowded, 2–3 lines broad, quite entire, dark blood-color.

Wholly *dark blood-color*, the stem when compressed pouring forth bloody juice. Odor of radish. Thinner than species nearest to it. The spores are ochraceous on a white ground, somewhat ferruginous on a black ground. *Fries.*

Spores 6x4µ *W.G.S.*

North Carolina, *Curtis;* Massachusetts, *Sprague, Farlow, Frost;* Connecticut, *Wright;* New York, *Peck*, 23d Rep.

Edible. *Leuba.*

C. cinnamo'meus Fr. Pileus 1–2½ in. across. **Flesh** thin, convexo-campanulate, umbonate, somewhat cinnamon color, silky squamulose with yellowish innate fibrils, becoming almost glabrous. **Gills** adnate, broad, crowded, shining, yellowish, then tawny-yellow. **Stem** 2–4 in. long, equal, yellow, as is also the flesh and the veil, hollow. **Spores** 7–8x4–5µ.

(Plate LXXXVII.)

CORTINARIUS CINNAMOMEUS.
Natural size.

A very common species, especially in mossy places in pine woods, occurring under many well defined forms, which can not be separated as species. Essential points common to all. (1) Stem everywhere equal, stuffed, then hollow, yellowish, fibrillose from the similarly colored veil. (2) Pileus thin, flattened and obtusely umbonate, silky with yellowish down, often glabrous when adult, and then bright cinnamon, but the color is variable. (3) Flesh splitting, yellowish. (4) Gills adnate, crowded, thin, broad, always shining. (5) Spores dark ochraceous, size and color very variable; pileus from ½–3–4 in. across; color of pileus changeable, depending on the more or less persistence of the down (fundamental color and veil constant in this species and its allies); gills varying through blood-red, reddish cinnamon, tawny saffron, golden and yellow. *Fries.*

322

Pileus thin, convex, obtuse or umbonate, dry, fibrillose at least when **Cortinarius.** young. **Flesh** yellowish. **Lamellæ** thin, close, adnate. **Stem** slender, equal, stuffed or hollow. **Spores** elliptical, 8μ long. *Peck*, 48th Rep. N. Y. State Bot.

Spores 7–8x4μ *Cooke.*

The Germans are said to be very fond of this species, which is generally stewed in butter and served with sauce for vegetables.

Catalogued by Dr. M. A. Curtis, North Carolina, as edible. **Edible.** *Cooke.*

Var. *semi-sanguin'eus* received from E. B. Sterling, Trenton, N. J., August, 1897. Juicy and good.

The species is common over the United States and plentiful in its numerous varieties from August to frost. It frequents mixed woods, borders and open and mossy places. The pine woods of New Jersey yield it in quantity, as do the hemlock forests of Eagle's Mere, Pa., and oak woods of West Virginia.

It has a smell and taste—mildly of radishes. Its flavor when cooked is decided but pleasant.

TELAMO'NIA. (Gr.—lint.)

C. armilla'tus Fr.—*armilla*, a ring. ([Color] Plate LXXXII, fig. 5.)

Pileus 3–5 in. broad, *red-brick color,* truly fleshy, but not very compact, at first cylindrical, soon campanulate, at length flattened, dry, at first smooth, soon innately fibrillose or squamulose, flesh dingy pallid. **Stem** 3–6 in. long, ½ in. thick, solid, firm, remarkably bulbous (bulb 1 in. thick, villous, whitish) and fibrillose at the base, when old striate and reddish-pallid, internally dirty yellow. Exterior veil woven, red, arranged *in 2–4 distant cinnabar zones encircling the stem;* partial veil continuous with the upper zone, arachnoid, reddish-white. **Gills**

(Plate LXXXVIII.)

CORTINARIUS ARMILLATUS.

adnate, slightly rounded, distant, at first pallid cinnamon, at length very broad (½ in.), dark ferruginous, almost bay-brown.

Cortinarius. Odor of radish. A very striking species. From the pileus not being hygrophanous, *at the first smooth* and at length torn into fibrils or squamulose, it might easily be taken for a species of Inoloma. The cortina itself is paler than the zones. It differs from all others in these zones. The rings are usually somewhat oblique. *Fries.*

Professor Peck in the 23d Rep. N. Y. State Cab. Nat. Hist., describes the American species as follows:

"**Pileus** fleshy, thick, convex or subcampanulate, then expanded, minutely squamulose, yellowish-red. **Lamellæ** not close, broad, slightly emarginate, whitish-ochraceous, then cinnamon. **Stipe** stout, solid, fibrillose, whitish, girt with one to four red bands, bulbous.

"Height 4–6 in., breadth of pileus 2–4 in., stipe 4–8 in. thick.

"Woods. North Elba. August.

"A large and noble species. The margin of the pileus is thin and sometimes uneven; the upper band on the stem is usually the brightest and most regular. The pileus is not distinctly hygrophanous."

Spores 10x6µ *Cooke.*

Edible. *Cooke.*

September 8, 1897, Mr. E. B. Sterling, Trenton, N. J., sent me several specimens new to me and remarkable in having two well-defined veils, the lower and thicker one of which left a dark zone upon the stem, the upper, fibrillose, was more persistent, but left a fainter impression. These veils are not mentioned in Professor Peck's description of the American species, but are prominently noted in that of Fries, as above. In a very young specimen both veils were present. Cap light brown, minutely squamulose, with a few small red spots; margin thin, involute, flesh thick, yellowish, firm; gills distant, rounded behind, slightly emarginate, alternate ones short, light brown inclined to cinereous on edge.

Spores brown. Small young specimens did not show bulbous stem as distinct as larger and older ones.

I afterward found several specimens at Mr. Gretna, Pa., September and October, 1897.

The flesh is excellent, closely resembling Pholiota subsquarrosa. The species seems to be rare. If found in quantity it will prove one of our very best edibles.

C. dis'tans Pk. **Pileus** thin except the disk, convex, squamulose, Cortinarius. bay-brown when moist, tawny when dry. **Lamellæ** broad, distant, thick, dark cinnamon-color. **Stipe** subequal, often a little tapering upward, solid, slightly fibrillose-scaly, concolorous.

Height 2–3 in., breadth of pileus 1–2 in., stipe 4–6 in. thick.

Grassy ground in pine woods. Greenbush. June.

The flesh is dull-yellowish. The pileus, when drying, has for a time a brown-marginal zone. *Peck*, 23d Rep. N. Y. State Bot.

New Jersey pines. Eagle's Mere, Pa., coniferous woods. August. Mt. Gretna, Pa., pines. August, September. *McIlvaine*.

Like most of the hygrophanous Cortinarii, the taste is more or less that of rotten wood. The flavor is flat and undesirable.

C. furfurel'lus Pk. **Pileus** thin, convex, furfuraceous with minute squamules, hygrophanous, watery-tawny when moist, pale ochraceous when dry. **Lamellæ** broad, thick, distant, adnate or slightly emarginate, tawny-yellow, then cinnamon. **Stem** equal, peronate, colored like the pileus, with a slight annulus near the top. **Spores** subelliptical, minutely rough, 8–10x6μ.

Plant 1–2 in. high. **Pileus** 1–2 in. broad. **Stem** 2–4 lines thick.

Moist ground in open places. Gansevoort. August. *Peck*, 32d Rep. N. Y. State Bot.

Haddonfield, N. J., Mt. Gretna, Pa. *McIlvaine*.

Strong woody flavor—like rotten wood. Not poisonous, but not desirable.

HYGROCYBE.

C. casta'neus Bull.—chestnut. **Pileus** fleshy, thin, campanulate or convex, then expanded, dark chestnut-color when moist, paler when dry. **Lamellæ** rather broad, violet-tinged, then cinnamon. **Stipe** fibrillose, stuffed or hollow, lilac tinged at the top, white below.

Height 2–3 in., breadth of pileus 1–2 in., stipe 3–4 lines thick.

Ground under spruce or balsam trees. Catskill mountains. October. Edible. *Peck*, 23d Rep. N. Y. State Bot.

Spores 8x5μ.

It is certainly a wholesome, esculent species, but a great number would be required to make a good dish. *M. C. Cooke*.

Catalogued by Rev. M. A. Curtis, North Carolina, as edible.

325

Cortinarius.   Eaten in Italy.   Inodorous, edible and agreeable.   *Cordier.*   More
than fair.   I have often eaten it.   *R. K. Macadam.*

PAXIL'LUS Fr.

Paxillus, a small stake.

Paxillus.   **Hymenophore** continuous with the stem, decurrent.   **Gills** membranaceous, somewhat branched, frequently anastomosing behind, *distinct
from the hymenophore and easily separable from it.*   **Spores** dingy-white
or ferruginous.

*Fleshy putrescent fungi, margin of pileus at first involute, then con-
tinually and gradually unfolding and expanding.   Fries.*

Pileus symmetrical or eccentric.   **Stem** central, eccentric or wanting.
Edge of gills entire, sharp.

The marked features of this genus are the strongly involute margin,
the soft, tough, decurrent gills, separating readily from the flesh, and
the color of the spores.

The members of this genus possess some of the characters of Boletus.
The gills separate easily from the hymenophore as do the tubes of the
latter, and their anastomosing tendency is in P. porosus so marked that
the hymenium consists of large angular tubes.   The gills of P. solidus
B. and C. form pores at the base, and its spores are elongated, both
features indicating an affinity with Boletus.

ANALYSIS OF TRIBES.

LEPISTA (a pan).   Page **327.**

Pileus entire, central.   Spores dingy-white, in P. panæolus somewhat
rust-color.   On the ground.

326

TAPINIA (to depress). Page 328

Pileus generally eccentric or resupinate. Spores **rust-color.** On the Paxillus. ground or on stumps.

So far as known the species of this genus are harmless. Many of them are large, fleshy and inviting in appearance, but their flesh is usually dry and coarse, and, though absorbent, is hard to cook tender. P. atrotomentosus, which seems to be rare, is an exception. The flesh of this species being firm in texture and readily made into a first-class dish.

LEPIS'TA.

P. lepis'ta Fr.—*lepista*, a pan. **Pileus** 2–4 in. broad, flat or depressed, dirty-white, smooth, sometimes minutely cracked near the margin which is thin, involute and often undulate. **Stem** very variable in length, 1–4 in., ½–¾ in. thick, dingy white or cream, solid, white inside, equal, with a cartilaginous cuticle passing between the gills and the flesh of the pileus, base blunt, villous, white. **Gills** very decurrent, crowded, 2–3 lines broad, slightly branched but not at the base, dingy-white becoming darker,

Spores reddish, becoming dingy brown. Broadly pyriforme 6x8μ *Massee.*

Pennsylvania. September, 1894. *McIlvaine.* Albion, N. Y., *Dr. Cushing,* 1898.

On ground in woods and margins of woods.

Flesh white. **Gills** narrow, crowded, brittle, decurrent, dingy-white or pale-buff, easily separating from cap. **Stem** solid, elastic, at length hollow, often short, an inch long, tapering downward, frequently up to four inches in length and equal, base villose.

Resembling Lactarius piperatus and some forms of Clitocybe. It is separated from the former by the absence of milk and from the latter by its involute margin. The Clitocybe resembling it are all edible.

Smell strong, like old oily nuts. Edible but coarse.

P. li'vidus Cke. **Pileus** 1–2 in. across, convex, at length slightly depressed at the disk, margin slightly arched and incurved, dingy-white, or livid ochraceous, opaque. **Gills** decurrent, arcuate, almost

Paxillus. crowded, 1½ line broad, white. **Stem** 3–4 in. long, ½ in. thick at the apex, attenuated downward, white, fibrillose, stuffed then hollow, usually rather flexuous. **Flesh** nearly white. **Spores** globose, 3–3.5μ diameter, nearly white.

In woods. Usually in small clusters. Closely allied to Paxillus revolutus, but distinguished by the absence of any tinge of violet on the pileus or stem, and by the persistently white gills. *Massee.*

Received from Katherine A. Hall, Danville, N. Y. October, 1898. Raw it tastes like a drug-store smell. Edible, pleasant.

TAPI'NIA.

P. involu'tus (Batsch) Fr.—*involutus*, rolled inward. **Pileus** 2–5 in. broad, fleshy, compact, convexo-plane then depressed, smooth, viscid when moist, shining when dry, yellowish or tawny-ochraceous, *strongly involute, margin densely downy*, flesh pallid. **Stem** 2–4 in. high, about ½ in. thick, solid, firm, paler than the pileus, central or eccentric. **Gills** 2–3 lines broad, crowded, branched, anastomosing, *forming pores behind*, whitish then yellowish or rusty, *spotting when bruised.*

Spores rust-color, ellipsoid or oblong-ellipsoid 8–16x6μ *K.;* 5x 6μ *W. G. S.* Elliptical, 8–10μ *Peck.*

(Plate XC.)

PAXILLUS INVOLUTUS.
One-half natural size.

It grows singly or in groups and likes damp mossy soil. Common in cool hemlock or spruce woods in the Adirondack mountains; not rare in the mixed woods of all our hilly districts. When growing on decayed stumps the stem is sometimes eccentric. August, November. *C. H. Peck.*

In open woods near Haddonfield, N. J., it grows to a large size and in quantity. In Angora woods near Philadelphia a complete ring of it 20 ft. in diameter was seen.

Considered edible throughout Europe and said to be highly esteemed in Russia. The flesh of the American plant is dry and coarse, does not cook tender and is rather tasteless.

P. a'tro-tomento'sus (Batsch.) Fr.—*ater*, black; *tomentum*, down. **Pileus** 3–6 in. broad, rust-color or reddish-brown, compactly fleshy, eccentric, convex then plane or depressed, margin thin, frequently· minutely rivulose, sometimes tomentose in the center. **Flesh** white. **Stem** 3–6 in. high, ½–1 in. thick, stout, solid, elastic, eccentric or lateral, unequal rooting, *covered with dense velvety down, very dark brown*. **Gills** adnate, 3 lines broad, close, anastomosing at the base, yellowish, interspaces venose.

Spores subhyaline 4–6x3–4μ *K*. Elliptical, pale-yellowish, 5x2.5–3μ *Massee*. Elliptical 5–6x4μ *Peck*.

Found near Philadelphia, gregarious in old woods. September. In New Jersey in pine woods on stumps and on the ground, probably growing from roots. *McIlvaine*.

Grows singly or cespitose, sometimes in large tufts, when the pileus is frequently irregular from compression. In wet weather the pileus is moist and sometimes obscurely mottled with dark spots. Occasionally it has an unpleasant dirt-like odor. *Peck*.

Cordier considers this species suspicious and Paulet inutile on account of its bad taste.

The flesh differs from most Paxilli in being very fine grained and cooked is of the consistency of a marshmallow. The taste is marked but pleasant.

Series IV. **PORPHYRO'SPORÆ** (Pratelli). *Gr.*—purple.

Spores typically black-purple or brownish-purple, more rarely dusky brown. (It is to be observed that the spores vary in color according to the color of the ground on which they are deposited.) There are sterile forms with the gills persistently white (A. obturatus, A. udus). Those species are more deceptive in which the gills continue for a long time white, and even begin to decay before they are discolored by the spores; these may be easily mistaken for Leucospori. *Fries.*

Pratelli is the name given by the early authors to this series, based upon the spore color; Porphyrosporæ is the name now used. The species within the group are closely allied to those having black spores without a tinge of purple or violet (Melanosporæ), but in none of the species do the gills deliquesce as in Coprinus, neither are there resupinate or lateral stemmed species.

There is a present tendency to do away with this series and include all dark-spored species in the Melanosporæ. Professor Atkinson and Bertha Stoneman, in their "Provisional Key to the Genera of Hymenomycetes," omit the series and give "Melanosporæ, Gill and Butz (Pratellæ and Coprinariæ in broadest sense). Spores dark brown, purplish-brown or black."

(Plate XCIII.)

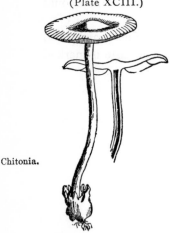

Chitonia.

CHITONIA RUBRICEPS.
Two-thirds natural size.

It is frequently difficult to determine by the spore-color of this series even to which series a specimen belongs. Many of our best edibles belong in this series. I know of none noxious.

CHITO'NIA Fr.

Universal veil distinct from the pileus, at maturity forming a distinct volva round the base of the ringless central stem. Gills free from the stem. Spores brownish-purple.

Analogous in structure with Volvaria and Amanitopsis. An exotic genus imported into this country.

No American species reported.

330

PLATE XCII.

PORPHYROSPORAE.

Hymenophore distinct from fleshy stem.

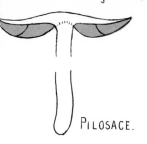

PILOSACE.

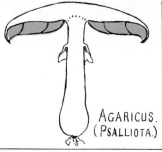

AGARICUS.
(PSALLIOTA.)

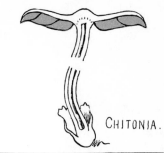

CHITONIA.

Hymenophore confluent and homogeneous with fleshy stem.

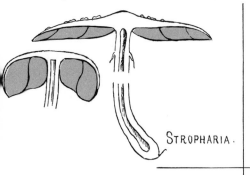

STROPHARIA.

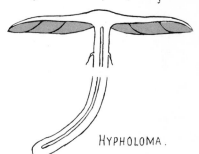

HYPHOLOMA.

Hymenophore confluent with, but heterogeneous from cartilaginous stem.

PSILOCYBE.

PSATHYRA.

DECONICA.

CHART OF GENERA IN PURPLE-SPORED SERIES—PORPHYROSPORAE,

AGAR'ICUS.

Agaricon, a Greek name for fungi, said to be derived from the name of a town, Agara.

Pileus fleshy, flesh of the stem different from that of the pileus, fur- Agaricus.
nished with a distinct ring. **Gills** at first enclosed by the veil, free,
rounded behind, at first white or whitish, in some species this stage lasting but a short time, then pink or reddish, at length dark purplish-
brown from the spores. **Spores** brown, brownish or reddish-purple.

On the ground, generally in pastures, meadows or manured ground,
a few species occur in woods.

Analogous with Lepiota of the white-spored series. Stropharia also
bears a ring and has similar colored spores, but is separated by the flesh
of stem and pileus being continuous and the gills being more or less
adnate.

Formerly in Agaricus as sub-genus Psalliota (*psallion, psalion*, in
poetry, a ring). When Psalliota was raised to generic rank it was given
the name of the great genus Agaricus as a mark of distinction on account of its including the most widely known and useful mushroom of
the world—Agaricus campester. The name Psalliota is not in modern
use.

Old Agaricus included many subgenera and consequently many more
species. Now it contains but few. All of them are highly flavored and of
marked excellence. Before the subgenera under Agaricus were promoted
to full generic standing it was customary to state the name of a species
thus: Agaricus (Psalliota) campester. Agaricus (Stropharia) semi-
orbicularis. This was lengthy and clumsy. In the older books this
form prevails. Often, however, the subgenus is omitted before the
name, which compels the student to look up the subgenus to which the
species belongs. The older books are therefore puzzling to modern
students, who find there simply the name Agaricus to guide them. The
present genus of a known species in old Agaricus can be easily found
by looking in the index for its specific name. The name of the genus
follows it in parentheses.

All of the genus can be cooked in any desired way.

331

ANALYSIS OF SPECIES.

Agaricus.   * Gills at first or very soon pink or rosy.
** Gills at first brownish or gray.
*** Gills at first white or whitish.

* *Gills at first or very soon pink or rosy.*

A. campes'ter Linn.—*campus*, a field.   ([Color] Plate XCI, fig. 4 (3 figs.) fig. 5, section.)   **Pileus** at first hemispherical or convex, then expanded with decurved margin or nearly plane, smooth, silky floccose or hairy squamulose, the margin extending beyond the lamellæ, the flesh rather thick, firm, white.   **Lamellæ** free, close, ventricose, *at first delicate pink or flesh color*, then blackish-brown, *subdeliquescent*.   **Stem** equal or slightly thickened toward the base, *stuffed*, white or whitish, nearly or quite smooth.   **Ring** at or near the middle, more or less lacerated, sometimes evanescent.   **Spores** elliptical, 6–8x4–5μ.

Plant 2–4 in. high.   **Pileus** 1.5–4 in. or more broad.   **Stem** 4–8 lines thick.   *Peck*, 36th Rep. N. Y. State Bot.

Spores spheroid-ellipsoid, 9x6μ *K.;* 6x8μ *W.G.S.*

The varieties of A. campester are numerous.   All of them are edible and vary but slightly in their excellence.

Var. *al'bus* Berk.—*albus*, white.   A very common wild form.   **Cap** 2–4 in. across, smooth or slightly fibrillose.   **Stem** 1½–3 in. long, ⅓–⅔ in. thick, white or whitish.   Spring to autumn, in rich grassy places.   Sometimes very large.   It is cultivated.

Var. *gri'seus* Pk.—*griseus*, gray.   **Cap** grayish, silky, shining.   **Ring** vanishing.   Reported from Virginia.

Var. *prati'cola* Vitt.—*pratum*, a meadow; *colo*, to inhabit.   Meadow variety.   **Cap** covered with reddish scales.   **Flesh** pinkish.   Parade ground, Mt. Gretna, Pa.

Var. *umbri'nus* Vitt.—*umber*, dark brown.   **Cap** brown, smooth. **Stem** short, minutely scaly.

"Var. *rufes'cens* Berk.—*rufescens*, becoming red.   **Pileus** reddish, minutely scaly.   **Gills** at first white.   **Stem** elongated.   **Flesh** turning bright red when cut or bruised.   This departs so decidedly from the ordinary characters of the type, especially in the white color of the young gills, that it seems to merit separation as a distinct species." *Peck*, 36th Rep.

Var. *villa'ticus* Brond.—belonging to a villa. Cap scaly. Stem scaly. Agaricus.

Var. *horten'sis* Cke.—growing in gardens. Cap brownish or yellow-ish-brown, covered with fibrils or minute hairs. This is a cultivated species.

"Var. *Bu'channi*. Cap white, smooth, depressed in center, the margin naked. Stem stout. Ring thin, lacerated. A rare variety, sometimes occurring in mushroom beds.

"Var. *elonga'ius*—elongated. Long-stemmed variety. Pileus small, smooth, convex, the margin adorned with the adherent remains of the lacerated veil. Stem long, slender, slightly thickened toward the base. Ring slight or evanescent. This is also a variety of mushroom beds.

"Var. *vapora'rius*. Green-house variety (A. vaporarius Vitt.) Pileus brownish, coated with long hairs or fibrils. Stem hairy-fibrillose, becoming transversely scaly. Conservatories, cellars, etc. Not differing greatly from Var. hortensis." *Peck*, 36th Rep. N. Y. State Bot.

The A. campester is known the world over as the common mushroom. It is cosmopolitan, appearing in pastures and rich places from spring and until long after severe frosts. It is the sweet morsel of gourmets. Indirectly it has done more damage than the assembled viciousness of all other toadstools. It is by mistaking the young button forms of the deadly Amanita for the button forms of the common mushroom that most cases of fatal toadstool poisoning are brought about. It is, also, usually the persons who think they know the mushroom, and can not be deceived, that get poisoned. If two rules are observed danger can be avoided. (1) Never eat a fungus gathered in the woods believing it to be the mushroom. The typical A. campester does not grow in the woods; species of Agaricus somewhat resembling it do. (2) Look at the gills; those of the mushroom are at first a light-pink which rapidly, as the plant matures, darken to a dark-brown, purplish-brown, or purplish-black. This is due to the ripening of the spores. Those of the Amanita are constantly white.

Pages could be written upon the mushroom and its culture, and recipes for the cooking of it would fill a volume. One important thing is omitted from them all—it is culinary heresy to peel a mushroom. Much of the flavor lies in the skin, as it does in that of apples, apricots, peaches, grapes, cherries and other fruits. The mushroom should be wiped with a coarse flannel or towel until the skin is clean. See chapter on cooking, etc.

Lafayette B. Mendel, in American Journal of Physiology, March, 1898, gives the following analysis of A. campester:

Two varieties of the common mushroom were collected in New Haven. Fifteen specimens of one variety weighed 1½ ounce, an average weight of 43 grains each. The analysis gave:

	a.	*b.*
Water	87.88%	92.20%
Total solids	12.12	7.80
Total nitrogen in dry substance	4.42	4.92
Ash in dry substance	11.66	17.18

A. comp'tulus Fr.—*comptus*, gaily adorned. **Pileus** 1–1½ in. broad, *yellowish-white*, slightly fleshy, convex then plane, obtuse, *adpressedly fibrilloso-silky*, becoming even. **Flesh** thin, soft, of the same color as the pileus. **Stem** 2 in. long, 2–3 lines thick, *hollow*, stuffed with floccules when young, *somewhat attenuated*, even, smooth, white, becoming somewhat light yellow. **Ring** medial, torn, *fugacious*, of the same color. **Gills** rounded-free behind, crowded, soft, broader in front, *flesh-color* then *rose*, not dingy-flesh-color except when old.

Closely allied to A. campestris, but constantly distinct in its more beautifully colored gills. *Fries.*

Cultivated ground. Menands. August. *Peck*, Rep. 41.

Closely allied to A. campestris, from which it may be separated by its smaller size, the yellowish hue of the dry plant and by the smaller spores. *Peck*, 41st Rep. N. Y. State Bot..

Mt. Gretna, Pa. Parade ground, with A. campester; Haddonfield, N. J. August to frost. *McIlvaine.*

A. comptulus appears frequently in the latitude of Philadelphia. It is a neat species, but not substantial in flesh. Here it usually grows close to the ground. The ring is very evanescent.

Its edible qualities are those of A. campester.

A. silvat'icus Schaeff.—belonging to woods. **Pileus** thin, at first convex or bell-shaped, then expanded, *gibbous or subumbonate*, fibrillose or variegated with a few thin tawny brownish or reddish-brown *spot-like adpressed scales*, whitish, brownish or smoky-gray, the disk sometimes tinged with red or reddish-brown, the flesh white or faintly reddish. **Lamellæ** thin, close, free, narrowed toward each end, reddish, then blackish-brown. **Stem** rather long, *equal or slightly taper-* Agaricus. *ing upward*, hollow, whitish. **Spores** elliptical, 5–6.5x4–5μ.

Plant 3–5 in. high. **Pileus** 2–4 in. broad. **Stem** 4–6 lines thick.

Woods. Summer and autumn. Not common. *Peck*, 36th Rep. N. Y. State Bot.

Massachusetts, *Farlow;* Minnesota, *Johnson;* California—edible, *H. and M.*

West Virginia, 1881–1885, New Jersey, Pennsylvania. August to frost. In pine and mixed woods. *McIlvaine.*

Edible, *Curtis.* Edible, *Peck.*

In taste and smell A. silvaticus resembles A. silvicola, but is stronger. It is a frequent but not common species in the localities where I have found it. Quantities of it have not occurred, but myself and friends have eaten it for years, knowing no distinction in effect between it and allied species. Its strong taste requires that it be well cooked. It does not lose its high flavor, which may be objectionable to some. I prefer using its juices as a flavoring.

A. dimīnuti'vus Pk.—diminutive. **Pileus** thin, fragile, at first convex, then plane or centrally depressed, sometimes slightly umbonate, whitish or yellowish, faintly spotted with small thin silky appressed brownish scales, the disk brownish or reddish-brown. **Lamellæ** close, thin, free, ventricose, brownish-pink becoming brown, blackish-brown or black. **Stem** equal or slightly tapering upward, stuffed or hollow, smooth, pallid. **Annulus** thin, persistent, white. **Spores** elliptical 5x4μ.

Plant 1.5–2 in. high. **Pileus** 1–1.5 in. broad. **Stem** 1–2 lines thick.

Woods. Croghan and Sandlake, N. Y. August. Autumn.

This is a small but symmetrical and beautiful Agaric. It is perhaps too closely related to the preceding species (A. silvaticus), of which it may possibly prove to be a mere variety or dwarf form. Its pileus is quite thin and fragile. Usually the darker or reddish hue of the disk gradually loses itself in the paler color of the margin, but sometimes the whole surface is tinged with red. *Peck*, 36th Rep. N. Y. State Bot.

Chester county; West Philadelphia, Pa., September; Mt. Gretna, Eagle's Mere, Pa., August. *McIlvaine.*

I have found A. diminutivus so intimately associated with A. sil-

335

Agaricus. vaticus that its being a dwarf form of the latter seemed more than probable. Its edible qualities are the same.

A. Rod′mani Pk. **Pileus** rather thick, firm, at first convex, then nearly or quite plane, with decurved margin, smooth or rarely slightly cracked into scales on the disk, white or whitish, becoming yellowish or subochraceous on the disk, the flesh white, unchangeable. **Lamellæ** close, *narrow*, rounded behind, free, reaching nearly or quite to the stem, *at first whitish then pink or reddish-pink*, finally blackish-brown. **Stem** short, subequal, solid, whitish, smooth below the ring, often scurfy or slightly mealy-squamulose above; ring variable, thick or thin, entire or lacerated, at or below the middle of the stem. **Spores** broadly elliptical or subglobose, generally uninucleate, 5–6x4–5µ.

Plant 2–3 in. high. **Pileus** 2–4 in. broad. **Stem** 6–10 lines thick.

Grassy ground and paved gutters. Astoria, L. I. *Rev. W. Rodman.* Washington Park, Albany. May to July.

This species is intermediate between A. campestris and A. arvensis, from both of which it may be distinguished by its narrow gills, solid stem and smaller, almost globose, spores. In size, shape of the pileus and general appearance it most resembles A. campestris, but in the whitish primary color of the gills and in the yellowish tints which the pileus often assumes, it approaches nearer to A. arvensis. * * * *Peck*, 36th Rep. N. Y. State Bot.

I can now add my own testimony to that of Mr. Rodman as to its edibility. Its flesh is firm but crisp, not tough, and its flavor, though not equal to that of the common mushroom, is nevertheless agreeable, and its use as food is perfectly safe. *Peck*, Rep. 49.

This species has grown freely for several years at Hull and Cohasset, Mass. It is usually found about June 1st, and is not seen again until early autumn. It is the handsomest mushroom I have seen, and its edible qualities are on a par with its appearance. *Macadam.*

A. hæmorrhoida′rius Shulzer. *Gr.*—discharging blood. **Pileus** 4 in. across, reddish-brown, fleshy, ovate then expanded, *covered with broad adpressed scales*, margin at first *bent inward*. **Flesh** when broken immediately blood-red. **Stem** 4 in. high, 1 in. thick, soon hollow, fibrillose, the solid base somewhat bulbous. **Ring** superior, large. **Gills** free, approximate, crowded, rosy-flesh-color, at length purple-umber.

Very striking, 3–4 in. high.   The pileus and the white stem become Agaricus. spotted blood-red when touched.   The stem when young is adpressedly squamulose below, when full grown mealy, becoming smooth.   *Fries.*

Spores purple-brown, 7–8x5μ *Massee;* brown, elliptical, 5–6x4μ *Peck.*

A rare or overlooked plant in United States, first recorded by Professor Peck, who found it but once, growing under a hemlock tree.   Rep. 45.

Nebraska, *Clements;* West Virginia; Eagle's Mere and Mt. Gretna, Pa.   In hemlock and mixed woods.   Autumn.   *McIlvaine.*

Cap 2–4 in. across.   **Stem** 3–4 in. long, up to ¾ in. thick.

Every part of the plant turns red and has a congested appearance when bruised.   The flesh is white but immediately becomes red when broken.

It is a frequent but not common species, growing singly, or in small clusters.

In flavor and substance it is equal to any mushroom.

A. mari'timus Pk.   **Pileus** very fleshy, firm, at first subglobose, then broadly convex or nearly plane, glabrous, sometimes slightly squamose with appressed spot-like scales, white becoming dingy or grayish-brown when old.   **Flesh** whitish, quickly reddening when cut, taste agreeable, odor distinct, suggestive of the odors of the seashore. **Lamellæ** narrow, close, free, pinkish becoming purplish-brown with age, the edge white.   **Stem** short, stout, firm, solid, equal, sometimes bulbous, white, the annulus delicate, slight and easily obliterated.   **Spores** broadly elliptic, purplish-brown, 7–8μ long, 5–6μ broad.

Pileus 2–8 in. broad.   **Stem** 1–2 in. long, .6 in. thick.

Sandy soil near salt water, Lynn, Mahant and Marblehead, Mass. June to December.   *R. F. Dearborn.*

This is a very interesting and an excellent mushroom.   Dr. Dearborn writes that he has used it on the table for fourteen years and that it is the only mushroom that he has ever eaten in which the stem is as good as the cap.   He considers it the most hearty and satisfying of all the numerous species that he has ever eaten.   Both its taste and odor is suggestive of the sea.   The latter is quite strong, and perceptible by one riding along the road by whose side the mushrooms are growing. They sometimes grow in semicircles and attain a larger size in warm weather than in the colder weather of autumn.   They are most abundant in August.   The flesh, when cut or broken, quickly assumes a pink

Agaricus. or reddish hue on the freshly-exposed surface. This is a very distinctive character and with the maritime habitat makes the species easy to recognize. Another species, Agaricus hæmorrhoidarius Kalchb. exhibits a similar change of color in its wounded flesh, but is of very rare occurrence with us, does not, so far as ascertained, grow near the sea, has a darker cap and a long hollow stem. The stem in the maritime mushroom is short and solid. Its collar is very slight and easily destroyed. *Peck*, Bull. Torr. Bot. Club, Vol. 26, No. 2, F. 1899.

A. Califor'nicus Pk.—**Pileus** at first subconical, becoming convex, minutely silky or fibrillose, whitish, tinged with purple or brownish-purple on the disk. **Flesh** whitish. **Gills** close, free, pink becoming purplish, then blackish-brown. **Stem** rather long, solid or stuffed, equal or tapering upward, distinctly and rather abruptly narrowed above the entire externally silky ring, pallid or brownish. **Spores** broadly elliptical, 5–6x4–5µ.

Pileus 1–3 in. broad. Stem 1.5–3 in. long, 2–4 lines thick.
Under oak trees. Pasadena. January. *McClatchie*.

This fungus is similar in size, shape and habitat to A. hemorrhoidarius, but it is unlike that species in color, in the adornment of the pileus and in its color not changing where bruised or broken. Bull. Torr. Bot. Club, 22–5 My. 95.

A. Elven'sis B. and Br.—Name from river Elwy, Wales, where first found. Tufted. **Pileus** 4–6 in. or more across, subglobose then hemispherical, fibrillose, broken up into large persistent brown scales, areolate in the center, margin very obtuse, thick, covered with pyramidal warts. **Stem** at first nearly equal, at length swollen in the center, and attenuated at the base, 4–6 in. high, 2 in. thick in the center, fibrillose and areolate below, nearly smooth within the pileus, solid, stuffed with delicate threads. **Ring** thick, very large, deflexed, broken here and there, warted in areas beneath. **Gills** rather crowded, ¼ in. broad, free, of a brownish flesh-color. **Spores** elliptic oblong, 8x4µ.

Under oak trees, etc. Edible, delicious eating. Flesh of pileus ¾ in. thick, red when cut. *Massee*.

California, *H. and M.*

Edible. *Cooke*, 1891.

338

A. fœdera'tus Berk. and Mont.—confederated. **Pileus** fleshy, thin, Agaricus. at first ovoid then bell-shaped, finally convex, somewhat umbilicate with the center slightly depressed, margin hanging down (when dry involute), fragments of the veil hanging from the margin, tawny, scaly with minute, scattered, white, persistent granules, 2–3 in. broad, ¾–1½ in. high. **Stem** stout, hollow, stuffed with fibers, gradually increasing in size to the base; below the ring rough from the ruptured bark, 4 in. high. **Ring** superior, broad, reflexed, torn, persistent. **Gills** linear, medium broad, at first pinkish-lilac, when adult brownish, edge white, pulverulent, adnate, gradually attenuated toward the margin. **Spores** dingy-brown, ovoid oblong, 10µ long. Somewhat cespitose. Elegant.

On the ground in pastures. July. Columbus, Ohio. *Sullivant,* Mont. Syll., p. 121.

Edibility not reported. I have not seen this species.

A. xylo'genus Mont. *Gr.*—produced on wood. **Pileus** membranaceous, at first ovoid, then conical, bell-shaped, umbonate, finally convexo-plane, smooth, pale-yellow, center brownish, margin split, striate when dry, 1½–2½ in. broad, 1¼ in. high. **Stem** cartilaginous, white, 3 in. high, ¼ in. thick, gradually thickened toward the base, hollow. **Ring** of medium size, inferior, erect or reflexed. **Gills** free, remote, lance-shaped, rounded behind, attenuated toward the margin, pink as in A. campester. **Spores** spherical, colorless, hyaline, 5–7.5µ.

On dead wood. August. Columbus, Ohio. *Sullivant.* Mont. Syll., p. 122.

Edibility not reported. I have not seen this species.

** *Gills at first brownish or gray.*

A. argen'teus Brændle—of silver. **Pileus** thin, convex becoming nearly plane, slightly silky or glabrous, pale grayish white or grayish brown, shining with a silvery luster when dry, the margin sometimes striate, at first incurved, often revolute when old. **Flesh** whitish, becoming blackish where cut. **Lamellæ** close, free, at first brownish becoming blackish brown or black with age. **Stem** short, glabrous, solid, often narrowed toward the base, the annulus slight, evanescent. **Spores** broadly elliptic, 7–10µ long, 6µ broad.

Agaricus. **Pileus** 1–2 in. broad. **Stem** 1–1 ½ in. long, ¼–⅜ in. thick.

Lawns and grassy places in rich soil. Often associated with Stropharia bilamellata Pk. After rains from April to November. Washington, D. C. *F. J. Brændle.*

This is a small mushroom, peculiar in having the young gills of a dark color and in the absence of any pink hues. The gills sometimes become moist and manifest a tendency to deliquesce. The drying specimens emit a strong but not unpleasant odor. Mr. Brændle says that their edible quality is excellent and that it is not impaired by drying. *Peck,* Bull. Torr. Bot. Club, Vol. 26, F. 1899.

A. praten′sis Schaeff.—a meadow. **Pileus** 2–3 ½ in. across, ovoid then expanded, becoming smooth or sometimes broken up into scales more or less concentrically arranged, whitish, then grayish. **Flesh** thick in the center, thin toward the margin, white. **Gills** free, rounded behind, about ¼ in. broad, grayish, then brown. **Stem** about 2 in. long, ½–⅔ in. thick, base thickened, smooth, whitish. **Ring** median, simple, usually deciduous. **Stem** becoming more or less hollow. **Spores** elliptical, apiculate, 6x3.5μ.

On pastures and woods. Distinguished by the grayish gills becoming brown without any intermediate pink or fleshy tinge, and in being rounded behind, the median deciduous ring, and the more or less hollow stem. *Massee.*

California. Common. Edible. *H. and M.* Not elsewhere reported.

A. achi′menes B. and C. *Gr.*—an amber-colored plant. **Pileus** 4–6 in. broad, pallid or yellowish-white, smooth like kid leather, but studded with warty excrescences especially toward the center. **Stem** 4–6 in. high, 3–4 lines thick, white, stuffed with floccose fibers, furnished toward the apex with a large deflexed ring. **Gills** broad, crowded at first, whitish then ash-colored and dingy-brown, free. **Spores** brownish, oval or ovate.

A splendid species allied to A. fabaceus, but differing in its paler spores, warty cap, ample ring, etc.

On the earth. Solitary. June. *S. C. Ravenel.* Am. Jour. Sci. and Arts, 1849.

I have not seen this species.

A. faba'ceus Berk.—relating to beans. **Pileus** 4–5 in. across, Agaricus. thin, almost submembranaceous, umbonate, conical when young, becoming nearly plane as it expands, white, viscid when moist; epidermis smooth, tough, feeling like fine kid leather, turning yellow when bruised. **Stem** 3–4 in. high, ⅓ in. thick, white, smooth, with the exception of a few fibrilla, equal except at the base. **Veil** large, at first covering the gills and connecting the margin with the stem, white, externally floccose. **Gills** crowded, very thin, not ventricose, free, brown when young, then darker brown, at length almost black like the dark part of a bean flower. A fine species allied to A. arvensis. When young it has a peculiar but not unpleasant smell. On the ground, amongst dead leaves in open woods. Waynesville, September 10, 1844. Hooker's London Jour. of Botany, 1847.

Described by Berkeley from specimens collected by Thomas G. Lea, in the vicinity of Cincinnati.

On ground among old leaves in woods. Common. **Pileus** 3–4 in. broad. **Stem** 3–4 in. high. **Spores** brown, nucleate on one side, small, 5.5μ long. *Morgan.*

This is among the most delicious species for the table. Fresh specimens have a distinct taste and odor of peach kernels or bitter almonds which is nearly lost in cooking. Am. Jour. Science and Arts, 1850. *Curtis.*

Ohio, *Lea, Morgan;* North Carolina, *Curtis;* South Carolina, *Ravenel;* Massachusetts, *Sprague.*

*** Gills at first whitish.

A. arven'sis Schaeff.—belonging to cultivated ground. Horse Mushroom, Plowed-Land Mushroom. (A. Georgii Sow., A. pratensis Scop., A. edulis Krombh., A. exquisitus Vitt.) **Pileus** at first convex or conical, bell-shaped then expanded, at first more or less floccose or mealy, then smooth white or yellowish. **Flesh** white. **Gills** close, free, generally broader toward stem, *at first whitish, then pinkish,* finally blackish-brown. **Stem** equal or slightly thickened toward the base, smooth, *hollow or stuffed* with a floccose pith; ring rather large, thick, the lower or exterior surface often cracked in a radiate manner.

Plant 2–5 in. high. **Pileus** 3–5 in. or more broad. **Stem** 4–10 lines thick.

341

Cultivated fields and pastures. Summer and autumn.

This species is so closely related to the common mushroom that it is regarded by some authors as a mere variety of it. Even the renowned Persoon is said to have written concerning it: ''It appears to be only a variety of A. campestris.'' Fries also says that it is commonly not distinguished from A. campestris, but that it is diverse in some respects; its white flesh being unchangeable, its gills never deliquescing, remaining a long time pale and not becoming dark-red in middle age. Berkeley says of it: ''A coarse but wholesome species, often turning yellow when bruised.'' *Peck*, 36th Rep. N. Y. State Bot.

Spores spheroid-elliptical, 9x6μ *K.;* 11x6μ *W.G.S.;* elliptical, 8–10 x5–6.5μ *Peck.*

Indiana, *H. I. Miller;* Minnesota, *B. L. Taylor;* West Virginia, North Carolina, New Jersey, Pennsylvania, *McIlvaine.*

Unless the numerical system of John Phœnix to express degrees of quality is adopted by a mycophagists' congress, and one species of fungus is chosen as the standard of excellence, the comparative excellence of species will never be settled. English epicures shun A. arvensis; the French prefer it. Berkeley says it is inferior to the common mushroom; Vittadini says it is very sapid and very nutritious. So opinion varies. Individual tastes must decide excellence. Comparison never will. Toadstools differ in substance, texture and taste as one meat or vegetable differs from another. Beef could not be chosen as the standard for meats, or cabbage as the standard for vegetables. Agaricus arvensis is good.

A. magni'ficus Pk.—magnificent. (Plate XCIV.) **Pileus** 5–15 cm. (2–6 in.) broad, fleshy, thick, convex, becoming nearly plane or centrally depressed, bare, often wavy and split on the margin, white or whitish, often brownish in the center. **Flesh** 1.5–2 cm. (½ in.) thick in the center, thin on the margin, white, unchangeable. **Gills** numerous, rather broad, close, free, ventricose, white becoming dark purplish brown with age, never pink. **Stem** 10–15 cm. long (4–6 in.), about 2.5 cm. thick (1 in.), firm, stuffed with cottony pith, bulbous or thickened at the base, fibrillose, striate, minutely furfuraceous (covered with scurf) toward the base, ringed, pallid or whitish, the ring thin, persistent, white. **Spores** small, elliptic, 5–6μ long, 3–4μ broad.

Gregarious or cespitose; thin woods, Mt. Gretna, Pa. August. Agaricus.
Charles McIlvaine.

A large fine species distinguished from its near allies by the absence of pink hues from the gills. Mr. McIlvaine remarks that it has an anise-like flavor and odor and that when young the whole fungus is tender and high flavored, but when full grown the caps only are edible. *Peck*, Bull. Torr. Bot. Club, Vol. 26, F. 1899.

A. silvic'ola Vitt.—*silva*, a wood; *colo*, to inhabit. ([Color] Plate XCI, fig. 2.) (A. arvensis, var. abruptus Pk.; now A. abruptus Pk.) **Pileus** convex or sub-bell-shaped, sometimes expanded or nearly plane, *smooth, shining*, white or yellowish. **Gills** close, thin, free, rounded behind, generally narrowed toward each end, *at first whitish, then pinkish*, finally blackish-brown. **Stem** *long*, cylindrical, stuffed or hollow, white, *bulbous;* ring either thick or thin, entire or lacerated. **Spores** elliptical, 6–8x4–5μ.

Plant 4–6 in. high. **Pileus** 3–6 in. broad. **Stem** 4–8 lines thick.

Woods, copses and groves or along their borders. Summer and autumn. *Peck*, 36th Rep. N. Y. State Bot.

Very good eating, though scarcely as highly flavored as the common mushroom. *Peck.*

West Virginia, Pennsylvania, New Jersey, June to frost. *McIlvaine.*

A. silvicola, by many authors considered a variety of A. campester, is, seemingly, becoming common. Professor Peck in 46th Rep. has made the abrupt bulb and its usual double veil distinctive marks which ally it to A. arvensis. He therefore calls it var. abruptus. As this book goes to press Professor Peck writes me that he concludes var. abruptus to be a good and distinct species. It is therefore given as such. While familiar with it since 1881, I never found it in quantity until 1898, at Mt. Gretna, Pa. There, among the straw and rubbish of abandoned camps on wood margins, it grew in great quantity; sometimes singly, at others in crowded clusters. When growing singly it exhibits all the characteristics of its description; when clustered, the stems are not always bulbous. The caps are thin but fleshy, brittle and bear a disproportionate width to the stem—like a plate on a pipe stem. The caps when mature are usually tinged with yellow and are spread flat; the ring is large, often double, yellowish, often torn, fragments of it frequently hang from the cap margin; the bulb when perfect is small, abrupt, as if it had once been round but the stem pushed into it. It has a strong spicy mushroom odor and taste, and makes a high-flavored dish. It is delicious with meats. It is the very best mushroom for catsup. Mixed with Russulæ or Lactarii or other species lacking in mushroom flavor, it enriches the entire dish. The stems, excepting of the very young, are tough.

Larvæ do not infest A. silvicola. Its habit of growth shows it to be cultivatable. It has but one draw-back. Growing as it does in woods and in the presence of the poisonous Amanita, it is possible for the careless collector to confound the two. The Amanitæ have larger bulbs, cups at the base, and *white gills;* the A. silvicola has no volva, has whitish gills when very young only, they become pinkish, then a marked blackish-brown.

A. creta′ceus Fr.—*creta*, chalk. **Pileus** 3 in. and more broad, wholly *white*, fleshy, lens-shaped-globose when young, then convexo-flattened, obtuse, dry, *sometimes even*, sometimes rivulose chiefly round the margin from the cuticle *separating into squamules*. **Flesh** thick, white, unchangeable. **Stem** 3 in. long, 3–6 lines and more thick, *hollow, stuffed with a spider-web pith*, firm, attenuated upward, even, smooth, not spotted, white. **Gills** free, then remote, ventricose but *very much narrowed toward the stem*, crowded, *remaining long white*, becoming dingy-brown only when old. *Fries.*

Spores 3x4µ *W.G.S.;* 5–6x3.5µ *Massee.*

Under certain conditions the spores are white. *M. J. B.*

In lawns and rich ground.

North Carolina, on earth and wood. Edible, *Curtis;* Minnesota, rare, *Johnson;* California, *H. and M.;* Ohio, *Lloyd;* Kentucky, *Lloyd*, Rep. 4; New York, *Peck*, Rep. 22.

A. subrufes′cens Pk.—*sub*, under; *rufescens*, becoming red. **Pileus** at first deeply hemispherical, becoming convex or broadly expanded, silky fibrillose and minutely or obscurely scaly, whitish, grayish or dull reddish-brown, usually smooth and darker on the disk. **Flesh** white, unchangeable. **Lamellæ** at first white or whitish, then pinkish, finally blackish-brown. **Stem** rather long, often somewhat thickened or bulbous at the base, at first stuffed, then hollow, white; the annulus flocculose or floccose-scaly on the lower surface; mycelium whitish,

forming slender branching root-like strings. **Spores** elliptical, 6–7µ Agaricus.
Peck, 48th Rep. N. Y. State Bot.

Indiana, *H. I. Miller*, 1898; Haddonfield, N. J., *McIlvaine*.

June 2, 1896, I found several specimens of a fungus new to me, and sent them to Professor Peck for identification. He pronounced it a dwarf form of his species A. subrufescens. The cluster grew on a florist's compost pile at Haddonfield, N. J. Its flesh has a flavor like that of almonds.

This species is now cultivated and has manifest advantages over the marketed species—it is easier to cultivate, very productive, produces in less time after planting the spawn, is free from attacks of insects, carries better and keeps longer.

Amateurs are likely to succeed in growing it, and to have goodly crops of mushrooms instead of disappointments.

A. placo′myces Pk. *Gr.*—a flat cake. ([Color] Plate XCI, fig. 3.) **Pileus** thin, at first convex, becoming flat with age, whitish, brown in the center and elsewhere adorned with minute brown scales. **Lamellæ** close, white, then pinkish, finally blackish-brown. **Stem** smooth, annulate, stuffed or hollow, bulbous, white or whitish, the bulb often stained with yellow. **Spores** elliptical, 5–6.5µ long.

Cap 2–4 in. broad. **Stem** 3–5 in. long, ¼ to nearly ½ in. thick.

It grows in the borders of hemlock woods or under hemlock trees from July to September. It has been eaten by Mr. C. L. Shear, who pronounces it very good. I have not found it in sufficient quantity to give it a trial. This mushroom is very closely related to the wood mushroom or silvan mushroom, Agaricus silvaticus, a species which is also recorded as edible, but which is apparently more rare in our state (New York) than even the flat-cap mushroom. This differs from the silvan mushroom in its paler color, in having the cap more minutely, persistently and regularly scaly, and in its being destitute of a prominent center. In the silvan mushroom the scales, when present, are few, and they disappear with age. *Peck*, 48th Rep. N. Y. State Bot.

Mrs. E. C. Anthony, Gouverneur, N. Y., June, 1898, writes: "In great abundance on lawn, tumbling over one another in their haste to make their appearance. One of the largest, which did not have half a chance to display its proportions, would probably measure 7 in., perhaps more. When mature they crack across the top, showing the white flesh. The gills are pink, stem white, solid and bulbous. There is no perceptible odor when fresh.''

Indiana, *H. I. Miller*, edible, good.

Specimens sent to me by Mrs. Anthony, though not fresh, were eaten by me. They very much resembled the common mushroom, but probably, owing to their condition, were not so tender.

I have not found the species. The illustration is after a painting by Mrs. E. C. Anthony,

A. varia'bilis Pk.—variable. ([Color] Plate XCI, fig. 1.) **Cap** 2–6 in. across, ovate, bell-shaped, irregularly convex and wavy, margin incurved but never striate, smooth, minutely fibrillose, with few remaining floccose scales; mature plant pure white, when young distinctly tinged with lilac and here and there with yellow when mature, slightly, broadly umbonate and depressed around umbo, cracks along gills. **Flesh** thick in center, very thin, even membranaceous toward. margin, spongy, unchangeable. **Gills** free, close, thin, flaccid, ventricose, narrow next stem, but few short, pure-white when young, then dark-umber without purple tinge. **Stem** equal, tubed, white, silky, smooth above ring, rippled and minutely furfuraceous (scurfy) below, flocculose-furfuraceous when young, densely hairy at base, and occasionally slightly expanding, but not bulbous, densely cespitose with a coarse, white, root-like mycelium. **Veil** heavy at first, mottled with yellow scales beneath; as cap expands veil becomes thin, like tissue paper, ruptures at both stem and margin leaving torn ring on stem and appendiculate fragments on edge of cap.

Spores shed in great quantity, rich dark umber-brown without shade of purple.

Taste strong like almond. **Smell** slightly of musk, like the running mycelium of A. campester.

Found at Mt. Gretna, Pa. *Charles McIlvaine.*

I have never found worms in this species. It is very prolific and its habitat shows that it can be cultivated. Its freedom from worms and lasting carrying quality will make it commercially valuable.

It grew in an old roofless stable from September until after several frosts, in enormous quantity, 25 or 30 pounds in a patch. It differs from A. subrufescens in not having a shade of red about it, in its very distinct light-lilac cap when full grown, and in its snow-white youth.

The young gills are pure white as are the caps. The stems sometimes Agaricus.
taper upward, but they are usually remarkably equal.

It is delicate when cooked and of excellent flavor.

A. tabula′ris Pk.—relating to boards. **Pileus** 5–10 cm. broad, very
thick, fleshy, firm, convex, deeply cracked in areas, whitish, flesh whitish, tinged with yellow, the areas pyramidal, truncate, the sides horizontally striate, their apices sometimes tomentose. **Lamellæ** narrow,
close, free, blackish-brown when mature. **Stem** short, thick, solid.
Spores broadly elliptical, 7.5–9µ long, 6–7.5µ broad, generally containing a single large nucleus.

In clay soil by roadsides. Craig, Colorado. August. *E. Bethel.*

This species is remarkable for the peculiar upper surface of the pileus
which is broken into pyramidal areas. The sides of these are marked
by parallel lines in such a way that they appear as if formed by small
tablets placed one upon another, each successive tablet being a little
smaller than the one immediately preceding it. Only dried and broken
specimens have been seen by me and the notes of the collector do not
give the color of the young lamellæ. There is a trace of a thick ring on
the broken stem of one specimen. *Peck*, Bull. Torr. Bot. Club, Vol.
25, No. 6, 1898.

Not elsewhere reported. Edible qualities not given.

PILOSACE Fr.

Pilosace. Hymenium differentiated from the stem. **Gills** free from the stem; general and partial veil both absent, hence there is no ring on the central stem. **Spores** purple-brown.

(Plate XCV.)

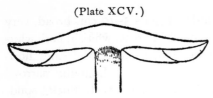

SECTION OF PILOSACE ALGERIENSIS.

A peculiar genus, with the habit of Agaricus, but without a trace of a ring. *Massee.*

P. eximius Pk., 24th Rep. N. Y. State Bot., is the only species thus far reported in America. Edible qualities unknown.

STROPHA'RIA.

Gr.—a sword-belt. (Referring to the ring.)

Stropharia. **Flesh** of stem and pileus *continuous*. **Veil** present, when ruptured forming a distinct ring on the stem. **Gills** more or less *adnate.*

On the ground or epiphytal.

Separated from all the genera of the purple-spored series but Agaricus by the presence of a distinct ring, and from that by the continuity of flesh in stem and pileus, and by the gills not being free. **Pileus** somewhat fleshy, sometimes viscid.

The species belonging to this genus are rather small, and from their habitats are frequently passed or overlooked. Yet many of them are common and plentiful. Those which have been tested are excellent and worth seeking in their season. The entire genus has been under a cloud. Writers upon it assert some of its members to be dangerously poisonous. So far as carefully tested by the writer no doubtful one has been encountered, and one—semiglobata—has been eaten by himself Stropharia. and friends since 1881, notwithstanding its dangerous reputation.

The division between this genus and Agaricus is not always sharply defined. S. æruginosa, S. semiglobata and S. stercoraria were formerly placed in Psalliota, now Agaricus.

ANALYSIS OF TRIBES.

A. VISCIPELLES (*viscum*, bird-lime; *pellis*, a skin). Page 349.
Pellicle of the pileus even or scaly, generally viscid.
* Mundi—*mundus*, clean. Not growing on dung.
** Merdarii—*merda*, dung. Ring often incomplete.

B. SPINTRIGERI (Stropharia spintriger).
Pileus without a pellicle, but fibrillose, not viscid. None known to be edible.

A. VISCIPELLES. Pellicle of the pileus even or scaly.

* Mun′di—*not growing on dung.*

S. ærugino′sa Curt. —*ærugo*, verdigris. **Pileus** fleshy, but not compact, convex-bell-shaped then flattened, somewhat umbonate (ob- tuse when larger), *with very viscid pellicle,* the ground color yellowish but *verdigris from the azure-blue slime* with which it is more or less covered over, becoming pale as the slime separates. **Stem** *hollow,* soft, equal, *at the first scaly* or fibrillose *below the ring, viscid, becoming* more or less *azure-blue green.* **Ring** distant. **Gills** adnate, plane, 2 lines and more broad, not crowded, soft, whitish then dusky, becoming somewhat pur- ple.

(Plate XCVI.)

STROPHARIA ÆRUGINOSA.
Natural size. (After Stevenson.)

The above are the essential marks of this species. Variable in form,

Stropharia. sometimes cespitose. The typical and handsomest form is gathered in soaking weather in later autumn in shaded woods; it is large (pileus and stem 3 in. and more), stem squarrose with white spreading scales, intensely verdigris or azure-blue-pelliculose and very glutinous. From this there is a long series of forms with the gluten more separating (on the separation of the gluten the pileus becomes yellow), and the scales alike of the pileus and stem rubbed off. Finally, a smaller form occurs in open meadows, stem scarcely 2 in. long, only 2 lines thick, becoming azure-blue-green and without scales, pileus 1–2 in. broad, pale verdigris soon light yellowish, less viscid. In this form the ring is incomplete, while in the typical form it is entire, spreading, and persistent.

In woods, meadows, etc. Common. July to November. *Stevenson.*

Spores ellipsoid or spheroid-ellipsoid, 8x4–5μ *K.;* 5x7μ *W.G.S.;* elliptical, 10x5μ *Massee.*

POISONOUS. *Stevenson.*

"There is a white variety, in which the pileus is perfectly white from the first." *Cooke.*

S. æruginosa has been noted here by Schweinitz in Pennsylvania, Curtis in North and South Carolina, Frost in Vermont and Massachusetts, Harkness and Moore, California, Morgan, Ohio. The qualities of the American representatives are not reported. I have not seen the species. As it is asserted to be poisonous by European writers it may be. M. C. Cooke says: "It has the reputation, which is somewhat general on the continent, of being poisonous, but probably this is only assumed from its disagreeable taste and repulsive appearance." Collectors are cautioned to look out for it, and not to eat of it carelessly.

I can find no case of poisoning by this species reported. It presents another case of "Not proven."

** Merda'rii—*ring often incomplete.*

S. stercora'ria Fr.—*stercus*, dung. **Pileus** 1 in. broad, yellow, fleshy, but thin at the margin, hemispherical then expanded, obtuse, orbicular, with a viscid pellicle, naked, smooth, even or at length slightly striate only at the margin. **Stem** 3 in. and more long, 2–3 lines thick, stuffed with a separate fibrous pith, equal, clothed to the ring (which is scarcely 1 in. distant from the pileus, viscous, narrow, but somewhat spreading) with the flocculose veil which is at the same time viscous (so that it

350

appears as if smooth), yellow. **Gills** adnate, very broad behind, 2 lines broad, somwhat crowded, dusky-umber or dusky-olivaceous, of one color, quite entire.

Stem silky-viscous when moist, when dry becoming even, shining and yellowish-white, and without a manifest veil. The gills are truncate and somewhat decurrent. *Fries.*

Spores 17x13μ *W.G.S.;* elliptical, 18–20x8–10μ *Massee.*

West Virginia, 1881–1885; Pennsylvania; New Jersey. June to November. *McIlvaine.*

I have enjoyed this species, which is common, since 1881. It is usually conspicuous upon droppings and manure piles. It also occurs on richly-manured ground, in wood and field, usually single; sometimes two or three are united.

Caps and stems are edible, but do not cook in the same time. It is better to cook the caps only. They are delicious.

S. semigloba'ta Batsch.—*semi*, half; *globus*, a ball. **Pileus** commonly ½ in. broad, *light-yellow*, slightly fleshy, hemispherical, not expanded, very obtuse, even, *viscous*. **Stem** about 3 in. long, 1 line thick, tubed, slender, firm and straight, equal, even, smooth, becoming yellow, paler at the apex, powdered with the spores, otherwise smeared with the glutinous veil which is abrupt above terminating in an *incomplete* (not membranaceous) viscous, distant *ring*. **Gills** adnate, *very broad*, plane, *clouded with black.*

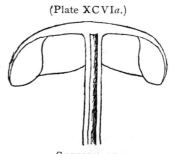

(Plate XCVI*a*.)

SECTION OF
STROPHARIA SEMIGLOBATA.
Natural size. (From Massee.)

Spores dusky-purple. *Stevenson.*

Spores blackish-purple, 13x8μ *W.G.S.;* elliptical, ends rather acute, 12x6μ *Massee.*

Grows on dung, rich lawns and pastures. April to November. A common, frequent, solitary species, easily recognized by its hemispherical cap, dark mottled gills. At first sight it resembles Naucoria semiorbicularis.

The caps are equal to any mushroom. I have eaten it since 1881. M. C. Cooke says: "It was Sowerby who drew attention to this species

Stropharia. as dangerous, and intimated that it had been fatal. **Since that period** we are not aware of any further evidence against it.

It is tender, good and harmless.

HYPHOLO'MA.

Gr.—a web; *Gr.*—a fringe.

Hypholoma. **Pileus** more or less fleshy, margin at first incurved. **Veil** *webby, adhering in fragments to the margin of the pileus*, not forming a distinct ring on the stem. **Stem** fleshy, similar in substance to that of the pileus with which it is continuous. **Gills** attached to the stem, sometimes with a notch at the juncture (emarginate), occasionally separating and then appearing to be free.

(Plate XCVIII.)

HYPHOLOMA FASCICULARIS.
Natural size.

Generally cespitose, mostly growing on wood above or under the ground.

Spores brownish - purple, sometimes intense-purple, almost black.

Corresponding to Tricholoma, Entoloma and Hebeloma.

352

ANALYSIS OF TRIBES.

FASCICULARES (H. fascicularis).   Page 354.

Pileus tough, smooth, bright colored, not hygrophanous.   Hypholoma.

VISCIDI (*viscidus*, viscid).

Pileus naked, viscid.   None known to be edible.

VELUTINI (*H. velutinus*).   Page 360.

Pileus silky or streaked with small fibers.

FLOCCULOSI (*floccus*, a lock of wool).

Pileus covered with superficial floccose scales, at length disappearing. (None reported edible.)

APPENDICULATI (*H. appendiculatus*).   Page 362.

Pileus smooth, hygrophanous.

Members of this purple-spored genus grow upon decayed wood, either standing or as roots in the ground, or from ground heavily laden with woody material.   They grow singly, in groups, or in densely-tufted or overlapping masses.   The several species vary in shades of yellow, red, orange, brick-color and brown; their caps are from 1–6 in. across; their stems are short or long, as the number in the cluster permits; when growing singly the stems are short and sturdy.   There is a floccose veil, or remnants of one, about the stem.   The gills are yellowish, greenish, olivaceous or greenish shades of yellow, gray, purple, almost black.   They are showy, easily recognized and are found from September until mid-winter.   I have gathered them when frozen hard. The flesh is solid, or spongy, flexible or fragile, white or yellowish; the tastes are sweet, nutty, bitter and saponaceous.   Patches of them—and they are frequent in almost every woods in the land—often yield several bushels.   Tons of them annually go to waste.

Old authors and some copyists say "the species are not edible, the tough ones being bitter, the fragile ones almost void of flesh."   Eighteen years of experience with them warrants my saying that there is not a single wild genus approaching it in economic value, and when its most prominent species are properly cooked, few equal it in consistency and flavor.   As a pickle the Hypholomas have no superior.

Half a dozen or more of the species are exceedingly difficult to separate. Professor Peck has happily made a new species, H. perplexum, which is well named. For all culinary purposes these affiliated species may be gathered under that convenient name; for botanic purposes his description covers several perplexing characteristics common to what have been written as separate species, and covers a composite species.

The occasional bitter taste of some species is not constant, and can not be relied upon as a distinguishing mark. In the same tufts some individuals may be mild, others bitter; some individuals in groups are in a position and of an age to absorb water; others are not. There will be a marked difference in their taste raw. A few in the same group may have been infested by insects; others not. Those infested are often intensely bitter, while their companions are of pleasant flavor. The same remarks apply to neighboring clusters and individuals. I am of the opinion, from long observation, that the bitter is largely due to the injury and excrement of larvæ. Changes of taste occur in toadstools in a most marked and rapid manner. Apples from the same tree, chestnuts from the same tree, acorns from the same oak, radishes from the same seed, blackberries from the same bush, differ widely in taste. Why not toadstools of the same species?

I have often seen species of this genus, described as having stems up to 5 in. long, stretch and twist their stems to over a foot in order to get their caps from the inside of, or from a crack in a decaying stump, out into the light; and I have seen stems of the same species stout, solid and sturdy when individuals grew upright and singly. But wherever and however they grow, Hypholomas are safe. I have eaten them indiscriminately since 1881, and as long ago as 1885 published their edibility.

FASCICULA'RES. Pileus smooth, etc.

H. perplex'um Pk.—*perplexus*, perplexed. Perplexing Hypholoma. ([Color] Plate XCVII, fig. 2.) **Pileus** convex or nearly plane, glabrous, sometimes broadly and slightly umbonate, reddish or brownish-red fading to yellow on the margin, the flesh white or whitish. **Lamellæ** thin, close, slightly rounded at the inner extremity, at first pale-yellow, then tinged with green, finally purplish-brown. **Stem** nearly equal, firm, hollow, slightly fibrillose, whitish or yellowish above, rusty-reddish or reddish-brown below. **Spores** elliptical, purplish-brown, $8 \times 4 \mu$.

The Perplexing hypholoma has received the name because it is one of a group of five or six very closely allied species, whose separation from each other is somewhat difficult and perplexing. Of these six species three have a decidedly bitter, unpleasant flavor, and three are mild, or not decidedly bitter, if we may rely on the published descriptions of them. The three bitter ones, also, have no purplish tints to the mature gills; but two of the mild ones have. By using these and other distinguishing characters the six species may be tabulated and their several peculiarities more clearly shown.

Taste bitter I

Taste mild, or not clearly bitter. 3

I. Stem solid or stuffed, flesh whitish, gills whitish, then
      sooty-olive. . . . . .sublateritium

I. Stem hollow, flesh yellow. 2

  2. Cap yellow or tinged with tawny, stem yellow, gills
        yellow, becoming greenish. . . . . .fasciculare

  2. Cap brick-red, stem ferruginous, gills green, becom-
        ing olive. . . . . .elæodes

3. Cap red or brick-red, with a yellow margin; gills yel-
      low, then greenish, finally purplish-brown. . . . . .perplexum

3. Cap yellow, or slightly tawny on the disk only. 4

  4. Gills gray, becoming purplish-brown. . . . . .capnoides

  4. Gills yellow, becoming gray, neither green nor pur-
      plish . . . . .epixanthum

Probably in general appearance the Perplexing hypholoma most nearly resembles the brick-red Hypholoma, H. sublateritium; but it has often been mistaken for the tufted Hypholoma, H. fasciculare. From this it may be separated by the more red cap, the whitish flesh, the purplish-brown color of the mature gills, and the mild flavor. From H. sublateritium it is distinguished by its usually smaller size, more slender hollow stem, the yellow greenish and purplish tints of the gills, and the absence of a bitter flavor. Some may prefer to consider it a variety of this fungus, rather than a distinct species.

Its cap is 1–3 in. broad, its stem 2–3 in. long and 2–4 lines thick. It commonly grows in clusters, though sometimes singly, on or about old stumps or prostrate trunks of trees, in woods or open places. The caps of the lower ones in a cluster are often defiled and apparently discolored by the spores that have lodged on them from the upper ones.

Hypholoma. It appears in autumn, and continues until freezing weather stops its growth. It is a very common species, as well as a late one, and may often be gathered in large quantity. Its flavor is not first quality, but with good preparation it makes a very acceptable dish. It has been tested by myself and correspondents several times, and has been proved harmless. *Peck*, 49th Rep. N. Y. State Bot.

West Virginia, 1881–1885; New Jersey, North Carolina, Pennsylvania, October to January. On stumps, roots, ground containing decayed woody matter. *McIlvaine.*

H. perplexum is abundant in most if not all the states. I have eaten it and its allied species since 1881; dried them, pickled them, and fed them to many. If the collector gets puzzled, as he will, over one or all of these species, because no description fits, he can whet his patience and appetite by calling it H. perplexum and graciously eating it.

H. capnoï'des Fr. *Gr.*—like smoke, from the color of the gills. **Pileus** 1 in. sometimes 3 in. broad, *ochraceous-yellowish*, fleshy, convex, then flattened, obtuse, dry, *smooth*. **Flesh** somewhat thin, white. **Stem** 2–3 in. long, 2–4 lines thick, growing together at the base, *hollow*, equal, often curved and flexuous, *becoming silky-even*, pallid, whitish at the apex, here and there striate, becoming rust-colored under the surface-. covering when old. Cortina appendiculate, white, then becoming brownish-purple. **Gills** adnate, easily separating, somewhat crowded, rather broad, arid, *at first bluish-gray then becoming brownish-purple.*

Cespitose, fasciculate; odor and taste mild. On pine-stumps. Uncommon. *Fries.*

Spores ellipsoid-spheroid, 7x5μ *K.;* elliptical, brownish-purple, 8x4μ *Massee.*

California, *H. and M.;* Minnesota, not necessarily in fir-woods, *Johnson;* New York, on or about stumps or decaying wood of spruce. *Peck,* 50th Rep.

Haddonfield, N. J., 1894. Pine roots and stumps, and on ground. Cespitose. September to frost. *McIlvaine.*

A pretty species with caps up to 1½ in. across. **Stem** 2–4 in. long, ¼–⅜ in. thick, growing together (connate). The taste and smell are pleasant. The basket is soon filled from its clusters. There is not a better Hypholoma. The slightly soapy taste which attaches to most of the abundant and better known species is absent in this.

Porphyrosporæ

H. fascicula'ris Huds.—*fasciculus*, a small bundle. (Plate XCVIII, Hypholoma.
p. 352.) **Pileus** about 2 in. broad, *light yellow*, the disk commonly
darker, fleshy, thin, convex, then flattened, somewhat umbonate or
obtuse, even, smooth, dry. **Flesh** *light yellow*. **Stem** very variable
in length, hollow, thin, incurved or flexuous, fibrillose, of the same
color as the pileus and flesh. **Gills** adnate, very crowded, linear, *somewhat deliquescent, sulphur-yellow then becoming green.*

It is very easily distinguished from the preceding species by its *bitter
odor and taste, light-yellow flesh, and somewhat deliquescent, sulphur-
yellow then green gills.* It forms also more crowded clusters. There
are many remarkable varieties; one *robustior* (more robust), stem
thickened at the base, another *nana* (dwarf), both on the ground.

Cespitose on old stumps and the ground. Extremely common.
Stevenson.

Spores elliptical, 7x4µ *Massee;* 6–7x4µ *K.;* 6x4µ *W.G.S.;* ferruginous purple, 6x4µ *Morgan.*

"It is very usual to regard this as a poisonous species, but possibly
it is not so in reality." *Cooke.*

West Virginia, 1881, Pennsylvania, New Jersey, North Carolina,
McIlvaine.

A very common species appearing in October and lasting until well
into the winter, growing in large, overlapping masses or in tufts from
old stumps or roots, and about trees where decay has begun. Sometimes solitary. It is then short-stemmed and sturdy. There are several closely allied species. To know the one from the other, a careful
study of the group is necessary. (See introduction to genus, H.
epixanthum, H. sublateritium, H. capnoides, H. elæodes, and H. perplexum.) Old authors give it as bitter and poisonous. The bitter is
not always present. Any there is disappears in cooking. It is not
poisonous, but one of our most valuable species. I have eaten it since
1881. A little lemon juice or sherry will cover the slightly saponaceous
taste sometimes present. The caps only are good. It makes a choice
pickle and a good catsup.

H. epixan'thum Fr. *Gr.—epixanthos*, yellowish-brown. **Pileus** 2–3
in. broad, light-yellow or becoming pale, the disk commonly darker,
fleshy, moderately thin, convexo-plane, obtuse or gibbous, even, *slightly
silky then becoming smooth.* **Flesh** white, becoming light-yellow. **Stem**

357

Hypholoma. about 8 in. long, 3–4 lines thick, *hollow*, attenuated from the thickened base or equal, *floccose-fibrillose, pale rust color* or becoming dingy-brown *below*, with a frosty bloom at the apex; veil hanging from margin of pileus, white.   Gills adnate, crowded, *at first light yellow-white*, *at length becoming ash-colored*, not deliquescent, and not becoming purple or green.

Strong smelling, odor acid; extremely variable in stature; not hygrophanous. *Fries.*

Spores elliptical, 7x4μ *Massee.*

West Virginia, Pennsylvania, New Jersey, North Carolina.   On oak, chestnut stumps and growing from tree roots in ground.   October to December. *McIlvaine.*

(See H. perplexum, H. sublateritium and compare descriptions.)

This species, in common with its allies, is extremely hard to determine.   When growing singly from roots or from ground heavily charged with decaying wood, it is a sturdy, solid plant; when in clusters the stem is longer, more flexible and the whole character of the plant is modified.   Except for botanic purposes there is no occasion to puzzle over it.   It is in every way an excellent and useful fungus.

H. disper′sus Fr.—*dispergo*, to scatter.   **Pileus** 1–1½ in. broad, *tawny-honey-color*, not hygrophanous, *slightly fleshy*, bell-shaped then convex, at length expanded, even, *superficially silky round the margin* with the veil, or squamulose, otherwise even and smooth.   **Flesh** thin, a little paler than the pileus.   **Stem** 2 in. or a little more long, 2 lines thick, tubed, equal, *tense and straight*, tough, *fibrilloso-silky*, somewhat rust-colored, becoming dingy-brown at the base, pale at the apex. **Gills** adnate, thin, *ventricose*, broad, 3–4 lines, *crowded, at first pallid-straw color, at length crowded*, obsoletely green.   *Fries.*

Gills broader than H. fascicularis, etc.   Solitary, scarcely ever cespitose.   On pine stumps and the ground.   April to November.

Spores elliptical, 7x3–4μ *Massee.*

North Carolina, in pine woods, *Curtis;* California, *H. and M.;* West Virginia, Pennsylvania, North Carolina, New Jersey, *McIlvaine.*

Difficult to distinguish from H. fascicularis when growing solitary. Its edible qualities are precisely the same.

H. elæo′des Fr. *Gr.*—an olive; *Gr.*—*eidos*, appearance.   **Pileus**

brick-red or tan, fleshy, rather plane, somewhat umbonate, *dry, smooth,* Hypholoma. opaque. **Flesh** yellow. **Stem** stuffed then hollow, equal, commonly slender, incurved or flexuous, fibrillose, of the same color as the pileus, becoming rust-color. **Gills** adnate, crowded, thin, *green then* pure olivaceous.

Cespitose. Odor bitter. On trunks and on the ground. *Fries.*

Cap 1–2 in. across. **Stem** 2–4 in. long, ¼–⅜ in. thick, stuffed then hollow.

West Virginia, 1881–1885, Haddonfield, N. J.; Pennsylvania. On stumps, roots and ground in woods, etc. *McIlvaine.* Not reported elsewhere.

Its habit is the same as H. fascicularis, to which it is closely allied, and to me seems but a form of this very variable species. It is equally good.

H. sublateri′tium Schaeff.—*sub* and *later,* a brick. ([Color] Plate XCVII, fig. 3.) **Pileus** 2–3 in. and more broad, tawny-brick-red, but paler round the margin and covered over with a superficial, somewhat silky, whitish cloudiness (arising from the veil), fleshy, convexo-plane, obtuse, *discoid, dry,* even, *becoming smooth.* **Flesh** *compact,* white, then becoming yellow. **Stem** 3–4 in. long, 3–5 lines thick, *stuffed, stout* and firm, commonly manifestly attenuated downward, rarely equal, *scaly-fibrillose,* fibrils pallid, rust-colored downward. **Cortina** superior, at first *white, at length becoming black.* **Gills** adnate, more or less crowded according to stature, narrow, at first *dingy-yellowish* and darker at the base, *then sooty,* and at length inclining to olivaceous.

Spores brownish purple. Somewhat cespitose. **Stem** incurved from position. There are many varieties: *B,* somewhat solitary, the pileus and stem, which is thickened at the base, of the same color, reddish. *C,* smaller, pileus light yellowish, the hollow stem equal. *Schaeff.*

Var. *squamo′sum,* Cooke. Pileus convex, bright brick-red, shading to yellow at the margin, spotted with superficial scales. Flesh very thick, yellowish. Gills narrowish, adnate. Stem elongated, stout, pale above, rust-colored below, hollow, veil hanging from the margin when young.

On trunks. A very beautiful variety, larger and more robust than the typical form. *Massee.*

Spores 6x3µ *W.G.S.;* elliptical, sooty-brown, 8x4µ *Massee*

Hypholoma. West Virginia, 1881–1885; Pennsylvania, New Jersey, densely cespitose on stumps and roots. October to long after frosts. *McIlvaine.*

Edible. *Dr. Taylor,* 1893. Dept. of Agr. Rep. No. 5.

H. sublateritium has many forms. Both Fries and Stevenson indicate this as a variable species and my own observation confirms the truth of this.

This is a very common autumnal species, lasting into the winter. Old authors give it as bitter and very poisonous. I tested it in 1881 and have been eating it, in common with all Hypholomas I have found, ever since. At times it is bitter. I believe this to be due to the passage of larvæ through the flesh. Unattacked specimens are slightly saponaceous to the taste while others in the same bunch are bitter.

VIS'CIDI. Pileus viscid, etc. (None known to be edible.)

Velutini. *Pileus silky, etc.*

H. veluti'nus Pers.—*vellus,* a fleece. Velvety. **Pileus** fleshy, thin, convex or expanded, brittle, minutely tomentose-scaly, becoming smooth, hygrophanous, yellow with the disk reddish. **Lamellæ** rather broad, attached, tapering toward the outer extremity, dark brown tinged with red, the edge whitish-beaded. **Stem** equal, rather slender, hollow, fibrillose, subconcolorous, white-mealy and slightly striate at the top. **Spores** black.

Height about 2 in., breadth of pileus 1–1.5 in.

Roadsides. Albany Cemetery. September. The pileus sometimes cracks transversely. *Peck,* 23d Rep. N. Y. State Bot.

Spores 6x8µ *W.G.S.;* elliptical, 10x5µ *Massee.*

Often used in catsup. Innocent and edible. *Cooke.*

West Virginia. 1881–1885, Pennsylvania, West Philadelphia, Bartram's Creek, 1887, *McIlvaine.*

Var. *leioceph'alus* B. and Br. (*Gr.*—smooth; *Gr.*—head, from its smooth pileus). **Pileus** hygrophanous, rugged, smooth except at the margin, where it is fibrillose, pallid as is the stem, whose apex is mealy.

Densely cespitose, much smaller than the common form, but apparently a mere variety, though a striking one from its smooth but very rugged disk. On old stumps. *Stevenson.*

New York, *Peck,* 23d Rep.; West Virginia, West Philadelphia, Bartram's Creek, Haddonfield, N. J., September to November. *McIlvaine.*

Quantities of var. leiocephalus grow in the West Virginia forests on Hypholoma. stumps and on the ground from decaying roots. 1 ½ in. is the limit of its width. Its frequent and dense clusters, its tenderness and delicacy of flavor make it a favorite.

H. aggrega′tum Pk.—*aggrego*, to grow together. Densely cespitose. **Pileus** thin, convex or subcampanulate, grayish-white, obscurely spotted with appressed brownish fibrils. **Lamellæ** subdistant, rounded behind, nearly free, at first whitish, then brown or blackish-brown with a whitish edge. **Stem** rather long, hollow, somewhat woolly or fibrillose, white. **Spores** brown, elliptical, 8x4–5μ.

Pileus about 1 in. broad. **Stem** 2–3 in. long, 1.5–2 lines thick.

At the base of trees and stumps in woods. Alcove. September.

The cespitose habit and obscurely spotted grayish-white pileus are marked features of this species. From H. silvestre the species may be distinguished by its smaller size, adnexed or nearly free lamellæ which have no rosy tint, and by its very cespitose mode of growth. *Peck*, 46th Rep. N. Y. State Bot.

Mt. Gretna, Pa., about trees and stumps. September to November, 1898–1899. *McIlvaine.* Not reported elsewhere.

The caps are oyster-color. Amateurs accustomed to the gayer colors of the autumnal Hypholomas will not suspect this of belonging to the genus, until the color of the spores is obtained.

The caps are fine.

H. lachrymabun′dum Fr.—*lachryma*, a tear. **Pileus** 2–3 in. broad, whitish when young, then dingy-brown, becoming pale around the margin, truly fleshy but not compact, convex, obtuse, scaly with hairs, the innate scales darker. **Flesh** white. **Stem** 2 in. long, 3–4 lines thick, hollow, somewhat thickened at the base, scaly with fibrils, becoming brownish-whitish. **Veil** separate, clothed with fibers, hanging from the pileus, white. **Gills** adnate, crowded, 3 lines broad, whitish then brownish-purple, edge whitish and distilling drops in wet weather.

Spores brownish-purple. From mutual pressure the caps are often irregular. Very cespitose, firm. *Fries.*

Spores brownish-purple, 9x4μ *Massee.*

On ground and on trunks. Truly cespitose. Smaller than H. velu-

Hypholoma. tinus, but firmer, truly fleshy, not hygrophanous.   Bushy pastures.
Bethlehem.   October.

Our specimens do not agree in all respects with the published description of the species.   The pileus is sometimes wholly destitute of
scales and sometimes densely clothed with hairy, erect ones.   The species
is manifestly variable.   *Peck*, 30th Rep. N. Y. State Bot.

" Like H. fascicularis in quality.   Intensely irritant.   It is bound with
the weight of its own guilt."   *Hay.*

This is a good specimen of Hay's comments.   H. fascicularis is never
irritant, is good eating, is innocent.

There is irony in the comment of Dr. Cooke: "This doubtful species is used by the smaller ketchup makers."

I have not seen this species.   When I do I shall eat it and expect to
live.

APPENDICULA'TI.   Pileus hygrophanous, smooth.

H. incer'tum Pk.   (Plate XCVIIa.)   **Pileus** fragile, convex or subcampanulate, then expanded, hygrophanous, often radiately wrinkled,
whitish with the disk yellowish, the thin margin sometimes purplish-
tinted, often wavy, adorned by fragments of the white flocculent fugacious veil.   **Lamellæ** close, narrow, whitish then rosy-brown, the edge
often uneven.   **Stem** equal, straight, hollow, easily splitting, whitish
with a frosty bloom or slightly scurfy at the top.   **Spores** elliptical,
purplish-brown, 8x5μ.

Plant gregarious or subcespitose, 2–3 in. high.   **Pileus** 1–2 in.
broad.   **Stem** 1–2 lines thick.

Ground among bushes.   Green Island and Sandlake.   June and July.

The veil is sometimes so strongly developed as to form an imperfect
ring.   The color is nearly white from the first.   *Peck*, 29th Rep. N. Y.
State Bot.

As the name indicates, I was uncertain whether this was a form of H.
Candolleanum, to which it is very closely related, but as Fries says of
that "Gills at first violaceous," and as our plant has them at first white
or whitish, I concluded to risk the uncertainty on a new species.

I have seen Central Park, New York, well covered with it in May.   It
is also common in the vicinity of Boston.   Of very agreeable flavor and
delicate substance.   The profusion of its growth compensates for its
small size.   *Macadam.*

362

HYPHOLOMA INCERTUM.

Photographed by Dr. J. R. Weist.

Indiana, *H. I. Miller;* Mt. Gretna, Pa., in great clusters between Hypholoma. railroad ties and beside track, *McIlvaine.*

Tender. One of the best.

H. appendicula'tum Bull.—a small appendage. From the veil adhering to margin of pileus. ([Color] Plate XCVII.) **Pileus** 2–3 in. broad, date-brown then tawny, becoming pale yellowish when dry, fleshy-membranaceous, thin, ovate then expanded, at length flattened, obtuse, smooth, when dry slightly wrinkled, somewhat sprinkled with atoms. **Stem** 3 in. long, 2–3 lines thick, fistulose, equal, smooth, white, *pruinate at the apex;* veil fringing the margin of the pileus, fugacious, white. **Gills** somewhat adnate, crowded, *dry, white* then flesh-colored, at length dingy-brown.

Densely cespitose, very fragile and hygrophanous. Much thinner and more fragile than A. Candolleanus. It may be safely distinguished from species which are nearest to it by the gills being whitish then brownish-flesh color.

Var. *lana'tum.* A curious form, densely woolly when young, traces of the woolly coat remaining at the apex when the pileus is fully expanded. Sibbertoft. B. and Br., 1876. *Stevenson.*

Spores ellipsoid, pellucid, 6–8x3–4μ *K.;* 4x6μ *W.G.S.;* elliptical, 5x2.5μ *Massee.*

Angora, West Philadelphia, October, November, December, 1897; Haddonfield, N. J., Mt. Gretna, Pa., cespitose and gregarious in woods about stumps. *McIlvaine.*

"It is very common and edible." *Farlow.*

At Mt. Gretna, Pa., October, 1898, in great abundance. When found it was gregarious in large patches and cespitose on stumps. My identification was confirmed by Professor Peck.

It dries well, and retains flavor and esculent qualities. Cooked it is among the best.

H. Candol'leanum Fr.—After De Candolle. **Pileus** 2–4 in. broad, date-brown then becoming white, the top somewhat yellowish, somewhat fleshy, acorn-shaped then bell-shaped, soon convex and at length flattened, obtuse and unequal, smooth, even. **Flesh** thin, white. **Stem** 3 in. long, 2–4 lines thick, fistulose, solid at the base, somewhat thickened, fibrillose, white, striate at the apex; veil in the form of a cortina,

Hypholoma. web-like, appendiculate (depending from the margin of the pileus), white, at length becoming dingy-brown. **Gills** rounded-adnexed, then separating, crowded, violaceous then brownish-cinnamon, the edge at first whitish.

Readily distinguished from neighboring species by the gills being at first beautifully dark violaceous, never flesh-colored. Densely cespitose, fragile, very hygrophanous. *Stevenson.*

Spores elliptical, 8x4μ *Massee.*

Edible, often used in catsup. *Cooke.*

A species variable in color with the weather. Its gills are cream-colored at first, then purplish, then very dark. After rain the fragile cap often turns up at the margin and splits.

It differs somewhat in texture from other Hypholomas, being more delicate in texture and substance. It is excellent.

H. suba'quilum Banning.—*aquilus*, brownish, tawny. **Pileus** brown, convex, smooth, hygrophanous, often shaded into ocher at margin, veil delicate, silk-like, encircling and covering the marginal extremities of the lamellæ but forming no ring on the stem. **Flesh** white, turning umber when cut. **Lamellæ** adnexed or nearly free, close, forked, umber. **Stem** cespitose, regular, hollow, silky, white, 2–3 in. long.

Spores brown, 4x5μ. *Banning* MS.

Druid Hill Park, Baltimore, *Miss Banning;* decaying wood, Adirondack mountains. August and September. New York. *Peck*, 45th Rep. N. Y. State Bot.

H. subaquilum is closely allied to H. appendiculatum, but is distinguished by its darker colored cap and gills.

Its edible qualities are the same. It is among the best.

PSILO'CYBE Fr.

Gr.—naked; head.

Pileus more or less fleshy, smooth, *margin at first incurved.* **Gills** Psilocybe. becoming brownish or purple. **Stem** somewhat cartilaginous, rigid or tough, tubular, hollow or stuffed, often rooting. **Veil** absent or rudimentary, never forming a membrane. **Spores** purple, purple-brown or slate-color.

Generally growing on the ground, gregarious, sometimes cespitose.

Psilocybe is analogous in form to Collybia, Leptonia and Naucoria, which are distinguished by their spore colors. Separated from Psathyra by the incurved margin of the pileus.

But one species of Psilocybe is herein given as edible. Of it, alone, the writer has had opportunity to eat meals. Several others of the species have been found by him and tested in small quantity. They are all of good texture, substance and flavor, though most are small. He is of the opinion that increased testing will prove the entire genus edible. Nothing can or should be prognosticated about a toadstool, but the indications are all in favor of Psilocybe.

P. spadi'cea Schaeff.—*spadiceus*, date-brown. **Pileus** thin, submembranaceous, hemispherical, then convex or expanded, smooth, hygrophanous, pale grayish-brown and striatulate when moist, white or yellowish when dry. **Gills** narrow, close, attached, easily separating from the stem, at first whitish, then brown, tinged with flesh-color. **Stem** straight, equal, hollow, smooth, white.

Height 1–2 in., breadth of pileus 1–1.5 in. **Stem** 1–2 lines thick.

Grassy ground in yards and fields. Albany. June. Gregarious or cespitose. The pileus is fragile, the spores are brown. *Peck,* 23d Rep. N. Y. State Bot.

(Plate XCIX.)

PSILOCYBE SPADICEA.
Two-thirds natural size.

Spores brown, 9x4μ *Massee;* purplish brown, 7.6x5.1μ *Morgan.*

Haddonfield, N. J., October, November, December, 1896. In large patches and where stumps had been taken from the ground. *McIlvaine.*

Psilocybe.   Var. *hygro'philus* Fr.   *Gr.*—moist; loving.

Pileus tawny, then clay-color.   Stem 4–6 in. long, rather fusiform, rooting.   Gills emarginate with a deeply decurrent line; at length umber-brown.

Var. *polyceph'alus* Fr.—*polus*, many; *cephale*, head.

Densely crowded.   Stem thinner, flexuous.   Gills nearly free, at length tawny-umber.

The plant is tender, cooks easily and is of fine flavor.

P. semilancea'ta Fr.—*semi*, half; *lancea*, a spear.   **Pileus** ½ in. high, not broad, various in color, becoming yellow, green, dingy-brown, somewhat membranaceous, *acutely conical*, almost *cuspidate*, never expanded, but the margin when young at first bent inward, *covered with a pellicle which is viscous and separable in wet weather*, slightly striate chiefly round the margin.   **Stem** as much as 3 in. long, scarcely 1 line thick, tubular and *containing a pith*, equal, more frequently *flexuous*, smooth, *capable of being* twisted round the finger, smooth, *becoming pale; furnished with a veil when young*.   **Gills** *ascending* into the summit of the cone, adnexed, almost *linear*, crowded, becoming purple-black.   *Fries.*

Gregarious, very tough.   Pastures and roadsides, etc.   Common. August to November.   *Stevenson.*

Spores ellipsoid, 9–16x4–9µ *K.;* 14x9µ *W.G.S.*

New York, *Peck*, Rep. 23; Novia Scotia, *Somers.*

Var. *cærules'cens* Cooke—becoming blue.   Base of stem turning indigo-blue.

Not common in America, but frequently found.   According to M. C. Cooke—a careful authority—P. semilanceata has a dangerous reputation.   It is said to have proved fatal to children when eaten raw.   It is not deleterious when cooked.

366

PSA'THYRA Fr.

Gr.—friable.

(Plate C.)   Psathyra.

Veil none or only universal, and floccoso-fibrillose. **Stem** somewhat cartilaginous, fistulose with a tube, polished, *fragile.* **Pileus** conical or bell-shaped, *membranaceous, the margin at the first straight and adpressed to the stem.* **Gills** becoming purple or brownish. *Slender, fragile, hygrophanous.*

Some of the last species of Hypholoma *and* Psilocybe *are very closely allied to them.* The Coprinarii are readily distinguished by the gills being white or ash-color, then black, *not dusky-brown nor becoming purple.*

Psathyra corresponds with Mycena, Nolanea, Galera and Psathyrella. All the species grow on the ground or on trunks. *Stevenson.*

But four American species reported.

PSATHYRA GYROFLEXA.
Natural size. (After Massee.)
Omitted from Index to Species.

Small and unimportant.

DECONICA.

Stem tough; margin of **Pileus** at first incurved. **Gills** subtriangularly decurrent. Corresponds with Omphalia, Eccilia, Tubaria.

Few American species. Small and unimportant.

 VARIOUS as are the spore colors in this series (in its broadest sense), there is an entire absence of brown and purple shades in the black spores of four of the genera belonging to this group or series. In Gomphidius the spores are dingy-olivaceous. It is an outsider affiliating with thoroughbreds because of more technical congeniality than other genera afford. Like comets in the universe, it has no home. The singular genus Montagnites (of which but one species has been found in America, and that in Texas) has the relationship of spore-color. Panæolus, Anellaria, Psathyrella, when young, have gills free from each other; Coprinus, in early life, presents them pressed tightly together; as the plants age and the spores ripen, the entire gill structure becomes black and dissolves into an inky fluid, the color of which is due to the spores.

The species are all of delicate body, and many of them add generously to table luxuries.

COPRI′NUS Pers.

Gr.—dung.

Coprinus. **Pileus** separate from the stem. **Gills** membranaceous, at first closely pressed together, cohering, at length melting into a black fluid. **Trama** obsolete. **Spores** oval, even, black.

The extreme closeness of the gills and their entire deliquescence into a fluid, black from the spores, sharply define this genus and separate it from all others. At first the form is oval or cylindrical; most are furnished with a downy or scurfy veil often adhering to the pileus, sometimes forming an adhering volva at the base of the stem. Nearly all are ephemeral, many completely disappearing in a day.

Cystidia (sterile cells) of large size are frequent on the gills of many species.

368

PLATE CI.

MELANOSPORAE.

Hymenophore distinct from fleshy stem.

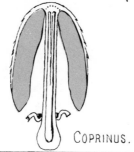

COPRINUS.

Hymenophore confluent and homogeneous with fleshy stem.

ANELLARIA.

PANAEOLUS.

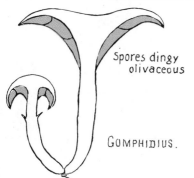

Spores dingy olivaceous

GOMPHIDIUS.

ymenophore confluent with, but heterogeneous from cartilaginous stem.

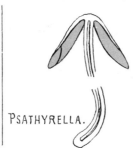

PSATHYRELLA.

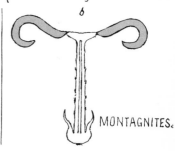

MONTAGNITES.

CHART OF GENERA IN BLACK-SPORED SERIES—MELANOSPORAE.

PLATE CIII.

Photographed by Dr. J. R. Weist.

COPRINUS COMATUS.

The majority grow on richly manured ground or dung, some on rotten Coprinus. wood and other materials. Bolbitius, the only ally, has the same ephemeral existence, and grows in similar situations, but the gills only soften (not melting) and the spores are somewhat rust-colored.

The blackening of the gills is not a process of decay, but is due to the growth of the spores, and the plant is still (before deliquescence) perfectly edible although not so inviting in appearance as before.

Species of Coprinus are very common and are easily recognized by the deliquescent gills which, when mature, stain the fingers black.

In "Once upon a Time," when country people made their own writing inks, the convenient Coprinus gave its juices for this purpose. A little corrosive sublimate added to the boiled and strained fluid prevented it from molding.

With few exceptions the species are small. They are tender, of real mushroom flavor and highly enjoyable. They make a thin, well flavored catsup, but are better used to give flavor to their less favored brethren.

They stew in from two to fifteen minutes, depending upon the solidity of the species.

ANALYSIS OF THE TRIBES.

A. PELLICULOSI (*pellicula*, a thin skin). Page 370.

Gills covered above with a fleshy or membranaceous skin, hence the pileus does not split along the lines of the gills, but becomes lacerated with the edges turned upward.

* Comati—*coma*, hair. Furnished with a ring formed from the free margin of the volva. The skin of the pileus torn into innate scales.

** Atramentarii—*atramentum*, ink. Ring imperfect. Volva absent. Pileus dotted with minute innate scales.

*** Picacei—*pica*, a magpie. Universal veil downy, at first continuous then broken up into superficial scales forming patches on the pileus.

**** Tomentosi—*tomentum*, down. Pileus at first covered with a loose hairy down, becoming torn into distinct scales, at length disappearing. Ring absent.

***** Micacei—*mico*, to glitter. Pileus at first covered with minute glistening scales, soon disappearing. Ring none.

****** Glabrati. Pileus smooth. Veil absent.

B. VELIFORMES (*velum*, a veil; *forma*, form). Page 380.

Coprinus. Pileus very thin without a skin, at length opening into furrows along the backs of the gills and becoming folded in furrows. Stem thin, hollow. Gills wasting away into thin lines.

* Cyclodei. *Gr.*—a circle; appearance. Stem with a ring or volva.

** Lanatuli—*lanatus*, woolly. Pileus covered with superficial woolly floccules, at length disappearing. Ringless.

*** Furfurelli—*furfureus*, branny. Pileus mealy or scurfy. Gills generally attached to a collar at the apex of the stem. Ringless.

· **** Hemerobii. *Gr.*—living a day. Pileus always smooth.

None known to be edible.

A. PELLICULO'SI. Cap becoming torn, edge turning upward, etc.

* Comati. *Furnished with a ring, etc.*

C. coma'tus Fr.—*coma*, hair. (Plate CIII.) **Pileus** 2–7 in. high, white, fleshy, at first oblong, becoming bell-shaped, seldom expanded, when in mature deliquescing state, splitting at the margin along the line of the gills, the cuticle, except upon the apex, separating into shaggy, often concentric scales, at times yellowish, at others tinged with purplish-black. **Gills** free from the stem, crowded and at first cohering, broad, white then tinged with pink or salmon color, then purple to black and dissolving into ink. **Stem** up to 10 in. long, up to ⅝ in. thick, attenuated upward, most part concealed within the cap, hollow, but with spider-web threads within, smooth or fibrillose, white or lilac-white, easily pulling out of cap, brittle. **Ring** thin, torn, sometimes entire and movable.

On rich soil, lawns, gardens, roads, dumps, especially where ashes have been placed. Solitary or in large dense clusters. August until after frost, but it is occasionally found during the spring months.

Spores elliptical, black, 13–18µ long *Peck*. Almost black, elliptical, 13–18x7–8µ *Massee;* 11–13x6–8µ *K.;* 15x8µ *W.G.S.*

Var. *brev'iceps* Pk. **Pileus** before expansion subovate, shorter and broader than in the typical form, 1.5–2.5 in. high. Dumping ground. Albany. November. *H. Neiman. Peck,* 49th Rep.

Coprinus comatus is common to the United States. In its perfection it is a stately and beautiful plant. I have seen it with the oblong cap eight inches long, but its usual height is from 2–4 in.   It occurs after Coprinus. hard rain and often in the most unexpected places.   It is a rather domestic species, usually in troops, but often in clusters of from five to fifty individuals.   I have seen it lift firmly sodded ground about railroad stations, and again, bulging the surface of gardens like mole-hills.

There are toadstools of higher flavor, but not one of greater delicacy. In this C. comatus is not excelled from its earliest stage until fully ripened.   It is everywhere commended.

Lafayette B. Mendel, in American Journal of Physiology, gives the following analysis:

The specimens were freshly gathered and had not yet turned "inky." They varied very widely in size, thirty-six mushrooms weighing 1485 grams, of which 980 grams belonged to the caps (pileus) and 505 grams to the stems.   The average weight of a fresh specimen was thus:

Pileus.. 27 grams
Stem 14

Total weight.................................... 41

A specimen which had attained the average growth weighed:

Pileus .... 43 grams
Stem 25

Total weight............................... 68

An analysis yielded the following results:

Water.. 92.19 per cent.
Total solids.. 7.81

The dry substance contained:

Total nitrogen... 5.79 per cent.
Extractive nitrogen.. 3.87
Protein nitrogen... 1.92
Ether extract... 3.3
Crude fiber.. 7.3
Ash 12.5
Material soluble in 85 per cent. alcohol.............. 56.3

C. soboli'ferus Fr.   **Pileus** 1½–2½ in. across, subcylindrical, then oval bell-shaped, lower half of pileus usually undulate but not furrowed or striate, disk obtuse, usually depressed, distinctly scaly, dingy white, toward the apex tinged with pale brown, scales darker.   **Flesh** very thin.   **Gills** free, tapering toward each end, ¼ in. or more broad, crowded, pale then blackish.   **Stem** 5–8 in. long, ¾ in. thick at the

Coprinus. base, slightly attenuated upward, silky-white, stuffed; toward the base there is a depressed zone caused by the edge of the pileus when young. **Ring** fugacious. **Spores** elliptical, 15x7μ.

Amongst grass near to trunks, buried wood, etc. A very large and beautiful species, distinguished from Coprinus atramentarius, its nearest ally, by the larger size of every part, the costate (ribbed) or waved lower portion of the pileus, the truncate, depressed disk, with distinct squamules, the whitish color of the pileus, and the imperfectly hollow or stuffed stem.

Spores elliptical, 15x7μ *Massee.*

Almshouse grounds, Philadelphia. On maple roots in grass-grown places, May, 1897–1898. *McIlvaine.* Not previously noted in United States.

C. soboliferus is a substantial food-giving species, very heavy for its size. It grows singly and in clusters and will immediately attract attention, wherever found. It is of fine flavor and substance. Cook at once.

C. ova'tus (Schaeff.) Fr.—*ovum*, an egg. **Pileus** white, somewhat membranaceous, *at the first egg-shaped* and *densely imbricated with thick spreading concentric scales*, covered with an even hood at the apex, then expanded, striate. **Stem** 3–4 in. long, solid at the base, rooting, otherwise hollow, with spider-web threads within, attenuated upward, downy, shining white. **Ring** not very conspicuous and soon vanishing. **Gills** free, remote, slightly ventricose, at the first somewhat naked and remaining long shining white, *at length umber-blackish*, never becoming purple.

Smaller, thinner, less handsome than C. comatus. For the most part solitary. *Fries.*

Spores 11–12x7–8μ *Massee.*

On rich ground, dumps, etc. Same habitat as C. comatus.

West Virginia, Pennsylvania, North Carolina, New Jersey. *McIlvaine.*

So closely allied to C. comatus that it is with difficulty distinguished from it. However, its edible qualities are the same, and into these the name does' not enter.

C. sterquili'nus Fr.—*sterquilinium*, a dunghill. **Pileus** about 2 in. across when expanded, conical, then expanded, sulcate more than half way from margin to disk, at first villous or silky, disk rather fleshy with Coprinus. rough scales, silvery-gray, tinged with brown at the apex. **Flesh** thin. **Gills** free, ventricose, about 2 lines broad, pale then umber-purple. **Stem** 4–6 in. high, slightly attenuated upward, white, fibrillose, hollow, thickened base solid, and booted for about an inch from the base, margin of sheath ending in a free border or ring.

On dung. A fine large species known by the scaly apex of the pileus, the basal portion of the stem surrounded by a volva-like, adnate structure with a free upper margin. The stem soon becomes black when bruised. Base of stem not rooting but abrupt, and furnished with a few white fibers. *Massee.*

Edible, *Cooke*, 1891; also *Leuba.*

Nova Scotia, *Dr. Somers.*

This species is not reported as found in the **United States.**

**Atramentarii. *Ring imperfect, etc.*

C. atramentarius (Bull.) Fr.—*atramentum*, ink. ([Color] Plate CII, fig. 1.) **Pileus** 1½–4 in. across, ovate, expanding, grayish, lead-color or grayish-brown, with occasionally a few obscure scales on disk, often covered with bloom; margin ribbed, sometimes notched, soft, tender. **Gills** free, ventricose, up to ½ in. broad, crowded and at first cohering and white with white floccose edges, then becoming black and dissolving into ink. **Stem** up to 5 in. long, up to ½ in. thick, smooth, whitish, hollow, at first spindle-shaped, then attenuated upward, with more or less distinct ring near base.

Spores subcylindrical, large cystidia numerous, 12x6μ *Massee;* 9–10 x6μ *K.;* 9x5μ *W.G.S.;* 8–10μ long *Peck.*

Indiana, *H. I. Miller;* Harrisburg, Pa., *Dr. J. H. Fager;* West Virginia, *McIlvaine.*

The stem is obscurely banded within, by which it may be recognized with certainty.

It grows singly or in clusters of many individuals on rich ground, whether lawns, gardens, gutter sides, or in woods, but not on dung. I know of a fine cluster growing year after year on a much-decayed pear-stump. Occasionally it appears in the spring months, but is common during the summer and autumn after rains, and from its first appearance

373

Coprinus. it occurs in successive crops until stopped by severe frost. It is common in Europe and over the United States.

The flavor is higher than that of C. comatus. It should be cooked as soon as gathered, and kept in a cool place until needed.

Analysis shows the following:

Two separate, freshly-gathered lots of this species were examined. The one (a) contained six young small specimens weighing 5.5 grams, or .9 gram each; the other (b)·contained eight mushrooms weighing 12 grams, or 1.5 grams each. An analysis gave:

	a.	*b.*
Water	92.31 per cent.	94.42 per cent.
Total solids	7.69	5.58
The dry substance contained:		
Total nitrogen	4.68	4.77
Ether extract	3.1	5.7
Crude fiber	9.3	......
Ash	16.8	20.1

Lafayette B. Mendel in American Journal of Physiology.

C. fusces′cens (Schaeff.) Fr.—*fuscus*, dark or swarthy. **Pileus** 1–1¼ in. across, submembranaceous, ovate, expanded, dull, disk rather fleshy, even or cracked into squamules, grayish-brown, disk reddish. **Gills** adfixed, blackish-umber. **Stem** 4–5 in. long, about ¼ in. thick, equal, fragile, hollow, subfibrillose. **Ring** indistinct or absent, whitish. *Massee.*

Smaller and more slender than Coprinus atramentarius. **Pileus** brownish-gray, disk becoming reddish, not sprinkled with micaceous particles, but at first covered with a mealy bloom. **Gills** adnexed, attenuated from the stem to the margin, deliquescent. *Fries.*

Spores elliptical, pointed at the ends, 10x6µ *Massee;* 10x5µ *W.G.S.* Solitary and in tufts. On stumps, trunks, etc. May to October. West Philadelphia, Pa., *McIlvaine.*

C. fuscescens is tender, delicate and of excellent flavor. In this it ranks with C. atramentarius

C. macro′sporus Pk. **Pileus** ovate, then expanded, rimose-striate (cracked in lines), obscurely floccose-squamulose, white, the small even brownish disk scaly. **Lamellæ** crowded, free, white then black. **Stem**

glabrous, white, with traces of an annulus (ring) near the thickened or Coprinus. subbulbous base.

Spores very large, elliptical, 20–20.5 long, 12–16μ broad.

Plant cespitose, 2–3 in. high. **Pileus** 1–2 in. broad. **Stem** 1 line thick.

Ground in open fields. Ticonderoga. August.

The prominent characters of this species are the cracked pileus, squamose disk, free lamellæ and large spores. In its early state it resembles some species of Lepiota. It seems to be intermediate between the sections Atramentarii and Micacei. *Peck,* 31st Rep. N. Y. State Bot.

Found in quantity at Mt. Gretna, Pa. August to September, 1898, growing among old stable bedding on parade ground.

(Plate CIV.)

COPRINUS MACROSPORUS.
Enlarged one-third.

C. macrosporus is an excellent species, higher in flavor than any other Coprinus.

***Pica'cei. Universal downy veil, etc.

C. pica'ceus (Bull.) Fr. **Pileus** 2–2½ in. across, membranaceous, ovato-bell-shaped, striate up to the disk, smoky-black, variegated with large, irregular, superficial white patches. **Gills** free, ½ in. or more broad, ventricose, grayish-black. **Stem** 5–6 in. long, base bulbous, abrupt, otherwise equal, ¼–⅔ in. thick, white, hollow, fragile, smooth. **Spores** elliptical, apiculate, 14x8μ; cystidia large, numerous. *Massee.*

Decaying trunks or branches of trees in woods. Lyndonville. June. *Fairman.*

The form here referred to this species differs somewhat from the description of the type in being smaller, in having no bulb to the stem and in having smaller spores. It is probably the "smaller variety growing on rotten wood" noticed by Stevenson in his British Fungi. I

375

Coprinus. have seen the true form of the species from Kansas. The New York plant seems to me to be worthy of distinctive designation, at least as a variety, and I call it

Var. *ebulbo'sus*. Plant smaller. **Stem** destitute of a bulb. **Spores** 8–10x5µ. *Peck*, 44th Rep. N. Y. State Bot.

Minnesota, *Johnson*, 1897; Kansas, *Cragin*, 1884; Wisconsin, *Bundy;* Nebraska, *Clements*.

Edible. *Leuba.*

Large quantities grew on rotting chestnut and oak rails at Mt. Gretna, Pa., from June to August, 1899. It is strong and unpleasant.

**** Tomento'si. *Pileus at first veiled with a loose hairy veil.*

C. fimeta'rius Fr.—*fimetum*, a dunghill. **Pileus** 1–2 in. across, membranaceous, thin, *at first cylindrical,* soon conical, *the edge at length revolute* and torn at the margin, *when young everywhere covered with floccose-squarrose white scales* (from the universal veil), which separate from the vertex toward the circumference, at length naked, longitudinally cracked, but not opening into furrows, the vertex which remains entire, livid. **Stem** about 3 in. long, 2–3 lines and more thick, hollow, fragile, *thickened and solid at the base*, attenuated upward, s h i n i n g w h i t e and downy with squamules of the same color. **Gills** free, reaching the stem, at first *ventricose, then linear, flexuous, black.* **Stem** when young curt and firmer. *Fries.*

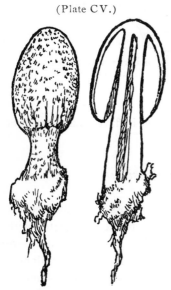

(Plate CV.)

COPRINUS FIMETARIUS.

Spores spheroid-ellipsoid, 15–18x9–12µ *K.;* 15x9µ *W.G.S.;* 12–14x7–8µ *Massee.*

Sometimes there is a root as long as the stem. *M.J.B.* Common on dung heaps in successive crops. Spring to autumn.

Var. *pulla'tus.* **Pileus** with adpressed scales and tomentose, soon naked, brownish, then blackish. **Stem** equal, becoming smooth.

On dung. Clustered. Stature of the type.

Var. *cine reus*. **Pileus** membranaceous, floccosely mealy, then naked, Coprinus.
ashy-gray. **Stem** subequal, rootless, hollow to the base, often twisted.
Spores 12–8μ.

On dung and rich soil.

Var. *macrorhi'za*. **Pileus** at first with feathery squamules. **Stem**
short, hairy, rooting, sometimes more or less marginately subbulbous.
Spores 13–14x8–9μ.

On dung. Pileus pale and smaller than in the typical form, stem
shorter, with a more or less elongated rooting base. *Berkeley*.

Of this very variable species there is a small form growing on decayed wood in woods. It has the spores rather smaller than in the
type, they being 10–11μ long, 8μ broad. It might be designated Var.
silvi'cola. *Peck*, 43d Rep. N. Y. State Bot.

West Virginia, 1881–1885, May to October. *McIlvaine*.

Common to the United States. Of excellent flavor and tender. It
must be cooked at once.

C. tomento'sus (Bull.) Fr.—*tomentum*, pubescence. **Pileus** very
thin, at first oblong-oval and floccose-scaly, soon bell-shaped, naked,
closely striate, grayish-brown or blackish-brown, often with a leaden
hue, finally expanded, the disk smooth, reddish or ochraceous-brown,
the margin turned upwards and much split or lacerated. **Lamellæ**
closely crowded, narrow, free, white then pinkish, finally black. **Stem**
white, tall, fragile, tapering upward, finely floccose-squamulose, hollow,
sometimes with a large tap root. **Plant** gregarious or cespitose.

Height 3–6 in., breadth of pileus 6–18 lines.

Very variable in size and color. The covering of the pileus is easily
rubbed off. It soon disappears and the plant quickly decays, seldom
continuing through the day. *Peck*, 23d Rep. N. Y. State Bot.

Mt. Gretna, Pa., about old picketing places in camp grounds. *Prof.
M. W. Easton*, July, 1898.

West Virginia, North Carolina, New Jersey, Pennsylvania, May to
September, on dung, rich ground, gardens and in woods. *McIlvaine*.

Very delicate; of strong mushroom flavor. It is common, and can
usually be collected in numbers. It is of little food value in itself, but
yields an excellent flavor to anything it is cooked with. It must be
cooked as soon as gathered.

377

Coprinus.   **C. ni′veus** Fr.—*nix*, snow.   **Pileus** white, 1–2 in. across, thin, ovate then bell-shaped, margin at length turned upward, split or covered with a dense white, mealy or downy covering, slightly pink.   **Gills** *adnexed*, narrow, crowded, at first cohering, white then pinkish, then black. **Stem** at first short, then up to 4 in., slender, attenuated upward, covered with white down, fragile, hollow.

Spores 16x11–13μ *Massee;* 10x12μ *W.G.S.*

Common on dung and dung heaps, clustered.   May to frost.

West Virginia, North Carolina, Pennsylvania, New Jersey.   *McIlvaine.*

Very variable in size, but clearly distinguished by its snow-white color and adnexed gills.   Like all of the thin, delicate species of this genus there is little substance left after cooking, but the savory flavor is imparted to the cooking medium.

***** Mica′cei.   *Pileus at first covered with minute, glistening scales, etc.*

C. mica′ceus (Bull.) Fr.—*mica*, grain, granular.   ([Color] Plate CII, fig. 2.)   **Pileus** thin, ovate, then bell-shaped, with the margin more or less revolute, wavy, splitting, closely striate, with a few minute scales and sparkling atoms, or naked, varying in color from whitish-ochraceous to livid-brown, generally darker when moist or old.   **Gills** rather narrow, crowded, white then pinkish, finally black.   **Stem** slender, fragile, easily splitting, slightly silky, white, hollow, often twisted. Plant mostly cespitose.

Height 2–4 in., breadth of pileus, 1–2 in.

Streets, yards and fields, on or about old stumps.   May to September.   *Peck*, 23d Rep. N. Y. State Bot.

Spores elliptical, blackish, 7–8x4–5μ *Massee;* 7x8μ *W.G.S.;* 10x5μ *W.P.;* elliptical, brown, 6–8μ *Peck.*

Var. *granula′ris.*   Pileus sprinkled with granules or furfuraceous scales.   New York.   August.   *Peck*, 47th Rep.

Indiana, *H. I. Miller;* West Virginia, North Carolina, Pennsylvania, New Jersey.   May to October.   *McIlvaine.*

Common from spring until frost.   This is the oval-capped toadstool found in clusters about trees, posts, along grassy sides of pavements, popping up, Brownie-like, from sodded places.   Although small and thin, its clusters soon fill baskets, and its continuous growth in some places, from Coprinus. month to month, year to year, makes it one to be depended upon. Stewed for ten minutes it makes a rich, luscious dish.   C. congregatus closely resembles it and is equally good.

****** Glabra'ti. *Pileus smooth, etc.*

C. deliques'cens (Bull.) Fr.   **Pileus** 3–4 in. broad, livid-fuliginous, membranaceous, bell-shaped then expanded, smooth, but *dotted with minute points on the disk*, never downy or split, the edge turning upward and striate, the striæ broad but not deep.   **Stem** 4 in. long, 2–4 lines thick, hollow, with a bark-like covering, equally attenuated upward, *smooth, shining white.* **Gills** free, *at length remote from the stem*, very crowded, flexuous, very narrow, only ½ line broad, lurid-blackish. *Fries.*

Frequent on stumps and among fallen leaves, sometimes in tufts. July to October.

Spores elliptical, obliquely apiculate, 8x5μ *Massee.*

Sometimes confounded with C. atramentarius.

West Virginia, Pennsylvania, New Jersey, *McIlvaine.*

C. deliquescens is of good size and quality.   The stems do not cook well with the caps.   The flavor is the same as C. atramentarius.

C. congrega'tus (Bull.) Fr.   **Pileus** ½–¾ in. high, cylindrical, then bell-shaped, finally expanded and split at the margin, smooth, viscid, margin slightly striate, ochraceous.   **Gills** about 1 line broad, slightly adnexed, white, finally becoming black.   **Stem** 1½ in. high, equal, smooth, hollow, whitish.

On the ground, also in hot-houses.   *Massee.*

Readily distinguished by the densely cespitose mode of growth, the small size, the viscid, ochraceous, glabrous pileus which remains elongato-cylindrical for some time, then becomes campanulate and finally expands and splits at the margin.

Densely cespitose, fragile, readily distinguished from C. digitalis by its much smaller size.   *Fries.*

Spores 7x8μ *W.G.S.;* 10x5μ *W.P.*

Fries and Cooke considered this a good species.

So closely allied to neighboring species that it is difficult to determine it.   Edible qualities are included in the alliance.

379

B. Veliformes. Pileus very thin, etc.

* Cyclodei. *Stem bearing ring, etc.*

Coprinus. None edible.

** Lanatuli. *Pileus with superficial downy covering, etc.*

C. lagopus Fr.—*Gr.*, a hare; a foot. **Pileus** 1 in. broad, whitish, disk livid, very tender, cylindrical then bell-shaped, when young beautifully downy then naked, flattened and split, radiately furrowed. **Stem** 5 in. and more long, 1 line thick, very weak, very fragile, slightly attenuated at both ends, everywhere white-woolly. **Gills** at length remote, narrow, black. *Fries.*

Fries distinguishes two forms. A, *nemorum.* **Stem** slender, 4–6 in. long. B, *viarum.* **Stem** 2–3 in. long. **Pileus** broader, livid. Both forms are inodorous. The pileus of the long-stemmed form is sometimes entirely clear brown, at others grayish with a brownish disk. **Stem** very weak, 5 in. and more in length, 1 line thick, attenuated at both ends. **Pileus** thin, expanded bell-shaped, about 1 in. across, when young elegantly flocculose, then furrowed, disk livid. **Gills** rather distant.

New York, *Peck*, 38th Rep.; Mt. Gretna, Pa., July, 1898, on rubbish about abandoned camp. *Prof. M. W. Easton.*

A strikingly beautiful species. Both forms were found in abundance, tested and eaten with enjoyment. They are extremely delicate, and of attractive but not high flavor.

C. Virgineus Banning. **Pileus** ovate, bell-shaped, or cylindrical, pale ocher, the margin thin, torn, downy. **Lamellæ** narrow, close, forked, at first white, turning dark but never black, adnexed. **Stem** 3½ in. long, stout, somewhat stuffed, attenuated where it meets the pileus, flattened, downy. **Spores** black.

Cespitose or gregarious at the roots of trees or about old stumps. Also found in Virginia.

The plant is not rapidly deliquescent, remaining perfect for some hours. *Banning* MS.

Maryland. Virginia. *Miss M. E. Banning* MS. *Peck*, 44th Rep.

Chester county, Pa. New Jersey, about pear trees and stumps. *McIlvaine.*

This little Coprinus is a valuable species when found.　A patch of it Coprinus. about a tree or stump is treasure trove.　Patches of it appear in July and bear until October.　The not-particular observer would mistake it for C. micaceus.

*** Furfurel′li.　*Pileus micaceous or scurfy, etc.*

C. domes′ticus (Pers.) Fr.—*domus*, a house.　**Pileus** 2 in. broad, fuliginous, disk date-brown, thin, ovate then bell-shaped, *covered with small branny scales*, then opening into furrows and flattened, *undulately sulcate*, disk obtuse, even.　**Stem** 2–3 in. long, 2–3 lines thick, fistulose, slightly firm, attenuated upward, *adpressedly silky*, becoming even, white.　**Gills** *adnexed*, at first crowded, distant when the pileus is split, linear, *white then reddish*, at length brownish-blackish.

A larger and more remarkable species than all the neighboring ones.　*Fries.*

Spores 14–16x7–8μ *Massee.*

On much decayed wood, damp carpets, in cellars, etc.　Often in clusters.

Mt. Gretna, Pa., *Prof. M. W. Easton*, July, 1898 ; West Virginia, New Jersey, Pennsylvania, *McIlvaine.*

(Plate C VI.)

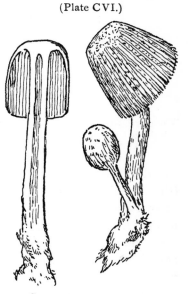

COPRINUS DOMESTICUS.
Natural size.

C. domesticus is the largest of its section and is sometimes of remarkable growth.　I have seen it start from under a board in a cellar and prolong its stems for over a foot to get its caps to air and light.　Under such conditions the stems are twisted in a confused mass.

It is very tender with a decided mushroom flavor.　Cook at once.

C. silvat′icus Pk.　**Pileus** membranaceous, with a thin fleshy disk, convex, striate in folds on the margin, dark-brown, the depressed striæ paler.　**Lamellæ** subdistant, narrow, attached to the stem, brownish.　**Stem** fragile, slender, smooth, hollow, white.　**Spores** gibbousovate, 12.7μ long.

Coprinus.   (Plate CVII.)

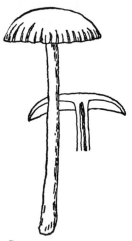

COPRINUS SILVATICUS.
Enlarged one-fourth.
(After Peck.)

Plant 2 in. high. **Pileus** 6–10 lines broad. **Stem** .5 lines thick. Ground in woods. Greig. September.

The striæ extend about half way up the pileus. Allied to C. plicatilis and C. ephemerus *Peck*, 24th Rep. N. Y. State Bot.

West Virginia, Pennsylvania, New Jersey. Frequent, but not common. On ground in woods, August to October. *McIlvaine*.

This pretty little fungus is frequently found. I have never been able to get it in quantity, but have often eaten it. Its flavor is musky, rather strong. It is edible, but is not obtainable in sufficient numbers to make it of much food value.

C. ephem′erus Fr. *Gr.*—lasting for a day. **Pileus** ½–¾ in. across, very thin, ovate, then bell-shaped, finally expanded and splitting, furrowed radiately, at first slightly scurfy, disk elevated, even, reddish. **Gills** slightly attached, linear, white, then brownish, at length blackish. **Stem** 1½–2½ in. high, 1 line or more thick, equal, glabrous, pellucid, hollow, whitish. **Spores** 16–17x9–10μ.

On dunghills, manured ground, etc. To the naked eye appearing almost glabrous, but under a lens seen to be distinctly scurfy. Known from Coprinus plicatilis by the disk of the pileus being prominent and not depressed. *Massee*.

Common dung and dung heaps. May to October. New York, *Peck*. 23d Rep.

Of such size and delicate substance as to be of little food value. But it has a strong mushroom flavor which is choice as a flavoring. It appears during the summer months on dung and dung heaps. It must be cooked as soon as gathered.

C. semilana′tus Pk. **Pileus** submembranaceous, broadly conical, then expanded and strongly revolute, and the margin sometimes split, covered with mealy atoms, finely and obscurely rimose-striate, pale grayish-brown. **Lamellæ** narrow, close, free. **Stem** elongated, fragile, hollow, slightly tapering upward, white, the lower half clothed with loose cottony flocci which rub off easily, the upper half smooth or Coprinus. slightly farinaceous. **Spores** broadly elliptical, 12.7μ long.

Plant very fragile, 4–6 in. high. **Pileus** 8–12 lines broad. **Stem** 1 line thick at the base. Rich ground and dung. Sandlake. August. (Plate IV, fig.15–18.) Allied to C. coopertus. *Peck*, 24th Rep. N. Y. State Bot.

West Virginia. 1881–1885, Mt. Gretna, Pa. July to October. *McIlvaine.*

I have seldom found it, though at times it was quite common about stables in West Virginia. It has good mushroom flavor and is edible. It is stately, attracting attention by its peculiar cap.

C. plica′tilis Fr.—*plico*, to fold. **Pileus** 1 in. broad, dusky-brown then bluish-gray-cinereous, *disk darker*, dusky-brown or reddish, oval-cylindrical then campanulate, soon expanded, opening into furrows, *sulcate-plicate*, for the most part *smooth, disk broad*, even, *at length depressed*. **Stem** 1–3 in. long, fistulose, thin, equal, even, *smooth*, pallid, *somewhat pellucid*. **Gills** *remote from the stem* and adnate to a *collar* which is formed from the dilated apex of the stem, distant, gray-blackish. *Fries.*

Very tender and fragile, but when scorched by the sun not melting into fluid. Very variable in stature and size. *Stevenson.*

Spores 12–14x8–10μ *Massee;* broadly elliptic, 5μ long, *M. J. B.;* 11–13μ *long*, 8–10μ broad *Peck*, Rep. 50.

Common in rich pastures, lawns, roadsides, etc. May to October.

West Virginia, Pennsylvania, New Jersey, *McIlvaine.*

A neat little fungus often found in great plenty. Though small it is nevertheless edible and must be written with its edible companions.

383

PANÆ'OLUS Fr.

Gr.—all; *Gr.*—variegated.

Panæolus. **Pileus** slightly fleshy, *not striate, margin exceeding the gills.* **Gills**
(Plate CVIII.) ascending in a conical manner, slate-gray, mottled with the black spores. **Stem** polished. **Veil** woven, often absent. **Spores** black.

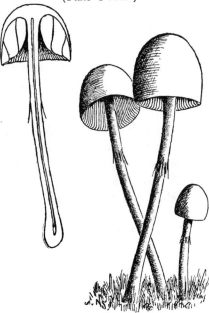

PANÆOLUS.

On the ground in rich earth, and on dung.

In the black-spored series Psathyrella is separated by the striate pileus, not exceeding the gills, Anellaria by the ring and Coprinus by the deliquescent gills.

Panæolus, in its entirety, has a precise looking membership. If the gills were cut from cardboard and fixed by machinery, they could not be more correct. Some of the species are among the earliest arrivals at toadstool lawn parties, and some are the last to leave. Several are culinary favorites, notably Panæolus solidipes. P. papilionaceus possesses intoxicating properties. P. campanulatus is reported to be a sedative.

The edible species are easily cooked and are exceptionally delicate and well flavored.

P. retiru'gis Fr.—*rete,* a net; *ruga,* a wrinkle. **Pileus** about 1 in. across, at first almost globose, then hemispherical, subumbonate, minutely mealy, opaque, moist, furnished with uniting raised ribs, pinkish tan-color; margin with irregular fragments of the veil attached. **Flesh** rather thick. **Gills** adnexed, ascending, 2 lines or more broad, grayish-black. **Stem** 2–4 in. long, about 2 lines thick, equal, pruinose, purplish flesh-color, hollow. *Fries.*

Spores elliptic-fusiform, 11–13x7µ *Massee.*

On dung. Distinguished among the species of **Panæolus by the** raised ribs on the pileus and its appendiculate margin. The pileus is Panæolus. sometimes grayish. Closely resembling, superficially, Psathyra corrugis, which is, however, distinguished by the violet-black gills.

Spores elliptical, shortly fusiform, 20µ *Q.;* 16x11µ *W.G.S.*

New York, *Peck,* 23d Rep. West Virginia, 1881–1885. Pennsylvania, New Jersey, frequent on dung. June to frost. *McIlvaine.*

P. retirugis is not a common species, and is a sparse grower, but is frequently found. It is seldom that a mess can be had at one time. It is an excellent species by itself and imparts a good flavor to others.

P. fimi'cola Fr.—*fimus,* dung; *cola,* to inhabit. **Pileus** ½–¾ in. across and high, slightly fleshy, convex bell-shaped, obtuse, glabrous, opaque, dingy-gray when moist, paler and yellowish when dry, with a narrow brown encircling zone near the margin. **Gills** adnate, 2 lines or more broad, gray, variegated with smoky-black. **Stem** 2–4 in. high, 1 line or more thick, equal, fragile, whitish, powdered with white meal upward, hollow. *Fries.*

Stem soft, fragile, obsoletely silky-striatulate, 2–4 in. long. **Pileus** when moist commonly smoky-gray, when dry grayish clay-color, sometimes discoid. **Gills** semi-ovate with a minute decurrent tooth. *Fries.*

West Virginia, Pennsylvania, New Jersey. Frequent. On dung and richly manured places. June to September. *McIlvaine.*

P. fimicola is neither as large nor heavy as P. solidipes, but in other respects equals it.

P. soli'dipes Pk.—*solidus,* solid; *pes,* a foot. ([Color] Plate CII, figs. 3, 4.) **Pileus** 2–3 in. across, firm, at first hemispherical, then subcampanulate or convex, smooth, whitish, the cuticle at length breaking up into dingy-yellowish, rather large, angular scales. **Gills** broad, slightly attached, whitish, becoming black. **Stem** 2–4 lines thick, firm, smooth, white, solid, slightly striate at the top. **Spores** very black with a bluish tint. Height of plant 5–8 in. Dung heaps. West Albany. June.

A large species, remarkable for its solid stem. The scales on the pileus are larger on the disk, becoming smaller toward the margin. The upper part of the stipe is sometimes beaded with drops of moisture. *Peck,* 23d Rep. N. Y State Bot.

West Virginia, 1881–1885. Pennsylvania, New Jersey, frequent on dung and dung heaps. May to frost. *McIlvaine.*

On mature plants, or after rains, the scales are not always present.

P. solidipes is a handsome, readily recognized species of good weight and substance. It is one of the best of toadstools.

P. campanula'tus Linn.—*campanula*, a little bell. **Pileus** oval, bell-shaped or obtusely conical, sometimes umbonate, smooth, somewhat shining, brownish, with a peculiar gray or lead-colored tint, sometimes becoming reddish-tinted, the margin, often scalloped or fringed with the appendiculate veil. **Lamellæ** not broad, attached, becoming grayish-black. **Stem** long, slender, hollow, reddish, pruinose and slightly striate at the top, at length dusted with the spores.

Height 4–6 in., breadth of pileus 6–12 lines.

On horse dung and rich soil. June and July. Common.

In very wet weather the cuticle of the pileus sometimes cracks into scales or areas. *Peck*, 23d Rep. N. Y. State Bot.

Spores subellipsoid, 16–18x10–13µ *K.;* 8–9x6µ *Massee.*

Mr. R. K. Macadam, Boston, Mass., informs me that he has information of a case of poisoning by this fungus. "The victim experienced dizziness, dimness of vision, trembling and loss of power and memory. He recovered after simple treatment and was well inside of 24 hours."

A full account of this case is in "The London Medical and Surgical Journal," Vol. 36, November, 1816. The poison acts as a sedative.

I have several times eaten of this fungus in small quantities, because larger could not be obtained, and with no other than pleasant effect. There does not appear to be any case of poisoning reported by it since 1816, which, considering the inquisitiveness of man, is singular. Caution is advised.

P. papiliona'ceus Fr.—*papilio*, a butterfly. **Pileus** subhemispherical, sometimes subumbonate, smooth, or with the cuticle breaking up into scales, whitish-gray, often tinged with yellow. **Lamellæ** very broad, attached, becoming black. **Stem** slender, firm, hollow, pruinose above, whitish, sometimes tinged with red or yellow, slightly striate at the top and generally stained by the spores.

Height 3–5 in., breadth of pileus 6–18 lines.

On dung and rich soil. Common. May and June.

A small form occurs with the pileus nearly white, scarcely half an inch in diameter, and the cuticle not cracking. *Peck*, 23d Rep. N. Y. State Bot.

The effects of P. papilionaceus are very uncertain. I have seen it produce hilarity in a few instances, and other mild symptoms of intoxication, which were soon over, and with little reaction. But I have seen, at table, the same effects from eating preserved peaches and preserved plums which had fermented. Many personal testings have been without effect. Testings upon others vary with the individuals. The fungus seems to contain a mild stimulant. It is not dangerous, but should be eaten with caution. Being of small size, and not a prolific species, quantities of it are difficult to obtain. Moderate quantities of it have no effect whatever.

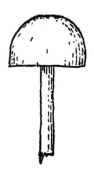

ANELLA′RIA Karst.

Anellus, a little ring.

Anellaria. **Pileus** slightly fleshy, smooth and even. **Gills** adnexed, dark slate-color, variegated with the black spores. **Stem** central, smooth, shining, rather firm. **Ring** present at first, either persistent or forming a zone around the stem.

The species of this genus were formerly included in Panæolus, from which this is separated by the presence of a ring, more or less definite. In other characters they are similar. As in Amanitopsis and Amanita.

A. separa′ta Karst.—*separatus*, distinct, separate. **Pileus** 1–1½ in.

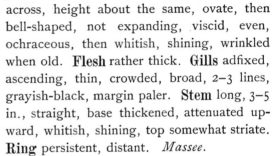

(Plate CIX.)

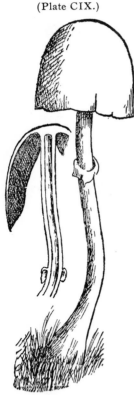

across, height about the same, ovate, then bell-shaped, not expanding, viscid, even, ochraceous, then whitish, shining, wrinkled when old. **Flesh** rather thick. **Gills** adfixed, ascending, thin, crowded, broad, 2–3 lines, grayish-black, margin paler. **Stem** long, 3–5 in., straight, base thickened, attenuated upward, whitish, shining, top somewhat striate. **Ring** persistent, distant. *Massee.*

On dung. Rather variable in size.

Pileus bell-shaped, but very obtuse at the summit, ½–1¼ in. from the base to the apex, not expanding at the base without cracking.

Spores broadly elliptic-fusiform, black, opaque, 10x7μ *Massee;* ellipsoid, 16–22x10–12μ *K.;* 16x11μ *W.G.S.*

West Virginia, 1881–1885, New Jersey, Mt. Gretna, Pa., July, 1898, on dung. *Mc-Ilvaine.*

A common, frequent species from May to October. It is substantial in flesh, excellent in substance and flavor. **Cook soon and not over fifteen minutes.**

ANELLARIA SEPARATA.
Natural size.

PSATHYREL′LA.

Gr.—fragile.

Pileus membranaceous, *striate*, margin straight, at first pressed to the stem, *not extending beyond the gills.* **Veil** inconspicuous. **Gills** sooty-black, *not variegated.* **Spores** black.

Closely resembling Psathyra in appearance, but separated by the spore color.

In the black-spored series Panæolus and Anellaria are distinguished by their pilei not being striate and Coprinus by its deliquescent gills.

The species are small and can seldom be gathered in quantity. But those tested have the full mushroom flavor and are valued for the flavor they give to less gifted species when cooked with them.

P. gra′cilis Fr.—slender. **Pileus** ½–1 in. broad, *sooty*, livid, etc., when dry, tan, rosy or whitish, hygrophanous, membranaceous, bell-shaped, obtuse, smooth, *even*, slightly and pellucidly-striate only round the margin. **Stem** 3 in. and more long, scarcely 1 line thick, tubular, *remarkably tense and straight*, equal, naked, smooth, whitish, *not rooted*, *white-villous at the base.* **Gills** wholly adnate, commonly *broader* behind (rarely linear), almost *distant*, distinct, at first whitish, then cinereous-blackish with the black spores, *edge rose-colored. Fries.*

When dry the pileus is soft to the touch. Gregarious, fragile. Very similar to A. corrugis, and there is a variety corrugated. *Stevenson.*

Spores ellipsoid, 13–14x7–8µ *K.;* 5x12µ *W.G.S.;* 7x3–3.5µ *Massee;* 14x8µ *Morgan.*

New York, *Peck*, Rep. 23; West Virginia, New Jersey, Pennsylvania, common, rich ground, June to October. *McIlvaine.*

A common and beautiful fungus, growing in patches on rich ground. It is decidedly prim. Its conical cap is regular as an extinguisher. It pays to gather it for flavoring other species. I have not seen the corrugated form mentioned by Fries. P. graciloides Pk. lacks the rosy-edged gills; gills are whitish.

Psathyrella. **P. graciloi'des** Pk.—slender. **Pileus** thin, conical or bell-shaped, glabrous, hygrophanous, brown and striatulate when moist, whitish and subrugulose when dry. **Lamellæ** ascending, rather broad, subdistant, brown, becoming blackish-brown, the edge whitish. **Stem** long, straight, fragile, hollow, smooth, white. **Spores** blackish, elliptical, 15–16.5x8–8.5μ.

(Plate CX.)

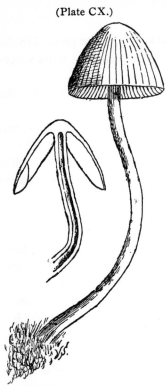

PSATHYRELLA GRACILOIDES.

Plant gregarious, 4–6 in. high. **Pileus** 1 in. broad. **Stem** 1 line thick.

Ground in an old dooryard. Maryland. September.

This is allied to A. gracilis Fr., but the edge of the gills is not rosy. When drying the moisture leaves the disk of the pileus first, the margin last. When dry the plant bears some resemblance to large forms of A. tener. Under a lens the texture of the surface of the pileus is seen to be composed of matted fibrils. *Peck*, 30th Rep. N. Y. State Bot.

Pennsylvania and New Jersey, on ground about houses and stables, often in barn yards, after they have been cleaned out and are empty for the summer. *McIlvaine.*

The whitish-edged gills with entire absence of rosiness on gill edges distinguish this species from P. gracilis Fr. It is frequent but not plentiful. Often a pint can be gathered. It has a fine mushroom flavor, resembling the delicate forms of Coprinus.

P. atoma'ta Fr.—*atomatus*, atomate. **Pileus** ½–1 in. broad, livid, when dry becoming pale tan or pale flesh-color, sometimes reddish, hygrophanous, membranaceous, bell-shaped, obtuse, *slightly striate*, when dry without striæ, slightly wrinkled, *sprinkled with shining atoms.* **Stem** 2 in. long, almost 1 line thick, tubular, equal, not rooted, *lax*, slightly bent (not tense and straight), *white and white pulverulent at* *the apex*. **Gills** adnate, broad, *ventricose, slightly distant*, distinct, Psathyrella. whitish, but cinereous-blackish with the black spores. *Fries.*

Solitary or gregarious. Pileus changing like A. gracilis from livid to whitish and rose-color, but more fragile. *Stevenson.*

Spores elliptical, 10x4µ *Massee;* 14x9µ *W.G.S.;* 11x8µ *Morgan.*

Chester county, Pa., June to September. *McIlvaine.*

Several specimens were eaten. In flavor they could not be distinguished from C. micaceus. The scarcity and small size of the species make it of little value, save as a flavoring.

P. dissemina′ta Pers.—*dissemino*, to scatter. Found everywhere.

Densely tufted. **Pileus** about ½ in. across, membranaceous, ovate, bell-shaped, at first scurfy, then naked, coarsely striate, margin entire, yellowish then gray. **Gills** adnate, narrow, whitish, then gray, finally blackish. **Stem** 1–1½ in. long, rather curved, mealy then smooth, fragile, hollow. *Massee.*

(Plate CXI.)

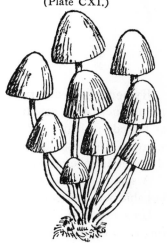

PSATHYRELLA DISSEMINATA.

Crowded. **Pileus** ovate, conical, at length bell-shaped, ⅓–½ in. from the base to the apex, striate and plicate, membranaceous, pale buff or reddish-brown, at length gray, becoming flaccid and dissolving. **Gills** distant, narrow, pale brown. **Stipes** 1–3 in. long, slender, weak, brittle, crooked, hollow, pale yellowish, whitish or grayish. Particularly partial to old willow trees, and when growing on a stump of a felled tree often covering nearly a square yard. *Grev.*

Spores 8x6µ *W.G.S.;* 7.6x5µ *Morgan.*

West Virginia, New Jersey, Mt. Gretna, Pa., about abandoned camp. Densely tufted. May to frost. *McIlvaine.*

Patches of it are very common on old trunks, about decaying trees, on ground. The caps rarely reach 1 in. in diameter. The plants cook

Psathyrella. away to almost nothing, but they are of fine flavor, which they impart to the cooking medium.

GOMPHI'DIUS Fr.

A wooden bolt or nail.

Gomphidius. **Hymenophore** decurrent. **Gills** distant, composed of a mucilaginous membrane, which can be readily separated into two plates, continuous at the edge which is acute and powdered with the blackish fusiform spores. **Veil** viscoso-floccose. Fleshy, putrescent, pileus at length the shape of an inverted cone.

A small genus with great difference among the species. Intermediate in habit between Cortinarius and Hygrophorus.

Universal **Veil** glutinous, at first terminating on the stem in a floccose ring soon disappearing. The **Gills** frequently admit of being detached and stretched out into a continuous membrane. *Fries.*

A genus possessing several well-marked characters. The very decurrent gills differ from all others in their soft mucilaginous consistency. The spores are larger than usual in the Agaricaceæ and have the elongated spindle-shape found in Boleti. The stem and pileus are of the same substance, and the pileus and veil are both glutinous when moist. The spores have been described as greenish-gray becoming black, and as dingy-olive.

I have had opportunity to see but two species of this small genus— G. rhodoxanthus and G. viscidus. Of these the spores are decidedly olivaceous. If the six other species recorded as found in the United States are as creditable, they are well worth hunting for. G. Oregonensis Pk. is reported as edible and as a valuable food species in Oregon.

The glutinous coatings to pileus and stem do not appear on the Ameri- Gomphidius. can form of G. rhodoxanthus in the localities I have found it in during fifteen years.

G. glutino′sus (Schaeff.) Fr.—*glutin*, glue. **Pileus** 2–5 in. broad, purple-brown, often mottled with black spots, fleshy, convex, obtuse, at length plane, even depressed, even, smooth, very glutinous. **Flesh** thick, about ½ in., soft, white. **Stem** 2–3 in. and more long, about ½ in. thick, solid, whitish, thickened and externally and internally yellow at the base, viscid with the veil, fibrillose or varying with black scales. **Cortina** often woven in the form of a ring, but soon fugacious. **Gills** deeply decurrent, distant, distinct, branched, quite entire, muci-laginous, 3–4 lines broad, *at first whitish, then cinereous*, clouded with the spores.

Trama none, wherefore the gills easily separate from the pileus. Taste watery, moldy. Odor not marked. *Stevenson.*

Spores 20μ *Cooke;* 18–23x6–8μ *K.;* 16–17x6μ *W.G.S.;* 18–20x6μ *Massee.*

Distinguished by the bright yellow base of stem.

Pine woods. July to November. Nova Scotia. *Somers.*

Edible. *Leuba.* Chiefly used for catsup. *Cooke.*

Var. *ro′seus.* **Pileus** rose-color. **Stem** white, attenuated and rosy flesh-color internally at the base. Very distinguished, always smaller.

Spores 20–22x6μ *K.*

Nova Scotia. Massachusetts. *Frost.*

I have not seen this species or its variety. Eminent authorities vouch for its edibility.

G. Oregonen′sis Pk. **Pileus** at first convex, becoming nearly plane or somewhat centrally depressed, viscid, brown or dark-brown, becom-ing black in drying, taste sweet and pleasant. **Lamellæ** numerous, rather close, adnate or slightly decurrent, blackish in the dried plant. **Stem** short, solid, equal or slightly tapering upward, colored like the pileus. **Spores** oblong, 10–12.5μ long, 4–5μ broad.

Pileus 5–10 cm. broad. **Stem** 2.5–5 cm. long, 4–10 mm. thick.

Fir woods. Oregon. September to December. *Lane.*

Dr. Lane writes that this species is edible and grows so abundantly in fir woods that it might be gathered by wagon loads and might be

Gomphidius. made a source of an abundant food supply. *Peck.* Torrey Bulletin, Vol. 25, No. 6, June, 1898.

G. vis′cidus Fr.—viscid.

(Plate CXII.)

GOMPHIDIUS VISCIDUS.
One-half natural size.

Pileus 2–3 in. and more broad, brownish-red, compact, at first bell-shaped, then expanded, umbonate, slightly viscous, shining when dry. **Flesh** yellowish. **Stem** 3–4 in. and more long, ½ in. thick, solid, equal or attenuated at the base which is rhubarb-colored internally, scaly-fibrillose, not very viscous, yellowish. **Cortina** very evidently floccose, not glutinous, woven in the form of a ring, but readily falling off. **Gills** deeply decurrent, distant, the shorter ones adnexed to the longer, not truly branched, at first paler, somewhat olive, at length brownish - purple, clouded with the spores. *Fries.*

Hymenophore descending between the gill plates. Odor not unpleasant. *Stevenson.*

Chiefly used in catsup. *Cooke.* Edible. *Leuba. Cooke.*

North Carolina, Massachusetts, *Frost.* Minnesota, California, Pennsylvania.

Many grew under pines at Mt. Gretna, Pa., September to November. The gills seemed branched, but were grown together. Taste and smell pleasant. The caps are good, but not equal to G. rhodoxanthus.

G. rhodoxan′thus Schw. ([Color] Plate XCVII, figs. 4, 5.) Solitary. **Pileus** 1–2 in. broad, cushion-shaped, reddish-yellow, sometimes with dusky hues. **Gills** arched, decurrent, orange-yellow. **Stem** attenuated, short, firm.

Spores oblong, 10–12.5μ in length. *Peck.* Olivaceous. *McIlvaine.* Solitary, gregarious or cespitose.

Among leaves and grass in shady places. August to October.

When the student has mastered the name and memorized the descrip-

tion, Gomphidius rhodoxanthus can not be mistaken for any other spe- Gomphidius.
cies.

It is not common in localities I have frequented, but its presence is pretty general in the United States, specimens having been sent to me from Georgia, Iowa, New York, New Jersey, etc., and I have found it in West Virginia, North Carolina, Woodland Cemetery, Philadelphia, and other places in Pennsylvania, from July to September, 1898, inclusive. Having enjoyed it in West Virginia in 1882, I was delighted to find it in generous quantity at Mt. Gretna, Pa., and to eat many meals of it. Its caps are not excelled by any edible fungus. They have solid, delicious substance and rich full flavor.

The plant is often cespitose. I have never found its cap viscid or glutinous. The cooked flesh has the latter consistency.

MONTAGNITES Fr.

After Montagne.   (Plate CI, fig. 6, p. 368.)

The universal veil forming a volva, persistent.   **Stem** dilated at the Montagnites.
apex into a plane round disk, even on both sides, *to the margin* of which are *adfixed the gills which are free, not joined by any membrane*, radiating, razor-shaped, persistent, obtuse at the edge. **Trama** cellulose. **Spores** oblong, even, black fuscous.   *Fries.*

A single species is reported from Texas.

FAMILY II.—POLYPORACEÆ.

Hymenophore inferior, facing the ground. Hymenium consisting of tubes with poriform mouths which are round or angular, sometimes sinuous or torn, lined with 4-spored sporophores and cystidia.

Fleshy, coriaceous or woody fungi, most abundant and luxuriant in warm countries. Intermediate between the Agaricaceæ and the Hydnaceæ, connected with the former by Dædalea and Lenzites, and with the latter by Fistulina and Irpex. *Fries.*

Within this large family are famed edible species, notably in Boletinus, Boletus and Fistulina. In the woody species the razor-strop man finds material for his strops (Polyporus celulinus); the surgeon styptics; the peasant punk to catch sparks from his flint, and the 4th of July urchin a fire-holder to light his pyrotechnics. The Chinese have placed some species in their fathomless materia medica, while the Polyporus of the locust tree is used in America as a medicine for horses. No fungoid growth is more universal. They are the ever active pruners of our trees and converters of forest debris. They begin the task in Nature's laboratory of changing decaying wood into assimilable shape as food to feed the very trees that dropped it. Some are of annual growth, others add to their substance year after year, often attaining enormous size. In summer and in winter they are ever present objects for interesting study.

SYNOPSIS OF GENERA.

BOLETINUS. Page 398.

Hymenium composed of broader radiating gills connected by very numerous more narrow anastomosing branches or partitions and forming large angular pores. Tubes somewhat tenacious, not easily separable from the hymenophore and from each other, adnate or subdecurrent, yellowish. *Peck.*

BOLETUS. Page 404.

Stratum of tubes easily separable from the hymenophore. Stem central.

STROBILOMYCES.   Page 475.

Tubes like Boletus, but pileus with large scales.   Stem central.

FISTULINA.   Page 477.

Fleshy, lateral, tubes crowded but distinct.

POLYPORUS.   Page 479.

Stratum of tubes distinct from hymenophore, but not separable, not stratose; fleshy and tough, stipitate or sessile.

FOMES.

Tubes as in Polyporus, often stratose; woody, sessile; dimidiate. (No edible species reported.)

POLYSTICTUS.

Tubes as in Polyporus, not stratose, generally developing from the center to the margin, at first shallow and punctiform, coriaceous or membranaceous.   (No edible species reported.)

PORIA.

Tubes as in Polyporus, not stratose; entirely resupinate. (No edible species reported.)

MUCRONOPORUS.

Tubes studded with reddish-brown spines, intermingled with the basidia, otherwise as in Polystictus (and also as in Polyporus and Fomes).   *Atkinson.*   (No edible species reported.)

TRAMETES.

Tubes immersed in flesh of pileus, of various depths, hence not forming a heterogeneous stratum, subcylindrical, not stratose; corky; sessile.

DÆDALEA.

Tubes as in Trametes, but sinuous and labyrinthiform; corky; not stratose; sessile.   (No edible species reported.)

397

HEXAGONIA.

Tubes from the first dilated in hexagonal channels, not stratose; plants corky, sessile. *Atkinson.* (No edible species reported.)

FAVOLUS.

Tubes large at first, radiating from a central stem, or from a lateral attachment in sessile or dimidiate forms; plants tough and fleshy. *Atkinson.* (No edible species reported.)

CYCLOMYCES.

Gills or tubes in concentric circles. Stem central, subcentral or none. *Atkinson.* (No edible species reported.)

MERULIUS. Page 490.

Subgelatinous. Tubes very shallow, formed by anastomosing wrinkles; resupinate.

———

BOLETI'NUS Kalchb.

(Plate CXIII, p. 402.)

Boletinus. **Hymenophore** not even (as in Boletus), but extended in blunt points descending like a trama among the tubes. **Tubes** not easily separable from the hymenophore and from each other. **Stem** ringed, hollow. **Spores** pale yellowish. Sylloge, Vol. VI, p. 51.

Professor Peck has for excellent reasons, given in his Boleti of the United States, emended the generic diagnosis of Fries thus: *Hymenium composed of broader radiating lamellæ connected by very numerous more narrow anastomosing branches or partitions and forming large angular pores. Tubes somewhat tenacious, not easily separable from the hymenophore and from each other, adnate or subdecurrent, yellowish.*

Professor Peck classifies Boletinus as follows:

Stem hollow..B. cavipes
Stem solid... I
1. Stem lateral or eccentric...........................B. porosus
1. Stem central... 2
  2. Pileus pale yellow, silky.........................B. decipiens
  2. Pileus red or adorned with red scales...................... 3

3. Pileus red.................................B. paluster Boletinus.

3. Pileus soon red-squamose.....................B. pictus

Boleti of the United States, p. 76.

There are six species given as found in the United States—B. cavipes Kalchb., B. pictus Pk., B. paluster Pk., B. decipiens Pk., B. porosus Pk., B. appendiculatus Pk.—of these I have found and eaten four. B. decipiens has, at this writing, not been seen by Professor Peck, but Professor Farlow, of Harvard, has informed him of authentic specimens. There is every probability of its being as edible as the others; a description of it is, therefore, given.

In consistency Boletinus is of the best, being rather like that of marshmallows, and the same as Boletus subaureus. The flavor is mild and pleasant.

Professor Peck mentions that the smell of B. porosus is sometimes unpleasant. I have been fortunate in not having had this experience.

B. ca′vipes Kalchb. **Pileus** broadly convex, rather tough, flexible, soft, subumbonate, fibrillose-scaly, tawny-brown, sometimes tinged with reddish or purplish. **Flesh** yellowish. **Tubes** slightly decurrent, at first pale-yellow, then darker and tinged with green, becoming dingy-ochraceous with age. **Stem** equal or slightly tapering upward, somewhat fibrillose or floccose, slightly ringed, *hollow*, tawny-brown or yellowish-brown, yellowish at the top and marked by the decurrent dissepiments of the tubes, white within. **Veil** whitish, partly adhering to the margin of the pileus, soon disappearing. **Spores** 8–10x4μ.

Pileus 1.5–4 in. broad. **Stem** 1.5–3 in. long, 3–6 lines thick. Swamps and damp mossy ground under or near tamarack trees. New York, *Peck;* New England, *Frost.*

The pileus is clothed with a fibrillose tomentum which becomes more or less united into floccose tufts or scales. The umbo is not always present and is generally small. The young stem may sometimes be stuffed, but, if so, it soon becomes hollow, though the cavity is irregular. The freshly shed spores have a greenish-yellow or olivaceous hue, but in time they assume a pale or yellowish-ochraceous hue. This species is apparently northern in its range. It loves cold sphagnous swamps in mountainous regions. *Peck*, Boleti of the U. S.

Boletinus.   West Virginia mountains under spruce trees.   Haddonfield, **N. J.**, among scrub pines.   Mt. Gretna, Pa., among pines.

It is of excellent consistency and of mild pleasant flavor.   **It is at its** best in patties, croquettes and escallops.

B. appendicula'tus Pk.   **Pileus** fleshy, convex, glabrous, ochraceous-yellow, the margin appendiculate with an incurved membranous veil.   **Flesh** pale-yellow, unchangeable.   **Tubes** rather small, yellow, their mouths angular, unequal, becoming darker or brownish where wounded.   **Stem** solid, slightly thickened at the base, yellow.   **Spores** pale-yellow, oblong, 10–12x4μ.   **Pileus** 4–8 in. broad.   **Stem** 2–3 in. long, 4–6 lines thick.

Under or near fir trees.   Washington.   September to December. *Yeomans.   Peck*, Bull. Torrey Bot. Club, Vol. 23, No. 10.

B. pic'tus Pk.   **Pileus** convex or nearly plane, at first covered with a *red fibrillose tomentum which soon divides into small scales revealing the yellow color of the pileus beneath.*   **Flesh** yellow, often slowly changing to dull pinkish or reddish tints where wounded.   **Tubes** tenacious, at first pale yellow, becoming darker or dingy ochraceous with age, sometimes changing to pinkish-brown where bruised, concealed in the young plant by the copious whitish webby veil.   **Stem** equal or nearly so, solid, *slightly* and somewhat evanescently annulate, clothed and colored like or a little paler than the pileus, yellowish at the top.   **Spores** ochraceous, 9–11x4–5μ.

Pileus 2–4 in. broad.   **Stem** 1.5–3 in. long, 3–6 lines thick.

Woods and mossy swamps.   New York, *Peck;* New England, *Frost;* North Carolina, *Curtis.   Peck*, Boleti of the U. S.

West Virginia mountains, 1882.   Haddonfield, N. J., Angora, West Philadelphia, Mt. Gretna, Pa.   August and September.   In mixed woods, principally oak.   Leominster, Mass.   *C. F. Nixon*, Ph. G.

It is sometimes found upon much decayed chestnut stumps.

The caps of some species are so cracked as to appear distinctly areolate.   The white webby veil is often persistent.   The fungus is one of the handsomest.   Its rich variegated colors impress it upon eye-memory.   It is óne of the very best edible species.

B. palus'ter Pk. — Pileus thin, broadly convex, plane or slightly depressed, sometimes with a small umbo, floccose-tomentose, *bright red.* **Tubes** very large, slightly decurrent, yellow, becoming ochraceous or dingy ochraceous. **Stem** slender, solid, subglabrous, red, yellowish at the top. **Spores** pinkish-brown, 8–9x4μ.

Pileus 1–2 in. broad. **Stem** 1–2 in. long, 2–3 lines thick.

Wet places and sphagnous mossy swamps. New York, *Peck.* Maine, *Harvey. Peck,* Boleti of the U. S.

Angora, West Philadelphia and Mt. Gretna, Pa. September. *McIlvaine.*

A few specimens found at Mt.

(Plate CXII*a.*)   Boletinus.

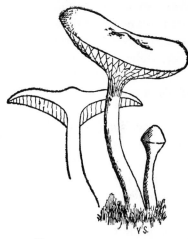

BOLETINUS PALUSTER.
Natural size. (After Peck.)

Gretna had stems slightly reticulated. Its taste is sweet, smell mild, and cooked it is of excellent body and flavor.

B. deci'piens (B. and C.) Pk. **Pileus** dry, minutely silky, *whitish-yellow or pale-buff,* flesh buff, one-third in. thick; hymenium plane or somewhat concave, yellow, consisting of large, unequal, flexuous radiating tubes resembling multiseptate lamellæ. **Stem** equal, solid but spongy. Veil floccose, evanescent, adhering for a time to the margin of the pileus. **Spores** rather minute, oblong, *ochraceo-ferruginous* (rusty yellow), 8–10x3.5–4μ.

Pileus 2 in. broad. **Stem** 2–2.5 in. long, 3–4 lines thick.

Thin woods. North and South Carolina. *M. A. Curtis.*

Specimens of this species have not been seen by me. The authors remark that its affinities are clearly with Boletinus flavidus and its allies, from which it is distinguished by its large radiating pores. They also say that when dry it is scarcely distinguishable from Paxillus porosus Berk., except by its spores. This would imply that its stem is eccentric or lateral, and I have been informed by Mr. Ravenel that it is sometimes so. But specimens of this kind, labeled Boletinus decipiens B. and C., have been received, which show by their spores that they are Paxillus porosus. Besides, Professor Farlow informs me that authentic

401

Boletinus. specimens of B. decipiens in the Curtisian Herbarium have only central stems, from which things I suspect that the two species have been confused. The spore dimensions here given are derived from a specimen in the Curtis Herbarium, through the kindness of Professor Farlow. *Peck*, Boleti of the U. S.

I have not recognized this Boletinus. Its affinities are with excellent edible species.

B. poro'sus (Berk.) Pk. (Plate CXIII.) **Pileus** fleshy, viscid when moist, shining, reddish-brown. **Flesh** 3–9 lines thick, the margin thin and even; hymenium porous, yellow, formed by radiating lamellæ a line to half a line distant, branching and connected by numerous irregular veins of less prominence and forming large angular pores. **Stem** lateral, tough, diffused into the pileus, reticulated at the top by the decurrent walls of the tubes, colored like the pileus. **Spores** semi-ovate.

Pileus 2–5 in. broad. **Stem** 6–16 lines long, 4–6 lines thick.

Var. *opa'cus* (Paxillus porosus Berk., Bull. N. Y. State Mus. 2, p. 32). **Pileus** dry, glabrous or subtomentose, not shining, brown or tawny-brown. **Spores** brownish-ochraceous, 9–11x6–8μ.

Damp ground in woods and open places. Ohio, *Lea, Morgan;* North Carolina, *Curtis;* New England, *Frost, Farlow;* Wisconsin, *Bundy;* New York, *Peck.*

This species is remarkable for its lateral or eccentric stem. There is often an emargination in the pileus on the side of the stem which gives it a kidney shape. In the typical form it is described as viscid when moist, and the Wisconsin plant is also described as viscid, but in all the New York specimens that I have seen it is dry and sometimes minutely tomentose. I have, therefore, separated these as a variety. The color of the pileus varies from yellowish-brown to reddish-brown or umber. A disagreeable odor is sometimes present. The tubes are rather short and tough and do not easily separate from the hymenophore and from each other. In the young plant they are not separable. They sometimes become slightly blue where wounded. As in other species they are pale yellow when young, but become darker or dingy-ochraceous with age. The spores have been described as bright yellow, but I do not find them so in the New York plant. The plant is incongruous among the Paxilli by reason of its wholly porous hymenium,

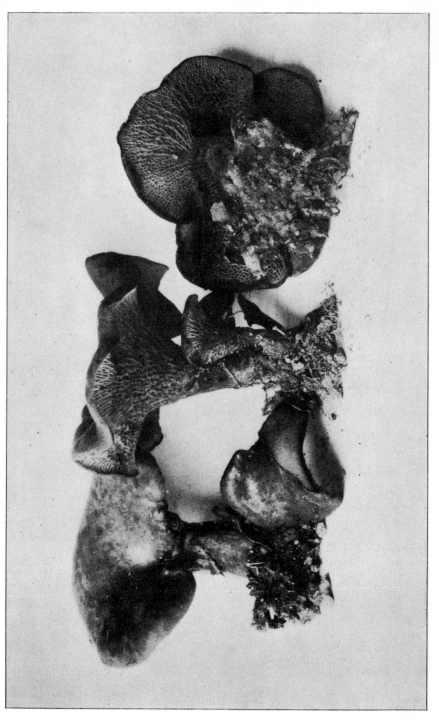

BOLETINUS POROSUS.

Photographed by Dr. J. R. Weist.

but in this place it seems to be among its true allies. *Peck*, Boleti of the U. S.

Fine specimens were sent to me by Mr. H. I. Miller, Terre Haute, and Dr. J. R. Weist, Richmond, Ind. They were in condition to be eaten and enjoyed. No disagreeable odor was perceptible.

B. borea'lis Pk. **Pileus** fleshy, convex, obtuse or subumbonate, brownish-yellow, obscurely and somewhat reticulately streaked with reddish-brown lines. **Pores** large, angular, unequal, slightly decurrent, brownish-yellow. **Stem** short, equal or slightly tapering upward, brownish-yellow with a whitish myceloid tomentum at the base. **Spores** oblong, 10–12.5x4–5μ.

Pileus 1–2 in. broad. **Stem** about 1 in. long.

Sandy soil. Capstan Island, Labrador. October. *Waghorne.*

The markings of the pileus appear as if due to the drying of a glutinous substance. The radiating lamellæ and the transverse partitions of the interspaces are very plainly shown. Described from two dried specimens. *Peck*, Bull. Torr. Bot. Club, Vol. 22, No. 5.

BOLETUS Dill.

Gr.—a clod.

Boletus.

THE name of a fungus considered a great delicacy among the Romans, derived from *bolos*, a clod, probably to denote the round figure of the plant.

Hymenium wholly composed of small tubes, connected together in a stratum, the surface of which is dotted with their poriform mouths, and which is distinct from the hymenophore on account of the latter not descending into a trama. **Tubes** packed close together, easily separating from the hymenophore and from one another. **Pores** or mouths of the tubes round or angular (in the subgenus Gyrodon sinuous or gyroso-plicate). **Spores** normally fusiform, rarely oval or somewhat round. *Growing on the ground, fleshy, putrescent, with central stems. Mostly edible, and of importance as articles of food; a few poisonous. Fries.*

No American species in Gyrodon. It is therefore omitted in synopsis of tribes. *C. M.*

This genus abounds in species and is related to Boletinus on one hand and to Polyporus on the other. From the latter it is distinguished by the absence of a trama and from both by the tubes being easily separable from the hymenophore and from each other. Some of the species are very variable, others are so closely allied that they appear to almost run together.

The species are generally terrestrial, but B. hemichrysus is habitually wood-growing, and others are occasionally so.

The spores vary so much in color in such closely related species that this character is scarcely available for general classification, but it is valuable as a specific character and should always be noted.

SYNOPSIS OF THE TRIBES.

Pileus and stem yellow-pulverulent, stem not reticulated
    with veins....................(p. 421.)   **Pulverulenti**
Pileus and stem not yellow-pulverulent, or if so then
    the stem reticulated with veins...................... I
   1. Tubes yellowish with reddish, or reddish-brown
    mouths ....(p. 453.)   **Luridi**

1. Tubes of one color, or mouths not reddish............... 2 Boletus.
2. Stem lacunose-reticulated and lacerated.(p. 436.) Laceripedes
2. Stem reticulated with veins, not lacerated............. 3
2. Stem not reticulated............................... 5
3. Tubes white, becoming flesh-colored...(p. 466.) Hyporhodii
3. Tubes not becoming flesh-colored...................... 4
4. Tubes free,or if adnate then stuffed when young.(p.444.) Edules
4. Tubes adnate, not stuffed when young.(p. 438.) Calopodes
5. Pileus viscid or glutinous when moist.................... 6
5. Pileus dry 7
6. Tubes adnate.....................(p. 406.) Viscipelles
6. Tubes free or nearly so, yellowish.......(p. 444.) Edules
6. Tubes free or nearly so, whitish.....(p. 459.) Versipelles
7. Stem solid... 8
7. Stem spongy within,soon cavernous or hollow..(p. 471.) Cariosi
8. Tubes becoming flesh-colored ......(p. 466.) Hyporhodii
8. Tubes not becoming flesh-colored.................... 9
9. Tubes adnate 10
9. Tubes free or nearly so..............................11
10. Pileus subtomentose..............(p. 430.) Subtomentosi
10. Pileus glabrous or pruinose........(p. 423.) Subpruinosi
11. Tubes yellowish or stuffed when young.....(p. 444.) Edules
11. Tubes whitish, not stuffed...........(p. 459.) Versipelles
Peck, Boleti of the U. S.

C. H. Peck, N. Y. State Botanist, has contributed to Mycological literature his careful arrangement and analysis of species of this genus, in his "Boleti of the United States." Species of the genus are found in every state of the Union. Several species are common to all the states. Comprehending, as do the states, all sorts of climates within their vast range of latitude, differences in appearance and structure in the same species must be expected, dependent largely, as they are in most fungi, upon habitat and environment. These variations will frequently suggest new species. Descriptions which are typical and which can be recognized as standard are most desirable. Professor Peck's are accepted by the writer as such, that there may be uniformity, and are quoted as fully as space will permit. Such variations as are attributable to locality will be noted.

Boletus.   Since 1882 the writer has given great attention to the edible qualities
of the Boleti.   He is convinced by many personal tests and those made
by his family and friends, that much, if not all, of the suspicion thrown
about Boleti is unjust and erroneous.   He is able to state positively that
change of color when bruised or broken; bitter and pepperiness have
nothing whatever to do with the edible qualities of species exhibiting
them, excepting in B. felleus, which exhibits an intense bitter, not lost
in cooking.   It is not poisonous.

The writer has the courage of his convictions, and has taken interest
in eating species with a bad reputation whenever opportunity afforded,
that their just dues might be given them.   He has never experienced
the slightest inconvenience.   But others may not be so fortunate.

Before cooking Boleti the stem, unless crisp and tender, should be
removed, as should the tubes unless young and fresh.   They broil, fry,
stew, make good soups and dry well.   See recipes.

It is believed that all species of Boleti up to this time found in
America are described in this volume.   When no remarks of the writer
follow the descriptions, he has not had an opportunity to test the edible
quality of the species.

VISCIPELLES—viscum, bird lime; pellis, a skin.

Pileus covered with a viscose pellicle.   **Stem** solid, neither bulbous,
lacerated nor reticulated with veins.   **Tubes** adnate, rarely sinuate, of
one color.

The first four and several of the final species here described recede
somewhat from the character of the central or typical species of the
group.

Stem with an annulus.................................... 1
Stem without an annulus................................ 9
1. Stem dotted both above and below the annulus............. 2
1. Stem dotted above the annulus.......................... 3
1. Stem not dotted....................................... 4
    2. Tubes salmon color......................B. salmonicolor
    2. Tubes yellowish..........................B. subluteus
  3. Annulus entirely viscose........................B. flavidus
  3. Annulus membranous, fugacious..................B. elegans
  3. Annulus membranous, persistent....................B. luteus

406

4. Pileus squamose.............................B. spectabilis Boletus.
4. Pileus not squamose................................ 5
5. Tubes whitish or grayish................................ 6
5. Tubes yellow or yellowish............................... 7
6. Flesh white, unchangeable....................B. Elbensis
6. Flesh white, changing to bluish...............B. serotinus
7. Spores globose or broadly elliptical...........B. sphærosporus
7. Spores much longer than broad........................ 8
8. Annulus fugacious............................B. flavus
8. Annulus persistent.......................B. Clintonianus
9. Stem dotted with glandules..........................10
9. Stem not dotted 16
10. Pileus some shade of yellow.......................11
10. Pileus some other color..........................15
11. Stem rhubarb color...........................B. punctipes
11. Stem some other color.............................12
12. Stem four lines or more thick.....................13
12. Stem less than four lines thick..............B. Americanus
13. Pileus adorned with tufts of hairs or fibrils.........B. hirtellus
13. Pileus glabrous....................................14
14. Stem yellow within..........................B. subaureus
14. Stem whitish or yellowish-white within........B. granulatus
15. Pileus white.............................B. albus
15. Pileus not white.......................B. granulatus
16. Stem squamulose.............................17
16. Stem not squamulose...........................18
17. Pileus dull red.............................B. dichrous
17. Pileus some other color......................B. collinitus
18. Pileous yellow................................19
18. Pileus bay-red or chestnut.......................20
18. Pileus some other color.........................21
19. Flesh pale-yellow...........................B. unicolor
19. Flesh white................................B. bovinus
20. Stem short, one inch or less................B. brevipes
20. Stem longer, two inches or more.................B. badius
21. Tubes olivaceous or golden-yellow..................B. mitis
21. Tubes ferruginous................................22
22. Taste mild...............................B. rubinellus

Boletus.   22. Taste acrid or peppery.........................B. piperatus
Peck, Boleti of the U. S., p. 83.

B. specta′bilis Pk.—*spectabilis*, distinguished. **Pileus** broadly con-

(Plate CXV.)

BOLETUS SPECTABILIS.
Natural size.

vex, *at first covered with a red tomentum, then scaly*, v i s c i d when
moist, *red*, the tomentose scales becoming grayish-red, brownish or yellowish. **Flesh** whitish or pale-yellow.
Tubes at first yellow and concealed
by a reddish glutinous membrane,
then ochraceous, convex, *large, angular, adnate*. **Stem** nearly equal, annulate, yellow above the annulus, red
or red with yellow stains below.
Spores *purplish-brown*, 13–15x6–7µ.
Pileus 2–5 in. broad. **Stem** 3–5
in. long, 4–6 lines thick.
Thin woods in swamps. New York,
Peck; Wisconsin, *Bundy*.

This is a rare and showy species
which inhabits the cold northern swamps of the country. It probably
extends into Canada. When cut, the flesh emits a strong, unpleasant
odor. Wounds of the flesh made by insects or other small animals have
a bright-yellow color. When young, the tomentose veil covers the
whole plant, but it soon parts into scales on the pileus and partly or
wholly disappears from the stem. *Peck*, Boleti of the U. S.
London, Can., *J. Dearness; Peck*, Rep. 44, N. Y. State Bot.

B. Elben′sis Pk. **Pileus** convex, glabrous, viscid when moist, dingy
gray or pinkish-gray inclining to brownish, obscurely spotted or streaked
as if with patches of innate fibrils. **Flesh** white. **Tubes** at first whitish,
becoming dingy or brownish-ochraceous, nearly plane, adnate or slightly
decurrent, rather large, angular. **Stem** nearly equal, annulate; *whitish
above the ring*, colored like the pileus below, sometimes slightly reticulated at the top. **Spores** *ferruginous*-brown, 10–12x4–5µ.
Pileus 2–4 in. broad. **Stem** 3–5 in. long, 4–6 lines thick. Thin
woods of tamarack, spruce and balsam. New York. *Peck*.

Its locality is thus far limited to the Adirondack region of this state. Boletus. *Peck,* Boleti of the U. S.

B. sero'tinus Frost.—late.   Bulletin Buffalo Soc. Nat. Sci., 1874.
Pileus flat or convex, viscid, sordid brown, streaked with the remnants
of the veil, especially near the margin, which is white, very thin, and
when partly grown singularly pendent. **Flesh** white, *changing to bluish.*
Tubes large, angular, unequal, slightly decurrent, at first sordid white
or gray, sometimes tinged with green near the stem, afterward cinna-
mon-yellow.   **Stem** reticulated above the ring which adheres partly to
it and partly to the margin of the pileus, white but stained by the
brownish spores and tinged with yellow at maturity.   **Spores** 10x6μ.
   Shaded grassy ground.   New England, *Frost.*
   Probably this is only a variety of the preceding species. *Peck,* Boleti
of the U. S.

B. salmoni'color Frost.   Bull. Buff. Soc. Nat. Sci., 1874.   **Pileus**
convex, soft, very glutinous, brownish or tawny-white with a faint tinge
of red, wine-color when dry, the margin thin.   **Flesh** *tinged with red.*
Tubes simple, even, angular, adnate, *pale salmon* color.   **Stem** small,
dotted above with bright ferruginous red, sordid below, annulus *dingy
salmon-color.*   **Spores** 8x2.5μ.
   Borders of pine woods.   New England.   *Frost.*
   Apparently a distinct species.   No specimens seen.   *Peck,* Boleti of
the U. S.

B. el'egans Schum.   **Pileus** convex or plane, viscose, *golden-yellow
or somewhat rust-color.*   **Flesh** pale-yellow.   **Tubes** decurrent, golden
or sulphur-yellow, the mouths minute, simple.   **Stem** unequal, firm,
golden or reddish, *dotted above the fugacious white or pale-yellowish
annulus.*
   Pileus 3–4.5 in. broad.   **Stem** 2–4 in. long.
   Woods, especially under or near larch trees. North Carolina, *Curtis;*
Wisconsin, *Bundy;* Minnesota, *Johnson.*   *Peck,* Boleti of the U. S.
   Cordier and Gillet give the species as edible though not delicate.
   West Philadelphia on lawns under larches, 1887–1891.   *McIlvaine.*
   The caps are of good flavor and consistency.   They are best fried or
broiled.

409

Boletus.   **B. Clin'tonianus** Pk. **Pileus** convex, very viscid or glutinous, glabrous, soft, shining, *golden-yellow, reddish-yellow or chestnut color*, the margin thin. **Flesh** pale yellow, becoming less bright or dingy on exposure to the air. **Tubes** nearly plane, adnate or subdecurrent, *small*, angular or subrotund, pale-yellow, becoming dingy-ochraceous with age, *changing to brown or purplish-brown where bruised.* **Stem** equal or slightly thickened toward the base, straight or flexuous, *yellow at the top*, reddish or reddish-brown below the annulus, sometimes varied with yellow stains, the annulus white or yellow, *persistent*, forming a thick band about the stem. **Spores** *brownish-ochraceous*, 10–11x4–5μ.

Pileus 2–5 in. broad. **Stem** 2–5 in. long, 4–9 lines thick.

Mossy or grassy ground in woods or open places, especially under or near tamarack trees. New York, *Peck;* New England, *Frost.*

This is apparently closely related to B. elegans, from which it differs in its thick persistent ring, in its stem which is not at all dotted and in its longer and darker-colored spores. Its smaller tubes and persistent ring separate it also from B. flavus. In the typical form the pileus is bay-red or chestnut color, but plants growing in open places generally have it yellowish or reddish-yellow. It is mild to the taste and I have eaten it sparingly. It sometimes grows in tufts. *Peck*, Boleti of the U. S.

B. inflex'us Pk.—curving. **Pileus** convex, glabrous, viscid, yellow, often red or reddish on the disk, the margin thin, inflexed, concealing the marginal tubes. **Flesh** whitish, not changing color where wounded. **Tubes** rather long, adnate, yellowish, becoming dingy-yellow with age, the mouths small, dotted with reddish glandules. **Stem** rather slender, not ringed, solid, viscid, dotted with livid-yellow glandules. **Spores** yellowish, 10–12x4–5μ.

Pileus about 1 in. broad. **Stem** about 2 in. long, 2–4 lines thick.

Open woods. Trexlertown. September. *Herbst.*

This Boletus belongs to the tribe Viscipelles. It is remarkable for and easily recognized by the inflexed margin of the pileus, which imitates to some extent the appendiculate veil of Boletus versipellis. It sometimes grows in tufts. The paper in which fresh specimens were wrapped was stained yellow. Boletus Braunii Bres. has an inflexed margin, but that is a much larger plant with a yellowish-brown pileus, a fibrillose stem and much smaller spores. *Peck*, Bull. Torr. Bot. Club, Vol. 22, No. 5.

B. fla'vus With. **Pileus** convex, compact, covered with a brownish separating gluten, *pale-yellow*. **Flesh** pale-yellow. **Tubes** large, angular, adnate, yellow. **Stem** yellow, becoming brownish, reticulated above the *membranous fugacious* dirty yellowish annulus. **Spores** 8–10x3–4μ.

Pileus 2–5 in. broad. **Stem** 2–3 in. long, 6–10 lines thick.

Woods. Minnesota, *Johnson;* Wisconsin, *Bundy.*

This is apparently a rare species in this country. I have not seen it. It is said to resemble B. luteus, from which it is separated by the large angular mouths of the tubes. In British Fungi the spores are described as "spindle-shaped, yellowish-brown;" in Sylloge, as "ovoid-oblong, acute at the base, granulose, pale ochraceous." *Peck*, Boleti of the U. S.

B. fistulo'sus Pk. **Pileus** convex, viscid, glabrous, yellow, the margin at first incurved or involute. **Flesh** yellow. **Tubes** plane or subventricose, medium size, round with thin walls, adnate or sometimes depressed around the stem, yellow. **Stem** rather slender, subequal, viscid, glabrous, hollow, yellow, with a white mycelioid tomentum at the base. **Spores** elliptical, 13x6μ.

Pileus about 1 in. broad. **Stem** 2–4 in. long, about 3 lines thick.

Grassy woods. Auburn, Ala. July. *Underwood.*

A small but pretty species of a yellow color throughout. It is remarkable for its hollow stem, which is suggestive of the specific name. It is referable to the tribe Viscipelles. *Peck*, Bull. Torrey Bot. Club, Vol. 24, No. 3.

B. sphæros'porus Pk.—globose-spored. (Bulletin Torrey Botanical Club, Vol. XII.) **Pileus** at first hemispherical, then convex, glabrous, viscid, creamy-yellow, becoming reddish-brown or chestnut color with age. **Flesh** pale yellowish-brown. **Tubes** adnate or slightly decurrent, large, angular, pale-yellow, becoming brown, sometimes tinged with green. **Stem** stout, equal, even or slightly reticulated at the top, the *membranous annulus persistent*, sometimes partly adhering to the margin of the pileus. **Spores** *globose or broadly elliptical*, 8–9μ long.

Pileus 3–8 in. broad. **Stem** 1–3 in. long, 6–12 lines thick.

Low ravines and sandy places. Wisconsin, *Trelease;* Iowa, *McBride.*

The spores easily serve to distinguish this species from its allies. The

Boletus. European B. sphærocephalus has ovoid spores, but its tube mouths are minute and rotund and its stem is densely squamose. *Peck*, Boleti of the U. S.

B. lu'teus L.—yellow. **Pileus** gibbous or convex, covered with a brownish separating gluten, becoming yellowish-brown and virgate-spotted. **Flesh** white. **Tubes** adnate, minute, simple, yellow, becoming darker with age. **Stem** *stout*, yellowish and *dotted above* the large membranous brownish-white annulus, brownish-white or yellowish below. **Spores** fusiform, yellowish-brown, 6–7x3–4μ.

Pileus 2–5 in. broad. **Stem** 1–2 in. long, 6–10 lines thick.

Pine woods and groves. New York, *Peck*.

B. luteus has an international reputation for edibility. I have found it at Waretown and Haddonfield, N. J.; in Bartram's Garden, West Philadelphia, always under pines. At Waretown it was gregarious. Pine needles, sand, anything through which it grows, adheres to the glutinous cap. It must be carefully cleaned before cooking. It is then of choice consistency and good flavor.

B. sublu'teus Pk.—luteus, yellow. **Pileus** convex or nearly plane, viscid or glutinous when moist, often obscurely virgate-spotted, dingy-yellowish, inclining to rusty-brown. **Flesh** whitish, varying to dull-yellowish. **Tubes** plane or convex, adnate, small, subrotund, yellow becoming ochraceous. **Stem** equal, *slender*, pallid or yellowish, *dotted both above and below* the ring with reddish or brownish glandules; ring submembranous, *glutinous*, at first concealing the tubes, then generally collapsing and forming a narrow whitish or brownish band around the stem. **Spores** subfusiform, ochraceo-ferruginous, 8–10x4–5μ. **Pileus** 1.5–3 in. broad. **Stem** 1.5–2.5 in. long, 2–4 lines thick.

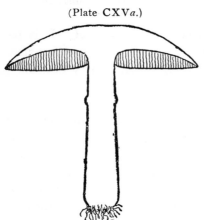

(Plate CXV*a*.)

SECTION OF BOLETUS SUBLUTEUS.

Sandy soil in pine woods. New York, *Peck, Clinton;* New England,
Frost.

The species is closely related to B. luteus, from which it differs in its smaller size, more slender stem and glutinous collapsing veil. *Peck,* Boleti of the U. S.

Found at Waretown, N. J., 1887, under pines and in same locality as B. luteus, for which it can be readily mistaken. It is usually covered with adherent sand or pine needles. Its flesh is tender with a pleasant glutinosity. Flavor good.

B. fla'vidus Fr.—light yellowish. **Pileus** thin, gibbous, then plane, viscose, livid, yellowish. **Flesh** pallid. **Tubes** decurrent, with *large angular compound mouths*, dirty yellowish. **Stem** *slender*, subequal, pallid, sprinkled with *fugacious glandules above the entirely viscose ring*. **Spores** oblong-ellipsoid, straight, subhyaline, 8–10x3–4µ.

Pileus 1–2 in. broad. **Stem** 2–3 in. long, 2–3 lines thick.

Pine woods and swamps. Pennsylvania, *Schweinitz;* North Carolina, *Curtis;* New England, *Frost;* California, *H. and M.;* Rhode Island, *Bennett.*

Fries says that this species is more slender than its allies, and differs from them all in its merely glutinous veil. *Peck,* Boleti of the U. S.

Dr. Curtis, of North Carolina, places it among edible species.

Many specimens were found by the writer near Waretown and Haddonfield, N. J., and a few at Mt. Gretna, Pa. The stems are thin and slightly spreading at the top. They are hard. The caps are excellent.

B. America'nus Pk. **Pileus** thin, convex or nearly plane, sometimes umbonate, soft, very viscid or glutinous when moist, *slightly tomentose on the margin when young*, soon glabrous or the margin sometimes remaining scaly, rarely scale-spotted from the drying of the gluten, yellow, becoming dingy or less bright with age, sometimes vaguely dotted or streaked with bright red. **Flesh** pale-yellow, less clear or pinkish-gray on exposure to the air. **Tubes** plane or convex, adnate, *rather large*, angular, pale-yellow, becoming sordid-ochraceous. **Stem** *slender*, equal or slightly tapering upward, firm, *not at all annulate*, yellow, often pallid or brownish toward the base, marked with *numerous brown or reddish-brown persistent glandular dots*, yellow within. **Spores** oblong or subfusiform, ochraceo-ferruginous, 9–11x4–5µ.

Boletus.   **Pileus** 1–3 in. broad.   **Stem** 1.5–2.5 in. long, 2–4 lines thick.

Woods, swamps and open places, especially under or near pine trees. New York, *Peck, Clinton;* Minnesota, *Arthur.*

A slight subacid odor is sometimes perceptible in our plant. It sometimes grows on much decayed wood. Its mycelium is white. *Peck,* Boleti of the U. S.

The caps, only, are good.

B. subau'reus Pk.—*sub* and *aureus,* golden.   ([Color] Plate CXIV, fig. 2.)   **Pileus** convex or nearly plane, viscose, pale-yellow, sometimes adorned with darker spots, the young margin slightly grayish-tomentose.   **Flesh** pale-yellow.   **Tubes** *small or medium,* somewhat angular, adnate or subdecurrent, pale-yellow becoming dingy-ochraceous.   **Stem** equal, *stout,* glandular-dotted, yellow *without and within.* **Spores** oblong or subfusiform, ochraceo-ferruginous, 8–10x4μ.

Pileus 2–4 in. broad.   **Stem** 1.5–2.5 in. long, 4–6 lines thick.

Thin woods and open places.   New York, *Peck;* North Carolina, *C. J. Curtis;* Massachusetts, Mississippi, *G. Survey* (Rep. 51).

This plant might almost be considered a stout variety of the preceding, but in addition to its thicker pileus and stouter stem, it has smaller tubes of a clearer yellow color, and the exuding drops are yellow, not whitish, as in that species.   In habit it appears more like B. granulatus, from which it is distinct in color.   *Peck,* Boleti of the U. S.

From early October, through heavy frosts and until long after November snows I found this species at Mt. Gretna, Pa., in 1897–1898. Specimens were sent to Professor Peck and identified as this species. It grew in grass on borders of woods, or gravelly ground, sometimes among pine needles.   Large troops of it were frequent, and tufts containing many individuals were common.

I regard B. subaureus as among the most valuable of our food species. Its plentifulness, lateness, excellent quality will commend it to all Mycophagists.   It can be cooked in any way.   The tubes need not be removed.

B. hirtel'lus Pk.—slightly hairy.   **Pileus** broadly convex, soft, viscose, golden-yellow, adorned with *small tufts of hairs or fibrils.* **Flesh** pale-yellow.   **Tubes** adnate, medium size, angular, becoming dingy-ochraceous. **Stem** subcespitose, equal, stout, glandular dotted, Boletus. yellow. **Spores** pale, *ochraceous-brown*, 9–10x4μ.

Pileus 2–4 in. broad. **Stem** 2–3 in. long, 4–6 lines thick.

Sandy soil under pine trees. New York, *Peck.*

This species is very rare and was formerly confused with the preceding from which it is separated by the hairy adornment of the pileus and the darker, more brown color of the spores. *Peck*, Boleti of the U. S.

B. punc′tipes Pk.—*punctum*, a dot; *pes*, a foot. **Pileus** convex or nearly plane, glutinous when moist, yellow, the thin margin at first minutely grayish-pulverulent, becoming recurved with age. **Tubes** short, nearly plane, adnate, small, subrotund, *at first brownish*, then sordid-ochraceous. **Stem** rather long, *tapering upward*, grandular-dotted, *rhubarb-yellow*. **Spores** 9–10x4–5μ.

Pileus 2–3 in. broad. **Stem** 2–3 in. long, 3–5 lines thick. Mixed woods. New York, *Peck.*

The rhubarb-colored stem and the brownish color of the young hymenium are the distinguishing features of this species. The glandules occur also on the tubes. The species is rare. *Peck*, Boleti of the U. S.

Not seen by Professor Peck since its discovery in 1878.

Spores when first dropped are olive-green on white paper, but the green hue soon changes to brownish-ochraceous. *Peck*, 44th Rep. N. Y. State Bot.

Ontario, *Prof. Dearness (Lloyd, R. 4).*

B. al′bus Pk.—white. **Pileus** convex, viscid when moist, *white*. **Flesh** white or yellowish. **Tubes** plane, small or medium, subrotund, adnate, whitish, becoming yellow or ochraceous. **Stem** equal or slightly tapering downward, both it and the tubes glandular-dotted, *white*, sometimes tinged with pink toward the base. **Spores** ochraceous, subfusiform, 8–9x4μ.

Pileus 1.5–3 in. broad. **Stem** 1.5–3 in. long, 3–5 lines thick.

Woods, especially of pine or hemlock. New York, *Peck;* New England, *Frost.*

This species is easily known by its white pileus, but its color is lost in drying. Sometimes the fresh plant emits a peculiar fetid odor. *Peck*, Boleti of the U. S.

Boletus. **B. granula′tus**—*granula*, a granule. **Pileus** convex or nearly plane, very viscid or glutinous and *rusty-brown* when moist, *yellowish* when dry. **Flesh** pale-yellowish. **Tubes** short, adnate, yellowish, their mouths simple, granulated. **Stem** dotted with glandules above, pale-yellowish. **Spores** spindle-shaped, yellowish-orange, 7.5–10x2–3µ.

Pileus 1.5–4 in. broad. **Stem** 1–2 in. long, 4–6 lines thick.

Woods, especially of pine and in open places under or near pine trees. Very common.

The plant is generally gregarious and sometimes grows in circles, whence the name B. circinans Pers. Occasionally it is cespitose. The pileus is very variable in color—pinkish-gray, reddish-brown, yellowish-gray, tawny-ferruginous or brownish—and is sometimes obscurely spotted by the drying gluten. The flesh is rather thick and often almost white, except near the tubes, where it is tinged with yellow. The tubes are small, at first almost white or very pale-yellow, but they become dingy-ochraceous with age. The stem is generally short, stout and firm, whitish-pallid or yellowish, and often dotted to the base, though the glandules are more numerous and distinct on the upper part. *Peck*, Boleti of the U. S.

B. granulatus is of frequent and general occurrence. I have found it in the pine woods of New Jersey, North Carolina, Pennsylvania and West Virginia, and in West Virginia and Pennsylvania in mixed woods.

It is a late-growing species, appearing in September and continuing until frost.

All authors, with one exception (Gillet), give the species as edible. From frequent and copious testings, the writer vouches for its edibility and excellence. It bears favorable comparison with any of the late Boleti.

B. bre′vipes Pk.—*brevis*, short; *pes*, foot. **Pileus** thick, convex, covered with a *thick, tough gluten* when young or moist, *dark chestnut color*, sometimes fading to dingy-tawny, the margin inflexed. **Flesh** white or tinged with yellow. **Tubes** short, nearly plane, adnate or slightly depressed around the stem, small, subrotund, at first whitish becoming dingy-ochraceous. **Stem** whitish, *not dotted or rarely with a few very minute inconspicuous dots at the apex, very short*. **Spores** sub-fusiform, 7.5x3µ.

Pileus 1.5–2.5 in. broad. **Stem** .5–1 in. long, 3–5 lines thick.

Sandy soil in pine groves and woods. New England, *Frost;* New Boletus.
York, *Peck*.

The species is closely related to B. granulatus, from which it differs especially in its darker colored pileus, more copious gluten, shorter stem and the almost entire absence of granules from the tube mouths and stem. In the rare instances in which these are present they are extremely minute and inconspicuous. The plant occurs very late in the season and the pileus appears as if enveloped in slime and resting stemless on the ground. *Peck*, Boleti of the U. S.

Specimens found in pine woods of New Jersey, identified by Professor Peck. Lambertville, N. J., *C. S. Ridgway;* Haddonfield, N. J., *T. J. Collins;* Pleasantville, *Isaac F. Shaner*.

B. brevipes is a disreputable, dirty, tramp-looking fungus, from which the collector would expect no good. Nevertheless, when it has had a good scrubbing it becomes respectable and is sweet, tender, good eating. When other species abound, it does not pay for the cleansing.

B. collini'tus Fr.—*collino*, to besmear. **Pileus** convex, even, *becoming pale when the brown gluten separates*. **Flesh** white. **Tubes** adnate, elongated, naked, *the mouths two-parted*, pallid, becoming yellow. **Stem** firm, often tapering downward, *somewhat reticulate with appressed squamules*, white, becoming brown.

Woods of pine or fir. North Carolina, *Curtis;* New England, *Frost*.

I have seen no specimens of this apparently rare species. It is said to be solitary in its mode of growth and to resemble B. luteus in size and color, but to be distinct from it by its ringless, dotless stem. Dr. Curtis records it as edible. *Peck*, Boleti of the U. S.

I found three specimens at Haddonfield, N. J., October, 1897, under scrub pines. Cap 2 ½ in. across, convex, gibbous; stem equal, 2 ½ in. long, ½ in. in diameter, slightly tapering at base. The two-parted mouths to the tubes were very distinct. The stems were tough, but the caps, washed and fried, were good.

B. di'chrous Ellis. **Pileus** convex, viscose, *dull red*. **Flesh** soft, dull, yellowish-white, *changing to greenish-blue* where wounded, finally yellow. **Tubes** subdepressed around the stem, large, unequal, straw-colored, changing color like the flesh where wounded. **Stem** thickened

417

Boletus. below, solid, covered with a *red scaly coat*, except at the yellow apex, yellow within. **Spores** elliptical, slightly bent at one end, 2μ long.

Pileus 2–3 in. broad. **Stem** 3 in. long, 6 lines thick.

Dry soil in oak and pine woods. New Jersey. *Ellis.*

I have seen no specimens of this species. From the description, its affinities appear to be with B. bicolor, but it is placed here because of its viscose pileus. *Peck*, Boleti of the U. S.

B. ba′dius Fr.—bay-brown. **Pileus** convex, even, soft, viscose or glutinous, shining when dry, *tawny-chestnut*. **Flesh** whitish, tinged with yellow, bluish next the tubes. **Tubes** large, angular, long, adnate or sinuate-depressed, whitish-yellow, becoming tinged with green. **Stem** subequal, even, solid, paler, *brown-pruinate*. **Spores** fusoid-oblong.

Pileus 2–3 in. broad. **Stem** 2–4 in. long, 3–5 lines thick.

Woods, especially of pine. New York, *Peck;* Minnesota, *Johnson;* Wisconsin, *Bundy;* Nova Scotia, *Somers.*

In the American plant the spores are 10–12x4–5μ.

Cordier classes it among the edible species. *Peck*, Boleti of the United States.

B. mi′tis Krombh.—mild. **Pileus** convex, then plane or depressed, firm, viscid, yellowish-flesh color, reddish-rust color when dry. **Flesh** pale, grayish-yellow. **Tubes** *short, olivaceous or golden-yellow*, their mouths compound, angular, unequal. **Stem** firm, short, even, narrowed toward the base, colored like the pileus. **Spores** 12–14x4μ.

Pileus 2–2.5 in. broad. **Stem** 2–2.5 in. long.

Mixed woods. New England, *Frost.*

This species is unknown to me and is recorded by Mr. Frost only. *Peck*, Boleti of the United States.

B. uni′color Frost MS. **Pileus** broadly convex or nearly plane, viscid when moist, even, sometimes streaked as if with minute innate brown fibrils, *pale-yellow*. **Flesh** *pale-yellow*. **Tubes** adnate or slightly decurrent, rather short, compound, *lemon-yellow*, becoming darker with age. **Stem** *even*, equal or narrowed toward the base, colored like the pileus. **Spores** reddish-yellow, 9–11x4μ.

Pileus 2–4 in. broad. **Stem** 2 in. long, 4–6 lines thick.

Pine woods and open sedgy places. New England, *Frost.*
Specimens not seen. The species seems too near B. bovinus, of which it may possibly be a variety, but its yellow flesh and the colors ascribed to the tubes and spores require its separation. Rev. C. J. Curtis sends notes of a species found by him in North Carolina, which agree with this in its characters so far as noted. *Peck,* Boleti of the U. S.

B. ignora′tus Pk. **Pileus** convex, viscid, bright lemon-color, marked with wrinkled lines of orange color, which are distributed over the pileus, giving it a streaked appearance. **Flesh** white, solid, does not change color when cut or broken; taste slightly acid. **Pores** lemon-color, moderately large, free, connected with the stem by web-like filaments. **Stem** larger at the apex, somewhat tapering toward the base, yellow, smooth, solid. **Spores** 4.5x11μ.

This closely approaches Boletus unicolor Fr., from which it scarcely differs except in its white flesh and free tubes. Fungi of Maryland, *Mary E. Banning. Peck,* 44th Rep. N. Y. State Bot.

B. bovi′nus—*bos,* an ox. **Pileus** nearly plane, glabrous, viscid, pale yellow. **Flesh** *white.* **Tubes** very short, subdecurrent, their mouths compound, pale yellow or grayish, becoming rust-colored. **Stem** equal, even, colored like the pileus. **Spores** fusiform, dingy greenish-ocher, 7.5–10x3–4μ.

Pileus 2–3 in. broad. **Stem** 1.5–2 in. long, sometimes cespitose.

Pine woods. North Carolina, *Schweinitz, Curtis;* Pennsylvania, *Schweinitz;* New England, *Frost, Palmer, Bennett, Sprague, Farlow;* California, *H. and M.*

The shallow tubes, 2–3 lines long, are said to resemble the pores of Merulius lacrymans. The species is recorded edible by Curtis, Gillet and Palmer. *Peck,* Boleti of the U. S.

West Virginia mountains under hemlocks, 1882–1885, and near Haddonfield, N. J., under pines. *McIlvaine,* 1892. Gregarious and in clusters. The pore surface was in some specimens broadly wrinkled.

Smell and taste pleasant. Cooked, the quality is of the best in Boleti.

B. rubinel′lus Pk.—dim. of *ruber,* red. **Pileus** broadly conical or convex, viscid when moist, subtomentose or slightly pubescent when dry, *red fading to yellow on the margin.* **Flesh** whitish or yellowish,

419

Boletus. taste *mild*. **Tubes** adnate or slightly depressed around the stem, dingy-reddish, becoming subferruginous. **Stem** equal, slender, even, colored like the tubes, *yellow within*, sometimes yellow at the base. **Spores** oblong-fusiform, ferruginous-brown, 12.5–15x4µ.

Pileus 1–2 in. broad. **Stem** 1–2 in. long, 1–3 lines thick.

Mixed woods or under or near coniferous trees in open places. New York, *Peck*. *Peck*, Boleti of the U. S.

B. pipera'tus Bull.—*piper*, pepper. **Pileus** convex or nearly plane, glabrous, *slightly viscid* when moist, *yellowish, cinnamon or subferruginous*. **Flesh** white or yellowish, taste *acrid, peppery*. **Tubes** rather long and large, angular, often unequal, plane or convex, adnate or subdecurrent, *reddish-rust color*. **Stem** slender, subequal, tawny-yellow, bright yellow at the base. **Spores** subfusiform, ferruginous-brown, 9–11x4µ.

Pileus 1–3 in. broad. **Stem** 1.5–3 in. long, 2–4 lines thick.

Woods and open places. Common and variable.

This species may easily be recognized by its peppery flavor. The pileus sometimes appears as if slightly tomentose, and both this and the preceding species recede from the character of the tribe by the slight viscidity of the pileus. This is sometimes cracked into areas and sometimes the margin is very obtuse by the elongation of the tubes. *Peck*, Boleti of the U. S.

Haddonfield, N. J., 1892. *McIlvaine*.

This fungus is reckoned poisonous by Stevenson. Massee gives its taste as very hot. The taste of the American plant is peppery but not offensively so. This pepperiness it loses in cooking. It has been eaten by the writer and his friends with enjoyment and without any discomfort.

B. subsanguin'eus Pk.—*sub* and *sanguineus*, bloody. ([Color] Plate CXVI, fig. 4.) **Pileus** convex or slightly depressed in the center, glabrous, viscid, bright-red or scarlet. **Flesh** thick, firm but flexible, white, slowly changing to a pale brownish-lilac on exposure to the air, taste slightly bitter. **Tubes** very short, 2–4 mm. long, adnate, but often separating from the stem with the expansion of the pileus, reddish, the mouths minute, stuffed at first, pinkish, then brownish-yellow, changing to a light-brown where wounded. **Stem** short, thick, uneven, often tapering downward, streaked with red, pale-yellow at the top, white at Boletus. the base, marked at the top by the decurrent walls of the tubes.

Pileus 2.5–10 cm. broad. **Stem** 2.5–5 cm. long, 2–4 cm. thick.

Solitary, gregarious or cespitose. Under beech trees. West Philadelphia, Pa. August. *C. McIlvaine.*

This is a very showy species, easily recognized by its bright-red viscid pileus and its short, thick and uneven or somewhat lacunose stem. It is closely related to the European B. sanguineus With., from which it is separated by its minute tubes, its uneven stem and the brownish hues assumed where wounded.

The spore characters of this and the four succeeding species are unknown, but the other characters are quite distinctive and apparently sufficient for the recognition of the species. The descriptions have been derived from colored figures and other data furnished by Mr. McIlvaine, who says all are edible. *Peck*, Bull. Torrey Bot. Club, No. 27.

When slowly stewed for thirty minutes, there is no better Boletus.

PULVERULENTI.

Pileus clothed with a yellow dust or a yellow powdery down. **Stem** more or less yellow powdered, neither bulbous nor distinctly reticulated.

The species which constitute this tribe are easily distinguished from all others by the sulphur-colored pulverulence which coats the pileus and stem like a universal veil. They appear thus far to be peculiar to this country. Though strongly resembling each other in the tribal character they are very diverse in other respects. One species, by its viscidity, connects with the preceding tribe; another by its differently colored tube mouths is related to the Luridi; and the third is peculiar in its ligneous habitat.

Plant growing on the ground...................................1
Plant growing on wood........................B. hemichrysus
1. Tubes adnate, of one color.......................B. Ravenelii
1. Tubes free, with red mouths..................B. auriflammeus
Peck, Boleti of the U. S., p. 103.

B. hemichry′sus B. and C.—half-golden. **Pileus** convex, at length plane or irregularly depressed, floccose-squamulose, covered with a yellow powder, sometimes cracked, bright golden-yellow. **Flesh** thick,

Boletus. *yellow.* **Tubes** adnate or decurrent, yellow, becoming reddish-brown, the mouths large, angular. **Stem** *short, irregular, narrowed below,* sprinkled with a yellow dust, yellowish tinged with red; mycelium yellow. **Spores** oblong, minute, dingy-ochraceous.

Var. *muta'bilis.* **Flesh** slightly changing to blue where wounded. **Stem** reddish, yellow within, sometimes eccentric. **Spores** oblong-elliptical, 7.5–9x3–4μ.

Pileus 1.5–2.5 in. broad. **Stem** about 1 in. long, 3–6 lines thick.

Roots of pine, *Pinus palustris.* The variety on stumps of *Pinus strobus.*

South Carolina, *Ravenel;* North Carolina, *Curtis;* New York, *Peck.* The species is remarkable for its habitat, which is lignicolous. The New York variety grew on a stump of white pine. By its eccentric stem it connects this genus with Boletinus, through Boletinus porosus. According to the authors of this species it resembles Boletus variegatus. *Peck,* Boleti of the U. S.

B. Ravenel'ii B. and C.—after Ravenel. **Pileus** convex or nearly plane, *slightly viscid when young or moist,* covered with a sulphur-yellow powdery down, becoming naked and dull-red on the disk. **Flesh** whitish. **Tubes** at first plane, *adnate,* pale-yellow, becoming yellowish-brown or umber, dingy-greenish where bruised, the mouths large or medium size, subrotund. **Stem** nearly equal, clothed and colored like the young pileus, yellow within, with a slight evanescent webby or tomentose ring. **Spores** ochraceous-brown, 10–12x5–6μ.

Pileus 1–3 in. broad. **Stem** 1.5–4 in. long, 3–6 lines thick.

Woods and copses. South Carolina, *Ravenel;* North Carolina, *Curtis;* New York, *Peck;* New England, *Frost.*

This is a very distinct and very beautiful species. Mr. Ravenel remarks in his notes that "this plant is not infested by larvæ and preserves more constant characters than any other Boletus with which I am acquainted." The webby powdered filaments constitute a universal veil which at first covers the whole plant and conceals the young tubes. As the pileus expands this generally disappears from the disk, and, separating between the margin and the stem, a part adheres to each. The flesh is sometimes stained with yellow. The tubes in some instances become convex and slightly depressed around the stem. They are almost white when young, and often exhibit brownish hues where wounded.

The plant is sometimes cespitose. I have observed a greenish tint to Boletus. the freshly shed spores, but it soon disappears. Boletus subchromeus Frost Ms. is this species. *Peck*, Boleti of the U. S.

B. auriflam'meus B. and C.—flaming yellow. **Pileus** convex, *dry*, powdered, bright golden-yellow. **Flesh** white, unchangeable. **Tubes** plane or convex, *free*, yellow, their broad angular *mouths scarlet*. **Stem** slightly tapering upward, powdered, colored like the pileus. **Spores** 10–12.5x5μ.

Pileus 8–12 lines broad. **Stem** 1–1.5 in. long.
Woods. North Carolina, *Curtis;* New York, *Peck.*

This is evidently a rare species and as beautiful as it is rare. The whole plant is bright-yellow except the tube mouths, and is sprinkled with yellow dust or minute yellow branny particles. In the New York specimen the scarlet color is wanting in the marginal tube mouths and the stem is marked with fine subreticulating elevated lines. In other respects it agrees well with the diagnosis of the species. *Peck*, Boleti of the U. S.

SUBPRUINOSI—sub, pruina, hoar frost.

Pileus glabrous, but more often pruinose. **Tubes** adnate, yellowish. **Stem** equal, even, neither bulbous nor reticulated.
The species of this tribe have the pileus neither viscid nor distinctly and permanently tomentose. Typically it is glabrous or merely pruinose, but Fries has admitted into the group one species with a pulverulent, and one with a silky pileus. The species are not sharply distinguished from those of the following tribes, and possibly some have been admitted here which might as well have been placed there. Some of the species are variable in color and their characters are not sufficiently well known.

Tubes bright-yellow, golden or subochraceous.................1
1. Tubes pale or whitish-yellow.............................6
1. Tubes changing to blue where wounded....................2
1. Tubes not changing to blue.............................3
  2. Stem pallid, with a circumscribing red line at the top..B. glabellus
  2. Stem yellow, sometimes with red stains....B. miniato-olivaceus
  2. Stem red, yellow at the top......................B. bicolor
3. Stem viscid or glutinous when moist..............B. auriporus

423

Peck, Boleti of the U. S.

B. minia′to-oliva′ceus Frost—olive-red. **Pileus** at first convex and firm, then nearly plane, soft and spongy, glabrous, vermilion, becoming olivaceous. **Flesh** pale-yellow, changing to blue where wounded. **Tubes** bright lemon-yellow, adnate or subdecurrent. **Stem** glabrous, enlarged at the top, pale-yellow, brighter within, sometimes lurid at the base. **Spores** 12.5x6μ.

Var. *sensi′bilis* (Boletus sensibilis Rep. 32, p. 33).

Pileus at first pruinose-tomentose, red, becoming glabrous and ochraceous-red with age. **Tubes** bright-yellow tinged with green, becoming sordid-yellow. **Stem** lemon-yellow with red or rhubarb stains at the base, contracted at the top when young, subcespitose. **Spores** 10–12.5 x4–5μ.

Pileus 2–6 in. broad. **Stem** 3–4 in. long, 3–6 lines thick.

Woods and their borders. New England, *Frost ;* New York, *Peck.*

Though the sensitive Boletus differs considerably in some respects from the olive-red Boletus, it is probably only a variety, and as such I have subjoined it here. In it every part of the plant quickly changes to blue where wounded, and even the pressure of the fingers in handling the fresh specimens is sufficient to induce this change of color. I have not found the typical plant in New York, but specimens received from Mr. Frost are not, in the dry state, distinguishable from the variety. *Peck*, Boleti of the U. S.

Indiana, *H. I. Miller*; West Virginia. Haddonfield, N. J. Cheltenham, Pa., *McIlvaine.*

Years ago I marked it edible and excellent when young. My friends have eaten it, and continue to do so. Yet Professor Peck (48th Rep., Boletus. p. 202) reports a case brought to his notice of an entire family being sickened by eating B. sensibilis. All recovered. It may, therefore, be one of those species which, while disagreeing with some persons, can be eaten by the majority. Clitocybe illudens, Lepiota Morgani and others of the Agaricaceæ are such species.

B. bi'color Pk.—two-color. ([Color] Plate CXVII, figs. 1, 2.) **Pileus** convex, glabrous or merely pruinose-tomentose, dark-red, firm, becoming soft, paler and sometimes spotted or stained with yellow when old. **Flesh** yellow, not at all or but slightly and slowly changing to blue where wounded. **Tubes** nearly plane, adnate, bright-yellow, becoming ochraceous, slowly changing to blue where wounded, their mouths small, angular or subrotund. **Stem** subequal, firm, solid, *red, generally yellow at the top..* **Spores** pale, ochraceous-brown, 10–12.5x4–5μ.

Pileus 2–4 in. broad. **Stem** 1–3 in. long, 4–6 lines thick.

Woods and open places. New York, *Peck;* Wisconsin, *Bundy*.

The color of this plant is somewhat variable. In the typical form the pileus and stem are dark red, approaching Indian red, but when old the color of the pileus fades and is often intermingled with yellow. The surface sometimes cracks and becomes cracked in areas. From the European B. Barlæ this species is separated by its solid stem; from B. versicolor by its small tube mouths and its red stem. *Peck*, Boleti of the U. S.

Plentiful at Mt. Gretna, Pa., July, August, September, 1898, in mixed woods. Very variable in shape and color. Identified by Professor Peck from painting and description.

Fine eating, one of the very best.

B. glabel'lus Pk.—smooth. **Pileus** fleshy, thick, broadly convex or nearly plane, soft, dry, subglabrous, *smoky-buff*. **Flesh** *white*, both it and the tubes changing to blue where wounded. **Tubes** nearly plane, adnate, ochraceous, tinged with green, their mouths small, subrotund. **Stem** subequal, glabrous, even, reddish toward the base, pallid above, with a *narrow reddish circumscribing zone or line at the top*. **Spores** oblong, brownish-ochraceous, tinged with green when fresh, 10–12.5x4μ.

Pileus 3–5 in. broad. **Stem** 1–3 in. long, 5–10 lines thick.

Boletus. Grassy ground under oaks. New York, *Peck*.

The species is well marked by the reddish band or line on the stem just below the tubes, but this disappears in drying. *Peck*, Boleti of the U. S.

B. aluta′ceus Morgan—yellowish. **Pileus** cushion-shaped, glabrous, *alutaceous* with a tinge of red. **Flesh** *white, inclining to reddish*. **Tubes** semifree, medium in size, unequal, angular, greenish-yellow. **Stem** nearly equal, striate, reticulate at the apex, colored like the pileus. **Spores** fusiform, brownish-olive, 12.5x5μ.

Pileus 3 in. broad.

Rocky woods of oak and chestnut. Kentucky, *Morgan*.

The general aspect of the figure of this species recalls some of the forms of Boletus subtomentosus. The tubes are nearly equal in length to the thickness of the flesh of the pileus. *Peck*, Boleti of the U. S.

Quite frequent at Mt. Gretna, Pa., in mixed woods, principally oak and chestnut.

Stem should be **removed**, and tubes when old. It cooks well and is especially good.

B. tenui′culus Frost—thin. **Pileus** nearly plane, *thin*, lurid-red on a yellow ground. **Flesh** unchangeable. **Tubes** short, adnate, small, *golden-yellow*. **Stem** *slender*, equal, colored like the pileus. **Spores** 10x6μ.

Pileus 1–2 in. broad. **Stem** 4–6 in. long.

Woods. New England. *Frost*.

The thin pileus and long slender stem readily **distinguish** this species. *Peck*, Boleti of the U. S.

B. auri′porus Pk.—golden-pore. **Pileus** convex or nearly plane, glabrous or merely pruinose-tomentose, grayish-brown, yellowish-brown, or reddish-brown. **Flesh** white, unchangeable. **Tubes** plane or slightly depressed around the stem, adnate or subdecurrent, *bright golden-yellow, retaining their color when dried*. **Stem** equal or slightly thickened at the base, *viscid or glutinous when moist*, especially toward the base, colored like or a little paler than the pileus. **Spores** 7.5–10x4–5μ.

Pileus 1–3 in. broad. **Stem** 1–3 in. long, 2–4 lines thick.

Thin woods and shaded banks. New York, *Peck;* New England,
Frost.

This species is remarkable for the rich yellow color of the tubes, which is retained unchanged in the dried specimens, and for the viscid stem. This character, however, is not noticeable in dry weather and was overlooked in the original specimens.

Boletus glutinipes Frost Ms. is not distinct. *Peck*, Boleti of the U. S.

Hopkins' Woods, Haddonfield, N. J. Grassy oak woods. 1891–1894. *McIlvaine.*

The caps are delicious.

B. innix′us Frost. **Pileus** convex or nearly plane, glabrous, yellowish-brown, slightly cracked in areas when old, yellow in the interstices. **Flesh** white. **Tubes** adnate, lemon-yellow, unchangeable. **Stem** slender, short, much thickened at the base in large specimens, yellowish, streaked with brown, brownish within. **Spores** 10x5μ.

Grassy woods. New England. *Frost.*

The whole plant often reclines as if for support, *Peck*, Boleti of the U. S.

B. parasi′ticus Bull.—a parasite. **Pileus** convex or nearly plane, dry, silky, becoming glabrous, *soon tessellately cracked*, grayish or dingy-yellow. **Tubes** decurrent, medium size, *golden yellow.* **Stem** equal, rigid, incurved, yellow without and within. **Spores** oblong-fusiform, pale-brown, 12.5–15x4μ.

Pileus 1–2 in. broad. **Stem** 1–2 in. long, 2–4 lines thick.

Parasitic on species of Scleroderma. New York, *Gerard;* New England, *Sprague, Bennett.*

This species is very rare in this country. It is remarkable for its peculiar habitat. *Peck*, Boleti of the U. S.

New York, *Lydia M. Patchen;* Westfield, on Scleroderma vulgare.

I found many specimens of this rare species during August, 1897, growing on Scleroderma vulgare.

Professor Peck, to whom I sent specimens, identified them as B. parasiticus. The tubes were large, unequal, dissepiments thin, decurrent. The Sclerodermas frequently appear to be parasitic upon the Boletus. I have seen the host plant thrown entirely free from the ground by the Boletus.

427

B. parasiticus is edible, but it is not of agreeable flavor.

B. dictyoceph'alus Pk.—reticulate. **Pileus** convex, glabrous, *reticulate with brown lines beneath the thin separable cuticle*, brownish-orange, darker in the center and there tinged with pink. **Flesh** white, unchangeable. **Tubes** nearly plane, slightly depressed around the stem, grayish-yellow, becoming brown where bruised. **Stem** equal or slightly tapering at the top, solid, rimose, dotted with scales, lemon-yellow, darker toward the base. **Spores** 15–20x6μ.

Pileus 2.5 in. broad. **Stem** 3–4 in. long, 5–6 lines thick.
Mixed woods. North Carolina. *C. J. Curtis.*

The description here given has been derived from a single dried specimen and from the notes kindly sent by Mr. Curtis. The species is apparently well marked and very distinct by the peculiar reticulations of the pileus. *Peck*, Boleti of the U. S.

B. subgla'bripes Pk.—rather smooth. **Pileus** convex or nearly plane, glabrous, reddish inclining to chestnut color. **Flesh** white, unchangeable. **Tubes** adnate, nearly plane in the mass, pale yellow, becoming convex and darker or greenish-yellow with age, the mouths small, subrotund. **Stem** equal, solid, scurfy, pale yellow. **Spores** oblong-fusiform, 12.5–15x4–5μ.

The smoothish-stemmed Boletus is well marked by its cylindric minutely scurfy stem which is colored like the tubes. Its cap is smooth and nearly always some shade of red or bay. Specimens occur occasionally in which it approaches grayish-brown or wood-brown. The flesh is white and unchangeable when cut or broken.

The tubes at first have a nearly plane surface, but this becomes somewhat convex with age, and slightly depressed around the stem. The tube mouths are small and nearly round. The color of the tubes is at first a beautiful pale yellow, but it becomes darker or slightly greenish-yellow with age.

The stem is colored very nearly like the tubes, but sometimes it has a slight reddish tint toward the base. Its peculiar feature consists of the minute, branny particles upon it. They are so small and pale that they are easily overlooked.

There is a variety in which the cap is corrugated or irregularly pitted and wrinkled. Its name is Boletus subglabripes corrugis Pk.

The cap is 1½–4 in. broad, the stem is 2–3 in. long and 4–8 lines Boletus. thick. The plants are found in woods in July and August. *Peck*, 51st Rep. N. Y. State Bot.

B. pal'lidus Frost—pale. ([Color] Plate CXVII, fig. 4.) **Pileus** convex, becoming plane or centrally depressed, soft, glabrous, pallid or brownish-white, sometimes tinged with red. **Flesh** white. **Tubes** plane or slightly depressed around the stem, nearly adnate, *very pale or whitish-yellow*, becoming darker with age, *changing to blue where wounded*, the mouths small. **Stem** equal or slightly thickened toward the base, rather long, glabrous, often flexuous, whitish, sometimes streaked with brown, often tinged with red within. **Spores** pale ochraceous-brown, 10–12x5–6μ.

Pileus 2–4 in. broad. **Stem** 3–5 lines long, 4–8 lines thick.

Woods. New England, *Frost;* New York, *Peck.*

The species is readily recognized by its dull pale color, rather long stem, and tubes changing to blue where wounded. *Peck*, Boleti of the U. S.

Common in West Virginia mountains, Angora, West Philadelphia, Mt. Gretna, Pa. Solitary, on ground in mixed woods.

The caps are tender and delicately flavored.

B. rubropunc'tus Pk.—red-dotted. ([Color] Plate CXVII, fig. 3.) **Pileus** convex, glabrous, reddish-brown. **Flesh** yellowish, unchangeable. **Tubes** nearly plane, depressed about the stem, their mouths small, round, bright golden-yellow, not changing color where bruised. **Stem** firm, solid, tapering upward, yellow, punctate with reddish dots or squamules. **Spores** olive-green, 12.5x4–5μ.

Pileus 1–2 in. broad. **Stem** 1–2 in. long, 3–6 lines thick.

Woods. Port Jefferson. July. Cold Spring Harbor, *H. C. Beardslee.*

This is a pretty Boletus, well marked by the red dots of the stem. It is apparently a very rare species. B. radicans is said to have the stem sprinkled with red particles, but that is a larger plant with the margin of the pileus persistently involute or incurved and with a radicating stem, characters which are not shown by our fungus. *Peck*, 50th Rep. N. Y. State Bot.

I found my specimens at Mt. Gretna, Pa., August-September, 1898.

429

Boletus. Identified for the writer by Professor Peck from painting and description.

Taste and smell slight. Cooks well and is pleasant to the taste. The tubes should be removed.

SUBTOMENTO'SI—*sub, tomentosus,* downy.

Pileus when young villose or subtomentose, rarely becoming glabrous with age, destitute of a viscid pellicle. **Tubes** of one color, adnate. **Stem** at first extended, neither bulbous nor reticulated with veins, wrinkled or striated in some species. **Flesh** in some changing color where wounded.

The tubes are generally yellow or greenish-yellow. In some species they are occasionally somewhat depressed around the stem, but they do not form a rounded free stratum, nor, with the exception of B. rubeus, are they stuffed when young as in most of the Edules. The species are scarcely separable from those of the preceding tribe except by the more evidently tomentose young pileus.

Tubes brown, becoming cinnamon..................B. variegatus
Tubes not having these colors................................1
1. Flesh or tubes changing to blue where wounded..............2
1. Flesh or tubes not changing to blue......... ....,5
    2. Stem glabrous.......................................3
    2. Stem not glabrous............................. 4
3. Flesh yellow under the cuticle.....................B. rubeus
3. Flesh red under the cuticle....................B. chrysenteron
    4. Stem velvety at the base.....................B. striæpes
    4. Stem with a reddish bloom or scurf...........B. radicans
    4. Stem with brown dot-like scales ....B. mutabilis
5. Tubes whitish, becoming yellow....................B. Roxanæ
5. Tubes yellow..6
    6. Tube mouths large and angular ....B. subtomentosus
    6. Tube mouths minute.....................B. spadiceus
Peck, Boleti of the U. S.

B. variega'tus Swartz. **Pileus** at first convex, then plane, obtuse, moist, sprinkled with *superficial bundled hairy squamules, dark-yellow,* the acute margin at first flocculose. **Flesh** yellow, here and there becoming blue. **Tubes** adnate, unequal, minute, *brown then cinnamon*. Boletus. **Stem** firm, equal, even, dark-yellow, sometimes reddish. **Spores** oblong-ellipsoid, hyaline or very pale-yellowish, 7.5–10x3–4μ.

Pileus 2–5 in. broad. **Stem** 2–3 in. long, 6 lines thick.

Woods, especially of pine. North Carolina, *Curtis, Schweinitz;* California, *Harkness, Moore;* Rhode Island, *Bennett. Peck*, Boleti of the U. S.

West Virginia mountains, 1882–1885. Haddonfield, N. J., *McIlvaine;* Doylestown, Pa., *Paschall.* Quite common on flat benches where hemlocks and spruces have grown.

When the caps are cooked they are sweet, nutty, excellent.

B. Roxa′næ Frost. **Pileus** broadly convex, at first subtomentose, then covered with red hairs in bundles, *yellowish-brown.* **Flesh** yellowish-white. **Tubes** at first *whitish, then light-yellow*, arcuate-adnate or slightly depressed around the stem, the mouths small. **Stem** enlarged toward the base, striate at the apex, yellowish or pale-cinnamon. **Spores** 10x4μ.

Var. *auri′color.* **Pileus** and subequal stem bright-yellow, the tomentum of the pileus yellow.

Pileus 1.5–3 in. broad. **Stem** 1–2 in. long, 3–5 lines thick.

Borders of woods. New England, *Frost;* New York, *Peck.*

Peck, Boleti of the U. S.

B. striæ′pes Secr.—striate stem. **Pileus** convex or plane, soft, silky, *olivaceous, the cuticle rust-color within.* **Flesh** white, yellow next the tubes, sparingly changing to blue. **Tubes** adnate, greenish, their mouths minute, angular, yellow. **Stem** firm, curved, marked with *brownish-black striations*, yellow, velvety and brownish-rufescent at the base. **Spores** 10–13x4μ.

Pine and oak woods. Minnesota, *Johnson.*

I have seen no specimens of this species, which is recorded from but one locality in our country. The character—flesh sparingly changing to blue—is given on the authority of Rev. M. J. Berkeley. *Peck*, Boleti of the U. S.

B. chrysen′teron Fr.—golden within. **Pileus** convex or plane, soft, floccose-squamulose, often cracked in areas, brown or brick-red. **Flesh** *yellow, red beneath the cuticle*, often slightly changing to blue where

431

Boletus. wounded. **Tubes** subadnate; greenish-yellow, *changing to blue where wounded;* their mouths rather large, angular, unequal. **Stem** subequal, rigid, fibrous-striate, red or pale-yellow. **Spores** fusiform, pale-brown, 11–12.5x4–5μ.

Pileus 1–3 in. broad. **Stem** 1–3 in. long, 3–6 lines thick.

Woods and mossy banks.

The species is common and very variable. The color of the pileus may be yellowish-brown, reddish-brown, brick-red, tawny or olivaceous. The subcutaneous reddish tint and the reddish chinks of the cracked pileus are distinguishing features. Wounds of the tubes sometimes become blue then greenish. Authors disagree concerning the edible qualities of this Boletus. Stevenson gives it as edible, but Cordier and Gillet say that it is regarded with suspicion. In one strongly marked form the tubes are decidedly depressed around the stem, in another the flesh is whitish tinged with red. It may be doubted whether these are varieties or distinct species. *Peck,* Boleti of the U. S.

I have found, and eaten plentifully of this species in West Virginia, North Carolina, New Jersey, Pennsylvania, from July until October. I have no hesitancy in recommending it in all of its varieties. Excepting from very young specimens the tubes and stems should be removed. The flesh is sweet, delicate and toothsome.

B. fumo′sipes Pk. **Pileus** convex or nearly plane, minutely tomentose, sometimes minutely rivulose, dark olive-brown. **Flesh** whitish. **Tubes** at first nearly plane, becoming convex with age, their mouths whitish when young, becoming yellowish-brown, changing to bluish-black where bruised. **Stem** equal, solid, smoky-brown, minutely scurfy under a lens. **Spores** purplish-brown, 12.5–15x5–6μ.

Pileus 1–2 in. broad. **Stem** 1–2 in. long, 3–4 lines thick.

Woods. Port Jefferson. July.

This species resembles small dark-colored forms of B. chrysenteron, and this resemblance is still more noticeable in those specimens in which the pileus cracks in areas, for in these the chinks become red as in that species. The different color of the stem and tubes will at once separate these species. *Peck,* 50th Rep. N. Y. State Bot.

B. ru′beus Frost—red. **Pileus** broadly convex, very finely appressed subtomentose, bright brick-red when young, becoming mottled with red and yellow, *yellow under the cuticle*, the thin margin at first inflexed, then horizontal, curved upward when old. **Flesh** pale-yellow, changing to blue where wounded. **Tubes** adnate or slightly depressed around the stem, lemon-yellow and *stuffed when young*, becoming yellow and sometimes red at the mouths. **Stem** small, often flexuous, colored like the pileus, reddish within, white-tomentose at the base. **Spores** 9–12.5 x4–5μ.

Pileus 2–4 in. broad. **Stem** 1–3 in. long, 3–5 lines thick.

Deep woods. Rare. New England, *Frost.*

This is apparently too closely related to B. chrysenteron, and it also resembles B. bicolor. *Peck*, Boleti of the U. S.

B. frater'nus Pk. **Pileus** convex, becoming plane or depressed, slightly tomentose, deep red when young, becoming dull red with age. **Flesh** yellow, slowly changing to greenish-blue where wounded. **Tubes** rather long, becoming ventricose, slightly depressed about the stem, their walls sometimes slightly decurrent, the mouths large, angular or irregular, sometimes compound, bright yellow, quickly changing to blue where wounded. **Stem** short, cespitose, often irregular, solid, subtomentose, slightly velvety at the base, pale reddish-yellow, paler above and below, yellow within, quickly changing to dark green where wounded. **Spores** 12.5x6μ.

Pileus 1–1.5 in. broad. **Stem** 1–1.5 in. long, 3–6 lines thick.

Shaded streets. Auburn, Alabama. July. Underwood.

The species is apparently allied to B. rubeus, but is very distinct by its small size, cespitose habit, color of the flesh of the stem and by the peculiar hues assumed where wounded. When the pileus cracks the chinks become yellow as in B. subtomentosus. The species belongs to the tribe Subtomentosi. *Peck*, Bull. Torrey Bot. Club, Vol. 24, No. 3.

B. subtomento'sus L.—*sub; tomentosus*, downy. **Pileus** convex or nearly plane, soft, dry, *villoso-tomentose, subolivaceous, concolorous beneath the cuticle*, often cracked in areas. **Flesh** white or pallid. **Tubes** adnate or somewhat depressed around the stem, yellow, their mouths large, angular. **Stem** stout, somewhat ribbed-sulcate, scabrous or scurfy with minute dots. **Spores** 10–12.5x4–5μ.

Pileus 1–4 in. broad. **Stem** 1–2.5 in. long, 2–5 lines thick.

Common and variable. The pileus is usually olivaceous or yellow-

Boletus. ish-brown, but it may be reddish-brown or tawny-red. When it cracks the chinks become yellow. The species, as I understand it, may be distinguished from its near relative, B. chrysenteron, by its paler flesh, the clearer yellow tubes not changing to blue where wounded, and by the chinks of the pileus becoming yellow. The species is recorded edible by Cordier, Curtis and Palmer. Gillet says it is only medium in quality. *Peck*, Boleti of the U. S.

Found and eaten in West Virginia, North Carolina, New Jersey, Pennsylvania. Specimens received from Indiana, Minnesota, Alabama. I have not seen any change of color in flesh or tubes. It is common in Woodland Cemetery and Fairmount Park, Philadelphia. If the tubes are not removed the dish is slimy. The B. chrysenteron also makes such a dish when stewed, but fried, and well done, both species are decidedly good.

B. cæspito'sus Pk.—cespitose. **Pileus** broadly convex or nearly plane, sometimes slightly concave by the elevation of the margin, even, brown or blackish-brown, the margin often a little paler or reddish-brown. **Flesh** slightly tinged with red. **Tubes** adnate or slightly decurrent, yellow, their mouths rather large, angular, concolorous. **Stem** short, even, solid, glabrous, tapering upward, brown or reddish-brown. **Spores** oblong-elliptic, 10μ long, 5μ broad.
Pileus 1–2.5 cm. broad. **Stem** 2–2.5 cm. long, 4–6 mm. thick.

Cespitose. Virginia. August. *R. S. Phifer*.

A small species growing in tufts and referable to the tribe Subtomentosi. The tubes retain their bright yellow color in the dried specimens. *Peck*, Bull. Torrey Bot. Club, January 27, 1900.

Edible qualities not stated.

B. spadi'ceus Schaeff.—nut brown. **Pileus** convex or plane, moderately compact, dry, tomentose, opaque, *date-brown*, irregularly cracked. **Flesh** white, unchangeable, brownish-red above. **Tubes** adnate, yellow, their mouths minute, subrotund. **Stem** firm, clavate, even, *woolly-scaled*, yellow or brownish, yellowish-white within. **Spores** 12x4μ.

Pileus 2–4 in. broad.

Woods. New England, *Frost*.

This species is admitted on the authority of Mr. Frost who alone has recorded it in this country. But specimens received from him under this name do not in my opinion belong to it, and its occurrence here is somewhat doubtful. *Peck*, Boleti of the U. S.

In oak woods near Bartram's Garden, West Philadelphia, in 1887–1888, I found several Boleti answering the description, exactly, of B. spadiceus. They proved to be good eating.

B. radi´cans Pers.—*radix*, a root. **Pileus** convex, dry, subtomentose, olivaceous-cinereus, becoming pale-yellowish, the margin thin, involute. **Flesh** pale-yellow, instantly changing to dark blue, taste bitterish. **Tubes** adnate, their mouths large, unequal, lemon-yellow. **Stem** even, *tapering downward and radicating, flocculose with. a reddish bloom*, pale-yellow, becoming naked and dark with a touch.

Pileus 2–3 in. broad. **Stem** 2 in. long, 6 lines thick.

Woods. Ohio, *Morgan*.

Of the American plant Mr. Morgan says that the pileus is quite firm and dry, becomes reddish or brownish-yellow and nearly glabrous, that the flesh is pale-yellow, but that he has not observed any bluish tinge, and that the spores are olive, fusiform, 10–12.5x5μ. Those of the European plant have been described as very pale ocher, almost white, 6μ long, 3μ broad. *Peck*, Boleti of the U. S.

Near Bryn Mawr, Pa. *W. C. Alderson*, 1894.

Several specimens brought to me were eaten. The change in color of flesh was instantaneous upon exposure to the air. Taste strong and raw rather than bitterish. The caps alone were cooked, and dish marked "fine."

B. muta´bilis Morg.—changeable. Jour. Cin. Soc. Nat. Sci., Vol. VII. **Pileus** convex, then plane or depressed, compact, dry, subtomentose, *brown*. **Flesh** bright-yellow, *promptly changing to blue where wounded*. **Tubes** adnate or subdecurrent, their mouths large, angular, unequal, some of them compound, yellow changing to greenish yellow and *quickly becoming blue where wounded*. **Stem** stout, solid, flexuous, subsulcate, yellowish beneath the *brown dot-like scales*, bright yellow within. **Spores** olive, fusiform, 12–13x5μ.

Pileus 2.5–4 in. broad. **Stem** 2–3 in. long, 6 lines thick.

Thick woods. Ohio, *Morgan*.

A shade of yellow sometimes appears beneath the brown of the pileus, and as the plants grow old the pileus becomes blackish, glabrous and

435

Boletus. shining.   The stem increases in thickness above and downward.   *Peck,* Boleti of the U. S.

B. badi′ceps Pk.—*badius,* bay and head.   ([Color] Plate CXVI.) **Pileus** firm, convex or somewhat centrally depressed when mature, dry, velvety, obliquely truncate on the margin, bay-red or dark-maroon color.   **Flesh** white unchangeable, taste and odor mild, sweet, suggestive of molasses.   **Tubes** plane, adnate, white or whitish, becoming dingy with age, the mouths minute.   **Stem** equal or slightly swollen in the middle, radicating, glabrous, solid, brownish.

Pileus 4–8 cm. broad.   **Stem** 4–5 cm. long, 1.5–3 cm. thick.

Oak woods. West Philadelphia, Pa. August and September. *Charles McIlvaine.*

The truncate or beveled margin of the pileus is a striking feature in this species.   It is about 4 mm. broad and as even as if cut with a knife.   Sometimes the surface of the stem ruptures transversely just below the top, the liberated shreds above curling upward against the tubes and those below curving outward and downward.   In mature plants brownish spots appear in the flesh of the pileus.   "When cooked it is of high flavor and tender as kidney," *C. McIlvaine.   Peck,* Bull. Torrey Bot. Club, January 27, 1900.

LACERI′PEDES—lacerated stem.

Stem elongated, coarsely pitted or deeply and lacunosely reticulated in small hollows, the ridges somewhat intumescent in wet weather and more or less lacerated, giving a rough or shaggy appearance to the stem.

The species of this tribe are few, very closely allied and so far as known are peculiar to this country.

Pileus viscid..I
Pileus dry.......................................B. Russelli
1. Stem red in the depressions, tubes tinged with green...B. Morgani
1. Stem pale-yellow, tubes not greenish.................B. Betula
Peck, Boleti of the U. S.

B. Rus′selli Frost—Russell's Boletus. ([Color] Plate CXVIII, fig. 2.) **Pileus** thick, hemispherical or convex, *dry, covered with downy scales or bundles of red hairs,* yellowish beneath the tomentum, often cracked in

436

areas. **Flesh** yellowish, unchangeable. **Tubes** subadnate, often de- Boletus. pressed around the stem, rather large, dingy-yellow or yellowish-green. **Stem** very long, equal or tapering upward, roughened by the lacerated margins of the reticular depressions, *red or brownish-red*. **Spores** olive-brown, 18–22x8–10μ.

Pileus 1.5–4 in. broad. **Stem** 3–7 in. long, 3–6 lines thick.

This is distinguished from the other species by the dry squamulose pileus and the color of the stem. The latter is sometimes curved at the base. *Peck*, Boleti of the U. S.

B. Russelli occurs in the West Virginia mountains, where I found and ate it in August, 1883. Though solitary in its method of growth, it is frequent in many parts of Pennsylvania, among leaves in mixed woods. August to October.

Taste when raw, sweet, mild. Cooked it is rather soft, tasty. Tubes and stem should be removed.

B. Mor'gani Pk. **Pileus** convex, soft, *glabrous viscid*, red or yellow, or red fading to yellow on the margin. **Flesh** whitish tinged with red and yellow, unchangeable. **Tubes** convex, depressed around the stem, rather long and large, bright-yellow becoming greenish-yellow. **Stem** elongated, tapering upward, pitted with long, narrow depressions, *yellow, red in the depressions*, colored within like the flesh of the pileus. **Spores** olive-brown, 18–22μ long, about half as broad.

Pileus 1.5–2.5 in. broad. **Stem** 3–5 in. long, 3–6 lines thick.

Rocky hillsides in woods of deciduous trees. Kentucky, *Morgan*.

In wet weather the anastomosing ridges of the stem swell and become broadly winged, thereby giving the stem a peculiar lacerated appearance. The glabrous viscid pileus and the coloration of the stem distinguish the species. *Peck*, Boleti of the U. S.

B. Morgani is found in like localities with B. Russelli. Excepting in its smooth, viscid cap and whitish flesh, it closely resembles the latter. The ridges in the stems of both species swell when moist.

Its edible qualities are the same as B. Russelli.

B. Be'tula Schw.—birch. **Pileus** convex, viscose and shining in wet weather, tessellately cracked and reticulated, orange-fawn color, rather small. **Flesh** yellowish-white. **Tubes** separating, rather large, *yellow*, almost like those of B. subtomentosus but *not greenish*. **Stem** long,

437

Boletus. *attenuated downward*, everywhere covered with a deciduous reticulated bark two lines high and separating like the bark of birches, *pale-yellow without and within.*

Pileus 1.5 in. broad. **Stem** 5–6 in. long.

Ligneous earth. North Carolina, *Schweinitz, Curtis;* Pennsylvania, *Schweinitz. Peck,* Boleti of the U. S.

During several seasons I found B. Betula in Woodland Cemetery, Philadelphia.

Edible qualities good.

CALO′PODES. *Gr.*—beautiful; *Gr.*—feet.

Stem stout, at first bulbous, typically venose-reticulated with veins. **Tubes** adnate, their mouths not reddish.

The reticulate stem and adnate tubes of one color distinguish the species of this tribe. In the Luridi the mouths of the tubes are differently colored, and in the closely related Edules the tubes are more or less depressed around the stem or sub-free, and their pores are commonly stuffed when young. Fries did not admit species with whitish tubes into this tribe, but we have done so in those cases in which this was the only character to exclude them.

Tubes yellow or yellowish.....................................I
Tubes white or whitish, at least when young...................7
1. Tubes or flesh changing to blue where wounded..............2
1. Tubes or flesh not changing to blue where wounded..........5
  2. Pileus red, at least when young..........................3
  2. Pileus some other color 4
3. Stem red ....B. Peckii
3. Stem yellow or reddish only at the base.............B. speciosus
  4. Tubes angular, pileus olivaceous.................B. calopus
  4. Tubes rotund, pileus not olivaceous..............B. pachypus
5. Pileus viscid.....................................B. Curtisii
5. Pileus pulverulent, stems cespitose..................B. retipes
5. Pileus neither viscid nor pulverulent........................6
  6. Stem yellow.....................................B. ornatipes
  6. Stem brown.....................................B. modestus
  6. Stem yellowish-white ....B. rimosellus
7. Pileus some shade of red....................................8

438

7. Pileus some shade of brown or gray........................9 Boletus.
8. Stem pallid or yellowish......................B. rubignosus
8. Stem dark-brown...........................B. ferrugineus
9. Pileus pale-brown, stem flexuous................B. flexuosipes
9. Pileus gray or grayish-black, stem straight...........B. griseus
Peck, Boleti of the U. S.

B. specio′sus Frost—handsome. **Pileus** at first very thick, subglobose, compact, then softer, convex, glabrous or nearly so, red. **Flesh** pale-yellow or bright lemon-yellow, changing to blue where wounded. **Tubes** adnate, small, subrotund, plane or but slightly depressed around the stem, bright lemon-yellow, becoming dingy-yellow with age, changing to blue where wounded. **Stem** stout, subequal or somewhat bulbous, reticulated, *bright lemon-yellow without and within*, sometimes reddish at the base. **Spores** oblong-fusiform, pale ochraceous-brown, 10–12.5x4–5µ.

Pileus 3–7 in. broad. Stem 2–4 in. long, 10–24 lines thick.
Thin woods. New England, *Frost;* New York, *Peck.*
This is a very beautiful Boletus. When young the whole plant except the surface of the pileus is of a vivid lemon-yellow color. Wounds quickly change to green, then to blue. The color of the pileus approaches closely to solferino. *Peck*, Boleti of the U. S.
Caps of specimens found in mixed woods at Mt. Gretna, Pa., were minutely areolate when old. Stems yellow at top and with purplish red over the bright yellow toward the bulbous base, solid, bright yellow within.
Stems and caps are edible and rank high in flavor and texture.

B. illu′dens Pk.—deceiving. ([Color] Plate CXVIII, fig. 3.) **Pileus** convex, dry, subglabrous, yellowish-brown or grayish-brown, sometimes tinged with red, especially in the center. **Flesh** pallid or yellowish. **Tubes** bright yellow, plane or somewhat convex when old, adnate, their mouths angular or subrotund, often larger near the stem. **Stem** nearly equal, sometimes abruptly pointed at the base, glabrous, pallid or yellowish, coarsely reticulated either wholly or at the top only. **Spores** oblong or subfusiform, yellowish-brown tinged with green, 11–12.5x4–5µ.
Pileus 1.5–3 in. broad. **Stem** 1.5–2.5 in. long, 3–5 lines thick.

439

Boletus. Woods and copses. Port Jefferson. July. *Peck*, 50th Rep. N. Y. State Bot.

Found in plenty at Mt. Gretna, Pa., September, 1898. On ground and old stumps in mixed woods. Identified by Professor Peck.

Taste and smell pleasant. Cooked as egg-plant it is one of the best. Remove tubes.

B. Peck′ii Frost—after C. H. Peck. **Pileus** convex, firm, dry, sub-glabrous, *red, fading to yellowish-red or buff-brown* with age, the margin usually retaining its red color longer than the disk. **Tubes** adnate or slightly decurrent, nearly plane, yellow, changing to blue where wounded. **Stem** equal or subventricose, reticulated, *red, yellow at the top*. **Spores** oblong, pale ochraceous-brown, 9–12x4–5μ.

Var. *læ′vipes*. **Stem** reticulated above, even below.

Pileus 2–3 in. broad. **Stem** 2–3 in. long, 3–6 lines thick.

Woods of frondose trees. New York, *Peck*. *Peck*, Boleti of the U. S.

B. cal′opus Fr. *Gr.*—beautiful; *Gr.*—foot. **Pileus** globose, then convex, unpolished, *subtomentose, olivaceous*. **Flesh** pallid, slightly changing to blue when wounded. **Tubes** adnate, their mouths minute, angular, yellow. **Stem** firm, conical, then elongated and subequal, reticulated, *wholly scarlet or at the apex only*, sometimes colored like the pileus toward the base. **Spores** fusiform, yellowish-brown, 7–8x 3–4μ.

Pileus 2–3 in. broad. **Stem** longer than the diameter of the pileus.

Woods. North Carolina, *Schweinitz, Curtis;* Pennsylvania, *Schweinitz;* New England, *Sprague, Bennett*. *Peck*, Boleti of the U. S.

B. orna′tipes Pk.—ornate-stem. (Boletus retipes, Rep. 23.) **Pileus** convex, firm, dry, glabrous or very minutely tomentose, *grayish-brown or yellowish-brown*. **Flesh** yellow or pale-yellow. **Tubes** adnate, plane, or concave, rarely convex, the mouths small or medium size, clear-yellow. **Stem** firm, subequal, distinctly and beautifully reticulated, yellow without and within. **Spores** oblong, *ochraceous-brown*, 12–16x4–5μ.

Pileus 2–5 in. broad. **Stem** 2–4 in. long, 4–6 lines thick.

Thin woods and open places. New York, *Peck*.

The color of the tubes becomes darker with age, but it does not change to blue where wounded. The species is related to the next following one with which it has sometimes been confused, but from which Boletus. it is clearly distinct. The color of the spores is quite dark and approaches snuff-brown. *Peck*, Boleti of the U. S.

Edible. Good.

B. re′tipes B. and C.—reticulate stem. **Pileus** convex, dry, *powdered with yellow*, sometimes rivulose or cracked in areas. **Tubes** adnate, yellow. **Stem** subequal, *cespitose*, reticulate to the base, *pulverulent below*. **Spores** *greenish-ochraceous*, 12–15x4–5μ.

Pileus 1.5–2 in. broad. **Stem** 2 in. long, 3–6 lines thick.

The tufted mode of growth, pulverulent pileus and paler-colored spores separate this species from the preceding one. *Peck*, Boleti of the U. S.

West Virginia, 1882–1885. Mt. Gretna, Pa. ; New Jersey, *McIlvaine*.

The caps, alone, of this species, are desirable, the stems not cooking well. Its way of bunching itself gratifies the collector, as do its flavor and quality.

B. pa′chypus Fr. *Gr.*—thick-footed. **Pileus** convex, subtomentose, brownish or pale tan-color. **Flesh** thick, whitish, changing slightly to blue. **Tubes** rather long, *somewhat depressed around the stem*, their mouths round, pale-yellow, at length tinged with green. **Stem** thick, firm, reticulated, at first ovate-bulbous, then elongated, equal, *variegated with red and pale-yellow*. **Spores** large, *ovate*, pale yellowish-ochraceous, 12.5–14x5–6μ.

Pileus 4–8 in. broad. **Stem** 2–4 in. long.

Woods, either of pine or beech.

This species is noted for its thick, stout stem, which sometimes attains a diameter of more than two inches. It approaches the Edules in habit, but according to Gillet it is poisonous, or at least to be suspected, has a penetrating unpleasant odor and a somewhat nauseous flavor. He also describes the pores as at first whitish. The stem is sometimes intensely blood-red. *Peck*, Boleti of the U. S.

A common species in West Virginia mountains, 1881–1885, in beech groves. August to frost. It is rare in the pines of New Jersey, though I have found it there. Like B. felleus, its size and attractiveness induce the finder to over and over again try cooking it, hoping the discovery of a successful way to rid it of its unpleasantness. I have never succeeded. It is not poisonous.

Boletus. **B. rimosel'lus** Pk.—cracked. **Pileus** broadly convex, flat or irregular, glabrous, *tessellately cracked*, dark-brown. **Flesh** whitish. **Tubes**
adnate or sinuately decurrent, somewhat depressed around the stem,
pale-yellow, becoming *darker or brownish* with age. **Stem** tapering upward, broadly reticulated with brown veins, *yellowish-white*. **Spores**
fusiform, 15–17.5x5–6μ.

 Pileus 3–5 in. broad. **Stem** 3–4 in. long, 6–9 lines thick.
 Mixed woods. North Carolina, *C. J. Curtis*.
 I have described this species from the notes and a single dried specimen sent me by Mr. Curtis. More extended observation may require
some modification of the description. The color of the spores is described as brown. They are remarkable for their size. *Peck*, Boleti
of the U. S.

 B. modes'tus Pk.—modest. **Pileus** convex or nearly plane, often
irregular, firm, dry, very minutely tomentose, *yellowish-brown*. **Flesh**
gray or pinkish-gray. **Tubes** nearly plane, adnate or subdecurrent,
the mouths angular, pale-ochraceous. **Stem** equal, reticulated, brown.
Spores elliptical, 10x5μ.

 Pileus 2–3 in. broad. **Stem** 1–2 in. long, 2–4 lines thick.
 Grassy ground in thin woods. New York, *Peck*.
 Miss Banning finds in Maryland what appears to be a form of this
species in which the part of the hymenium near the stem consists of
lamellæ, the rest of tubes. The species needs further investigation.
Peck, Boleti of the U. S.

 B. Cur'tisii Berk.—after Dr. Curtis. **Pileus** hemispherical or convex, *viscose, golden-yellow*. **Tubes** depressed around the stem, nearly
free, their mouths umber, at length tawny. **Stem** slender, attenuated
upward, polished, reticulated, straw-colored. **Spores** ferruginous, subelliptical, slightly attenuated at each end.

 Pileus 1 in. or more broad. **Stem** 2 in. long, 2–3 lines thick.
 Pine woods. North and South Carolina, *Curtis*.
 In the original description the stem of this species is said to be hol-
low. *Peck*, Boleti of the U. S.

 B. gri'seus Frost—gray. **Pileus** broadly convex, firm, dry, subglabrous, *gray or grayish-black*. **Flesh** whitish or gray. **Tubes** adnate

or slightly depressed around the stem, nearly plane, their mouths small, Boletus. subrotund, *white or whitish*. **Stem** equal or slightly tapering upward, distinctly reticulated, *whitish or yellowish*, sometimes reddish toward the base. **Spores** ochraceous-brown, 10–14x4–5μ.

Pileus 2–4 in. broad. **Stem** 2–4 in. long, 3–6 lines thick.
Thin woods and open places. New York, *Peck*.
Peck, Boleti of the U. S.

B. flexuos'ipes Pk.—flexuous stem. **Pileus** convex or plane, even, subtomentose, *pale-brown*. **Flesh** white, unchangeable, the cuticle separable. **Tubes** long, convex, *decurrent*, white or whitish, becoming brownish with age. **Stem** *flexuous*, solid, reticulated, whitish or pallid, *changing to brown where bruised*. **Spores** 7.5–10x4μ.

Pileus 3–4 in. broad. **Stem** 4–6 in. long, 8–15 lines thick.
Mixed woods. North Carolina, *C. J. Curtis*. *Peck*, Boleti of the U. S.

B. ferrugi'neus Frost—rust color. **Pileus** convex, soft, subto-mentose, dark reddish-brown. **Flesh** white, unchangeable. **Tubes** generally adnate, dingy-white, their mouths stained brown by the spores. **Stem** short, reticulated, dark-brown. **Spores** 10–13x6μ.

Pileus 3–6 in. broad.
Borders of woods. New England, *Frost*. *Peck*, Boleti of the U. S.
Alabama, 1897.

B. rubigino'sus Fr.—rusty. **Pileus** convex, soft, pubescent, soon bare, *brownish-rust color*. **Flesh** subspongy, white, unchangeable. **Tubes** *adnate*, their mouths unequal, white. **Stem** firm, stout, reticulated, at first *whitish or pallid*, then *yellowish*, subcinereous or yellowish-olivaceous where touched.

Pileus 2–4 in. broad. **Stem** 2–3 in. long, 1 in. thick.
Woods. North Carolina, *Curtis*.
Although apparently distinct, this and the two preceding species are not sufficiently well known. *Peck*, Boleti of the U. S.

B. tabaci'nus Und. **Pileus** fleshy, convex or nearly plane, subglabrous, often cracked in areas, tawny-brown. **Flesh** at maturity soft and similarly colored. **Tubes** concave or nearly plane, depressed around

Boletus. the stem, their mouths small, angular, colored like the pileus. **Stem** subequal, solid, reticulated, concolorous. **Spores** oblong or subfusiform, 12.5–14x5μ. **Pileus** 2.5–5 in. broad. **Stem** 1.5–3 in. long, 6–10 lines thick.

Along road-sides. Alabama. May. *Underwood.*

The species is referable to the section Calopodes, but the tubes are more or less depressed about the stem. *Peck*, Bull. Torrey Bot. Club, Vol. 23, No. 10.

EDU′LES—*edulis*, edible.

Tubes subfree, rounded-depressed around the stem, their mouths not at first reddish, but commonly white-stuffed. **Stem** stout, bulbous as in the Luridi but not, with a few exceptions, reticulate nor dotted with pointed scales nor red. **Flesh** scarcely changeable. **Taste** pleasant.

This tribe is not sharply limited but partakes to some extent of the characters of Calopodes and Luridi. From the former its nearly free and at first white-stuffed tubes and its generally even stem separate it, from the latter its tubes with concolorous mouths or at least with mouths not red or reddish when young will distinguish it. The species are generally of large or medium size and noted for their esculent qualities.

Stem brownish-lilac or chocolate color...........................1
Stem some other color..2
1. Stem reticulated.................................B. separans
1. Stem not reticulated, furfuraceous..................B. eximius
  2. Pileus viscid...................................B. limatulus
  2. Pileus not viscid...3
3. Tubes yellow with no tinge of green........................4
3. Tubes tinged with green or becoming green where bruised.......6
  4. Pileus whitish...............................B. æstivalis
  4. Pileus not whitish...5
5. Stem glabrous.................................B. affinis
5. Stem pubescent.................................B. impolitus
  6. Pileus becoming white-spotted where bruised....... B. leprosus
  6. Pileus not becoming spotted..............................7
7. Pileus glabrous.................................B. edulis
7. Pileus not glabrous...8
  8. Stem reticulated, whitish or pallid..................B. variipes
  8. Stem even, brownish-red.........................B. decorus
Peck, Boleti of the U. S.

B. sep′arans Pk. ([Color] Plate CXVIII, fig. 1.) **Pileus** convex, thick, glabrous, subshining, often pitted, pitted or corrugated, brownish-red or dull-lilac, sometimes fading to yellowish on the margin. **Flesh** white, unchangeable. **Tubes** at first nearly plane, adnate, white and stuffed, then convex, depressed around the stem, ochraceous-yellow or brownish-yellow and sometimes separating from the stem by the expansion of the pileus. **Stem** equal or slightly tapering upward, reticulated either wholly or in the upper part only, colored like the pileus or a little paler, sometimes slightly furfuraceous. **Spores** subfusiform, brownish-ochraceous, 12–15x5–6μ.

Pileus 3–6 in. broad. Stem 2–4 in. long, 6–12 lines thick.

Thin grassy woods. New York, *Peck*. *Peck*, Boleti of the U. S.

West Virginia. September, 1881. New Jersey and Pennsylvania. October, 1887, *McIlvaine*. Indiana, October, 1898. *Dr. J. R. Weist, H. I. Miller*.

One of the handsomest of Boleti. It varies greatly in size and color, but traces of purple or lilac are always detectable. The reticulations upon the stem are often obscure, especially in young specimens.

It is pleasant when raw, and quite equal to any Boletus when cooked.

B. edu′lis Bull.—*edulis*, edible. ([Color] Plate CXVIII, fig. 5.)

Pileus convex or nearly plane, *glabrous*, moist, at first compact, then soft, variable in color, grayish-red, brownish-red or tawny-brown, often paler on the margin. **Flesh** white or yellowish, reddish beneath the cuticle. **Tubes** convex, nearly free, long, minute, round, *white, then yellow and greenish*. **Stem** short or long, straight or flexuous, subequal or bulbous, stout, more or less reticulate, especially above, whitish, pallid or brownish. **Spores** oblong-fusiform, 12–15x4–5μ.

(Plate CXIX.)

1, BOLETUS EDULIS, VAR. CLAVIPES.
2, 3, BOLETUS EDULIS.

Var. *cla′vipes*. Plate CXIX. **Stem** tapering upward from an enlarged base, everywhere reticulated.

Pileus 4–6 in. broad. Stem 2–6 in. long, 6–18 lines thick.

Woods and open places. Not rare. *Peck*, Boleti of the U. S.

Boletus. Indiana, *H. I. Miller, Dr. J. R. Weist;* New Jersey, Pennsylvania, West Virginia, *McIlvaine.*

Some species of fungi appear to have that prize of Fairyland—the Wishing Cap—and by its power be able to take on any form they please. Boletus edulis is one of them. Its variableness is puzzling. It is eaten everywhere where found and is a favorite. Carefully sliced, dried and kept where safe from mold it may be prepared for the table at any season.

B. edulis Bull.—Var. *clavipes* Pk. (Plate CXIX, fig. 1, p. 445.) **Pileus** fleshy, convex, glabrous, grayish-red, bay-red or chestnut-color. **Flesh** white, unchangeable. **Tubes** at first concave or nearly plane, white and stuffed, then convex, slightly depressed around the stem, ochraceous yellow. **Stem** mostly obclavate (inversely club-shaped) and reticulate to the base. **Spores** oblong-fusiform, 12–15x4–5µ.

The club-stemmed Boletus is so closely related to the edible Boletus and so closely connected by the intermediate forms that it seems to be only a variety of it, but one worthy of illustration. It differs in the more uniform color of the cap, in having the tubes less depressed around the stem and less tinted with green when mature, and in having the stem more club-shape and commonly reticulated to the base. The lower reticulations are usually coarser but less permanent than the upper. The cap is more highly colored when young and is apt to become paler with age, but the margin does not become paler than the central part, as it so often does in the edible Boletus. Individuals sometimes occur in which the stem is nearly cylindric and reticulated only on the upper part. These connect so closely with the edible Boletus that we have considered this to be a mere variety of it. In size and in edible qualities it is very similar to that species. *Peck*, 51st Rep. N. Y. State Bot.

Same in quality as B. edulis.

B. vari'ipes Pk.—variable stem. **Pileus** convex or nearly plane, thick, soft, dry, *scaly, pointed scaly or minutely tomentose*, grayish or pale grayish-brown, sometimes tinged with yellow or ochraceous. **Flesh** white, unchangeable. **Tubes** convex or nearly plane, slightly depressed around the stem, at first white, then greenish-yellow, their mouths small, subrotund, *ochraceous*, stuffed when young. **Stem** firm, reticulated, whitish or pallid. **Spores** oblong-fusiform, ochraceous-brown **Boletus.** tinged with green, 12–15x5µ. *Peck*, Boleti of the U. S.

Mt. Gretna, Pa. August, 1898. Stem slightly reticulated at top, indistinctly striate below. ˙Smell and taste strong, like B. felleus, but sweetish, not bitter. When tubes are removed and cap fried it is excellent.

Var. *al'bipes*. **Stem** whitish, wholly reticulated, the reticulations coarser near the base. *Peck*, Boleti of the U. S.

Mt. Gretna, Pa. August, 1898. Taste slightly acrid, smell slight. Excellent.

Var. *pallid'ipes*. **Stem** pallid, slightly furfuraceous, even or obscurely reticulated toward the base, distinctly reticulated above. *Peck*, Boleti of the U. S.

Satiny, shining. Taste slightly acrid, smell slight. Excellent.

Var. *tenu'ipes*. **Stem** slender, elongated. *Peck*, Boleti of the U. S.

Mt. Gretna, Pa. August, 1898, on decaying chestnut stump and on ground. Excellent. *McIlvaine.*

This species, with its varieties, grows in mixed woods, the density of which has much to do with its general appearance. Individuals growing where the sun plays upon them, show the reticulations plainer than those maturing in the shade. The tubes should be removed before cooking. The caps are best fried.

B. exi'mius Pk.—select. **Pileus** at first very compact, subglobose or hemispherical, subpruinose, *purplish-brown or chocolate color*, sometimes with a faint tinge of lilac, becoming convex, soft, smoky-red or pale-chestnut. **Flesh** grayish or reddish-white. **Tubes** at first concave or nearly plane, stuffed, colored nearly like the pileus, becoming paler with age and depressed around the stem, their mouths minute, rotund. **Stem** stout, generally short, equal or tapering upward, abruptly narrowed at the base, *minutely branny*, colored like or a little paler than the pileus, purplish-gray within. **Spores** subferruginous, 12.5–15x5–6µ.

Pileus 3–10 in. broad. **Stem** 2–4 in. long, 6–12 lines thick.

Woods and their borders. New England, *Frost;* New York, *Peck. Peck*, Boleti of the U. S.

In mixed woods and in new clearings near Bartram's Garden, Philadelphia, Pa. *McIlvaine.*

Boletus.   A patch of it is treasure trove.

B. lepro'sus Pk.—leprous. **Pileus** very convex, glabrous, soft like kid, cinereous-yellowish-drab or pale-brown, *slowly changing to whitish where bruised*, the cuticle separable. **Flesh** *white, changing to yellowish.* **Tubes** yellow or brownish-yellow, *changing to greenish where wounded,* plane, depressed around the stem, short, small, stuffed when young. **Stem** solid, enlarged at the top, *lemon-yellow.* **Spores** oblong-fusiform, 12.5–15x5μ.

Pileus 4–6 in. broad. Stem 2 in. long, 1 in. thick.

Mixed woods. North Carolina, *C. J. Curtis.*

This plant is remarkable for the whitish or leprous spots which the pileus assumes, even from being handled, and for the change in the color of the flesh and tubes. The stem is very thick at the top but tapers downward. *Peck*, Boleti of the U. S.

B. affi'nis Pk.—related. **Pileus** convex above or nearly plane, subglabrous, reddish-brown or chestnut color fading to tawny or dingy-ochraceous with age. **Flesh** white. **Tubes** plane or convex, adnate or slightly depressed around the stem, at first white and stuffed, then glaucous-yellow or subochraceous, changing to rusty-ochraceous where wounded. **Stem** subequal, even, glabrous, colored like or paler than the pileus. **Spores** rusty-ochraceous, 9–12x4–5μ.

(Plate CXX.)

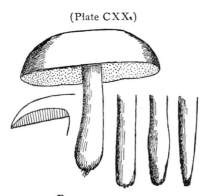

BOLETUS AFFINIS.

The Related boletus belongs to the tribe of Boleti known as Edules because of their especially esculent character, but it differs from the general character of the tribe in having its tubes not at all or but slightly shortened around the stem and in its stem not being thickened or bulbous at the base. The species is quite variable in the color of the cap, which is generally darker in young plants, paler in old ones. It may be brown, reddish-brown or blackish-brown when young, but is more or less tinged with tawny or ochraceous when old. It is smooth and even or minutely tomentose and sometimes slightly rugose. In wet weather

448

the margin of the cap sometimes curves upward, giving a very convex Boletus. surface to the tubes. Sometimes the wounded flesh slowly assumes a yellowish hue. The peculiar rusty-ochraceous hue of the spores is also seen sometimes in the tubes of old specimens. As in many species, the flesh of old plants is more soft than that of young ones. The stem is quite variable and is often narrowed downward. It is sometimes very obscurely reticulated at the top.

The cap is generally 2–4 in. broad, the stem 1.5–3 in. long, 4–8 lines thick. The plants are found in thin woods or in bushy places in July and August.

Var. *maculo'sus* Pk. differs from the type simply in having a few yellowish spots scattered over the cap.

While not as high flavored as some Boleti this is, nevertheless, a fairly good and perfectly safe one. *Peck*, 49th Rep. N. Y. State Bot.

Very open timber in Woodlands Cemetery, Philadelphia. August, 1898. *McIlvaine*.

A solitary species which does not appear to be plentiful. The whole fungus is edible, but the stems and tubes are of different texture from the caps and do not cook well with them.

B. æstiva'lis Fr.—pertaining to summer. **Pileus** convex or nearly plane, even, *glabrous, whitish,* granulose in dry weather. **Flesh** yellow below, white above. **Tubes** nearly free, the mouths minute, equal, yellow. **Stem** very thick, bulbous, even, glabrous, pale yellow, reddish within at the base. **Spores** elongated-oval, greenish-brown, rather dark, 11x4–5μ.

Pileus 4–6 in. broad. **Stem** 4–5 in. long.

Woods and woodland pastures. Minnesota, *Johnson;* California, *H. and M.*

A large species, recorded as edible and said to be pleasant and delicate in flavor. I have seen no specimens of this. *Peck*, Boleti of the United States.

West Virginia mountains, 1882, Haddonfield, N. J., 1894, *McIlvaine*, on grassy margin of woods.

The flesh is sweet, nutty. Remove stems and tubes when old.

B. impoli'tus Fr.—unpolished. **Pileus** convex, dilated, *flocculose,* at length grained in lines, unpolished, *tawny-brown.* **Flesh** white or

449

Boletus. whitish, unchangeable, yellowish under the cuticle. **Tubes** free, their mouths minute, yellow. **Stem** stout, subbulbous, even, *pubescent, pale-yellow*, sometimes with a reddish zone near the top. **Spores** oval or fusiform, pale greenish-brown, 7.5–10x5μ.

Pileus 4–6 in. broad. **Stem** 2 in. long.

Oak woods. California, *Harkness and Moore*.

This species is recorded as edible and said to be among the most delicious. It is evidently rare in this country. According to Quelet the spores are ellipsoid, papillate, 15–18μ long. *Peck*, Boleti of the U. S.

Near Bartram's Garden, West Philadelphia, Pa., 1885. Thin mixed woods. *McIlvaine*.

That this species is edible and delicious is vouched for by many. I can add my own pleasurable experience.

B. deco'rus Frost.—decorous. **Pileus** convex, rather firm, tomentose, brownish tinged with red, the margin often darker colored. **Flesh** white, unchangeable. **Tubes** becoming free, yellow, *changing to green where wounded.* **Stem** bulbous, minutely branny, *brownish-red*, the bulb sometimes white and attenuated at the base. **Spores** 13x5μ.

Rich woods. New England, *Frost*. *Peck*, Boleti of the U. S.

Leominster, Mass., *C. F. Nixon*, August, 1897; Woodland Cemetery, Philadelphia, Pa., August, 1897, *McIlvaine*.

Cap 2–3 in. broad. **Stem** 2–2½ in. high, but variable in size. Its edible qualities are excellent.

B. lima'tulus Frost—polished. **Pileus** nearly flat, thin, glabrous, *viscid when moist*, somewhat polished and shining when dry, rich yellowish-brown. **Flesh** *reddish in the pileus*, darker in the stem. **Tubes** depressed around the stem, greenish-yellow, their mouths yellowish-brown. **Stem** small, subbulbous, colored like the pileus. **Spores** 12–15x4–5μ.

Pileus 1–2.5 in. broad.

Woods. New England, *Frost*.

By the differently-colored tube mouths, this species approaches those of the next following tribe, but it is placed here because these are not red or reddish. *Peck*, Boleti of the U. S.

B. au'ripes Pk.—yellow-stem. **Pileus** convex, subglabrous, yellowish-brown, sometimes cracking in areas when old. **Flesh** yellow, fading to whitish with age. **Tubes** nearly plane, their mouths small, subro- Boletus. tund, at first stuffed, yellow. **Stem** nearly equal, solid, even or slightly reticulated at the top, bright yellow, a little paler within. **Spores** ochraceous-brown tinged with green, 12x5μ.

Pileus 3–6 in. broad.   **Stem** 3–5 in. long, 8–12 lines thick.

Under mountain laurel, *Kalmia latifolia*.   Port Jefferson.   July.

The whole plant, except the upper surface of the pileus, is of a beautiful yellow color.   The stem is sometimes more highly colored than the tubes.   The species is referable to the tribe Edules.   *Peck*, 50th Rep. N. Y. State Bot.

Mt. Gretna, Pa.   August, September, 1898.   *McIlvaine*.

In mixed woods in which *Kalmia latifolia* is plentiful.   The specimens found were in its vicinity.   The caps are excellent.

B. leptoceph′alus Pk. *Gr.*—thin; *Gr.*—head. **Pileus** thin, broadly convex or nearly plane, dry, minutely cracked, especially near the margin, light tawny-brown, sometimes tinged with reddish-brown. **Flesh** yellowish-white, taste at first mild, then slightly acrid. **Tubes** subventricose, depressed about the stem, nearly free, dingy olive-yellow, the mouths small, subrotund. **Stem** nearly equal, enlarged at the top, solid, glabrous or slightly pruinose-mealy, reticulated above, colored like the pileus, white within, with a white mycelium at the base. **Spores** greenish-olivaceous, fusiform, 12.5–17.5μ long, 5–6μ broad.

Pileus 10–12.5 cm. broad.   **Stem** 10–12.5 cm. long, 1.2–1.6 cm. thick.

Dry, open woods.   July.   *Earle*.

The reticulation of the upper part of the stem appears to be formed by the decurrent walls of the tubes.   The species belongs to the tribe Edules.   *Peck*, Bull. Torr. Bot. Club, Vol. 25.

Edible.

B. fra′grans Vitt.—fragrant. Fasciculate or solitary. **Pileus** 1–4 in. across, convex, dark-brown or umber-brown, often wavy, slightly tomentose, margin incurved. **Flesh** very thick, yellowish, sometimes unchangeable, at others changing to green or blue, and finally becoming reddish when broken. **Tubes** shortened around the stem and almost free, ½ in. or more long, openings small, roundish, yellow then greenish. **Stem** at first stout, ovate, usually tapering at the base, then length-

Boletus. ening and becoming thinner upward, even, variegated with yellow and red, solid. **Spores** pale-olive, elongato-fusiform, 10–12x4μ.

In woods, under oaks, etc. **Pileus** bronze-brown, sometimes with purple shades. Often grows in dense clusters, and in this particular differing from any other British species. Very good for eating. *Massee.*

Haddonfield, N. J. Oak woods. August to September, 1894. Mt. Gretna, Pa., 1898. *McIlvaine.*

Solitary. A handsome valuable species which appears to be rare in the United States. Shade a beautiful bronze. Cap 3–4 in. across. A dozen or more individuals were found and eaten. Excellent.

B. frustulo′sus Pk.—*frustulum*, a small bit. **Pileus** thick, convex or nearly plane, subglabrous, cracked in areas, white or whitish. **Flesh** whitish. **Tubes** equal to or a little longer than the thickness of the flesh of the pileus, depressed about the stem, whitish, becoming pale brown. **Stem** equal, solid, whitish, reticulated above. **Spores** 15–17x5–6μ.

Pileus 3–5 in. broad. **Stem** 1–2 in. long, 6–10 lines thick.

Open grounds and clay banks. Ocean Springs, Mississippi and Akron, Alabama. May and June. *Underwood.*

The deeply cracked surface of the pileus is the most notable feature of this species. This sometimes is seen even in quite young plants. The cracked areas are quite unequal in size. The deep chinks with sloping sides cause them to appear like frusta of polygonal pyramids. In some specimens the reticulations of the stem extend nearly or quite to its base, and make the place of the species ambiguous between the Calopodes and Edules. *Peck*, Bull. Torrey Bot. Club, Vol. 24, No. 3.

Mt. Gretna, Pa., September, 1898, on soil over red conglomerate and on road-sides. *McIlvaine.*

The deep cracks in the cap readily distinguish this species. After rains the caps are frequently slightly dished and widely cracked at margin. The exposed flesh dries with a fine silky gloss. The caps are excellent. The tubes and stem should be removed.

B. cras′sipes Pk.—thick-footed. ([Color] Plate CXVI, fig. 5.) **Pileus** convex or centrally depressed, firm, dry, velvety, brown tinged with yellow, the wavy or lobed involute margin extending beyond the tubes. **Flesh** lemon-yellow, unchangeable, taste sweet, odor like that of yeast. **Tubes** rather short, depressed around the stem, almost free, Boletus. yellowish mottled with brown, the mouths minute, stuffed when young. **Stem** stout, thick, sometimes swollen in the middle and sometimes bulbous, beautifully reticulated but the reticulations sometimes disappearing with age, orange-yellow tinged with brown. **Flesh** of a brighter yellow than that of the pileus.

Pileus 5–10 cm. broad. **Stem** 6–8 cm. long, 2.5–3.5 cm. thick.

Oak woods. Mt. Gretna, Pa. August and September. *McIlvaine.*

The thick, beautifully reticulated stem, the deep velvety brown color of the pileus and the yellow color of the flesh serve to distinguish this species. *Peck,* Bull. Torr. Bot. Club, Vol. 27, January, 1900.

It is one of the best edible mushrooms. I have also found it in New Jersey.

LU'RIDI.

Stratum of tubes rounded toward the stem and free, their mouths at first closed and red. **Pileus** compact, then soft, cushion-shaped, the flesh juicy, changeable. **Stem** stout, at first short, bulbiform, then elongated and subequal, subreticulated or dotted.

Growing especially in frondose woods. Very poisonous.

In this tribe the tubes and their mouths are differently colored, the latter being red or some shade of red. By this character the species are easily distinguished from those of other tribes.

Flesh distinctly changing color where wounded..................1
Flesh not at all or scarcely changing color where wounded.......7
1. Flesh white or whitish.....................................2
1. Flesh yellow or yellowish.................................5
  2. Flesh changing to red or violet...................B. Satanus
  2. Flesh changing to blue................................3
3. Stem roughened..............................B. alveolatus
3. Stem even...4
  4. Stem hairy at the base.....................B. subvelutipes
  4. Stem not hairy at the base ....B. vermiculosus
5. Stem red..................................B. luridus
5. Stem yellow or reddish only at the base......................6
  6. Pileus purplish-red.........................B. purpureus
  6. Pileus gray................................B. firmus
  6. Pileus yellow or yellowish................B. magnisporus

453

Boletus. 7. Pileus blood-red......................................B. Frostii
     7. Pileus reddish-tawny or brown.....................B. Sullivantii
     Peck, Boleti of the U. S.

All authors, up to this date, agree in stating that the species within this series are poisonous. Experiments made by Smiedeberg and Koppe with Boletus Satanus developed symptoms closely resembling poisoning by Amanitæ. Kobert, who made analysis of B. luridus, shows that it contains muscarine, which is one of the most deadly poisons. Such a mass of evidence commands respect. It is urged upon finders of these species to either leave them alone or test them in minute quantities until they have established their ability to eat them without injury.

I have taken special pains to establish the edibility of B. Satanus and B. luridus. For fifteen years I have eaten them in quantity when opportunity afforded, in West Virginia, New Jersey and Pennsylvania. My family, and my friends in widely separated localities, have partaken freely of them many times and without discomfort. They are remarkably fine eating. The same can be said of B. alveolatus, B. purpureus, B. subvelutipes. I have not seen the other species of this tribe.

I have determined so many of the reputed poisonous species to be edible, that unless positively authenticated, I do not accept repute as truth, but carefully test suspicious species upon myself. When sure there is no danger, I as carefully have them tested by my numerous under-tasters—male and female.

B. Sa'tanus Lenz.—Satanic. **Pileus** convex, *glabrous*, somewhat gluey, *brownish-yellow or whitish*. **Flesh** whitish, becoming *reddish or violaceous* where wounded. **Tubes** free, yellow, their mouths bright red becoming orange-colored with age. **Stem** thick, ovate-ventricose, marked above with red reticulations. **Spores** 12x5μ.

Pileus 3–8 in. broad. **Stem** 2–3 in. long.

Woods. Rare. North Carolina, *Curtis;* New York, *Peck;* California, *H. and M., N. J. Ellis.*

Though mild to the taste, this Boletus is said to be very poisonous, a character suggestive of the specific name. Fries describes the color of the spores as earthy-yellow; Smith as rich brown. *Peck*, Boleti of the U. S.

West Virginia, New Jersey, Pennsylvania, *McIlvaine*.

Boletus Satanus is sometimes plentiful in spots. Where it luxuriates it is a rich decoration to the ground, and earth upon upturned-roots upon which it often grows. It does not live long after reaching maturity, but decomposes into a putrescent mass.

Its reputation rivals that of the original possessor of its name. But old proverb sayeth that even "The Devil is not as black as he is painted." See remarks heading Luridi.

B. alveola′tus B. and C. **Pileus** convex, glabrous, shining, bright crimson or maroon-color, sometimes paler and varied with patches of yellow. **Flesh** firm, white, changing to blue where wounded. **Tubes** *adnate, subdecurrent*, yellow with maroon-colored mouths, the hymenial surface *uneven with irregular alveolar depressions*. **Stem** very *rough with the margins of rather coarse subreticular depressions*, the reticulations bright-red above with yellow stains. **Spores** yellowish-brown, 12.5–15x4–5μ.

Pileus 3–6 in. broad. **Stem** 3–4 in. long, 9 lines thick.

Damp woods. New England, *Frost*. *Peck*, Boleti of the U. S.

West Virginia mountains, New Jersey, Pennsylvania, in mixed woods and on banks of streams. *McIlvaine*.

B. alveolatus appears to be more generally distributed than B. Satanus. It is not as clannish, though occasionally three or four are found growing together. When growing from the banks of creeks, or between the roots of beech and other trees in low places, it is often deformed in cap and stem. The texture is firm, close and the taste is very pleasant. It botanically takes its place in this suspected series. I consider it one of the best Boleti. See remarks heading Luridi.

B. lu′ridus Schaeff.—lurid in color. **Pileus** convex, tomentose, *brown-olivaceous*, then *somewhat viscose*, sooty. **Flesh** yellow, changing to blue where wounded. **Tubes** free, yellow, becoming greenish, their mouths round, vermilion, *becoming orange*. **Stem** stout, vermilion, somewhat orange at the top, *reticulate or punctate*. **Spores** greenish-gray, 15x9.

Pileus 2–4 in. broad. **Stem** 2–3 in. long.

The lurid Boletus, though pleasant to the taste, is reputed very poisonous. *Boletus rubeolarius* Pers., having a short bulbous scarcely reticu-

455

Boletus. lated stem, is regarded as a variety of this species. The red-stemmed Boletus, *B. erythropus* Pers., is also indicated as a variety of it by Fries. It is smaller than B. luridus, has a brown or reddish-brown pileus and a slender cylindrical stem, not reticulated, but dotted with squamules. It has been reported from California by Harkness and Moore. *Peck*, Boleti of the U. S.

Var. *erythropus* received from Dr. J. W. Harshberger, Philadelphia, May, 1896.

Often shining as if varnished and very handsome. I frequently found it in West Virginia, New Jersey and Pennsylvania in mixed woods among leaves. Its reputation is bad. It is undoubtedly edible by many, and is delicious. The caution heading Luridi should be carefully observed.

B. purpu'reus Fr.—purple. **Pileus** convex, opaque, dry, *somewhat velvety, purplish-red.* **Flesh** in the young plant only becoming blue, then dark-yellow. **Tubes** nearly free, yellow or greenish-yellow, their mouths minute, *purple-orange,* changing to blue where wounded. **Stem** stout, firm, adorned with purple veins or dots, sometimes reticulated at the apex only, yellow, reddish within, especially at the base. **Spores** greenish-brown, 10–12x5–6µ.

Pileus 2–4 in. broad. **Stem** 2–4 in. long, 6–8 lines thick.

Woods. North Carolina, *Curtis;* New York, *Peck;* Minnesota, *Johnson. Peck*, Boleti of the U. S.

West Virginia, Mt. Gretna, Pa., *McIlvaine.*

At Mt. Gretna, Pa., 1897–1898, B. purpureus was common in oak and chestnut woods. It is a showy species, easily distinguished by its velvety cap. In young specimens the stem is robust, then tapering upward. When old the cap loses its rich color toward the margin, becoming yellowish. The flesh is thick, firm and of excellent flavor. It undoubtedly proved itself delicious and harmless to many eating it.

B. vermiculo'sus Pk.—wormy. **Pileus** broadly convex, thick, firm, *dry,* glabrous, or very minutely tomentose, brown, yellowish-brown or grayish-brown, sometimes tinged with red. **Flesh** white or whitish, quickly changing to blue where wounded. **Tubes** plane or slightly convex, nearly free, yellow, their mouths small, round, brownish-orange, becoming darker or blackish with age, changing promptly to blue where wounded. **Stem** subequal, firm, *even*, paler than the pileus. Boletus. **Spores** ochraceous-brown, 10–12x4–5μ.

Var. *Spra'guei.* (Boletus Spraguei Frost, Bull. Buff. Soc., p. 102.) **Stem** yellow above, minutely velvety below.

Pileus 3–5 in. broad. **Stem** 2–4 in. long, 4–10 lines thick.

Woods. New York, *Peck;* Ohio, *Morgan;* New England, *Frost.*

The species is separated from B. luridus by its dry pileus, white flesh, even stem, which is neither reticulated nor dotted, and by its smaller spores. I can not distinguish specimens of B. Spraguei received from Mr. Frost, from this species. The name is scarcely appropriate, for specimens are not always infested by larvæ. *Peck*, Boleti of the U. S.

I have not seen this species, therefore, have not tested it. CAUTION.

B. subvelu'tipes Pk.—velvety-stem. **Pileus** convex, firm, subglabrous, yellowish-brown or reddish-brown. **Flesh** whitish, both it and the tubes changing to blue where wounded. **Tubes** plane or slightly convex, nearly free, yellowish, their mouths small, brownish-red. **Stem** equal or slightly tapering upward, firm, even, somewhat pruinose above, *velvety with a hairy tomentum toward the base*, yellow at the top, reddish-brown below, varied with red and yellow within. **Spores** 15–18x 5–6μ.

Pileus 2–3 in. broad. **Stem** 2–3 in. long, 4–6 lines thick.

Woods. New York, *Peck.*

This species resembles the preceding one in general appearance, but it is very distinct by its much longer spores and by the velvety hairiness toward the base of the stem. *Peck*, Boleti of the U. S.

Boletus subvelutipes is common in some localities in Pennsylvania, especially on the Springton Hills, in chestnut and oak woods. I have frequently eaten it and found it excellent. Others should carefully test it.

B. fir'mus Frost—firm. **Pileus** convex, *very firm*, slightly tomentose, gray, often pitted. **Flesh** *yellowish or deep-yellow*, changing to blue where wounded. **Tubes** *adnate*, deeply arcuate, unequal, yellow, their mouths *tinged with red*. **Stem** solid, hard, *very finely reticulated*, yellowish, reddish at the base. **Spores** 13x3μ.

Pileus 2.5–4 in. broad. **Stem** 2–4 in. long.

Rich moist wood. New England, *Frost.*

Boletus. Apparently a well-marked and very distinct species. According to the author, it is readily distinguished by its tenacity and generally distorted growth. I have not seen it nor the next. *Peck*, Boleti of the U. S.

Professor Peck's measurement of spores, 50th Report, New York State Botanist, is 13μ long, 6μ wide.

B. magnis'porus Frost. **Pileus** convex, firm, tomentose, *golden-yellow;* tubes *scarcely adnate*, even, greenish-yellow, their mouths light cinnabar-red. **Stem** long, slender, yellow above, red below. **Spores** 15–18x6μ.

Pileus 2.5 to 3.5 in. broad.

Woods and thickets. New England, *Frost;* Ohio, *Morgan.* *Peck*, Boleti of the U. S.

I have not recognized it. CAUTION.

B. Fros'tii Russell. **Pileus** convex, polished, shining, *blood-red*, the margin thin. **Flesh** scarcely changing to blue. **Tubes** nearly free, greenish-yellow, becoming yellowish-brown with age, their mouths blood-red or cinnabar. **Stem** equal or tapering upward, distinctly reticulated, firm, blood-red. **Spores** 12.5–15x5μ.

Pileus 3–4 in. broad. **Stem** 2–4 in. long, 3–6 lines thick.

Grassy places under trees or in thin woods. New England, *Frost;* New York, *Peck;* New Jersey, *Ellis.*

This is a highly colored, beautiful Boletus, but it is not common. The stem sometimes fades with age, and both it and the tubes are apt to lose their color in drying. *Peck*, Boleti of the U. S.

I have not recognized it. CAUTION.

B. Sullivan'tii B. and M. **Pileus** hemispherical, glabrous, reddish-tawny or brown, brownish when dry, cracked in squares. **Tubes** free, convex, medium size, angular, longer toward the margin, their mouths reddish. **Stem** solid, violaceous at the thickened base, red-reticulated at the apex, expanded into the pileus. **Spores** pallid ochraceous, oblong-fusiform, 10–20μ long.

Pileus 3–4 in. broad. **Stem** 1.5–3 in. long.

Compact soil. Ohio. *Sullivant.*

The species is said to be intermediate between Boletus scaber and B.

edulis. From the former it differs in its reticulated stem, from the Boletus. latter, in its larger tubes and from both in its stratum of tubes being remote from the stem. I have not seen it. *Peck*, Boleti of the U. S.

B. Un'derwoodii Pk. **Pileus** rather thin, convex, becoming nearly plane, slightly velvety, bright brownish-red, becoming paler with age. **Flesh** yellow, changing to greenish-blue where wounded. **Tubes** adnate or slightly decurrent, greenish-yellow, becoming bluish where wounded, their mouths very small, round, cinnabar red, becoming brownish-orange. **Stem** equal or slightly tapering upward, somewhat irregular, solid, yellow without and within. **Spores** $10-12 \times 5\mu$.
  Pileus 2–3 in. broad. **Stem** 3–4 in. long, 4–6 lines thick.
  Grassy woods. Auburn, Alabama. July. *Underwood.*
  This species is remarkable for its adnate or subdecurrent tubes, in which it departs from the character of the tribe to which it belongs according to the colors of the tubes. *Peck*, Bull. Torrey Bot. Club, Vol. 24, No. 3.

B. par'vus Pk.—*parvus*, small. **Pileus** convex, becoming plane, often slightly umbonate, subtomentose, reddish. **Flesh** yellowish-white, slowly changing to pinkish where wounded. **Tubes** nearly plane, adnate, their mouths rather large, angular, at first bright red, becoming reddish-brown. **Stem** equal or slightly thickened below, red. **Spores** oblong, $12.5 \times 4\mu$. **Pileus** 1–2 in. broad. **Stem** 1–2 in. long, 2–3 lines thick.
  Grassy woods. Auburn, Ala. July. *Underwood. Peck*, Bull. Torrey Bot. Club, Vol. 24, No. 3.

VERSIPEL'LES—*verto*, to change; *pellis*, a skin.

Tubes at first white or whitish, minute, round, equal, forming a convex stratum free from the stem.
  Stem black . . . . .B. alboater
  Stem some other color. 1
1. Stem yellow at the base. . . . . .B. chromapes
1. Stem not yellow at the base. 2
  2. Margin of the pileus appendiculate . . . . .B. versipellis
  2. Margin not appendiculate. 3

Boletus. 3. Stem scabrous or punctate-squamulose................B. scaber
       3. Stem even..4
          4. Pileus white or whitish.........................B. albellus
          4. Pileus dark-brown ....B. sordidus
       Peck, Boleti of the U. S.

B. alboa'ter Schw.—black and white. Pileus convex, subtomentose-velvety, black. Tubes free, their mouths rather small, white. Stem black.

Pileus 3 in. broad. Stem 2 in. long.

Moist woods. Frequent. North Carolina and Pennsylvania, *Schweinitz*.

In Epicrisis, p. 424, Fries adds to the description here quoted, that the stem is flocculose-veiled. He subjoins to this as a subspecies, Boletus floccosus Schw.; but in Syn. N. A. Fung., Schweinitz makes this a synonym of Boletus floccopus. The species does not appear to have been recognized by recent collectors, which seems strange unless there is some error concerning it. Can it be a black variety of Boletus scaber? *Peck*, Boleti of the U. S.

Mt. Gretna, Pa. Gravelly woods. *McIlvaine.*

Cap 1½–4 in. across, convex, slightly depressed, *margin involute when young*, black, densely velvety in youth and age—beautifully so. Flesh firm, thick, solid, white changing to grayish. Tubes white, stuffed, sometimes blackish when young, excepting a grayish-white circle around stem, becoming yellowish-white when matured, rotund, minute, up to ½ in. long, plane when young; when caps expand tubes draw away from stem leaving a deep white depression. This drawing away apparently elongates many dissepiments, creating a gill-like effect, decurrent upon stem. Stem 2–3 in. long, swollen toward base when young, equal, expanding into cap and tapering to a point at base; ¾–1 in. thick, slightly compressible, hard, sooty-black, velvety near base, satiny and glossy upward, has the appearance of having been blackened with burnt cork, usually with narrow white band next to the tubes, no trace of veil, composed of rather hard waved fibers, white when split, but changing to sooty black toward base, lighter upward.

Smell like common mushroom; taste nutty.

Gregarious in sandy-conglomerate soil in mixed woods, among moss and leaves. Mt. Gretna, Pa.

Differs from B. alboater Schw., in having densely tomentose cap, Boletus. tubes widely separated from stem in age.

A young specimen of apparently same species in same patch had very short, decurrent tubes (not over 1 line) which were sooty-black. Delicious.

B. sor'didus Frost—sordid. **Pileus** convex, subtomentose, dirty dark-brown. **Flesh** white, slightly tinged with green. **Tubes** long, nearly free, at first white, changing to bluish-green. **Stem** smaller at the top, brownish, marked with darker streaks, generally greenish above. **Spores** 10–13x5μ.

Pileus about 2 in. broad.

Recent excavations in woods. New England, *Frost;* Ohio, *Morgan.*

The Ohio plant occurs in damp woods, has the flesh sometimes tinged with red and green, the tubes white, then sordid, but changing to bluish-green when bruised, their mouths large and angular, the stem somewhat flexuous and striate and the spores fusiform and dirty-brown *Peck*, Boleti of the U. S.

B. versipel'lis Fr. **Pileus** convex, *dry*, at first compact and minutely *tomentose*, then squamose or smooth, reddish or orange-red, the margin *appendiculate* with the inflexed remains of the membranous veil. **Flesh** white or grayish. **Tubes** at first concave or nearly plane, almost or quite free, minute, sordid-white, their mouths gray. **Stem** equal or tapering upward, solid, wrinkled-scaly, whitish or pallid. **Spores** oblong-fusiform, 14–18x4–6μ.

Pileus 2–6 in. broad. **Stem** 3–5 in. long, 4–10 lines thick.

Woods and open places, especially in sandy soil. North Carolina, *Curtis;* New England, *Frost;* New York, *Peck;* California, *H. and M. Peck*, Boleti of the U. S.

West Virginia, New Jersey, Pennsylvania. *McIlvaine.*

The caps are good cooked in any way.

B. sca'ber Fr.—*scaber*, rough.   ([Color] Plate CXVIII, fig. 4.) **Pileus** convex, *glabrous, viscid when moist*, at length wrinkled or lined. **Tubes** free, convex, white, then sordid, their mouths minute, rotund. **Stem** solid, attenuated above, *roughened with fibrous scales*. **Spores** oblong-fusiform, snuff-brown, 14–18x4–6μ.

461

Boletus. **Pileus** 1–5 in. broad. **Stem** 3–5 in. long, 3–8 lines thick.

(Plate CXXI.)

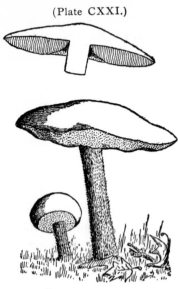

BOLETUS SCABER.
One-half natural size.

Woods, swamps and open places. Very common and appearing through summer and autumn.

This may fairly be called our most common and variable species. It is recorded in nearly every local list of fungi. The pileus is convex, hemispherical or even subconical. It may be glabrous, minutely tomentose, subvelvety or squamulose. The flesh is white or whitish and sometimes slightly changeable where wounded. The tubes are generally rather long and with a rounded or convex surface. The stem is distinctly scabrous or roughened with small blackish-brown or reddish dots or scales, the ground color generally being whitish, grayish or pallid. The spores have been described as pale-brown and light-yellowish. When caught in a mass on white paper they appear to me to approach snuff-brown. The viscidity of the pileus is not always clearly discernible. Indeed the pileus is often quite as *dry as in* B. versipellis. When moistened by heavy rains it sometimes is smooth and clammy to the touch but scarcely viscid. Several varieties have been indicated which are expressive of the variations in the color of the pileus.

Var. *testa'ceus*. **Pileus** brick-red.

Var. *auranti'acus*. **Pileus** orange or orange-red.

These appear to connect this species and B. versipellis.

Var. *aluta'ceus*. **Pileus** yellowish-tan color.

Var. *fuligin'eus*. **Pileus** fuliginous or cinereous-fuliginous.

Var. *fus'cus*. **Pileus** brown or dark-brown.

Var. *oliva'ceus*. **Pileus** olivaceous.

Var. *ni'veus*. **Pileus** white, when old sometimes stained with blue or livid-blue.

To these might be added:

Var. *areola'tus.* **Pileus** rimose-areolate. ([Color] Plate CXVIII, Boletus. fig. 4.)

Var. *mutab'ilis.* **Flesh** changing slightly to brown or pinkish where wounded.

Var. *graci'lipes.* **Stem** very slender, 2–3 in. long, 2–3 lines thick. **Pileus** thin, translucent when held toward the light.

This Boletus is classed among the edible species, but it is said to be less agreeable than B. edulis. *Peck*, Boleti of the U. S.

West Virginia, North Carolina, New Jersey, Pennsylvania, *McIlvaine*.

The numerous varieties with their peculiarities here given by Professor Peck will enable the finder of a Boletus with a distinctly scabrous stem —*roughened with scales, not reticulate*—to select its name. For the mycophagist it is enough to know that he has Boletus scaber. In all of its varieties it is edible. The stems, often the tubes, unless young, should be discarded, as they do not cook in the same time as the caps. The comparative excellence of the species rests with the devourer. It deserves a high place.

B. scaber, var. areolatus, Plate CXVIII, fig. 4, has slight flavor, but is of pleasing consistency.

B. durius'culus Schulz—somewhat hard. **Pileus** 2–5 in. across, hemispherical, minutely velvety, viscid when moist, varying in color from pale-brown, through dingy-chestnut, to umber-brown, often becoming cracked in areas when dry, interstices paler. **Flesh** thick, white or tinged yellow, when cut becoming reddish copper-color. **Tubes** ½– ¾ in. long, shortened round the stem and free, openings about ⅔ mm. across, often compound, irregularly angular, bright-yellow. **Stem** 4–7 in. long, fusiform, thickest part 1½–2 in. across, situated below the middle, yellowish, rough with blackish points, which are sometimes arranged in a subreticulate manner, apex sometimes more or less grooved, solid, flesh of upper part becoming coppery like the pileus. **Spores** elongato-cylindrical, pale-umber, 14–16x5–6μ.

In woods. Esculent and very delicious. Allied to Boletus scaber, but distinguished by the bright-yellow tubes and the very firm flesh, which turns coppery-red when exposed to the air; this color eventually changes to a dingy grayish-violet. Also allied to Boletus porphyrosporus. *Massee.*

Snow Hill, N. J. Gravelly soil, mixed woods, 1892. *McIlvaine.* The stem and tubes should be removed. The caps are very fine.

B. albel'lus Pk.—whitish.

Pileus convex or gibbous, soft, glabrous, whitish. **Flesh** white, unchangeable. **Tubes** convex, free, or nearly so, small, subrotund, whitish, unchangeable. **Stem** *glabrous or minutely branny*, substriate, *bulbous or thickened at the base*, whitish. **Spores** brownish-ochraceous, 14–16x5–6μ.

Pileus 1–2 in. broad. **Stem** 1–2 in. long, 3–6 lines thick.

Woods. New York, *Peck*.

This is closely related to B. scaber, of which it may possibly prove to be a dwarf form; but it is easily distinguished by its smooth or only slightly scurfy and subbulbous stem. It presents no appearance of the colored dot-like squamules which are a constant and characteristic feature of that species. *Peck*, Boleti of the U. S.

West Virginia. Woodland Cemetery, Philadelphia. *McIlvaine.*

Specimens found at Mt. Gretna, Pa., had a satiny, glossy stem, beautifully furfuraceous, and stem *not* thickened at base. Professor Peck, to whom specimens were sent, writes: ''Stem is a little more furfuraceous, and not thickened at the base, otherwise the agreement is very good.'' It is good fried.

B. chro'mapes Frost.

Pileus convex or nearly plane, slightly and sometimes fasciculately tomentose, pale-red. **Flesh** white, unchangeable. **Tubes** subadnate, more or less depressed around the stem, white or whitish, becoming brown. **Stem** equal or slightly tapering upward, *rough-spotted*, whitish or pallid, *chrome-yellow at the base both without and within*, sometimes reddish above. **Spores** oblong, 12–14x4–5.

Pileus 2–4 in. broad. **Stem** 2–4 in. long, 4–6 lines thick.

Woods. New England, *Frost;* New York, *Peck*.

The yellow base of the stem appears to be a peculiar and constant character by which the species may easily be recognized. It imitates Boletus piperatus in this respect, but in everything else it is very distinct from that plant. Sometimes the stem is so badly infested by larvæ that it is difficult to procure a sound specimen. The spores have a subferruginous color with a slight incarnate tint, but the rough-dotted stem indicates a relationship with B. scaber. Through this species, Boletus conicus and B. gracilis, the Versipelles and the Hyporhodii ap-

pear to run together. In the Catalogue of Plants of Amherst the specific name is "chromapus." It would be more in accordance with
present custom to write it "chromopus." *Peck*, Boleti of the U. S.

A dozen or more specimens referable to this species were found by
me at Mt. Gretna, Pa., August, 1897, in mixed woods. The caps were
eaten and were excellent.

B. nebulo'sus Pk. **Pileus** convex, dry, snuff-brown or smoky-brown.
Flesh white, unchangeable. **Tubes** convex, depressed around the stem,
pallid or brownish, becoming purplish-brown where wounded, the
mouths small, rotund. **Stem** enlarged toward the base, solid, scurfy,
colored like the pileus. **Spores** 12.5–15x6µ.

Pileus 2–4 in. broad. **Stem** 3–4 in. long, 4–6 lines thick.

Shaded banks by road-side. Raybrook. August.

No young or immature specimens were seen, and the description is
to that extent incomplete. *Peck*, 51st Rep. N. Y. State Bot.

By a painting made by the writer September, 1885, Professor Peck
identified the species of which it is a picture as B. nebulosus Pk. The
following notes accompany it, which have been verified many times
since their writing:

Oak woods. West Philadelphia, Pa., September. Mt. Gretna, Pa.,
September.

Pileus chestnut-brown and darker, covered with small, low, black
spots; convex, often depressed in center, sharp on margin. **Flesh**
white, thick, solid, unchangeable. **Tubes** very small, and light pinkish-brown. When touched they change to a deeper hue. **Stem** same
color as pileus, but a shade lighter, solid, scurfy, having a striate appearance, enlarging toward base.

Taste sweet and pleasant. Cooked it is juicy, meaty and very fine.

B. ful'vus Pk.—brownish-yellow. ([Color] Plate CXVI, fig. 3.)
Pileus thick, convex or subcampanulate, dry, glabrous, rimose-areolate,
tawny-yellow, the extreme margin dark-brown. **Flesh** spongy, tough,
white, slowly assuming a reddish tint upon exposure to the air. **Tubes**
rather long, ventricose, depressed around the stem and free or nearly
so, greenish-yellow, the mouths small, tawny-yellow. **Stem** rather long,
often narrowed and striate at the top, dotted with brownish-orange gran-

Boletus. ules or points, radicating, tough, stuffed with greenish-yellow fibers, colored like the pileus. **Spores** unknown.

Pileus 2–3 in. broad. **Stem** 4–5 in. long, 4–8 lines thick.

Cespitose on decaying stumps. West Philadelphia, Pa. **August.** *McIlvaine.*

Mr. McIlvaine says that there were between twenty and thirty specimens on and about an old stump and that they were as attractive to the eye as a cluster of Clitocybe illudens. *Peck*, Bull. Torrey Bot. Club, Vol. 27, January, 1900.

Excellent in flavor, rather spongy, but fine.

HYPORHO'DII.   Gr.—somewhat rose-colored.

Tubes adnate to the stem, whitish, then white-incarnate **from the** rosy spores.

In this tribe the tubes are at first whitish, but with the development of the spores they usually assume a pinkish or flesh-colored hue. Wounds of the tubes in some species cause a change in color but not to blue, nor are the tube mouths differently colored as in the Luridi. The stem in some is more or less reticulated but this is scarcely a constant or reliable character in these species. Typically the spores are rosy or flesh-colored, but I have admitted species in which they incline to rust-colored, giving more weight to the color of the tubes than to that of the spores.

```
Pileus black or blackish...........................B. nigrellus
Pileus some other color ................................1
1. Stem more than four lines thick .....................2
1. Stem slender, generally less than four lines thick.......B. gracilis
  2. Stem not reticulated...............................3
  2. Stem more or less reticulated......................4
3. Tubes angular, flesh-colored....................B. conicus
3. Tubes round, white............................B. alutarius
  4. Taste mild..................................B. indecisus
  4. Taste bitter ...............................B. felleus
```
Peck, Boleti of the **U. S.**

B. con'icus Rav.—conical. **Pileus** convex or *subconical*, clothed with bundled appressed *yellowish flocci*. **Flesh** white, unchangeable,

466

tasteless.   **Tubes** ventricose, flesh-colored, becoming darker from the Boletus. spores, the mouths small, angular, slightly fringed.   **Stem** glabrous, tapering upward, pale-yellow.   **Spores** fusiform, subferruginous.

Pileus 1–2 in. broad.   **Stem** 2 in. long, 6 lines thick.

Damp pine woods.   South Carolina, *Ravenel*.

The species is compared to Boletus scaber, from which it differs in its smaller tubes and smooth stem, and from both this and B. albellus it differs in the color of the tubes and in the yellowish flocci of the pileus. I have seen no specimens, but on account of the color of the tubes I have placed the species with the Hyporhodii. *Peck*, Boleti of the U. S.

B. gra cilis Pk.—slender.   ([Color] Plate CXIV, fig. 1.)   **Pileus** convex, glabrous or minutely tomentose, rarely squamulose, ochraceous-brown, tawny-brown or reddish-brown.   **Flesh** white.   **Tubes** plane or convex, depressed around the stem, nearly free, whitish, becoming pale flesh-colored, their mouths subrotund.   **Stem** *long, slender*, equal or slightly tapering upward, pruinose or minutely branny, even or marked by slender elevated anastomosing lines which form long narrow reticulations.   **Spores** subferruginous, 12.5–17.5x5–6μ.

Var. *læ'vipes*.   **Stem** even.

Pileus 1–2 in. broad.   **Stem** 3–5 in. long, 2–4 lines thick.

Woods.   New York, *Peck;* New England, *Frost;* Ohio, *Morgan.*

The slender habit separates this species from all the others here included in this tribe.   Its spores are not a clear incarnate in color, but incline to dull-ferruginous, and by this character this and the preceding species connect this tribe with Versipelles.   In color B. gracilis resembles some forms of B. felleus, but in size, habit and color of spores it is easily distinct.   The tomentum of the pileus sometimes breaks into tufts or squamules.   This is Boletus vinaceus, Frost MS.   *Peck*, Boleti of the U. S.

B. gracilis, var. lævipes, was found by the writer in Woodland Cemetery, West Philadelphia, August, 1897, and at Mt. Gretna, Pa., September, 1898.   The stem of some specimens spreads at the top.   The pileus is often cracked on the margin, and the upturning of the margin often exposes the tubes.   Painting, as of this species, identified by Professor Peck.

The taste is at first sweet, then bitter.   The bitterness is lost in cooking.   Edible, good.

Boletus. **B. indeci'sus** Pk.—undecided. ([Color] Plate CXXII, fig. 1.) **Pileus** convex or nearly plane, dry, slightly tomentose, ochraceous-brown, often wavy or irregular on the margin. **Flesh** white, unchangeable; taste mild. **Tubes** nearly plane or convex, *adnate*, grayish becoming tinged with flesh color when mature, changing to brownish where wounded, their mouths small, subrotund. **Stem** minutely furfuraceous, straight, or flexuous, *reticulated above*, pallid without and within. **Spores** oblong, *brownish flesh color*, 12.5–15x4μ.

> **Pileus** 3–4 in. broad. **Stem** 2–4 in. long, 4–6 lines thick.
> Thin oak woods. New York, *Peck*.

The mild taste and darker colored spores will separate this Boletus from any form of B. felleus. Its stem reticulated above distinguishes it from B. alutarius. It resembles B. modestus in some respects, but its tubes are not at all yellow. *Peck*, Boleti of the U. S.

> Kentucky, *Lloyd*, Rep. 4.
> Woodland Cemetery, Philadelphia, July, 1897, *McIlvaine;* Trenton, N. J., August, 1897, *Sterling*. In open mixed woods.

Boletus indecisus so closely resembles B. felleus in some of its forms that until the color of the spores is ascertained, the sweet taste, without trace of bitter, is the only thing that will enable the finder to discriminate between them. Young B. felleus are at first pleasant to the taste and do not, at once, develop their intense bitter in the mouth. They may readily be taken for B. indecisus. If, by mistake, a single B. felleus is cooked with mild species, the dish will be spoiled. Specimens believed to be B. indecisus should be tested. A minute will perfectly satisfy anyone.

The B. indecisus is delicious.

B. aluta'rius Fr.—*aluta*, tanned leather. **Pileus** convex, then nearly plane, soft, *velvety*, becoming glabrous, *brownish tan color*. **Flesh** almost unchangeable, taste *mild, watery*. **Tubes** depressed around the stem, plane, short, round, white, becoming brownish where wounded. **Stem** solid, bulbous, nearly even, *small, irregular prominences at the top*. **Spores** 14x4μ.

> **Pileus** 3–4 in. broad. **Stem** 4–5 in. long.
> Grassy woods. Minnesota, *Johnson*. *Peck*, Boleti of the U. S.
> West Virginia mountains, 1882–1885. Margins of woods. Chelten-

ham, Pa.   Margins of woods, 1888–1889, grassy woods and margins. Boletus.
McIlvaine.

Common in West Virginia mountains where it grows with B. felleus, from which it is impossible to distinguish it without tasting.   It is delicious when cooked.   But I long ago ceased collecting for the table any Boletus questionable for B. felleus.   I have been deceived so many times—taken the bitter for the sweet—that, preferring the sweet, I take no chances for the bitter.

B. fel'leus Bull.—*fel*, gall.   Bitter.   ([Color] Plate CXXII, figs. 2, 3, 4.)   **Pileus** convex or nearly plane, firm, becoming soft, *glabrous*, even, variable in color, pale-yellowish, grayish-brown, yellowish-brown, reddish-brown or chestnut.   **Flesh** white, often changing to flesh color where wounded, taste *bitter*.   **Tubes** adnate, long, convex, depressed around the stem, their mouths angular, white, becoming tinged with flesh-color.   **Stem** variable, equal or tapering upward, short or long, sometimes bulbous or enlarged at the base, subglabrous, generally reticulated above, colored like or a little paler than the pileus.   **Spores** oblong-fusiform, flesh-colored, 12.5–17.5x4–5μ.

Var. *obe'sus.*   **Pileus** large.   **Stem** thick, coarsely and distinctly reticulated nearly or quite to the base.

Pileus 3–8 in. broad.   **Stem** 2–4 in. long, 6–12 lines thick.

The variety is large and solitary in its mode of growth.   It is remarkable for the coarse reticulations of the stem which extend nearly or quite to the base.   After heavy rains the pileus is viscid.   It may prove to be a distinct species.

The flesh in the American plant does not always assume incarnate hues where wounded.   The color of the fresh tubes often changes to a deeper tint where wounded.   *Peck*, Boleti of the U. S.

West Virginia, Pennsylvania, New Jersey, North Carolina, *McIlvaine;* Indiana, *H. I. Miller.*

A very common species in woods and on thin margins, on open grassy places, and about decayed stumps.   I saw hundreds of plants, var. obesus, some a foot in diameter, in a wheat stubble near oak woods.

One of the most attractive of Boleti.   Its cap resembles a handsomely browned cake.   Its solidity is inviting; its flesh, generous in quantity, excites appetite.   Until one experiences its intense lasting bitter, one clings to it with hope. Even after tasting, it is thrown away with regret. It is not poisonous, but a small piece of one will embitter a whole dish. *McIlvaine*, Bull. Phila. Myc. Center. July, 1898.

B. nigrel'lus Pk.—blackish. **Pileus** broadly convex or nearly plane, dry, subglabrous, *blackish*. **Flesh** soft, white, unchangeable. **Tubes** plane or convex, adnate, sometimes slightly depressed around the stem, their mouths small, subrotund, whitish becoming flesh-colored, slowly changing to *brown or blackish where wounded*. **Stem** equal, short, *even*, colored like or a little paler than the pileus. **Spores** dull flesh-colored, 10–12x5–6μ.

Pileus 3–6 in. broad. **Stem** 1.5–2.5 in. long, 6–12 lines thick.
Woods and copses. New York, *Peck*.
The blackish color of the pileus and stem distinguishes this species. From Boletus alboater Schw., the adnate, flesh-colored tubes will separate it. The surface of the pileus sometimes becomes cracked in areas. *Peck*, Boleti of the U. S.
Mt. Gretna, Pa., August, 1898. *McIlvaine*.
Another distinguishing mark from B. alboater is the velvety pileus of the latter. B. nigrellus is mild in taste and smell and an excellent species for the table.

B. eccen'tricus Pk.—eccentric. ([Color] Plate CXVI, fig. 1.)
Pileus thick, firm, convex, irregular, glabrous, more or less lobed or wavy on the involute margin, gray or yellowish-gray. **Flesh** white, close-grained, elastic, unchangeable, taste and odor farinaceous. **Tubes** convex, depressed around the stem, not reaching the margin of the pileus, somewhat uneven and pitted on the surface, yellowish-brown, the mouths subangular, at first concolorous, becoming reddish or reddish-purple. **Stem** eccentric, tapering downward, solid, uneven with short irregular shallow grooves or obscure reticulations, tinged with red at the top, grayish below, tinged with red or purple within at the base.

Pileus 5–10 cm. broad. **Stem** 4–5 cm. long, 3–4 cm. thick at the top.
Sandy soil in grassy places in woods. Mt. Gretna, Pa. August and September.
The species is well marked by its eccentric stem, thick irregular pileus and the reddish or reddish-purple mouths of the mature tubes. Mr.

McIlvaine remarks that when it is cooked it is delicate and savory. Boletus. *Peck*, Bull. Torrey Bot. Club, No. 27.

In commenting upon this new species to the writer, Professor Peck says: "I suspect that the spores of this (B. eccentricus) are pinkish or rosy. If so, it belongs here (in Hyporhodii). If not, it may have to go in the Luridi, or possibly may be made the type of a new tribe.

Cario'si—*caries*, rottenness.

Stem never reticulated, stuffed with a spongy pith, at length commonly excavated. Tubes at first white, then often yellowish, their mouths minute, round.

Fries adds to these characters, "spores white." But in our species the spores are pale-yellow when shed in a mass on white paper. They are more elliptical in outline than the spores of most Boleti. The character of the stem is peculiar and easily distinguishes the tribe. The exterior is firm, the interior soft and spongy, becoming irregularly hollow or cavernous in the typical species.

Flesh unchangeable. I
Flesh quickly changing to blue where wounded. . . . . . . .B. cyanescens
1. Pileus minutely velvety-tomentose. . . . . .B. castaneus
1. Pileus granulated. . . . . .B. Murrayi
Peck, Boleti of the U. S.

B. cyanes'cens Bull.—*cyaneus*, deep-blue. **Pileus** convex or nearly plane, opaque, floccose-scaly or covered with an appressed tomentum, pale-buff, grayish-yellow, yellowish or somewhat brown. **Flesh** rigid, white, *quickly changing to blue* where wounded. **Tubes** free, white, becoming yellowish, the mouths minute, round, changing color like the flesh. **Stem** ventricose, hoary with fine hairs, stuffed, becoming cavernous, contracted and even at the top, colored like the pileus. **Spores** subelliptical, 10–12.5x6–7.5µ.

Pileus 2–5 in. broad. **Stem** 2–4 in. long, 8–18 lines thick.

Woods and open places. New York, *Peck;* New England, *Frost, Bennett;* Minnesota, *Johnson;* Wisconsin, *Bundy. Peck*, Boleti of the U. S.

High ground in woods. Solitary. West Virginia mountains, Springton Hills, Pa., Kingsessing, Philadelphia, Mt. Gretna, Pa., *McIlvaine.*

Boletus.   Boletus cyanescens is a sparse grower.   The quality of the juice varies.   That of young specimens stains the fingers blue, that of old, brown. The caps are firm and make an excellent dish cooked in any way.

B. casta′neus Bull.—chestnut. ([Color] Plate CXIV, fig. 3.) **Pileus** convex, nearly plane or depressed, firm, even, dry, minutely *velvety-tomentose, cinnamon or reddish-brown*. **Flesh** white, unchangeable **Tubes** free, short, small, white becoming yellow. **Stem** equal or tapering upward, even, stuffed or hollow, clothed and colored like the pileus. **Spores** 10–12.5x6–7.5µ.

Pileus 1.5–3 in. broad.   **Stem** 1–2.5 in. long, 3–5 lines thick.

Woods and open places.   Rather common and wide spread. *Peck*, Boleti of the U. S.

Boletus castaneus is one of the neatest looking of fungi.   The prevailing color is cinnamon, that of the tubes white or very light yellow, spotted with brown wherever insects have touched them.   The pore surface of mature specimens is usually irregular.   Whoever has seen the stalagmites of Luray Cave will recognize their color on the stems of B. castaneus.   These are brittle, snapping like pipe stems, with a small tube in center.

The fungus is common from June until September.   It is gregarious, occasionally three or four individuals form a group.   Either raw or cooked the caps are edible and will become favorites.

B. Mur′rayi B. and C.   **Pileus** hemispherical, *granulated, vivid red*. **Flesh** yellow.   **Tubes** decurrent, about 1 line deep, yellow.   **Stem** clavate, even, pale-yellow.   **Spores** pale-yellow.

Pileus 2–3 in. broad, nearly 1.5 thick.

New England, *Murray*.

On account of the color of the spores this species has been placed with the Cariosi.   The description does not mention the character of the interior of the stem, and the decurrent tubes depart from the character of the typical species so that its true position is uncertain.   The species seems well marked by the character of the pileus. *Peck*, Boleti of the U. S.

B. isabelli′nus Pk.   **Pileus** convex, firm, minutely tomentose, whitish, becoming ᐧdarker and smoother with age.   **Flesh** isabelline.

472

Tubes adnate, minute, sometimes larger near the stem, nearly round, Boletus.
whitish. Stem nearly equal, subglabrous, hollow, whitish. Spores
subelliptical, 7.5–9x5–6μ. Pileus 2–3 in. broad. Stem 1–2 in. long,
4–6 lines thick.

Woods. Ocean Springs, Miss. June. *Underwood*.

The species belongs to the Cariosi. *Peck*, in Bull. Torrey Bot. Club,
Vol. 24, No. 3.

APPENDIX (Boletus).

The descriptions of the following species are scarcely sufficient to
permit of the satisfactory reference of the species to their places in the
tribes. It is to be hoped that these plants may again be found and
their proper relations be ascertained.

B. Ana'nas Curt. Pileus pulvinate, thickly and rigidly floccose-ver-
rucose, yellow, flocci white above, flesh-colored beneath, the margin
thin, membranous, lacerated; hymenium plane, depressed around the
stem, yellow or tawny-yellow, becoming greenish where wounded, their
mouths medium size, obtusely angular. Stem even, solid, somewhat
enlarged at the base, white. Spores ferruginous.

Pileus 3–4 in. broad. Stem 3–4 in. long, 6–9 lines thick.

Under prostrate trunks of pine trees.'

South Carolina, *Ravenel;* North Carolina, *Curtis*.

This is said to approach S. strobilaceus in habitat, but to be otherwise very different. It is placed among the Subtomentosi in Sylloge,
but from these it recedes by its floccose wart-like scales. *Peck*, Boleti
of the U. S.

B. radico'sus Bundy. Pileus thin, wide, recurved, yellow tinged
with brown, the cuticle easily removed. Flesh pale-yellowish tinged
with pink, not changing color when bruised. Tubes decurrent, large,
uneven-mouthed, compound, angular, tinged with brown. Stem flexuous, yellow above, whitish below, rough with dark appressed scales,
fibrous-rooted.

Pileus 4 in. broad. Stem 3–4 in. long, 5 lines thick.

Wisconsin, *Bundy*.

The pileus is not described as viscid, but in other respects the spe-

473

Boletus. cies appears to belong to the Viscipelles and to be related to Boletus collinitus. *Peck*, Boleti of the U. S.

B. Po'cono Schw. **Pileus** pulvinate, cervine (dun color), minutely covered with bundles of tomentum on the closely-inflexed margin. **Tubes** rather large, somewhat prominently angular, concolorous. **Stem** subattenuated, thickened toward the base, pallid-striate at the apex, elsewhere spadiceous, subfurfuraceous.

Pileus 1 in. broad. **Stem** 2–3 in. long.

Beech woods. Pennsylvania, *Schweinitz*.

STROBILO'MYCES Berk.

Gr.—a pine cone; a fungus.

(Plate CXXIV.)

Hymenophore even. **Tubes** not easily separable from it, large, equal. **Pileus** and **stem** distinctly rough-scaled, the **flesh** tough. Syl. Fung., Vol. VI, p. 49.

I have given Professor Saccardo's emended diagnosis of this genus, because it expresses what appears to me to be the most important generic character, that is, tubes not easily separable from the hymenophore. By this character and by the tough substance the transition between Boletus and Polyporus is made.

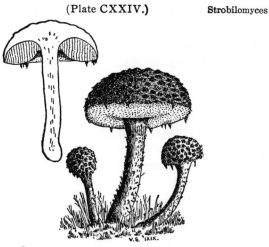

STROBILOMYCES STROBILACEUS.
Two-thirds natural size.

Tubes nearly equal in length..... ....S. strobilaceus
Tubes shortened around the stem.....................S. floccopus
   Peck, Boleti of the U. S.

S. strobila'ceus Berk. *Gr.*—cone-like. (Plate CXXIV.) **Pileus** hemispherical or convex, dry, covered with thick floccose projecting blackish or blackish-brown scales, the margin somewhat appendiculate with scales and fragments of the veil. **Flesh** whitish, changing to reddish and then to blackish where wounded. **Tubes** adnate, whitish, becoming brown or blackish with age; their mouths large, angular, changing color like the flesh. **Stem** equal or tapering upward, sulcate at the top, floccose-tomentose, colored like the pileus. **Spores** subglobose, rough, blackish-brown, $10-12.5\mu$.

Pileus 2–4 in. broad. **Stem** 3–5 in. long, 4–10 lines thick. *Peck*, Boleti of the U. S.

West Virginia mountains, Pennsylvania, *McIlvaine;* Indiana, *H. I. Miller.*

Common in woods and their margins, under the overhanging sods of washes and road-cuts. Often in troops, occasionally cespitose. The

Strobilomyces. rough fuzzy cap reminds of short fur that has been wet and dried. Its appearance is unique among Boleti. Before cooking the stem and tubes should be removed, unless the latter are very firm and fresh. The squamules must be cut away or the dish will be rough.

With many this Boletus is a prime favorite. It has a strong woody taste, sometimes musky, sometimes faintly of anisette. It cooks well by any method.

S. floc′copus Vahl.—floccose-stemmed.

Pileus convex, soft, covered with areas of bunched rough, scaly tomentum, cinereous, at length blackish, appendiculate with the silky, thick annular veil. **Tubes** *shortened behind*, their mouths large, whitish-gray. **Stem** stout, pitted above, umber-tomentose below. **Spores** perfectly globose, brown, 9µ broad.

Pileus 4–5 in. broad. **Stem** 4–5 in. long, 1 in. thick.

Woods. North Carolina and Pennsylvania, *Schweinitz;* Ohio, *Morgan;* New York, *Peck.*

According to Fries this is a larger and firmer species than S. strobilaceus but manifestly related to it. The New York specimens which I have referred to it differ from S. strobilaceus in no respect, except in the tubes being depressed around the stem. Unless there are other differences in the European plant, it scarcely seems to me to be worthy of specific distinction. Boletus floccopus, Rost. tab. 40, is referred to Boletus scaber, as is B. holopus, Rost. tab. 48. *Peck,* Boleti of the U. S.

I agree with Professor Peck that this species is not worthy of specific distinction. During 1898 I found a bunch containing eight individuals which varied through all botanic characteristics given to both species. The largest individual was 4½ in. across cap, the smallest 1½ in. On some the tubes were adnate, on others shortened behind. **There was no difference in flavor excepting that due to age.**

476

FISTULI'NA Bull.

Fistula, a pipe.

Hymenium formed on the under surface of a fleshy hymenophore, at Fistulina. first warted, the warts developing into cylindrical tubes that remain distinct and free from each other, producing in their interior cellular processes each bearing four spores. Conidia are produced in cavities of the old hymenophore.

With the outward appearance of a Polyporus, but separated by the tubes being free from each other.

A small genus of which F. hepatica is the principal species. This is known and valued in Europe and wherever found in this country. Unfortunately it is rare or unknown in many localities. A new species has recently been found in the United States—Fistulina firma, by Mrs. A. M. Hadley, Manchester, N. H.—a white-flesh species whose edibility is not reported. Torrey Bull., 1899. F. pallida B. and Rav.; F. radicata, Schw.; F. spathulata B. and C., are reported from Alabama. Edible qualities not stated. The writer has not seen them or he surely would have tested them. The spread and cultivation of F. hepatica is possible. Experiments in this line are desirable.

F. hepat'ica (Huds.) Fr. *Gr.*—resembling the liver. ([Color] Plate CXXV, fig. 1.) Juicy-fleshy, not rooting. **Pileus** entire, blood-red. **Flesh** thick, soft, viscid above, transversed with tenacious fibers, hence variegated-red. **Tubes** at first pallid.

Changeable in form, sessile or extended into a lateral stem. *Fries.*

Spores salmon-color, nearly round with an oblique apiculus, 3μ *W. G.S.;* broadly elliptical, $5-6 \times 3-4\mu$; conidia, $6-10 \times 5\mu$ *Massee;* yellowish, elliptical, $5-6.5\mu$ long *Peck.*

West Virginia, New Jersey, Pennsylvania. August to frost. *McIlvaine.*

Small specimens may be confounded with F. pallida, which follows.

Fistulina hepatica is celebrated in most countries, and known usually as the Beefsteak fungus. It grows from decaying crevices in oak, chestnut and other trees and stumps, but those named are its favorites. July, August, September are its months, and after rains. In some localities and years it is rare. At Mt. Gretna, in 1898, a hundred pounds of it could be gathered almost any day.

Fistulina. August, 1899, at Mt. Gretna, Pa., I found several specimens in vicinity which, though evidently F. hepatica, were remarkable for their structure—2–4 in. across, irregularly cylindrical, with spore surface covering the entire fungus. Stem curt, eccentric, almost central. Specimens were sent Professor Peck, who writes:

"The sample of Fistulina which you send is a singular thing. Saccardo has noted a somewhat similar form but without pore surface. Yours has pore surface, but I do not find spores developed in it. I am inclined to think it a monstrosity, as you do, but as you say you have found several of them I think it would be well to put it on record and I will enter it in my record as Fistulina hepatica monstrosa n. var. and indicate its characters." Letter from Professor Peck, August 28, 1899.

I have partially succeeded in transplanting the mycelium of F. hepatica. Experiments in this direction, I feel confident, will introduce this valuable fungus to localities where it is not now found, or is rare. Experiments with the spores have not been as yet successful.

F. hepatica monstrosa n. var. Pk. Subglobose, supported on a short stem or stem-like base, the external surface entirely covered with tubules 2–4 mm. long.

Pennsylvania. *C. McIlvaine.* In color and texture resembling the common form, but Mr. McIlvaine informs me that there is nothing in the position or place of growth of the specimens to account for their peculiar character. They are 2–4 in. in diameter. *Peck*, Bull. Torrey Bot. Club, 27, January, 1900.

Excellent.

F. pal'lida B. and Rav.—*pallidus*, pale. **Pileus** kidney-shaped, pallid-red, pulverulent, 1–2 in. broad, about 1 in. long, margin inflexed. **Tubes** more or less decurrent. **Stem** lateral, striate, when dry, 1 ½ in. long, ⅓ in. thick.

Mountains of South Carolina on the ground. *Ravenel.* Alabama, base of stumps of white oak. Peters. Grev., Vol. 1, No. 5. New Jersey, *Ellis.*

PLATE CXXVI.

No. 1. POLYPORUS FUMOSUS.
2. POLYSTICTUS VERSICOLOR.
3. MERULIUS CORIUM.
4. POLYPORUS PERENNIS AND SECTION.
} About natural size.
5. DÆDALEA QUERCINA.
6. FOMES IGNIARIUS.
7. TRAMETES GIBBOSA.
} Reduced in size.

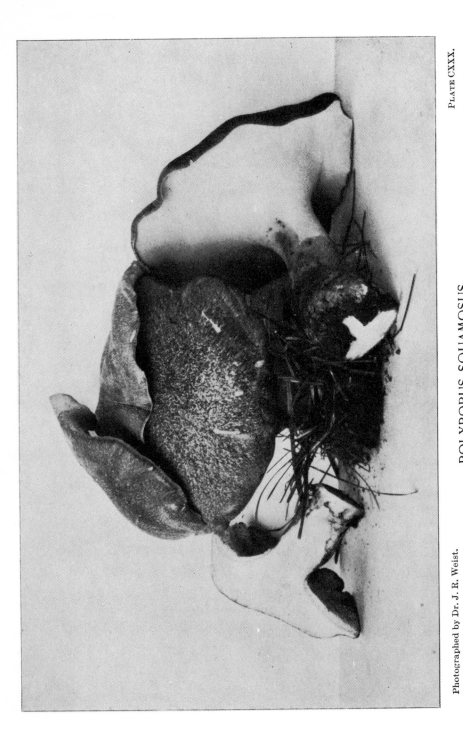

PLATE CXXX.

POLYPORUS SQUAMOSUS.

Photographed by Dr. J. R. Weist.

POLY'PORUS Fr.

Gr.—many; a passage, pore.

Pileus fleshy, moist, tough, becoming harder in age, internally com- Polyporus.
posed of radiating fibers; the spore-bearing surface is within passages
or pores which are made by the descending substance of the pileus forming the dissepiments or separating walls, hence they are not easily
separable from the pileus or from one another.   The pores not appearing at first, then becoming rounded, angular or torn.   They form a distinct strata.   **Stem** central, eccentric, lateral or absent.

With few exceptions growing from wood.   Section Merisma contains
species which are conspicuous among fungi for their size and beauty.

The majority of this genus are unedible, because of their being woody,
tough or bitter.   Few of the edible species are of the first class.

Excellent dishes are made by stewing the species well, serving them
in patties or in croquettes.   The cooking of P. intybaceus is a guide to
all.

ANALYSIS OF TRIBES.

I.—MESOPUS. (*Gr.*—middle; a foot.) Page 479.

Stem single, distinct, central or eccentric; not black at the base.

II.—PLEUROPUS. (*Gr.*—the side; a foot.) Page 480.

Stem single, lateral or eccentric; base black.

III.—MERISMA. (*Gr.*—to divide.) Page 482.

Divided into numerous pileoli, borne on a simple or much-branched
stem, or a short, thick tubercle.

IV.—APUS. (*Gr.*—without; a foot.) Page 488.

Stem wanting; pileus attached by the side or spread on the matrix.

V.—RESUPINATI. (Lying on the back.) Page 489.

The pores being placed directly upon the wood or on the mycelium,
the pileus proper is absent.

I.—ME'SOPUS.

P. ovi'nus Schaeff.—relating to sheep. **Pileus** 2–4 in. broad, fleshy,
thick, fragile, irregular in shape, becoming scaly, whitish. **Stem** short,

Polyporus. thick, 1 in. or more in length, white. **Pores** minute, equal, round, white then citron-color.

On the ground. Autumn.

North Carolina, *Curtis;* Massachusetts, *Frost;* Ohio, *Morgan;* New York, ground in pine woods. Bethlehem. September, *Peck,* 22d Rep.

Cordier says it possesses an agreeable odor of almonds and that Fries and his companions ate it raw in their mycological excursions.

Edible. *Peck, Curtis.*

P. leuco'melas (Pers.) Fr.—*leucos*, white; *melas*, black. **Pileus** 2–4 in. broad, fleshy, somewhat fragile, irregularly-shaped, silky, sooty-black. . **Flesh** soft, reddish when broken. **Stem** 1–3 in. in length, stout, unequal, somewhat tomentose, sooty-black, becoming black internally. Pileus and stem becoming black in places. **Pores** rather large, unequal, ashy or whitish, becoming black in drying.

Spores pale brown, 10–12x4–5µ. *Massee.*

North Carolina, edible, *Curtis;* Ohio, a curious esculent. *Morgan.*

P. circina'tus Fr.—round. **Pileus** 3–4 in. broad, compact, thick, round, plane, zoneless, velvety, reddish-brown. **Flesh** the same color. It forms duplicate strata of pilei, the inferior contiguous with the stem and corky; the superior compact, soft, floccose. **Stem** 1 in. thick and high, bearing a reddish-brown tomentum. **Pores** decurrent, entire, dusky-gray.

In fir woods.

A noble species, memorable for the stratified duplicate pilei.

Var. *prolif'erus.* Like the typical form but having one or more pilei developed from the upper surface of the first one. Fulton Chain August. *Peck,* 46th Rep. N. Y. State Bot.

New York. On ground in borders of woods. September. *Peck,* 32d, 46th Rep.

On ground in oak woods, West Philadelphia. *McIlvaine.*

When young the soft pilei are good.

II.—PLEUROPUS.

P. squamo'sus Fr.—*squama*, a scale. (Plate CXXX, p. 479) **Pi-**leus 3 in.–1 ½ ft. broad, somewhat ochraceous, *variegated with a broad,* *adpressed, spot-like*, centrifugal, *darker scales*, fleshy-pliant, fan-shaped, <small>Polyporus.</small> flattened. **Stem** excentric and lateral, obese, *reticulated* at the apex, blackish at the base. **Pores** thin, variable (at first minute), then large, angular and torn, pallid. *Fries.*

Handsome, commonly very large, somewhat central and umbilicate when young, at length lateral, very variable in shape.

On trunks and stumps, chiefly ash. Common. May to November. *Stevenson.*

Spores oval, white, 14x6µ *W.G.S.;* elliptical, colorless, 12x5µ *Massee.*

Massachusetts, *Sprague;* Iowa, *Macbride;* New York. Trunk of elm. May. *Peck*, 27th Rep.; West Virginia, New Jersey, Pennsylvania. On fallen trunks and on stumps. May to November. *McIlvaine.*

This species does not seem to be common in America, but is found throughout Europe. It varies in size from 3 in. to over 3 feet. It has been known to attain the circumference of 7 ft. 5 in., and the weight of 40 lbs. Dr. Badham says that it can not be masticated and that its expressed juice is very disagreeable. The fact, however, remains that it is eaten, and is recorded as edible by most authors. It is undoubtedly tough, but cut fine and stewed slowly for half an hour it is quite as tender as the muscle of an oyster and has a pleasant flavor.

P. pi'cipes Fr.—*pix*, pitch; *pes*, a foot. Pallid then chestnut, commonly pale yellowish-livid, with the disk chestnut. **Pileus** fleshy-coriaceous, then rigid, tough, even, smooth, depressed at the disk or behind. **Flesh** white. **Stem** excentric and lateral, equal, firm, at first velvety, then naked, dotted, black up to the pores. **Pores** decurrent, round, very small, rather slender, white, then slightly pale yellowish. *Fries.*

Imbricated, odor somewhat sweet. The pileus is depressed behind, commonly emarginate, funnel-shaped with lobes all round.

On trunks, especially willow. Frequent. July to December. *Stevenson.*

Many young plants, in tufts upon a decaying oak log, were found by me at Mt. Gretna, August, 1899. They were oyster-color, the very thin caps translucent, 2–6 in. across, $\frac{1}{16}$ in. thick; pores not visible to the naked eye. The black dots upon the stems developed some time after gathering.

They were pleasantly crisp when stewed and of fine flavor. Older specimens were bitter and tough.

III.—MERISMA.

P. umbella′tus Fr.—*umbella*, a sun-shade. Very much branched, fibrous-fleshy, toughish. **Pileoli** very numerous, ½–1½ in. broad, sooty, dull-red or pallid light-yellow, *entire, umbilicate*. **Stems** elongated, separate, united at the base, white. **Pores** minute, white.

The pileoli have occurred white. *Fries.*

Edible. *Fries.*

New York, *Peck*, Rep. 51; Richmond, Ind., *Dr. J. R. Weist;* Gouverneur, N. Y., *Mrs. E. C. Anthony;* West Virginia, New Jersey, Pennsylvania. On decaying roots in ground and on stumps. May to November. *McIlvaine.*

Tufts dense, branches spreading from a center. The pilei up to 2 in. across, connected at base. The dense spreading tufts, up to a foot across and half as high, are very noticeable. The flesh is soft and of good flavor. Cook like P. intybaceus.

P. a′nax Berk. Fleshy, fibrous, rather tough, dusky-gray, branching out from a thick, single stem at the base and forming a large head of branches and pileoli 10–20 lines in diameter; the branches terminate in numerous large pileoli of various forms and size, imbricating, confluent and recurved. **Flesh** and **pores** white. **Stems** thick, growing together, white. **Pores** large, unequal, angular, white. **Spores** white, subelliptic, 7–8µ long.

Ohio, at the base of oak trees and stumps. Autumn. *Morgan.*

This species has apparently been confused by some American mycologists with P. intybaceus. I have received specimens of it bearing that name. The spores of that species are described as elliptic or ovoid. The spores of Polyporus anax, as shown by our specimens, are globose. *Peck*, 51st Rep. N. Y. State Bot.

Ohio, *Morgan;* New York, *Peck*, 51st Rep.; New Jersey, *Sterling;* Angora, West Philadelphia, growing on rotting stump. September, 1897, *McIlvaine.*

Edible when young and fresh.

Photographed by Dr. J. R. Weist.

POLYPORUS FRONDOSUS.

PLATE CXXVIII.

P. frondo'sus Fr.—*frons*, a leafy branch. (Plate CXXVIII, p. 482.) Polyporus.
Tuft ½–1 ft. broad, very much branched, fibrous-fleshy, toughish.
Pileoli very numerous, ½–2 in., sooty-gray, *dimidiate, wrinkled*, lobed,
intricately recurved. **Flesh** white. **Stems** growing into each other,
white. **Pores** *rather tender, very small, acute*, white.

Pores commonly round, but in an oblique position, gaping open and
torn. *Fries.*

North Carolina, *Curtis;* Iowa, *Macbride;* New York, *Peck*, 24th
Rep.; West Virginia, 1881–1885, Chester county, Angora, Philadel-
phia, Pa. On stumps, roots, etc. Rare. September to frost, *McIlvaine.*
Edible. *Curtis.* Sold in the Roman market.

Tufts up to 12 in. across; the branches very numerous, up to 2 in.
wide. The plant is tender when young and grows tough as it matures.
When young it is of good flavor and edible—older it makes a well-
flavored gravy, or is edible if chopped fine and very well cooked.

P. intyba'ceus Fr.—succory-like. Very much branched, fleshy,
somewhat fragile. **Pileoli** *very nu-
merous*, pale-yellowish inclining to
fuscous, *dimidiate, stretched out*, sinuate, at length spathulate. **Stems**
connate in a very short trunk. **Pores**
firm, obtuse, white, inclining to dingy-
brown. *Fries.*

About same size as P. frondosus
and larger. *Stevenson.*

Spores colorless, elliptical, 7x3.5µ
Massee; 6x3µ *W.G.S.*

(Plate CXXIX.)

POLYPORUS INTYBACEUS.

Indiana, *H. I. Miller.* Base of
living trees. Woodland Cemetery, West Philadelphia, Pa., Mt. Gretna,
Pa., West Virginia, New Jersey. Large tufts growing from oak roots
in ground and at base of oak trees. *McIlvaine.*

Edible. *Stevenson.* Paulet says: In place of its being heavy upon
the stomach, *he* will feel all the lighter who sups upon it.

The people of the Vosges call it the Hen-of-the-Woods.

The words of the old song—

"So very much depends upon
The way in which it's done,"

483

apply with exceptional force to the cooking of P. intybaceus. **If it is** cut in thin slices across the grain and slowly stewed for half an hour it will be tender and of good flavor. It can then be served in that way, or made into patties or croquettes.

P. crista′tus Fr.—*crista*, a crest. Branched, firmly fleshy, fragile. **Pileoli** about 3 in. broad, *reddish-green*, entire and dimidiate, imbricated, *depressed, somewhat pulverulent-villous, then cracked into scales.* **Stems** connate, irregularly shaped, white. **Pores** minute, angular and torn, whitish. *Fries.*

Very changeable in form, sometimes simple with an undulato-lobed, central pileus.

Edible. *Curtis.*

Mt. Gretna, Pa., Woodland Cemetery, Philadephia, West Virginia. On ground over roots, open woods and grassy places. September, October. *McIlvaine.*

Variable in form, but usually in rose-shaped clusters, which are slightly greenish at times; oftener shades of yellow. The substance is the same in texture as P. intybaceus. Cook in same manner.

P. con′fluens Fr.—stems confluent; adherent. **Pilei** branched, fleshy, fragile, thick, dimidiate, imbricated, confluent, smooth, fleshy-yellow becoming obscure, slightly scaly. **Stem** short. **Pores** short, minute, pallid-white.

Eaten about Nice; savor a little sharp. *Cordier;* North Carolina, superior eating. *Curtis.* Pine woods. New Scotland. September.

Our specimens are not at all squamulose, and this character is not attributed to the species by all authors. It is probable that it is not uniform in this respect. *Peck*, 39th Rep. N. Y. State Bot.

P. Berk′eleyi Fr. Very much branched. **Pileoli** very large, subzonate, finally tomentose, yellowish, fleshy, tough becoming corky and hard. **Stem** short or none, arising from a long and thick common base growing out of the ground usually near trees or stumps. **Pores** rather large, irregular, angular, pale yellowish.

A magnificent specimen found near Boston a dozen years ago and exhibited in the window of Doyle, the florist, was fully four feet high and from two to three feet broad, containing very many pileoli.

North Carolina, edible, *Curtis;* Iowa, *Bessey;* Ohio, *Morgan;* Mt. Polyporus. Gretna, Pa., very large specimens, 20 in. across. *McIlvaine.*
Edible when young.

P. gigante'us Fr.—*gigas*, a giant. Tuft 1–2 ft. and more broad, in many imbricated layers, fleshy-pliant then somewhat coriaceous. **Pilei** *date-brown*, dimidiate, very broad, flaccid, somewhat zoned, rivulose, depressed behind. **Stems** connato-branched from a common tuber. **Pores** *minute, somewhat round, pallid*, at length torn.

The rigid cuticle separates into granules or fibrillose squamules. Pores becoming dark when touched. *Fries.*

Edible, *Curtis.* Esculent when young. On the continent its esculent qualities are known and appreciated. *Cooke.*

West Virginia, Chester county, Pa., Eagle's Mere, Pa. On decaying stumps and roots. *McIlvaine.*

It is well marked by its spore-surface becoming black to the touch. When young and fresh it stews to a pleasant, edible consistency, but is tough if not well cooked or too old. The flavor of a gravy from it is at all times good.

P. sulphu'reus Fr.—*sulphur*, brimstone. ([Color] Plate CXXV, fig. 2.) In many cespitose layers, 1–2 ft. and more, *juicy-cheesy.* **Pilei** 8 in. or more broad, *reddish-yellow*, imbricated, undulated, rather smooth. **Flesh** light yellowish, then white, splitting open and not hardened when old. **Pores** minute, plane, *sulphur-yellow. Fries.*

Soon becoming pale. Commonly sessile, but varying with a stem, lateral on standing trees, but expanded on all sides on fallen ones; also club-shaped, porous throughout. *Sow.* In its fullest vigor it is filled with sulphur-yellow milk.

On living trees and stumps. Frequent. August to October. *Stevenson.*

Spores oval, white, minutely papillose, 8x5μ *W.G.S.;* elliptical, hyaline, slightly papillose, 7–8x4–5μ *Massee.*

Edible. *Stevenson, Curtis.*

Maryland, *Miss Banning;* Indiana, *H. I. Miller;* West Virginia, New Jersey, Pennsylvania. On willow, apple, cherry, maple, hickory, etc. Frequent. August to November. *McIlvaine.*

Frequently in large masses. Commonly broadly attached, but some-

Polyporus. times with a short stem. Very occasionally a single pileus will protrude from a tree like a giant yellow tongue shaded with reddish-orange. Usually the pilei are in clusters united in a solid base, white-fleshed and rich in color. I have seen clusters two feet across. On an old willow at Mt. Gretna, a cluster 18 in. across afforded a dozen meals. Whenever a meal was wanted a pound or two was broken off. It lasted until January. If P. sulphureus is cooked properly it is a delicious fungus. Cut fine, stew slowly and well, season, add butter, milk with a little thickening.

P. macula'tus Pk.—having *maculæ*-spots. **Pileus** of a cheesy consistence, broad, flattened, sometimes confluent, sessile or narrowed into a short stem, slightly uneven, white or yellowish-white, marked with darker zones and watery spots. **Pores** minute, subangular, short, whitish, sometimes tinged with brown. **Flesh** white.

Pileus 4–6 in. broad, 6–8 lines thick.

Prostrate trunks of trees in woods. Worcester. July.

In texture and shape this species is related to P. sulphureus, but the pores are smaller than in that species. The plants are sometimes cespitose, sometimes single. The spots in the dried specimens have a smooth depressed appearance. *Peck*, 26th Rep.

Angora, West Philadelphia. September, 1896. Mt. Gretna, Pa., September, 1897–1898. On white oak trunks. *McIlvaine.*

Several specimens of different ages proved good eating. Like P. sulphureus it must be well cooked.

P. hetero'clitus Fr. *Gr.*—one of two; *Gr.*, to lean. In many cespitose layers, coriaceous. **Pilei** 2½ in. broad, *orange, sessile, expanded on all sides from a radical tubercle*, lobed, villous, zoneless. **Pores** irregularly shaped and elongated, golden-yellow. *Fries.*

On the ground under oak. Rare.

The flat pilei extend horizontally from the tubercle. Irregular, eccentric. *Stevenson.*

Minnesota, *Johnson.*

Haddonfield, N. J., Hopkin's woods. June to July, 1890-1896. *McIlvaine.*

Of all fungoid growth this is the most showy. Its clusters, often a foot and a half in diameter and spread like mammoth dahlias, are gor-

geous in color and conspicuous in design. Resting upon the ground or Polyporus.
reared against the base of tree or stump, they deceive by their likeness
to gaudy bouquets, left by foreign picnickers. In quality it is the same
as P. sulphureus. It does not, however, retain its edibility. As it ages
it becomes offensive.

P. por'ipes Fr.—porous-stemmed. **Pileus** 1.5–3 in. broad, rather
fleshy, sinuately repand, smooth, grayish-brown. **Stem** central or excentric, firm, smooth, 1.5–3 in. long, 4–6 lines thick, punctuated by
the whitish decurrent pores.

On earth in hilly regions.

Cap 2 in. across, light drab, smooth, slightly furfuraceous toward
center, broken into minute appressed squamules, zoned. **Flesh** fibrous,
white-pliable. **Tubes** very shallow, round mouths with obtuse divisions,
china-white, running down to base of stem. **Stem** eccentric, almost
lateral, entirely surrounded by pores, connate at base, ⅛ in. thick.

·**Smell** pleasant.

New York. Ground. August, *Peck*, Rep. 24; Mt. Gretna, Pa.,
August to November, *McIlvaine*. A large tufted species growing on
the ground in woods, August to November, *McIlvaine*.

When raw tastes like the best chestnuts or filberts, but rather too dry
cooked. *Curtis.*

It must be chopped fine and slowly cooked.

P. immi'tis Pk.—wide, rude. **Pilei** cespitose-imbricated, broad,
slightly convex or flattened, more or less rough or uneven, radiately-
wrinkled, tuberculose or fibrous-bristled, zoneless, white, becoming
tinged with yellow or alutaceous in drying. **Flesh** white, slightly fibrous,
soft and moist when fresh, cheesy when dry, with a subacid odor. **Pores**
minute, angular or even subflexuous, about equal in length to the thickness of the pileus, the dissepiments thin, white, often at length dentate
or lacerate on the edge. **Spores** minute, white, elliptical, $3-4 \times 18-20\mu$.

Pilei 2–4 in. broad, the flesh commonly 3–4 lines thick.

Decaying ash trunks. East Berne. August.

The species is apparently related to P. cæsareus, but the character of
the pores is quite different in the two species. *Peck*, 35th Rep. N. Y.
State Bot.

Mt. Gretna, Pa. On dead black oak. August to November, 1898.

Polyporus. Several clusters grew on dead black oaks. The pilei overlap and the wrinkled corrugated margins curve downward, giving them the semblance of shells. From a distance a group looks like Pleurotus ostreatus. The substance is juicy; while cooking it is at first bitter, but this disappears. It becomes tender and well flavored.

P. alliga′tus Fr.—*alligo*, to bind to. In many cespitose layers, fibrous-fleshy, rigid-fragile. **Pilei** tan-isabelline, imbricated, unequal, *zoneless, villous*. **Pores** minute, soft, white, readily becoming stopped up with flocci.

Often clavate when young. Commonly wrapping round stipules and grasses. *Fries.*

Spores elliptical, pale, 6x7μ *Massee*.

Woodland Cemetery, Philadelphia. Among oak trees on grassy ground. July, August, September. *McIlvaine.*

Tufts frequently weigh two pounds. When young the plant cooks well, is tender and of sweet, pleasant flavor. When old it has a sour unpleasant odor.

IV.—APUS.

P. chio′neus Fr. *Gr.*—snow. White **pileus** 1 in. and more broad, fleshy, *soft, becoming even, smooth*, zoneless, often extended behind, margin inflexed. **Pores** curt, very small, round, equal, quite entire. *Fries.*

Always soft, fragile, hyaline-white when moist, shining white when dry. Odor acid. Without a cuticle. *Stevenson.*

Spores white, oval, 21x3μ *W.G.S.*

New York. Decaying wood of frondose trees. *Peck*, 33d Rep.

Angora, Philadelphia, Mt. Gretna, Pa. On standing and fallen timber. June to September. *McIlvaine.*

This snow-white Polyporus is too conspicuous to be passed unseen. One does not expect to find snow-balls stuck against trees in August. At a distance it resembles one. When young and fresh it is good.

P. betuli′nus Fr.—*betula*, birch. **Pileus** fleshy, then corky, hoof-shaped, obtuse, zoneless, smooth, *the oblique vertex in the form of an umbo*, pellicle thin, separating. **Pores** late of being developed, curt, minute, unequal, at length separating. *Fries.*

488

On living and dead birch.   Common.   May to December.

Pileus 3–6 in. broad.   The pileus is at first pale, then acquiring a brownish tinge.   The edge is always very obtuse.   *Stevenson.*

The lower surface or hymenium is frequently rough with numerous acicular projections, making the plant look like a Hydnum when viewed horizontally.   *Peck*, 24th Rep. N. Y. State Bot.

Massachusetts, Kansas, New York.   *Peck*, Rep. 24.

Wherever the birch grows this neat, white-fleshed Polyporus abounds.

When young it is eaten by deer.   Dried it burns with a white flame, or holds fire as well as the best punk.   It is a valuable fuel, already prepared for the stove.   In the birch forests near Eagle's Mere, Pa., tons of it can be seen protruding from tree and log.

When very young it is fair.   Unpleasant when old.

V.—Resupinati.

P. sinuo'sus Fr.—full of folds.   Broadly effused, adnate, dry, the evanescent mycelium somewhat rooting, white then yellowish.   **Pores** large, surface flexuous, acute, lacerated.   Odor of licorice.

New York.   Decaying wood of maple.   *Peck*, 40th Rep.

Mt. Gretna, Pa.   *McIlvaine.*

Of but little food value.   Collected carefully and boiled, it yields a pleasantly flavored liquor.

MERU'LIUS Hall.

(Plate CXXVI, fig. 3, p. 478.)

Merulius.   Hymenophore resting on a loose mold-like mycelium, covered with the soft, waxy, continuous hymenium, having its surface variously plicate or wrinkled, the folds forming irregular pores, sometimes obsoletely toothed.

Generally on wood.

I have tasted, raw, every species I have found. They are all more or less woody in flavor, and I believe them to be edible. At the best Merulius would be an emergency genus. M. tremellosus is substantial, as is M. rubellus Pk.

M. tremello'sus Schrad.—*tremellosus*, trembling. Resupinate; margin becoming free and more or less reflexed, usually radiately-toothed, gelatinoso-cartilaginous; hymenium variously wrinkled and porous; whitish and subtranslucent looking, becoming tinged brown in the center. **Spores** cylindrical, curved, about $4 \times 1\mu$.

On wood. From 1–3 in. across, remaining pale when growing in dark places. Margin sometimes tinged rose, radiating when well developed. *Massee.*

Spores cylindrical, curved, hyaline, $4 \times 1\mu$ *K.*

New York. Old logs, stumps, Catskill mountains. *Peck*, 22d Rep. N. Y. State Bot.

Mt. Gretna, Pa. Common, both rose-colored and translucent brown species, numerous on decaying wood. October to November, 1898–1899. *McIlvaine.*

M. tremellosus is a common species and rather attractive looking. In substance it approaches Tremella and Peziza. The spore-bearing surface is superior (turned upward) and then sometimes turned in at the margin which frequently is bright rose color, sometimes yellowish-rose. It is rather tasteless—slightly woody in flavor, rather tough. An emergency species.

M. rubel'lus Pk.—*rubellus*, dim. of *ruber*, reddish. Generally cespitose, imbricated, sessile, dimidiate, soft, tenacious, tomentose, evenly red, pale when dry; margin mostly undulately inflexed; hymenium

white or flesh-color; folds branching, forming anastomosing pores. Merulius.
Spores elliptical, hyaline, minute, 4–5x2.5–3μ.

Pileus 2–3 in. long, 1.5 in. broad.

Somewhat related to M. tremellosus.

On trunks of beech in woods.

Ohio, *Morgan;* Indiana, *Dr. J. R. Weist;* Mt. Gretna, Pa., November. *McIlvaine.* Specimens identified by Professor Peck.

Tough, but edible.

FAMILY III.—**HYDNA'CEÆ.**

Hymenium inferior or amphigenous (not confined to one surface), from the first definitely protuberant, spread over persistent spines, bristles, teeth, tubercles or papillæ. *Fries.*

While the highest members of this family possess the general form of the mushroom, others, lacking a stem, recline on the back (resupinate); the lowest, without even the appearance of a distinct pileus, seem to be simply spread over the supporting body (effused). In the highest class the spines or other spore-bearing surface are inferior, *i. e.*, below the pileus; in the others they are of course superior, *i. e.*, above the pileus.

Of the eleven genera but two contain species of food value. Hydnum, characterized by its acute spines, embraces species which are eaten as delicacies, and Irpex, distinguished by its somewhat acute teeth growing from a ridgy hymenium, contains those which may furnish sustenance in time of need. In Caldesia, bearing spines, the texture is floccose not fleshy. Sistotrema has a pileus and a central stem, but instead of spines bears irregular flattened teeth. The remaining genera are separated by the tubercles, granules, folds, etc., which take the place of spines or teeth.

Several species of Hydnum are common to earth and wood, others are distinct in their habitats.

SYNOPSIS OF THE GENERA.

HYDNUM. Page 494.

Sporophore fleshy, with a central stem or entirely resupinate, texture compact, spines acute, distinct at the base.

CALDESIELLA.

Resupinate; texture floccose, spines acute; spores muriculate. (No edible species reported.)

SISTOTREMA.

Pileate; fleshy, central-stemmed, teeth flattened, irregular, inferior. (No edible species reported.)

IRPEX. Page 504.

Resupinate; teeth rather acute, springing from folds or ridges that often anastomose irregularly.

RADULUM.

Resupinate; tubercles coarse, deformed, subcylindrical, obtuse. (No edible species reported.)

PHLEBIA.

Resupinate; hymenium covered with folds or wrinkles, having the edge entire or corrugated. (No edible species reported.)

GRANDINIA.

Resupinate; hymenium with crowded, globose, persistent, hemi spherical, minute granules, having their apices more or less excavated. (No edible species reported.)

POROTHELIUM.

Resupinate; hymenium with scattered wart-like granules, which become more or less elongated and excavated at the apices. (No edible species reported.)

ODONTIA.

Resupinate; hymenium densely covered with small granules that are divided at the apices in a penicillate manner. (No edible species reported.)

KNEIFFIA.

Resupinate; hymenium covered with very minute, barren, acute spinules. (No edible species reported.)

MUCRONELLA.

Spines slender, elongated, acute, not springing from a sporophore or subiculum. (No edible species reported.)

493

HYD'NUM.

Gr.—name for some edible fungus.

Hydnum.

HYMENIUM inferior, bearing awl-shaped Spines, distinct at the base. *Fries.*

In this genus the spines proceed from an even surface, not folded or wrinkled, and are covered with the spore-bearing surface.

The forms are extremely variable, the type of the first section, H. repandum, being easily mistaken for one of the Agaricaceæ until examined, the stem being nearly central and upright, while in other forms it is lateral or absent. Some are dimidiate (as if part of the pileus had been removed and the plant attached by the remaining portion); the lower forms are resupinate.

ANALYSIS OF TRIBES.

MESOPUS (*Gr.*—middle, a foot). **Page 495.**

Entire, simple, stem central.
On the ground, mostly in pine woods.

PLEUROPUS (*Gr.*—the side; a foot).

Stem lateral.
None known to be edible.

MERISMA (*Gr.*—to divide). **Page 501.**

Very much branched or of an irregular form without a distinct margin.

APUS (*Gr.*—without; a foot). **Page 503.**

Stemless, dimidiate, margin distinct.

RESUPINATI (*resupino,* to throw on the back).

Without stem or distinct pileus.
None known to be edible.

494

MES'OPUS. *Gr.*—middle; a foot.

(Entire, simple, stem central. On the ground, mostly in pine woods.)

H. imbrica'tum L.—*imbrex*, a tile. **Pileus** about 2–5 in. broad, Hydnum. *umber*, zoneless, fleshy, rather plane, somewhat umbilicate, *floccose*, tessulato-scaly. **Flesh** dingy whitish. **Stem** curt, 1–3 in. long, 1–2 in. thick, even. **Spines** 4–6 lines long, decurrent, ashy-white.

There are two forms; one with the pileus plane and with thick persistent scales, another with the pileus somewhat infundibuliform, and with thinner, at length separating scales. *Stevenson.*

Spores pale yellow brown, rough, 6–7x5μ *Massee;* 6x5μ *W.G.S.*

Fleshy. The numerous scales over lapping toward the center. The surface of the cap often cracks in a tesselated manner. Flesh dingy, buffish or reddish. **Spines** short, blunt, grayish-white and mostly of equal length.

In pine and mixed woods. Autumn.

Of delicate taste. *Cordier.* Edible. *Curtis.*

Fine specimens grew at Mt. Gretna, Pa., from September to November. Until closely examined the cap may be mistaken for that of H. zonatum. The zones of the latter and the pervading rust-color will distinguish it. Both are edible, though H. zonatum is much tougher. H. imbricatum is slightly bitter, raw. It must be sliced thin and well cooked.

H. læviga'tum Swartz—*lævis*, smooth. **Pileus** 4–6 in. broad, *umber*, fleshy, *compact*, firm, regular, plane, *even, very smooth*, margin circinate (not repand). **Flesh** whitish, compact, but by no means fibrous, soft when fresh, pliant when dry. **Stem** short, thick, even, pallid-brown. **Spines** thin, pallid-brown.

Its size is that of H. imbricatum, but it occurs twice as large, with the pileus minutely rimuloso-rivulose, by no means scaly. The stem varies curt and unequal or longer and equal. Quite distinct from H. fragile. *Stevenson.*

Spores 10–15μ long, *Massee;* globose, warted, pale lemon-yellow, 7μ *Q.*

In pine woods. August to October.

Edible. *Curtis;* edible, *Leuba.* "Eaten in Alpine districts." *Barla.*

495

Hydnum.

H. scabro'sum Fr.—*scabrosus*, rough. **Pileus** about 1 ½ –4 in. broad,
brownish-yellow, compactly fleshy,
at first top-shaped, then plane above,
very convex beneath, at first tomentose, then rough with flocci which
are fasciculate in the form of minute
crowded squamules, slightly repand
at the margin. **Flesh** very thick,
white, descending into the stem.
Stem very curt, 1 in. long, and
equally thick, round or compressed,
dotted with the rudiments of spines
decurrent upon it, ash-color, attenuated downward, roundish and blackish at the base. **Spines** 4 lines long,
equal, awl-shaped, dingy-rust color, whitish at the apex, at first sight
grayish-brown. *Fries.*

(Plate CXXXI.)

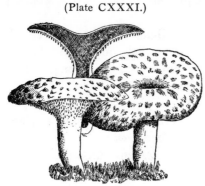

HYDNUM SCABROSUM.
Natural size.

Spores 4–5µ diameter. *Massee.*

Hydnum scabrosum is frequently found in Pennsylvania, among pines
and in mixed woods where pines grow. It occurs at Mt. Gretna, Pa.,
and on Springton Hills under hemlocks.

The caps are soft, fleshy, and equal to H. repandum in quality.

H. squamo'sum Schaeff.—*squama*, a scale. **Pileus** 1 ½ –3 in. across,
reddish-brown, fleshy, irregular, depressed, *smooth*, breaking up into
irregular scales. **Flesh** whitish. **Stem** curt, attenuated downward,
white. **Spines** grayish-brown, whitish at the apex. *Stevenson.*

Spores subglobose, 5–6µ diameter. *Massee.*

Pileus smooth and even when young. **Flesh** whitish. **Spores** grayish-brown. **Spines** whitish, giving the lower surface a much lighter appearance than the upper.

Under hemlock and spruce in West Virginia, 1884. *McIlvaine.*

Caps are good when sliced thin and well cooked.

H. subsquamo'sum Batsch. **Pileus** fleshy, somewhat convex, subumbilicate, brownish-rust color, superficial scales soon dropping off;
spotted with brown. **Stem** stout, unequal, smooth. **Spines** whitish,
becoming brown, apex remaining whitish.

North Carolina, *Curtis;* Alabama, *Peters;* Massachusetts, *Sprague.*
Edible. *Curtis.* Edible. *Cordier.*

H. repan'dum L.—*repandus*, bent backward (of the cap, upward).
Pileus 2–6 in. broad, *pallid*, etc.,
fleshy, fragile, *somewhat repand*,
rather smooth. **Stem** 2–5 in. long,
½–1½ in. thick, irregularly shaped,
pallid. **Spines** 4 lines long, unequal, of the same color. *Stevenson.*
   Spores pointed, 5–8μ *Massee.*

(Plate CXXXII.)

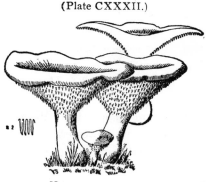

HYDNUM REPANDUM.

   Pileus sometimes depressed, often
turned upward at margin, often
waved, sometimes tomentose. Color
variable—light-buff, brown, pinkish,
reddish. **Flesh** whitish, compact,
fragile. **Spines** conical, up to ¼ in.
in length, whitish but rich creamy shades, mostly pointed, but sometimes appearing to be hollow. **Stem** central or eccentric, sometimes
covered with white down, thick, uneven, usually crooked, solid, fleshy,
light in color.
   July to November.
   Edible. *Curtis.*
   Common to most countries, and, although given as a ground-growing
species, it is rather indiscriminate in its habitats. Woods, fields, leaf-
covered or bare places, much decayed wood and stumps are its living
places. Dr. Cooke thinks it irreproachable. Popularly it goes by the
name of the Hedgehog mushroom.
   H. repandum varies greatly in shape, color and texture. In the open
it is usually symmetrical and tough; when clustered it is irregular, often
fanciful and quite brittle—tender.
   When sliced thin an hour's slow cooking is sufficient. All writers
commend it, and properly.

   H. rufes'cens Pers.—*rufus*, red. **Pileus** 2–3 in. across, thin, fragile,
usually regular, pubescent, reddish. **Spines** 1–3 lines long, regular.
Stem 1–3 in. long, commonly thin, nearly equal, reddish.

497

The whole plant is reddish. In all other respects it resembles H. repandum. Usually more regular.

Commonly found in woods. New York, *Peck;* North Carolina, *Curtis, Schweinitz.*

Edible, *Curtis.* Edible, *Leuba.*

Fries considered H. rufescens a variety of H. repandum, and the writer agrees with him. It is given distinct place here because Massee and Stevenson—books in the hands of many students of fungi—give it importance.

It is quite as good as H. repandum.

H. ferrugi'neum Fr. **Pileus** 1–4 in. across, corky, soft, convex, then plane or depressed, irregularly pitted, ferruginous, at first with whitish tomentum. **Flesh** ferruginous. **Spines** thin, acute, about 2 lines long, rusty-brown. **Stem** firm, 2–3 in. long, unequal, rusty-brown. **Spores** subglobose, 4μ diameter.

In fir woods. Often gregarious; soft when young, corky and dry at maturity. *Massee.*

Mt. Gretna, Pa. November to December, 1898. Among pine leaves.

Taste mild, mealy. Tough, but when young it cooks tender.

H. zona'tum Batsch. Ferruginous. **Pileus** 1–2 in. broad, *equally coriaceous*, thin, expanded, somewhat infundibuliform, *zoned, becoming smooth, radiately-wrinkled*, the paler margin sterile beneath. **Stem** ½–¾ in. long, 2–3 lines thick, slender, somewhat equal, floccose, base tuberous. **Spines** 1–1½ lines long, slender, pallid, then rust-color. *Stevenson.*

Spores rough, globose, pale watery brown, 4μ diameter *Massee.*

New York, *Peck*, 24th Rep. Mt. Gretna, Pa. Abundant among hemlocks; West Virginia. *McIlvaine.*

Coriaceous. Edible. It will not cook tender, but yields a pleasant flavor to a gravy made of its juices.

H. albo'nigrum Pk. **Pileus** convex or nearly plane, broadly ob-conical, tough but soft and densely tomentose on the upper surface, buff-brown or smoky brown, often wholly covered with a whitish downy tomentum, sometimes on the margin only, substance within soft tomen-

tose and buff-brown in the upper stratum, the lower half hard and black. Hydnum.
Spines short, at first white, then whitish or grayish. **Stem** short, often irregular, compressed or growing together, blackish when moist, buff-brown when dry, covered with a thick dense tomentum, which is frequently more abundant toward the base, hard and black within. **Spores** white, globose, 4–5μ.

Pileus 1–3 in. broad, sometimes 2 or 3 confluent. Stem 1–2 in. long.

Ground in mixed woods. Gansevoort. August. *Peck,* 50th Rep. N. Y. State Bot.

Specimens from pine woods New Jersey, *T. J. Collins,* September, 1897. 1½ in. across. Frequent at Mt. Gretna, Pa.

Edible. Good flavor, but tough.

H. velle'reum Pk. This species appears to be very much like the preceding one (H. albonigrum Pk.) from which it is separated by its smaller size and the paler brownish or rusty-brown substance of its pileus and stem. *Peck,* 50th Rep. N. Y. State Bot.

At Mt. Gretna, Pa., the species grows with H. albonigrum. In quality it is the same.

H. al'bidum Pk. **Pileus** fleshy, thin, broadly convex or nearly plane, subpruinose, white. **Flesh** white. **Spines** short, white. **Stem** short, solid, central or eccentric, white. **Spores** subglobose, 4–5μ broad.

(Plate CXXXIII.)

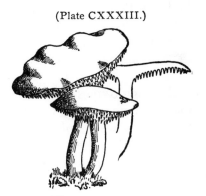

HYDNUM ALBIDUM.

The whitish Hydnum is uniformly colored in all parts. It grows in groups or in clusters. In the latter case the caps are sometimes irregular because of the crowded mode of growth and the stems are occasionally eccentric. It is a small species not liable to be mistaken for any other except possibly for very small pale forms of the spreading Hydnum. But wholly white examples of this species have never been seen by me.

The caps are 1–2 in. broad and the stems are generally about 1 in. long and 3–5 lines thick.

499

Hydnum. The plants grow in thin woods or in open bushy places and appear in June and July. It is not a common species, and though well flavored it is not of very great importance as an edible mushroom, because of its scarcity and small size. *Peck*, 51st Rep. N. Y. State Bot.

Port Jefferson. July. This fungus has been tested and found to be edible. *Peck*, 50th Rep. N. Y. State Bot.

Mt. Gretna, Pa., 1897. Specimens identified by Professor Peck. *McIlvaine*.

The caps are edible and superior to H. repandum.

H. fen'nicum Karst. **Cap** fleshy, fragile, unequal, at first scaly, at length breaking up, reddish-brick color becoming darker, margin undulately lobed, 2–4 in. broad. **Flesh** white. **Stem** sufficiently stout, unequal below, attenuated, flexuous or curved, smooth, of the same color as the cap, base acute, light white tomentum outside, inside light pale-blue or dark-gray (wood-ash), 1–3 in. long, .4–1 in. thick. **Teeth** decurrent, equal, pointed, from white dusky, about 4 mm. long. **Spores** ellipso-spheroidical or sub-spheroidical, rough, dusky, 4–6μ long, 3–5μ broad.

Found in gravelly or sandy soil in woods.

Found at Angora near Philadelphia. Top cracked. Identified by Professor Peck.

Occurs frequently at Mt. Gretna, Pa., ground in mixed woods. August to September. The taste and smell are at first inviting, but the extreme bitter which develops destroys all desire to eat it.

H. spongio'sipes Pk. **Pileus** convex, soft, spongy-tomentose, but tough in texture, rusty-brown, the lower stratum more firm and fibrous. but concolorous. **Spines** slender, 1–2 lines long, rusty-brown, becoming darker with age. **Stem** hard and corky within, externally spongy-tomentose, colored like the pileus, the central substance often transversely zoned especially near the top. **Spores** subglobose, nodulose, purplish-brown, 4–6μ broad.

Pileus 1.5–4 in. broad. **Stem** 1.5–3 in. long, 4–8 lines **thick**.

Woods. Rensselaer and Saratoga counties. August.

This plant was formerly referred to Hydnum ferrugineum Fr. *Peck*, 50th Rep. N. Y. State Bot.

Found in pine woods, near Haddonfield, N. J., by T. J. Col-

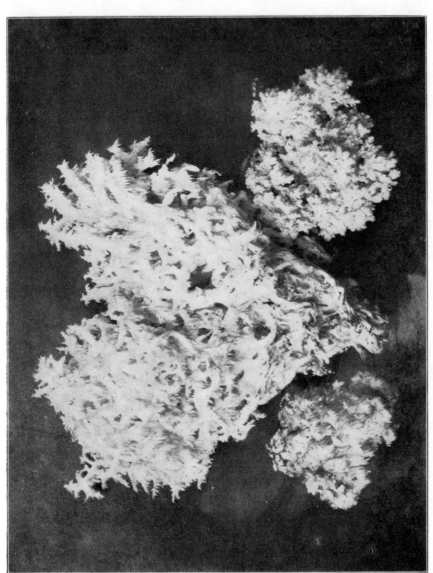

Plate CXXXIV.

Photographed by Dr. J. R. Weist.   HYDNUM CORALLOIDES.

 PLATE CXXXV.

HYDNUM CAPUT-MEDUSÆ.

lins, September, 1897. **Cap** and **stem** dark brown. **Spines** darker. Hydnum.
Stem swelling toward base, which then tapers in a long rooting way.
Cap umbilicate. Specimens identified by Professor Peck.

Edible but tough and uninviting. Yields a good fungoid flavor to
the water in which it is boiled.

H. gelatino sum Scop. Transferred to Tremelledon as T. gelatinosum, under which heading it is described and its edible qualities noted.

MERIS'MA. *Gr.*—a division.

(Very much branched or of an irregular form without a distinct margin.)

H. coralloi'des Scop. (Plate CXXXIV.) 6–18 in. across. Tufts
on wood. Pure shining white growing yellow with age, composed
wholly of attenuated interlacing branches ⅛ in. at base, tapering to a
point. **Spines** growing from one side of the branches, 3–4 lines in
length, awl-shaped.

Spores globose, 4–6µ diameter *Massee.*

Peck, Rep. 22; Indiana, *H. I. Miller;* Massachusetts, *Sprague;* California; West Virginia, New Jersey, Pennsylvania, *McIlvaine.*

Edible. *Curtis.*

Grows upon standing and fallen timber which is attacked by decay.
Fir, oak, beech, ash, birch, hickory and other trees are inhabited by it.
August to frost.

This beautiful species can not be mistaken for any other. Its name
is the best guide to its identification. Dame Nature has made many
exquisite decorations for herself and this is one of them.

It is generally eaten, but is rare. Professor Peck speaks affectionately
of it as a gratuitous adjunct to his bill of fare when on botanical tramps
in the Adirondacks.

H. caput-ur'si Fr.—bear-head. 6–8 in. high, 6–8 in. across. Tufts
usually pendulous, compact, white, becoming yellow and brownish.
Spines up to 1 in. long, round, pointed. **Branches** in every direction,
short.

Closely resembling H. coralloides and in small forms with shorter
spines easily mistaken for it. Position of growth has much to do with

501

Hydnum. its shape and appearance.   On fallen timber the branchlets and spines may be erect.

New York, *Peck*, 44th Rep.; North Carolina, *Curtis;* West Virginia, New Jersey, Pennsylvania, *McIlvaine.*

Edible. *Curtis.* Edible. *Peck.*

Hydnum caput-ursi is common in West Virginia forests.   It is conspicuous on standing oaks, and at a distance a puzzling object to one not familiar with such excrescences.   It grows on standing oaks near Haddonfield, N. J., and sparsely at Mt. Gretna, Pa.

It is more compact, and is tougher than H. coralloides and H. Medusæ, but cooks tender and is very good.

H. caput-Medu'sæ Bull.—head of Medusæ.   (Plate CXXXV.) 3–18 in. across, 2–8 in. high.   Tufts pendulous.   White then grayish. Body compact, tapering to a solid base, more or less stem-like. **Spines** covering entire surface.   Those upon top are long, thin, straight or distorted, growing shorter around and to the under side where they are short and straight.   The wavy appearance of the slender spines remind of the snaky locks of Medusa, hence the name.

Edible. *Curtis.* Edible. *Leuba.*

On elms at Haddonfield, N. J.; on oaks at Mt. Gretna, Pa., and in Woodland Cemetery, and on elms in Washington Square, Philadelphia, Pa. *McIlvaine.*

Commonly eaten in Italy and parts of Austria; rare elsewhere in Europe.   Occurring over the United States.   Specimens eighteen inches across were seen by the writer in the West Virginia mountains.

Mr. H. I. Miller, Terre Haute, Ind., sent me a fine specimen weighing 10½ pounds.

The American species, as far as seen by the writer, changes to a light yellow when ageing.   The entire fungus is edible and excellent, but the tender spines and more delicate parts make a dish equaled by few fungi.

H. erina'ceum Bull.—*erinaceus*, a hedgehog.   2–8 in. and more across.   Tufts pendulous.   White and yellowish-white becoming yellow-brownish, fleshy, elastic, tough, sometimes emarginate (broadly attached as if tuft was cut in two, sliced off where attached), a mass of latticed branches and fibrils. **Spines** 1½–4 in. long, crowded, straight, equal, pendulous. **Stem** sometimes rudimentary.

On trunks of oak, beech, etc.   July to October.

Spores subglobose, 5–6µ diameter *Massee;* white, plain, 5x6µ *W.G.S.*

Alabama, *Miss K. Skehan;* Pennsylvania, *McIlvaine;* Massachusetts, *Sprague;* New York, *Peck*, Rep. 22.

Eaten in Germany and France.   *Cooke.*

A dead beech trunk at Eagle's Mere, Pa., in August, 1898, bore at least fifty pounds of it.   It draped one side of the tree from root to top with yellowish, pendulous tufts, with spines up to 3 in. long, which waved in the wind.   The spines and tender parts were stewed, and enjoyed by many.   It shrinks very much in drying, becoming sour.

A'PUS.   *Gr.*—without; a foot.

(Stemless, dimidiate, margin distinct.)

H. septentriona'le Fr.—Northern.   Fleshy-fibrous, becoming pale, imbricated.   **Pilei** not numerous, growing one above the other, plane, behind thick, consolidated, margin straight, whole.   **Spines** very crowded, slender, equal.

The largest known Hydnum.

Received from E. B. Sterling, Trenton, N. J., September, 1897. The specimens formed part of a dense fasciculate mass weighing over 20 pounds, growing on a beech stump.   Edges of the young plant are edible, but have little taste.

503

IR'PEX Fr.

A harrow.

Irpex.   Hymenium inferior, toothed from the first. Teeth firm, somewhat coriaceous, acute, concrete with the pileus, arranged in rows or like network, connected at the base by folds, which are gill-like (in sessile species) or resemble honeycomb (in resupinate ones). Sporophores 4-spored. Growing on wood, somewhat growing from the side or upon the back, approaching Lenzites and Dædaleæ.

Irpex differs from Hydnum in having the spines connected at the base, and in their being less awl-shaped and pointed.

It is reported as found well up in the northern States, but its species prefer warm climates. Irpex contains no choice species, but all I have tested can be eaten.

I. obli'quus Fr.—oblique. White, inclining to pale, effused (spread), forming an adnate crust, circumference flaxy. Teeth *extended from a base resembling honeycomb, compressed, unequal, incised*, oblique, 2–3 lines long.

At first abundantly porous, but toothed from the first, at length quite as in Hydna.

On stumps and dead branches. November to February. *Stevenson*.

This spreads in irregular patches on the surface of decaying wood. The pores for a small space round the margin are round and distinct, but toward the center are greatly lengthened out, lying one upon another in an imbricated manner. The color is white at first, when old it changes to a yellow-brown, and at last to a dirty fuscous black. *Bolton*.

At first it looks more like a small white orbicular resupinate Polyporus than an Irpex. *Peck*.

The species is common and can be collected at most times of the year. When fresh and moist it can be shaved from its host plant. Goodly quantities can thus be obtained. It stews to a firm gelatinous mass of pleasant flavor. The lost hunter need not die of starvation in any woods if he will but study the tree-growing fungi, and especially the small species, hitherto insignificant in food circles.

I. car'neus Fr.—resembling the color of flesh. Reddish, effused, 1–

504

3 in. long, *cartilaginous-gelatinous*, membranaceous, adnate.   Teeth _{Irpex.} obtuse and awl-shaped, entire, united at the base.

It inclines to Radula and Phlebia.   *Stevenson.*

On tulip poplar, Haddonfield, N. J., September, 1892; on hickory, Angora, Philadelphia, September, 1897.   *McIlvaine.*

The entire fungus is good, cooking like a Hydnum.

I. defor'mis Fr.—deformed.   White, effused, crustaceous, thin, cir cumference pubescent, somewhat flaxy.   Teeth *extended in awl-shape from a minutely porous base, thin*, somewhat digitato-incised (cut in finger-shape), 1–2 lines long.   *Fries.*

It approaches the Polypori.   Grows on wood.   *Stevenson.*

North Carolina, *Schweinitz*, *Curtis;* Massachusetts, *Frost.*

Common on stumps and trees.   The awl-shaped teeth, which have the appearance of shreds, can be scraped from the fresh plant, or if dried plants are moistened, the teeth are detachable, and are food-giving.

I. fusco-viola'ceus Fr.—*fuscus*, brown; *violaceous*, violet.   **Pileus** 2 in. long, more than 1 in. broad, *white inclining to hoary*, effuso-reflexed, coriaceous, silky, zoned.   Teeth in rows in the form of plates, *brownish-violet*, incised at the apex.   *Fries.*

On pine trunks.   *Stevenson.*

Decaying trunks of spruce, abies nigra.   Adirondack mountains. July.

Our specimens are not "silky," as required by the description, but villose or tomentose-villose as in Polyporus hirsutus and P. abietinus, the latter of which this species closely resembles.   The hymenium, however, is coarser, more highly colored and lamellated to such an extent that young specimens might easily be taken for a Lenzites. *Peck*, 30th Rep. N. Y. State Bot.

Found in West Virginia, New Jersey, Pennsylvania, and elsewhere. *McIlvaine.*

Very common on logs of coniferous trees.   It is difficult to collect it entirely free from resin, which as a seasoning is not recommended.

FAMILY IV.—THELEPHORA'CEÆ Fr.

Gr.—a teat; *Gr.*—to bear.

Sporophore erect and stipitate, with a central stem, effused, with the upper portion free and bent backward, or entirely resupinate. **Hymenium** perfectly even or radiately wrinkled, glabrous or minutely bristled with projecting cystidia; basidia normally 4-spored. **Spores** without a division, colorless or colored. *Massee.*

In Thelephoraceæ are shapes closely resembling those found in Hydnaceæ, Polyporaceæ and Agaricaceæ. The genus Craterellus is closely allied to Cantharellus, and, though the spore surface is much less wrinkled or veined, resembles it in several of its species. Other types show likeness to Merulius in Polyporaceæ; others to Tremellineæ and Clavariaceæ. Many puzzles are presented by its species, but the solving is interesting.

Though populous it contains but few edibles. The best of them is Craterellus cornucopoides.

SYNOPSIS OF GENERA.

A. Spores Colored.

Spores smooth.

Coniophora.

Resupinate, **dry** and pulverulent. (No edible species reported.)

Aldridgea.

Resupinate, **soft** and subgelatinous. (No edible species reported.)

Spores warted or echinulate..

Thelephora.

Dry and fibrous, hymenium rugulose. (No edible species reported.)

Soppittiella.

Subgelatinous, effused or variously incrusting, hymenium even. (No edible species reported.)

506

B. SPORES COLORLESS.

Parasitic on living leaves or stems.

EXOBASIDIUM.

Saprophytes growing on dead wood, branches, etc. Hymenium minutely setulose with projecting cystidia.

PENIOPHORA.

Cystidia colorless, rough at the tip with particles of lime. (No edible species reported.)

HYMENOCHÆTE.

Cystidia brown, smooth. (No edible species reported.)

Hymenium glabrous.

CORTICIUM.

Entirely resupinate, hymenium usually cracked when dry. (No edible species reported.)

STEREUM.

Effuso-reflexed, pileus silky or strigose, hymenium even. (No edible species reported.)

CLADODERRIS.

Horizontal and attached by a narrow point behind, hymenium radiato-rugulose. (No edible species reported.)

CRATERELLUS. Page 508.

Large, erect, funnel-shaped.

CYPHELLA.

Minute, cup-shaped, mouth open. (No edible species reported.)

SOLENIA.

Minute, cylindrical, gregarious or crowded, tubular, mouth contracted. (No edible species reported.)

CRATEREL'LUS Fr.

Crater, a bowl.

Craterellus. **Hymenium** waxy-membranaceous, distinct but adnate to the hymeno-phore, inferior, continuous, smooth, even or wrinkled. **Spores** white. *Fries*.

This, the only genus of Thelephoraceæ containing edible fungi, has the form and general appearance of Cantharellus to which it is allied, but it is distinguished by its nearly even hymenium, which in Canthar-ellus has the form of gills, fold-like and thick but still distinctly gills. The species vary from fleshy to membranaceous, all having a funnel-shaped pileus and stem merging into it. On the ground. Autumn. The slightly veined surface where the spores are borne, and the spores themselves, when a microscope is brought to bear upon them, distin-guish this genus from Cantharellus; and its thin flesh and funnel-shape from the large forms of Pistillaria. Several of the species are edible. It is probable that all are.

Toadstools, despite their name, are more popularly associated with fairies than with toads. "Fairy rings," "Fairy Bread" and "Fairy Clubs" are titles belonging to them, and these link us to the pretty be-lief of childhood—a belief we often do not outgrow. A group of C. lutescens or C. cornucopoides may well be likened to fairy trumpets, or to a tiny orchestrion thrusting its horns through wood earth where roots of stumps abound.

C. cantharel'lus Schw. ([Color] Plate XLVI, fig. 3.) **Cap** 1–3 in. across, convex, often becoming depressed and funnel-shaped, glabrous, yellowish or pinkish-yellow. **Flesh** white, tough, elastic. **Hymenium** slightly wrinkled, yellow or faint salmon color. **Stem** 1–3 in. high, 3–5 lines thick, glabrous, solid, yellow. **Spores** on white paper yel-lowish or pale salmon.

Spores 7.5–10x5–6µ *Peck*.

West Virginia, *McIlvaine*.

No one not looking for minute botanic details would separate this species from Cantharellus cibarius, especially if found growing near or with it. The pinkish tinge sometimes present in C. cantharellus I have never observed in C. cibarius. The present species is of equal excel-lence.

C. cornucopoi′des Pers.—*cornu* and *copiæ*, horn of plenty. ([Color] Craterellus. Plate CXXXVI, fig. 8.) **Cap** dark sooty shades of gray or brown —shades of well-worn velveteen—1–2 in. across, whole plant from 2–4 in. high, trumpet-shaped, or like a funnel with its open mouth, plane, wavy, split or in folds. Substance very thin and either brittle or tough. The inside is sometimes minutely scaly, the opening extending to the base; outside, where the spores are borne, it has neither gills, pores nor protuberances, but a slightly uneven surface varying little in color. **Stem** obsolete or seldom noticeable. **Odor** slight.

Spores pointed, 11–12x7–8µ *Massee.*

Grows single, clustered or in troops along shaded roads, or from leaf mold and ground in woods. July to frost.

Large patches, clustered, grow near stumps in moist places on Botanic Creek, West Philadelphia. It is plentiful near Haddonfield, N. J., at Mt. Gretna, Pa., and many other places in the United States.

It is not pleasant to look upon, because of its peculiar color, but when one gets used to it it has an attractiveness of its own. Its graceful shape, even its funereal hue and name—Trompet du Morte—are alluring.

It dries well, and when moistened expands to its normal size. It is a first-class edible fungus. It should be stewed slowly until tender.

C. clava′tus Fr.—*clava*, a club. **Pileus** 2 in. broad, somewhat light-yellowish, fleshy, *top-shape, truncate* or depressed, flexuous, unpolished, *attenuated into the solid stem.* **Flesh** thick, white. **Hymenium** even, then corrugated, purplish then changing color. *Fries.*

Spores elliptical, pale-yellow, 10–12x4–5µ *Massee.*

Professor Peck notes that the species so closely resembles Cantharellus cibarius that it might easily be mistaken for a deformed condition of it.

The resemblance to the yellow forms of Clavaria pistillaria is marked.

Massachusetts, *Sprague, Farlow;* New York, *Peck,* Rep. 32; West Virginia, Pennsylvania, *McIlvaine.*

An excellent species. Its scarcity is regrettable.

C. du′bius Pk. **Pileus** infundibuliform, subfibrillose, lurid-brown, pervious to the base, the margin generally wavy and lobed. **Hymenium** dark cinereous, rugose when moist, the minute crowded irregular folds abundantly anastomosing, nearly even when dry. **Stem** short. **Spores** broadly elliptical or subglobose, 6–7.5µ long.

509

Craterellus. Plant simple or cespitose, 2–3 in. high. Pileus 1–2 in. broad.
Ground under spruce trees. Adirondack mountains. August.

In color this species bears some resemblance to Cantharellus cinereus.
From Craterellus sinuosus it is separated by its pervious stem, and from
C. cornucopoides by its more cespitose habit, paler color and smaller
spores. *Peck*, 31st Rep. N. Y. State Bot.

West Virginia, Pennsylvania, *McIlvaine*.

Its edible qualities are in every way equal to those of C. cornucopoides.

C. sinuo'sus Fr.—*sinus*, a curve. Strong scented. Pileus funnel-
shaped, downy, grayish-brown, margin undulated. Stem pale yellow,
elongated, stuffed. Hymenium with anastomosing ribs, grayish.
Spores elliptical, pale yellow, 8–9x5μ.

In woods. Pileus ½–1 in. high and broad. Stem about 1 in. high,
sometimes very short. Smell strong, musky. Hymenium becoming
tan-color when dry. Pileus more or less villose. *Massee.*

The above description is given so that Var. crispus which follows
may be compared with it. Fries considered var. crispus a good species.

Var. *cris'pus*—*crispus*, curled. ([Color] Plate CXXXVI, fig. *7.*)
Margin of hymenium sinuous and crisped. Pileus pervious. Stem
stuffed at base only. Hymenium almost even. *Massee.*

Solitary and cespitose in mixed woods.

Found by *Dr. S. C. Schmucker* near West Chester, Pa., 1896; *Wm.
H. Rorer*, Mt. Gretna, Pa., August, 1897.

Cap varies in color from dark to light brownish-gray. Gills brownish-gray, almost even. Stem hollow, dark yellow. Smell strong,
musky, much like A. silvicola.

Substance tender and of markedly high and pleasant flavor.

510

FAMILY V.—CLAVARIA'CEÆ.

Hymenium not distinct from the hymenophore, covering entire outer surface. Somewhat fleshy, not coriaceous, vertical, simple or branched. *Fries.*

For the most part growing upon the ground.

In this family there is no separation into stem and pileus, with the spore-bearing surface restricted to gills or tubes, but the substance of the plant is continuous, and the spores are produced on the clubs or branches.

But three genera—Clavaria, Sparassis and Pistillaria—include species of food value. They are easily recognized.

The genus Calcocera resembles Clavaria in form, but is very different in material, being a jelly-like viscid, cartilaginous substance, horny when dry, resembling that of Tremella.

SYNOPSIS OF GENERA.

SPARASSIS. Page 512.

Very much branched, branches compressed, plate-like, crisped.

TYPHULA.

Simple or club-shaped, with a thread-like stem.

CLAVARIA. Page 513.

Fleshy, simple or branched, branches typically round, some forms club-shaped.

PISTILLARIA.

Club-shaped, simple, rigid when dry; usually minute.

PTERULA.

Branches numerous, slender, forming a tuft, or single, leathery, round or compressed.

511

SPARAS'SIS Fr.

Gr.—to tear in pieces.

Sparassis. Fleshy, branched, with flat leaf-like branches, composed of two plates, fertile on both sides, with four-spored sporophores. *Fries.*

Very beautiful plants of striking appearance.

Unfortunately they are not common, although they generally occur yearly in the same locality.

S. Herb'stii Pk. Plants much branched, forming tufts 4–5 in. high and 5–6 in. broad, whitish, inclining to creamy-yellow, tough, moist, the branches numerous, thin, flattened, concrescent, dilated above and spatulate or fan-shaped, often somewhat longitudinally curved or wavy, mostly uniformly colored, rarely with a few indistinct, nearly concolorous, transverse zones near the broad, entire apices.

Spores subglobose or broadly elliptical, 5–6x4–5μ.

Trexlertown. August.

Closely allied to S. spathulata Schw., but differs in its paler color with no rufescent hues, more branching habit and absence of any distinct zones.

Four specimens were found at Mt. Gretna, Pa., during August, 1898. These were not as symmetrical as S. crispa, which they closely resembled in fold and texture. They were of equal excellence cooked.

S. lamino'sa Fr.—a thin plate. **Base** branching, straw-color. **Branches** erect, crowded, growing together, straight at the top, zoneless, entire.

North Carolina, *Curtis.* On oak log.

Edible, *Curtis.* "Deliciosa," *Fries.*

S. cris'pa Fr.—*crispus*, curly. (Plate CXXXVII.) **Height** 3–12 in., width 4–24 in. Tufts very handsome, whitish, oyster color or pale-yellow, very much branched. **Branches** flat, leaf-like. Spore surface on both sides, sometimes crimped on edges. Compacted into a round mass, ending below in a solid rooting base.

Spores pale-ochraceous, 5–6x3–4μ *Massee.*

Very variable in size. On ground in woods and grassy places in open woods. Summer, autumn.

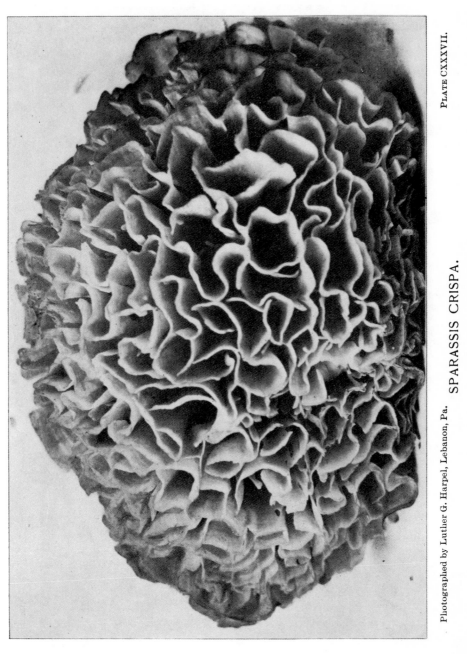

Photographed by Luther G. Harpel, Lebanon, Pa. SPARASSIS CRISPA.

CLAVARIA CRISTATA.

Photographed by Dr. J. R. Weist.

North Carolina, *Curtis;* West Virginia, New Jersey, Pennsylvania, ₛₚₐᵣₐₛₛᵢₛ. *McIlvaine.*

Have seen it 2 ft. across. ''Delicosissima.'' *Fries.*

A perfect specimen of S. crispa resembles a huge rosette, round and many-folded in tortuous design. The folds are wide, flattened branches springing from a common base, thin, semi-transparent, not unlike damp sheets of gelatine although thicker. Surfaces of the leaves are dull, like the flattened seaweeds and the light-colored sea-rock mosses. S. crispa may be easily dried, and though shrinking much in size, retains its shape, forming a very pretty ornament for the desk of the mycologist. It is not common. Where it has chosen a habitat several tufts may be found during the moderate season. The writer found three specimens ranging from 6–12 in. in diameter near Haddonfield, N. J., others, not as large, in West Virginia and in Chester county, Pa.

It has long been known as edible. It makes an ever-to-be-remembered dish.

CLAVA'RIA L.
Clava, a club.

Fleshy, branched or simple, somewhat round, without a distinct stem. Clavaria. **Hymenium** continuous, dry, homogeneous. *For the most part growing on ground. Fries.*

The members of this genus vary greatly in form, which in some is that of a club growing singly or cespitose, while others present a more or less bush-like appearance, being slightly or excessively branched.

The color of the plant covers a wide range, as it may be white, red, yellow, violet or their various shades, and to be in harmony the spores do not confine themselves to one color, but are white, ochraceous or cinnamon. In cases where the plant is not otherwise well defined the spore colors will be found a valuable aid in placing it.

ANALYSIS OF TRIBES.

RAMARIA (*ramus*, a branch). Page 514.
Branched, branches attenuated upward.

A. SPORES WHITE OR PALLID.
* Plant, color bright, red, yellow or violet.
** Plant white, gray or yellowish.

Clavaria.   * Plant yellow or dingy ochraceous.
** Growing on wood.

Syncoryne (*Gr.*—together ; a club).   Page 523.

Clubs almost simple, tufted at the base.

Holocoryne (*Gr.*—entire ; a club).   Page 524.

Clubs almost simple, distinct at the base.

Excepting to toadstool hunters the Clavaria, though numerous, are not known to those who "Know a toadstool when they see it." They bear no semblance to the stereotyped toadstool. They seem to possess an imitative faculty. Those growing among grasses harmonize with the faded stalks under debris or the bleached surfaces of blades famishing for sunlight; those of the woods take on the color of the leaf mat or of the lichens, and shapes of club and deer-horn mosses, or assemble in groves as pigmy trees, boled and sturdy-branched in mimicry of their giant protectors towering above them. In their forms many are delicate, graceful, beautiful, others are intricate. There is fascination for eye and brain in looking through the vistas and labyrinths of their branches.

A few species are tough as shoe-strings; a few bitter; one, C. dichotoma, on the authority of Leuba, contains a minor poison. The genus is plentiful and reliable. Many individuals are of marked excellence. In soups, stews, patties, they remind one of noodles; sometimes of macaroni. The hard parts of the stem should be removed, the branches broken or cut in ½ in. lengths. If stewed, they require time and slow cooking; if fried in butter they are crisp, choice bits.

Rama'ria—*ramus,* a branch.

Branched, branches attenuated upward.

A. Spores White or Pallid.

* *Plant, color bright, red, yellow or violet.*

C. fla'va Schaeff.—yellow. Fragile, trunk thick, fleshy, white, very much branched. **Branches** even, round, fastigiate, obtuse, yellow. *Fries.*

514

Height 2–4 in., 2–4 in. across; pale-yellow, dingy-yellow. **Stem** Clavaria. or trunk short, robust, whitish. **Branches** very numerous, dense, fragile, erect, straight, lighter than the yellow tips (fading with age) which are toothed. **Flesh** white. **Spores** white. Taste and odor pleasant.

Woods and open places. June to frost.

Indiana, *H. I. Miller;* West Virginia, New Jersey, Pennsylvania, *McIlvaine.*

The C. flava and C. botrytes have long been noted edible species, liberally commended abroad and in the United States. Variations in their structure are interchangeable; variations in their quality are due to environment. There is a slight difference in the measurement of their spores, but the difference is not so great as between spores of the same specimen. Specific differences may exhibit themselves in young plants, yet disappear with age.

Plants for the table should be young and fresh. When aged or when the ravages of insects appear, they should not be used, as they then have an unpleasant taste which will effect a whole dish.

They should be cut into small pieces and stewed slowly for fully thirty minutes. They can be seasoned and eaten as a stew or made into patties.

C. botry'tes Pers. *Gr.*—a cluster of grapes (from shape). **Height** 3–4 in., 3–6 in. across, white, yellow, pinkish, dingy in shades of these colors. **Base** thick, short, fleshy, unequal. **Branches** many, swollen, thick, crowded, unequal, enlarged at the ends and divided into several small branchlets which are sometimes reddish at tips. **Flesh** white.

Spores ellipsoid, sub-transparent, white, 8x5μ *Massee.*

On wood earth. Common.

New York, *Peck*, Rep. 24; West Virginia, New Jersey, *McIlvaine.* A general favorite and highly esteemed in Europe. Edible. *Curtis.*

"When old the branches both of this species and of C. flava become elongated, obtuse, very fragile, and of a uniform color. The yellow tips of the latter and the red ones of the former species wholly disappear." *Peck*, 32d Rep.

Excepting when young (not always then) the red tips to the branchlets can not be relied upon as distinctive features of this species. The place of its growth and the character of the soil have very much to do

Clavaria. with its size, and the color and quality of its flesh. A well-shaded thin-soiled spot will, after a rain, grow pale, spindling, tender bunches, having but a tinge of red upon the points; perhaps not any. A rich, better lighted spot will produce more robust and highly colored plants. The same can be said of C. flava. C. botrytes is plentiful in Pennsylvania, New Jersey, West Virginia and like latitudes. It must be well cooked.

C. amethys'tina Bull.—amethyst in color. ([Color] Plate CXXXIX, fig. 1.) **Height** ½–3 in. **Color** violet, very much branched or almost simple. **Branches** round, even, fragile, smooth, obtuse, known by its color.

Spores elliptical, pale ochraceous, sub-transparent, 10–12 x 6–7µ *Massee*.

Common in open woods and grassy places.

New York, *Peck* 30th Rep.; West Virginia, Pennsylvania, New Jersey, *McIlvaine*. August, September.

Eaten in Europe, and by some preferred to any other.

A handsome species, very brittle, and though large, delicate.

C. fastigia'ta—*fastigium*, the top. **Height** 1–2 in., tufted, yellow. **Branches** numerous, flexible, tough, equal, fastigiate (branches pointing upward), sometimes short and simple, when higher very much branched.

Spores white, irregularly globose, 4–6µ *Massee*.

In pastures and grassy places, during warm months.

North Carolina, *Curtis;* California, West Virginia, New Jersey, Pennsylvania, *McIlvaine*.

Commonly eaten throughout Europe. In Germany they call it Ziegenbart—goat's beard.

This is one of the species that has to be looked for. Grass tufts hide it. Its yellowish stools are not unlike them in color. It is freely found, and, though not of the best, well rewards the seeker.

C. muscoi'des—*muscus*, moss. **Height** 1–1½ in., slightly tufted, yellow. **Stem** slender, tomentose at base, becoming two or three times forked. **Branchlets** thin, tapering, crescent-shaped, acute.

Spores white, subglobose, 5–6µ *Massee*.

In pastures.

516

North Carolina, *Schweinitz, Curtis;* Ohio; New York, *Peck,* 47th Clavaria. Rep.

Edible. *Curtis.*

C. Her'veyi Pk. Gregarious or subcespitose, simple or with a few branches, often compressed or irregular, scarcely 1 in. high, golden-yellow, sometimes brownish at the apex. **Flesh** white. **Branches** when present, short, simple or terminating in few or many more or less acute denticles. **Spores** globose, 7.5µ broad, minutely roughened; mycelium white.

Ground under hemlock trees. Orono, Me. September. *F. L. Hervey.*

Allied to C. fastigiata and C. muscoides, but distinct from both by its more irregular and less branching character and by its larger spores. *Peck,* 45th Rep. N. Y. State Bot.

Near Haddonfield, N. J., August, 1890, among scrub pines and spruce. A pretty species of medium flavor.

* Plant white, gray or yellowish.

C. coralloi'des Linn. **Height** 2–4 in., usually tufted, growing into each other, white. **Trunk** thick, short, much branched. **Branches** repeatedly forked, compressed, hollow within, fragile, dilated upward, tips crowded acute.

Occasionally the branches do not develop entirely and are obtuse; they then somewhat resemble in shape C. rugosa, but are not wrinkled.

Spores pale-ochraceous, pointed, 10x8µ *Massee.*

Indiana, *H. I. Miller;* West Virginia, New Jersey, Pennsylvania, *McIlvaine.*

A common edible species in Europe. Common in United States.

The writer has eaten it for many years. It is not always tender. It should be young, fresh, and the branches alone cooked. It requires slow, patient cooking if at all old. It does dry well, as stated by some writers, but it does not wet well again.

C. cine'rea Bull.—*cinis,* ashes. (Plate CXL.) **Height** 1–3 in., gregarious or tufted, sometimes in rows. Gray. **Stem** either thin or thick, short, lighter than branches. **Branches** very numerous com-

517

Clavaria. (Plate CXL.)

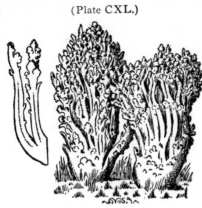

CLAVARIA CINEREA.
Two-thirds natural size.

pressed, wrinkled, irregular, somewhat obtuse or flattened and divided into slender points.

Its gray color easily distinguishes it from others. It is variable in its mode of growth and in its shape.

On ground in woods. Common. June to frost.

Eatable, but injurious in quantities. *Cordier.* Edible, but provokes indigestion in delicate stomachs. *Leuba.*

Eaten generally in Europe. In France it is called *pied de coq.*

Plentiful in United States, in mixed woods. June to frost.

The writer and his friends have eaten it for fifteen years, and know of no Clavaria equalling it.

C. tetrago'na Schw.—Four-angled. Very fragile, deep orange-yellow, twice forked. **Stem** and **branches** quadrangular, 1–1 ½ in. tall.

Moist shady places.

New York. Ground in shaded places. August and September. Poughkeepsie, *Gerard, Peck,* 24th Rep.; North Carolina, *Schweinitz, Curtis;* Pennsylvania, *Schweinitz.*

Edible. *Curtis.*

C. crista'ta Pers.—*crista,* a crest. (Plate CXLI, p. 513.) **Height** 1–5 in., whitish, tufts of broad flattened branches cut on margins or crested. **Base** short, stout. **Branches** numerous, irregular, flattened upward and divided like moose horns, tough, stuffed, dingy. This peculiarity distinguishes it and separates it from C. coralloides.

Spores pale ochraceous, pointed, 10x8µ *Massee.*

Woods. Common. Summer and autumn. Indiana, *H. I. Miller;* West Virginia, Pennsylvania, New Jersey, *McIlvaine.*

Edible. *Curtis.*

After a summer rain the crested Clavaria is usually abundant where there is good encouragement of mossy beds or mats of rich wood-soil in woods where leaves and mold accumulate. It is not as tender as

518

many other species, but chopped fine and stewed slowly for an hour it Clavaria. will be eaten with enjoyment.

C. rugo'sa Bull.—*ruga*, a wrinkle. White or dingy, simple or tufted, 2–4 in. high, branched from the base with irregular blunt branches wrinkled lengthwise, sometimes thickened upward.

Distinguished by the distinct, irregular, longitudinal wrinkles.

Spores white, irregularly globose, 8–10µ *Massee.*

In woods, solitary or gregarious. August to November.

North Carolina, *Schweinitz, Curtis.* Pennsylvania, Ohio.

It is reported edible by Dr. Curtis, M. C. Cooke and Dr. Badham.

C. pyxida'ta Pers.—*pyxis*, a small box. Tufted, light tan-color, shaded with red, 1–3 in. high. Stem or trunk thin, smooth, variable in length, dividing into many erect forked branches, which are cup-shaped at the tips. The margins of these tips have slender branchlets issuing from them (proliforme).

Distinguished by the cup-like tips. Spores white, 4x3µ *Massee.*

On rotten wood, on rotten roots in ground. June and into the autumn.

North Carolina, *Schweinitz, Curtis;* Pennsylvania, *McIlvaine.*

Specimen sent by writer to Prof. Peck, June, 1897, and identified by him. Not tested by writer, but is in Dr. Curtis' list of edible species.

C. subtil'is Pers. Scattered, slender, subtenaceous, pallid-white, bases smooth and of equal thickness, branches few, forked, subfastigiate.

North Carolina, *Schweinitz, Curtis;* Pennsylvania.

Edible. *Curtis.*

C. den'sa Pk. Tufts 2–4 in. high, nearly as broad, whitish or creamy-yellow, branching from the base. Branches very numerous, nearly parallel, crowded, terete, somewhat wrinkled when dry, the tips dentate, concolorous. Spores slightly colored, elliptical, 7.5–10x5–8.5µ.

Ground in woods. Selkirk. August.

Apparently closely allied to C. condensata, but differing decidedly in color. *Peck*, 41st Rep. N. Y. State Bot.

Specimens identified by Professor Peck.

Clavaria. Large masses of it grew at Mt. Gretna, Pa., July, August and September, 1898, in mixed woods.

Brittle; when young it is very compact. **It is without much** flavor, but stews tender and makes a good dish.

B. SPORES OCHRACEOUS OR CINNAMON.
Plant yellow or dingy ochraceous.

C. au'rea Schaeff.—*aurum*, gold. ([Color] Plate CXXXIX, fig. 2.) **Trunk** thick, elastic, pallid. **Flesh** white, dividing into numerous thick branches that become repeatedly divided in a dichotomous manner upward, and terminate in slender, erect, round, yellow branchlets. **Spores** pale ochraceous, elliptical, 10–11x5–6u.

In woods. Forming large tufts 2–3 in. high, colorless or almost so below, tips yellow. *Massee.*

North Carolina, *Curtis;* Ohio, Alabama. Found in West Virginia, 1882; Devon, Angora, Eagle's Mere, Mt. Gretna, Pa.; Haddonfield, N. J. August and September. *McIlvaine.*

Eaten in Europe. Edible. *Curtis.*

In structure it reminds one of a miniature cropped Lombardy poplar. The color is not bright, but dingy-yellow. Resembles C. flava; distinguished by different color of spores. The branches (not stem) are tender and good.

Var. *rufes'cens* Schaeff.

This plant occurs after heavy rains. It sometimes grows in continuous rows several feet in extent. The pinkish-red tips of the branches fade with age. The axils are rounded and the plant is quite fragile. Fries considers it a variety of C. aurea. *Peck,* 25th Rep. N. Y. State Bot.

Found at Springton, Chester county, Pa., August, 1887. It is edible and good. The plant is tender and easily cooked.

C. formo'sa Pers.—*formosus*, finely formed. ([Color] Plate CXXXIX, fig. 3.) **Height** 2–4 in. **Trunk** 1 in. and more thick, whitish or yellowish, elastic. **Branches** numerous, crowded, elongated, divided at ends into yellow branchlets which are thin, straight, obtuse or toothed.

Spores ochraceous 9x3–4u *Massee;* elongated, oval, rough, 16x8u *W.G.S.*

On ground in woods, in large tufts, frequently in rows several feet Clavaria. long.

North Carolina, *Schweinitz, Curtis;* Pennsylvania, New Jersey, *Mc- Ilvaine*.

Esteemed in Europe. Edible. *Dr. Curtis.*

Common in the United States in woods. Variable. An orange-rose color is sometimes prominent on the tips. The tenderer portions of the plant are excellent, but must be well cooked.

C. spinulo′sa Pers.—spined. **Height** 2–3 in. high. **Stem** ½–1 in. thick. **Trunk** stout, short, whitish. **Branches** numerous, crowded, erect, tense, elongated, tapering upward. **Color** cinnamon-brown or darker.

Spores ochraceous, elliptical, 11–13x5–6µ *Massee.*

On ground in pine woods. August to October.

New York, *Peck*, 24th Rep.; New Jersey, *Sterling;* Pennsylvania, *McIlvaine.*

Of same edible quality as C. aurea, which it resembles, excepting that it is darker and less abrupt in the ending of its clusters.

C. flac′cida Fr.—*flaccidus*, flaccid. **Height** 1–3 in., bright ochrace- ous, slender. **Stem** short, smooth, sometimes wanting, thin, 1–2 lines thick, repeatedly branched. **Branches** crowded, unequal, flaccid, upper ones forcep-shaped, pointed. Does not turn green when bruised like C. abietina. The whitish mycelium creeps over the leaves on which it grows. Brittle, tender, flesh white.

Spores ochraceous, broadly elliptical 4–5x3µ *K.*

Received from E. B. Sterling, Trenton, N. J.

Two specimens eaten. These were quite dry. After soaking they were tender and had good flavor.

C. cir′cinans Pk.—*circino*, to make round. (Plate CXLII.) **Stem** short, solid, dichotomously or subverticillately branched. **Branches** slightly diverging or nearly parallel, nearly equal in length, the ultimate ones terminating in two or more short acute concolorous ramuli. **Spores** ochraceous.

Plant 1–2 in. high, obconic in outline, flat-topped, appearing almost

Clavaria.   (Plate CXLII.)

CLAVARIA CIRCINANS.
(After Peck.)

as if truncated, pallid or almost whitish in color, generally growing in imperfect circles or curved lines.

Under spruce and balsam trees. Adirondack mountains. August. *Peck*, 39th Rep. N. Y. State Bot.

Where pines have grown, but where now oak and chestnut trees make rather open woods, it grows at Mt. Gretna, Pa. A stumpy fungus impressing one as stunted. Its texture is solid. It does not cook tender, but yields a fungus flavor to the cooking medium.

** *Growing on wood.*

C. stric′ta Pers.— *stringo*, to draw tight. **Height** 2–3 in. **Color** pale dull-yellow becoming brown when bruised. **Stem** distinct, thick, short. **Branches** numerous, repeatedly forked, straight, closely pressed, tips pointed.

Spores dark cinnamon, *Fries;* creamy yellow 4x6µ *W.G.S.*

Var. *fu′mida.* The whole plant is a dingy, smoky-brownish hue, otherwise of the typical form. Catskill mountains. September. In the fresh state the specimens appear very unlike the ordinary form, but in the dried state they are scarcely to be distinguished. *Peck*, 41st Rep. N. Y. State Bot.

Eaten in Germany.

This form occurs in West Virginia mountains and at Mt. Gretna, Pa., Trenton, N. J., in August and September, among leaves in mixed woods. It compares favorably with the ordinary run of Clavaria.

C. dicho′toma God.—dividing by pairs. Cespitose, white; branches regularly dividing by pairs, elongated, flexuous, diverging, somewhat compressed, extremities obtuse, rounded at or just below the apex broadly compressed.

On the ground, under beeches.

"Notwithstanding its beauty this is dangerous. In 1883, when it was very plentiful, I saw entire families sick from it and in 1888 there was a repetition with new victims.

"It produces nausea, vertigo and violent diarrhea." *Leuba.*
I have not seen the plant.

SYNCO'RYNE. *Gr.*—together, a club.

Clubs almost simple, tufted at the base.

C. fusifor'mis Sow.—*fusus*, a spindle. ([Color] Plate CXXXVIII, fig. 1.) *Yellow*, cespitoso-connate, slightly firm, soon hollow. **Clubs** somewhat fusiform, simple and toothed, even, attenuated to the base which is of the same color. *Stevenson.*

Spores pale yellow, globose, 4–5μ *Massee.*

Closely resembles C. inæqualis Fl. Dan.

Woods and pastures. August to November.

Received from *E. B. Sterling*, Trenton, N. J., August, 1897.

The clubs are $\frac{1}{16}$ in. through, 4 in. high, light clear yellow, translucent, clustered in groups of four or five united at the base.

Tender, well flavored, cooks easily.

C. auran'tio-cinnabari'no Schw.—*aurantius*, orange; *cinnabaris*, vermilion. Orange-red; base white with a sub-hairy powder; clubs simple, flexuous, fleshy, somewhat tenacious, fasciculate, thickened in the middle and attenuated toward either end, at first cylindrical then compressed, 6–7 mm. thick, 2–4 in. high.

Pennsylvania. On the ground among rhododendrons.

Received from E. B. Sterling, Trenton, N. J.

The plant when fresh is a beautiful rose color, inclining to orange at the tips. It reminds one of the peach-blow vase color in some of its shades. The single clubs, growing in cluster, to the height of four inches, graceful in outline, exquisitely shaded, are a sight one lingers over. While they invite the mycophagist to eat them, his voracity is checked by their beauty. They are tender and delicious. It is regrettable that thus far it has not been reported in quantity.

C. inæqual'is Fl. Dan.—unequal. **Height** 2–3 in. club-shaped, yellow, gregarious, single or in loose tufts, fragile, *stuffed*. **Clubs** club-shaped or almost equal, simple, sometimes forked or variously cut at tip, one color.

Spores colorless, elliptical, 9–10x5μ *Massee.*

523

Clavaria. Woods and pastures. August to October.

Distinguished from C. fusiformis by the tips not being sharp-pointed and colored.

North Carolina, *Schweinitz;* New Jersey, *Sterling*.

This Clavaria is quite common in New Jersey. Its clusters are clear bright yellow and conspicuously pretty. The clubs are translucent and smooth. Excepting in color it resembles C. aurantio-cinnabarino. In the many specimens seen there was nothing to suggest the propriety of the name, excepting height of clubs.

A dish of it is a delicacy.

C. vermicula'ris Scop.—*vermis*, a worm. **Height** 1–2 ½ in., white, tufted. **Clubs** simple, quill-shaped, stuffed, awl-shaped, brittle, pointed.

Spores white, elliptical, 4x3µ *Massee*.

New York, North Carolina, Pennsylvania, Ohio. Thin grassy woods and among grass. July to October.

Edible. *Cordier*.

Common in southern New Jersey, and in warm soils from June to frost. When growing among grass it is not conspicuous and is often missed unless specially sought for. Its purity, its choice of refreshing abode, its excellent qualities, make it select among Clavaria.

HOLOCO'RYNE. *Gr.*—entire; *Gr.*—a club.

Clubs almost simple, distinct at the base.

C. pistillar'is L.—*pistillum*, a pestle. ([Color] Plate CXXXVIII, figs. 2, 3.) **Height** 2–12 in., up to 1 in. and more thick, color light yellow, ochraceous, brownish, chocolate. **Clubs** Indian-club shape, ovate-rounded, puckered at top, simple, fleshy, white within, spongy, exterior smooth or more or less wrinkled, usually with smooth base.

Spores white, 10x5µ *W.G.S.;* 9–11x5–6µ *Massee*.

Mixed woods, moss and grassy places. August until November.

North Carolina, Pennsylvania, California, Alabama.

Eaten in Poland, Russia and Germany.

The writer first found this truly club-like species in West Virginia in 1882, and ate it. But few specimens were found, and those of a dark chocolate color. At Mount Gretna in 1897 and 1898 the yellow va-

riety grew in considerable quantity from July until after frost. The Clavaria.
largest specimen found measured 5 ½ in. and was 1 in. in diameter at
its thickest part.   The average height is 2 ½ in.   Both varieties grew
in mixed woods from the leaf-covered ground.   They are often clustered, four or five together, and of different sizes.   The surface, especially of the dark variety, is regularly, vertically wrinkled, truncated in
few places, very much resembling that of the Craterellus cantharellus.
The stems of both are white.   The apex of the clubs is folded inward
as though pulled by drawing-strings.

The flesh is soft, white, fine grained. A slight bitter is present in the
dark variety, when raw, which entirely disappears upon cooking.   This
is one of the best of Clavariæ.

C. clava′ta Pk.   Simple, straight, clavate, obtuse, smooth, not hollow, yellow when fresh, rugose-wrinkled and orange-colored when dry,
4–6 lines high.

Damp shaded banks by road-sides.   Sandlake.   June.   *Peck*, 25th
Rep. N. Y. State Bot.

Patches of it are conspicuous—golden-hued upon somber background.   They are seen at Eagle's Mere, Mt. Gretna, and on the
Springton Hills, Pa., along wooded road-sides.   Raw, they have a mild,
pleasant flavor, and have the same when cooked.   A small species seldom found in sufficient quantity to make a comforting dish.

FAMILY VI.—**TREMELLA'CEÆ** Fr.

Whole fungus homogeneous, gelatinous, shrivelling when dry, reviving when moistened, pervaded internally with branched filaments, terminating toward the surface all round in sporophores. Spores transparent, from globose to sausage-shape and curved, sometimes septate. *Fries*.

The Tremellaceæ, as their name signifies, tremble, because jelly-like when moist. They are hard, tough, horny when dry, but swell and become gelatinous when wet. In the typical genus, Tremella, there is often but little consistency. Whoever has climbed an old rail fence on a rainy day has had the doubtful pleasure of acquaintance with some of them. Sections for the microscope are obtainable by hardening them in alcohol.

There are several edible species in the family. They are good in soups, giving them flavor and body, and some are excellent when stewed.

SYNOPSIS OF GENERA.

Sub-Family—**Auricularieæ.** Page 528.

AURICULARIA.

Broadly attached, margin free and reflexed. (No edible species reported.)

HIRNEOLA. Page 528.

Cartilaginous, ear-shaped, attached by a point.

Sub-Family—**Tremellineæ.** Page 529.

EXIDIA.

Cup-shaped, truncate, or irregularly lobed; spores reniform, producing curved sporidiola on germination. (No edible species reported.)

ULOCOLLA.

Pulvinate and gyrose; spores reniform, producing rod-shaped sporidiola on germination. (No edible species reported.)

TREMELLA.   Page 529.

Brain-like or lobed; spores globose or ovoid.

NÆMATELIA.

Firm, convex, with a central hard nucleus.   (No edible species reported.)

GYROCEPHALUS.

Erect, spathulate.   (No edible species reported.)

TREMELLEDON.   Page 533.

Gelatinous, tremelloid, fan-shaped, fleshy; hymenium with distinct spines.

Sub-Family—**Dacryomyceteæ.**

DACRYOMYCES.

Small, pulvinate and gyrose.   (No edible species reported.)

GUEPINIA.

Irregularly cup-shaped, hymenium on one surface only.   (No edible species reported.)

DACRYOPSIS.

Hymenium at the apex of a short stem, bearing conidia and spores. (No edible species reported.)

DITIOLA.

Stem distinct, bearing the hymenium at its expanded apex. (No edible species reported.)

APYRENIUM.

Subglobose or lobed, hollow.   (No edible species reported.)

CALOCERA.

Subcylindrical and erect, simple or branched.   (No edible species reported.)

Sub-Family—**Auricularieæ.**

HIRNE'OLA Fr.

Hirnea, a small jug.

Hirneola.   Gelatinous, **rather** cartilaginous, soft and tremulous when moist, but not distended with jelly, horny when dry, becoming somewhat cartilaginous when moistened.   The hard skin forming the hymenium, which covers the cup-shaped cavity and is of a different color, can be separated entire after a thorough soaking in water.   **Sporophores** (spore-bearing processes) not involved in jelly.   **Spores** oblong, curved.   *Fries.*

A very peculiar and distinct genus separated from the neighboring genera by its disk-like, somewhat cup-shaped cavity and by its not being distended with jelly.

H. auri'cula–Jude'a (Linn.) Berk.—Jew's ear.   1–4 in. across, thin, and flexible when moist, hard when dry,

(Plate CXLIII.)

HIRNEOLA AURICULA-JUDEA.
About two-thirds nat. size.

date-brown or blackish.   **Hymenium** veno-so-plicate (vein-plaited), forming irregular depressions such as are in the ear, yellowish-gray or grayish beneath and hairy.   The large depressions or corrugations branch from smaller ones near the center of the plant.

Spores 20–25x7–9µ *Massee.*

H. auricula-Judea is not very particular in the trees it patronizes.   Elm, maple, hickory, balsam-fir, spruce, alder bear it.   When the plant grows on upright timber it usually turns upward.   It is not generally reported in the United States.

Ohio, Maryland, *Miss Banning;* Indiana, *H. I. Miller;* New York, *Peck;* New Jersey, Pennsylvania, West Virginia, *McIlvaine.*   Extensively used in China, where eating it probably antedates all European records by several thousand years.   It is brought there dried from Tahiti in great quantities and made into soup.

The writer has found and eaten several specimens of it.   It is not as tender as other gelatinous species, but it is an oddity that pleases.

528

Sub-Family—**Tremellineæ.**

TREMEL'LA Dill.

Tremo—to tremble.

Distended with jelly when moist, tremulous, without a defined mar- Tremella.
gin and without nipple-like elevations. Spore-bearing processes globose,
becoming divided into four parts, each division producing an elongated
free point terminating in a simple spore. *Fries.*

Distinguished by its peculiarly convoluted habit and jelly-like substance, which is more or less inclined to be cartilaginous.

Exidia, similar in form, is separated by possessing minute nipple-like
elevations and Hirneola by its distinct difference in form.

Generally growing on dead wood; some species are found on trees
and others on the ground, etc.

Old tradition, in many countries, attests that the Tremellas are Fairy
bread, and T. albida the choicest baking. Pretty, indeed, must have
been the feasts when piles of such purity filled the board, and the brilliant Pezizae were wassail cups.

They are better suited to Fairy appetites than to those of mortals;
being watery their nutritive value is small. Nevertheless they have
dainty flavor.

So far as tested no suspicion rests upon Tremellæ.

ANALYSIS OF TRIBES.

MESENTERIFOR'MES (*Gr.*—the mesentery). Page **530.**

Gelatinous inclining to cartilaginous, foliaceous, naked.

CEREBRINÆ (*cerebrum*, the brain). Page 530.

Firm, then pulpy, somewhat pruinose with the spores.

CRUSTA'CEÆ (*crusta*, a crust).

Diffused, becoming plane.

TUBERCULIFOR'MES (*tuberculum*, a little **tuber**).

Small, somewhat erumpent.

I.—MESENTERIFOR'MES. Gelatinous, inclining to cartilaginous.

Tremella. **T. fimbria'ta** Pers.—*fimbriæ,* fringe. Olivaceous inclining to black, cespitose, clusters 2–3 in. high and even broader, *erect, corrugated; lobes* flaccid, incised at the margin, *undulately fringed.*

When soaked with water it has a dark tawny tinge. *Stevenson.*

Spores subpyriform.

On roots, dead branches, stumps, rails, etc.

From July to December, 1898, tufts five inches in diameter grew from an oak stump close by the writer's cottage at Mt. Gretna, Pa. These tufts dried, and revived after rain into a gelatinous condition. They were nibbled at raw, and several were cooked. Tufts were found elsewhere in the same woods and eaten by others. They were unanimously approved. The species dries hard, like thin glue, but is darker. A dried piece swells in the mouth, grows tough, and has but little taste. Flavor develops in cooking.

T. lutes'cens Pers.—*luteus,* yellow. *Yellowish,* cespitose, small, cluster ½–1 in. broad, very soft, circling in wavy, undulating folds; lobes entire, naked.

Inclining to be fluid. Whitish when young. *Stevenson.*

Spores subglobose, 12–16μ diameter *Massee.*

(Plate CXLIV.)

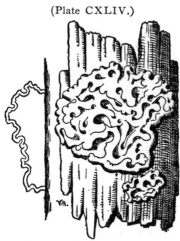

TREMELLA MESENTERICA.
Natural size.

North Carolina, common. *Curtis.* On decaying branches, stumps, etc. July to February.

It dries and revives, or swells with moisture, very soft and tremulous.

Edible. *Leuba.*

II.—CEREBRINÆ. Firm then pulpy, etc.

T. mesenter'ica Retz. *Gr.* — the mesentary. Gelatinous but firm, bright orange-yellow, variously contorted; lobes short, smooth, pruinose with the white spores at maturity. **Spores** broadly elliptical, 6–9μ diameter; conidia 1–1.5μ diameter.

On dead branches. Very variable in

form but known by the bright orange color. From ½–2 in. across. Tremella. *Massee.*

North Carolina. Common, edible. *Curtis;* California, Ohio, West Virginia, New Jersey, Pennsylvania. *McIlvaine.* Dr. J. R. Weist, Richmond, Ind., November, 1898, sent me fine specimens.

Very common as an apparent exudation from sticks, branches and rails. It can usually be collected in quantity from June until far into the winter. It can be found in every month in the year.

During the civil war the writer's first attempt at making a dish of cornstarch resulted in getting it *into knots.* T. mesenterica, when stewed, very much resembles these same knots. It has a mild, woody flavor, slightly sweet, and is good.

T. myceto'phila Pk. (Plate CXLIVa.) Suborbicular, depressed, circling in folds, tremelloid-fleshy, slightly pruinose, yellowish or pallid, 4–8 lines broad. *Peck,* 28th Rep. N. Y. State Bot.

Haddonfield, N. J., August, 1895. *McIlvaine.*

Professor Peck notes it as found parasitic upon Collybia dryophila.

I found T. mycetophila growing parasitic upon Marasmius oreades, August, 1894. The mass was 2 in. in diameter. Separating them was taking the host from the parasite. Cooked it is glutinous, tender—like calf's head. Rather tasteless.

T. al'bida Huds.—*albidus,* whitish. *Whitish,* becoming dingy-brown when dry, 1 in. broad, ascending, tough, expanded, undulated, somewhat circling in folds, *powdered. Stevenson.*

(Plate **CXLIVa.**)

TREMELLA MYCETOPHILA on COLLYBIA DRYOPHILA. (After Peck.)

Spores oblong, obtuse, curved, 2-guttate, subhyaline, 12–14x4–5μ *K.*

Where birch, sugar-maple, hickory are in abundance the T. albida will be found. At Eagle's Mere and Springton, Pa., and other wooded places, it is common during the warm months. It has slight taste, sweet, woody, but makes a pleasant dish.

531

T. intumes'cens Eng. Bot.—*intumesco,* to swell up. Gelatinous; sub-cespitose, rounded, broken up into numerous tortuous lobes, brown, shining, obscurely dotted, becoming darker when dry. Spores oblong, slightly curved, 12–14x3–4µ.

From 1–2 in. across. *Massee.*

Entire year, but dried or frozen during winter, swelling in wet weather.

North Carolina. Common. *Curtis.* West Virginia, Pennsylvania, New Jersey, *McIlvaine.*

T. intumescens is not rare in West Virginia, or where beech logs are in plenty, though it does not confine itself to beech. It occurs on maples and some other woods.

It resembles the T. mesenterica in taste, but is sweeter. It is not as large, but is equally good.

TREMEL'LODON Pers.

Tremo, to tremble.

Gelatinous, pileate, prickly below, spines awl-shaped, equal. *Fries.* Tremellodon.
The members of this genus resemble in form the section Mesopus of
Hydnum and have the same awl-shaped spines, but differ in their gela-
tinous consistency and fructification.

T. gelatino'sum Pers.—*gelatina*, jelly. **Pileus** covered with a greenish-brown bloom, *gelatinous*, tremu-
lous, dimidiate, somewhat stipitate,
covered with small pimples. **Spines**
soft, glaucous.

(Plate CXLV.)

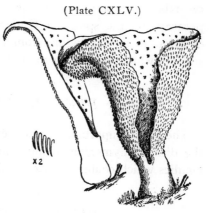

On fir, trunks and sawdust.
September to October. *Stevenson.*

Of singular beauty, almost translucent with steel-blue tints shading
into violet, while the spines are of a
pure soft white.

Spores round, somewhat irregular, white, 2µ *W.G.S.*

Can not be confounded with any.
The only gelatinous spiny fungus.

TREMELLODON GELATINOSUM.

North Carolina, *Schweinitz, Curtis;* Pennsylvania, Massachusetts, *Farlow, Frost;* New York, *Peck*, Rep. 22. T. gelatinosum is well distributed
over the United States but is not reported in quantity. It is an autumnal
grower, lasting well into the winter. The writer found specimens near
Haddonfield, N. J., in February, 1894, and sent them to Professor
Peck. It is delicious when slowly stewed.

Sub-Class ASCOMYCETES.

The reproductive bodies consisting of sporidia mostly definite, contained in asci—mother cells or sacs—springing from a naked or enclosed stratum of fructifying cells and forming a hymenium or nucleus. The sporidia are often accompanied by simple or branched threads, which are abortive asci, called paraphyses.

In Hymenomycetes the spores are entirely unenclosed and are borne on stalk-like processes on the gills of Agaricaceæ, in the tubes of Polyporaceæ, on the spines of Hydnaceæ, etc. In Ascomycetes they are enclosed in sacs springing from the external layer of the fruit-bearing surface, which may be on the outer surface of the plant or enclosed.

Cohort *DISCOMYCETES.*   *Gr.*—a sac; *Gr.*—a fungus.

The most important distinctive feature of Discomycetes consists in the disk or hymenium being fully exposed at maturity. It includes families which contain choice edible species.

FAMILY.—HELVELLA'CEÆ.

Fleshy, waxy or gelatinous; hymenium or sac-bearing surface exposed at first, or at length more or less exposed. Where a distinct stem is present it is surmounted by a more or less definite pileus or the stem is expanded into a club-like head. In Peziza the definite stem is absent and the plant is seated on the supporting surface.

Many more genera than are noted below are included in Helvellaceæ, but are not known to contain edible species.

SYNOPSIS OF THE GENERA.

*Margin only or whole of pileus free from sides of stem.

HELVELLA.   Page 536.

Pileus drooping, irregularly waved and lobed.

534

VERPA.   Page 539.

Pileus drooping, regular, margin entire, thimble-shaped.

LEOTIA.   Page 540.

Pileus fleshy, discoid.

** *Pileus adnate throughout to the stem.*

MORCHELLA.   Page 541.

Surface of pileus furnished with stout, anastomosing ribs bounding deep irregular pits.

GYROMITRA.   Page 546.

Surface of pileus covered with rounded, variously contorted folds.

MITRULA.   Page 548.

Pileus subglobose or clavate, surface even.

SPATHULARIA.   Page 549.

Pileus flattened, running down the stem for some distance on opposite sides.

GEOGLOSSUM.   Page 550.

HELVEL'LA Linn.

A small pot herb.

Helvella. **Stem** of medium thickness. **Pileus** hanging loosely over the stem, more or less folded, but not into pits. Hymenium on the upper side only.

Helvella esculenta is now Gyromitra esculenta, and is in bad repute.

Meanings of the unfamiliar words are too lengthy to give in the descriptions of species. They are in the Glossary.

Dr. Badham says: ''All Helvellæ are esculent, have an agreeable odor, and bear a general resemblance in flavor to the Morell.''

H. cri'spa Fr.—curled. **Pileus** deflexed, lobed or variously contorted, white or whitish. **Stem** equal or slightly swollen at the base, deeply and uninterruptedly grooved, white or whitish. **Spores** elliptical, 18x22μ long. *Peck*, 48th Rep. N. Y. State Bot.

(Plate CXLVI.)

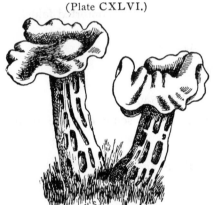

HELVELLA CRISPA.
Natural size.

Distinguished from all other species by the stout, costate, lacunose, hollow stem; entirely glabrous, fragile and with a semi-transparent look. Color variable, included under the following forms:

Var. *al'ba*. Pileus whitish.

Var. *Grevil'lei*. Under surface of the pileus reddish; stem white.

Var. *incarna'ta*. Pileus and stem flesh-color.

Var. *ful'va*. Pileus yellowish or tawny. *Massee*.

Pileus whitish, flesh-colored or yellowish, deflexed, lobed, at length free, crisped. **Stem** hollow, ribbed outside forming deep **pits, 3–5** in. high, snowy white.

Edible. *Badham, Cordier, Cooke, Berkeley, Peck.*

West Virginia, Pennsylvania, New Jersey, *McIlvaine.*

H. crispa is white and variable in shape of cap. In its color it differs from all others of its genus. It is found in the woods only, from July until frost. It is not usually abundant. It is an esculent species and good of its kind.

536

H. Califor'nica Phillips. **Pileus** bell-shaped or saddle-shaped, de- Helvella.
flexed, sublobate, free, veined beneath, purplish-brown. **Stem** longitudinally pitted between ridges, rosy-pink. **Asci** cylindrical, narrowed
toward the base. **Sporidia** 8, elliptical, binucleate, $17 \times 9\mu$; paraphyses
linear, clavate and brown at the apices.

2–6 in. in diameter. **Stem** 2–6 in. high, .75–1.5 in. in diameter.

On the earth in dense forests near rocks. Sierra Nevada **mountains;**
California, *Harkness.*

Edible. *Harkness.*

It presents characters essentially different from those of any species
hitherto described. Its nearest ally is H. crispa, from which it differs
in the color of the hymenium and stem and in being a larger species.

H. lacuno'sa Afzel.—uneven, pitted. **Pileus** inflated, lobed, cinere-
ous-black, lobes deflexed, adnate. **Stem** white or dusky, hollow, exterior ribbed, forming intervening cavities; asci cylindrical, stemmed;
sporidia ovate, hyaline.

Solitary or gregarious.; very variable in size.

North Carolina, *Curtis;* Massachusetts, *Sprague, Frost;* White
mountains, *Farlow;* Rhode Island, *Bennett;* California, *H. and M.*

Edible. *Cordier, Berkeley, Badham, Cooke, Curtis.*

H. sulca'ta Afzel.—furrowed. **Pileus** deflexed, equally 2–3 lobed,
even, compressed, darker when dry. **Stem** 2 in. long, 4–5 lines thick,
stuffed, equal, longitudinally furrowed. **Spores** very broadly elliptic,
with a single large globose nucleus, $15–18\mu$ long *B. and Br.*

Solitary, rarely gregarious.

Var. *mi"nor* Clem. Bot. Surv. of Neb. Univ. of Neb. Pileus .8–1.2
in., rarely 3.2 in. wide, .8–2 in. high. Stem .8–1.2 in., rarely 4 in.
high, .6–1.4 in. wide; sporidia $15 \times 10\mu$.

On shady ground. Otowanie woods, Lancaster county.

The prominent character in this species, as indicated by the name, is
the sulcate stem. The furrows are very deep, and extend, without
interruption, the entire length of the stem. The whole stem, as shown
by a cross-section, is made up of the costæ intervening between these
furrows. I do not find the stem "stuffed," as required by the description in Syst. Myc., Vol. II, p. 15. The pileus is generally darker than
that of H. crispa. *Peck*, 31st Rep. N. Y. State Bot.

Helvella.   On decaying wood, stumps, trunks.   Spring until autumn. Known to be edible. *Peck.*

H. elas'tica Bull.—elastic.   **Pileus** free from the stem, drooping, 2–3 lobed, center depressed, even, whitish, brownish or sooty, almost smooth underneath, about 2 cm. broad.   **Stem** 2–3.5 in. high, 3–5 lines thick at the inflated base; tapering upward, elastic, even or often more or less pitted, colored like the pileus, minutely velvety or furfuraceous, at first solid, then hollow.   **Spores** hyaline, smooth, continuous, elliptical, ends obtuse, often 1-guttulate, 18–20x10–11μ; 1-seriate; paraphyses septate, clavate. *Massee.*

It is not uncommon to find the pileus attached in one or two points to the stem. *Peck*, 32d Rep.

Var. *al'ba* (Pers.) Sacc.

On decaying wood.   August to frost.

Massachusetts, *Frost;* Rhode Island, *Bennett;* Nebraska, *Clements;* New York, *Peck*, Rep. 24, 32, 51.

Edible.   *Unger, Cordier.*   Known to be edible.   *Peck.*

H. in'fula Schaeff.—a head dress.   **Pileus** hooded, in 2–4 irregular, drooping lobes, at length undulate, strongly adherent to the sides of the stem, reddish-brown or cinnamon more or less deep in color, whitish and downy underneath, 1.5–3 in. broad.   **Stem** 1½–2½ in. long, ½ in. and more thick, usually smooth and even, sometimes compressed and irregularly pitted, pallid or tinged with red, covered with a white meal or down, solid when young but becoming hollow with age; asci cylindrical, apex somewhat truncate, 8-spored.   **Spores** hyaline, smooth, continuous, elliptical, ends obtuse, 21–23x11–12μ *Massee.*

West Virginia, Pennsylvania.   Decaying trunks, stumps and roots. *McIlvaine.*

Edible.   *Cooke, Curtis, Peck.*

Equal to any Helvella.

VER'PA Swartz.

Verpa, a rod.

Ascophore stipitate, campanulate, attached to the tip of the stem and Verpa. hanging down like a bell, surrounding but free from the side of the stem, regular, smooth or slightly wrinkled but not ribbed, persistent, thin, excipulum formed of interwoven, septate hyphæ, hymenium entirely covering the outer surface of the ascophore; asci cylindrical, 8-spored. **Spores** elliptical, continuous, hyaline or nearly so, 1-seriate; paraphyses septate. **Stem** elongated, stuffed.

Very closely allied to Helvella; distinguished by the ascophore being more regular in form, and more evidently deflexed round the apex of the stem, which it surrounds like a thimble on a finger, and is quite free from the stem except at the apex.

The species grow on the ground, in spring. *Massee.*

V. digitalifor'mis Pers.—*digitus*, a finger. **Pileus** at first nearly even, olivaceous-umber, dark at the apex. **Stem** obese, furnished at the base with a few reddish radicles, white with a slight rufous tinge, marked with transverse reddish spots; smooth to the naked eye, but under a lens clothed with fine adpressed flocci, the rupture of which gives rise to the spots, which are, in fact, minute scales. In the mature plant the pileus is ¾ in. high, bell-shaped, finger-form, or subglobose, more or less closely pressed to the stem, but always free, the edge sometimes inflexed so as to form a white border, wrinkled, but not reticulated, under side slightly pubescent; sporidia yellowish, elliptic. **Stem** 3 in. high, ½ in. or more thick, slightly attenuated downward, loosely stuffed, by no means hollow. *Berkeley.*

Minnesota, *Johnson;* California, *H. and M.;* New York, Buffalo, *Clinton;* Oneida, *Warne*, May. *Peck*, 30th, 32d Rep.

1t. Gretna, July, 1897. Road-side bank. *McIlvaine.*

Sold in Italy. Vittadini. Not to be despised when one can not get better nor to be eaten when one can. *Badham.*

The substance of this fungus is the same as that of Helvella. It is pleasant but rather tasteless.

LEOTIA Hill.

Leotia. Ascophore stipitate, substance fleshy, soft and somewhat gelatinous. **Pileus** orbicular, spreading; margin drooping or incurved free from the stem, glabrous, hymenium entirely covering the upper surface. **Stem** central, elongated; asci cylindric-clavate, apex narrowed, 8-spored. **Spores** hyaline, continuous or 1-septate, elongated and narrowly elliptical, obliquely 1–2 seriate; paraphyses present.

Growing on the ground, or on decaying wood. *Hill.* Emended. *Massee.*

Stem long. **Pileus** flattened, margin incurved, covered everywhere with the smooth, somewhat viscid hymenium.

L. chloroceph'ala Schw.—*chloros*, green; *kephalos*, a head. Cespitose, stipitate. **Pileus** 4–6 lines across, depresso-globose, somewhat translucent, more or less wavy, margin incurved, dark verdigris-green to blackish-green. **Stem** 1–1 ½ in. long, almost equal, green but often paler than the pileus, pulverulent, often twisted; asci cylindric-clavate, apex rather narrowed, 8-spored. **Spores** smooth, hyaline, narrowly elliptical, ends acute, often slightly curved, usually 2–3-guttulate, 17–20x5μ, irregularly 2-seriate; paraphyses slender, hyaline.

On the ground.

Distinguished from L. lubrica by the green stem. *Massee.*

North Carolina, *Curtis;* West Virginia, New Jersey, Pennsylvania. Cespitose. In mixed woods, moist ground. July until long after frosts. *McIlvaine.*

A small clustered plant having a green gelatinous appearance. Quarts of it can frequently be gathered after rains. Both it and L. lubrica have less flavor than the larger Helvellaceæ, but they make a palatable dish.

L. lu'brica Pers.—slippery. Gregarious or in small clusters, stipitate, somewhat gelatinous. **Pileus** irregularly hemispherical, inflated, wavy, margin very obtuse, yellowish olive-green, 6–8 lines across. **Stem** 1.5–2 in. high, nearly equal or more or less inflated at the base, pulpy within then hollow, externally yellowish and covered with minute white granules; asci cylindrical, apex slightly narrowed, 8-spored. **Spores** obliquely 1-seriate, hyaline, continuous, smooth, often guttulate, narrowly elliptical, straight or very slightly curved, 22–25x5–6μ; paraphyses slender, cylindrical, hyaline.

On the ground in woods. *Massee.*

North Carolina, *Curtis;* Massachusetts, *Frost;* Minnesota, *Johnson;* New York, *Ellis.*

New York, *Peck*, 23d Rep.; Trenton, N. J. Cespitose on damp ground in woods. Forty specimens, July, 1898. *E. B. Sterling;* New Jersey; Pennsylvania. Gregarious and cespitose in several localities. July to frost. *Mc-Ilvaine.*

Irregular in appearance. Helvella-like but with a very soft gelatinous stem, yellow. The color of the stem distinguishes it from L. chlorocephala, which has a green stem. It is a small plant, but of good food value. Where it occurs there is often a goodly quantity.

(Plate CXLVII.)   **Leotia.**

LEOTIA LUBRICA.
Natural size.

MORCHEL'LA Dill.

Gr.—a mushroom.

Stipitate or subsessile. **Pileus** globose or ovate, adnate throughout **Morchella.** its length to the sides of the stem, remaining closed at the apex, hollow and continuous with the cavity of the stem; externally furnished with stout, branched and anastomosing ribs or plates, every part bearing the hymenium. **Stem** stout, stuffed or hollow; asci cylindrical, 2–4–8-spored. **Spores** 1-seriate, continuous, hyaline, elliptical; paraphyses septate, clavate.

Most nearly allied to Gyromitra; differs in the ribs of the pileus being deep and plate-like, and anastomosing to form elongated or irregularly polygonal deep pits.

Growing on the ground in the spring. *Massee.*

Stem stout; pileus ovoid or conical, deeply folded into pits, resembling honeycomb.

Notwithstanding Dill, the author of the genus, describes the caps as adnate throughout their length to the stem, such is not the case. Pro-

Morchella. fessor Peck arranges the genus into two groups, "in one of which the margin of the cap is wholly attached to the stem, in the other it is free." In the latter group are M. bispora and M. semilibera.

The species are so much alike that botanical descriptions are omitted of all but M. esculenta and Professor Peck's species.

Not one of the Morells is even suspicious. They are favorites wherever found. The Morell is one of the few species known to the settler and to the farmer. It loves old apple orchards, probably because ashes have been used about the trees; ashes and cinders are its choice fertilizers. In Germany peasants formerly burned forests to insure a bountiful crop. Mr. Moore, of San Francisco, Cal., says: "We find it in profusion on burnt hillsides all along the Pacific coast."

But it does not confine its habitat to burned surfaces. It grows in thin open woods or on borders of woods. It grows under pine, ash, oaks and other trees. Strange to say it grows under the walnut tree where very few fungi of any kind grow. Especially does it love the white walnut or butternut.

Morchella dry well and keep well for winter use.

M. esculen'ta Pers.—esculent. ([Color] Plate XLVI, fig. 2.) **Pileus** globose, ovate or oblong, adnate to the stem at the base, hollow, ribs stout, forming irregular, polygonal, deep pits, pale dingy yellow, buff or tawny, 1.25–2.5 in. high and broad. **Stem** stout, whitish, almost even, hollow or stuffed, 1.25–2.5 in. high, .8 in. and more thick; asci cylindrical, 8-spored. **Spores** continuous, smooth, hyaline, elliptical, ends obtuse, 19–20x10μ, paraphyses rather slender, slightly thickened upward.

On the ground. Spring and early summer. Edible.

Variable in form, size and color, but distinguished by the pileus being adnate to the stem at the base, and the stout ribs anastomosing to form irregular, polygonal pits of about equal size, and not elongated. *Massee.*

Common over the states, West Virginia, Pennsylvania, New Jersey. In orchards, on ashes and cinders, under walnut, pine and oak trees. May and June. *McIlvaine.*

The common Morell varies in size, 2–4 in. high, sometimes larger. The cap, usually broader than it is long, oval, at times tapering to a rounded top. The cavities resemble those of a weather-beaten honey-

comb, and are whitish, or grayish or brownish. The stem is about ⅓ Morchella.
in. in diameter. It is an easily recognized species. Edible. Choice.
Total nitrogen, according to Lafayette B. Mendel, 4.66 per cent.

M. cras′sipes Pers.—*crassus*, thick; *pes*, a foot. Agreeing with M.
esculenta in having the pits of the pileus irregular in form, not much,
if at all, longer than broad, and in not having a main series of more or
less parallel and vertical ribs; differing in the stout stem being much
longer than the pileus. *Massee.*
Attains a height of 9 in. or more.
Not rare in May. Kansas, *Cragin;* Minnesota, *Johnson.*
Esculent. *Cooke.*

M. delicio′sa Fr. The Delicious morell is easily known by the shape
of its cap, which is cylindrical or nearly so. Sometimes it is slightly
narrowed toward the top and occasionally curved, as in the preceding
species, but its long narrow shape and blunt apex is quite strongly contrasted with that species. It is usually two or three times as long as it
is broad, and generally it is longer than the stem. Specimens also occur in which the cap is slightly more narrow in the middle than it is
above and below, and rarely it is slightly pointed at the apex. The
pits on its surface are rather narrow and mostly longer than broad. The
stem is often rather short.
The plant varies from 1½–3 in. high. *Peck*, 48th Rep. N. Y. State
Bot.
Its name gives it esculent properties.

M. con′ica Pers.—conical. The Conical morell has the cap conical
or oblong-conical, as its name indicates. The longitudinal ridges on
its surface run more regularly from top to base than in the Common
morell. They are connected by short transverse ridges which are so
distant from each other or so incomplete that the resulting pits or depressions are generally longer than broad, and sometimes rather irregular. The color in the young plant is a beautiful buff-yellow or very
pale ochraceous, but it becomes darker with age.
The plants are generally 3–5 in. high, with the cap 1½–2 in. thick
in its broadest part, and distinctly broader than the stem. *Peck*, 48th
Rep. N. Y. State Bot.
Kansas; California; Rhode Island; Ohio, *Lloyd;* New York; Indiana,

Morchella. *H. I. Miller*, orchards, thin woods; New Jersey, Pennsylvania, West Virginia, *McIlvaine*.

The conical form distinguishes M. conica from M. esculenta, if they are really different species, as some writers doubt. For the table there is not any difference.

M. bi'spora Sor.—Two-spored. The Two-spored morell is very similar to the Half-free morell in external appearance. It is distinguishable by its cap, which is free from the stem almost or quite to the top. The stem of the European plant has been described as stuffed, but in our plant it is hollow, though possibly in very young plants it may be stuffed. The remarkable and very distinctive character which gives name to the species can only be seen by the aid of a microscope. In this species there are only two spores in each ascus or sack and these are much larger than the spores of the other species. They are two or three times longer and sometimes slightly curved. The spores of the other species are eight in an ascus and are very much alike in size and shape, and do not furnish decided specific characters; but in this species their importance can not be overlooked. Their length is about 60μ, while in the others it is 20–25μ.

This is probably our rarest species. I am not aware that it has been found in but one locality in our state. A few years ago Mr. H. A. Warne detected it growing among fallen leaves in a ravine near Oneida. I have not tested its edible qualities, but would have no hesitation in eating it if opportunity should be afforded. *Peck*, 48th Rep. N. Y. State Bot.

Var. *trunca'ta*. **Pileus** broadly rounded or truncate, its costæ slightly prominent, the margin often a little recurved; paraphyses numerous. **Stem** long.

Michigan. May. *Hicks*. *Peck*, 46th Rep. N. Y. State Bot.

M. angus'ticeps Pk.—*angustus*, narrow; *caput*, head. **Pileus** oblong-conical and subobtuse or narrowly conical and acute, adnate to the stem, 1–2 in. high, and about half as broad at the base, ribs longitudinal, here and there anastomosing or connected by transverse veins. **Stem** subequal, hollow, whitish, furfuraceous without and within, even or rarely rough with irregular longitudinal furrows; asci cylindrical. **Spores** elliptical, whitish tinged with ocher, 20–25x12.5–18μ; paraphyses short, clavate, with one or two septa near the base.

544

Sandy soil in the borders of woods and in open places. West Albany Morchella. and Center. April and May.

Two forms occur, one with the pileus oblong-conical, rather obtuse, often tipped with a slight umbo or papilla, and with a diameter a little surpassing that of the stem from which the base is separated by a slight groove; the other with the pileus narrowly conical, rather acute, scarcely exceeding the stem in diameter and without any separating groove. The stem and fruit are alike in both forms. The stem is usually about equal in length to the pileus. The species is related to M. conica and M. elata, but may be separated from both by the size of the spores and the character of the paraphyses. In our plant I have never seen these as long as the asci. Large forms appear also to approach M. rimosipes, but that species has the margin of the pileus more free, the stem proportionately longer, and the paraphyses as long as the asci, if we may rely upon the figure of it. Our plant is edible. *Peck*, 32d Rep. N. Y. State Bot.

The plants are commonly 2–3 in. high, with the cap generally less than an inch broad in its widest part, but sometimes much larger specimens occur. (Plate CXLIX.) *Peck*, 48th Rep. N. Y. State Bot.

M. semilib′era D.C.—half-free. The Half-free morell has a conical cap, the lower half of which is free from the stem. It rarely exceeds 1 in. or 1 ½ in. in length, and is usually much shorter than its stem. The pits on its surface are longer than broad. Deformed specimens occur in which the cap is hemispherical and very blunt or obtuse at the apex; in others it is abruptly narrowed above and pointed.

The plants are 2–4 in. high. The species is rare with us. *Peck*, 48th Rep. N.Y. State Bot.

Spores pale-yellow.

Odor feeble, becomes stronger in drying. Much less sapid than M. esculenta. Neither of these funguses should be gathered after rain, as they are then insipid and soon spoil.

MORCHELLA SEMILIBERA.

Badham.

545

GYROMI'TRA Fr.

Gyro, to turn; *mitra*, a head-covering.

Gyromitra.   Ascophore stipitate; hymenophore subglobose, inflated and more or less hollow, or cavernous, variously gyrose and convolute at the surface, which is everywhere covered with the hymenium; substance fleshy; asci cylindrical, 8-spored. **Spores** uniseriate, elongated, hyaline or nearly so, continuous; paraphyses present.

Helvella of old authors.

Distinguished from Morchella by the thick, brain-like folds of the hymenophore not anastomosing to form irregularly polygonal depressions; and from Helvella in the hymenophore not being free from the stem at the base.

Growing on the ground. *Massee.*

G. esculen'ta Fr.   ([Color] Plate VI, fig. 6.)   **Pileus** rounded, lobed, irregular, gyrose-convolute, glabrous, bay-red.
(Plate CXLVIIIa.)   **Stem** stout, stuffed or hollow, whitish, often irregular. **Spores** elliptical, binucleate, yellowish, 20–22μ long.

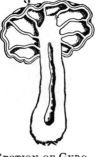

The Edible gyromitra, formerly known as Helvella esculenta, is easily recognized by its chestnut-red irregularly rounded and lobed cap with its brain-like convolutions. The margin of the cap is attached to the stem in two or three places. When cut through it is found to be hollow, whitish within and uneven, with a few prominent irregular ribs or ridges. The stem is whitish, slightly scurfy, and when mature, hollow. In large specimens it sometimes appears as if formed by the union of two or more smaller ones.

SECTION OF GYRO-
MITRA ESCULENTA.

The plant is 2–4 in. high and the cap commonly 2–3 in. broad. Specimens sometimes occur weighing a pound each. It is fond of sandy soil and is found in May and June. It grows chiefly in wet weather or in wet ravines or springy places in the vicinity of pine groves or pine trees. *Peck*, 48th Rep. N. Y. State Bot.

G. esculenta crispa n. var.   Whole surface of the pileus finely reticulated with anastomosing costæ (ribs or veins).

Under evergreens. North Elba. June. *Peck*, 51st Rep. N. Y. State Bot.

546

Since 1882 myself and friends have repeatedly eaten it. In no instance Gyromitra. was the slightest discomfort felt from it. It was always enjoyed. Mr. Charles H. Allen, San Jose, Cal., writes to me that G. esculenta grows plentifully in his region, and that it is not only edible, but he has found it one of the best. But the species, though long ago esteemed highly in Europe and by many in America, now rests under decided suspicion. It is not probable that in our great food-giving country anyone will be narrowed to G. esculenta for a meal. Until such an emergency arrives, the species would be better let alone.

G. cur'tipes Fr.—*curtus*, short; *pes*, a foot. **Pileus** inflated, gyrosely undulated, oblong, rotund, at first pallid then brownish; margin of pileus closely adnexed to the stem. **Stem** irregular, short or almost absent. **Asci** cylindrical. **Sporidia** .30x9µ fusiform, uninucleate. Paraphyses clavate.

On the ground. Spring. Readily distinguished from other species by the almost obliterated stem. Fries commends it highly as an esculent.

Separated from G. esculenta by paler color, shorter stem and different spores.

G. Carolinia'na (Bosc.) Fr. **Pileus** rotund, base free, surface woven into deep irregular undulating folds. **Stem** conical, sulcate. **Asci** cylindrical. **Sporidia** 3-3.2x1µ; somewhat fusiform; paraphyses thickened toward the top.

In woods. Esculent.
Massachusetts. *Sprague.*

(Plate CXLVIII.)

G. brun'nea Underwood—*brunneus*, brown. A stout, fleshy, stipitate plant, 3-5 in. high, bearing a broad, much contorted, brown ascoma. **Stem** ¾-1.5 in. thick, more or less enlarged and spongy, solid at the base, hollow below, rarely slightly fluted, clear white; receptacle 2-4 in. across in the widest direction, the two diameters usually considerably unequal, irregularly lobed and plicate, in places faintly marked into areas by indistinct anastomosing ridges, closely cohering with the stem in the various

GYROMITRA BRUN-
NEA.

547

Gyromitra. parts, rich chocolate-brown or somewhat lighter if much covered with the leaves among which it grows, whitish underneath; asci 8-spored. **Spores** oval, 28–30μ long, by about 14μ wide, hyaline, somewhat roughened-tuberculate, usually nucleate, the highly refractive nucleus spherical or oval, 11μ or, if oval, 14x11μ in diameter; paraphyses slender, enlarged at the apex, faintly septate.

In rich woods, mostly in beech-leaf mold. Putnam county, Ind., May, 1892, 1893 and 1894. First found by Dr. W. V. Brown.

The plant is esculent, tender and possesses a fine flavor. Often as many as 8 or 10 plants would be found in one small area, but the plant appears to be local and never very abundant. Some single plants would weigh nearly half a pound.

MI'TRULA Fr.

(Emended, *Massee.*)

Mitrula. Ascophore stipitate, fleshy. **Head** subglobose, ovate, or clavate, even, glabrous, everywhere covered with the hymenium, adnate throughout to the more or less elongated stem; asci cylindric-clavate, 8-spored. **Spores** narrowly elliptic-fusiform, hyaline, continuous or septate, irregularly 1–2-seriate; paraphyses present. *Fries.*

(Plate CL.)

MITRULLA VITELLINA.

M. vitelli'na Sacc., var. *irregularis* Pk.— *vitellus*, egg-yolk. **Pileus** clavate, often irregular or compressed and somewhat lobed, obtuse, glabrous, yellow, tapering below into the short, rather distinct, yellowish or whitish stem. **Spores** narrowly elliptical, 8–10μ long.

When the Irregular mitrula is well grown and symmetrical it closely resembles the typical European plant, but usually the clubs or caps are curved, twisted, compressed or lobed in such a way that it is difficult to find two plants just alike. The plants are usually only one or two inches high, so that they would scarcely be thought of any importance as an edible species. But

548

sometimes it grows in considerable profusion in wet mossy places in Mitrula. woods, so that it would not be difficult to gather a pint of them in a short time. Its beautiful bright yellow color makes it a very attractive object. It is our largest species of Mitrula and occurs in autumn.

It was first reported as an edible species in the forty-second report. Its flesh is tender and its flavor delicate and agreeable. *Peck*, 48th Rep.

Ontario, *Dearness* (Ll. R. 4). West Virginia, New Jersey, Pennsylvania. Common, gregarious in moist woods. September to November. *McIlvaine*.

Those fortunate enough to find this species will hunt for it again assiduously. Even raw, when cut in strips, it makes a picturesque and delicious salad.

SPATHULA′RIA Pers.

A spatula.

Receptacle erect, spathulate, compressed, hollow, adnate to the stem, Spathularia. down which it runs for some distance on opposite sides, everywhere covered with the hymenium. **Stem** subcylindrical, hollow; asci clavate, apex narrowed, 8-spored. **Spores** elongated, cylindric-clavate, multiseptate at maturity, arranged in a parallel fascicle in the ascus; paraphyses filiform, septate.

Distinguished by the broad, flattened ascophore running down opposite sides of the stem.

Growing on pine leaves or on the ground among moss. *Massee.*

Resembling a spatha, an instrument for stirring a liquid, shaped like an apothecary's spatula.

Pileus irregular, compressed, folded, running down into the stem on either side.

S. clava′ta (Schaeff.) Sacc.—club-shaped. S. flavida Pers. Elvela clavata Schaeff. ([Color] Plate CXXXVI.) **Head** spathulate or broadly clavate, obtuse or sometimes more or less divided at the apex, hollow, much compressed, running down the stem for some distance on opposite sides, glabrous, margin crisped or undulated, surface wavy or slightly lacunose, yellow, rarely tinged red, .8–1.2 in. high, .6–1 in. broad. **Stem** white then tinged yellow, 1.2–2.4 in. long, .2–3 lines thick,

Spathularia. hollow, cylindrical or slightly compressed; asci clavate, apex narrowed, 8-spored. **Spores** arranged in a parallel fascicle, hyaline, linear-clavate, usually very slightly bent, multiguttulate then multiseptate, 50–60x3.5–4µ; paraphyses filiform, septate, often branched, tips not thickened, wavy. *Massee.*

New York. Woods in hilly and mountainous districts. Common. *Peck*, 22d Rep.

Professor Peck gives S. rugosa, which has the club wrinkled.

This odd, pretty little plant was found by me in great numbers at Eagle's Mere, Pa., August, 1897, growing among mosses. The contrast of its bright yellow paddle-shapes against the moss-green is very pleasing to one who loves choice bits of color. Its consistency when stewed is tenacious but tender, and its flavor is delicate.

GEOGLOS'SUM Pers.

(Emended.)

Geoglossum. (Plate CLI.)

GEOGLOSSUM
GLUTINOSUM.
About nat. size.

Entire fungus more or less clavate, erect, the apical, thickened portion everywhere covered with the hymenium; glabrous or hairy, often viscid; asci clavate, apex narrowed, 8-spored. **Spores** elongated, arranged in a parallel fascicle, cylindrical or very slightly thickened above the middle, and inclined to become cylindric-clavate, brown, septate, usually slightly curved; paraphyses septate, brown at the tips, often longer than the asci.

Distinguished among the clavate species by the long, narrow, brown, septate spores. The entire plant is black in all British species.

Growing on the ground, among grass, etc. *Massee.*

G. glutino'sum Pers. Ascophore 1.5–2 in. high, black, glabrous; ascigerous portion about ⅓ of the entire length, oblong, lanceolate, up to .4 in. broad, obtuse, slightly viscid, more or less compressed, passing imperceptibly into the somewhat slender, cylindrical, viscid, brownish-black stem; asci clavate, taper-

550

ing downward into a long, slender pedicel. **Spores** 8, arranged more Geoglossum. or less parallel near the apex of the ascus, cylindrical, ends obtuse, 3-septate and clear-brown at maturity, straight or very slightly curved, 65–75x5–6μ; paraphyses numerous, distinctly septate, about 2μ thick, pale-brown, apex broadly pyriform and filled with dark-brown coloring matter.

On the ground among grass, etc.

The most important features of the present species are 3-septate brown spores and compressed ascophore. *Massee.*

New Jersey, *E. B. Sterling.* Mt. Gretna, Pa., August, 1899, gregarious in wet ground. Over a quart found in one patch. *McIlvaine.*

Stewed it is delicious.

FAMILY.—PEZIZÆ.

PEZI'ZA Linn.

Pezizæ, a sort of mushroom without root or stalk, mentioned by Pliny.

Peziza.   Ascophore sessile, but sometimes narrowed to a short, stem-like base, fleshy and brittle, closed at first, then expanding until cup-shaped, saucer-shaped, or in some species quite plane or even convex; disk even, nodulose or veined; externally warted, scurfy, or rarely almost glabrous; cortical cells irregularly polygonal; asci cylindrical, 8-spored. **Spores** obliquely 1-seriate, continuous, hyaline (rarely tinged brown), elliptical, epispore smooth or rough; paraphyses present. *Dill.* Emended. *Massee.*

The genus is large. Professor Peck reports 150 American species. Some are large, others require the microscope to find them.

They are rather indiscriminate in their habitats; some are eccentric; these grow on damp walls, on dung, in cellars and cisterns, on spent hops and on old fungi. One or two species grow on sticks under water, an unusual place for fungi of any kind. Minute species grow upon stems of herbaceous plants; nine or ten upon the nettle. Two species contain a milky fluid, P. succosa and P. saniosa. Many are known in Europe which have not been found in America. European authors differ as to their qualities; some call them insipid, some speak of them with kindly respect. Much depends upon their cooking. They are, as a rule, tenacious in texture. To cook them properly requires time and slow stewing. They then become soft and rather glutinous. Their flavor is slight but pleasant, and their consistency agreeable.

ANALYSIS OF TRIBES.

I.—ALEURIA. Page 553.

Externally powdered or with a woolly scurf.

II.—LACHNEA. Page 558.

Externally hairy or downy.

III.—PHIALEA.

Externally almost naked, smooth. No edible species reported.

I.—Aleuria Fr.

Fleshy or fleshy-membranaceous, externally powdered or with a Peziza. woolly scurf.

* Macropodes—*macros*, long; *podes*, feet. Stem firm, elongated, furrowed.

** Cochleata—*cochleatus*, spiral. Subsessile, oblique or twisted.

*** Cupulares. Subsessile, regular.

**** Humaria. Small, somewhat fleshy, margin downy. (None known to be edible.)

***** Encœlia. More or less coriaceous. (None known to be edible.)

* Macropodes. Stem firm, elongated, etc.

P. aceta′bulum Linn.—a cup. **Ascophore** stipitate, cup-shaped, fleshy, rather tough, disk dark umber-brown, externally paler and minutely scurfy or flocculose; mouth somewhat contracted; 1.2–2 in. broad, 1.2–1.4 in. high. **Stem** .4–.6 in. high, often .4 in. thick, imperfectly hollow, with parallel or anastomosing ribs, which continue for some distance up the ascophore as branching veins, pale umber; cells of the cortex give off short, rather closely septate hyphæ in groups; asci cylindrical, 8-spored. **Spores** obliquely 1-seriate, hyaline, smooth, broadly elliptical, ends obtuse, with a very large oil-globule, 18–22x 12–14μ; paraphyses straight, septate, the brownish, clavate tip 5–6μ thick.

The fluted stem and veined outside of the excipulum mark the pres ent species. The colorless hypothecium is composed of very densely and compactly interwoven hyphæ. *Massee.*

Season spring.

North Carolina, *Curtis;* New Jersey, *Ellis;* Massachusetts, *Frost;* Rhode Island, *Bennett;* Ohio, *Lloyd,* R. 4.

Esculent. *Cordier, Cooke.*

P. ma′cropus Pers.—*macros*, long; *pous*, a foot. Solitary, 1–3 in. high, cups 1–2 in. broad. The cups become expanded, and sometimes reflexed; the exterior is ash-colored and clothed with little hairy or villous warts, the hairs consisting of concatenate cells, their extremities free. The stem is enlarged downward, often pitted, occasionally becoming hollow with age. *Phillips.*

Peziza.   Asci cylindrical, 8-spored. **Spores** 1-seriate, smooth, hyaline, elliptical, $28–33\times11–13\mu$; paraphyses straight, tips brownish and thickened in a clavate manner up to $8–10\mu$ *Massee*.

On the ground in shady places. Summer and autumn.

North Carolina, *Curtis;* New Jersey, *Ellis;* Minnesota, *Johnson;* Massachusetts, *Frost;* New York, *Peck*, Rep. 22.

Esculent. *Cordier.*

** Cochlea'ta. *Subsessile, oblique, etc.*

P. veno'sa Pers.—*venosus*, full of veins. Smell strong, nitrous; sessile or contracted into a short, stout, stem-like base; cup-shaped and with the margin incurved when young, then expanding and the margin becoming more or less split or lobed and wavy, 1.2–2 in. across; disk umber-brown, externally whitish, minutely granular, and furnished with rather stout, anastomosing ribs which radiate from the base; excipulum pseudoparenchymatous, cells largest at the periphery, where some run out as clavate, free tips; asci cylindrical, 8-spored. **Spores** obliquely 1-seriate, smooth, hyaline, often with 1 large oil-globule, elliptical, ends obtuse, wall rather thick, $18–24\times11–13\mu$; paraphyses septate, tips clavate, brownish. On the ground. Spring. *Massee.*

Massachusetts, *Frost;* California, *H. and M.;* Rhode Island, *Bennett;* New York, *Peck*, Rep. 24.

Edible. Has a most decided nitrous odor and also fungoid flavor. *Cooke.*

P. ba'dia Pers.—of a brown or bay color. (Plate CLII, p. 554.) Gregarious or cespitose, sessile or narrowed into a very short, stout, stem-like base and often more or less lacunose; subglobose and closed at first, then cup-shaped or more expanded, margin entire or nearly so, the entire cup often wavy, rather thick, 1.2–2 in. across; disk dark-brown, externally paler-brown and minutely granular, often with a purple tinge; hypothecium and excipulum formed of stout, septate, irregularly inflated hyphæ, hypothecium compact, excipulum spongy and cavernous; cortex compact, the hyphæ running out in irregular lumps to form the external granulations; asci cylindrical, apex truncate, 8-spored. **Spores** obliquely 1-seriate, hyaline, continuous, elliptical, with one large oil-

Plate CLII.

PEZIZA BADIA.

Photographed by C. G. Lloyd, Cincinnati, O.

globule, minutely warted at maturity, 15–19x9–10µ; paraphyses septate, tips slightly clavate.

On the ground among grass, etc., also on scorched places.

Readily distinguished by the bay or umber-brown disk, and the mi-
nutely-warted spores. *Massee.*

North Carolina, *Curtis;* California, *H. and M.;* Minnesota, *Johnson;*
Nebraska, *Clements;* New York, *Peck,* Rep. 25.

Alabama. On ground, Alabama Bull. No. 80, West Virginia, New
Jersey, Pennsylvania. On ground. Frequent. July to October. *Mc-
Ilvaine.*

Esculent. *Cordier.*

P. badia is frequent on bare ground, along wood roads, etc. In the
West Virginia mountains it occurs where there have been brush fires.
It is a meaty plant, without much flavor. It must be cut fine and slowly
cooked if stewed, or can be quickly fried in a hot buttered pan. It has
more flavor fried crisp than stewed.

P. cochlea′ta—spiral. Sessile, cespitose, variously contorted and
plicate, fleshy, brittle, disk umber-brown, externally paler and pruinose,
sometimes altogether paler and leather-color or pale dingy-ochraceous,
2–3.2 in. diameter; when solitary or almost so, at first globose, then
expanding with the margin involute, finally spreading and irregularly
plicate; excipulum spongy and cavernous, due to the loose weft formed
by interlacing, hyaline, thin-walled, flaccid, septate hyphæ, cortex compact, running out into irregular groups of cells that form the scurfy exterior; asci cylindrical, apex slightly truncate, 8-spored. **Spores** obliquely 1-seriate, hyaline, continuous, smooth, usually 2-guttulate,
16–18x7–8µ; paraphyses slender, septate; tip slightly clavate, often
curved and sometimes branched.

The entire substance is brittle and rather watery, and usually assumes
a yellowish tint when bruised. Smell and taste almost none.

Sometimes the ascophores are closely crowded, hence irregular and
much contorted, and resembling a foliaceous Tremella or a small specimen of Sparassis crispa. *Massee.*

New York. Ground in woods. Helderberg mountains and Greenbush. June. *Peck,* Rep. 23; Alabama, *Peters,* Ala. Bull. No. 80;
North Carolina, *Curtis;* Massachusetts, *Frost;* Ohio, *Lloyd,* Rep. 4.

Peziza. This species is quite insipid and somewhat leathery, but Mr. Berkeley has seen it offered for sale under the name of Morell. *Badham.* Esculent. *Cordier, Cooke.*

P. lepori'na Batsch.—*lepus*, a hare. **Cup** 1–3 in. high, 1–3 in. broad, gregarious, often cespitose; margin involute, divided to the base on one side; disk even or rarely wrinkled, a shade darker than the exterior; paraphyses slender, hardly thickened at the summits, but almost invariably crooked. This fine species grows as large as O. onotica at times, but is not so brightly colored, being throughout of a sober tan-color, resembling common wash leather used for cleaning plate. *Phillips.*

Asci cylindrical, 8-spored. **Spores** obliquely uniseriate, hyaline, smooth, continuous, 1–2 guttulate, elliptical, 12–15x7–8μ; paraphyses filiform, septate, apex slightly swollen, and usually strongly curved.

On the ground in woods, among leaves, etc. *Massee.*

California, edible, *H. and M.*

Esculent. *Cordier.*

P. onotica Pers. Very variable in form, usually elongated on one side and ear-shaped, but sometimes almost equal-sided and entire, 1–3 in. high, up to 2 in. wide, becoming narrowed to a more or less wrinkled, short stem-like base; disk pale orange, usually with a rosy tinge, externally pale tawny-orange. Asci elongated, narrowly cylindrical, 8-spored. **Spores** obliquely 1-seriate, hyaline, smooth, colorless, ends obtuse, 1–2-guttulate, 14–15x8–9μ; paraphyses straight, septate, apex clavate.

On the ground in woods, among leaves, etc. *Massee.*

North Carolina, *Curtis;* Iowa, *Fitzpatrick* (Ll. R. 4); New York, *Peck*, Rep. 28.

Esculent. *Cordier.*

P. unici'sa Pk.—implying one incision. **Cup** large, thin, split on one side to the base, sessile or with a short stem, externally wrinkled, minutely pulverulent under a lens, yellow, within pale-yellow slightly tinged with pink. **Spores** elliptical, usually containing two nuclei, 12–15μ.

Ground in woods Croghan. September.

The cups are about two inches broad. The species is related to P. Peziza. onotica. *Peck*, 26th Rep. N. Y. State Bot.

Minnesota, *Johnson;* Mt. Gretna, Pa. On ground in mixed woods, gravelly ground. September to October. *McIlvaine.*

Many specimens were found scattered and in patches, and were eaten. They were of slight flavor but good.

P. auran'tia Pers. ([Color] Plate CXXXVI, fig. 3.) Sessile or protracted into a very short stem-like base, cespitose and irregular, or growing singly and then circular in outline and regular, becoming almost plane; thin, brittle, disk clear, deep orange or sometimes orange-red, externally much paler, or sometimes almost white, with a pink tinge, delicately tomentose, due to the presence of short, stout, blunt, 1–2-septate hyaline hairs; varying from ½–3.2 in. broad. **Spores** 15–16x7–8μ.

On the ground, often near stumps or among chips.

Sometimes crowded, large, with the margin raised and very much waved and more or less incised, at others scattered, smaller, almost or quite even and finally spread flat on the ground. Easily recognized by the large size, bright orange disk, pale, downy exterior, and the broadly elliptical spores covered with a delicate net-work of raised lines at maturity. *Massee.*

Massachusetts, *Frost;* Rhode Island, *Bennett;* Minnesota, *Johnson;* California, *H. and M.;* Alabama, *Peters;* New York, October, *Peck,* 23, 24 Rep.; Indiana, Richmond, November, *Dr. J. R. Weist;* West Virginia, New Jersey, Pennsylvania. On ground. September to October. *McIlvaine.*

Esculent. *Cordier.*

At Mt. Gretna, Pa., patches of it twenty feet long, made the ground along a road on the margin of a woods golden with its clusters. The plants grew from sand mixed with leaf-mold. I have eaten it for fifteen years. Fair flavor.

*** Cupulares. *Subsessile, etc.*

P. repan'da Wahlenb.—bent backward. Clustered or scattered, sub-sessile, contracted into a short, stout, stem-like base, which is often rooting; saucer-shaped, then quite expanded and the margin more or less split and wavy, sometimes drooping and revolute, extreme edge often crenate; 1.6–4 in. across; disk pale or dark brown or umber, more or less wrinkled toward the center, externally whitish, minutely granular. **Spores** obliquely 1-seriate, hyaline, smooth, continuous, elliptical, ends obtuse, 18–22x11–12µ; paraphyses septate, clavate and brownish at the tips. *Massee.*

On the ground, often in beech-woods; also on decayed trunks.

New York, *Ellis;* Minnesota, *Johnson;* Ohio, *Lloyd*, R. 4. New York. Ground and decaying wood. Croghan. September. *Peck*, 28th Rep.

Specimens sent to the writer by Dr. W. B. Miller, Altoona, Pa., were 3½ in. across, and a beautiful velvety brown. Cooked they had a mushroom flavor.

P. vesiculo′sa Bull.—full of bladders. Clustered, often distorted from mutual pressure, sessile but more or less narrowed at the base, globose and closed at first, then expanding, but the margin usually remaining more or less incurved and somewhat notched; disk pale brown, externally brownish and coarsely granular from the presence of minute, irregular warts, 1.2–3 in. across. **Spores** obliquely 1-seriate, smooth, hyaline, continuous, elliptical, ends obtuse, 21–24x11–12µ; paraphyses slender, septate, clavate.

Var. *ce′rea* Rehm. Similar in size, habit and general structure to the typical form; differing in the wax-yellow color, the more distinct stemlike base, and the slightly smaller spores, 18–19x10µ; very brittle. *Massee.*

North Carolina, *Curtis;* California, *H. and M.;* Massachusetts, *Frost;* New Jersey, *Ellis;* Ohio, *Lloyd*, Rep. 4; var. minor, *Sacc.;* Nebraska, *Clements;* New York, *Peck*, Rep. 25.

Esculent. *Cordier.*

II.—LACHNEA.

P. odora′ta Pk. **Cups** .5–3 in. broad, gregarious or scattered, thin, sessile, rather brittle when fresh, shallow, expanded or even convex from the decurving of the margin, at first brownish, then white or whitish, the hymenium ochraceous-brown; asci cylindrical, opening by a lid, .01–.012 in. long, .0006–.0008 in. broad, paraphyses filiform, obscurely septate, slightly thickened at the tips. Spores elliptical, even, 20–22.2 Peziza. x10–12.5µ.

Ground in cellar. Maine. June. *F. L. Harvey.*

The plant when fresh has the peculiar fungoid flavor suggestive of that of chestnut blossoms. The species is apparently allied to P. Petersii, from which it may be distinguished by its larger spores and distinct but peculiar odor. The spores also are not binucleate, as in that species. In drying, the hymenium is apt to become blackish. *Peck,* Bull. Torrey Bot. Club, Vol. 23, No. 10.

A cluster 4 inches across, in general appearance resembling P. repanda, was found by the writer at Mt. Gretna, Pa., June, 1898, growing from between the staves of an empty flour barrel which was exposed to the weather. The margin instead of being revolute, turned inward (involute) until it touched the short stem. The cluster was eaten and had the flavor of P. repanda. In June, 1899, several pounds grew on and around the same barrel. Professor Peck recognized it as P. odorata.

P. cocci'nea Jacq.—scarlet or crimson. Geopyxis coccinea Mass. ([Color] Plate CXXXVI, fig. 2.) Scattered or in groups of 2–3 specimens, stipitate; at first closed, then expanding and becoming shallowly cup-shaped, margin entire, .8–1.6 in. across; disk clear and deep carmine, externally whitish or pinkish, delicately tomentose, due to the presence of wavy, usually aseptate, hyaline, cylindrical hyphæ, 5–6µ thick. Stem .4–.8 in. long, 1.2–2 in. thick, whitish and tomentose. Spores 1-seriate, elliptic-oblong, ends obtuse, hyaline, wall rather thick and forming a hyaline border, straight, 25–30x8–9µ; paraphyses very slender, hardly thickened at the tips.

On rotten branches lying on the ground. Spring.

Readily distinguished among the large, stipitate Pezizæ by the deep rose-red or carmine disk and the whitish, tomentose exterior. The stem varies considerably in length; when the fungus springs from the underside of a branch the stem is often elongated and curved. The base of the stem is attached to the branch by a mass of whitish, tomentose mycelium. *Massee.*

New York. Half-buried sticks. April and May. *Peck,* 23d Rep.; New Jersey, *E. B. Sterling;* Mt. Gretna, Pa., New Jersey. On sticks on ground. Spring. *McIlvaine.*

This brilliant fungus is one of the beauties of the woods. Though

Peziza. small it attracts the eye by its deep carmine in striking contrast with the somber carpeting. It is frequent when in season. A half pint of it may be gathered from a few acres. Its substance is tenacious, taste pleasant. Mr. Massee mentions that it is abundant in some of the woods near Scarboro, England, and is regularly collected and sold along with moss for decorative purposes. Exquisite effects may be produced by arranging the brightly colored fungi among moss and leaves. "Fairy Cups," they are called. Rosy must be the lips that do not pale beside them.

P. calyci'na Schum.—resembling a bud. Ascophores cespitose, gregarious or scattered, narrowed into a short, stout, stem-like base, rather fleshy, 1–3 mm. broad; disk orange-yellow, externally white and villose, hairs rather wavy, cylindrical, obtuse, colorless, minutely rough, $100–150x4–5\mu$; asci subcylindrical, apex obtuse, 8-spored. Spores 1-seriate or inclined to be 2-seriate above, hyaline, elliptic-fusiform, continuous, $18–25x6–8\mu$; paraphyses slender, hyaline, cylindrical.

On bark of larch and Scotch fir. *Massee.*

North Carolina, *Curtis;* Massachusetts, *Frost;* **New York.** Gum spots on spruce, bark of pines, *Peck*, 22d Rep.

Esculent. *Cooke.*

Cohort *PYRENOMYCETES.*

FAMILY.—HYPOCREACEÆ.

HYPO'MYCES Fr.

Gr.—under; *Gr.*—fungus.

Perithecia (the hollow narrow-mouthed cases which contain the
spores) gregarious, with a cottony stroma in which they are more or
less immersed. Mostly parasitic on various Hymenomycetes or Discomycetes; bright colored, with papilliform (nipple-shaped) or slightly
elongated ostiola (apertures). Asci mostly cylindrical, 8-spored, without paraphyses. Sporidia oblong or fusoid, uniseptate, hyaline. Co-
nidial stage represented by Asterophora, etc.

This parasite attacks several species of fungi, and so alters their
structure and appearance that it is difficult to distinguish the host-plant.
The attack is made in the extreme youth of the plant. The writer is
fully satisfied from his own observation that H. lactifluorum and H.
purpurea infest Lactarius piperatus. The milk cells are so changed by
H. lactifluorum that they yield no milk. When attacked by H. purpurea the milk is a beautiful purple. In both cases the pepperiness of
the host-plant is destroyed. I have seen the same host plant attacked
by both forms of the Hypomyces. After the host-plant of Hypomyces
lactifluorum is fully grown, and infested, it is frequently attacked by
Hypomyces purpureus. Purple spots appear, which gradually spread
until the entire plant is covered. This Hypomyces seems to affect the
milk cells. A beautiful, profuse, purple fluid results.

The parasite is proving itself an enemy to fungi, but a friend to man.
Upon L. piperatus and upon Amanita rubescens it very much adds to
the weight of the plants, and improves the texture and edible qualities.
The same may be said of L. volemus, but not to such a degree. Prof.
M. W. Easton in August, 1899, found this species at Mt. Gretna, Pa.,
attacked by a parasitic fungus in such a manner as to destroy its milk-

561

giving quality and completely transform its gills into a corrugated, granular surface.

Professor Peck, to whom I submitted the parasite, thinks it a new species and calls it H. volemi.

Further study of Hypomyces and its effect upon fungi, and of the particular host-plants is desirable.

H. lac'tifluorum (Schw.) Tulasne—*lac*, milk; *fluorum*, flowing. ([Color] Plate CXXXVI, fig. 5.)

Asci long and slender, sporidia in one row, spindle-shaped, straight or slightly curved, rough, hyaline, uniseptate, cuspidate-pointed at the ends, 30–38x6–8μ.

The general appearance is much the same as that of H. aurantius (Pers.) Tul., but the sporidia are larger, rough and warted, and the felt-like mycelium is wanting.

In the affected species of Lactarius the gills are entirely obliterated, so that the hymenium of the agaric presents an even, orange-colored surface on which the subglobose perithecia are thickly bedded, with only their slightly prominent reddish ostiola visible. In decay the color changes to a purplish-red.

On Lactarius, especially L. piperatus.

New Jersey, *Ellis;* Alabama, *U. and E.;* Minnesota, *Arthur;* Nova Scotia, *Dr. Somers;* on various species of Lactarius, 1895, Ala. Bull.; South Carolina, *Ravenel;* Pennsylvania, *Everhart.*

West Virginia, 1881–1882; Haddonfield, N. J., Mt. Gretna, Pa., August to October. *McIlvaine.*

This fungus puzzled me for many years. August, 1896, I sent several specimens to Professor Peck of different colors—orange, red, whitish and purple.

Professor Peck kindly identified the specimens and wrote: "In one the matrix of the host-plant has not been so completely changed or transformed as in the other. It would be interesting to know what species of Lactarius it is that Hypomyces attacks. I have never been able to ascertain, and have sometimes thought it might be Cantharellus cibarius, but this specimen of yours indicates, rather, a Lactarius."

Of the purple specimen he wrote: "This is a beautiful thing, and as I find nothing like it described I have given it a name—Hypomyces purpureus Peck."

Well cooked, in small pieces, it is one of the very best.

H. purpu′reus Pk.—*purpureus*, purple.   Subiculum effused, purple, permeating, transforming and discoloring the matrix; perithecia minute, sunk in the subiculum, the ostiola emergent, black; asci cylindrical; spores fusiform, uniseptate, purple, with a cusp-like point at each end, $35–40\mu$ long, 7.5μ broad, oozing out and forming beautiful purple masses or patches on the surface of the matrix.

Pennsylvania.   August.   *Charles McIlvaine.*

The species is similar in all respects to H. lactifluorum, except in color.   It is apparently parasitic on some species of Lactarius, but the host plant is so transformed and discolored that the species is not recognizable.   *Peck*, Bull. Torrey Bot. Club, Vol. 25, No. 6.

H. purpureus Pk. was sent by the writer to Professor Peck in August, 1897, who wrote: "This is a beautiful thing and as I find nothing like it described, I have given it a name."

Of itself H. purpureus is a minute parasitic fungus as above described. But it possesses the power of so altering the structure—changing form, shape and appearance—of the fungus upon which it has taken its abode that the host-plant, be it Cantharellus cibarius, Craterellus cantharellus or one of the Lactari, or whatever the species, becomes difficult to recognize, so that it is not yet certain upon which species it is parasitic. It may be upon many.

The present plant seems to be parasitic upon one of the Lactarii. It therefore becomes necessary to describe the host as it appears when possessed by the parasite.   The plant is variable in shape from an irregular nodule to a distorted-capped, short-stemmed mass, 2–4 in. across, 1–3 in. high, hard, brittle, coarse in appearance and rough to the touch; deep orange, wholly or in part stained with a beautiful purple.   The purple juice exudes and dyes everything with which it comes in contact.   The growth is very heavy for its size.

To all appearances it is the same host as is attacked by Hypomyces lactifluorum, resembling it in every particular excepting in the purple stain and juice.

It is frequent in open oak and chestnut woods, but prefers oak.   It grows from among leaves or from grassy spots.   August to October.

While it is beautiful in its coloring it is not inviting in appearance as an edible.   Yet sliced, cut small and stewed for twenty minutes it is one of the very best fungi I have eaten.

Hypomyces. **H. vole'mi** Pk. Subiculum very thin, whitish or isabelline; perithecia minute, brown, nestling in the subiculum; asci very slender, 100–125μ long, sporiferous part 4μ broad. **Spores** oblong-fusiform, 12–15μ long, 4μ broad, commonly binucleate.

Parasitic on the hymenium of Lactarius volemus. Pennsylvania. *Charles McIlvaine.*

The hymenium of the host plant is changed in appearance by the parasite, but the stem and upper surface of the pileus remain unchanged. *Peck*, Bull. Torrey Bot. Club, 27, January, 1900.

The edible qualities are in nowise affected by the parasite.

FAMILY.—**TUBERA'CEÆ.**

Subterranean; ascophore irregularly globose, usually large, not rup- Tuber. turing.

To this family belongs the Truffle of commerce renowned for its flavoring qualities. It has not yet been found in America, though several fungi are ignorantly bought in our markets under that name; notably Coprinus comatus or maned mushroom. The writer has frequently been informed with all the logical force of genuine marketwomen that this was the real Truffle, because they raised it themselves.

Until quite recently but one species of Truffle has been reported

(Plate CLIII.)

× 820

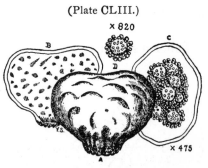

× 475

TUBER NIVEUM or TERFEZIA LEONIS.
By courtesy Rev. A. B. Langlois.
A. Plant. B. Interior (section). C. Asci.
D. Spore.

as growing in America. This, Tuber niveum Desf. or Terfezia leonis Tul. was found by Rev. A. B. Langlois, St. Martinville, La. He reported it as growing plentifully, buried or nearly so in the red sand land along the Red river near Natchitoches. He writes me: "The people where it is found are looking for it with great care and are eating it with great relish. I had occasion to eat it once and I found it delicious." He kindly sent the original illustration from which the accompanying drawing was made. It was taken from Jour. Myc., January, 1887, J. B. Ellis, who first published a description of the American representative of the species. He describes it as "subglobose, up to full two inches in diameter, strongly plicate or furrowed below, nearly smooth and pale reddish-brown outside, marbled-white within and of compact texture much like a potato, but softer. When first dug from the ground the color is pure white, the reddish tint being due to exposure to the air. The asci obovate or subglobose, 75–80x60–70μ. Each contains eight globose spores, thickly clothed with obtuse, elongated, wart-like tubercles and about 20μ in diameter. The home of the white Truffle is

565

Tuber. said to be in Northern Africa, though it is not uncommon in Southern Europe, where its growth is favored by mild winters.''

It is probable that the Truffle will be found in other southern states. Perhaps in the north, as Fries reports that two specimens were found near Linkoping, Sweden, and Mr. H. W. Harkness reports Tubers in the Sierras at the height of 7,000 feet. It is worth hunting for.

It is possible that the common Truffle—Tuber æstivum—will be found in America. Fame awaits the finder. A description of it with illustration is therefore given.

Tuber æsti'vum Vitt.

(Plate CLIV.)

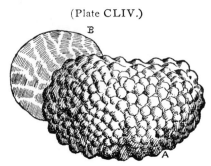

TUBER ÆSTIVUM.
(Common Truffle.)
A. Plant. B. Section showing interior.

Peridium warty, of a blackish-brown color, the warts polygonal and striate; flesh transversed by numerous veins; asci 4–6-spored; spores elliptical, reticulated.

This plant, the common Truffle of our markets, is abundant in Wiltshire and some other parts of England, and probably occurs in many places where it escapes observation from its subterranean habit. *Badham.*

It is cultivated largely in France. ''Perigord Truffles'' are a costly delicacy. The Truffle is of subterranean habit, growing under various kinds of trees and from 12–48 in. under ground. As it does not manifest its presence above ground, dogs and pigs are trained to find it by scent. An interesting chapter on Truffles will be found in British Edible Fungi, M. C. Cooke, 1891. Any plant of similar habit, when found, should be immediately sent by the finder to a known expert for identification.

Thirteen species of Tuber and several Terfeziæ are reported in California, and are described and beautifully illustrated in ''California Hypogæous Fungi'' by H. W. Harkness, ''Proceedings of the California Academy of Sciences,'' 1899.

Terfezia spinosa Harkness closely resembles T. leonis Tul., and T. (sphærotuber) Californicum n. sp., found under oaks beneath vegetable humus in Alameda county, Cal., Professor Harkness remarks, is nearly

566

identical with an edible species found in Italy.   All species found in Tuber.
California are said to be edible, but to be too rare to be of food value.

There is a well known growth, found from New Jersey south to the
Gulf and west to Kansas, called Tucka-
hoe (Pachyma cocos).   (Plate CLV),
an Indian name meaning a round loaf or
cake, and famed for its edible qualities.
Its exact place in plant growth has been
variously determined.   It is now con-
ceded that it is the sclerotium or cellular
reservoir of reserve material of some
fungus.   It is usually found attached to
the roots of trees, in low marshy places.
It grows several feet below the surface,
and to the size of a man's head.   It
varies in shape, being oblong or round,
having a coarse brown covering, looking
like a cocoanut.   Its interior is white,
compact, without cellular structure; it
has no mycelium or trace of fructifica-
tion.   It contains as high as 77 per cent. of pectose and is therefore
highly nutritious.

(Plate CLV.)

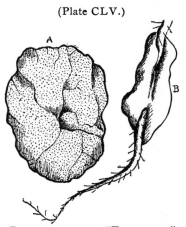

PACHYMA COCOS—"TUCKAHOE."
After Century Dictionary.
A. Mass of Tuckahoe. B. Showing
method of growing around a root.

For full accounts see Torrey Bulletin, October, 1882; Smithsonian
Inst. Rep., 1881, p. 693; article by Professor J. Howard Gore; also
Garden and Forest, IX, p. 302.

The illustration is after that in the Century Dictionary, " Tuckahoe."

Sub-Class BASIDIOMYCETES.

Cohort *GASTROMYCETES.* *Gr.—gasteron,* a sac, etc.

(Plate CLVI.)

1

I.

A. Exterior skin, bark, rind, cortex, scurf, warts, spines, bristles—peridium. Plants with long spines—echinate.

B. Inner rind or true peridium. [A. B.— peridia (plural of peridium).]

C. Columella—those filaments springing from the base and rising, which do not unite freely with those issuing from the inner peridium. This mass of threads is usually conical, but sometimes globose.

D. Capillitium—a soft mass of cottony threads interspersed with minute dust-like spores; the space occupied is called the gleba.

E. Coarse empty, sterile cells. The space they occupy is called the subgleba.

F. Echinate spores magnified.

G. Spines (magnified) which fall off and leave the inner peridium exposed.

2.

A. Lycoperdon echinatum.

B. Spines (magnified) which fall off and leave tesselated inner peridium exposed.   (After Morgan.)

As has been stated, the two Cohorts in which a hymenium or spore-bearing surface is present are called Hymenomycetes and Gastromycetes. In the first the hymenium is exposed, as in the common mushroom. In the second—Gastromycetes—the hymenium is at first enclosed in a sac or peridium, as in the common puff-ball.

The botanical description of Gastromycetes, given by M. C. Cooke, is: "Hymenium more or less permanently concealed, consisting in most cases of closely-packed cells, of which the fertile ones bear naked spores on distinct spicules, exposed only by the rupture or decay of the insisting coat or peridium.'

The Gastromycetes are usually large, ground-growing fungi. A few grow upon wood. The peridium is of dense structure, usually globose

568

and of considerable thickness. It commonly consists of two layers. These form the sac holding the spore-bearing structure, which is called the gleba. The gleba consists of innumerable chambers or cells, curved and branched, and only to be distinguished by magnifying. The primary structure is retained in some species throughout the life of the plants, excepting changes due to growth and maturing, or in others these cells or chambers are large and few, and form distinct peridiola, which contain the spores.

The maturing of the plant and the consequent changes in the gleba is accompanied by various transformations of the peridium.''

It is impossible within the scope of this book to even name all the genera of Gastromycetes. Professor Morgan's table of the families and table of the genera of Lycoperdaceæ are here given. The orders are defined as are some of the genera, and the edible species are described.

TABLE OF FAMILIES OF GASTROMYCETES.

A. TERRESTRIAL.

(a) Peridium double.

I.—Phalloi'deæ. Page 570.

Peridium becoming transformed into a receptacle of various shape, with a volva at its base. Gleba becoming dissolved into a dark green mass of jelly.

II.—Lycoperda'ceæ. Page 577.

Peridium sessile, usually with a more or less thickened base or sometimes stipitate, at maturity filled with a dusty mass of mingled threads and spores.

(b) Peridium single.

III.—Scleroderma'ceæ. Page 615.

Peridium discrete from the gleba, often with a columella; cells of the gleba subpersistent.

IV.—Hymenogastra'ceæ.

Peridium concrete with the gleba, indehiscent; cells of the gleba persistent. (No edible species reported. C. McIlvaine.)

B. Epiphytal.

V.—**Nidularia′ceæ.**

Peridium cyathiform, open at the top, containing one or more distinct peridiola. *Morgan.*

(Small. No species reported edible. *C. McIlvaine.*)

A. Terrestrial.

(*a*) *Peridium double.*

FAMILY I.—**PHALLOIDEÆ.**

Receptacle and **gleba** at first enclosed in a universal volva composed of three distinct layers, the central one being gelatinous at maturity. **Spores** minute, elliptic-oblong, smooth, when mature involved in mucus. *Massee.*

Spores 3–5μ in length. *Morgan.*

There are but few edible species within the family, and those edible only when very young. The family embraces the very offensive fungi known as stink-horns.

TABLE OF GENERA.

I.—**PHALLEÆ.**

Receptacle consisting of an elongated stipe bearing the gleba on a conical pileus at its apex.

1. PHALLUS. Page 571.

Pileus attached only to the apex of the stipe, dependent free all around below.

2. MUTINUS. Page 575.

Pileus wholly adnate to the summit of the stipe.

II.—**CLATHREÆ.**

Receptacle a hollow clathrate body, with the gleba attached to the upper part of the inner surface.

3. CLATHRUS.

Receptacle composed of obliquely anastomosing bars and sessile.

4. SIMBLUM.

Receptacle composed of obliquely anastomosing bars and stipitate.

5. LATERNEA.

Receptacle composed of a few vertical columns and sessile.

Morgan.

I.—PHAL'LEÆ.

Receptacle consisting of an elongated stem bearing the gleba on a conical pileus at its apex. **Stem** cylindric, hollow, composed of one to several layers of round-celled tissue; the gleba occupying the outer surface of the pileus.

GENUS I.—PHAL'LUS Mich.

Stem hollow within, the wall composed of several layers of round- Phallus. celled tissue. **Pileus** attached only to the apex of the stipe, dependent free all around below, the gleba occupying its outer surface. *Morgan.*

The following synoptical tables will exhibit the prominent distinctive features of the species of Phallus of this state (New York) and the United States, so far as I am able to get them from the published descriptions and the specimens at my command.

NEW YORK SPECIES OF PHALLUS.

Denuded pileus reticulate with coarse deep pits or cells.
 Veil exposed, reticulate with small perforations.......

P. Dæmonum Rumph.

 Veil none...................... ....P. impudicus L.
Denuded pileus porous, veil not perforate, concealed....

P. Ravenelii B. and C.

UNITED STATES SPECIES OF PHALLUS.

Denuded pileus reticulate with coarse deep pits or cells.
 Veil exposed.

571

Phallus. Large and reticulate with large perforations.....P. indusiatus Vent.
Smaller and reticulate with small perforations.P. Dæmonum Rumph.
Smaller and plicate........................P. duplicatus Bosc.
Veil none ....P. impudicus L.
Denuded pileus even or merely porous.
Veil short, concealed beneath the pileus.....P. Ravenelii B. and C.
Veil none.................................P. rubicundus Bosc.
Peck, 32d Rep. N. Y. State Bot.

I.—HYMENOPHAL'LUS.

An indusium or veil surrounding the stipe and dependent from its apex beneath the pileus.

a. *Veil reticulate, hanging below the pileus.*

P. Dæ'monum Rumph. **Volva** globose, not very thick, pinkish; segments 3 or 4, irregular. **Stem** cylindric, tapering at each end, cellulose; the veil reticulate, somewhat expanded and bell-shaped, hanging nearly to the middle of the stem. **Pileus** bell-shaped, somewhat oblique; the surface reticulate-pitted after deliquescence; the apex truncate, smooth, perforate. **Spores** elliptic-oblong, 4x2µ.

Plant 9 in. high. **Volva** 2 in. in diameter. **Stem** 1⅛ in. thick at the middle. **Pileus** 2 in. in height; the lower edge of the veil hangs about 4 in. from the apex of the stem. The short veil and the smooth ring at the apex will distinguish this species from the next. *Morgan.*

Growing on the ground in woods.

Ohio, *Morgan, Lea;* Maryland, *Miss Banning;* New York, *Peck.*

Mt. Gretna, ground in mixed woods, August, 1899. *McIlvaine.*

Several specimens were found; but two in the early or ovum stage. In this condition the species is edible. Quality same as P. impudicus.

P. duplica'tus Bosc. **Volva** depressed globose, thick, flabby white; segments 3–5, acute. **Stipe** fusiform-cylindric cellulose; the veil reticulate, hanging down to the volva, sometimes much expanded, often torn and shreddy with pieces adherent to the stipe. **Pileus** campanulate, reticulate-pitted after deliquescence; the apex acute, not regularly perforate. **Spores** elliptic-oblong, 4x2µ.

Photographed by Dr. J. R. Weist.

PHALLUS IMPUDICUS.

PLATE CLVIII.

Plant 6–8 in. high. **Volva** 2½ in. in diameter. **Stipe** 1¼ in. thick in the middle. **Pileus** 2 in. in height. The long veil usually clings close to the stipe though sometimes swinging free and much expanded. In this species the gleba extends over the apex and there is no thick smooth ring encircling the perforation as in the preceding species (P. Dæmonum). *Morgan.*

Growing in woods about old stumps and rotten logs. West Virginia, in woods, along mountain trails; Mt. Gretna, Pa., in mixed woods, summer. *McIlvaine.*

In the forests of the West Virginia mountains, P. duplicatus is frequent. Before rupture of the volva the plant is a semi-gelatinous mass, tenacious and elastic. It has little taste or smell. Cut in slices and fried, or stewed, it is a tender, agreeable food.

b. Veil not reticulate, concealed beneath the pileus.

P. Ravenel'ii B. and C. **Volva** subglobose or ovoid, pinkish; with an inner membrane, the lower half of the veil surrounding the base of the stem; segments 2 or 3. **Stem** cylindric, tapering at each end, cellulose; the veil membranous, scarcely half as long as the pileus and concealed beneath it. **Pileus** conico-bell-shaped; the surface not reticulate-pitted after deliquescence; the apex smooth and closed or finally perforate. **Spores** elliptic-oblong, $4 \times 5 – 2\mu$.

Plant 5–7 in. high. **Volva** 1½–2 in. in diameter. **Stem** nearly 1 in. thick. **Pileus** 1½ in. in height. This species vitiates the genus Dictyophora and it can not very well be placed in Ithyphallus.

Growing in woods and fields about rotting stumps and logs. *Morgan.* South Carolina, *Ravenel;* Ohio, *Morgan;* New York, *Peck.*

II.—ITHYPHAL'LUS Fischer. (*Gr.*—erect; *Gr.*—phallus.)

Stipe without an indusium or veil dependent from its apex. *Morgan.*

P. impudi'cus Linn. (Plate CLVIII.) **Volva** globose or ovoid, white or pinkish; segments 2 or 3. **Stem** cylindric, tapering at each end, cellulose, without a veil. **Pileus** conic-campanulate; the surface reticulate pitted after deliquescence, the apex smooth, at first closed, at length perforate. **Spores** elliptic-oblong, $4–5 \times 2\mu$.

573

Phallus.   Growing on the ground in woods.

Plant 6–8 in. high.   **Volva** 2 in. in diameter.   **Stem** 1¼ in. thick. **Pileus** 2 in. in height.

By the elongation of the stem the thin membrane which separates the stipe from the pileus is torn into shreds and the pileus is thus liberated from the stipe except at the apex.   *Morgan.*

West Virginia, New Jersey, Pennsylvania.   Summer and autumn. *McIlvaine.*

P. impudicus makes itself known wherever it grows.   The stench of the full-grown plant is aggravatingly offensive, attracting blow-flies in quantities, and the carrion beetle Necrophorus Americanus.   It is common over the United States, in woods, open fence corners, along road-sides, but a favorite abode is in kitchen yards and under wooden steps, where, when mature, it will compel the household to seek it in self-defense.   It is a beautiful plant.

When in the egg-shape it is white or light dull-green, semi-gelatinous, tenacious and elastic.   As many as a dozen sometimes grow in a bunch, each from a peculiar white, cord-like root or mycelium.   They look, when young, like bubbles of some thick substance.   In this condition they are very good when fried.   They demand to be eaten at this time, if at any.

GENUS II.—MUTI'NUS Fr.

Stipe hollow within, the wall composed of a single layer of round- Mutinus. celled tissue. **Pileus** wholly adnate to the summit of the stipe, the gleba occupying its outer surface. *Morgan.*

Distinguished from Ithyphallus by the cap being adnate to the receptacle.

M. cani'nus Fr. (*Phallus caninus* Berk.; *Phallus inodorus* Sow.)

Receptacle elongato-fusiform, cellular, white or rosy. **Pileus** short, subacute, rugulose, red. **Spores** cylindrical, involved in green mucus, 3–5x2µ.

(Plate CLIX.)

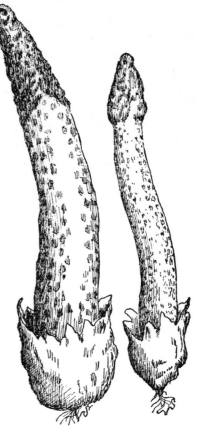

In woods and bushy places. Sporophore from ½–¾ in. before the volva is ruptured. When fully evolved 3–4 in. high. Sometimes scentless, at others with a distinct odor, but never so strong and disagreeable as in Ithyphallus impudicus. *Massee.*

Spores elliptic, 6x4µ *Morgan;* 3–5x2µ *Massee.*

New England, *Frost;* New York, *Warne;* West Virginia, New Jersey, Pennsylvania, *McIlvaine.*

This species is common. Few old woods are without it. It is conspicuous in color among the brown of the forest carpet. The plant has not the mal-odor of its relations, but is not pleasant. In the egg shape it is gelatinous, tenacious, rather firm, edible and good when sliced and fried.

MUTINUS CANINUS.
(After Massee and Morgan.)

M. bovi'nus Morg. **Volva** oblong-ovoid, pinkish, segments 2 or 3.

575

Mutinus. **Stem** cylindric, tapering gradually to the apex, white or pinkish below, bright red above. **Pileus** indeterminate, conic-acuminate, perforate at the apex. **Spores** elliptic-oblong, 4–5x2μ.

Plant 4–7 in. in height, the stem ¾ of an in. in thickness, the volva not much thicker and 1–1½ in. in height; the pileus occupies 1–2 in. of the pointed apex, but is not definitely limited below. This plant has the strong disagreeable odor of other Phalloids.

Growing in rich soil in cultivated grounds and in woods. *Morgan.*

Common in mixed woods, West Virginia, Pennsylvania. Smell strong, but not so offensive as P. impudicus. Edible in the egg-shape.

M. brevis B. and C.—short. **Volva** globose or ovoid, segments 2 or 3. **Stem** bright red, coarsely cribrose, attenuated below. **Pileus** somewhat broadly clavate, sometimes conical, but always more or less obtuse, perforate at the apex.

Plant 2–3 in. high. **Stem** 4–5 lines thick, the volva ¾ of an in. in diameter, the pileus sometimes half as long as the stem.

Growing on the ground in fields and gardens. *Morgan.*

North Carolina, *Curtis;* South Carolina, *Ravenel;* New England, *Wright;* New York, *Howe, Gerard, Peck.*

In the remaining genera, Clathrus, Simblum, Laternea, no species have been reported as tested.

FAMILY II.—**LYCOPERDA'CEÆ.**

Peridium sessile, usually with a more or less thickened base or sometimes stipitate, at maturity filled with a dusty mass of mingled threads and spores.

This order contains many of our most delicious and important food species. The characteristics of all genera are given. In several of them no species are reported edible, but it is more than probable that all are. The genera are therefore given in this table, but are omitted in place to save room. The omitted genera are Nos. 1, Polyplocium; 2, Batarrea; 3, Myriostoma; 5, Astreus; 6, Mitremyces.

TABLE OF GENERA OF LYCOPERDACEÆ.

I.—**VOLVATÆ.**

Outer peridium a thick, firm, persistent coat, bursting irregularly or splitting from the apex downward into segments.

(*a*) *Inner peridium stipitate, the outer remaining as a volva at the base of the stipe.*

1. POLYPLOCIUM.

Inner peridium pileate, with aculeiform processes underneath; threads of the capillitium slender, hyaline, scarcely branched.

2. BATARREA.

Inner peridium circumscissile, the upper part coming off like a lid; threads of the capillitium with spiral markings.

(*b*) *Inner peridium sessile, the outer splitting into segments which become reflexed.*

3. MYRIOSTOMA.

Inner peridium dehiscent above by many mouths; columella ——; threads of the capillitium simple, tapering to each extremity.

4. GEASTER. Page 580.

Inner peridium dehiscent at the apex by a single mouth; columella present; threads of the capillitium simple, tapering to each extremity.

5. ASTRÆUS.

Inner peridium membranaceous; dehiscent at the apex by a single mouth; columella none; threads of the capillitium very long, much branched and interwoven.

6. MITREMYCES.

Inner peridium cartilaginous, dehiscent at the apex by a stellate fissure; columella none; threads of the capillitium very long, much branched and interwoven.

II.—CORTICATÆ.

Outer peridium (cortex) a soft, fragile, more or less deciduous layer, often with external projections in the shape of warts, spines or scales.

(c) *Peridium stipitate.*

7. TYLOSTOMA. Page 582.

Peridium membranaceous, dehiscent by a regular apical mouth; threads of the capillitium very long, much branched and interwoven.

(d) *Peridium sessile, but with a more or less thickened base.*

8. CALVATIA. Page 582.

Peridium large, globose or turbinate, breaking up into fragments from above downward, and gradually falling away; threads of the capillitium very long, much branched and interwoven.

9. LYCOPERDON. Page 589.

Peridium small, globose, obovoid or turbinate, membranaceous, dehiscent by a regular apical mouth, threads of the capillitium long, slender, simple or branched.

10. BOVISTELLA. Page 608.

Peridium subglobose, membranaceous, dehiscent by a regular apical

mouth; threads of the capillitium free, short, several times dichotomously branched.

(e) Peridium sessile, without any thickened base.

11. CATASTOMA.   Page 609.

Peridium globose, subcoriaceous, dehiscent by a basal aperture;
threads of the capillitium free, short, simple, or scarcely branched.

12. BOVISTA.   Page 610.

Peridium subglobose, membranaceous, dehiscent by an apical mouth,
or opening irregularly; threads of the capillitium free, short, several
times dichotomously branched.

13. MYCENASTRUM.   Page 613.

Peridium subglobose, very thick, coriaceous, the upper part finally
breaking up into irregular lobes or fragments; threads of the capillitium
free, short, with a few short branches and scattered prickles. *Morgan.*

GENUS IV.—GEA'STER Mich.

Gr.—the earth; Gr.—star.

Mycelium filamentous or fibrous, much branched and interwoven with Geaster.
the soil.   Peridium subglobose, composed of two distinct persistent
coats; outer peridium thick, fleshy-coriaceous, at first closely investing
the inner, but discrete (distinct) at maturity splitting from the apex downward into several segments which become reflexed; inner peridium thin,
membranaceous then papyraceous (like parchment), sessile or with a

Geaster. short pedicel, dehiscent at the apex by a single mouth. Capillitium tak-

(Plate CLX.)

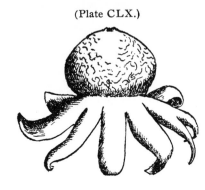

Geaster hygrometricus.
Natural size. (After Morgan.)

ing its origin from the inner surface of the peridium and also from a distinct central columella, which arises from its base; threads simple, long, slender, thickest in the middle and tapering to each extremity, fixed at one end and free at the other. Spores small, globose, minutely warted, brown. *Morgan.*

G. minimus, when found by the writer, was not tested because not found in condition. It is a plant beautiful in its oddity. Its seven to nine outer segments of skin loosen at the bottom, spring up, raising the oval body of the plant with them, turn their points down and balance on the lower points, and look, in miniature, just as would two sectional orange peels spread at their loose points if one was rested, point to point, upon the other. This hoisting of the spore-bearing part aloft, that it may better eject its spores to the wind, does not seem to have been noted by Professor Morgan. Specimens sent to Professor Peck by the writer beautifully illustrated this enterprise of the plant.

G. hygrome'tricus Pers. (Plate CLX, 2 figs., p. 580.) Peridium depressed-globose, the cuticle deciduous with the mycelium; outer peridium deeply parted, the segments 7–20, strongly hygrometric, acute at the apex; inner peridium depressed-globose, sessile, reticulate, pitted, whitish becoming gray or brownish; the mouth an irregularly lacerate aperture. Threads of the capillitium rather thinner than the spores, hyaline. Spores globose, minutely warted, brown, 8–11μ in. in diameter.

Growing in fields and woods in sandy soil. A very common species found everywhere in the world. Inner peridium ¾–1 in. in diameter, the segments expanding to a breadth of 2–3 in. The inner layer of the outer peridium is cartilaginous-gelatinous, hard and rigid when dry, swelling greatly and flexible when wet; though constantly becoming more and more cracked and fissured, it retains its hygroscopic qualities a long time, and the outer peridium remains lying on the soil, stellate in shape, spreading out its rays in moist weather and bending them inward in dry. *Morgan.*

Mr. Morgan made a new genus—Astræus, in which he placed this species. It is so widely known as Geaster hygrometricus that to avoid confusion it is placed in its old genus.

This natural barometer, spreading its stellate covering on the soil about it when the air is laden with moisture, and closing it around its puffy body when humidity is absent, is odd and interesting. The entire genus is more or less gifted with this weather-wise quality. The species is very common, but seldom found in number. Once, in the West Virginia mountains, 1882, I found a large patch of it, and was able to collect from it enough young ones to test its edibility. It is difficult to find before it bursts its outer coat. When young it is, when cooked, soft and creamy inside. The outer part is tough and semi-glutinous but of pleasant texture. It has not a marked flavor, but makes a succulent dish.

II.—CORTICATÆ.

GENUS VII.—TYLOS'TOMA Pers.

Tylostoma.   (Plate CLXI.)

Gr.—a knob.

TYLOSTOMA
MEYENIANUM.
(After Morgan.)

Plants growing on the ground, oftenest in dry and sandy regions. The genus is readily distinguished from all others of the Lycoperdaceæ by the entire peridium being mounted upon the apex of the stem. *Morgan.*

The genus contains but few species. Those I have found were not in condition to test. There is no report upon the edibility of any.

GENUS VIII.—CALVA'TIA Fr.

Calvatia.   Mycelium fibrous, usually thick and cord-like, rooting from the base. **Peridium** large, globose and nearly sessile, or turbinate with a well-developed base; cortex a very thin adherent layer, often smooth and continuous, sometimes composed of minute spinules or granules; inner peridium a loosely woven and very fragile covering, after maturity breaking up into fragments from above downward and gradually falling away. Subgleba cellulose, mostly definitely limited and concave above, persistent; mass of spores and capillitium dense, compact, persistent a long time and slowly dissipating after the fracture of the peridium; the threads very long, slender, much branched and interwoven. **Spores** small, globose, usually sessile or with only a minute pedicel. *Morgan.*

Puffballs of the largest size, growing on the ground in fields and woods. *Morgan.*

I.—SESSILES.

Peridium very large, without a distinct base; subgleba nearly obsolete, the mass of spores and capillitium quite filling the interior.

C. gigantea Batsch.—gigantic. (*L. bovista* Linn.; *L. maximum* Schaeff.; *L. giganteum* Batsch.) Very large, 10–20 in. in diameter, obconic or depressed-globose, nearly or quite sessile, white or whitish, becoming discolored by age, smooth or slightly roughened by weak spinose or minute floccose warts, sometimes cracking in areas; capillitium and spores yellowish-green to dingy-olive. **Spores** smooth, 4μ in. in diameter. Edible. *Peck*, 32d Rep. N. Y. State Bot.

Spores globose, even or sometimes minutely warted, 3.5–4.5 in. in diameter, often with a minute pedicel. *Morgan.*

Common over the states. Growing on the ground in grassy places in fields and woods. August to October.

As the name implies, this species is gigantic. It is the largest of all fungi. It has attained the diameter of three feet in this country, but is reported larger in Europe. I have found it in West Virginia weighing nine pounds, but one is reported as found in Gordon Park weighing forty-seven pounds. I have often followed the advice of Vittadini and sliced a meal for my family from growing individuals. The cut surface contracts and dries. The plant seems to be deprived of its power to further ripen. It can thus be cut for many days. It has other than food uses in its dry form—as a sponge, as tinder, as a color, as a styptic in hemorrhage; the Finns make a remedy of it for diarrhea in calves, and it is burned under bee-hives to stupefy bees.

It, as well as L. cyathiforme, is an admirable and delicate fungus.

C. pachyder'ma Pk. *Gr.*—thick-skinned. **Peridium** very large, globose or obovoid, often irregular, with a thick cord-like root; cortex thin, smooth, whitish, persistent, drying up into polygonal areolæ which are white in the center with a brown border; inner peridium very thick but fragile, with a separable membranaceous lining, after maturity gradually breaking up into fragments and falling away. Subgleba obsolete; mass of spores and capillitium greenish-yellow then olive-brown; the threads very long, occasionally septate, branched, mostly thinner than the spores. **Spores** globose, distinctly warted, $5-6\mu$ in diameter, sometimes with a minute pedicel.

Growing on the ground. Arizona, *Pringle;* Dakota, *Miss Nellie Crouch*. **Peridium** 4–8 in. in diameter. Remarkable for its thick peridium, which becomes white spotted and areolate. *Morgan.*

I have not seen this species.

II.—CYATHIFORMES.

Calvatia. Peridium large, top-shaped, with a stout thick base; subgleba limited and concave above, persistent.

C. cyathifor'mis Bosc.—cup-shaped. (*L. cyathiforme* Bosc.) (Plate CLXII, p. 584.) **Peridium** 3–6 in. in diameter, globose or depressed-globose, smooth or minutely floccose or scaly, whitish cinereous brown or pinkish brown, often cracking into areas in the upper part, commonly with a short, thick, stem-like base; capillitium and spores purple-brown, these and the upper part of the peridium falling away and disappearing when old, leaving a cup-shaped base with a ragged margin. **Spores** globose, rough, purple-brown, 5–6.5μ broad. *Peck*, 48th Rep. N. Y. State Bot.

Common over United States. Indiana, *H. I. Miller;* West Virginia, New Jersey. On open grassy ground. July to October. *McIlvaine.*

Often a queer, ragged, cup-shaped, purplish mass is noticed protruding from the ground, looking as if the upper half had been cut off horizontally. This is the mature C. cyathiformis, or rather, what is left of it. The upper half has blown away and is spreading its spores elsewhere.

A first-class Lycoperdon, meaty and of excellent flavor. When it occurs, it is usually in plenty. On the great parade ground at Mt. Gretna, Pa., it annually appears in large quantities. Cows are fond of it, and it is this fungus which is currently believed among farmers to affect milk. I have watched cows pawing it to pieces and eating portions of it.

C. fra'gilis Vitt.—fragile. **Peridium** obovoid, plicate below, with a short-pointed base and a cord-like root. Cortex a smooth continuous layer, very thin and fragile, separable, white or grayish, becoming brownish and tinged with violet and purple, commonly areolate above; inner peridium thin, violet to purple, velvety, extremely fragile, after maturity the upper part soon breaking up into fragments and falling away. Subgleba occupying but a small portion of the peridium, cup-shaped above, persistent; mass of spores and capillitium from violet to pale purple; the threads very long, mostly thinner than the spores, scarcely branched. **Spores** globose, minutely warted, 4–5.5μ in diameter, sessile.

CALVATIA CYATHIFORMIS.
(Lycoperdon Cyathiforme.)

Photographed by Dr. J. R. Weist.

Growing on the open prairies.   Wisconsin, *Brown;* Iowa, *McBride;* Calvatia. Nebraska, Wyoming, *Webber;* Kansas, *Cragin;* California, *Harkness.* **Peridium** 1½–3 in. in diameter.   *Morgan.*

Not seen by writer.   Doubtless edible.

C. sigilla′ta Cragin—adorned with figures.   **Peridium** large, depressed above, narrowed below into a stem-like base.   Cortex very thin and fragile, white, easily abraded; inner peridium subcoriaceous, with a fragile rust-color brown lining, marked off above into polygonal areas by lines of depression, at length breaking up into fragments and falling away.   Mass of spores and capillitium violet to dark-purple.   **Spores** globose, even, 3.5–4.5μ in diameter, with a long pedicel.

Growing on the open prairie.   Kansas, *Cragin.*   **Peridium** 4–5 in. in diameter.   The species is well marked by the even pedicellate spores. *Morgan.*

C. cæla′ta Bull.—carved in relief.   (*L. cæla′tum* Bull.; *L. bovista* Pers.)   **Peridium** large, obovoid or top-shaped, depressed above, with a stout thick base and a cord-like root.   Cortex a thickish floccose layer, with coarse warts or spines above, whitish then ochraceous or finally brown, at length breaking up into areola which are more or less persistent; inner peridium thick but fragile, thinner about the apex, where it finally ruptures, forming a large irregular lacerate aperture. Subgleba occupying nearly half the peridium, cup-shaped above and a long time persistent; mass of spores and capillitium compact, farinaceous, greenish-yellow or olivaceous, becoming pale to dark-brown; the threads very much branched, the primary branches two or three times as thick as the spores, very brittle, soon breaking up into fragments. **Spores** globose, even, 4–4.5μ in diameter, sessile or sometimes with a short or minute pedicel.

Growing on the ground in fields and woods.

Peridium 3–5 in. in diameter, sometimes larger.   *Morgan.*

Wisconsin, *Brown;* Minnesota, *Johnson;* Kansas, *Kellerman;* L. cælatum, New York, edible, *Peck,* 23d Rep.; Indiana, good, *H. I. Miller.*   Common, West Virginia, wooded lanes, *McIlvaine.*

An excellent species.

C. hiema′lis Bull.—belonging to winter.   **Peridium** obovoid or top-shaped, depressed above, with a stout thick base and a cord-like root.

Calvatia. Cortex a thin furfuraceous coat, with stout convergent spines above, whitish or gray, becoming yellowish and reddish, after maturity gradually falling away from the upper part; inner peridium thin, submembranaceous, pallid or brownish, dehiscent at the apex by an irregular lacerate mouth. Subgleba occupying nearly half the peridium, cup-shaped above and a long time persistent; mass of spores and capillitium soft, lax, greenish-yellow then brownish-olivaceous; the threads very long, much branched, the primary branches about as thick as the spores, the ultimate ones long, slender and tapering. Spores globose, even, 3.5–4.5μ in diameter, with a short or minute pedicel.

Growing on the ground in fields and pastures. Peridium 2–4 in. in diameter and 3–5 in. in height. I find this species referred to North America in Saccardo's Sylloge. It is Lycoperdon cælatum of Fries S. M. Possibly the L. cælatum of Curtis's catalogue may be this species. *Morgan.*

Not seen by writer.

C. craniifor'mis Schw.—*cranion*, a skull. Peridium very large, obovoid or turbinate, depressed above, the base thick and stout, with a cordlike root. Cortex a smooth continuous layer, very thin and fragile, easily peeling off, pallid or grayish, sometimes with a reddish tinge, often becoming folded in areas; the inner peridium thin, ochraceous to bright brown, velvety, extremely fragile, after maturity the upper part breaking up into fragments and falling away. Subgleba occupying about one-half of the peridium, cup-shaped above and a long time persistent; mass of spores and capillitium greenish-yellow then ochraceous or dirty olivaceous; the threads very long, about as thick as the spores, branched. Spores globose, even, 3–3.5μ in diameter, with a minute pedicel.

(Plate CLXIII.)

CALVATIA CRANIIFORMIS.
(After Morgan.)

Growing on the ground in woods. Peridium commonly 3–6 in. in diameter and 4–5 in. in height, but much larger specimens are some- Calvatia. times met with. This species abounds in the woods of southern Ohio, growing in great patches of numerous individuals. I do not know that the edible qualities of this species have been tested. *Morgan.*

Chester county, Pa. Springton Hills. On ground in mixed woods. August to October. *McIlvaine.*

Not a frequent species with us. I have seen it only in the locality named. The substance is very like that of L. pyriforme. When white it has a strong but pleasant odor, and in this condition it is an excellent fungus. The slightest change to yellow makes it bitter.

C. rubro-fla′va Cragin—reddish-yellow. **Peridium** obconic, tapering gradually downward to the rooting mycelium. Cortex a very thin furfuraceous or granulose coat, with a few short, scattered spinules above; inner peridium thin and fragile, at first whitish, soon becoming orange-red to orange-brown in color, after maturity the upper part breaking up into fragments and falling away. Subgleba occupying about a third part of the peridium; mass of spores and capillitium reddish-ocher then olivaceous-orange; the threads very long, rather thicker than the spores, branched. **Spores** globose, even, 3–3.5μ in diameter, sometimes with a minute pedicel.

Growing on the ground. Kansas, *Cragin, Kellerman.* **Peridium** 1½–3 in. in height with a breach of 1–2 in. The peculiar orange or rather reddish-ocher color with which the whole plant is pervaded at maturity is very remarkable. *Morgan.*

III.—STIPITATÆ.

Peridium depressed, globose above, abruptly contracted below into a long stem-like base; subgleba not definitely limited above, continuous with the capillitium, persistent.

C. sacca′ta (Vahl.) Fr.—*saccus*, a bag or pouch. Medium size, 2–4 in. high, 1–2 in. broad. **Peridium** depressed-globose or somewhat lentiform, supported by a long stem-like base, furfuraceous with minute persistent mealy or granular warts or spinules, often plicate beneath, white or creamy-white, at maturity becoming brown or olive-brown, subshining and very thin or membranous, breaking up into irregular fragments which sometimes adhere to the capillitium for a considerable time, the stem-like base cylindrical or narrowed downward, sometimes thick;

587

Calvatia. capillitium rather dense, subpersistent, and with the spores dingy-olive or dingy-brown, sometimes verging toward purplish-brown. **Spores** rough, 4–5μ in diameter. Edible.

Low mossy grounds and bushy swamps, especially under alders. Sandlake, Center and Adirondack mountains. August to October. *Peck*, 32d Rep. N. Y. State Bot.

West Virginia, 1881–1885; Pennsylvania, New Jersey, North Carolina. Frequent, thin moist woods. July to November. *McIlvaine*.

C. saccata, the long-stemmed puff-ball, is a common and pleasing species. Shape, color, feel, combine to make it attractive. It is one of the very best we have. When white inside and otherwise in good condition it is delicious.

C. ela'ta Massee.

(Plate CLXIV.)

CALVATIA ELATA.
(After Morgan.)

Peridium globose or depressed-globose above, plicate below and abruptly contracted into a long stem-like base; the base slender, cylindric or tapering downward, sometimes pitted; mycelium fibrous and filamentous. Cortex a very thin coat of minute persistent spinules or granules; inner peridium white or cream-colored, becoming brown or olivaceous, very thin and fragile, after maturity the upper part soon breaking up into fragments and falling away. Subgleba occupying the stem-like base, a long time persistent; mass of spores and capillitium brown or brownish-olivaceous; the threads very long, branched, the main stem as thick as the spores, the branches more slender. **Spores** globose, even or very minutely warted, 4–5μ in diameter with a short or minute pedicel.

Growing among mosses in low grounds and bushy places. New England, *Humphrey;* New York, *Peck*. **Peridium** 1–2 in. in diameter and 3–6 in. in height, the stem-like base ½–¾ of an inch in thickness. This American form of Lycoperdon saccatum has lately been separated from it, and named, figured and described as Lycoperdon elatum by George Massee. *Morgan*.

Edible.

588

GENUS IX.—LYCOPER'DON Tourn.

Mycelium fibrous, rooting from the base. **Peridium** small, globose, Lycoperdon. obovoid or turbinate, with a more or less thickened base; cortex a subpersistent coat of soft spines, scales, warts or granules; inner peridium thin, membranaceous becoming papyraceous, dehiscent by a regular apical mouth. *Morgan.*

When the plant sits (without stem) directly upon the ground or wood it is *sessile.* The outer layer of the two parts of its covering is the *exterior peridium* (sometimes spoken of as *cortex*). This frequently breaks up into scales, spines, bristles, minute flocculent or powdery masses, and these vary in size and in many species disappear as the plant matures. These are of determining value in several species of Lycoperdaceæ. Plants with coarse, long spines are *echinate* because they bristle. When the spines incline together and form a point they are *stellate.* Various formations of this outside covering are also called *warts.* The inner rind or skin is the true *peridium.*

The mass of thread-like filaments which fills the interior of the plant is called the *capillitium.* The filaments are deftly interlaced. At times filaments springing from the base do not interlace with the others; these are called *columellæ.* These filaments bear the spores—the dust which puffs out in such quantity and gives the common name to the plant— puff-ball—and its Mephistophelian one—The Devil's Snuff-box. In some species the filaments at the base of the plant are *sterile*—they do not bear spores. These filaments are more contracted and form the neck, stem or *subgleba.* The *gleba* is the upper interior of the plant, in which the spores are contained. See plate CLVI.

Dehiscent is said of an organ which opens of itself at maturity. A plant is dehiscent at the discharging point of its spores. If this is at the summit it is *apically dehiscent.*

The descriptions herein given of American representatives of European species are in many instances those of A. P. Morgan, who has made special study of this genus, and those of Professor C. H. Peck, whose interstate experience acquaints him with every varying form. Mr. Morgan has kindly given permission to use his text and drawings.

No one has yet had reason to doubt the harmlessness of any puffball. There are a few I have not eaten, but believing that these will be proven edible, descriptions of all species occurring in America are given.

Lycoperdon.   There are first and second-class puff-balls. Usually the small species are slightly strong, and if a shade of yellow appears upon breaking any puff-ball, it will be more or less bitter and will spoil a whole dish. The larger species are milder. The flavor of puff-balls appears to be issued to them as a ration. It is all there in a little fellow, and in a big one it is simply spread through more substance.

Lafayette B. Mendel in Am. Jour. of Physiology, March, 1898, gives the nitrogenous compounds in L. bovista as:

Nitrogen soluble in gastric juice	3.13
Digestible protein nitrogen	3.13
Indigestible protein nitrogen	2.70
Protein nitrogen	5.79
Extractive nitrogen	2.40
Total nitrogen	8.19

TABLE OF THE SPECIES OF LYCOPERDON.

I.—PURPLE-SPORED SERIES.

Mature spores purplish-brown.

a. Cortex consisting of very long convergent spines. Page 591.
b. Cortex composed of long slender convergent spines. Page 592.
c. Cortex composed of minute spinules. Page 594.
d. Cortex a furfuraceous persistent coat. Page 595.
e. Cortex a smooth, continuous layer, becoming areolate. Page 597.

II.—OLIVE-SPORED SERIES.

Mature spores usually brownish-olivaceous.

A. PERIDIUM OBOVOID OR TURBINATE, THE SUBGLEBA WELL DEVELOPED.

f. Cortex of long spines mingled with shorter ones, the former at length fall away, leaving a reticulate surface to the inner peridium. Page 598.
g. Cortex of stout spines which fall away and leave a tomentose or furfuraceous surface to the inner peridium. Page 599.
h. Cortex of long spines, curved and convergent at the apex, which fall away and leave a smooth surface to the inner peridium. Page 600.
i. Cortex of minute spinules and granules or furfuraceous scales. Terrestrial. Page 602.

k. Cortex of minute spinules, scales or granules. Lignatile. Page 603. <small>Lycoperdon.</small>

B. Peridium Very Small, Globose, the Subgleba Nearly Obsolete.

l. Cortex a thin coat of minute spinules, scales or granules. Page 604. *Morgan.*

I.—Purple-spored Series.

(a) Cortex consisting of very long convergent spines; denuded peridium smooth.

L. echina′tum Pers.—prickly. (*L. Peck′ii* Morg.) (Plate CLVI, fig. 2, p. 568.) **Peridium** ¾–1½ in. broad, subglobose, generally narrowed below into a short stem-like base, whitish brownish or pinkish-brown, echinate above with rather stout spines, which at length fall off and leave the surface smooth; toward the base spinulose or furfuraceous; capillitium and spores dingy-olive. **Spores** minutely rough, 4µ in diameter.

Ground and decaying wood in woods. Albany, Forestburg and Adirondack mountains. August to October.

The whole plant is generally obovate, pyriform or turbinate, and the spines are larger and more or less curved at and near the apex, diminishing in size toward the base where they are more persistent. In the immature condition it is difficult to distinguish it from L. pedicellatum; but when mature its smooth peridium and spores destitute of pedicels separate it. *Peck,* 32d Rep. N. Y. State Bot.

L. echinatum appears to be common to all the states. August to frost. It is frequent but not abundant. Raw the taste is slight. Cooked it is tender and of good flavor.

L. pulcher′rimum B. and C. (*L. Frost′ii* Pk.) **Peridium** usually obovoid, sometimes subturbinate, with a short stout base; the mycelium forming a thick cord-like root. Cortex consisting of very long white spines, converging and often coherent at the apex; the spines at length fall away from the upper part of the peridium, leaving the inner peridium with a smooth purplish-brown shining surface, sometimes faintly reticulated. Subgleba occupying about a third part of the peridium; mass of spores and capillitium at first olivaceous, then brownish-purple; the threads much branched, the main stem thicker than the spores, the

Lycoperdon. branches long, slender and tapering. **Spores** globose, minutely warted, 4.5–5.5μ in diameter.

Peridium 1–2½ in. in diameter and 1–2 in. in height.

The fresh specimens of this plant have a strong and not unpleasant fragrance.

Growing in low grounds, in fields and woods. September, October.

L. pulcherrimum is frequent, but not abundant. It ranks with second-class puff-balls. It is good when young and fresh.

L. constella′tum Fr.—grouped. **Peridium** subglobose or ob-ovate, sometimes depressed, 10–18 lines broad, echinate with rather long stout crowded brown spines which are either straight curved or stellately united and which at length fall off and leave the surface reticulate with brown lines; capillitium and spores brown or purplish-brown, columella present. **Spores** rough, 5–6.5μ in diameter.

(Plate CLXV.)

LYCOPERDON CONSTEL-LATUM.
(After Peck.)

Ground in dense shades and groves. Oneida, *Warne*. Rare. Autumn. *Peck*, 32d Rep. N. Y. State Bot.

(b) Cortex composed of long, slender convergent spines; denuded peridium smooth.

L. hirtum Mart.—hairy. **Peridium** broadly turbinate, depressed above, contracted below into a short, thick, tapering or pointed base, with a cord-like root. Cortex a dense coat of soft spines, long, slender and convergent above, becoming shorter downward, gray or brownish in color; these finally fall away, leaving the inner peridium with a brown or purplish-brown, smooth, shining surface. Subgleba occupying from one-third to one-half of the peridium; mass of spores and capillitium olivaceous, then brownish-purple; the threads branched, the main stem about as thick as the spores, with slender, tapering branches. **Spores** globose, distinctly warted, 5–6μ in diameter.

Growing on the ground in woods. **Peridium** 1–2½ in. in diameter and 1½–2 in. in height. This species in this country heretofore has been included with L. atropurpureum. I have followed Mr. Massee in

keeping them separate. This is perhaps L. bicolor W. and C., of the Lycoperdon.
Pacific Coast Catalogue. *Morgan.*

New York, *Peck*, 46th Rep.; West Virginia, New Jersey, Pennsylvania. Ground in woods. August to October. *McIlvaine.*

It is edible. Good when young and fresh.

L. atropurpur'eum Vitt.—*ater*, black; *purpureus*, purple—of the
spores. **Peridium** globose depressed-globose or obovate, 6–30 lines
broad, generally narrowed below into a short stem-like base, white cinereous or brownish, mealy-spinulose, hairy-spinulose, echinate or stellately echinate, when denuded smooth and subshining; capillitium and
spores finally purplish-brown, columella present. **Spores** rough, 5–6µ in
diameter.

Sandy pastures, woods and bushy places. Common. August to
October.

This appears to be one of the most polymorphous species we have.
It is so variable that I have been obliged to modify the usual description
very much, in order to include forms which are quite diverse, yet which
appear to me to run together in such a way that I am unable to draw
any satisfactory line of distinction between them.

There are three principal varieties which I have referred to this
species. The first is usually 1–2 in. broad, sessile, or with a very short
stem, nearly smooth, being mealy or pruinose, and having a few minute,
weak, scattered spinules or scales. Its color is generally whitish or
white slightly clouded with brown. It grows in sandy pastures and
cleared lands, and is probably the nearest of the three in its resemblance
to the type.

I regard the second and third as worthy of a name and designate and
define them as follows:

Var. *hirtel'lum*. **Peridium** hairy-spinulose with erect or curved
sometimes stellately united spinules, which are often of a blackish color.

Ground and decaying vegetable matter in woods.

Var. *stella're*. **Peridium** echinate or stellately echinate with rather
stout easily deciduous spines.

Ground in woods and bushy places.

In this species the capillitium and spores are at first greenish-yellow,
olive-tinted or brownish; but when fully mature they are purple-tinted.
Some care will, therefore, be necessary lest the last variety be confused

Lycoperdon. with the Echinate Puff-ball, L. echinatum. *Peck*, 32d Rep. N. Y. State Bot.

Spores 6–7μ *Massee;* globose, distinctly warted, 5.5–6μ *Morgan.*

Ohio, West Virginia, New Jersey, Pennsylvania. On ground in woods. August to October. *McIlvaine.*

L. atropurpureum is frequent, not abundant. It is edible, good.

(c) Cortex composed of minute spinules; denuded peridium smooth.

L. cu'pricum Bon.—coppery. **Peridium** obconic, depressed above and tapering downward, the base plicate, with a fibrous mycelium. Cortex gray or flesh-color, composed of minute spinules circularly arranged and convergent and coherent at the apex; these dry up, becoming dark purplish in color, and finally fall away from the smooth, shining, copper-colored surface of the inner peridium. Subgleba occupying nearly a third part of the peridium; mass of spores and capillitium, at length purplish-brown; the threads branched, the main stem thinner than the spores, with long, tapering branches. **Spores** globose, distinctly warted, 6–7μ in diameter.

Growing in sandy soil in woods. New Jersey, *Ellis.*

Peridium about 1 in. in diameter and an inch or more in height. The microscopic features are given from specimens received from Mr. Ellis. *Morgan.*

Near Haddonfield, N. J., 1891–1896. Sandy woods. *McIlvaine.*

Not frequent. Those found upon several occasions were eaten and found good.

L. asterosper'mum D. and M.—*aster*, star; *sperma*, seed. **Peridium** obovoid or pyriform (pear-shaped), the base short and pointed, with a slender fibrous mycelium. Cortex a thin coat of minute spinules with intermingled granules, gray or brownish above, paler below; these dry up and are a long time persistent, but they finally fall away, leaving the inner peridium with a pale brown, smooth, shining surface. Subgleba obconical, occupying nearly a third part of the peridium; mass of spores and capillitium olivaceous, then brownish-purple; the threads about as thick as the spores, with slender tapering branches. **Spores** globose, distinctly warted, 5.5–6.5μ in diameter.

Peridium 1–1 ½ in. in diameter. A very pretty species of regular Lycoperdon. form; its glossy cortex is quite persistent. *Morgan.*

Growing on the ground in open woods. Ohio, *Morgan;* Nebraska, *Webber.*

New York, *Peck,* 46th, 51st Rep.; Ohio, *Lloyd.*

L. delica′tum Berk. **Peridium** subglobose, plicate underneath, with a fibrous mycelium. Cortex a thin coat of minute spinules and gran-ules, gray or brownish above, whitish below, finally falling away from the smooth, shining, pale or brownish surface of the inner peridium. Subgleba very small or quite obsolete; mass of spores and capillitium olivaceous, then pale or brownish-purplish; the threads rather thinner than the spores, with slender tapering branches. **Spores** globose, dis-tinctly warted, 5–6μ in diameter.

Peridium 1–2 in. in diameter. *Morgan.*

Growing on the ground. Pennsylvania, *Gentry;* Missouri, *Professor Trelease* (*Peck,* Rep. 40); Louisiana, *Langlois.*

(*d*) *Cortex a furfuraceous persistent coat.*

L. glabel′lum Pk.—smooth, bare. Subglobose or subturbinate, 1–1.5 in. broad, sometimes narrowed below into a short stem-like base, furfuraceous with very minute nearly uniform persistent warts, which appear to the naked eye like minute granules or papillæ, yellow, opening by a small aper-ture; inner mass purplish-brown, capillitium with a central columella. **Spores** purplish-brown, globose, rough, 5–6.5μ in diameter.

(Plate CLXVI.)

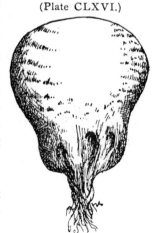

Ground in copses and in pine woods. West Albany and North Greenbush. Autumn. *Peck,* 31st Rep. N. Y. State Bot.

Ohio, *Morgan;* Wisconsin, *Trelease;* West Virginia, Pennsylvania, New Jersey. Fre-quent. Autumn. *McIlvaine.*

One of the prettiest Lycoperdons. Sym-metrical, and otherwise attractive. Sweet, firm, solid. It is not high in flavor, but is delicate.

LYCOPERDON GLABELLUM.
(After Morgan.)

Lycoperdon.

L. elonga′tum Berk.—elongated. **Peridium** globose above, contracted below into a stout thick base, more or less elongated and cylindric or tapering downward; mycelium composed of thick fibers. Cortex a loose flocculose white or yellowish coat, drying up into a mealy or furfuraceous persistent layer, which scarcely reveals the pale shining surface of the inner peridium. Subgleba occupying more than half the interior of the peridium; mass of spores and capillitium pale olivaceous, then pale brown or finally purplish; the threads much branched, the main stem much thicker than the spores, the branches tapering. **Spores** globose, distinctly warted, 5.5–6.5μ in diameter.

Growing on the ground in damp woods. Ohio, *Morgan*. **Peridium** 1–2 in. in diameter and 2–3 in. in height, the base ¾–1 in. in thickness. In form it somewhat resembles L. gemmatum, but it has a cortex like that of L. glabellum. *Morgan.*

New York, *Peck*, 49th Rep. Closely allied to L. glabellum. Its stout elongated base serves as a mark of distinction.

Ohio, *Morgan;* Pennsylvania, Washington, Pa., Myc. Club.

Not common. Sometimes tufted, three or four together. Edible, good.

L. el′egans Morgan—elegant. **Peridium** large, depressed globose, plicate underneath and sometimes with a narrow umboniform base, which is continuous with the thick root. Cortex at first flocculose, white or yellowish, drying up into a dense furfuraceous persistent coat, which becomes ochraceous or brownish in color, and sometimes obscurely areolate. Subgleba broad, convex above, occupying a third part or more of the peridium; mass of spores and capillitium olivaceous, then pale-brown or finally purplish-brown; the threads much branched, the main stem thicker than the spores, the branches long and tapering. **Spores** globose, distinctly warted, 5–6μ in diameter.

Growing on rich soil on the open prairie about Iowa City, Ia., *Prof. T. H. McBride*. **Peridium** 1½–3 in. in diameter. In form and size this species somewhat resembles Calvatia fragilis, but the threads are arranged in two sets as in Lycoperdon; the cortex is similar to that of L. glabellum; the mycelium forms a remarkably thick root. *Morgan.*

(*e*) *Cortex a smooth, continuous layer, becoming areolate.*

L. rimula′tum Pk.—*rimula*, a small chink. **Peridium** depressed— Lycoperdon.
globose or broadly obovoid, plicate under-
neath with a slender fibrous mycelium.
Cortex at first a thin, smooth, continuous
fibrillose layer, gray or bluish-gray, sometimes with a purplish tinge; this at length
breaks into a network of fine lines or fis-
sures, gradually dries up into minute thin
adnate scales, and finally falls away from
the smooth grayish or purplish-brown
surface of the inner peridium. Subgleba
broad, but distinct, plane above, occupying about a fourth part of the peridium;
mass of spores and capillitium purplish-
gray, then brownish-purple; the threads

(Plate CLXVII.)

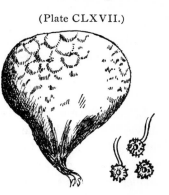

LYCOPERDON RIMULATUM.
With magnified spores.
(After Morgan.)

simple or scarcely branched, variable in thickness, but always thinner
than the spores. **Spores** globose, distinctly warted, 6–7μ in diameter,
often pedicellate.

Growing on the ground in fields and open woods. New York, *Peck;*
South Carolina, *Atkinson;* Ohio, *Morgan;* Wisconsin, *Trelease.* **Peridium** ¾–1 ½ in. in diameter, scarcely an inch in height. *Morgan.*

New Jersey, *T. J. Collins;* Pennsylvania. Autumn. *McIlvaine.*

A pretty species, generally in groups. Frequent. It is not common,
but occasionally generous patches of it are found. Edible, good.

L. vela′tum Vitt.—*velatus*, having a velum. **Peridium** globose or
obovoid, with a cord-like root. Cortex white or yellowish, at first a
thickish continuous layer, then breaking up into circular or irregular
persistent patches with fimbriate margins. Subgleba occupying about
a third part of the peridium; mass of spores and capillitium olivaceous,
then purplish-brown; the threads branched, the main stem nearly as
thick as the spores, the branches long and tapering. **Spores** globose,
distinctly warted, 5–6μ in diameter.

Growing on the ground in woods. South Carolina, *Ravenel.*
Peridium 1–2 in. in diameter. *Morgan.*

597

Lycoperdon. New Jersey, *T. J. Collins;* Chester county, Pa., sometimes clustered, *McIlvaine.*

Good.

II.—OLIVE-SPORED SERIES.

A. PERIDIUM TOP-SHAPED, THE SUBGLEBA WELL DEVELOPED.

(*f*) *Cortex of long spines, etc.*

L. gemma'tum Batsch.—gemmed.   **Peridium** turbinate, depressed above, the base short and obconic or more elongated and tapering or subcylindric, arising from a fibrous mycelium.   Cortex consisting of long, thick, erect spines or warts of irregular shape, with intervening smaller ones, whitish or gray in color, sometimes with a tinge of red or brown; the larger spines first fall away, leaving pale spots on the surface, and giving it a reticulate appearance.   Subgleba variable in amount, usually more than half the peridium; mass of spores and capillitium greenish-yellow, then pale-brown; threads simple or scarcely branched, about as thick as the spores.   **Spores** globose, even or very minutely warted, 3.5–4.5μ in diameter.

Peridium 1–2 in. in diameter and 1–3 in. in height.   This species is distinguished from all others by the peculiar large erect terete spines or warts, the so-called gems which stud its upper surface.   *Morgan.*

Growing on the ground and sometimes on rotten trunks in woods, often cespitose.   *Frost.*   New York, *Peck.*

Found in every part of the world.

New York, *Peck*, Rep. 22; Indiana, *H. I. Miller;* West Virginia, New Jersey, Pennsylvania.   On the ground and on logs.   *McIlvaine.*

Edible, but not pleasant.   *Peck.*   Edible, *H. I. Miller.*

Professor Peck gives two varieties:

Var. *hir'tum.*   Turbinate, subsessile, hairy with soft, slender warts which generally become blackish.

Var. *papilla'tum.*   Subrotund, sessile, papillose, furfuraceous-pulverulent.

Very common and known in all countries.   It is, to my thinking, our prettiest puff-ball.   Its beautifully studded surface, reminding of exquisite settings, is in itself worth studying for the designs.   It is usually solitary or in small groups, but at times these groups contain scores of individuals. It grows in the open on the ground or from both ground Lycoperdon. and wood, in woods.

I think it equal to any other puff-ball. But great care must be taken to examine each specimen before putting it into the pan. A single one, which has turned yellow in the slightest degree, will spoil a whole dish. And this is the case with any of the small puff-balls. One ageing L. pyriforme will embitter a hundred.

L. perla′tum Pers.—*perfero*, to endure. (Enduring through winter.) **Peridium** turbinate, broad and depressed above, plicate underneath and contracted into a short and pointed or sometimes elongated and tapering base; mycelium fibrous. Cortex of long slender spines, mingled with smaller spinules and warts, gray brown or blackish in color; the longer spines first fall away, leaving a reticulate surface to the inner peridium. Subgleba occupying one-third to one-half of the peridium; mass of spores and capillitium greenish-yellow, then brownish-olivaceous; the threads mostly simple, some of them thicker than the spores. **Spores** globose, even or very minutely warted, $3.5-4.5\mu$ in diameter.

Growing on the ground in woods. **Peridium** 1–2 in. in diameter and 1–2 in. in height. This is *L. gemmatum*, var. *hirtum*, of Peck's United States species of Lycoperdon. *Morgan.*

New York, *Peck*, 46th Rep.; Maryland, *James;* West Virginia, New Jersey. Occasional. On ground and decaying wood. *McIlvaine.*

Edible. Same habit and quality as L. gemmatum.

(*g*) *Cortex of stout spines which fall away, etc.*

L. excipulifor′me Scop.—*excipula*, a receptacle. **Peridium** turbinate, depressed above, plicate below and contracted into a more or less elongated base. Cortex of large stout spines, convergent above, becoming smaller downward, which at length fall away, leaving a tomentose surface to the inner peridium. Subgleba occupying one-half or more of the peridium; mass of spores and capillitium greenish-yellow, then brownish-olivaceous; the threads about as thick as the spores, scarcely branched. **Spores** globose, minutely warted, $4-5\mu$ in diameter.

Peridium 1–2 in. in diameter and 1–4 in. in height. *Morgan.*

Lycoperdon.   Growing on the ground in meadows and woods.   Pennsylvania, North Carolina, *Schweinitz;* Canada, *Saccardo.*

(*h*) *Cortex of long spines, etc., which fall away, etc.*

L. pedicella′tum Pk.—*pediculus*, a little foot.   **Peridium** ¾–1 ½ in.

(Plate CLXIX.)

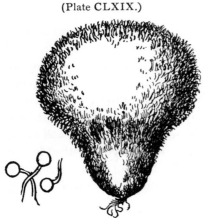

LYCOPERDON PEDICELLATUM
With magnified spores. (After Morgan.)

in diameter, globose or depressed-globose, sessile or narrowed below into a stem-like base, whitish or cinereous, becoming dingy or smoky-brown with age, echinate with rather dense spines which are either straight, curved or stellately united and which at length fall off and leave impressions or obscure reticulations on the surface; capillitium and spores greenish-yellow, then dingy-olive, columella present. **Spores** smooth, pedicellate, 4–4.5μ in diameter, the pedicel three to five times as long.   Ground and decaying wood in woods and bushy places.   Croghan, Center, Brewerton and Catskill mountains.   Autumn.   Oneida.   *Warne.*

The pedicellate spores constitute the peculiar feature of this species. It is one which suggests the name and which enables the species to be easily distinguished from all its allies.   The spore is terminally and persistently attached to the pedicel, as in some species of Bovista.   The plant is sometimes sessile, but usually it is narrowed below into a stem-like base.   In the immature state it has a rough, shaggy appearance, but the spines shrivel with age so that it appears less rough when old. The pitted surface of the denuded peridium affords a mark of distinction from L. echinatum.   L. pulcherrimum B. and C. is evidently the same species, but the name here adopted has priority of publication. *Peck*, 32d Rep. N. Y. State Bot.

Growing on the ground and on rotten wood in woods.   New York, *Peck;* Alabama, *Atkinson;* Ohio, *Morgan;* Wisconsin, *Trelease.*

L. exi'mium Morgan—*eximius*, excellent. **Peridium** obovoid, with Lycoperdon. a fibrous mycelium. Cortex white or brownish, composed of long slender spines, often curved and convergent at the apex, which at length fall away from above downward, leaving a pale smooth surface to the inner peridium. Subgleba small, occupying scarcely more than a fourth part of the peridium; mass of spores and capillitium greenish-yellow, then brownish-olivaceous; the threads mostly thinner than the spores, much branched. **Spores** oval, even, 5-6x4–4.5μ, usually furnished with a short pedicel.

(Plate CLXX.)

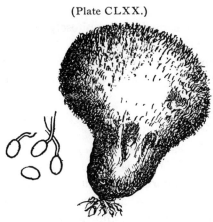

LYCOPERDON EXIMIUM.
With magnified spores. (After Morgan.)

Peridium ¾–1½ in. in diameter and about 1 in. in height. This species is readily distinguished by its large oval spores. *Morgan.*

Growing on the ground in sandy soil. South Carolina, *Prof. Geo. F. Atkinson;* Alabama *U. and E.*

L. Cur'tisii Berk.—in honor of Rev. M. A. Curtis. **Peridium** globose, with a very short rooting base and a slender fibrous mycelium. Cortex consisting of a pale yellowish farinaceous layer, covered by a coat of soft, fragile white spines, curved and convergent at the apex; after maturity it soon disappears, leaving a pale smooth surface to the inner peridium. Subgleba small, but distinct, convex above and definitely limited; mass of spores and capillitium greenish-yellow, then pale olivaceous; the threads long, simple, hyaline, two to three times as thick as the spores. **Spores** globose, even, 3.5–4μ in diameter.

(Plate CLXX*a*.)

LYCOPERDON CURTISII.
(After Morgan.)

Peridium ⅜–¾ of an inch in diameter. This is *L. Wrightii*, var. *typicum*, of Peck's U. S. species of Lycoperdon. The peculiar charac-

Lycoperdon. teristic of the species is the hyaline threads of the capillitium; although they are of large diameter, yet the walls are very thin and the threads collapse in drying. *Morgan.*

General. Growing gregariously and sometimes cespitosely on the ground, in meadows, pastures and even in cultivated fields.

This being L. Wrightii, var. typicum Pk., and being edible, it is hardly necessary to repeat the old axiom: Things which are equal to the same thing are equal to one another.

(*i*) *Cortex of minute spinules, granules, etc.*

L. molle Pers.—*mollis*, soft. (*L. muscorum* Morg.) **Peridium** 6–16 lines broad, globose or depressed-globose, narrowed below into a stem-like base, furfuraceous with nearly uniform persistent minute weak spinules or granular warts, sometimes with a few larger papilliform ones toward the apex, whitish, sometimes tinged with yellow, when mature brownish or olive-brown, nearly smooth, subshining; capillitium and spores dingy-olive; columella present. **Spores** minutely rough, 4–4.5μ in diameter.

Among mosses, especially Polytrichum, in old meadows and pastures. Albany, Summit and South Corinth. Autumn.

Peck, 32d Rep. N. Y. State Bot.

West Virginia, Pennsylvania. On ground in woods and grassy places in the open. *McIlvaine.*

(Plate CLXX*b*.)

LYCOPERDON TURNERI
With magnified spores.
(After Morgan.)

L. molle is of frequent occurrence but not abundant. Though exceedingly soft, it holds its body in cooking and is well flavored.

L. Tur'neri E. and E. **Peridium** obovoid, somewhat depressed above, plicate underneath, with a mycelium of rooting fibers. Cortex white, often gray or brownish above, consisting of minute spinules with intermingled granules; these after maturity dry up and are quite persistent, forming a minutely scabrous coat on the olive-brown shining surface of the inner peridium. Subgleba broad and shallow, scarcely occupying more than a fourth part of the peridium; mass of

602

Plate CLXXI.

Photographed by Dr. J. R. Weist.

LYCOPERDON PYRIFORME.

spores and capillitium greenish-yellow, then brownish-olivaceous; the Lycoperdon. threads with the main stem about as thick as the spores, and long tapering branches. **Spores** globose, minutely warted, 4–5μ in diameter, mostly with a short pedicel.

Peridium 1–2 in. in diameter and 1–2 in. in height. A very pretty puff-ball with a silky shining coat. *Morgan.*

New York, *Peck*, 49th Rep.; West Virginia, Pennsylvania, New Jersey. Ground in woods. August to October. *McIlvaine.*

Not frequent though general. It is good but must be young.

L. calyptrifor'me Berk.—hood-shaped. **Peridium** about 6 lines high, 3–4 in. broad, ovate or subconical, sessile, whitish, furfuraceous with minute warts or spinules; capillitium and spores olivaceous or yellowish-olivaceous. **Spores** smooth, 4μ in diameter.

Moss-covered rocks. Very rare. Adirondack mountains. August.

I have met with this very small and rare species but once, and then but two specimens were found. In these the apex was compressed or laterally flattened, instead of papilliform, as required by the original description of the species; but in all other respects they agree well with the specific characters. The plant is very distinct from all our other species by its small size and ovate or conical shape. *Peck*, 32d Rep. N. Y. State Bot.

(k) Cortex of minute spinules, scales or granules. Lignatile.

L. pyrifor'me Schaeff.—pear-shape. (Plate CLXXI, p. 602.) Plant 6–15 lines broad, 10–20 lines high, generally cespitose, obovate, pyriform or turbinate, sessile or with a short stem-like base, radicating with white branching and creeping root-like fibers, subumbonate, covered with very minute subpersistent, nearly uniform warts or scales, often with a few slender scattered deciduous spinules intermingled, pallid dingy-whitish or brownish; capillitium and spores greenish-yellow, then dingy-olivaceous, columella present. **Spores** smooth, 4μ in diameter. Edible, but not well-flavored.

Decaying wood and ground both in woods and cleared fields. Very common. July to October. *Peck*, 32d Rep. N. Y. State Bot.

Common the world over. Growing on logs, stumps, ground containing decaying woody matter. So dense in its clusters at times as to present an impervious surface. It is slightly acrid to taste and smell when raw.

Edible. Tender and of second-class flavor when young; white

603

Lycoperdon. inside; intensely bitter when slightest tinge of yellow is visible. One too old will embitter a whole dish. A little lemon juice or sherry improves it.

L. subincarna'tum Pk.—pale

(Plate CLXXII.)

Lycoperdon subincarnatum
With spines and pits magnified.
(After Morgan.)

flesh-color. **Peridium** 6–12 lines broad, globose, rarely either depressed or obovate, gregarious or cespitose, sessile, with but little cellular tissue at the base, covered with minute nearly uniform pyramidal or subspinulose at length deciduous warts, pinkish-brown, the denuded peridium whitish or cinereous, minutely reticulate-pitted; capillitium and spores greenish-yellow, then dingy-olivaceous, columella present. **Spores** minutely roughened, 4–5μ in diameter.

Prostrate trunks, old stumps, etc., in woods. Common. August to October. *Peck*, 32d Rep. N. Y. State Bot.

New York, *Peck*, Rep. 24th, 32d; Pennsylvania, *Gentry;* Ohio, *Morgan;* Wisconsin, *Brown.*

B. Peridium Very Small, Globose, Etc.

(*l*) *Cortex a thin coat of minute spinules, etc.*

(Plate CLXVIII.)

L. Wright'ii B. and C.—in honor

of Charles Wright. **Peridium** globose, depressed-globose or lentiform, 6–24 lines in diameter, generally sessile, white or whitish, echinate with deciduous sometimes crowded stellate spines or pyramidal warts, when denuded smooth or minutely velvety; capillitium and spores dingy-olive, columella present. **Spores** smooth, 4μ in diameter. Edible.

Ground in pastures and grassy places. Very common. July to October.

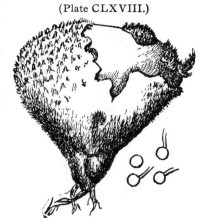

Lycoperdon separans
With magnified spores.
(After Morgan.)

604

This is another very variable species. The typical form is a small one, Lycoperdon. minutely echinate and having the denuded peridium smooth. The plant often occurs much larger and more coarsely echinate with stout angular spines or pyramidal warts, which fall off and generally leave the surface of the peridium velvety.

Var. *typ'icum.* Small, 6–9 lines broad, globose, minutely echinate, the warts quickly falling off and leaving the peridium smooth. (L. Wrightii B. and C.)

Var. *sep'arans.* Larger, 10–24 lines broad, subglobose or lentiform, echinate with coarse substellate spines or pyramidal warts, which at length fall off and leave the peridium smooth or velvety. (L. separans Pk.)

Var. *atropunc'tum.* Larger, 10–15 lines broad, subglobose, pure white, warts or coarse spines brown or blackish at the tips.

This species is generally gregarious, but sometimes it forms tufts of several individuals closely crowded together. It sometimes occurs in cultivated grounds and stubble fields. The under surface is occasionally plicate as in the long-stemmed puff-ball. In the var. separans the warts or spines are crowded at their thickened bases and slightly attached to each other, so that they come off at maturity in flakes or patches. When the denuded surface of the peridium is velvety, it is usually of a darker color than when smooth, being subcinnamon, reddish-brown or dark-brown. *Peck*, 32d Rep. N. Y. State Bot.

Ground in grassy places. July to frost. *McIlvaine.*

I have found var. separans in December, under snow.

Edible. *Peck*, Rep. 32.

The edible qualities of L. Wrightii and varieties are good.

L. calves'cens B. and C.—*calvesco*, to become bald. Subglobose, at first rough with warts which soon disappear, leaving the surface slightly velvety, 1¼ in. broad, bearing short rootlets at the base. **Spores** globose, smooth, having at first only a slight stalk (pedicel), dingy-ochraceous, 3–4μ.

Nearly related to L. Wrightii.

Connecticut, *Wright*, New York, ground in open woods. Bethlehem, *Peck*, 22d Rep. N. Y. State Bot.

L. pusil'lum (Batsch.) Fr.—small. **Peridium** ¼–1 in. broad, globose, scattered or cespitose, sessile, radicating, with but little cellular tissue

Lycoperdon.

(Plate CLXXIII.)

LYCOPERDON PUSILLUM.
(After Morgan.)

at the base, white or whitish, brownish when old, rimose-squamulose or slightly roughened with minute floccose or furfuraceous persistent warts; capillitium and spores greenish-yellow, then dingy-olivaceous. **Spores** smooth, 4μ in diameter.

Ground in grassy places and pastures. Common. June to October. *Peck*, 32d Rep. N. Y. State Bot.

West Virginia, Pennsylvania, New Jersey, North Carolina. Common. Spring to autumn on ground in grassy places. *McIlvaine.*

Grows where almost nothing else will, and where I have despaired of finding a meal of fungi, I could always find the ubiquitous L. pusillum.

L. oblongi'sporum B. and C.—oblong-spored. **Peridium** subglobose, with a slender mycelial cord. Cortex a thin, whitish, furfuraceous coat, drying up into minute persistent granules on the pale-brown surface of the inner peridium. Subgleba nearly obsolete; mass of spores and capillitium olivaceous, then brown; threads much branched, the main stem about as thick as the spores, the branches tapering. **Spores** elliptic, even, 5–6x3–4μ, sometimes with a minute pedicel.

Growing on the ground in dense woods. Wisconsin, *Trelease.* **Peridium** ⅜–1 in. in diameter. This pretty species, previously known only from Cuba, is indistinguishable from L. pusillum when immature, the spores affording the only really characteristic feature. *Morgan.*

L. cepæsfor'me Bull.—onion-shaped. **Peridium** globose or depressed-globose, plicate underneath, with a cordlike root. Cortex at first a thin, white, minutely furfuraceous coat, this soon becomes rimulose and at length breaks up into small scales and patches, which finally disappear from the pale or pale-brown surface of the inner peridium. Subgleba nearly obsolete; mass of spores and capillitium greenish-yellow, then pale-olivaceous; the threads very much branched, the main stem thicker than the spores, the branches long and tapering. **Spores** globose, even, 3.5–4μ in diameter, often with a minute pedicel.

(Plate CLXXIV.)

LYCOPERDON
CEPÆSFORME.
(After Morgan.)

606

Peridium ½–1 in. in diameter.   Lycoperdon.
Growing on the ground in meadows and pastures.
New York, *Peck*, 51st Rep.
Good.

L. colora'tum Pk.—colored. **Peridium** 5–10 lines broad, globose
or obovate, subsessile, radicating, yellow or reddish-yellow, brownish when old, slightly rough-   (Plate CLXXV.)
ened with minute granular or furfuraceous per-
sistent warts; capillitium and spores at first pale,
inclining to sulphur-color, then dingy-olive.
Spores subglobose, smooth, about 4µ in diameter.
Ground in thin woods and bushy places. Sandlake and Catskill mountains.   July and August.
Peck, 32d Rep. N. Y. State Bot.
New York, *Peck*, 29th Rep.; New England,   LYCOPERDON COL-
Morgan; Ohio, *Morgan;* Wisconsin, *Trelease*.   ORATUM.
                                                        (After Morgan.

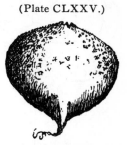

L. acumina'tum Bosc.—pointed. **Peridium** globose, then ovoid,
with a mycelium of fine white fibers.

(Plate **CLXXVI**.)

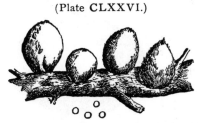

LYCOPERDON ACUMINATUM
With spores. (From Morgan.)

Cortex a white soft delicate continuous coat, drying up into a thin furfuraceous persistent layer on the surface of the inner peridium. Subgleba
obsolete; mass of spores and capillitium pale-olivaceous then dirty-gray;
threads simple, hyaline, two to three
times as thick as the spores. **Spores**
globose, even, 3µ in diameter.   Plate
II, fig. 8.   **Peridium** ¼–½ of an inch in height.
Growing on the mosses of old logs and about the base of living trees.
New York, *Peck;* North Carolina, *Curtis;* South Carolina, *Ravenel,*
Atkinson; Ohio, *Morgan;* Costa Rica, *Oersted*.

GENUS X.—BOVISTEL'LA Morg.

Bovistella.

(Plate CLXXVIa.)

SECTION BOVISTELLA
OHIENSIS.
Showing cellulose and definitely limited subgleba and the free threads of the capillitium. (From Morgan.)

Mycelium cord-like, rooting from the base. Peridium subglobose, with a well-developed base; cortex a dense floccose subpersistent coat; inner peridium thin, membranaceous, dehiscent by a regular apical mouth. Subgleba cellulose, cup-shaped above and definitely limited, persistent; capillitium originating within the tissue of the gleba; the threads free, short, several times dichotomously (two-forked) branched, the main stem thicker than the diameter of the spores, the branches tapering. Spores small, globose or oval, even, pedicellate.

A puff-ball of moderate size, growing in fields and open woods. *Morgan.*

B. Ohien'sis Ellis and Morg. **Peridium** globose or broadly obovoid, sometimes much depressed, plicate underneath, with a thick cord-like root. Cortex a dense floccose coat, sometimes segregated into soft warts or spines, white or grayish in color; this dries up into a thick buff-colored or dirty ochraceous layer, which gradually falls away, leaving a smooth, shining, pale-brown or yellowish surface to the inner peridium. Subgleba broad, ample, occupying one-half the peridium, a long time persistent; mass of spores and capillitium lax, friable, clay-color to pale-brown; the threads .6–.8 mm. in extent, three to five times branched, the main stem 6–8μ in thickness, the branches

(Plate CLXXVII.)

BOVISTELLA OHIENSIS.
Natural size.

tapering. Spores globose or oval, even, 4–5μ in length by 3.5–4μ in breadth, with long hyaline persistent pedicels.

Growing on the ground in old pastures, in fields and open woods. *Morgan.*

This species of puff-ball is made the type of the new genus Bovistella by Mr. Morgan.

GENUS XI.—CATAS'TOMA Morg.

Puff-balls growing just beneath the surface of the ground and con- *Catastoma.* nected immediately with it by filamentous threads, which issue from every part of the cortex; after maturity, when the peridium breaks away, the lower part of the outer coat is held fast by the soil, while the upper portion which has attained the surface remains, covering the inner peridium like a cap or inverted cup; consequently the apparent apex at which the mouth is situated is the actual base of the plant as it grows. The capillitium threads are similar to the densely interwoven hyphæ, which form the inner peridium and are evidently branches of them radiating from the interior. It is plain that the affinities of these plants are closest with Tylostoma and Astræus, but the needs of a systematic arrangement, according to more obvious characters, causes us to place them next to Bovista. *Morgan.*

C. circumscis'sum B. and C. (Plate CLXXVIII.) **Peridium** subglobose, more or less depressed and often quite irregular; cortex thickish, fragile, usually rough and uneven from the adhering soil, after maturity torn away, leaving the lower two-thirds or more in the ground; inner peridium depressed - globose, subcoriaceous, rather thin, pallid, becoming gray, minutely furfuraceous, with a small regular

(Plate CLXXVIII.)

CATASTOMA CIRCUMSCISSUM.
Showing method of growth, breaking away and turning over. Section of same showing origin of the threads of the capillitium.
(After Morgan.)

basal mouth. Mass of spores and capillitium soft, compact, then friable, olivaceous, changing to pale brown; the pieces of the threads short, unequal in length, flexuous, hyaline, 3–4µ in thickness. **Spores** globose, minutely warted, 4–5µ in diameter, often with a minute pedicel.

Growing in heavy clay soil in old lanes and pastures, especially along the hard-trodden paths.

Maine, *Blake;* Ohio, *Morgan;* Kansas, *Kellerman;* Nebraska, *Webber.* Inner peridium ½–¾ in. in diameter.

This is Bovista circumscissa B. and C., of Berkeley's Notices of

Catastoma. North America Fungi. It grows in great abundance with us some seasons, right in the hard-trodden barn-yard, and along the lane to the cattle pasture. Arachnion album Schw. usually keeps it company. *Morgan.*

I have not seen this acrobatic species. Study of its unique habit suggests the query: Is not the turning over of its spore-filled portion a substitute for an original but lost power of growing right side up?

GENUS XII.—BOVIS'TA Dill.

Bovista. **Mycelium** fibrous or sometimes filamentous. Peridium subglobose,

(Plate CLXXIX.)

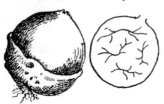

BOVISTA MINOR AND SECTION.
(From Morgan.)

without a thickened base; cortex a thin fragile continuous layer, shelling off or disappearing at maturity, except sometimes a small portion about the base; inner peridium thin, membranaceous, becoming papyraceous, dehiscent by an apical mouth or opening irregularly. Capillitium originating within the tissue of the gleba; the threads free, short, several times dichotomously branched, the main stem much thicker than the diameter of the spores, the branches tapering. **Spores** small, globose, or oval, even, brown. *Morgan.*

Small puff-balls growing upon the ground in fields and woods. One grows underground.

B. pi'la B. and C.—a ball. **Peridium** globose or obovoid, with a stout, cord-like root. Cortex a thin, white, smooth, continuous coat, breaking up at maturity into minute scales, which soon disappear; inner peridium thickish, tough, rigid, becoming brown or purplish-brown, smooth and shining, a long time persistent, and finally with age often fading to silvery-gray; dehiscence taking place at length by an irregular, torn aperture at or about the apex. Mass of spores and capillitium very firm, compact and persistent, at first clay-colored, pale brown or olivaceous, at length dark or purplish-brown; the threads rather small, .6–.8 mm. in extent, three to five times branched, 12–15μ thick, the ultimate branches rigid, nearly straight, tapering to a fine point. **Spores**

globose, even, 4–5µ in diameter, sessile or with only a minute Bovista. pedicel.

Growing on the ground in woods.

Peridium 1 ½–2 ½ in. in diameter.

This Bovista is remarkably tough, it maintains its shape firmly and persists a long time; it breaks away from its root and rolls about over the old leaves before the wind, even till the following season. *Morgan.*

West Virginia, Pennsylvania, New Jersey, in fields and woods on ground. June to October. *McIlvaine.*

Edible. *Trelease.*

When young and fresh it is excellent.

B. Monta′na Morg. Peridium subglobose with a cord-like root. Cortex a thin white continuous layer, breaking up at maturity into a mealy or furfuraceous coat, which soon falls away; inner peridium thin, flaccid, becoming brown, smooth and shining, dehiscent by an irregular torn aperture about the apex. Mass of spores and capillitium soft, lax, at first ochraceous or pale-brown, finally purplish-brown; the threads curled and flexuous, very large, with an expanse of 1.25–1.75 mm., four to seven times branched, the main stem 15–20µ in thickness, the ultimate branches long and tapering. **Spores** globose, even, 4.5–5.5µ in diameter, often with a minute pedicel.

Growing on the ground. Rocky mountains. *Jones.* Peridium 1 ½–2′in. in diameter. This differs from B. pila in being soft, flaccid, and soon collapsing; it, no doubt, is not so persistent. Microscopically it is readily distinguished by its much larger threads. *Morgan.*

B. nigres′cens (Vitt.) Pers.—blackish. Peridium subglobose, with a fibrous mycelium. Cortex a thin, smooth, white continuous layer, at maturity breaking up into scales, which soon disappear; inner peridium thin, flaccid, becoming dark-brown, smooth and shining, dehiscent at the apex by a lacerate mouth. Mass of spores and capillitium soft, lax, at first ochraceous or olivaceous, at length purplish-brown; the threads flexuous, about 1 mm. in extent, three to five times branched, the main stem 12–18µ thick, the ultimate branches tapering. **Spores** globose or oval, even, 5–6µ in diameter, with long hyaline pedicels.

Growing in old pastures, in fields and woods. Canada, *Saccardo;* Pennsylvania, *Schweinitz;* North Carolina, *Curtis;* Ohio, *Lea;* California, *Harkness.*

Bovista.   Peridium 1–2 in. in diameter.   I have never succeeded in obtaining an American specimen of this species; my description is drawn up from European specimens. *Morgan.*

Pennsylvania, *McIlvaine.*   B. nigrescens is a first-class puff-ball.

B. plum'bea Pers.—lead-colored.   Peridium ¾–1¼ in. in diameter, depressed-globose, with a fibrous mycelium.   Cortex a thin, smooth, white continuous coat, loosening at maturity and shelling off, except sometimes a small portion about the base; inner peridium thin, tough, smooth, lead-colored, dehiscent at the apex by a round or oblong aperture.   Mass of spores and capillitium soft, lax, ochraceous or olivaceous, then purplish-brown, the threads .8–1.0 mm. in extent, three to five times branched, the main stem 12–16μ thick, the ultimate branches long, straight and tapering to a fine point.   Spores oval, even, 6–7x5–6μ, with long hyaline pedicels.

Growing on the ground in meadows and pastures. *Morgan.*

Indiana, in abandoned brick-yard, *H. I. Miller;* West Virginia, New Jersey, Pennsylvania.   Common on ground in open places.   Solitary or in groups.   Spring to autumn; after rains, *McIlvaine.*

Edible. *Trelease, Badham.*

The botanic difference between a Lycoperdon and a Bovista does not affect the Mycophagist.   He can not distinguish the difference when cooked.   B. plumbea is given in Cooke and in Massee as Lycoperdon plumbeum.   Bovista plumbea is a first-class edible.

B. mi'nor Morg.   (Plate CLXXIX, p. 610.)   Peridium subglobose, deeply sunk in the soil and connected with it by a filamentous mycelium, which issues from every part of the surface.   Cortex thickish, rough and irregular from the adherent soil, fragile, falling away at maturity, except sometimes a small portion about the base; inner peridium thin, smooth, flaccid, reddish-brown, dehiscent by a regular apical mouth.   Mass of spores and capillitium olivaceous, then reddish-brown; the threads curled and flexuous, with an expanse of 1.0–1.5 mm., two to four times branched, the main stem 10–15μ thick, the ultimate branches very long and tapering to a fine point.   Spores globose or slightly oval, even, 3.5–4.5μ in diameter, with long hyaline pedicels.

Growing in damp shaded situations.   Ohio, *Morgan;* Nebraska, *Webber.*   Peridium ½–¾ of an inch in diameter.   A species well marked by its peculiar habit. The curled and flexuous threads are Bovista. interesting microscopic objects. *Morgan.*

GENUS XIII.—MYCENAS'TRUM Desv.

Mycelium funicular, rooting from the base. **Peridium** subglobose, Mycenastrum. without a thickened base; cortex a smooth continuous layer, at first closely adnate to the inner peridium, after maturity gradually breaking up and falling away; inner peridium thick, tough, coriaceous, becoming hard, rigid and corky, the upper part finally breaking up into irregular lobes or fragments. Capillitium originating within the tissue of the gleba; the threads free, short, thick, with a few short branches, acutely pointed and with scattered prickles. **Spores** large, globose, s e s s i l e, brown.

(Plate CLXXXII.)

Mycenastrum spinulosum Pk.
(After Morgan.)

Puff-balls of considerable size, growing in the sandy soil of dry regions. A very distinct genus, in no way related to Scleroderma, and resembling it only in its thick, corky, inner peridium. The threads of the capillitium originate within the tissue of the gleba, along with the spores, and are set free by deliquescence, the same as in Bovista. *Morgan.*

M. spinulo'sum Pk. **Peridium** globose, depressed globose, sometimes elongated and often irregular, with a thick, cord-like root. Cortex at first a thickish, white, smooth, continuous layer; after maturity it cracks or becomes furrowed into large polygonal areas, and at length falls away in large flakes or scales; inner peridium very thick, at first white and coriaceous, becoming hard, dry, brown and rigid, the upper part finally breaking up into irregular lobes or fragments. Mass of spores and capillitium compact then friable, at first olivaceous, then

613

Mycenastrum. dark purplish-brown; the threads bent, curved and flexuous, subhyaline, .2–.7 mm. in length, about the same thickness as the spores, with a few short branches, and with scattered prickles, which are most abundant toward the acute extremities. **Spores** globose, very minutely warted, opaque, 9–12μ in. diameter, often with a minute or slender hyaline pedicel.

Growing on the sandy soil of the western prairies. Wisconsin, *Brown;* Dakota, *Ellis;* Nebraska, *Webber;* Colorado, *Trelease;* Kansas, *Kellerman, Cragin;* New Mexico, *Irish.*

Peridium 2–4 in. in diameter. The plants are said to grow together in groups, sometimes of many individuals; after maturity they are easily loosened from their place of growth and are then rolled about by the wind. *Morgan.*

No report upon edibility. Probably good.

FAMILY III.—SCLERODERMA'CEÆ.

Peridium discrete from the gleba, often with a columella; cells of the gleba subpersistent. *Morgan.*

GENUS I.—SCLERODER'MA Pers.

Scleros, hard; *derma,* skin.

Skin firm with an innate bark, bursting irregularly; woolly threads Scleroderma. adhering on all sides to the bark and forming distinct veins in the central mass. Base sterile, usually becoming elongated into a stem-like structure. **Spores** large, granulated.

Scleroderma vulgare and verrucosum are general and very common over the United States. S. bovista and S. geaster have the same range but are not so common. They much resemble puff-balls, but are more pudgy, solid-looking. All are edible. Their qualities are noted under their descriptions.

S. vulga're Fr.—*vulgaris,* **common.** (Plate CLXXX.) Subsessile, irregular; bark corky, hard, opening indefinitely; inner mass in which the spores are collected into little heaps separated by a few grayish woolly threads, bluish-black. **Spores** dingy; in the mass blackish with purple tinge, globose, warted, 9–11μ *Massee.*

(Plate CLXXX.)

SCLERODERMA VULGARE.
A–B. Firm when young and remain nearly so when mature.

The larger form is generally of a yellowish or brownish hue, surface warty or covered with rough scales; the smaller, stemless minutely warty, bright brown.

Under trees, etc. Often cespitose, 1–3 in. across. Peridium vari-

Scleroderma. able, white or pale-brown, often becoming pink when cut. Dehiscing by decay of upper portion of peridium. *Massee.*

Scleroderma vulgare is one of our most common and plentiful toad-stools. Its hard, rough, warty, light brown knobs, single or clustered, growing along brook-banks or under trees, generally choosing hard ground, are known to all who observe Nature's curiosities. When quite young they are white inside. As they enlarge the center darkens and this purplish color finally develops into a grayish-purplish-black which extends throughout the interior and gives it a granular appearance. The fungus is solid, cutting like a potato. Its smell is strong; also its taste when raw. Sliced and well-cooked the species is good, even after it has become purplish, but if a single one is wilted it will embitter a whole dish. Or if it is not very well stewed or fried it remains strong. In no condition is it injurious. Specimens must be pared, and the base well cut away.

S. bovis'ta Fr. Subsessile, often irregular, peridium thin, pliant, almost smooth; tramal walls floccose, *yellow*, mass of spores olive-brown, spores globose, warted, 10–13μ.

Sandy soil under trees, etc. From 1–2 in. across. Distinguished by the thin, almost smooth peridium, and the yellow tramal walls. *Massee.*

West Virginia, New Jersey, Pennsylvania. On ground under trees. June to November. *McIlvaine.*

Not rough like S. vulgare and S. verrucosum, nor as solid. Same habit, same edible qualities when young. It is not good after it begins to change color.

S. verruco'sum Pers.—*verrucosus*, covered with warts (*verrucæ*). Peridium thin above, ochraceous or dingy brown, covered with minute warts, subglobose, continued downward as a more or less elongated stem-like base. **Spores** umber in the mass; trama whitish.

Spores globose, warted, 10–13μ.

On the ground, under trees, etc. Peridium 1–3 in. across. **Stem** ½–2 in. long, thick, flatly pitted, sometimes almost sessile, when it approaches S. vulgare, but is distinguished by the thin peridium and absence of purple tinge in the immature spore mass. *Massee.*

West Virginia, New Jersey, Pennsylvania. June to October. On Scleroderma. ground under trees. Same habit as S. vulgare. *McIlvaine.*

S. verrucosum closely resembles S. vulgare. The distinctions are noted in the description. It must be young, fresh and white inside, or it is bitter. It is not of as good quality as S. vulgare.

S. geas'ter Fr.—resembling genus Geaster in its manner of opening. Subglobose, sessile, peridium thick, rigid, almost smooth, splitting in an irregularly stellate manner at the apex.

Spores warted, 12–16μ.

Sandy places. Known by the peridium dehiscing in a stellate manner; from 1–2 in. across. *Massee.*

New Jersey, August. In sandy woods. *McIlvaine.*

I have found but few specimens. Those were edible and good.

GENUS II.—POLYSAC'CUM De C.

Polus, many; *saccus*, a sack.

Polysaccum.

Peridium irregularly globose, thick, attenuated downward into a stem-like base, opening by disintegration of its upper portion; internal mass (gleba) divided into distinct sack-like cells.

(Plate CLXXXI.)

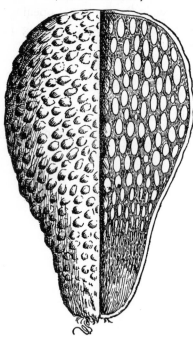

POLYSACCUM PISOCARPIUM.
Natural size.

Allied to Scleroderma and distinguished by the cavities of the gleba containing distinct peridiola. *Massee*.

P. pisocar'pium Fr. *Gr.*—a pea; *Gr.*—fruited. **Peridium** irregularly globose, indistinctly nodulose, passing downward into a stout stem-like base; peridiola irregularly angular, 4–5x2–3μ, yellow. **Spores** globose, warted, coffee-color, 9–13μ *Massee*.

P. pisocarpium was quite common at Mt. Gretna, Pa., from August to October, 1898, in open pine and mixed woods, growing from sandy ground. The height reached 5 in. and diameter 2 in. The shapes were usually those of inverted pears, more or less flattened along their lengths. Skin hard, polished, olivaceous-black with dull yellow mottlings, not unlike rattlesnake skin. When broken the peridiola (small ovate cylinders which bear the spores within) are very distinct, often over ⅛ in. long. The entire interior is dark when mature, and the rupture of the plant is irregular and by disintegration of the upper part. They often dry without rupturing. Search as I would, I could not find a young one, or one in edible condition. The plant is here given because interesting and one the student will wish to identify. It is so odd that it is not surprising to find it employed as a medicine in China.

TOADSTOOL POISONING AND ITS TREATMENT

By W. S. CARTER, M. D.

Professsor of Physiology and Hygiene, University of Texas, Galveston, Texas.

THE poisonous mushrooms, or so-called toadstools, may be grouped in two classes: (1) Those containing minor or irritant poisons, which act locally on the gastro-intestinal tract, such as the Clitocybe illudens, Lepiota Morgani and others, and (2) those containing major poisons which act on the nerve centers after absorption, causing symptoms to appear a long time after the poison has been taken and very often terminating fatally. This group includes the Amanita muscaria, the Amanita bulbosa or Am. verna and the Amanita phalloides.

From the prompt way in which vomiting and purging begin after eating the toadstools of the first group there seems to be no doubt of the local irritant action on the alimentary canal. Grave symptoms from any constitutional effect or any serious disturbances of the circulation do not occur. Although intensely disagreeable, such poisoning terminates in recovery and may not be regarded as dangerous unless the poison be taken in enormous quantity or by one in poor health.

In poisoning by the deadly toxic Amanitæ vomiting and purging may also occur as prominent symptoms, but generally only occur late—ten to fifteen hours after eating the toadstool—and are due to the action of the poison on the nerve centers. This is clear from the fact that these symptoms appear when the poison is given either hypodermatically or intravenously to animals.

It is exceedingly unfortunate that these deadly poisonous toadstools do not give some warning either in an unpleasant taste or contain an irritant which would act locally to cause emesis and purgation, for in that case the patient would get rid of the poison before such large quantities were absorbed and fatal poisoning would be less frequent. They are not at all unpalatable and sometimes large quantities are eaten by mistake.

Formerly frequent allusion was made to the possibility of poisoning by decomposition products from putrefactive changes in mushrooms. Not a single case has come to the writer's notice where this form of poisoning has *actually* occurred. In many reported cases of poisoning it is distinctly stated they were eaten soon after gathering; in none does the reporter mention any suspicion of poisoning of this nature.

At one time very many mushrooms were supposed to be poisonous. We now know that there are only a few dangerous ones, and where serious poisoning occurs it seems clearly to be due to some definite and constant poisons contained in certain fungi. We shall not deal here with the irritant poisons, as they are not dangerous and usually Nature gets rid of them easily, but shall consider the poisons of the Amanita muscaria or "Fly mushroom," the A. bulbosus vernus Bull. or A. verna, and the A. phalloides or "death cup." The writer has not had any personal experience with poisoning by these in man, but bases his observations upon over one hundred experiments made upon lower animals.* This is a distinct advantage in reaching any conclusion, as the facts are definite without any doubt as to the *kind* or *quantity* of the poison taken. In many of the reports of poisoning by mushrooms in man no mention of the species is made. In all these cases many kinds have been mixed together in preparing them for the table and it has never been known how many of the poisonous and how many of the edible ones have been eaten by any one individual partaking of the dish. Of course the fluid in which they are cooked contains some of the poison. This explains why some patients recover after having eaten several mushrooms while others die from a part of one only. (See report of six cases by Dr. G. E. Caglieri, New York Med. Record, August 28, 1897; also Dr. Berry's cases reported by Prentiss, Philadelphia Med. Journal, September 24, 1898.) Then, too, different poisonous species may be mixed together. The symptoms produced by the different Amanitæ poisons are quite different. Those containing irritant poisons may be taken with those containing deadly poisons. This accounts for the great variety of symptoms presented in cases reported.

---

*All of the toadstools used have been very kindly furnished and identified by Capt. Chas. McIlvaine. Unfortunately only fifteen experiments could be made with the fresh fungi while the writer was at the University of Pennsylvania. Since leaving there it has been impossible to get any in a fresh state, and the other experiments have been made either with dried fungi or alcoholic or glycerine extracts of the fresh.

POISONING BY AMANITA MUSCARIA.

The symptoms of poisoning by this fungus usually appear from eight to twelve hours after it has been eaten, unless it has been taken in enormous quantities, as in the cases reported by Prentiss (Phila. Med. Jour., September 24, 1898), where they came on in half an hour.

These begin with cramp-like pains in the extremities, colicky pains in the abdomen, burning thirst, vomiting and purging. The pulse may be very slow and strong at first, but later becomes rapid, small and feeble. The blood pressure is low and, as a result of this disturbance of the circulation, faintness is a common early symptom. Extreme pallor is often noticed. The secretions are increased, and the sweat and the saliva may be secreted in abnormal quantities.

The pupils are strongly contracted and dullness of the vision or double vision may be noticed early.

The respirations are slow and become shallow and stertorous when the poisoning is severe.

The mental state may be clear at first, but becomes dull, deepening into unconsciousness and deep coma if a large quantity has been taken.

Convulsions are reported to have occurred in some cases from poisoning by this toadstool in man. The dried Amanita muscaria or extracts of the fresh fail to produce convulsions in the lower animals, even in frogs, which are very susceptible. Either there is a considerable difference in the susceptibility to this poison or there is some poison present in the fresh fungus which is lost by drying.

Small amounts of the dried Amanita muscaria are said to be used by inhabitants of northern Asia for the stimulating effect upon the nervous system, producing, like other narcotic poisons, a dreamy state of intoxication, deepening into sleep (Von Boeck in Ziemssen's Cyclopedia of Medicine, Vol. VII).

In animals the most striking effect is upon the circulation. When injected intravenously it causes tremendous inhibition of the heart's action —a moderate amount causing the heart to beat slowly and powerfully; a large amount causing complete arrest. Even with the partial inhibition there is an enormous fall of pressure. The slowing of the heart soon passes off, and when a moderate amount has been injected, the circulation quickly returns to normal.

623

In one of my experiments on a dog, the heart stopped for 1¾ minutes and then began beating again, the circulation soon recovering.

Late in the poisoning the heart beats may be rapid and feeble and the blood pressure low. The lowered blood pressure is largely due to dilatation of the small blood vessels resulting from a loss of control over them by the nerve center which normally keeps the arterioles in a state of partial contraction.

The inhibition of the heart is due to the action of the well-known alkaloid *muscarine* upon nerve ganglia in the heart. The contraction of the pupil and the increased secretory activity of the glands are also due to this substance which was discovered by Schmideberg and Koppe in 1869.

It was soon found that although dogs recovered from the *immediate* or *early effects* (*i. e.*, from the muscarine) of enormous quantities of toadstools, they succumbed from the *late effects of much smaller quantities.* Atropine fails to avert this result from the late effect, whether given before the poison, with it, or after it. The inhibition of the heart passes off long before death occurs. Late death does not appear to be due to muscarine.

All these facts put together point to the existence of some other poison or poisons in the Amanita muscaria to which atropine is *not* an antidote.

This peculiar poisoning causing death so late will be discussed again after considering the other poisonous mushrooms as they act similarly.

Gastro-intestinal symptoms were not as common in my experiments with Amanita muscaria as with the Amanita phalloides. Vomiting and purging occasionally occurred early, but much more frequently late in the poisoning and often not at all.

Convulsions did not occur in any of the animals poisoned by this fungus. Convulsions are recorded in some cases of poisoning in man, but not so constantly as with the A. phalloides and A. verna. Where they occurred either a large amount had been taken (as in Prentiss' case) or there is some doubt about the Amanita muscaria having been the only toadstool eaten (as in Caglieri's cases). Frogs are very easily thrown into spasms, but no spasms were observed, even in fatal poisoning of them by this toadstool.

Regarding cerebral symptoms, little can be said except that unconsciousness and coma may come on early and persist till death. In cases terminating fatally the animal seemed to be conscious, but so depressed that it was unable to stand or even move when called.

Concerning differences in the susceptibility of different animals to the poisons of Amanita muscaria, cats seemed to be more susceptible than dogs in the earlier experiments with extracts of the fresh fungus, but more numerous experiments with the dried fungus failed to show any greater difference than can be observed between different animals of the same kind.

As to the nature of the poisons very little can be stated from the experiments, as they were undertaken as a preliminary step to chemical studies to be carried on later. The alkaloid muscarine is one of our best known poisons and nothing can be added to what is already known about it. The poisons are extracted by distilled water as well as by a solution of sodium chloride; they are soluble in glycerine and in alcohol and very little difference can be seen in the action of these extracts, unless the alcoholic extract contains more of the muscarine, while the glycerine extract contains more of the other poisons.

It is stated that muscarine is not poisonous to flies; that the Amanita muscaria contains a volatile poison which is poisonous to flies (hence the name "Fly mushroom"), and which is lost by drying; that inhabitants of northern Asia use the *dried* fungus (after the volatile poison has been lost) for producing intoxication (Von Boeck in Ziemssen's Cyclopedia, Vol. VII, p. 927). My experiments have been entirely with mammals and frogs, and unfortunately those performed with the fresh toadstools were not numerous enough to enable me to draw positive conclusions as to any loss of toxicity by drying. A single experiment with a cat seemed to indicate that boiling of the fungus lessened the toxicity but subsequent experiments indicated that a boiled solution was no less toxic than one not boiled.

One thing we can state definitely; that boiling the dried A. muscaria does not destroy its toxicity. This indicates that the poison is not of an albuminous nature, which would be coagulated by heat.

Whether or not any volatile poison is lost by boiling a solution of the fresh fungus or by drying at 40° C. can not be stated definitely as the experiments made with the fresh fungus were few in number on account of the extreme difficulty in getting them perfectly fresh.

The average of six observations in which it was possible to weigh the toadstools before and after drying at 40° C. showed a loss of 84.4

per cent. of water. In other words, 1 gram of the dried equals 6.4 grams of the fresh.

Comparing the lethal doses of the *dried* with the lethal doses of the *fresh* extracted by glycerine and alcohol, it does not appear that there is any great loss of the toxicity by drying as is shown by the following: Lethal dose of dried in Experiment 31 was .085 gram. per kilo of body weight; in Experiment 55, .033 gram. per kilo caused *early* death, while .223 gram. of dried per kilo and .120 gram. per kilo caused death from *late* effects (Exps. 32 and 57). The lethal doses of the *fresh* were .91 gram. per kilo (Exp. 29) and 1.055 gram. per kilo (Exp. 36) when a glycerine extract of the fresh growth was used, while 1.222 gram. per kilo (Exp. 16) made from an alcoholic extract failed to kill.

It may be well to introduce here the results of an experiment which shows there is no highly poisonous volatile material given off from the A. phalloides. This is rather an important fact to determine, as the opinion is held by some that there is a volatile poison, and most of my experiments were made with the dried fungus. A 1 per cent. solution of fresh A. phalloides was distilled until three-fourths of the fluid had passed over as distillate. The latter was injected into the vein of a dog and found not at all toxic. The opportunity has not been afforded me of repeating this experiment personally, but Dr. J. P. Arnold has kindly repeated it for me, injecting the distillate into rabbits and frogs and failed to find it toxic. Certainly if there is any volatile poison in the A. phalloides it must be either in very minute quantity or very slightly toxic.

ANTIDOTAL VALUE OF ATROPINE.

In arriving at any conclusion we must bear in mind the variation of different animals in their susceptibility to poisons. Thus, to give the greatest difference observed, .085 gram. dried Amanita muscaria per kilo of body weight killed one dog in an hour, while in another dog .223 grams. of the same preparation per kilo only killed after 24 hours, the cardiac inhibition having disappeared one-half hour after the poison was injected. However, an average of six (6) experiments on cats and dogs with dried A. muscaria in which no antidote was given shows the lethal dose to be .103 gram. per kilo of body weight. The average of four (4) experiments, in which the fungus, dried in the same way, was used but *atropine was given as an antidote*, gives the lethal dose of .335 gram. per kilo and death only occurred *late* in each case.

There can be no doubt, therefore, of the antidotal value of atropine for poisoning by Amanita muscaria.

It should be borne in mind, however, that it is not an infallible antidote even when given early, and that it does not prevent death from the *late effects* in severe cases, although given in large doses. In some experiments atropine was administered at the same time the poison was given and in others before it.

The important practical lesson is that too much reliance should not be placed upon atropine. It will be shown later that it has little value as an antidote to A. verna and A. phalloides. Probably these fungi contain less muscarine than A. muscaria. Although there is no drug so antagonistic in its physiological action to the poison of the A. muscaria as atropine, the use of other remedies should not be neglected. The symptoms have to be treated as they arise. Strychnia, alcohol in moderate amounts and suprarenal extract could all be used to advantage in restoring the circulation, especially late in the poisoning. Atropine merely removes the inhibition of the heart which occurs as an early symptom.

External heat should be applied if the body temperature is subnormal. The treatment of gastro-intestinal symptoms will depend upon the conditions of each individual case. The injection of a large amount of warm physiological salt solution (.6–.7 per cent. sodium chloride) into the subcutaneous tissues should also be tried in severe cases seen late in the poisoning.

POISONING BY AMANITA VERNA OR A. BULBOSUS VERNA BULL.

The symptoms appear from six to fifteen hours after the ingestion of the poison and may be largely choleraic in nature, *i. e.*, vomiting and purging, the discharges from the bowel being watery with small flakes suspended and sometimes containing blood.

The disturbance of the circulation is somewhat similar to that caused by A. muscaria, viz., slow, strong pulse early, but rapid and weak later. Dizziness and faintness may be early symptoms. Sometimes the skin is pale and covered with cold, clammy sweat; at others there is great cyanosis. The body temperature is subnormal, unless nervous symptoms are very severe. Very prominent among the symptoms are tetanic convulsions, which may appear comparatively early and persist until the end.

In animals the effect of this toadstool is entirely different from that of A. muscaria. Perhaps the most striking difference is the frequency with which convulsions appear. Convulsions occurred repeatedly in mammals and in nearly every frog to which the toadstool was given. This fungus seems to contain some poison that acts upon the spinal cord very much as strychnia does, though less powerfully, of course.

The circulatory conditions are also different. The inhibition of the heart may be pronounced as an early condition, but the pressure does not return to the normal after this disappears, either from giving atropine or from cutting the pneumogastric nerves. Section of these nerves removes the cardiac inhibition much more completely than after poisoning by the A. muscaria. There is often a fall of pressure without cardiac inhibition. In other words, there is a much greater permanent fall of blood-pressure due to paralysis of the nerve center controlling the blood vessels (vaso-motor center). This condition will last a long time and does not show the same tendency to disappear as after A. muscaria. Moreover it is produced by comparatively small amounts of the A. verna.

The respirations are very slow. The blood is poorly oxygenated and this probably causes the cyanosis sometimes observed in men poisoned by this fungus.

Bloody fluid is sometimes vomited or comes from the nose. It may also occur in the discharge from the bowel.

Retching and purging occurred more frequently as early symptoms than in animals poisoned by A. muscaria.

Coma appeared early and continued until death. The administration of atropine soon after giving the poison when cardiac inhibition was present, caused a slight temporary rise of blood pressure but did not affect the dilated condition of the blood vessels. The pressure continued low notwithstanding the atropine. Although the experiments with this fungus were not as numerous as with the A. muscaria because of difficulty in obtaining it, yet it seems clear that atropine is of very little value as an antidote. Death very rarely resulted from the cardiac inhibition occurring early but usually came on late after that condition had disappeared. The lethal dose was no larger when atropine was given than when no antidote was used.

Amanita verna is very much more toxic than A. muscaria, the average of four experiments in which the former was given without an antidote

628

being .034 gram. (dried) per kilo of body weight, while .103 gram. (dried) per kilo, was the average for the latter fungus.

POISONING BY AMANITA PHALLOIDES.

The symptoms described in man are very similar to those caused by the A. verna, except that the convulsions are less constant and cyanosis is not mentioned. In some cases vomiting and purging are prominent symptoms. There is dizziness and fainting, extreme ashy pallor, cold skin covered with sweat, subnormal temperature, muscular twitchings and occasional convulsions and somnolence which deepens into coma and lasts until death, which usually occurs two or three days after eating the poison. Sometimes the gastro-intestinal symptoms are less severe or may be absent, though they are usually present; in that case the nervous symptoms are more prominent, particularly the convulsions and circulatory disturbance.

In experiments upon animals the convulsions were not observed so constantly as with the A. verna. Out of twenty-five dogs poisoned by the Amanita phalloides, convulsions only occurred twice, while twelve frogs injected with different preparations (dried toadstool and glycerine and alcoholic extracts of the fresh) failed to show a convulsive seizure in a single instance. It seems to be difficult for mycologists to draw a sharp line between the A. verna and the A. phalloides and say to which of these two certain fungi belong. This may explain why convulsions are recorded more frequently in persons poisoned by this toadstool than in animals poisoned by it. Frogs are very susceptible to poisons acting upon the spinal cord, and all of those poisoned by lethal doses of A. verna had convulsions, while none of those poisoned by the A. phalloides had any. It would therefore appear from this striking difference in the physiological actions that the two are separate and distinct.

The circulatory and gastro-intestinal symptoms were quite similar to those caused by the A. verna.

A. phalloides is less toxic than the A. verna, but more so than the A. muscaria, the average lethal dose of the dried fungus (eight experiments) for dogs, where no antidote was used, being .117 gram. per kilo.

The antidotal value of atropine is very slight, if indeed it has any action other than removing the temporary cardiac inhibition. The animals very seldom died from this, but mostly from the late effects after

629

the inhibition had disappeared. In four experiments on dogs in which atropine was given either at the same time as the poison or before it, the average lethal dose was .198 gram. of the dried fungus per kilo. Two dogs were killed by .1 gram. per kilo without atropine; another was given the same amount and was given atropine hypodermatically a number of times and recovered, though very ill for two days.

Transfusion of physiological salt solution (.6 per cent. table salt) was practiced in three dogs. Although death occurred in all of these and the lethal dose was not unusually high, the pressure was restored for a time at least. It should be employed in treating poisoning in man, and not be depended upon as the *only* procedure, but used in conjunction with other remedies. This will be referred to again in describing treatment.

It will be seen from the above that poisoning by the A. verna and A. phalloides present symptoms in the lower animals which are quite different from those caused by the A. muscaria, and that in either case poisoning is far more serious than by the latter fungus. This is not only because they are so much more toxic, but also because there is no decidedly antagonistic action by atropine, and hence its value as an antidote is much less.

In treating a case of poisoning by either A. verna or A. phalloides the only thing that can be done is to meet the indications in the individual case. If the heart is beating slowly, atropine should be given in liberal doses. This will not overcome the chief disturbance of the circulation, viz., the tremendous dilation of the blood vessels. Strychnia will do this to a certain extent, but its use may be contra-indicated by twitchings or convulsions from the toadstools. If it can be used it is exceedingly valuable, as it stimulates not only the vaso-motor center, but the respiration and heart as well. Caffein or strong coffee may also be used to this end if the stomach will retain it. Suprarenal extract should also be given hypodermatically, as it will restore the blood pressure more nearly to normal than any other drug, according to our experiments. It has the advantage of not increasing the excitability of the spinal cord as strychnia does, and hence would not be contra-indicated by nervous symptoms.

Perhaps the most rational treatment to meet the most serious condition of the poisoning by these toadstools is the transfusion of normal saline solution (.6–.7 per cent. solution of table salt) into the subcu-

630

taneous tissues. This should, of course, only be given by a physician, as great care is required in sterilizing the syringe. It can be given with a fountain syringe and aspirating needle beneath the skin of the thigh. Large quantities should be used—at least a quart (1000 cc.) or more. The fluid is rapidly absorbed by the lymphatics and gets into the blood vessels. It restores the blood pressure by increasing the fluid in the vessels and also doubtless aids the organs of excretion in eliminating the poison; at the same time it would relieve the intense thirst patients complain of. Clinicians who have observed cases of poisoning by the A. phalloides in man have suggested this procedure as the most rational one to meet the symptoms presented. From the condition produced in animals poisoned by this toadstool the writer was led to the same conclusion. In two experiments upon dogs, when transfusion of warm physiological salt solution was made directly into the vein after poisoning by the A. phalloides, death occurred in both cases and the lethal dose was not unusually large, although the amount transferred was equal to the estimated volume of the blood of the animal in one case and half that amount in another. In another animal atropine was given before the poison and the pressure had been reduced by the latter to one-fifth of the normal, the transfusion of an amount of normal salt solution equivalent to two-thirds of the bulk of blood restored the pressure to three-fourths of normal in about 15 minutes, but further injection of the poison caused late death.

Although the rise of pressure is not so great from transfusion as from suprarenal extract in large doses, it is more *permanent*. Transfusion (or transfusion into the subcutaneous tissues by hypodermoclysis which amounts to the same thing) has the additional advantage of increasing the flow of urine, which is often suppressed in these cases. Even if it does no good it can do no harm if done antiseptically and should be tried but *always in conjunction with other remedies*.

A remarkable case of recovery after the injection of a large amount of normal saline solution has been reported by Delobel (Presse medicale September 30, 1899). A man aged fifty-two ate some A. phalloides; he was seen four hours afterward. The skin was covered with cold, clammy sweat; body temperature was sub-normal; shivering and tremors present; had not vomited or purged; urine suppressed; respiration stertorous; pulse 28 per minute and so feeble that it was almost imperceptible. Two full doses of atropine were given hypodermatically

631

as well as 10 cc. of ether and 200 cc. of strong coffee with 20 cc. of rum were given by the mouth and hot bottles applied externally. In spite of all this the symptoms became worse and the patient sank into a condition of profound collapse, the pulse dropping to 24 per minute and the tremors ceased. One liter (1 quart) of normal saline solution was injected hypodermatically and improvement began in 15 minutes after the injection. The respiration lost the Cheyne-Stokes character; the pulse improved in tension and in an hour was 60 per minute; the skin improved and the temperature returned to normal and the patient went to work next day.

The circulatory symptoms are most prominent and demand most attention. Vomiting and purging have to be treated according to the conditions in the individual case and no rule can be followed. As the peripheral vessels are dilated the body temperature is usually subnormal. This should be overcome by applying hot bottles externally.

The suppression of urine should receive attention, and the activity of the kidneys be stimulated as much as possible. It is probable the suppression is largely due to the tremendous fall of blood pressure. If the urine is secreted but retained in the bladder it should be drawn off.

Just as there is no simple way of detecting the presence of poisonous mushrooms in a mixture of mushrooms, so there is no simple way of destroying or removing the poisons. Pouchet stated that boiling destroyed the poison and Chestnut has stated the poison of A. phalloides is a toxic albumen. If this were the case boiling would destroy it. In our experiments, however, boiling has not diminished the toxicity at all and it can be definitely stated that the poison is *not* an albumen.

There is also a popular impression that vinegar will remove the poison and numerous observers claim to have removed the poison of A. muscaria completely by soaking the fungus in vinegar. We have not had the opportunity of trying this with fresh A. muscaria, but in one experiment in which the A. verna was soaked over night in vinegar it failed to get rid of the poison—any more than would have dissolved in that amount of water.

Toadstool poisoning differs from most poisonings in the long time elapsing before death in fatal cases. The only inorganic poisons causing death after such a long interval produce profound tissue changes. Husemann believed death from poisonous mushrooms to be due to fatty degeneration of the various organs. We have examined microscopically

the tissue of dogs and cats dying from the *late* effects of the A. muscaria and A. phalloides and found them to be perfectly normal.

Mr. V. K. Chestnut, in a bulletin published by the United States
Department of Agriculture (Circular No. 13, p. 23), states that death
from the A. phalloides is due to a destruction of the red-blood corpuscles. Upon what authority this assertion is made is not stated. The
conclusion has probably been based upon the venosity of the blood in
cases of poisoning resulting from the disturbance of the respiration and
circulation. The blood corpuscles of animals poisoned by all three of
the Amanitæ studied have been counted repeatedly in our experiments
and in *none of them has there been any appreciable reduction.*

It can be positively stated that death is *not* due to a destruction of
the red blood cells.

Further, the coloring matter of the blood (hæmoglobin), which carries oxygen to the tissues, has been examined with the spectroscope to
see if any new compound had been formed which would prevent it from
carrying oxygen. No such compound has been found—no alteration
could be detected in the hæmoglobin. It is quite evident that these
toadstools do not kill by their action on the blood, for in a number of
experiments the blood was examined a very short time before death.

Thinking that they might act upon the nerve cells of the brain and
spinal cord very much as certain toxins of infectious diseases do, those
structures were examined by special staining methods (silver impregna-
tion), but no greater variation than is normal could be detected in any
of those examined.

No statement can be made as to the cause of this late death, but it
would appear to be due to some disturbance of nutrition.

Late death occurs not only in animals, but in most of the cases of
poisoning in man recorded in medical literature.

The contrast between the early and late symptoms is not so great in
poisoning by A. phalloides and A. verna as in the case of poisoning by
A. muscaria. In the first two the serious symptoms appear early and
continue till the end; in the last the early effects of the muscarine soon
passes off or can be removed by atropine, but the late symptoms, strikingly in contrast with the early ones, still appear, and continue till death.

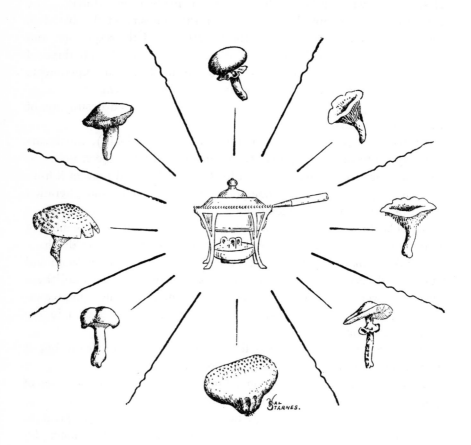

RECIPES

FOR

COOKING AND PREPARING FOR THE TABLE

PREPARING toadstools for the table should begin while collecting them. Have a soft brush, a knife, half a dozen one or two-pound paper bags and an open-topped, roomy, shallow basket. As edible species are found, cut them loose well above their attachment. Keep the spore surface down until the top is brushed clean and every particle of dirt removed from the stem. This prevents dirt from getting upon the spore surface, from which it is very hard to dislodge. Never clean a toadstool over other toadstools. If the stem is hard, tough or wormy, remove it.

Collecting. Cleansing.

Having cleaned the plant, place it in one of the paper bags, spore surface down. Write its name on the bag. Place but one kind in the same bag, unless species of about the same texture and flavor are found and mixing is not objectionable. Where another species is found, give it a bag to itself.

Select fresh, inviting plants only. Do all possible cleaning in the field. Plants keep clean, pack better, and more of them can be carried. A careless jumble is gritty, bruised and disappointing.

Selection.

If not ready to cook the find, place the bags in the ice chest. It is best to cook fungi as soon as possible. Cooked, they can be kept much longer than when uncooked.

When ready to cook, wash the plants by throwing them into a deep pan of water. Pass the fingers quietly through them upward; let stand a moment for the dirt to settle, then gather them from the water with the fingers as a drain. Remove any scurf or adhering dirt with a coarse flannel or a cloth. Wash in this way through two or three waters. Lay to drain. By experience in draining, exactly the amount of water necessary to cook a particular species

Washing.

can be allowed to remain within its spore surface, if it is a gilled species. To other kinds, water must usually be added.

The removal of the skin of any toadstool is seldom justifiable. As with the apple and most fruits, the largest amount of flavor is in the skin.

By the consistency of the species in hand, decide upon the best method of cooking it and the time and medium required. If it is thin, juicy, tender, from five to ten minutes' slow stewing will be ample; if it is thick, dry, tough, from thirty to forty minutes will be required. After any species is cooked tender, it may be seasoned to one's liking and served as one chooses.

Many species, which absolutely refuse to become tender after prolonged stewing, quickly succumb in the frying pan and make crisp, delicate morsels. Edible kinds which dry well, or are hard when found, often grate or powder easily, and are excellent (after soaking) made into soups, fritters or pâtés.

Hundreds upon hundreds of recipes for cooking the common mushroom and the few other fungi heretofore eaten, are at hand. The simpler methods—those which retain the natural flavor of the species cooked—are the best.

When a species has good body, and but little flavor, it may be made delicious by cooking with it another species of higher flavor.

The most concise instruction is: Cook in any way you can cook an oyster.

The writer's best and long-tried recipes are here given. Mrs. Sarah Tyson Rorer has kindly contributed some of her own choice methods; Mrs. Emma P. Ewing, of culinary celebrity, is represented; and that every recipe needed may be found herein, the most sensible of English and French recipes are given.

TO COOK MUSHROOMS.

Gather mushrooms whenever they can be found. That is the best time of the day to collect them. The gills grow darker and the flavor improves as the spores ripen. They are in good condition up to the time the gills begin to grow moist and to soften.

Cut off the extreme butt of the stem, holding the gills downward. Rub off the cap and stem with a rough towel or flannel. Do not peel. Wash in cold water. Drain well, gills downward.

The English method is to scald them, but there is more of custom than use in it.

Mushrooms may be preserved temporarily by boiling them in salt and water for five minutes, draining and wiping dry. A better way is to cook them, place in ice-chest, and reheat when wanted. *McIlvaine.*

To Broil.—Use well-spread caps only. Use double iron broiler. Place the caps on it, gills down, and broil two minutes, turn and broil two minutes more. While hot, season with salt and pepper, and butter well, especially upon the gill side. Serve upon toast. *Mrs. S. T. Rorer.*

BAKED MUSHROOMS ON TOAST.

Bake under a glass or basin, on toast along with scalded or clotted cream or a little melted butter, and salt and pepper to taste. They take about a quarter of an hour in a gentle oven or before a fire; when they are taken up, do not remove glass for a few minutes; by that time the vapor will have condensed and gone into the toast. *Stevens.*

CRUSTS OF MUSHROOMS.

Cut into small, even-sized squares a pint of the selected toadstool; stew in a little water until done; add two ounces butter and one teaspoonful of salt, one-half teaspoonful of pepper. Wet a teaspoonful of flour with two gills of cream and mix with the beaten yolks of two eggs. Add, and mix well with the toadstool.

Cut the upper crust from some small French rolls. Scoop out the inside of both upper and lower part, brush them with melted butter and brown in the oven; fill them, put on the top. Serve.

Or, when cooked as directed, serve in paper cases, or pastry shells.

TO DRY MUSHROOMS. (English method.)

Take those neither very young nor very old. Remove the butts only. Then slice, string or skewer the slices lightly, and expose to a current of warm dry air. A warm oven, with the door open, is a good place. When quite dry and shrivelled, pack in tins, with spice at top and bottom. When wanted for use, soak the slices in tepid water for some hours. Then cook. *Hay.*

STEWED MUSHROOMS ON TOAST.

Cut the mushrooms—caps and stems—into pieces of equal size. Place in a covered saucepan. To each pint add one ounce (two level table-spoonfuls) of butter. Enough water will have been retained by the gills after washing to make sufficient liquor. Stew slowly twenty minutes; season to taste with pepper and salt. Place upon toast. *McIlvaine.*

TO FRY MUSHROOMS.

Lay them in a frying pan in which butter has been heated boiling hot. After frying five minutes serve on a hot dish—pouring over them the sauce made by thickening the butter with a little flour. This is as delicious as more elaborate ways of cooking and retains the mushroom's distinctive flavor in full perfection.

FRICASSE OF MUSHROOMS.

Wash, put them into a chafing dish, sprinkle over a teaspoonful of salt, a quarter of a teaspoonful of black pepper, cover and cook slowly for five minutes. Moisten a tablespoonful of flour in a gill of milk, strain this into the mushrooms, bring to boiling point, add the yolks of two eggs slightly beaten, a tablespoonful of chopped parsley, and serve at once. *Mrs. S. T. Rorer.*

MUSHROOM PICKLES.

One-half peck of either Agaricus campester, Lepiota procera, Hypholoma fascicularis, Hypholoma perplexum, Clitocybe multiceps, Russula virescens. Select sound specimens, cut off ends of stems (entire stem of fascicularis or procerus), rub the tops with flannel dipped in salt. Throw them into milk and water (one-fourth milk). Drain and put them into a stew pan. Sprinkle the layers with salt—one-half gill to one-half peck mushrooms. Cover them close and put them over a gentle fire for five minutes to draw out the water. Then put them on a coarse cloth and drain until cold (or put on mosquito netting in a colander).

To prepare a pickle for them: Take one-half gallon vinegar (if strong dilute with water), two ounces mace, one-fourth ounce cloves, one-half pound salt (Worcester), one teaspoonful red pepper, one nutmeg cut in slices.

Put in a jar covered with a wet cloth and keep the cloth wet. Place over a very slow fire, cook as long as the acid is prominent *and no longer.*

Take small wide-mouthed bottles, fill with the mushrooms, pour on the pickle until the bottle is filled. Tie down tight. (To slice a nutmeg, boil it in vinegar—slice while hot. Makes of salt vary in strength; the "Worcester" is a strong salt.)

N. B.—When H. fascicularis is used, wipe the tops with a wet cloth.

McIlvaine.

TO PICKLE MUSHROOMS. (English style.)

Take buttons and remove butts only. Put into jars and cover with cold, spiced pickling vinegar. Add a few peppercorns and mustard seeds and seal hermetically. *Hay.*

MUSHROOM CATCHUP.

Take the opened toadstools, clean carefully, bruise them. Put a layer on the bottom of an earthen pan, strew salt over them (two tablespoonfuls to one-fourth peck), another layer, more salt and so on. One-half ounce cloves, one-half ounce mace, one-half ounce allspice, one-half ounce whole pepper. Let stand six days. Stir every day. Then put in gentle oven, cover pan with wet cloth, keep wet, and heat for four hours. Strain through a fine cloth or sieve. To every gallon of liquor add one quart red wine. Salt to taste. Add a race or two of ginger cut small. Strain; let catchup get cold. Pour it from the settlings. Bottle. Cork tight. *McIlvaine.*

MUSHROOM CATCHUP.

A catchup superior to that sold in the stores may be made at home. Break the toadstools into bits and place in a stone jar, with an ounce and a half of salt for every quart of plants. Let stand in a cool place for three days, stirring several times a day. On the third day put over the fire, in a porcelain kettle, and heat slowly. In about half an hour the juice will flow freely, when strain through a hair sieve, return to the fire and boil twenty minutes. Measure the liquid and to each quart allow an ounce of ginger root, a blade of mace, a bay leaf, a pinch of cayenne, and an ounce each of allspice and black pepper; boil down to one-half the quantity, add a teaspoonful of the best brandy to each half-pint. Bottle. Cork and seal with wax or rosin. *Anon.*

MUSHROOM CATCHUP. (English style.)

Remove the butts. Sprinkle all with salt. Pile in a bowl. Let them remain so for three days, stirring occasionally. Then squeeze out all the liquor. To each gallon of it add cloves and mustard seed, crushed, of each half an ounce; allspice, peppercorns and ginger, crushed, of each one ounce. Heat slowly up to boiling point in a covered vessel. Set aside in a warm place for a fortnight. Then strain and bottle. If the catchup shows signs of not keeping, add more salt and spice, heat and proceed as before.   *Hay.*

CANNED MUSHROOM SAUCE.

Cook together, until a light brown color, two tablespoonfuls each of butter and flour, add a can of button mushrooms, with the water it contains, and a cupful of water or broth. Simmer five minutes, stirring meanwhile, season and serve. The flavor of the mushroom is more distinct and pronounced if the sauce is seasoned only with salt and mixed pepper. If broth is used in the preparation of mushroom sauce instead of water, it should be the broth of such meat as the sauce is to be served with—for instance, chicken broth when to be served with chicken, beef broth when to be served with beef, etc.

Mrs. Emma P. Ewing.

FRESH MUSHROOM SAUCE.

Put in a graniteware or porcelain-lined saucepan two tablespoonfuls of butter. When hot add two cups of fresh, prepared mushrooms, cover closely, and cook briskly two or three minutes. Season to taste with salt and pepper, and serve with broiled beefsteak, birds, or sweetbreads.

Mrs. E. P. Ewing.

TO COOK BOLETI.

Remove the stems, and the tubes unless they are compact and young, or the dish will be slimy from the tubes. Wipe the caps clean.

To BROIL.—Put on wire broiler or in a hot buttered pan. Cook well. Add butter, pepper and salt.

To STEW.—After cutting the caps in pieces of similar size, stew in a covered saucepan for twenty minutes. Do not use much water. When done, add butter, or cream, pepper and salt. Some persons may prefer to add a little lemon juice or sherry.

To Bake.—Bake for half an hour in covered dish, add oil or butter, a little parsley, and garlic if liked, pepper and salt.

To Fry.—Remove the tubes from all the caps, excepting of very young or very short-tubed species. Slice the caps as you would egg-plant. Fry in butter, oil or fat, or dip in batter or in egg crumbs.

McIlvaine.

B. Edulis Soup (as made in Hungary).—Having dried some Boleti in an oven, soak them in tepid water, thickening with toast bread, till the whole be of the consistency of a puree, then rub them through a sieve, throw in some stewed Boleti, boil together, and serve with the usual condiments. *Paulet.*

To Dry Boleti (English method).—Gather in dry weather. Remove stems and tubes. Wipe clean with a damp cloth. Slice. String the slices. Hang up in a warm place for two days. Then give them a minute in a moderately warm oven. Pack in tins with spice. When wanted steep the slices in tepid water for some hours, till they swell. Then proceed to dress as for fresh Bolets. The Russians retain the stems and dry their Bolets whole, stringing them up the stem and through the center of the cap. *Hay.*

TO COOK CANTHARELLUS CIBARIUS.

Cut the mushrooms across and remove the stems; put them into a closely-covered saucepan with a little fresh butter, and sweat them until tender, at the lowest possible temperature. A great heat always destroys the flavor. *Mrs. Hussey.*

Wash, cut into pieces and put into boiling water; then stew with fresh butter, a little olive oil, chopped tarragon, pepper, salt and a little lemon peel; when cooked simmer over a slow fire for twenty minutes, moistening from time to time with beef gravy or cream. When ready to serve thicken with the yolk of egg. *M. C. Cooke.*

To Fry.—Dip caps in egg and bread crumbs, season with pepper and salt and fry in hot butter or oil.

To Stew.—Cut the whole plant into small pieces across the grain, stew slowly in a covered saucepan for forty minutes. Add thickened cream or milk. Use freely of butter and season to taste.

To Roast.—Place in a hot dry pan over a slow fire, shake and turn until the plants are crisp. Butter and season with pepper and salt. A fine camp dish.

To Preserve for Winter Use.—Pull into strips one-half inch wide, spread on a piece of mosquito netting and place in the sun or current of warm air. When dry hang up in small bags or mosquito netting in a dry place. *McIlvaine.*

TO COOK CLAVARIA.

Fry in hot butter, oil or fat until well done; or stew, covered with a little water, over a slow fire for half an hour. When done add cream or milk, a little flour, plenty of butter and season with pepper and salt. Salt last, always, or it will harden the plants. *McIlvaine.*

To Pickle (English recipe).—Put the tender parts into jars with peppercorns, mustard seeds and nasturtium seeds. Pour on them cold white wine vinegar. Fill up and cork hermetically. *Hay.*

TO COOK CLITOCYBE MULTICEPS.

To Bake.—Wash caps, remove stems, let drain for a few minutes; place gills upward in a pan; place on gills a small-sized lump of butter; season with pepper and salt; grate cheese over each layer, cover pan, and place in hot oven to bake for one-half hour.

An exceptionally fine dish. They are excellent fried.

Other species of similar consistency may be cooked in the same way. See Toadstools with Cheese. *McIlvaine.*

TO COOK COPRINUS.

"In regard to the C. micaceus I find that they are better cooked after the following recipe:

"Trim the stems, wash the toadstools carefully through several waters, then drain them in a colander. Spread them out in a long baking pan, dust lightly with salt, pepper, put over a few bits of butter, cover with another pan and bake in a moderate oven for twenty-five minutes. Add four tablespoonfuls of cream, bring to boiling point; dish on toast.

"The C. atramentarius may be cooked in precisely the same manner. I find that all these inky mushrooms are better cooked in a very slow heat in the oven, and they must be covered or they lose their flavor." *Mrs. S. T. Rorer.*

C. comatus, or any other Coprinus, may be treated in the same manner; or they may be stewed slowly in a covered dish for from five to ten minutes. *McIlvaine.*

CROQUETTES.

To one pint of any well-cooked toadstool of meaty species, add two hard-boiled eggs, a sprig of parsley; pepper and salt to taste; chop all very fine, then take two level tablespoonfuls of butter and one of flour, put over the fire with the toadstools and eggs; mix thoroughly together, set aside to cool. When cold, shape, dip in egg and bread crumbs, and fry in hot oil, butter or fat. *McIlvaine.*

DEVILED TOADSTOOLS.

For deviled toadstools prepare the meat as for patties, adding the yolks of two hard-boiled eggs to each pint of meat, a pinch of red pepper and a little chopped parsley. Serve hot or cold in halves of egg shells, nested among green. *McIlvaine.*

TO COOK FISTULINA HEPATICA.

Mrs. Hussey says of it: "If it is not beef itself, it is sauce for it."

It can be sliced thin and dressed as a salad with mayonnaise dressing or otherwise.

The simplest and best way is to cut the fungus into slices as one would egg-plant. If it is small, slice it into two parts, fry in hot butter, season with pepper and salt.

Another favorite way is to slice the plant across the grain, cut into squares of one-half inch and cook very slowly in a covered pan for twenty minutes. Add a little water, and plenty of butter. Season with pepper and salt.

The F. hepatica always has a slightly acid taste, which is very acceptable to most persons, but objectionable to a few. *McIlvaine.*

SALAD.—Cut in thin slices and rub them with garlic. Mingle with lettuce or other green salad. Dress with oil, vinegar, pepper, mustard and salt. Serve. *Hay.*

TO BROIL ANY CAPPED FUNGUS.

Select those that are spread open and keep the unopened for other styles of serving. Cut off the stems close to the tops. Baste well with melted butter and sprinkle lightly with pepper and salt. Heat the broiler very hot, lay the caps upon it with the gills up and broil over a clear fire, turning the broiler first on one side and then on the other. As soon as tender, which will be in about five minutes, open the broiler, remove the caps with care, and place on well-buttered slices of the toast which have been previously prepared. Pour over the whole a sauce made of drawn butter, or hot water thickened with flour to the consistency of cream.

FRIED TOADSTOOLS.

Take the caps only—one pint—well drained and carefully seasoned with one teaspoonful of salt, one-half teaspoonful black pepper. Place in a pan with one ounce of butter (a lump the size of a small egg). Fry slowly for ten minutes. Add a little milk or cream thickened with flour. Serve on hot toast.

TO COOK HYDNUM.

In cooking Hydnei care must be taken to cook slowly and well.

Use the tender parts only of stems and caps of the capped species, and soft, fresh parts of the maned species; cut into small pieces of similar size, stew slowly in covered saucepan for from thirty to forty minutes, season with butter, pepper and salt. Serve.

Or, after stewing for forty minutes as above, drain off the water, chop fine, make into croquettes or into pâtés.

A HUNTER'S TOAST.

Carry a vial of olive oil or a small can of butter, some pepper and salt mixed. An edible toadstool found, collect a few dry twigs, fire them. Split a green stick (sassafras, birch or spice-wood best) at one end; put the toadstool in the cleft, hold it over the fire; oil or butter, season. Eat from the stick. *McIlvaine.*

TO COOK HYPHOLOMAS.

To STEW.—Wash the caps, stew slowly in the water which the gills retain, for half an hour, keeping dish covered. Add plenty of butter, pepper and salt to taste, add cream or milk with a little thickening.

The Hypholomas have a slightly bitter taste, of which most persons become very fond; if it is objectionable, add a small amount of lemon juice or sherry. *McIlvaine.*

H. PERPLEXUM.—Put one dessertspoonful of vinegar in a quart of water. Soak the caps in this mixture twenty minutes. Then take them out and stew slowly for half an hour in a covered vessel, adding butter, pepper and salt to suit the taste. A small quantity of onion is thought by some to improve the flavor, and a thickening of flour and milk just before serving is an improvement. *Prof. C. H. Peck.*

The above is given as recipe for cooking H. perplexum. It answers equally well for the many Hypholomas resembling it. *McIlvaine.*

To BAKE.—Wash caps, remove stems, let drain for a few minutes, place gills upward in a pan, place on gills a small-sized lump of butter, season with pepper and salt, cover pan, and place in oven to bake for one-half hour. *McIlvaine.*

TO COOK LACTARII.

The rich juices of the Lactarii are best retained by baking. The species grow hard and granular if cooked rapidly. Baked they are excellent. This method is preferable to stewing, but no one will despise a properly made stew of them.

TO COOK MARASMIUS OREADES.

Remove the stems, wash the caps, place in a covered saucepan and simmer for thirty minutes, adding sufficient water to prevent scorching; add a little milk or cream, butter and season with salt and pepper to taste.

Or, simply fry in butter, make a gravy and season to taste.

To DRY.—String the caps on threads and loop up in a dry place, and when thoroughly dry place in tight glass jars or tin cases.

TO COOK THE MORELL—MORCHELLA (from Persoon).

1. Having washed and cleansed them from the earth which is apt to collect between the plaits or hollows of the plant, dry thoroughly in a napkin and put them into a saucepan with pepper, salt and parsley, adding or not a piece of ham. Stew for an hour, pouring in occasionally a little broth to prevent burning. When sufficiently done, bind with the yolks of two or three eggs and serve on buttered toast.

2. MORELLES A L'ITALIENNE.—Having washed and dried, divide them across, put them on the fire with some parsley, scallion, chevril, burnet, tarragon, chives, a little salt, and two spoonfuls of fine oil. Stew till the juices run out, then thicken with a little flour; serve with bread crumbs and a squeeze of lemon juice.

PÂTÉS.

The toadstools good for croquettes and pâtés are such as the Puff-balls, Lactarii, Gomphidius rhodoxanthus, Fistulina hepatica, Tricholoma personatum and many others of the meaty kinds. Cut the toadstools into small pieces, cook slowly until tender, adding butter, pepper and salt. Let them cook almost dry, then add cream or milk and thickening. Fill pastry shells and serve.　　　　　*McIlvaine.*

A pretty effect is produced by dipping the rims of the shells in partially beaten white of egg, then in finely-chopped parsley before filling.

TO COOK PLEUROTUS OSTREATUS.

Remove tough stem-part, if any, and use only such parts of the plants as seem fresh and tender.

TO STEW.—Cut in small pieces across the grain. Stew twenty minutes over slow fire with a little water. Add cream or milk with a little thickening; season with butter, pepper, salt.

TO FRY.—Cut into pieces about the size of a medium-sized oyster, dip in egg and bread crumbs, and fry in hot butter or oil, as oysters are fried.　　　　　*McIlvaine.*

WITH CHEESE—*au gratin.*—Cut into medium-sized pieces. Stew slowly, rather dry, for fifteen minutes. Pour off liquor; save it. Place in baking dish (or in individual dishes, clam shells, etc.) a layer of ostreatus, buttering and seasoning each layer, sprinkle with bread crumbs and grated cheese and so on until dish is filled, placing cheese on top. Pour liquor over the dish. Place in slow oven and bake until well browned.

This manner of cooking is a favorite. Any toadstool may be cooked in this way.　　　　　*McIlvaine.*

Wash. Put them into a chafing dish with one ounce of butter to each half pound of plant. Sprinkle over half a teaspoonful of salt, cover the dish and cook slowly for five minutes. Beat the yolks of two eggs with one gill of good milk, lift the lid, add the mixture of eggs and milk; when smoking hot serve. Do not allow the mixture to boil or the eggs will become curdled.

RECIPE NO. 2.—Wash, Dust with salt and pepper, dip in egg, then in bread crumbs and fry quickly in smoking hot olive oil.

The following recipe was given me by a chef:

"Put into the saucepan a tablespoonful of butter, add a clove of garlic,

646

a thin slice of onion, stir until slightly brown and add a tablespoonful of flour. Mix carefully, add a quarter of a teaspoonful of beef extract dissolved in half a cup of water and the same quantity of cream. Bring to boiling point, add a tablespoonful of chopped carrot, a bay leaf, and a blade of mace. Stand the mixture on the back part of the stove where it will scarcely boil, for ten minutes. Strain and add half a pound of ostreatus. Cover and cook for ten minutes. Serve on toast.''

Mrs. S. T. Rorer.

TO COOK PUFF-BALLS.

To FRY.—Remove the thin outer rind, slice, dip in egg and bread crumbs, and fry as egg-plant; serve without tomato sauce.

To STEW.—Cut in dice-shaped pieces, stew for fifteen minutes in a little water, pour off the water, dust with a little flour, add a small quantity of milk or cream, butter, pepper and salt and a little parsley. Stew slowly for five minutes. Serve. These also may be served in pâtés. When these are broken open they should be perfectly white inside; any stains or yellow part should be removed, otherwise they will be bitter. *McIlvaine.*

SALADS.—Cut into strips, mingle with mustard and cress, or with blanched dandelions, scallions and hard-boiled egg, and dress as ordinarily for a salad. Or, amalgamate with potato salad a l'Allemande.

Hay.

TO COOK TRICHOLOMA PERSONATUM.

To STEW.—Wash and cut into small pieces. Stew for thirty minutes. Pour off the water, add milk slightly thickened, butter, pepper, salt and a little chopped parsley. *McIlvaine.*

TO STEW THE TOUGHER TOADSTOOLS.
(Hydnaceæ, Polyporaceæ, Etc.)

Cut into small pieces of even size. Soak for half an hour in tepid water. Remove from water, do not drain; place in covered pan and simmer for forty minutes. Add proper proportion of thickened milk or cream, butter, pepper, salt.

Those who like may add parsley or nutmeg, or beef gravy; in fact, any flavoring. *McIlvaine.*

647

SALADS.

Many species of fungi make good salads. The best of these are, Russulæ when young, fresh and firm; either sliced raw or stewed and drained; Clitocybe multiceps stewed and drained; Tricholoma personatum, raw or stewed; Clitopilus prunulus, raw or stewed; Coprinus comatus, C. micaceus, atramentarius, raw; Clavaria, fresh, young, brittle, either raw or stewed; Fistulina hepatica, raw; any of the edible Polyporaceæ, after stewing; any of the edible Hydnaceæ after stewing; the puff-balls, raw or stewed. Any favorite species will make a salad.

After cooking allow to drain and cool; then mix with mayonnaise dressing, or make a dressing to taste of oil, vinegar, salt and pepper. Serve on lettuce.

SOUP.

Dame Nature never made a soup. Soup is a human invention of more or less distinctiveness. Usually it is a successful disguise or covering of invisibility for something which furnishes the name.

To make two quarts of a distinctly fungoid soup take one quart of any edible toadstools, carefully cleaned. Put in a well-covered boiler with three pints of water, and boil slowly for one hour. Rub the whole through a colander. Reject that which does not rub through readily. Add one-half pint of milk thickened with one tablespoonful of flour, one ounce of butter, a dessertspoonful of salt, a teaspoonful of pepper. Bring to a boil. Serve.

Any chosen thing or things may be added to the above—the toadstools can not resent it. *McIlvaine.*

TOADSTOOLS WITH CHEESE.

Several varieties of fungi are delicious when baked with a small quantity of cheese grated upon them; notably Clitocybe multiceps, the Hypholomas, Armillarias, Pleurotus ulmarius and ostreatus, Lentinus lepideus and many Boleti. See recipe for baking. When several layers of plants compose the dish, cheese should be grated on each layer.

McIlvaine.

BAKED TOADSTOOLS OF ANY GILLED KIND.

Wash, place the caps in a tightly covered dish or pan after dipping them in bread crumbs. Lay them in layers, with a small piece of butter on each toadstool, as well as the proper amount of pepper and salt. Bake from twenty to forty minutes as suits the consistency of the species. Serve on toast.

Or, the caps prepared as above, may be laid upon pieces of toast and placed in the pan. If this plan is adopted the lower pieces of toast become saturated with the liquor; therefore, in serving, cut from top to bottom of dish.

See To Cook Clitocybe Multiceps. *McIlvaine.*

A CAMP BAKE.

Cover the bottom of a tin plate with caps, spore surface up. Sprinkle with salt and pepper, place a bit of butter on each. Put another tin plate on top. Set on coals or a heated stone for fifteen minutes. Eat. No better baking will result in the best oven. *McIlvaine.*

GLOSSARY

A-, prefixed signifies absence; as *aseptate*, without septa.

ABBREVIATIONS:
$\qquad$ cm. = centimeter.
$\qquad$ mm. = millimeter.
$\qquad$ μ = micron.
$\qquad$ in. = inch.
$\qquad$ ′ = inch or inches.
$\qquad$ ″ = line ($\frac{1}{12}$ inch) or lines.
$\qquad$ nov. gen. = new genus.
$\qquad$ n. sp. = new species.
$\qquad$ x *between* two figures signifies by; 2 x 4 = 2 by 4.
$\qquad$ - between two figures = to; 2–4 = from 2 to 4.

ABER′RANT (*aberran(t-)s*, ppr. of *aberrare*, stray from, <*ab*, from, + *errare*, to stray), differing in some of its characters from the group in which it is placed, said of a plant, species, genus.

ABJEC′TION (*abjectio(n-)*, act of casting away, *abicere, abjicere*, <*ab*, away, + *jacere*, throw), throwing off with force, as spores or seeds; expulsion.

ABJOINT′ (*ab*, from, + *junctus*, adjoining), to joint off or delimit by septa or partitions.

ABNOR′MAL (*abnormis*, deviating from a fixed rule, irregular, <*ab*, from, + *norma*, a rule), not conforming to the usual type; irregular, unnatural.

ABOR′TIVE (*abortivus*, born prematurely), imperfect or wanting.

ABRUPT′ (*abruptus*, steep, disconnected, <*ab*, off, + *rumpere*, break), terminating suddenly.

ABSTRIC′TION (*abstrictus*, ppr. of *abstringere*, <*abs*, from, + *stringere*, bind), separation of one part from another by constriction, especially of spores from their hyphæ.

ACAULES′CENT, ACAU′LINE, ACAU′LOSE, ACAU′LOUS (*caulis*, a stem or stalk of a plant), having a very short stem or none; stemless.

AC′EROSE (*acerosus*, chaffy), narrow, stiff and pointed like spruce needles; intermediate in form between acicular and subulate.

ACETAB′ULIFORM (*acetabulum*, a cup-shaped vessel; *forma*, form), cup-shaped, having the form of a shallow bowl.

ACH′ROOUS (*Gr.*—priv. + *Gr.*—color), colorless, achromatic.

ACIC′ULA (pl. ACICULÆ) (a needle, a small pin, dim. of *acus*, a needle), a needle-shaped spine, prickle or other body.

ACIC′ULAR, ACIC′ULATE, AC′IFORM (*acicula*, a small pin or needle), needle-shaped, having a sharp point like a needle, as pine leaves.

AC′ROGEN (*Gr.*—at the top, + *Gr.*—born, produced), a cryptogam which increases by development [of an apical cell] at the summit of an axis, having a true stem, leaf-like appendages, etc., as ferns, mosses, etc.

ACROG′ENOUS (as *acrogen* + *ous*), (a) produced at the apex, as some spores from the apex of a hyphal branch; (b) of the nature of or pertaining to acrogens.

ACROP'ETAL (*Gr.*—the top, + *L.*—*petere*, seek), developing from below upward, or from the base toward the apex.

ACU'LEATE, ACU'LEATED (*aculeatus*, furnished with prickles or stings), slender-pointed.

ACU'LEUS (pl. ACU'LEI) (a sting, prickle, spine, dim. of *acus*, a needle), a prickle.

ACU'MINATE (*acumen*, a point or extremity), terminating in a long drawn point.

ACUTE' (*acutus*, sharp), sharp, applied to gills having sharp edges or pointed at either end.

AD'NATE (*adnatus*, grown to, pp. *adnasci*, to grow to), growing into or fast to; of gills, *e. g.* closely attached to the stem.

ADNEXED' (*adnexus*, connected), of gills attached to the stem, but not adnate to it.

ADPRESSED' (*adpressus*, pp. of *adprimere*, <*ad*, to; *premere*, to press), pressed in close contact but not adherent.

ADVENTI'TIOUS (*adventitius*, prop. *adventicius*, coming from abroad, <*adventus*, pp. of *advenire*, come to, arrive at), appearing casually, or in an abnormal or unusual position or place.

ÆRU'GINOSE, ÆRU'GINOUS (*æruginosus*, <*ærugo*, rust of copper), verdigris-green.

AFFIN'ITY (*affinita(t-)s*, <*affinis*, neighboring, related by marriage), morphological relationship; resemblance in general plan of structure.

AGAM'IC, AG'AMOUS (*Gr.*—unmarried + *ic*), sexless.

AG'AMOGEN'ESIS (*Gr.*—unmarried, + *Gr.*—production), non-sexual reproduction.

AGAM'OSPORE (*Gr.*—unmarried, + *spora*, spore), spore formed without fertilization.

AG'ARIC (*agaricum*, a kind of tree-fungus used as tinder, named, according to Dioscorides, from the country of the Agari in Sarmatia, where this fungus abounded), any gill-bearing fungus; formerly applied only to members of the genus Agaricus.

AGAR'ICIOID, of the nature of an agaric; mushroom-like.

AGGLOM'ERATE (*agglomeratus*, pp. of *agglomerare*, *adglomerare*, wind into a ball, <*ad*, to, + *glomerare*, wind into a ball), (a) clustered densely, but not connected together; (b) gathered into a rounded mass or into a compacted heap or pile.

AGGLU'TINATED (*agglutinatus*, pp. *adglutinare*, paste to), glued to a surface; grown together fast; applied to fungi that are firmly attached to matrix

AG'GREGATE, AG'GREGATED (*aggregatus*, pp. *adgregare*, lead to a flock; add to), collected together but not cohering.

ALBU'MINOID (*albumen* + *oid*), an organic substance containing nitrogen in its composition, as proteids.

ALLAN'TOID (*Gr.*—a sausage, + *Gr.*—form), sausage-shaped; narrowly oblong.

ALLIA'CEOUS (*allium*, garlic, + *aceous*), having the odor of onions.

ALUTA'CEOUS (*alutacius*, <*aluta*, soft leather), having the quality or color of tanned leather; leathery.

ALVEO'LATE (*alveolatus*, hollowed out, <*alveolus*, a small hollow), with small depressions like a shallow honeycomb, pitted.

AMOR'PHOUS (*Gr.*—without form, shapeless, misshapen), without definite form, structure or position.

AM'PHIGEN (*Gr.*—around, + *Gr.*—produce), a thallogen; a name applied to a cryptogam which increases by development of cellular tissue in all directions and not at the summit of a distinct axis. See ACROGEN.

AMPHIG'ENOUS (*Gr.*—about; *Gr.*—to beget), not confined to one surface, growing all around; *e. g.*, hymenium of Clavaria.

AMYG'DALINE (*amygdalinus*, <*amygdala*, almond), resembling the almond.

AMYLA′CEOUS (*amylum*, starch, + aceous), composed of, containing or resembling starch.

AM′YLUM (*Gr.*—starch), starch.

ANAL′OGY (*Gr.*—equality of ratios, proportion), superficial or general resemblance, without structural agreement; physiologically or functionally alike, morphologically unlike.

ANAS′TOMOSING (*Gr.*—an opening, outlet, discharge), united by running together irregularly; intercommunication of vessels, lines, gills or veins with each other.

ANGIOCAR′POUS (*Gr.*—a capsule, case, vessel of any kind), having the hymenium developed in a closed receptacle.

ANGUS′TATE (*angustatus*, pp. of *angustere*, straiten, narrow, <*angustus*, narrow), narrow.

AN′NUAL (*annualis*, a year old, <*annus*, a year), completing growth in one year or season.

AN′NULAR (*annularis*, relating to a ring, <*annulus*, a ring), ring-shaped.

AN′NULATE (See ANNULAR), having a ring.

AN′NULUS (See ANNULAR), the ring on the stem of a mushroom formed by the separation of the veil from the margin of the cap.

ANOM′ALOUS (*anomalus*, irregular, uneven), deviating from a general rule, method or analogy.

ANOM′ALY (*anomalia*, irregularity, unevenness), any deviation from the usual character.

ANTE′RIOR (as if from *anterus*, <*ante*, before), in front; denotes a position on the under side of the pileus adjacent to the margin; thus the end of a lamellæ next the margin is called the anterior end.

APARAPH′YSATE (a + *Gr.*—an offshoot), without paraphyses.

A′PEX (pl. API′CES) (*apex*, the extreme end), in mushrooms the extremity of the stem nearest the gill; the end furthest from the base or point of attachment.

AP′ICAL (*apex*, the extreme end, point), relating to the apex or top.

AP′ICES, plural of apex.

APIC′ULATE (*apiculatus*, dim. of *apex*, a point), terminating in a short, abrupt point.

APIC′ULUS (pl. APIC′ULI) (dim. of *apex* (apic-), a point), a short, sharp point.

APOTHE′CIUM (pl. APOTHE′CIA), (*Gr.*—a storehouse), in Ascomycetes, an open cup-shaped fructification with the hymenium on its upper concave surface; cup.

APPENDIC′ULATE (*appendiculatus* <*appendicula*, appendix, an appendage), hanging in small fragments; having an unusual appendage.

AP′PLANATE (*applanatus* <*ad*, to, + *planus*, flat), flattened out or horizontally expanded.

APPRESSED′ (*appressus*, *adpressus*, pp. of *adprimere*, press to, <*ad*, to, + *premere*, press), applied closely to the surface or to each other; adpressed.

APPROX′IMATE (*approximatus*, pp. *approximare* <*ad*, to; *proximare*, to approach), of gills which approach but do not reach the stem.

A′QUEOUS (as if *aqueus*, <*aqua*, water), watery; nearly colorless; hyaline.

ARACH′NOID (*Gr.*—a spider's web; + *forma*, form), like a cobweb.

ARBO′REAL, ARBOR′ICAL (*arboreus*, pertaining to trees), tree-inhabiting.

AR′CUATE (*arcuatus*, pp. *arcuare*, to bend like a bow, <*arcus*, a bow), bow-shaped.

ARENA′CEOUS, ARENA′RIOUS, ARE′NOSE (*arenaceus*, *harenaceus*, sandy, <*arena*, *harena*, sand), sandy; growing in sandy places.

ARE′OLATE (*areola*, dim. of *area*, a plot), divided into little areas or patches.

ARGILLA′CEOUS (*argillaceus* <*argilla*, white clay), resembling or like clay.

AR′ID (*aridus*, dry, <*arere*, be dry), dry.

ARIS′TATE (*aristatus* <*arista*, awn or beard), having a pointed beard-like process as in barley.

ARMIL′LA (*armilla*, a bracelet, armlet, hoop, ring, dim. prob. of *armus*, shoulder), a plaited frill hanging from the apex of the stem.

ARTE′RIOLE (*arteriola*, dim. of *arteria*, artery), a small artery.

ARTIC′ULATE (*articulatus*, pp. of *articulare*, divide into joints or members, <*articulus*, a joint, etc.), jointed.

ASCEND′ING (*ad*, to, + *scandere*, to climb), inclining or growing upward; applied to a lamella where its edge forms a line ascending in the direction from the margin of pileus toward the apex of the stipe; as in conical shaped pilei; applied to the partial veil when in the young stage its stem-attachment is below the level of its marginal one; in this case a ring formed from it is called inferior.

AS′CI (*Gr.*—a leathern bag, bladder), spore cases of certain mushrooms, in which a definite number of spores are enclosed in a sac.

ASCIF′EROUS, ASCOPH′OROUS (*ascus* + *ferre*, bear), ascus-bearing.

ASCIG′EROUS (*ascus* + *gerere*, bear), bearing asci.

AS′COCARP (*Gr.*—a bag, + a fruit), in Ascomycetes, sporocarp producing asci and ascospores.

ASCOG′ENOUS (*Gr.*—a bag, + producing), producing asci.

ASCOMYCE′TES (*Gr.*—a bag, + *Gr.*—a mushroom), group of fungi in which the spores are produced within little sack-like cells, called asci.

ASCOMYCE′TOUS, of or pertaining to the ascomycetes.

AS′COPHORE (*Gr.*—bearing wine-skins; *Gr.*—a bag), sporophore bearing an ascus or asci.

ASCOPH′OROUS, bearing an ascus or asci.

AS′COSPORES (*Gr.*—a bag, + *Gr.*—seed), one of a number of spores formed within an ascus.

AS′CUS (pl. AS′CI) (*Gr.*—a leather bag, bottle, bladder, etc.), microscopic sack-like cells in which spores, generally eight in number, are developed.

ASEP′TATE (*Gr.*—without, + *L.*—*septum*, a fence), without partitions or septa, said of hyphæ and spores.

ASH COLOR (See CINEREOUS).

AS′PERATE, AS′PERATED (*asperatus*, pp. of *asperare*, roughen, <*asper*, rough), having a rough, uneven surface.

AS′TICHOUS (astichus, <*a* + *Gr.*—row), not arranged in rows.

ASTO′MATOUS (*astomatus*, mouthless), without a mouth or aperture; without stomata.

AS′TOMOUS (*astomus*, mouthless), without a stoma or mouth.

AT′OMATE (*Gr.*—an atom), sprinkled with atoms or minute particles.

A′TRO (*ater*, black), in composition "black," or "dark."

A′TROPURPU′REOUS (*ater*, black, + *purpura*, purple dye, + ous), dark purple.

A′TROSANGUIN′EOUS (*ater*, black; *sanguineus*, blood, bloody), dark purple; dark blood color.

ATTEN′UATE (*attenuatus*, pp. of *attenuare*, make thin, weaken, lessen, <*ad*, to, + *tenuare*, make thin), becoming gradually narrowed or smaller.

AURANTI′ACEOUS (*aurantium*, an orange), orange-colored.

AUR′EOUS (*aureus*, of gold, golden, <*aurum*, gold), golden-yellow; yellow with a slight tinge of red.

Auric′ulate, Aur′iform (*auriculatus*, <*auricula*, the external ear), ear-shaped.

Auto-basid′ium (*actus*, an act, dim. of *Gr.*—a base, + basidium), an unseptated basidium giving rise at the apex to four slender sterigmata (sometimes fewer, sometimes more), each bearing a spore.

Auton′omous (*Gr.*—independent; of one's own free will; *Gr.*—self, + *Gr.*—hold sway), said of plants that are perfect and complete in themselves; not forming part of a cycle; independent.

Ax′is (*axle*, axis, pole of the earth), the central line of growth; stipe, stalk, etc.

Azo′nate (*Gr.*—without, + *L.*—*zona*, a zone), without zones or circular bands of different color.

Ba′dious (*badius*, bay), bay; reddish-brown; chestnut color.

Band, a broad bar of color.

Banded, marked with bands.

Barbed (*barba*, beard), furnished with barbs, fibrils or hairs.

Base (*bassus*, low, short, thick), the extremity opposite to the apex; the part of an organ nearest its point of attachment; applied to lamellæ; (a) the line of attachment to the pileus (as connected by veins at the base); (b) sometimes used to define the end attached to the stipe (broad or reticulate at the base).

Basid′iogenet′ic (*Gr.*—a base + genesis), produced upon a basidium.

Basid′iomyce′tes (*basidium* + *Gr.*—a mushroom), group of fungi which has its spores produced upon basidia.

Basid′iophore (*basidium* + *Gr.*—to bear), a sporophore bearing basidia.

Basid′iospore (*basidium* + *Gr.*—*spora*, spore; seed), spore acrogenously abjointed upon a basidium.

Basid′ium (pl. Basid′ia), mother cells in the hymenium of basidiomycetes formed on the end of a hyphal branch and abstricting spores; the spores are generally four in number, each on a sterigma, but sometimes more, sometimes fewer, and sometimes sessile. See Auto-basidium and Proto-basidium.

Basip′etal (*basis*, a base, + *petere*, seek, + al), in the direction of the base.

Bay (*badius*), a very rich dark-reddish chestnut; badious.

Bi-, prefix, meaning twice.

Bib′ulous (*bibulus*, <*bibere*, drink), having the quality of absorbing or imbibing moisture.

Bicip′etal, Bicip′itous (*biceps* (*bicipit-*), two-headed, + al), in botany divided into two parts at the top or bottom.

Bi′fid (*bifidus*, forked, <*bi*, two, + *findere*, cleave, divide), cleft or divided into two parts.

Bifur′cated (*bifurcus*, two-forked), divided into two forks or branches as in the gills of certain Agarics.

Biloc′ular (*bi*, two, + *loculus*, a cell, <*locus*, a place), two-celled.

Biog′enous (*bi*, two, + *genus*, <*gena*, born), growing on living organisms.

Bise′riate, Bise′rial (*bi*, two, + seriate), arranged in two rows.

Bis′tre (*fuligineus*), a dark brown color somewhat more reddish than sepia, but much less so than burnt umber.

Boot′ed, applied to the stem of a mushroom when enclosed in a sheath or volva; peronate.

Boss, a knob or short rounded protuberance; umbo.

Bossed, Bull′ate (*bulla*, a bubble), furnished with a boss, stud or umbo.

Glossary

BRANCHED (*brancha*, claw), dividing from the sides; also styled furcate and forked; ramifying, diverging.

BRICK, trade-term for a mass of mushroom spawn, in dimensions the size of a brick of masonry.

BRICK RED (*testaceus, lateritius, rutilus*), a dull brownish-red color like the color of burnt bricks.

BROAD, wide or deep vertically, not narrow.

BROCCOLI COLOR, the color of a variety of cabbage.

BUFF (*luteus, luteolus*), a light dull brownish-yellow, like the color of dressed buckskin or chamois.

BUL'BOUS (*bulbosus,* <*bulbus*, bulb), said of the stem of a mushroom when it has a bulb-like swelling at the base.

BYSSA'CEOUS, BYS'SOID (as if *byssaceus,* < *byssus*), resembling or consisting of fine filaments like the flax or cotton.

BYS'SUS (*Gr.*—originally a fine yellowish flax), an old name for the filamentous mycelium of certain fungi.

CÆRU'LEUS, CERU'LEOS (*cœruleus*, dark-blue, dark-green, dark colored), light blue; sky-blue.

CÆ'SIOUS (*cœsius*, bluish-gray), pale, bluish-gray; lavender colored.

CÆS'PITOSE, CÆS'PITOUS, CES'PITOSE (*cœspitosus,* <*cœsposus*, a clump of turf), growing in tufts or clumps.

CALCA'REOUS (*calcarius*, pertaining to lime, <*calyx*, lime), chalky, chalk-like.

CALLOS'ITY, CAL'LUS (*callosita,* <*callosus*, callous), a hard or thickened spot or protuberance.

CALYP'TRA (*Gr.*—a veil, hence *calyptra*, a hood), applied *e. g.* to the portion of the volva covering the pileus.

CAMPAN'ULATE (*campana*, a bell), bell-shaped.

CANALIC'ULATE (*canaliculus*, a little channel), channeled, furrowed.

CAN'CELLATE (*cancellatus*, pp. of *cancellare*, make like or provide with a lattice), latticed, marked both longitudinally and transversely with an open network.

CAN'DIDOUS (*candidus*), shining white.

CANES'CENT (*canescen(t-)s*, pp. of *canescere,* <*canus*, white or hoary), having whitish, grayish or hoary pubescence.

CAP, pileus; the expanded, umbrella-like receptacle of the common mushroom.

CAP'ILLARY (*capillaris*, pertaining to the hair, <*capillus*, the hair), pertaining to or resembling hair.

CAPIL'LIFORM (*capillus*, hair, + *forma*, form), in the shape or form of a hair.

CAPILLIT'IUM (*capillus*, hair), spore-bearing threads, filling as a packing material the fruiting part of certain fungi, variable in thickness and color, sometimes continuous with the sterile base, sometimes free, dense, persistent or lax and evanescent, often branched; found in the Lycoperdons.

CAP'ITATE (*capitatus*, having a head, <*caput*, head), having a head, or the form of a head.

CAPIT'ULUM (*capitulum*, a small head, <*caput*, head), a small head.

CAP'SULE (*capsula*, a small box or chest, dim. of *capsa*, a box), an enclosing envelope usually thin and membranous.

CARBONA'CEOUS (carbon + aceous), rigid, blackish and brittle; like or composed of carbon or coaly matter.

Car′diac (*cardiacus*, heart), of or pertaining to the heart; pertaining to the esophageal portion of the stomach, opposed to pyloric.

Ca′rious (*cariosus*, <*caries*, decay), decayed.

Carmine (*carmineus, coccineus*), a very pure and intense crimson, the purest of the cochineal colors.

Car′neous (*carneus*, <*caro*, flesh), fleshy; flesh-colored.

Car′nose (*carnosus*, fleshy, <*caro*, flesh), fleshy.

Cartilag′inous (*cartilaginosus*, <*cartilago*, gristle), firm and tough; gristly.

Casta′neous (*castaneus*), chestnut-colored; chestnut color. (Burnt umber + vermilion.)

Cau′date (*caudatus*, <*cauda*, a tail), having a tail-like appendage.

Caulic′olous (dim. of *caulis*, a stalk), growing on herbaceous stems.

Cell (*cella*, a small room, barn, etc.), (a) a small cavity, compartment or hollow place; (b) a mass of protoplasm of various size and shape, generally microscopic, with or without a nucleus and enclosing wall, the fundamental form-element of every organized body.

Cell′ular (*cellula*, dim. of a cell, + ar), composed of cells.

Cell′ulose (*cellula*, a cell), the essential constituent of the primary wall-membrane of cells, a secretion from the contained protoplasm; allied to starch, sugar and inulin. Chemical formula, $C_6H_{10}O_5$.

Centimeter, cm. (*centum*, a hundred, + metre, meter), in the metric system a measure of length, the hundredth part of a meter, equal to 0.3937 of an English inch.

Centrif′ugally (*centrum*, the center, + *fugere*, flee), from the center outwards.

Centrip′etally (*centrum*, the center, + *petere*, seek, move toward), from the circumference toward the center.

Ce′pæform (*cepa*, an onion; *forma*, form), onion-shaped.

Cera′ceous (*ceraceus*, <*cera*, wax), wax-like, waxy.

Cereb′riform (*cerebrum*, the brain, + *forma*, form), brain-shaped.

Cer′vine (*cervinus*, <*cervus*, deer), of a deep tawny or fawn color.

Chan′neled (*canalis*, a water-pipe, canal), hollowed out like a gutter; canaliculate

Charta′ceous (*chartaceus*, <*charta*, paper), like paper.

Chestnut Color (*castaneus, spadiceus*), a rich dark reddish-brown of a slightly purplish cast. (Vermilion + burnt umber.)

Chlam′ydospores (*Gr.*—mantle, + *Gr.*—seed) (encased spores), one of a number of thick-walled resting spores usually formed in rows from the breaking up of the hyphæ into spherical bead-like cells; on germination they may develop sporangia or conidiophores.

Chlo′rophyll (*chlorophyllum*, *Gr.*—yellowish-green; *Gr.=L.*, *folium*, a leaf), the green coloring matter of plants.

Chloro′sis (*Gr.*—greenness, paleness), loss of color, etiolation.

Chocolate-Brown (*chocolatinus*), a rich dark reddish-brown color, like the exterior glazed surface of a cake of chocolate.

Chrome-Green (*chromium-viridis*), a dull green color, nearly intermediate between malachite green and sage green.

Chrome-Yellow, a deep yellow.

Cil′ia (pl. of Cili′um), (*cilium*, an eye-lid), marginal hair-like processes.

Cil′iate (*cilium*, an eye-lid), fringed with hair-like processes.

Glossary

CINE′REOUS, CINERA′CEOUS (*cinereus, cineraceus*), ash-gray; a light bluish-gray color, lighter than plumbeous.

CIN′NABARINE (*cinnabar*, vermilion, + ine), cinnabar-colored; bright red; vermilion.

CINNAMO′MEOUS, CINNAMON (*cinnamomeus, cinnamominus*), a light reddish-brown color, like the inner surface of cinnamon bark.

CINNAMON-RUFOUS (*cinnamomeo-rufus*), rufous with a tinge of cinnamon. (Burnt sienna + raw umber + light red + white.)

CIR′CINATE (*circinatus*, pp. *circinare*, to make round), disposed in a circle; circular, coiled like a shepherd's crook.

CIRCUMSCIS′SILE (*circumscissus*, pp. of *circumscindere*, cut about), opening or dividing by a transverse circular line; applied to a mode of dehiscence in some fruits.

CIT′RINE, CIT′REOUS, CIT′RINOUS (*citrus*, a lemon or citron), lemon-yellow colored.

CLATH′RATE, CLATH′ROID (*clathratus*, *Gr.*—a lattice), latticed.

CLA′VATE, CLAV′IFORM (*clavatus*, <*clava*, a club), club-shaped, gradually thickened towards the top.

CLAY COLOR (*lutescens, luteolus, lutosus, argillaceus*), a dull light brownish-yellow color, nearly intermediate between yellow ocher and Isabella color.

CLEIS′TOCARP, CLIS′TOCARP (*Gr.*—that can be closed, + fruit), an ascocarp which is entirely closed, and from which the spores escape by its final rupture.

CLOSE, packed closely side by side; said of lamellæ when they are close together; also styled crowded.

COALES′CENT (*coalescens*, ppr. of *coalescere*, grow together), growing together of similar parts; coherent.

COCH′LEATE, COCHLEAR′IFORM (*cochleatus, cocleatus*, spiral, <*cochlea, coclea*, a snail's shell), shaped like a snail shell.

COHE′RENT (*coherens*, ppr. of *cohærere*, stick together, cohere), sticking together of similar parts; sometimes used in the sense of connate.

COLLEN′CHYMA (*Gr.*—glue, + *Gr.*—an infusion), in Geaster, etc., a cartilaginous-gelatinous tissue, hygroscopic and with great capacity for swelling, forming one of the inner layers of the peridium; its swelling at maturity causes the outer peridium to burst outward in a stellate manner.

COLLIC′ULOSE (*colliculus*, a little hill, dim. of *collis*, a hill), covered with little hill-like elevations.

COL′LOID (*Gr.*—glue, + semblance), like glue or jelly.

COLUMEL′LA (*columella*, a little column), a sterile tissue rising column-like in the midst of the capillitium, serving as a point of insertion for the threads which connect it with the peridium in the form of a network. (In Lycoperdaceæ.)

CO′MATE, CO′MOSE, CO′MOUS (*comatus*, hairy, <*coma*, a hair), furnished with a tuft of silky hairs; hairy.

COM′PLANATE (*complanatus*, pp. of *complanare*, make plane or plain), flattened vertically to a level surface above and below.

COMPRESSED′ (*compressa*, fem. of *compressus*, pp. of *comprimere*, compress), flattened laterally.

CONCAT′ENATE (*con*, together, + *catenare*, link, chain, <*catena*, a chain), linked together in a chain.

CONCAVE′ (*concavus*, hollow, arched, vaulted; *com*, together, + *cavus*, hollow), having a rounded, incurved surface.

CONCEN′TRIC (*con*, together; *centrum*, center), having a common center, as a series of rings, one within another.

CONCEP'TACLE (*conceptaculum*, <*concipere*, pp. *conceptus*, contain, conceive), a closed sporiferous body.

CON'CHIFORM (*concha*, a shell, + *forma*, shape), shell-shaped, resembling a clam-shell in shape.

CONCOLORED, CONCOL'OROUS (*concolor*, of one color), of a uniform color.

CONCRES'CENT (*concrescentia*, <*concrescere*, grow together), growing together.

CON'CRETE (*concretus*, grown together, solid), coalescent; united in a coagulated, condensed or solid mass; grown together.

CONFER'VOID (*conferva*, a name applied to certain of the Algæ, sea-weeds; + *Gr.*—form), like a Conferva, from the finely branched threads; loose and filamentous.

CON'FLUENT (*con*, together, + *Gr.—fluere*, flow), blended into one.

CON'GENER (*congener*, of the same race, <*con*, together, + *genus* (*gener*), race), of the same genus or kind.

CONGENER'IC, CONGENER'ICAL, CONGEN'EROUS (*congener*, of the same race, + ic, ous), belonging to or nearly allied to the same genus.

CONGENET'IC (*con*, together, + *Gr.*—generation, seed), produced at the same time or by the same cause; alike in origin.

CONGLOM'ERATE (*conglomeratus*, pp. of *conglomerare*, roll together, heap together, <*com*, together, + *glomerare*, gather into a ball), densely clustered; gathered into a round mass; composed of heterogeneous materials.

CONGLU'TINATE (*conglutinatus*, pp. of *conglutinare*, glue together, <*com*, together, + *glutinare*, glue), as if glued together.

CONID'IAL (conidium + al), pertaining to or of the nature of a conidium or conidia; characterized by the formation of conidia; bearing conidia.

CONIDIIF'EROUS, CONIDIOPH'OROUS (*conidium*, dust, + *ferre*, bear), bearing conidia.

CONID'IOPHORE (*conidium*, + *Gr.*—bearing), a hypha from which are abstricted conidia.

CONID'IUM (pl. CONIDIA), a non-sexual spore formed singly or in chains by abstriction from the ends of hyphæ or hyphal branches. See under SPORE.

CO'NIFER (*conifer*, cone-bearing, <*conus*, a cone, + *ferre*, bear), a cone-bearing tree.

CONJUGA'TION (*conjugatio(n-)*, a joining, entomological relationship, <*conjugare*, pp. *conjugatus*, join), union of two cells to form a spore.

CON'NATE (*connatus*, pp. *connasci*, <*con*, together; *nascor*, to be born), united by growing together from the first.

CON'NIVENT (*conniven(t-)s*, *coniven(t-)s*, ppr. of *connivere*, *conivere*, wink at; over-look), having an inward direction, converging, coming in contact, said of a cup whose sides curve inward and meet at the margin.

CON'STANT (*constan(t-)s*, steady, firm, <*com*, together, + *stare*), always present or always in the same condition.

CONSTRIC'TED (constrict + ed), contracted so as to be smaller in one or more places than in others.

CON'TEXT (*contextus*, pp. of *contexere*, join or weave together), texture; substance.

CONTIG'UOUS (*contiguus*, touching), near, or in contact.

CONTIN'UOUS (*continuus*, joined, <*continere*, hold together), without a break; applied to spores or hyphæ that have no septa.

CONTOR'TED (*contortus*, pp. of *contorquere*, twist, <*com*, together; *torquere*, twist), distorted, twisted, crooked or deformed.

CON'VEX (*convexus*, vaulted, arched, convex, concave), elevated and regularly rounded; forming the segment of a sphere or nearly so.

Glossary

Convex′o-Plane, between convex and flat.

Con′volute (*convolutus*, pp. of *convolvere*, roll together), covered with irregular convexities and depressions resembling the convolutions of the brain.

Cor′date (*cordatus*, heart-shaped), heart-shaped.

Coria′ceous (*coriaceus*, <*corium*, leather), of a leathery texture.

Cor′neous (*corneus*, horny), of a horny texture.

Cor′rugated (*corrugatus*, pp. *corrugare*, <*con*, together, + *rugare*, to wrinkle), wrinkled; contracted; puckered; having a wrinkled appearance.

Cor′tex (*cortex*, cork), literally bark; a covering of cells enclosing the axis; cortical layer; the outer rind-like layer or layers of some fungus bodies.

Cor′tical (*cortex* (*cortic*), bark, rind, + al), of or pertaining to the cortex.

Cor′ticate, Cor′ticated (*corticatus*, pp. adj., <*cortex*, bark), furnished with bark-like covering; having a rind.

Corti′na (*cortina*, a veil of spider-web structure rupturing at or near the stem; applied to the peculiar veil of the genus Cortinarius.

Cor′tinate (*cortinatus*, <*cortina*, a curtain), provided with or pertaining to a cortina.

Cos′tate (*costatus*, ribbed, <*costa*, rib), having a ridge or ridges as if ribbed.

Costæ (pl. of *costa*, a rib, a side), ribs or primary veins (as in a leaf).

Crate′ra (*crater*, a bowl), a cup-shaped receptacle.

Crater′iform (*crater*, a crater, + *forma*, shape), basin or saucer-shaped; having the form of a crater.

Cream Color (*cremeus*), a light pinkish-yellow color like cream.

Cre′nate (*crenatus*, <*crena*, a notch), notched at the edge, indented, scalloped: The notches are blunt or rounded, not sharp as in a serrated edge.

Cren′ulate, Cren′ulated (*crenulatus*), same as crenate.

Creta′ceous (*cretaceus*, chalky, <*creta*, chalk), chalky; of the color of chalk.

Crib′rate, Crib′riform (*cribrum*, a sieve, + *forma*, shape), sieve-like; perforated with small holes.

Crib′rose (*cribrosus*, <*cribrum*, a sieve), pierced with holes; perforated.

Crimson (*carmineus, sanguineus, sanguineo-ruber*), blood-red, the color of the cruder sorts of carmine.

Cri′nite (*crinitus*, haired, pp. of *crinire*, provide with hair, <*crinis*, hair), having a tuft of long, weak hairs.

Crisp, Crisped, Cris′pate (*crispus*, curled, wavy, uneven, tremulous), having the surface, especially near the margin, strongly and finely undulate, as the leaves of the Savoy cabbage.

Cris′tate (*cristatus*, <*crista*, a crest), crested; bearing a ridge, mane or tuft on the top.

Cru′ciate, Cru′ciform (*cruciatus*, pp. of *cruciare*, torture), having the form of a cross with equal arms.

Crusta′ceous (*crusta*, a crust, + aceous), of hard and brittle texture.

Cryp′togam (*Gr.*—hidden, + marriage), a plant of the order Cryptogamia.

Cryptoga′mia (*Gr.*—hidden, + marriage), flowerless plants propagated by spores.

Cryptog′amy (*Gr.*—hidden, + marriage), obscure fructification as in plants of the class Cryptogamia.

Culm (*culmus*, a stalk), the stem of grasses.

Cu′neate, Cune′iform (*cuneatus*, pp. of *cuneare*, wedge, make wedge-shaped, <*cuneus*, a wedge), wedge-shaped.

CUP (*cupa*, a tub, cask, vat), the concave fruiting body of angiocarpous lichens and discomycetous fungi; the peridium of a clustering fungus. See APOTHECIUM.

CU′PREOUS (*cupreus*, of copper, <*cuprum*, copper), copper-colored.

CU′PULAR, CU′PULATE (a little cup, dim. of *cupa*, a cup), cup-shaped.

CU′PULE, CU′PULA (a little cup, dim. of *cupa*, a cup), a receptacle shaped like a little cup, as in Peziza.

CURLED, same as CRISP.

CURT (*curtus*, clipped, broken, shortened), short.

CUR′TAIN (*cortina*, a small croft, screen, etc.), same as cortina.

CUS′PIDATE (*cuspidatus*, pp. *cuspidare*, <*cuspis*, a point, spear), with a sharp spear-like point.

CU′TICLE (*cuticula*, dim. of *cutis*, the skin), a distinct skin-like layer; cutis, cuticle, pellicle and epidermis have been used indiscriminately to describe the separable or inseparable skin-like layer sometimes present on the outer surface of the pileus and stem; of these terms, cuticle is used most commonly.

CU′TIS. See CUTICLE.

CYA′NEOUS (*cyaneus*, dark blue), bright blue; azure; lapis-lazuli blue.

CYANO′SIS (*Gr.*—dark-blue, + osis), in pathology a blue or more or less livid color of the surface of the body, due to imperfect circulation and oxygenation of the blood.

CY′ATHIFORM (*cyathus*, a cup; *forma*, form), cup-shaped, shape of a drinking glass slightly widened at the top.

CYLIN′DRIC, CYLIN′DRICAL (*cylindricus*, cylinder), cylinder-shaped; applied to a branch or stem having the same or nearly the same diameter throughout, and its cross-section circular.

CYM′BÆFORM, CYM′BIFORM (*cymba*, a boat, + *forma*, shape), boat-shape.

CYST (*cystis*, the bladder, bag, pouch), a bladder-like cell or cavity.

CYSTID′IUM (pl. CYSTID′IA) (*Gr.*—the bladder, + the dim. termination), sterile bladder cells of the hymenium, generally larger than the basidia cells between which and with which they are formed.

DASH, -, between two figures = to; from 2 to 4.

DAUGHTER-CELL, any cell when mentioned in relation to the one (mother-cell) from which it is derived.

DEAL′BATE (*dealbatus*, pp. of *dealbare*, whiten, white-wash, etc., <*de* + *albare*, whiten), as if white-washed; covered with very white opaque powder.

DECID′UOUS (*deciduus*, that falls down; <*decidere*, <*de*, down, + *cedere*, to fall), falling off at maturity or at the end of the season, not permanent; losing the foliage every year.

DECOR′TICATE, DECOR′TICATED (*decorticatus*, pp. of *decorticare*, <*de*, from, + *cortex*, bark), denuded of bark; destitute of a cortex or cortical layer.

DECUM′BENT (*decumben(t-)s*, ppr. of *decumbere*, lie down, <*de*, down, + *cumbere*, lie) applied to a stem having the lower part resting on the ground.

DECUR′RENT (*decurren(t-)s*, ppr. *decurrere*, run down), applied to lamellæ (gills) which are prolonged down the stem.

DECURVED′ (decurve + ed, after *decurvatus*, curved back), curved downward; opposed to recurved.

DEFLEXED′ (*deflexus*, pp. *deflectere*, turn aside), bent or turned down.

Glossary

DEHIS'CENCE (*dehiscen(t-)s*, dehiscent), the spontaneous opening of a peridium at maturity to discharge the spores.

DEHIS'CENT (*dehiscere*, gape, open), a closed organ opening of itself at maturity or when it has attained a certain development.

DELIQUES'CENT (*deliquescere*, melt away), relating to mushrooms which at maturity become liquid or melt down.

DELIMITA'TION (*delimitare*, mark out the limits, <*de* + *limitare*, limit, bound), the marking, fixing or prescribing the limits or boundaries.

DEN'DROID, DEN'DRIFORM (*Gr.*—a tree, + *L.*—*forma*, form), tree-shaped.

DEN'TATE (*dentatus*, tooth), toothed with a concave serrature.

DENTIC'ULATE (*denticulatus*, <*denticulus*, a small tooth), finely dentate.

DENU'DATE (*denudatus*, pp. of *denudare*, make bare, strip), naked; exposed, not immersed.

DEPRESSED' (*depressus*, pp. of *deprimere*, <*de*, down, + *premere*, press), as if pressed down or flattened; sunk below the level of the surrounding margin.

DERMINI, a group of fungi with brown or rust-colored spores.

DESCEND'ING (*descindere*, pp. *descensus*, come down, fall, <*de*, down, *scandere*, climb), applied to a marginal veil when, in the young stage, its marginal attachment is below the level of its stem-attachment; a ring formed from it is called superior; turned downward.

DES'ICCATE, DES'ICCATED (*desiccatus*, pp. of *desiccare*, dry up, <*de*, intensive, <*siccare*, dry), dried.

DETER'MINATE (*determinatus*, pp. *determinare*, fix, limit), ending definitely; having a distinctly defined outline.

DETERMINA'TION (*determinatio(n)*, boundary, conclusion, end, *determinare*, pp. *determinatus*, bound, determine), assignment to the proper place in a classification or series.

DIAGNO'SIS (*diagnosis*, a distinguishing), scientific discrimination of any kind; a short distinctive description, as of a plant.

DIAPH'ANOUS (*Gr.*—through, + to appear), of a transparent texture; permitting the passage of light.

DICHOT'OMOUS (*Gr.*—in two, + to cut), dividing into two; regularly forked.

DICHOT'OMY (*Gr.*—a cutting in two), a mode of branching by constant forking or dividing in pairs.

DID'YMOUS (*Gr.*—double, twofold, twin), double; of two equal parts.

DIFFEREN'TIATED (*differentia*, difference), exhibiting differentiation.

DIFFERENTIA'TION (*differentia*, difference, + ation), (a) discrimination between by observing or describing the differences; (b) the evolutionary process or results by which originally different parts or organs become differentiated or specialized in either form or function; specialization.

DIF'FLUENT (*diffluen(t-)s*, ppr. of *diffluere*, <*dis*, away, apart, + *fluere*, flow), readily dissolving.

DIF'FORM, DIFFORMED' (*deformis*, deformed), irregular in form, not uniform.

DIFFUSE' (*diffusus*, pp. of *diffundere*, pour in different directions, pour out, <*dis*, away, + *fundere*, flow), spreading widely, loosely and irregularly.

DIG'ITATE (*digitatus*, having fingers or toes, <*digitus*, finger), furnished with fingers; dividing like the fingers of the hand.

DILA'TED (*dilatare*, spread out; extend), expanded; enlarged.

662

DIMID′IATE (*dimidiatus*, <*dimidiare*, halve), halved; *e. g.* of gills which reach half-way to the stem; also of pileus when it is semi-circular in outline or nearly so; as many Polyporei.

DIMOR′PHIC, DIMOR′PHOUS (*dimorphus*, having two forms), existing in two distinct forms.

DIMOR′PHISM (*dimorphus*, having two forms), the property of existing under two distinct forms.

DISC, DISK (*discus*, a disk, trencher), (a) any flat circular disk-like growth; (b) the central portion of the upper surface of a pileus; the cup-shaped or otherwise variously shaped hymenial surface of a Discomycete.

DIS′CIFORM, DIS′COID, DIS′COIDAL (*discoides*, disk-shaped), of a circular, flat form; disk-shaped.

DIS′COCARP (*Gr.*—a disk, + *Gr.*—fruit), ascocarp in which the hymenium or disk lies exposed while the asci are maturing as in Peziza, Morchella, etc.

DISCOMYCE′TES (*Gr.*—a disk, + *Gr.*—fungus), a group of ascomycetous fungi in which the hymenium is exposed; the fruiting body is cupular, discoid or clavate, and sometimes convoluted.

DISCRETE′ (*discretus*, distinguished, separated), distinct, not coalescent.

DISSEC′TED (*dissectus*, pp. of *dissecare*, cut asunder, <*dis*, asunder, + *secare*, cut), cut deeply into many lobes or divisions.

DISSEP′IMENTS (*dissepimentum*, a partition) dividing walls; partitions.

DIS′TAL (*dist(ance)* + al), pertaining to the apex or outer extremity.

DIS′TANT, (*distans*, ppr. *distare*, stand apart), far apart; of gills which have a wide distance between them.

DIS′TICHOUS (*Gr.*—having two rows), disposed in two rows.

DIVAR′ICATE (*divaricatus*, pp. *divaricare*, spread asunder), separating at an obtuse angle; diverging widely.

DOR′SAL (*dorsalis*, <*dorsum*, the back), pertaining to the back, literally on the upper side.

DOWN, fine, soft pubescence.

E or EX·, prefix signifying "destitute of," "outside of," or "away from."

EBE′NEOUS (*ebeneus*, of ebony, <*ebenus*, ebony), black like ebony.

EBUR′NEOUS (*eburneus*, of ivory, <*ebur*, ivory), ivory-white.

ECCEN′TRIC (*Gr.*—out of the center), excentric.

ECH′INATE (*echinatus*, set with bristles, prickly), furnished with stiff bristles.

ECHIN′ULATE (*echinulus*, dim. of *echinus*, a hedgehog), beset with short bristles.

ECTO- (*Gr.*—without, outside), prefix signifying "outside."

ECTOBASID′IA (*Gr.*—outside, + basidium), basidia placed on an exposed surface; not enclosed.

EDEN′TATE (*edentatus*, toothless, pp. of *edentare*, render toothless), without teeth.

EFFUSED′ (*effusus*, pp. *effundere*, pour cut), spread over without regular form.

EFFU′SO-REFLEXED′, effused with upper margin reflected forming a pileus.

EGG (*ovum*, an egg), a young plan tbefore rupture of the volva in Phalloids, Amanitas, etc.

EGUTT′ULATE, not containing guttulæ.

ELLIP′SOID (*Gr.*—ellipse, + *Gr.*—form), a solid figure all plane, sections of which are ellipses or circles.

ELLIPSOI′DAL, shaped like an ellipsoid.

E<small>LLIP</small>′<small>TIC</small>, E<small>LLIP</small>′<small>TICAL</small> (*Gr.*—ellipse), elongate-ovate; more than twice as long as broad; parallel-sided in the middle and rounded at both ends.

E<small>MAR</small>′<small>GINATE</small> (*emarginatus*, pp. *emarginare*, <*e*, out of; *margo*, the margin), notched at the end; of gills with a sudden scoop, as if scooped out at the point of attachment to the stem.

E<small>MBOSSED</small>′, in botany projecting in the center like the boss or umbo of a round shield.

E<small>M</small>′<small>BRYO</small>, the mushroom before leaving its volva, also an early stage of mushrooms which have no volva.

E<small>NCRUST</small>′<small>ING</small> (*incrustare*, cover with a rind or crust, <*in*, on, + *crusta*, a crust).

E<small>NDEM</small>′<small>IC</small> (*Gr.*—native), peculiar to and characteristic of a locality or region; indigenous in some region and not elsewhere.

E<small>NDO</small>-, E<small>NTO</small>- (*Gr.*—"in," "within"), prefix signifying "within," "inside."

E<small>NDOBASID</small>′<small>IA</small> (*Gr.*—within, + basidium), basidia enclosed in a dehiscent or indehiscent conceptacle.

E<small>NDOCAU</small>′<small>LOUS</small> (*Gr.*—within, + *caulis*, a stalk), growing in the substance of herbaceous stems.

E<small>N</small>′<small>DOGEN</small> (*Gr.*—within, + producing). See M<small>ONOCOTYLEDON</small>.

E<small>NDOG</small>′<small>ENOUS</small> (*Gr.*—within, + *Gr.*—producing, + ous), produced within another body; of or pertaining to the class of endogens.

E<small>NDOPERID</small>′<small>IUM</small> (*Gr.*—within, + peridium), inner layer of the peridium.

E<small>N</small>′<small>DOPHYTE</small> (*Gr.*—within, + a plant), a plant growing within an animal or another plant, usually as a parasite; entophyte.

E<small>N</small>′<small>DOSPORE</small>, E<small>NDOSPO</small>′<small>RIUM</small> (*Gr.*—within, + *Gr.*—seed), (a) the inner coat of a spore; (b) spore which is produced within a sporangium or spore-sac as the ascospores.

E<small>N</small>′<small>SIFORM</small> (*ensis*, a sword, + *forma*, shape), sword-shaped.

E<small>NTIRE</small>′ (*integer*, <*integrum*, whole), the edge quite devoid of serrature or notch; continuous.

E<small>NTOMOG</small>′<small>ENOUS</small> (*Gr.*—an insect, + produced), growing upon or in insects.

E<small>NTOMOPH</small>′<small>YTOUS</small> (*Gr.*—within, + grow), growing upon or in insects.

E<small>N</small>′<small>TOPHYTE</small>, endophyte.

E<small>PIDER</small>′<small>MIS</small> (*Gr.*—the outer skin), the external or outer layer of the plant.

E<small>PIG</small>′<small>ENOUS</small> (*Gr.*—growing after or late), growing upon the surface of a part; often limited to growth upon the upper surface, in distinction from hypogenous.

E<small>PIGÆ</small>′<small>OUS</small>, E<small>PIGE</small>′<small>OUS</small> (*Gr.*—on or of the earth; on the ground), growing on or in the ground.

E<small>P</small>′<small>IPHRAGM</small> (*Gr.*—a covering; lid), a delicate membrane closing the cup-like receptacle of the Nidulariaceæ.

E<small>P</small>′<small>IPHYTAL</small>, E<small>PIPHYT</small>′<small>IC</small>, E<small>PIPHYT</small>′<small>ICAL</small> (*Gr.*—upon, + a plant), of the nature of an epiphyte.

E<small>P</small>′<small>IPHYTE</small> (*Gr.*—upon, + a plant), growing upon the outside of another plant; either parasitic or not.

E<small>P</small>′<small>ISPORE</small>, E<small>PISPO</small>′<small>RIUM</small> (*Gr.*—upon, + seed), the outer coat of a spore; same as exosporium.

E<small>PITHE</small>′<small>CIUM</small> (*Gr.*—upon, + a case), the layer sometimes formed above the asci by the concrescent tips of the paraphyses.

E<small>PIX</small>′<small>YLOUS</small> (*Gr.*—upon, + wood + ous), growing upon wood.

E′QUAL (*æqualis*, equal, like), all gills of the same, or nearly the same, length from back to front; stem of uniform thickness.

ERO′DED (*erodere*, gnaw off), the edge ragged as if torn.

ERUM′PENT (*erumpen(t-)s*, ppr. of *erumpere*, break out), prominent; originating beneath and bursting through the surface of the matrix.

E′TIOLATE, E′TIOLATED (*stipula*, straw), whitened, blanched by exclusion of the sun's rays or by disease.

EVANES′CENT (*evanescen(t-)s*, ppr. of *evanescere*, vanish away), fleeting; vanishing; soon disappearing.

E′VEN, of a surface which is quite plane as contrasted e. g. with one which is striate, pitted, etc. Distinguished from smooth. A surface may not be smooth and yet be even.

EX-, prefix. See "E-."

EXCEN′TRIC (*Gr.*—out of the center), not central; the stems of some mushrooms are always excentric.

EXCIP′ULUM (*excipulum*, a vessel for receiving liquids, <*excipere*, take out, receive), outer layer of an apothecium or cup developed as part of the receptacle.

EXO-, prefix signifying "outside."

EX′OGEN (*Gr.*—outside, + producing), a plant in which the growth of the stem is in successive concentric layers.

EXOG′ENOUS, growing by additions on the outside; belonging to or characteristic of the class of exogens; produced on the outside, as the spores of hyphomycetous and many other fungi.

EXOPERID′IUM (*Gr.*—outside, + peridium), outer layer of the peridium.

EX′OSPORES (*exosporium*), spores which are free, not produced within a sporangium, as basidio-spores.

EXOSPO′RIUM (*Gr.*—outside, + seed), the outer coat of a spore; same as episporium.

EXOT′IC (*exoticus*, foreign, alien), foreign, not native.

EXPAN′DED (*expandere*, pp. *expansus*, spread out, <*ex*, out, + *pandere*, spread), spread out, as a pileus from convex to plane.

EX′PLANATE (*explanatus*, flattened, spread out), flattened, expanded; applied usually to a part which has been rolled or folded.

EXSER′TED (*exsertus*, thrust out, pp. of *exsere*, stretch out), projecting; standing out.

EXSICCA′TI (*exsiccatus*, pp. of *exsiccare*, dry up), dried specimens; especially those published in sets and distributed.

FAC′ULTATIVE (*faculta(t-)s*, faculty), capability, etc., having a faculty or power, but exercising it only occasionally or incidentally; optional or contingent.

FAC′ULTATIVE-PAR′ASITE, an organism which normally lives throughout as a saprophyte, but which may also go through its course either wholly or in part as a parasite.

FAC′ULTATIVE-SAP′ROPHYTE, an organism which normally is parasitic, but which can vegetate at certain stages as a saprophyte.

FAL′CATE, FAL′CIFORM (*falcatus*, bent, curved, hooked, sickle-shaped, <*falx*, a sickle), hooked, curved like a scythe or sickle.

FAMILY (*familia*, household establishment, <*famulus*, a servant), a systematic group in a scientific classification embracing a greater or less number of genera which agree in certain characters not shared by others of the same order.

FARC′TATE (*farctus*, <*farcio*, to stuff), stuffed; without vacuities; opposed to fistulose.

Farina′ceous (*farinaceus,<farina*, meal), mealy.

Far′inose (*farinosus*, mealy), covered with a white mealy powder.

Fas′cia (*fascis*, a bundle), a band or bar.

Fas′ciate, Fas′ciated (*fascia*, a band or girth), having broad parallel bands or stripes; banded or compacted together; exhibiting fasciation.

Fascia′tion (*fascia*, a band), the act or manner of binding with fasciæ, a monstrous flattened expansion of the stem; condition of being bound or compacted together.

Fas′cicle, Fascic′ulus (*fasciculus*, a small bundle; packet, etc.), a close cluster; a small bundle.

Fascic′ulate (*fasciculus*, a small bundle), growing in small bundles or fascicles.

Fastig′iate (*fastigiatus*, sloping, <*fastigium*, the top of a gable, slope), with branches erect and close together; sloping upward to a summit, point or edge.

Favose′ (*favosus*, <*favus*, a honey-comb), honey-combed; resembling a honey-comb.

Fawn-Color (*cervinus, cervineus*), a light warm-brown color.

Ferru′gineous, Ferru′ginous (*ferrugineus*), rust-red or the color of iron rust.

Fi′brillar, Fi′brillate, Fi′brillose, Fi′brillous (*fibrilla*, a fiber), appearing to be covered or composed of minute fibers.

Fi′brous (*fibrosus*, <*fibra*, a fiber), clothed with small fibers.

Fi′brous-Myce′lium, Fibrillose-Mycelium, elongated branching mycelial strands, formed by the union of hyphæ.

Fig′urate (*figuratus*, pp. *figuare*, <*figura*, a form, shape), of a certain determinate form or shape.

Fil′ament (*filum*, thread), a separate fiber or fibril of any animal or vegetable tissue, as a filament of silk, wool, etc.

Filamen′tous, like a thread; composed of threads or filaments.

Filamen′tous-Myce′lium, Floc′cose-Myce′lium, mycelium of free hyphæ which are at most loosely interwoven, but without forming bodies of definite shape and outline.

Fil′aceous, Fil′iform (*filum*, a thread, + aceous), like a thread or filament.

Fim′briate, Fim′briated (*fimbriatus*, <*fimbriæ*, a fringe), fringed; cut jaggedly.

Fis′sile (*fissilis*, cleft, <*fissus*, pp. *findere*, split), capable of being split, cleft or divided in layers.

Fis′sured (*fissura*, a cleft, chink, fissure), cleft or split.

Fis′tular, Fis′tulose (*fistularis*, like a pipe, <*fistula*, a pipe), tubular, hollow in the center like a pipe.

Fixed, said of lamellæ or spines not readily detached from the underlying tissue.

Flabel′late, Flabel′liform (*flabellum*, a fan), fan-shaped.

Flac′cid (*flaccidus*, flabby, pendulous), soft and limber; flabby; without firmness or elasticity.

Flaves′cent (*flavescens*, ppr. *flavescere*, become yellow, <*flavus*, yellow), yellowish or turning yellow.

Fla′vous (*flavus*, golden-yellow, reddish-yellow), yellow.

Flesh, inner substance of a fungus-body as distinguished from the cortical and hymenial layers.

Flesh-Color (*carneus, incarnatus*), a pinkish-color like that observable in the cheeks of a person of fair complexion; carnation.

Flesh′y, succulent; composed of juicy cellular tissue.

Flex′uose, Flex′uous (*flexuosus*, <*flexus*, a bending, winding), wavy.

Floc′ci (pl. of Floc′cus) (*floccus*, a lock of wool), woolly locks.

Floc′cose (*floccosus*, <*floccus*, a lock of wool), downy, woolly; composed of or bearing flocci.

Floc′culose (*flocculosus*, <*flocculus*, dim. of *floccus*, a lock of wool), covered with flocci; composed of or bearing minute flocci.

Folia′ceous (*foliaceus*, leafy, of leaves, <*folium*, a leaf), leaf-like; bearing leaves.

Fo′veate (*foveatus*, <*fovea*, a small pit, pitfall), marked with pits or depressions.

Fov′eolate (*foveolatus*, <*foveola*, dim. of *fovea*, a small pit), marked with minute pits or depressions.

Free, said of gills which are not attached to the stem; said of any part not attached to another; of spores not inclosed in a special envelope.

Frill, same as Armilla.

Front, same as Anterior.

Fringe (*fimbria*, a border), a lacerated, marginal membrane.

Fruc′tification (*fructificare*, bear fruit), reproducing power of a plant; fruiting; also the organs concerned.

Fuga′cious (*fugar*, <*fugere*, flee), fleeting, transitory; falling or fading early.

Fu′gitive (*fugitivus*, fleeing away; a fugitive), quickly disappearing; evanescent.

Fulig′ineous, Fulig′inous (*fuliginosus*), sooty-brown or dark smoke-color.

Fulves′cent (*fulvescens*), inclining to a fulvous color.

Ful′vous (*fulvus*), a rather indefinite brownish-yellow or yellowish-brown tint, like tanned leather; tawny.

Fu′mose, Fu′mous (*fumosus*, full of smoke, <*fumus*, smoke, steam), smoke-colored, fuliginous.

Fun′goid (*fungus*, mushroom, + *Gr.*—form), of, or pertaining to fungi.

Fungol′ogy (*fungus*, mushroom, + *Gr.*—speak), mycology.

Fun′gus (pl. Fun′gi) (*fungus*, a mushroom), a thallophyte characterized by the absence of chlorophyl and deriving its sustenance from living or dead organic matter.

Funic′ular (*funiculus*, a small cord), having the character of a funicle or small cord.

Funic′ulate (*funiculus*, a small cord), having a funicle.

Funic′ulus (*L.*—a small rope), in Nidulariaceæ the cord of hyphæ attaching a peridiolum to the inner wall of the peridium.

Fur′cate (*furcatus*, <*furca*, a fork), forked.

Furfura′ceous (*furfuraceus*, <*furfur*, bran), with branny scales or scurf.

Fusces′cent (*fuscus*, dark, dusky, + escent), somewhat fuscous.

Fus′cous (*fuscus*, dusky), brownish in color; brown or brown tinged with gray; dingy, not pure.

Fu′siform, Fu′soid (*fusus*, a spindle; *forma*, form), spindle-shaped.

Gamogen′esis (*Gr.*—marriage, + generation), sexual reproduction.

Gas′teromyce′tes, Gas′tromyce′tes (*Gr.*—stomach, + mushroom), a group of Basidiomycetes in which the hymenium is enclosed in a sack-like envelope called the peridium.

Gelat′inous (*gelatinosus*, <*gelatina*, gelatine), jelly-like.

Gener′ic (*genus*, race, sort), pertaining to, of the nature of, or forming a mark of a genus; having the rank or classificatory value of a genus.

Genet′ic (*Gr.*—generation), of or pertaining to origin or mode of production.

Glossary

GE′NUS (pl. GEN′ERA) (*L.*—race, birth, origin, kind), a group of species having one or more characteristics in common; the union of several genera presenting the same features constitute a tribe.

GIB′BOUS (*gibbus*, hump-backed), in the form of a swelling; of a pileus *e. g.* which is more convex or tumid on one side than the other.

GILLS, the plates of an agaric on which the hymenium is situated; the lamellæ.

GIL′VOUS, isabelline; color of sole-leather.

GLA′BROUS (*glaber*, smooth), smooth, devoid of pubescence; a surface may be glabrous or smooth, and not even, or vice versa.

GLAIR (*clarus*, clear), any viscous transparent substance resembling white of an egg.

GLANDS, GLAN′DULES (*glans*, an acorn, dim. *glandula*, a gland), moist or sticky dots resembling the glands on the epidermis of phenogams.

GLAN′DULAR, bearing glands.

GLAUCES′CENT (*glaucescen(t-)s*, <*glaucus*, silvery, gleaming), inclining to glaucous.

GLAU′COUS (*glaucus*, silvery, gleaming), covered with a whitish-green bloom or very fine white powder easily rubbed off. Somewhat like that of cabbage.

GLE′BA (*gleba*, a clod), in Gastromycetes, spore-bearing tissue composed of chambers lined with the hymenium and enclosed by the sack-like peridium, as in puff-balls, etc.; in phalloids the peridium or volva ruptures and the gleba is carried up on the stem-like or clathrate receptacle.

GLO′BOSE, GLOB′ULAR, GLOB′ULOSE (*globosus*, round as a ball), nearly spherical.

GLU′TINOSE, GLU′TINOUS (*glutinosus*, gluey, viscous, <*gluten*, glue), covered with a sticky exudation; viscous; glue-like.

GONID′IUM (*Gr.*—generation; seed), same as conidium; also preferably applied to the algal element of lichens.

GRAN′ULAR, GRAN′ULATE, GRAN′ULOSE (*granula*, dim. of *granum*, grain), covered with or composed of granules.

GRAN′ULE (*granula*, dim. of *granum*, grain), a little grain; a fine particle; a sporule found in all cryptogamic plants.

GRAY (*griseus; cœsius; cinereus; canus; leucophœus*), a color produced by the mixture of black and white. Various shades depending upon varying relative proportions of the components.

GREAVED (*greve*, the shin-bone), of a stem clothed like a leg in armor.

GREGA′RIOUS (*gregarius*, of a flock), of mushrooms not solitary but growing together in numbers in the same locality; in groups but not in a tufted manner.

GRU′MOUS (*grumosus*, <*grumus*, a little heap), clotted; of flesh *e. g.* composed of little clustered grains.

GUT′TATE (*guttatus*, <*gutta*, a tear), marked with tear-like spots or drops.

GUTT′ULA (pl. GUTTULÆ) (dim. of *gutta*, a drop), a small drop or drop-like particle; the oil-globule in some spores resembling a nucleus.

GUTT′ULATE, finely guttate; also, containing or composed of fine drops or drop-like particles; said of spores containing an oily nucleus-like globule or guttula.

GYMNOCAR′POUS (*Gr.*—naked, + *Gr.*—fruit), having the hymenium exposed when the spores are maturing.

GY′RATE, GY′ROSE (*Gr.*—a circle), circling in wavy folds; having folds resembling the convolutions of the brain.

HAB′ITAT (*habitat*, it dwells), natural abode of a vegetable species.

HAUSTO′RIUM (pl. HAUSTO′RIA) (*haustor*, a drawer, <*haurire*, pp. *haustus*, draw), special branch of filamentous mycelium, which serves as an organ of adhesion and suction.

HEMIANGIOCAR′POUS (*hemi*, half, + *Gr.*—a vessel, a case), partly angiocarpous as those agarics where the hymenium is at first enclosed by a veil or otherwise and later becomes exposed.

HEPAT′IC (*hepaticus*, of the liver), pertaining to the liver, hence liver-colored; brownish-red.

HERBIC′OLOUS, growing on herbaceous plants.

HETEROGE′NEOUS (*Gr.*—one of two), of a structure which is different from adjacent ones.

HIBERNAC′ULUM (pl. HIBERNAC′ULA) (winter residence, <*hibernare*, pass the winter), applied to bodies which are the forms in which certain fungi (*e. g.* Typhulæ) pass the winter.

HIRSUTE′ (*hirsutus*, rough, shaggy, bristly), hairy with stiff hairs.

HIR′TO-VER′RUCOSE, bearing hairs grouped in wart-like masses.

HIS′PID (*hispidus*, rough, shaggy, bristly), having strong hairs or bristles; bristly.

HOAR′Y, covered with short dense grayish-white hairs; canescent.

HOLO-, (*Gr.*—entire, complete in all parts), a prefix signifying entire; whole.

HOMOGE′NEOUS (*Gr.*—one and the same, + kind), similar in structure; of the same character.

HOST, the name given to any plant or animal supporting a parasitic fungus.

HOMOL′OGOUS (*Gr.*—agreeing, correspondent), having the same relative position, proportion, value or structure; having correspondence or likeness.

HU′MUS (earth, ground, soil), vegetable mold; woody fiber in a state of decay.

HY′ALINE (*Gr.*—clear), colorless; transparent; clear like glass.

HYGROMET′RIC (*Gr.*—wet, moist, + a measure, + ic), readily absorbing and retaining moisture.

HYGROPH′ANOUS (*Gr.*—moist; *Gr.*—to show), of a watery appearance when moist and opaque when dry.

HYGROSCOP′IC, having the property of absorbing moisture from the atmosphere; sensitive to moisture.

HYME′NIUM (*hymenial*, belonging to the hymenium; *Gr.*—a membrane), the fruit-bearing surface; *e. g.* covering intimately each side of the gills of an Agaric.

HY′MENOMYCE′TES (*Gr.*—a mushroom, + *Gr.*—a membrane), a group of Basidiomycetes having the hymenium on the free, exposed surface of the sporophore.

HY′MENOPHORE, HYMENOPH′ORUM (*Gr.*—a membrane, + to bear), the structure which bears the hymenium; in Agarics *e. g.* the under surface of the pileus to which the gills are attached.

HY′PHA (pl. HY′PHÆ), the elementary filament or thread of a fungus; a cylindric thread-like branched body developing by apical growth, and usually becoming transversely septate.

HY′PHAL, of or pertaining to the hypha.

HYPOCRATER′IFORM (*Gr.*—the stand of a crater, + *forma*, form), having the shape of a cylindrical cup the margin of which turns outward; salver-shaped.

HYPOGÆ′OUS, HYPOGE′AL, HYPOGE′OUS (*hypogœous*, underground), subterranean; forming below the surface of the ground.

HYPOG′ENOUS (*Gr.*—under, + produced, + ous), growing on the under surface.

HYPOPHYL′LOUS (*folium*, a leaf, + ous), growing on the under side of a leaf.

Glossary

HYPOTHE′CIUM (*Gr.*—under, a case), layer of hyphal tissue immediately beneath a hymenium.

IDENTIFICATION, the determination of a genus and species to which a given specimen belongs.

IM′BRICATE, IM′BRICATED (*imbricatus*, pp. of *imbricare*, cover with gutter tiles; form like a gutter tile), to lay or lap one over another, like shingles.

IMMAR′GINATE (*in*, negative; *marginatus*, marginate), without a well-defined margin.

IMMERSED′ (*immersus*, pp. of *immergere*, dip or plunge into), sunk into the matrix; originating beneath the surface of the matrix or of the ground; growing wholly under water.

IMPER′FORATE (*in*, not, + *perforatus*, pp. of *perforare*, perforate), without any aperture.

INCANES′CENT (*incanescen*(*t-*)*s*, ppr. of *incanescere*, become gray or hoary), somewhat or slightly canescent.

INCAR′NATE (*in*, in, on, + *caro* (*carn*), flesh), flesh-colored.

INCISED′ (*incisus*, pp. *incindere*, cut into), appearing as if cut into; having marginal slits or notches.

INCRAS′SATED (*incrassatus*, pp. of *incrassare*, <*in*, in; *crassare*, make thick), becoming thicker by degrees, swelling or swollen.

INCRUS′TING (*incrustare*, cover with a rind or crust, <*in*, on, + *crusta*, a crust), forming a crust-like coating.

INDEHIS′CENT, applied to a peridium which does not open spontaneously at maturity; the spores within it becoming freed by its decay.

INDIF′FERENT, primitive, homogenous, not developed into parts or organs of different structure or function.

INDIG′ENOUS (*indigena*, a native), native of a country.

INDIGO BLUE, a dark blue-color like the indigo of commerce.

IN′DURATED (*induratus*, pp. of *indurare*, harden, <*in*, in, + *durare*, harden), hardened.

INDU′SIUM (*L.*—a tunic, <*induere*, put on), in certain phalloids, an appendage or veil hanging from the apex of the stem beneath the pileus.

INFE′RIOR (*inferior*, lower), growing below some other part; of the ring of an Agaric which is far down on the stem.

INFLA′TED, swollen like a bladder.

INFLEXED′ (*inflexus*, pp. *inflexere*, bent), bent inward.

INFUNDIB′ULIFORM (*infundibulum*, a funnel; *forma*, form), funnel-shaped.

INHIBIT′ION (*inhibitio*(*n-*), a restraining, <*inhibere*, restrain), the lowering of the action of a nervous mechanism by nervous impulses reaching it from a connected mechanism.

IN′NATE (*innatus*, pp. of *innasci*, <*in*, into; *nascor*, to be born), originating within the substance of the plant or matrix; appearing to be within or blending with the substance of a part.

INORGAN′IC, not produced by vital processes; not organic.

ISABELLA COLOR (*alutaceus*), a light grayish-cinnamon color, or light buff-brown.

INSER′TED (*insertus*, pp. *inserere*, to insert), growing like a graft from its stock; attached to or growing out of some other part.

INSITI′TIOUS (*insitio*, an ingrafting, <*inserere*, pp. *insitus*, sow or plant, ingraft), inserted.

INTER- (*L.*—in the midst, between, among, during), prefix signifying "between" or "among" or "during."

INTER´CALARY, INTER´CALATED (*intercalatus*, pp. of *intercalare*, < *inter*, between, + *calare*, call), interposed; inserted between.

INTERCELL´ULAR (*inter*, between, + *cellula*, cellule, + ar), situated between the cells.

INTERRUPT´ED, said of any surface or series the continuity of which is broken.

INTER´STICES, spaces between any surfaces or things.

INTRA- (*intra*, within), prefix signifying "within."

INTRACELL´ULAR, situated within a cell or cells.

INTRALAM´ELLAR, situated within or between the plates of the lamellæ (gills).

INTRAVENAL (*intra*, within, + *vena*, vein), situated or occurring within veins.

INTRODUCED´, applied to plants brought from another country and growing spontaneously.

INTUMES´CENT (*intumescens*, ppr. of *intumescere*, swell up), swelling up, becoming tumid.

INVAG´INATED (*in*, in, + *vagina*, a sheath), sheathed.

IN´VOLUTE (*involutus*, pp. *involvere*, to roll up), rolled inwards.

ISABEL´LINE, of the color of soiled linen or sole leather; alutaceous; brownish-yellow, yellowish-gray.

LA´BIATE (*labiatus*, lipped, < *labium*, lip), said of an aperture with distinct lip-like borders.

LABYRIN´THINE, LABYRIN´THIFORM (*labyrinthus*, labyrinth), characterized by intricate and sinuous lines; like a labyrinth.

LAC´CATE (*laccatus*, < *lacca*, lac), as if varnished or covered with a coat like sealing wax.

LAC´ERATE, LAC´ERATED, as if torn.

LACIN´IATE (*laciniatus*, < *lacinia*, a lappet), divided into flaps; irregularly cut into jagged edges, more regular and larger than fimbriate.

LACTES´CENT (*lactescere*, turn to milk), milk-bearing, provided with a milky juice.

LACU´NA (pl. LACU´NÆ) (*lacuna*, a pit, hollow, cavity, etc.), a pit or hollow, a gap; a vacancy caused by the admission, loss or obliteration of something necessary to continuity or completeness.

LACU´NOSE, LACU´NOUS (*lacunosus*, full of hollows, < *lacuna*, a pit), marked with small hollows, pitted; having or full of lac.

LAMEL´LA (pl. LAMEL´LÆ) (*lamella*, a thin piece of metal, wood), a gill or gills of mushrooms, on which the hymenium is extended.

LA´NATE (*lanatus*, woolly, < *lana*, wool), woolly; covered with a wool-like pubescence.

LAN´CEOLATE (*lanceolatus*, < *lanceola*, a little spear), lance-shaped; tapering to both ends.

LAT´ERAL, attached to or by one side.

LATERIC´EOUS, LATERIT´IOUS (*latericeus, lateritius*, consisting of bricks, < *later*, a brick), brick-colored.

LA´TEX (*latex*, liquid; *ferre*, bear), thick milky juice.

LATICIF´EROUS (*latex*, liquid, + *ferre*, bear), applied to the tubes containing latex, as in the Lactarii.

LAT´TICED, formed by interlacing and crossing lines or columns which leave open spaces between.

Glossary

Lavender (*lavendulaceus*), a very pale purplish color, paler and more delicate than lilac.

Lax (*laxus*, loose, slack), not compact, limber, flaccid.

Lead-Color (*plumbeus*), same as Plumbeous.

Lemon-Yellow (*citreus, citrinus*), a very pure light-yellow color, much like gamboge, but purer and richer.

Lentic'ular, Len'tiform (*lenticularis*, lentil-shaped, <*lenticula*, a lentil), shaped like a double convex lens; lentil-shaped.

Lep'idote (*Gr.*—scaly), scurfy with minute scales.

Leucos'poræ (*Gr.*—white; *Gr.*—seed), a group of fungi having white spores, hence *leucospore*, a white spore; *leucosporous*, having spores of a white color.

Lev'igate (*levigatus*, pp. of *levigare*, make smooth), having a polished surface.

Lig'natile (*lignatilis*, <*lignum*, wood), growing on wood.

Lig'neous (*ligneus*, wooden), of woody texture.

Lig'ulate (*ligula*, a tongue, strap, etc., + ate), strap-shaped, flattened like a strap.

Lilac, Lilaceous (*lilacinus, lilaceus*), a light-purple color, like the flowers of the lilac.

Lin'ear (*linearis*, <*linea*, a line), narrow and straight, slender.

Ling'uiform, Ling'ulate (*lingua*, tongue), tongue-shaped.

Livid (*lividus*, black-and-blue), bluish-black, like the black and blue of a bruise.

Lobed, Lo'bate (*Gr.*—the lobe of an ear), having divisions which are large and rounded.

Lob'ulate, having small lobes.

Loc'ular, Loc'ulate, Loc'ulose, Loc'ulous (*loculus*, a box, cell), divided by internal partitions into loculi or cells.

Loc'ulus (pl. Loc'uli), a little chamber or cell.

Lu'cid (*lucidus*, light, bright, clear), clear, transparent, bright.

Lu'men (*lumen*, a window-light), the internal cavity or spaces in a cell or any tubular organ.

Lu'rid (*luridus*, pale-yellow, wan, etc.), a color between purple, yellow and gray; livid.

Lu'teous (*luteus*), yellowish; more or less like buff or clay color.

Lutes'cent (*lutescen(t-)s*, ppr. of *lutescere*, turn to mud, <*lutum*, mud), yellowish.

Macro- (*Gr.*—long), in composition "large" or "long."

Mac'ulate, Mac'ular, Mac'ulose (*maculatus*, pp. of *maculare*, spot, speckle), spotted.

Mam'miform (*mamma*, a breast; *forma*, form), breast-shaped; mastoid; teat-like.

Mar'ginal Veil, a horizontal membrane extending from the margin of the pileus to the stem; found in Hymenomycetes.

Mar'ginate, having a well-defined border.

Maroon (*atro-purpureus, atro-coccineus*), a rich brownish-crimson, nearly like the pigment called purple madder; claret color.

Ma'trix (*matrix*, a womb), the substance upon or in which a fungus grows.

Mauve (*malvaceus, malvinus*), a light tint of violet. (Aniline violet + white.)

Medial (*medialis*, <*medius*, the middle), applied to ring when situated about at the middle of stem.

Medul'la (*medulla*, marrow, pith, kernel,<*medius*, middle), pith, marrow, kernel; inner substance as distinguished from outer or cortical layer or layers.

MED′ULLARY, composed of or pertaining to a medulla.

MEGA- (*Gr.*—great, large), prefix signifying "great."

MELANOS′PORÆ (*Gr.*—black, —seed), a group of fungi having black spores.

MEMBRANA′CEOUS (*membranaceus*, of skin or membrane, <*membrana*, skin, membrane), pertaining to, or of the nature of, skin, membrane; membranaceous; thin, rather soft and pliable.

MERIS′MOID (*merisma*, from *Gr.*—to divide, + —form), applied to pileus which is subdivided into many smaller pilei; resembling a Merisma; having a branched or laciniate pileus.

MES′OPOD (*Gr.*—middle, + —foot), plant having a central stem.

MICA′CEOUS (*micaceus*, <*mica*, mica), covered with glistening mica-like particles.

MI′CRON, MI′KRON, μ, microscopic unit of measure; $\frac{1}{1000}$ of a millimeter; nearly .00004 inch; to convert inches to microns, approximately, divide by .00004; represented by the Greek letter μ, following the number.

MILLIMETER, MM. (*mille*, a thousand, + meter), the thousandth part of a meter, equal to 0.03937 inch or nearly $\frac{1}{25}$ inch. It is denoted by mm., as 25.4 mm. is 1 inch.

MIN′IATE (*miniatus*, pp. of *miniare*, color with red lead, <*minium*, red lead), vermilion-colored; of a bright, vivid red color.

MI′TRATE, MIT′RIFORM (*mitra*, a miter), miter-shaped, bonnet-shaped.

MOLD, MOULD, fine soft earth; a general term to describe certain fungus growths of a low type.

MONIL′IFORM (*monile*, necklace; *forma*, form), contracted at intervals in the length like a string of beads.

MONOS′TICHOUS (*Gr.*—single, + —a line), arranged in one row.

MON′STROUS, of unnatural formation; deviating greatly from the natural form or structure (has no reference to size).

MORPHOLOG′IC, MORPHOLOG′ICAL, of or pertaining to morphology.

MORPHOL′OGY (*Gr.*—form, + —speak), the science of organic form; the science of of outer form and internal structure.

MOTHER-CELL, a cell from which another is derived.

MOUSE-GRAY (*murino-griseus*; *murinus*). (Lamp-black + white + sepia.)

MOV′ABLE, applied to a ring which has separated from the stem and can be moved up and down.

MUCED′INOUS (*mucedo*, mucus), having the character of or resembling mold or mildew.

MU′CID (*mucidus*, moldy, <*mucere*, be moldy or musty, <*mucus*, mucus), musty, moldy, slimy.

MUCILAG′INOUS (*mucilago*, a moldy, musty juice, + ous), slimy, ropy, slightly viscid, soft, moist.

MU′COUS (*mucosus*, slimy, <*mucus*, slime), pertaining to mucus, or resembling it; slimy, ropy, lubricous.

MU′CRO (*L.*—a sharp point, esp. of a sword), a short and abrupt point of a leaf or other organ.

MU′CRONATE (*mucronatus*, pointed, <*mucro*, a sharp point), tipped with an abrupt, sharp short point.

MULTI- (*multus*, much, many), in composition "many."

MUL′TIFID (*multifidus*, many-cleft), having many divisions.

MULTIPAR′TITE (*multipartitus*, much divided), divided into many parts.

MULTISEP′TATE (as if *multiseptatus*, <*multus*, many, + *septum*, a partition), divided by many partitions.

Mu′ricate (*muricatus*, pointed), rough with short hard points.

Muric′ulate (*muriculatus*, dim. of *muricatus*, pointed), finely muricate.

Mu′riform (*murus*, wall, + *forma*, shape), resembling the arrangement of the bricks in the walls of a house; said of spores having septa at right angles to each other.

Mu′rine, Mu′rinous (*murinus*, of a mouse), mouse-colored.

Mush′room, a cryptogamic plant of the class fungi: applied in a general sense to almost any of the larger, conspicuous fungi, such as toadstools, puff-balls, hydnei, etc., but more particularly to the agaricoid fungi and especially to the edible forms.

Mu′tualism, symbiosis of two organisms living together and mutually helping and supporting each other.

Myc, Mycet, Myceto, Myco, prefix signifying "fungus."

Myce′lial, of or pertaining to mycelium.

Myce′lium (*Gr.*—a fungus, + an excrescence), spawn of fungi resulting from the germination of spores; in agarics *e. g.* forming root-like threads; the weft of threads from which the mushroom arises.

Myce′lioid, like mycelium.

Myc′eloid, like a fungus.

Mycetol′ogy (*Gr.*—a fungus, + —speak), mycology.

Mycolog′ical, relating to fungi.

Mycol′ogist, one who is versed in mycology.

Mycol′ogy (*Gr.*—a fungus, + —speak), the science of fungi, their structure, classification, etc.

Mycoph′agist, one who eats fungi.

Mycoph′agy (*Gr.*—a fungus, + —eat), the eating of fungi.

Na′ked, bare; without covering of any kind, as of an enveloping membrane, pruinose, farinaceous or furfuraceous particles, tomentum, fragments of volva or veil, etc.

Nap′iform (*napus*, a turnip, + *forma*, form), turnip-shaped.

Narrow, of a very slight vertical width.

Nas′cent (*nascen(t)s*, ppr. of *nasci*, be born), in the earliest rudimentary condition; beginning to exist or to grow.

Nat′uralized, said of a plant of foreign origin which thrives as if indigenous.

Navic′ular, Naviculoid (*navicula*, a small ship or boat), boat-shaped; scaphoid.

Netted, covered with projecting, reticulated lines.

Nigres′cent, Nig′ricant (*nigrescen(t-)s*, ppr. of *nigrescere*, become black, grow dark), becoming black, also blackish, dusky, fuscous.

Nit′id, Nit′idus (*nitidus*, shining, bright, <*nitere*, to shine), lustrous, shining, polished.

Ni′veous (*niveus*, snowy, <*nix*, snow), snow-white.

Nod′ule (*nodulus*, a little knot, dim. of *nodus*, a knot), a little knot or lump.

Nod′ulose, Nod′ulous (*nodulosus*, <*nodulus*, a little knot), having little swellings, knotty.

Non-, not; prefix giving a negative sense to words.

Nu′cleate, Nu′cleated (*nucleatus*, having a kernel), having a nucleus or nuclei.

Nucle′olus (pl. Nuclei) (*nucleolus*, dim. of *nucleus*, a little nut), sharply defined point often seen in the nucleus.

Nu′cleus (pl. Nu′clei) (*nucleus*, a little nut, kernel, stone of a fruit), the central, highly differentiated mass of protoplasm in a spore or other cells of a fungus, controlling cell division and reproduction, functionally the most important portion of a cell, for in it the process of cell division begins; sometimes improperly applied to the oil globules or guttulæ and the vacuoles within some spores.

Ob-, in composition "inversely."

Obcla′vate (as if *obclavatus*, <*ob*, from; *clava*, a club), inversely club-shaped.

Obcon′ic, Obcon′ical (*ob*; *conus*, a cone), inversely conical.

Obcor′date (*ob*; *cordatus*, heart-shaped), like an inverted heart.

Obese′ (*obesus*, fat), stout, plump.

Ob′ligate-Parasite, can only grow as a parasite; see facultative parasite.

Ob′ligate-Saprophyte, can only grow as a saprophyte. See Facultative Saprophyte.

Ob′long, two or three times longer than broad, with nearly parallel sides.

Obo′vate, inversely ovate, having the broad end upward or toward the apex.

Obpyr′iform (*ob* + *pyriform*, pear-shaped), inversely pear-shaped.

Ob′solete, indistinct, very imperfectly developed; hardly perceptible.

Obtuse,′ blunt or rounded.

Ochra′ceous, O′cherous, O′chreous, O′chroid, O′chry, O′chrous, O′chery (*ochre*, ocher, + aceous), ocher-yellow; brownish-yellow.

Ochra′ceous-Rufous (*ochraceo-rufus*). (Yellow ocher + burnt sienna + light red.)

Ochre Yellow (*ochraceo-flavus*). The color of the pigment called yellow ocher.

Ochros′poræ (*Gr.*—pale yellow), a group of fungi having ocher or brown-colored spores.

Oleag′inous (*oleum*, oil), oily or oil-like.

Oliva′ceous, Olive (*olivaceus*, *olivinus*), a greenish-brown color like that of olives (Sepia + light zinnober-green.)

Olive-Buff (*olivaceo-luteus*). (Yellow ocher + cobalt-blue + white.)

Olive-Green (*olivaceo-viridis*), a peculiar color, produced by the mixture of yellow and gray, resulting in a tint somewhat between olive and dull yellowish-green.

Olive-Yellow (*olivaceo-flavus*). (Light-cadmium + black + white.)

Opaque′, Opake′, mostly used in the sense of dull, not shining.

Oper′culum (*operculum*, a lid, cover, <*operire*, cover, shut, conceal), a lid-like cover.

Orange (*aurantius*), a deep reddish-yellow like the rind of an orange.

Orange-Rufous (*aurantio-rufus*). (Neutral-orange or cadmium-orange + light-red.)

Orange-Yellow (*aurantio-flavus*), a color intermediate between orange and yellow.

Orbic′ular (*orbicularis*, <*orbiculus*, a little disk), having the form of an orb; having the shape of a flat body nearly circular in outline.

Order, the most important unit of classification above the genus.

Organ′ic, pertaining to either living or dead animal or vegetable organism.

Os′mose (*osmosis*, thrust, push, impel), the impulse or tendency of fluids to pass through membranes and mix or become diffused through each other.

Osmo′sis (*Gr.*—impulsion, pushing), the diffusion of fluids through membranes; see osmose.

Osmot′ic, of or pertaining to or characterized by osmose.

Os′tiole, Osti′olum (*ostiolum*, a little door), mouth of the perithecium; orifice through which the spores are discharged.

675

O′vate (*ovatus*, egg-shaped), egg-shaped; having a figure the shape of a longitudinal section of an egg.

O′void (*ovum*, egg, + *Gr.*—form), egg-shaped; used to describe solids.

Pales′cent, inclining to paleness; becoming pallid.

Pal′lid, pale, undecided color.

Pal′udine, Palu′dinous, Pal′udose, Palus′trine (*palus* (*palud-*), a swamp), growing in marshes or swamps.

Papil′iona′ceous (*papilio(n-)*, butterfly), variegated; mottled; marked with different colors; as the lamellæ of some species of Panæolus mottled with black spores.

Papil′la (pl. Papil′læ) (*papilla*, a nipple, a teat, also a bud, pimple, dim. of *papula*, a pustule), a small nipple-shaped elevation.

Pap′illate (*papillatus*, <*papilla*, a nipple), furnished with one or more nipple-like elevations.

Papil′liform, Papil′læform, shaped like a papilla.

Papyra′ceous (*papyraceus*, <*papyrus*, paper), parchment; resembling the material covering a hornet's nest; pergamentous.

Paraph′ysis (pl. Paraph′yses) (*Gr.*—an off-shoot), slender, thread-like bodies growing with the asci; sterile cells usually club-shaped found with the reproductive cells of some plants.

Par′asite (*Gr.*—one who eats at another's table, a guest), a plant growing on or in another living body from which it derives all or part of its nourishment.

Parasit′ic, growing on and deriving support from another plant.

Paren′chyma (*Gr.*—the peculiar tissue of the lungs, liver, kidney and spleen), the fundamental cellular tissue of plants composed of thin walled, approximately isodiametric cells; absent in fungi. See Pseudoparenchyma and Prosenchyma.

Parenchym′atous, pertaining to, containing, consisting of or resembling parenchyma.

Pa′ries (pl. Pari′eties), wall of a cavity or capsule.

Pari′etal (*parietalis*, belonging to walls, <*paries* (*pariet-*), a wall), pertaining to or arising from a wall.

Par′tial (*partialis*, divisible, solitary, <*pars*, a part), secondary; of a veil clothing the stem and reaching to the edge of the pileus, but not extending beyond it; marginal.

Pat′ellate (*patella*, a small pan or dish, a plate), shaped like a dish.

Patell′iform (*patella*, a pan, dish; *forma*, form), having the shape of a patella or knee pan.

Patent (*paten(t)s*, ppr. *patere*, lie open), spreading, diverging widely.

Pea Green, a pale, dull green color like the color of green pea pods.

Pearl Blue, a very pale, purplish-blue color.

Pearl Gray (*margaritaceus*), a very pale, delicate, blue-gray color.

Pec′tinate (*pectinatus*, comb-like, pp. *pectinare*, <*pecten*, a comb), with narrow teeth, arranged as in a comb.

Ped′icel (*pediculus*, a little foot), foot stalk; any short, very small, stem-like stalk.

Ped′icellate (*pedicellus*, dim. of *pediculus*, a little foot), having a pedicel or little foot stalk.

Pel′licle (*pellicula*, a small skin, dim. of *pellis*, skin), a little or thin skin, a cuticle; same as cortical layer and cuticle.

Pellic′ulose (*pelliculosus,* < *pellicula,* dim. of *pellis,* skin), furnished with a pellicle or distinct skin.

Pellu′cid (*pellucidus, perlucidus,* transparent, < *pellucere, perlucere,* shine through, be transparent), admitting the passage of light, transparent, translucent.

Pel′tate (*peltatus,* armed with a light shield, < *pelta,* a light shield), formed like a shield and fixed to the stalk by the center, or by some point distinctly within the margin.

Pen′ciled (*pencillum,* a painter's brush), marked with fine lines; with pencil-like hairs either on the tip or border.

Pen′dulous, hanging down.

Pen′icillate (*penicillus,* a pencil), pencil-shaped; having a tuft of short hairs resembling a camel's-hair brush.

Peren′nial (*perennis,* lasting the year through, < *per,* through, + *annus,* year), continuing growth from year to year.

Pergame′neous (*pergamena,* parchment, + eous), like parchment.

Pericli′nal (*Gr.*—sloping on all sides + al), said of wall cells or any lines when parallel with the outer surface.

Peridi′olum (dim. of peridium), a secondary or interior peridium containing a hymenium.

Perid′ium (pl. Perid′ia) (*Gr.*—a pouch, wallet), the outer enveloping coat of the sporophore in angiocarpous fungi, as in puff-balls.

Periph′eral, of, belonging to or situated on the periphery.

Periph′ery (*peripheria,* the line around a circle, circumference, part of a circle), the exterior surface of any body.

Per′istome (*peristomium,* around a mouth), toothed or variously shaped ring around the mouth or orifice for discharge of spores in a peridium.

Perithe′cium (pl. Perithe′cia) (*Gr.*—a lid), cup-shaped ascocarp with the margin incurved so as to form a narrow, mouthed cavity; the case or hollow shell which contains the spores.

Per′onate (*peronatus,* < *pero,* a kind of high boot), sheathed, booted; said of the stem when it has a boot-like or stocking-like covering.

Persis′tent, enduring, continuing without withering, decaying or falling off.

Per′sonate (*personatus,* masked, < *persona,* mask), masked or disguised in any way.

Per′vious (*pervius,* passable; < *per,* through; *via,* a way), having an open tube-like passage.

Pezi′zoid, resembling a Peziza.

Pi′leate (*pileatus,* capped, bonneted), having a cap or pileus.

Pile′olus (pl. Pileoli) (*pileolus,* dim. of *pileus,* a hat), secondary pileus, arising from the division of a primary pileus; a little pileus.

Pi′leus (pl. Pi′lei) (*pileus,* a hat), a part of the receptacle of a fungus, *e. g.* the cap-like heads of agarics; it may be stipitate, sessile, dimidiate, regular or irregular in form.

Pilif′erous, Pilig′erous, Pi′lose, Pi′lous (*pilus,* hair, + *ferre,* bear), covered with hair, especially with fine or soft hair.

Pi′lose (*pilosus,* < *pilus,* a hair), covered with hairs; furry.

Pink (*caryophyllaceus*), a dilute, rose-red color.

Pinkish-Buff (*caryophyllaceo-luteus*). (Yellow-ocher + light-red + white.)

Pinkish-Vinaceous (*caryophyllaceo-vinaceus*). (Indian-red + white.)

Pip-Shaped, the shape of an apple seed.

677

Glossary

Pɪ′sɪꜰᴏʀᴍ (*pisum*, a pea, + *forma*, form), pea-shaped.

Pɪᴛʜ (pit, marrow, kernel), central stuffing in some stems.

Pɪᴛs (*puteus*, a well, a pit), depressions in tubes or cells resembling pores; applied also to hollow depressions in the surface of the cap of the Morell.

Pɪᴛ′ᴛᴇᴅ, covered with pits or small depressions.

Pʟᴀᴄᴇɴ′ᴛɪꜰᴏʀᴍ (*placenta*, placenta, + *forma*, form), in the form of a thickened circular disk depressed in the middle, both above and below.

Pʟᴀɴᴇ, having a flat surface.

Pʟɪ′ᴄᴀᴛᴇ (*plicatus*, fold, bend), folded like a fan; plaited.

Pʟᴜᴍ′ʙᴇᴏᴜs (*plumbeus*), a deep, bluish-gray color like tarnished lead; lead-color. (Lamp-black + intense blue + white.)

Pʟᴜ′ᴍᴏsᴇ, Pʟᴜ′ᴍᴏᴜs (*plumosus*, full of feathers or down), feathery or feathered.

Pʟᴜʀɪ-, prefixed has the significance of "many."

Pɴᴇᴜᴍᴏɢᴀs′ᴛʀɪᴄ (*Gr.*—lung, + —stomach), pertaining to the lungs and stomach, or to the functions of respiration and digestion; in anatomy noting several nervous structures.

Pᴏᴄ′ᴜʟɪꜰᴏʀᴍ (*poculum*, cup, + *forma*, form), cup-shaped.

Pᴏʟʏ-, a prefix meaning "many."

Pᴏʟʏɢ′ᴏɴᴀʟ, having many angles.

Pᴏʟʏᴍᴏʀ′ᴘʜɪsᴍ, Pᴏʟ′ʏᴍᴏʀᴘʜʏ, existence in or exhibition by the same species or group of different types of structure.

Pᴏʟʏᴍᴏʀ′ᴘʜᴏᴜs (*Gr.*—multiform, manifold), varying much in appearance, form or structure in the same species or group; characterized by polymorphism.

Pᴏʟʏs′ᴛɪᴄʜᴏᴜs (*Gr.*—many, + —row, line), arranged in many rows.

Pᴏʀᴇ (*porus*, a pore), in Pyrenomycetes same as ostiole; in Hymenomycetes same as tubulus or tube, as the tubules of Polypores; also the mouth of a tubulus.

Pᴏ′ʀɪꜰᴏʀᴍ (*porus*, a pore; *forma*, form), in the form of pores.

Pᴏ′ʀᴏsᴇ, Pᴏ′ʀᴏᴜs (*porosus*, <*porus*, a pore), furnished with pores or tubules; pierced with small holes.

Pᴏʀᴘʜʏʀᴏs′ᴘᴏʀᴁ (*Gr.*—purple, + —seed), a group of fungi having purple spores.

Pᴏʀʀᴇᴄᴛ′ (*porrectus*, pp. of *porrigere*, stretch out before, reach out, extend, <*por*, forth, + *regere*, stretch, direct), extended forward; stretched forth horizontally.

Pᴏsᴛᴇ′ʀɪᴏʀ (*posterior*, compar. of *posterus*, coming after, etc.), denotes a position or under side of the pileus adjacent to the stem; the end of a lamella next the stem is the posterior end.

Pʀᴇᴍᴏʀsᴇ′ (*premorsus*, pp. of *premordere*, bite in front or at the end), having the apex irregularly truncate as if bitten or broken off.

Pʀɪᴍᴏʀ′ᴅɪᴀʟ (*primordium*, pl. *primordia*, origin, beginning), first formed; existing from the beginning.

Pʀɪᴍᴏʀ′ᴅɪᴜᴍ (*L.*—commonly in pl. *primordia*, the beginnings, <*primus*, first, + *ordiri*, begin), first beginning of any structure.

Pʀɪᴍʀᴏsᴇ-Yᴇʟʟᴏᴡ (*primulaceo-flavus*), a very delicate pale-yellow, of a more creamy tint than sulphur-yellow. (Pale cadmium + white.)

Pʀᴏᴄ′ᴇss, an outgrowth or projection from a surface.

Pʀᴏᴄᴜᴍ′ʙᴇɴᴛ (*procumben(t-)s*, ppr. of *procumbere*, fall forward or prostrate, <*pro*, forward, + *cumbere*, *cubare*, lie), prostrate; unable to support itself, therefore lying on the ground.

Pʀᴏᴊᴇᴄ′ᴛɪɴɢ, the anterior end jutting out beyond the margin.

678

PROLIF'EROUS (*proles*, offspring; *fero*, to bear), applied to an organ which gives rise to secondary ones of the same kind.

PRO'TEAN (*Gr.*—the name of a sea-god), exceedingly variable; changeable in form.

PRO'TEID, albuminoid.

PROTO- (*Gr.*—first), an element in compound words of Greek origin meaning "first" and denoting precedence in time, rank and degree.

PROTOBASID'IUM, basidium divided by transverse septa into four cells, each giving rise to a spore from a laterally inserted sterigma, or a basidium divided longitudinally by septa intersecting each other at right angles into four cells terminating in a long, tubular sterigma.

PRO'TOPLASM (*Gr.*—first, + anything formed or molded), the nitrogenous fluid of variable composition found in living cells; it is the vital substance into which all food is assimilated, and from which all parts of the plant are formed.

PROX'IMAL (*proximus*, nearest), pertaining to the base or extremity of attachment.

PRU'INATE, PRU'INOSE (as if *pruinatus*, < *pruina*, hoar-frost), covered with a bloom or powder so as to appear as if frosted.

PRUN'IFORM (*prunum*, a plum; *forma*, form), plum-shaped.

PSEUDO (*Gr.*—false, counterfeit, etc.), prefix signifying "false" or "spurious."

PSEUDO-PAREN'CHYMA, a fungus tissue formed of closely woven and felted hyphal threads, which on section has the appearance of the cellular structure of true parenchyma.

PUBES'CENCE (pubescen(t-) + ce), general term to describe hairyness; specifically covered with short, soft, downy hairs.

PUBES'CENT (*pubes*, of mature age), covered with soft, short hairs, downy; hairy.

PULLULA'TION (*pullulare*, pp. *pullulatus*, pullulate), a mode of cell multiplication in which a cell forms a protuberance on one side which enlarges to size of parent cell and is cut-off from it by a dividing wall; sprouting; budding.

PULVERA'CEOUS, PULVER'ULENT (*pulvis* (*pulver*), dust, powder), covered as if with powder or dust.

PUL'VINATE (*pulvinatus*, < *pulvinus*, a cushion), cushion-shaped.

PUNC'TATE (*punctatus*, < *punctus*, a point), dotted with points.

PUNC'TIFORM (*punctum*, point; *forma*, form), like a point or dot.

PUS'TULAR, PUS'TULATE (*pustulatus*, pp. of *pustulare*, to blister, < *pustula*, a blister, pimple), having low elevations shaped like blisters or pustules.

PUTRES'CENT, soon decaying.

PYR'IFORM (*pyrum*, a pear; *forma*, form), pear-shaped.

QUAD-, QUADRI- (*quadru*, four-cornered, square, fourfold, < *quattuor*, four), prefix signifying "four."

QUAD'RATE (*quadratus*, square, pp. of *quadrare*, make four-cornered, square, < *quadra*, a square), square; sometimes used to mean "of four equal parts."

QUATER'NATE (*quaternatus*, < *quaterni*, four each), arranged in groups of four.

RA'DIATE, RA'DIATING (*radiatus*, pp. of *radiare*, furnish with spokes, give out rays, shine), arranged like the spokes of a wheel.

RAD'ICATING (*radicatus*, pp. of *radicare*, take root), rooting; having root-like strands which penetrate the matrix.

RAD'ICLE (*radicula*, dim. of *radix*, a root), a rootlet.

RA'MEAL, RA'MEOUS (*rameus*, a branch), growing on twigs or branches.

679

Glossary

RAMIC′ULOUS, growing on branches.

RAM′IFICATION (*ramiflcare*, ramify), branching, or the manner of branching.

RAM′IFY (*ramus*, a branch, + *flcare*, < *facere*, make), to form branches.

RA′MOSE, RA′MOUS (*ramosus*, full of branches), having many small branches.

RECEP′TACLE, RECEPTAC′ULUM (*receptaculum*, place to receive things in), a part of the mushroom extremely varied in form, consistency and size, enclosing the organs of reproduction; usually implying a hollowed-out body containing other bodies; same as STROMA; same as SPOROPHORE; in Phalloids the stem, stem and pileus, or the clathrate body which supports the gleba.

REFLEXED′, REFLEC′TED (*reflexus*, pp. of *reflectere*, reflect), turned or bent back.

REMOTE′ (*remotus*, pp. *removere*, remove), of gills which do not reach the stem, but leave a free space between them and it.

REN′IFORM (*ren*, the kidney; *forma*, form), kidney-shaped.

REPAND′ (*repandus*, bent backward), bent or turned up or back; having a slightly undulating or sinuous margin.

REP′LICATE (*replicatus*, pp. of *replicare*, fold or bend back), folded back upon itself as when the margin of a cup turns outward and downward.

RESU′PINATE (*resupinatus*, pp. of *resupinare*, throw on the back), attached to the matrix by the back, the hymenium facing outward; said of fungi spread over the matrix without any stem and with the hymenium upwards.

RETIC′ULATE, RE′TIFORM (*reticulatus*, < *reticulum*, a little net), marked with crossed lines like the meshes of a net.

REV′OLUTE (*revolutus*, pp. of *revolvere*, revolved), rolled backwards or upwards; of the margin of a pileus *e. g.* the opposite of involute.

RHI′ZINES, RHI′ZOIDS (*Gr.*—root, +), delicate filiform hyphal branches which serve to attach the sporophore to the substratum and supply nourishment.

RI′MOSE, RI′MOUS (*rimosus*, < *rima*, a crack), cracked, full of clefts.

RHI′ZOMORPHS (*Gr.*—root, + *L.*—*forma*, form), long, branching or anastomosing, rigid, root-like cords of mycelium with a dark or black exterior, often growing between the bark and timber or about and penetrating the roots of dead and living trees, produced by Agaricus melleus and various other fungi.

RHIZOMOR′PHOID (*Gr.*—root, + form), root-like in form.

RHODOSPO′RÆ (*Gr.*—rose, + seed), rose or pink spores.

RIM′ULOSE, RIM′ULOUS (*rimula*, a little crack), covered with small cracks.

RIND, cortex; bark.

RING, a part of the veil adhering in the form of a ring to the stem of an agaric; same as annulus.

RI′VOSE (*rivus*, a stream, channel, groove), marked with furrows which do not run in parallel directions.

RIV′ULOSE (*rivulosus*, < *rivula*, a little stream), marked with lines like rivulets.

ROOT′ING, same as radicating.

ROSACEOUS (*rosaceus*, *pallidoroseus*, *caryophyllaceus*), a very pure purplish-pink color, like some varieties of roses.

ROSE-RED (*roseus*, *rosaceo-ruber*), the purest possible purplish-red color.

ROS′TRATE (*rostratus*, having a beak, hook or crooked point, < *rostrum*, a beak), beaked; having a process resembling the beak of a bird.

ROTUND′, round or nearly so.

RUBES′CENT (*rubescens*, ppr. of *rubescere*, become red, < *rubere*, be red), tending to a red-color.

Rubig′inous (*rubiginosus*, <*rubigo*, rust), rust-colored.

Rufes′cent (*rufescere*, to become reddish), tending to rufous or a dull red color.

Ruf′fled, very strongly undulate.

Ru′fous (*rufus*), a brownish-red color like the pigment called Venetian-red, light red, Indian-red, red chalk, etc., which represents various shades of rufous. The typical shade is light red.

Ru′gose (*rugosus*, <*ruga*, a wrinkle), wrinkled.

Ru′gulose (*rugula*, dim. of *ruga*, a wrinkle), minutely rugose.

Run′cinate (*runcina*, a plane), irregularly saw-toothed, the divisions or teeth hooked backward.

Russet (*russatus*), a bright tawny-brown color with a tinge of rusty.

Sab′uline, Sab′ulose (*sabulum*, sand, + *ine*), growing in sandy places.

Sac′cate (*saccus*, a bag), in the form of a sack or pouch.

Sac′charine (*saccharon*, sugar), of or resembling sugar, covered with shining grains like those of sugar.

Sac′cule, Sac′culus (*sacculus*, dim. of *saccus*, a bag), a small sack or pouch.

Salmon-Color (*salmonaceus*) (*carneus*), a color intermediate between flesh color and orange, like the flesh of the salmon. (Saturn red or orange chrome + white.)

Sanguin′eous (*sanguineus*, of blood, bloody), blood-colored; of a deep, somewhat brownish-red color; like the color of clotted blood.

Sap′id (*sapidus*, having taste, savory, <*sapere*, have a taste), agreeable to the taste.

Saprog′enous (*Gr.*—rotten, + *Gr.*—producing), growing in decaying or decomposing animal or vegetable matter.

Sap′rophyte (*Gr.*—rotten, + *Gr.*—a plant), a plant that lives on decaying vegetable or animal matter.

Saprophyt′ic, living upon and deriving its sustenance from dead organic matter.

Sca′brate, Sca′brous (*scabrosus*, <*scaber*, rough), rough on the surface; rugged.

Scalar′iform (*scalaria*, a flight of steps; *forma*, form), in the form of a ladder.

Scaph′oid (*Gr.*—like a bowl or boat, + *Gr.*—form), boat-shaped.

Sca′riose, Sca′rious (*scariosus*, from *scaria*, a thorny shrub), thin, dry, membranaceous; applied to a shriveled membrane.

Scis′sile (*scissilis*, to cleave), capable of being easily split or cleft; said of gills which can easily be split into two plates.

Sclerit′ic, Scle′roid, Scle′rose, Scle′rosed (*Gr.*—hard, rough, harsh), having a hard texture.

Sclero′tioid (*Gr.*—hard, + resemblance), in the form of a sclerotium; a form assumed by the mycelium of certain fungi.

Sclero′tium (pl. Sclerotia) (*Gr.*—hard), hard, black, compact, mostly tuber-like body, which is the resting stage of certain fungi, as in Peziza tuberosa; it remains dormant for a time and then sends up shoots, which develop into sporophores at the expense of the reserve material.

Scrobic′ulate (*scrobiculatus*, <*scrobiculus*, dim. of *scrobis*, a trench), marked with small pits; furrowed.

Scrupose, rough with small irregular prominences.

Scu′tellate (*scutellatus*, <*scutella*, a salver, dish), shaped like a plate or platter.

Section, a cutting, cutting off, excision, amputation, etc.

Semi-, prefix meaning "half" or "partial."

SEP′ARABLE (*separabilis*, that can be separated, <*separare*, separate), capable of being detached.

SEP′ARATING, becoming detached, as lamellæ from the stem, or resupinate fungi from the matrix.

SE′PIA, a deep, dark-brown color, with a little red in its composition. The pigment called sepia is a carbonaceous matter, prepared from the natural ink of a species of cuttle-fish.

SEP′TATE (*septum*, a fence), having partitions.

SEP′TUM (pl. SEPTA) (*septum*, a fence), partition.

SE′RIATE (*seriatus*, pp. of *seriare*, arrange in a series), arranged in rows.

SERIC′EOUS (*sericum*, silk), silky.

SER′RATE (*serratus*, saw-shaped), having marginal teeth shaped like saw teeth.

SER′RULATE (*serrulatus*, <*serrula*, dim. of *serra*, a saw), minutely serrate.

SES′SILE (*sessilis*, <*sessus*, pp. *sedere*, sit), attached by the base; having no stem or support.

SE′TA (pl. SE′TÆ) (*seta*, a bristle), a stiff-bristle-like hair.

SETA′CEOUS, SETIG′EROUS, SE′TOSE (*seta*, bristle), beset with bristles.

SE′TOSE (*setosus*, abounding in bristles), bristly.

SET′ULOSE (*setula* + *ose*), finely setose; covered with setules.

SIG′MOID (*Gr.*—of the shape of a sigma), said of an elongated spore having the ends bent slightly in opposite directions; S-shaped.

SIMPLE, in botany not formed by a union of similar parts or groups of parts; a simple stem or trunk is one not divided at the base.

SIN′UATE, SIN′UOSE, SIN′UOUS (*sinuatus*, pp. *sinuare*, <*sinus*, a curve), waved; serpentine; applied to an edge the outline of which is alternately concave and convex; a sinuate lamella has a sudden wave or sinus in its edge near the stem.

SI′NUS (*sinus*, the fold of a garment, a curve, hollow), a rounded inward curve between two projecting lobès.

SLATE-COLOR (*schistaceus*), a dark gray or blackish gray color, less bluish in tint than plumbeous or lead color.

SMOKE-GRAY (*fumidio-canus*). (Black + white + raw umber.)

SMOOTH, glabrous; applied to a surface which is destitute of hairs; a surface may be uneven and yet smooth.

SOR′DID (*sordidus*, dirty, filthy, mean, <*sordere*, be dirty), of a dingy, dirty hue.

SPADIC′EOUS (*spadiceus*, <*spadix*, a palm branch), date-brown, duller and darker than bay-brown.

SPATH′ULATE, SPAT′ULATE (*spathula*, dim. of *spatha*, a broad, flat instrument for stirring liquids), shaped like a spathula or spoon; oblong or rounded and flattened at the top with a long, narrow, attenuate base.

SPE′CIES, an individual, or collectively those individuals which differ specifically from all other members of a genus and which do not differ from each other except within narrow limits of variability, and which produce by propagation other individuals of the same kind.

SPECIF′IC (*species*, kind, + *ficus*, <*facere*, make), of, pertaining to, constituting, peculiar to, characteristic of, designating species or a species; not generic, not of wider application than to a species.

SPHAG′NUM (*Gr.*—a kind of moss), peat or bog moss.

SPHER′ICAL, SPHE′ROID, of the shape of a ball or globe or nearly so.

SPIC′ULAR, SPIC′ULATE, SPIC′ULOUS (*spicule* + ar), covered with spicules.

SPIC′ULE (*spicula*, a little sharp point), in Hymenomycetes one of the small projections on the basidia which bear the spores.

SPIN′ULE (*spinula*, dim. of *spina*, a thorn), a small spine or prickle.

SPORAN′GIOPHORE (sporangium + *Gr.*—bear), special mycelial branch bearing a sporangium.

SPORAN′GIUM (pl. SPORANGIA) (*spora*, a spore, + *Gr.*—vessels), sac producing spores endogenously.

SPORE (*Gr.*—a sowing, seed time, seed, etc.), the reproductive body of cryptogams analogous to the seed of phenogams; the terms spores, sporidia, sporules and conidia have been applied somewhat indiscriminately to all spore bodies.

SPORIDIF′ERA, a class of fungi in which the spores are enclosed in asci.

SPORIDIF′EROUS, SPORIDIIF′EROUS (*sporidium*, + *ferre*, bear), bearing sporidia; applied to a fungus of the class Sporidifera.

SPORID′IUM (pl. SPORID′IA) (dim. of *Gr.*—spore), an ascospore or endospore. See SPORE.

SPORIF′ERA, a class of fungi in which the spores are free, naked or soon exposed.

SPORIF′EROUS (*spora*, spore + *ferre*, bear), bearing spores; applied to a fungus of the class Sporifera.

SPO′ROCARP (*spora*, spore, + *Gr.*—fruit), in Ascomycetes the entire fruit, composed of the ascophore and the asci.

SPOROG′ENOUS (*spora*, spore, + *Gr.*—producing), producing spores.

SPO′ROPHORE (*spora*, spore), branch or portion of thallus which bears spores or spore-mother-cells; said to be simple or filamentous when consisting of a single hypha or branch of a hypha; compound, when formed by the cohesion of the ramifications of separate hyphal branches (the common mushroom is a compound sporophore.)

SPORT, an animal or plant, or any part of one that varies suddenly or singularly from the normal type of structure, and is usually of transient character or not perpetuated; not so much deformed as "monster."

SPO′RULE, see under SPORE.

SQUA′MA (pl. SQUA′MÆ) (*squama*, a scale), a scale or scale-like appendage.

SQUA′MOSE, SQUA′MOUS (*squamosus*, <*squama*, a scale), covered with appressed scales; scale-like.

SQUAM′ULA, SQUAM′ULE (dim. of *squama*, a scale), a small squama.

SQUAM′ULOSE (*squamulosus*, <*squamula*, dim. of *squama*, a scale), covered with small scales.

SQUAR′ROSE (*squarrosus*, scaly), rough with scales; roughened with projecting points.

STALK, stipe; any stem-like supporting organ.

STEL′LATE (*stellatus*, pp. of *stellare*, set or cover with stars, <*stella*, star), star-shaped.

STERIG′MA (pl. STERIGMATA) (*Gr.*—a prop, support), stalk-like branch of a basidium bearing a spore.

STER′ILE, not fertile; producing no spores.

STIPE (*L.*—a stock, trunk, post, etc.), stalk of a mushroom.

STIP′ITATE (*stipitatus*, <*stipes*, a stalk), stemmed, elevated on a stipe.

STO′MA (pl. STO′MATA) (*Gr.*—the mouth, opening, entrance, out-let, etc.), a mouth or aperture; little orifices in the epidermis of leaves, etc., opening into air cavities or intercellular spaces.

STRAIGHT, applied to margin of pileus when not involute.

STRAMIN′EOUS (*stramineus*, made of straw, <*stramen*, straw), straw-colored.

STRA′TOSE (*stratum*, a layer), arranged in distinct layers or strata,

STRA′TUM (pl. STRA′TA) (*L.*—coverlet, bed, pavement, etc.), a layer.

STRAW-COLOR, STRAW-YELLOW (*stramineus*), a very light impure yellow, like cured straw.

STRI′A (pl. STRI′Æ) (*stria*, a channel, furrow, hollow), parallel or radiating lines or markings.

STRI′ATE (*striatus*, pp. of *striare*, <*stria*, a channel, flute of a column), marked with striæ.

STRI′GOSE (*strigosus*, <*striga*, a swath), rough with stiff hairs.

STROBIL′IFORM (*strobilus* + *forma*, form), resembling a pine cone.

STRO′MA (pl. STRO′MATA) (*stroma*, a covering, coverlet), a mass in which another object is imbedded; a compact mass of mycelium in the form of a cushion, crust, club or branched expansion upon or in which perithecia or other organs of fructification are borne.

STUFFED, of a stem filled with material of a different texture from its walls.

SUB- (sub, under, before, near), prefixed signifies "somewhat," "almost" or "under."

SU′BERIZED (*suber*, cork, + ized), transformed into suberin or cork.

SUB-EROSE′ (*sub*, under, + *erosus*, pp. of *erodere*, gnaw off or away, consume), slightly erose; appearing as if eaten or gnawed on the margin.

SU′BEROSE (*suber*, cork), corky.

SUBGLE′BA (*sub*, under; *gleba*, a clod), basal portion of the gleba.

SUBIC′ULUM (*subiculum*, an under layer), a more or less thin and dense felt of hyphæ covering the matrix; upon its surface is spread the hymenium, or from it arise stalks supporting sporophores.

SUBSTRA′TUM (*substratum*, neut. of *substratus*, spread under), sometimes used in the sense of matrix.

SUBTERRA′NEAN, under ground.

SU′BULATE, SU′BULIFORM (*subulatus*, <*subula*, an awl, <*suere*, sew), awl-shaped.

SUC′CULENT (*succulentus*, full of juice, sappy, <*succus*, prop. *sucus*, juice), fleshy, juicy.

SUL′CATE (*sulcatus*, pp. *sulcare*, <*sulcus*, a furrow), marked with furrows; grooved.

SUL′CUS (pl. SUL′CI) (*sulcus*, a furrow, trench), groove or furrow.

SULPHU′REOUS, SULFU′REOUS (*sulfureus, sulphureus*, of or like sulphur, <*sulfur*, sulphur), sulphur-colored.

SULPHUR YELLOW (*sulphureus*), a very pale pure yellow color, less orange in tint than dilute gamboge or lemon yellow.

SUPER-, SUPRA- (*super*, over, above, beyond), prefix meaning "above" in position or degree.

SUPERFIC′IAL (*superficialis*, of or pertaining to the surface, situated on or close to the surface.

SUPE′RIOR (*superus*, <*super*, above), the upper surface; or applied to a ring when it is near the apex of the stem.

SUPRAVENAL (*super*, over, above; *vena*, vein), situated or occurring above veins.

SYM′BION, SYM′BIONT (*Gr.*—live together with), an organism which lives in a state of symbiosis.

SYMBIO′SIS (*Gr.*—a living together), the co-existence in more or less mutual interdependence of two different organisms; mutualism; mutual parasitism; commensalism; consortism; with some authors commensalism implies an association less necessary or mutually helpful than symbiosis.

SYMBIOT'IC, living in that kind of consociation called symbiosis.

SYN'ONYM (*Gr.*—a word having the same name with another), a discarded name for a species or genus; either of two or more names for the same species or genus.

SYNON'YMOUS, expressing the same idea; equivalent in meaning; having the character of a synonym.

TAPE'SIUM (*tapesium*, tapestry, carpet), a carpet or layer of mycelium on which the receptacle is situated.

TAWNY (*fulvus, fulvescens, alutaceus*), the color of tanned leather. (Nearly synonymous with fulvous.) (Neutral orange + raw sienna.)

TAWNY OCHRACEOUS (*fulvo-ochraceus*). (Yellow ocher + burnt sienna + raw umber.)

TAWNY OLIVE (*fulvo-olivaceus*). (Yellow ocher + raw umber.)

TENA'CEOUS (*tenax (tenac-)*, holding fast, <*tenere*, hold), tough.

TE'RETE (*teret*, round, smooth), cylindrical or nearly so, having a circular, transverse section; top-shaped.

TERRES'TRIAL (*terrestris*, of or belonging to the earth + al), growing on the ground.

TES'SELATED (*tessellatus*, made of small square stones, checkered + ed), arranged in small squares; checkered or reticulated in a regular manner.

TESTA'CEOUS (*testaceus*, consisting of tiles or sherds, <*testa*, tile, shell), same as brick-red.

TETAN'IC (*tetanicus*, affected with tetanus), pertaining to or characterized by tetanus; tetanic spasm; tonic spasm of the voluntary muscles as seen in tetanus, strychnic poisoning, etc.

TETRA- (*quatuor*, four), prefix signifying "four."

TET'RASPORE (*Gr.*—four, + seed), four spores forming one.

THALA'MIUM (*Gr.*—inner chamber, bedroom, bed), synonym for hymenium.

THAL'LOGEN (*Gr.*—a young shoot, + *Gr.*—producing), same as thallophyte.

THAL'LOPHYTE (*Gr.*—a young shoot, + *Gr.*—a plant), one of the so-called "lower cryptogams," plants in which the vegetative body usually consists of a thallus.

THAL'LUS (*Gr.*—a young shoot or twig), a vegetative body which is not differentiated into a true root, stem and leaf, has no true vessels or woody fiber; in fungi it is the whole body of the plant not serving directly as an organ of reproduction, *i. e.*: mycelium, if any, and sporophore but not including the hymenial layer.

THE'CA (*Gr.*—a case, box, receptacle, + put, place), a sac or case, generally used in the sense of capsule.

THE'CASPORE, the spore thus enclosed by the wall of the sac; an ascospore.

TIS'SUE (*texere*, weave), the cellular fabric out of which plant structures are built up.

TOAD'STOOL (toad + stool), a general name applicable to any form of visible fungus; usually applied to fleshy fungi as distinguished from the molds, smuts, etc. Mushroom is a name given to a few species of toadstools known to commerce, and wrongly to other edible species, of which there are many.

TOMEN'TOSE, TOMENTOUS (*tomentum*, wool, etc.), densely pubescent with matted wool or tomentum.

TOMEN'TUM (*tomentum*, a stuffing of wool, hair, feathers, etc.), a species of pubescence consisting of longish, soft, entangled hairs pressed close to the surface.

TORN, said of pores which are superficially rough and jagged as if torn.

To′ROSE, TOR′ULOSE (*torosus*, full of muscle or flesh, <*torus*, a bulging, protuberance), swollen at intervals.

TOR′SION (*torquere*, pp. *tortus*, twist, wring), the state of being twisted spirally.

TOR′SIVE (*torsus*, pp. *torquere*, twist), spirally twisted.

TOR′TUOUS, bending or turning in various directions.

TOX′IC (*toxicum*, poison), poisonous.

TRA′MA (*trama*, the weft or filling of a web), the substance proceeding from the
hymenophore, between the plates of (central in) the gills in Agarics, and between the double membranes of which the dissepiments of the pores are composed in Polyporei; the hyphal plates forming the walls of the chambers of the
gleba, in Gasteromycetes.

TRANSLU′CENT (*translucen*(*t*)*s*, ppr. of *translucere*, shine across or through, <*trans*,
over, + *lucere*, shine), transmitting rays of light without being transparent.

TRANS′VERSE (*transversus*, lying across), from side to side.

TREM′ELLOID, TREM′ELLOSE (*tremo*, to tremble), of a gelatinous or jelly-like consistency; resembling Tremella.

TRI- (*tri*, three), prefix signifying "three."

TRI′FID (*trifidus*, <*tres* (*tri-*) three, + *findere*, cleave), divided half way into three
parts by linear sinuses with straight margins; three-cleft.

TRIQUET′ROUS (*triquetrus*, three-cornered, triangular), having three acute angles
with concave faces; triangular; applied to the vertical radial section of some
dimidiate pilei; three-edged.

TRUN′CATE (*truncatus*, pp. *truncare*, cut off), ending abruptly as if cut short; cut
squarely off.

TU′BÆFORM, TU′BIFORM (*tuba*, a tube; *forma*, form), trumpet-shaped, tubular.

TUBE, TU′BULE (*tubus*, a pipe, tube), in polypores, tube lined with hymenium; same
as pore.

TU′BER (*tuber*, a bump, swelling, knob on plant, etc.), fleshy body, usually of a
rounded or oblong form, produced on underground stems, as the potato or arti-
choke; a genus of underground fungi.

TU′BERCLE (*tuberculum*, dim. of *tuber*, a swelling), a small, wart-like excrescence; a
small swelling.

TUBER′CULAR, TUBER′CULATE, TUBER′CULOSE (*tuberculum*, tubercle), having or covered with tubercles; formed like or forming a tubercle.

TUBER′CULIFORM (*tuberculum*, tubercle, + *forma*, form), shaped like a tubercle.

TU′BEROUS (*tuberosus*, full of lumps or protuberances, <*tuber*, a knob, lump), rounded and swollen; resembling a tuber.

TU′BULAR (*tubulus*, a small pipe), hollow and cylindrical.

TU′BULUS (pl. TUBULI) (*tubulus*, tube), same as tube; pore.

TU′MID (*tumidus*, swollen, swelling, <*tumere*, swell), swollen, slightly inflated.

TUR′BINATE (*turbinatus*, shaped like a top or cone), top-shaped; shape of an inverted cone.

TUR′GID (*turgidus*, swollen, <*turgere*, swell out), thickened as if swollen; distended
with liquid.

TUR′GOR (*turgere*, swell), the state of being turgid; a state of distension and tension
of plant cells and parts by reason of their fullness of liquid.

TYPE, a perfect specimen or individual exemplifying the essential characters of the
species to which it belongs; the original specimen from which a species was
described.

Typ'ical, agreeing closely with the characters assigned to a group or species.

Ulig'inose, Ulig'inous (*uliginosus*, full of moisture, damp, <*uligo*, moisture, marshiness), growing in marshes or swamps.

Ul'timate (*ultimatus*, farthest, last, pp. of *ultimare*, come to an end, <*ultimus*, last, finish), farthest, last.

Um'ber, Um'brinous (*umbra*, shade, shadow), the color of the pigment called raw umber.

Umbil'icate (*umbilicatus*, <*umbilicus*, navel), with a central depression or rounded pit; having a navel-like depression.

Umbili'cus (*umbilicus*, navel), a navel-like depression.

Umbo (boss of a shield), applied to the central elevation of the cap of some mushrooms.

Um'bonate (*umbonatus*, <*umbo*, the boss of a shield), with a central boss-like elevation.

Un'cinate (*uncinatus*, <*uncinus*, a hook), hooked; forming a hook.

Un'dulate, Un'date (*undatus*, pp. of *undare*, rise in waves, <*unda*, a wave), having the surface near the margin alternately concave and convex; waved.

Une'qual, applied to gills when of unequal lengths; to a stem not of uniform thickness.

Une'ven, said of surfaces that are irregular, striate, sulcate, etc.

Un'gulate, Un'gulous (*ungulatus*, having claws or hoofs, <*ungula*, claw, talon, hoof), hoof-shaped.

Uni-, prefix signifying "one."

Unicol'orous (*unicolor*, having one color, + ous), of a uniform color; of the same color.

Unise'riate (*unus*, one, + *series*, series), arranged in one row.

Univer'sal (*universus*, whole), said of the veil or volva which entirely envelopes the fungus when young.

Unsep'tate (*un*, not; *septum*, a fence), having no partitions.

Ur'ceolate (*urceolus*, a little pitcher, + ate), shaped like a pitcher with a contracted mouth.

U'terus (*uterus*, the womb, belly), same as peridium in Gastromycetes.

U'tricle (*utriculus*, a little leather bag or bottle, etc.), any thin bladder-like or bottle-like body.

Vac'uolate, Vac'uolated (*vacuole* + ate), provided with vacuoles.

Vac'uole (*vacuolum*, dim. of *vacuum*, an empty space), a cavity of greater or less size within the protoplasmic mass of active vegetable cells filled with water or cell-sap, as it is called.

Vag'inate (*vagina*, a sheath), furnished with or contained in a sheath; sheathed.

Vague, indefinite, indistinct.

Vandyke Brown, a rich deep brown, very similar to burnt umber, but rather less reddish.

Va'riable (*variabilis*, changeable, <*variare*, change), said of a species which embraces many individuals which depart more or less from the type of the group.

Va'riegated, marked with different colors; mottled; same as Papilionaceous.

Vari'ety, a subdivision of a species with minor characteristics uniformly varying from the type; an incipient species.

Vas′cular (*vasculum*, a small vessel), consisting of, relating to or furnished with vessels or ducts.

Vaul′ted, arched like the roof of the mouth.

Veil, Ve′lum (*velum*, a veil), a covering of various texture more or less completely enwrapping a fungus; occurring chiefly among the Agaricini; *partial or marginal veil*, a special envelope extending from the margin of the pileus to the stem enclosing the gills; *universal veil or volva*, a special envelope enclosing the entire plant in the young state, either concrete with the cuticle of the pileus as in Lepiota or discrete as in Amanita, ultimately ruptured by the expanding pileus, a membranaceous or fibrous or granulose coating stretched over the mouth of an apothecium or cup soon breaking into fragments.

Veins, swollen wrinkles on the sides of, and at the base between the gills, often connected to form cross partitions, (b) so-called, the rounded, obtuse-edged gills found upon Cantharellus, Craterellus, (c) the vein-like protuberances upon the surface of some fungi.

Vel′iform, Velamen′tous (*velum*, covering; *forma*, form), resembling or serving as a veil; of a thin veil-like covering.

Ve′lum, veil.

Velu′tine, Velu′tinous (*velutum*, velvet, + ine), velvety.

Ve′nate, Veined, Ve′nose, Ve′nous (*vena*, vein, artery), intersected by swollen wrinkles below and on the sides.

Ven′tral (*ventralis*, of or pertaining to the belly, stomach, <*venter*, belly, stomach), applied to the under side of pileus; opposite to "dorsal."

Ven′tricose (*venter*, the belly), swollen in the middle; bellied.

Vermic′ular, Vermic′ulate (*vermiculus*, a worm), worm-shaped.

Vermilion (*cinnabarinus*, *cinnabarino-ruber*), a very fine red color, lighter and less rosy than carmine, and not so pure or rich as scarlet.

Ver′nal (*vernalis*, of the spring, vernal, <*ver*, spring), of or pertaining to the spring.

Ver′nicose (*vernix*, varnish), appearing as if varnished.

Verru′ca (pl. Verru′cæ) (*verruca*, a wart, steep place or height), wart.

Ver′rucose (*verrucosus*, full of warts), covered with warts or glandular elevations.

Verru′ciform (*verruca*, a wart, + *forma*, form), warty, resembling a wart in appearance.

Verru′culose (*verrucula*, a little eminence. a little wart, dim. of *verruca*, a wart, + ose), minutely verrucose.

Ver′tex (*vertex*, vortex (*tic*-), a whirl, eddy, highest point, etc.), the upper extremity.

Verticil′late (*verticillus*, a whirl), whorled.

Ves′cicle (*vesicula*, a little blister, a vesicle, dim. of <*vesica*, bladder, blister), a minute bladder-like cell or cavity.

Vesic′ular, Vesi′culate, Vesic′ulose, Vesic′ulous (*vesicula*, vescicle), composed of or like vescicles.

Vil′lose, Vil′lous (*villosus*, <*villus*, a tuft of hair), downy with soft weak hairs.

Vina′ceous, a brownish-pink or delicate brownish-purple color like wine dregs; a soft, delicate wine-colored pink or purple.

Vina′ceous-Buff (*vinaceo-luteus*). (Indian-red + yellow ocher + white.)

Vina′ceous-Cinnamon (*vinaceo-cinnamomeus*). (Burnt umber + burnt sienna + white.)

Vina′ceous-Pink (*vinaceo-caryophyllaceus*). (Madder-carmine + light-red + white.)

VINA′CEOUS-RU′FOUS (*vinaceo-rufus*).   (Indian-red + light-red + white.)

VI′NOUS (*vinosus*, <*vinum*, wine), wine-colored; vinaceous.

VIOLET, VIOLA′CEOUS, a purplish-blue color, like the petals of a violet.   (Aniline-violet or mauve.)

VIRES′CENT (*virescere*, grow green, greenish), green or becoming green.

VIR′GATE (*virgatus*, <*virga*, a twig, rod), streaked; having an erect, slender shape like a rod.

VIRIDES′CENT (*viridescens*, ppr. of *viridescere*, be green, <*viridis*, green), slightly green; greenish.

VIS′CID (*viscum*, bird-lime, anything sticky), moist and sticky, glutinous, clammy, adhesive; covered with a shiny liquid which adheres to the fingers when touched.

VIS′COSE, VIS′COUS (*viscosus*, <*viscum*, bird-lime), glutinous, clammy, adhesive.

VITEL′LINE (*vitellus*, yolk of egg, + ine), egg-yellow color; luteous.

VOLUTE′ (*voluta*, a spiral scroll), rolled up in any direction.

VOL′VA (*volva*, a wrapper), wrapper; same as universal veil; the name is often applied to that portion of a discrete volva which is left after rupturing, either attached in fragments to, or forming a distinct membranous sheath about, the base of the stem, the peridium in phalloids analogous to the volva in Amanitæ.

VOL′VIFORM (*volva*, wrapper, + *forma*, form), having the form of a volva.

WART (*verruca*, wart, excrescence), a wart-like excrescence found on the pileus of some mushrooms; the remains of the volva in form of irregular or polygonal excrescences, more or less adherent, numerous and persistent.

WAVED, WA′VY.   See UNDULATE.

WAX-YELLOW (*ceraceus*), a deep but dull yellow, resembling the color of fresh bees′ wax.

WHORLED, having parts arranged in a circle around an axis; verticillate.

WINE-PURPLE (*vinaceo-purpureus*), a clear reddish-purple of a slightly brownish cast.

WOOD-BROWN, a light brown color like some varieties of wood.   (Raw umber+burnt sienna+white.)

YELLOW-OCHER, a bright yellowish-ochraceous or ocher-yellow color.

ZO′NATE, ZONED (*zona*, a zone or girdle), marked with concentric bands of color.

ZONES (*zona*, a zone or girdle), circular bands of color.

INDEX TO GENERA

INDEX TO SPECIES

Index to Species

Index to Species

INDEX TO RECIPES FOR COOKING AND PREPARING FOR THE TABLE

INDEX TO GENERAL CONTENTS

SUPPLEMENT

ONE THOUSAND AMERICAN FUNGI

PREFATORY

THE first edition of "One Thousand American Fungi" so fully embodied the species known to be edible, that the field for fresh investigation has been confined principally to newly discovered species. In the eighteen months elapsing since the publication of the first edition, Professor Charles H. Peck—the American authority upon fungi—has reported several. These, with his descriptions, are named in the supplement.

The many requests made of the author for information upon the raising of mushrooms show a prevalent interest in the industry. What he knows is stated herein; what he does not know, and what is not known upon the subject, would furnish the matter for a volume.

Interest in the study of fungi is well established and is rapidly increasing. This department of botany has been made a specialty in many colleges and schools. Its importance is everywhere recognized.

The author and publishers feel a just pride in the success of "One Thousand American Fungi." The prompt sale of the first edition, and immediate demand for the second, warrant it. Their thanks are due to the many who have kindly interested themselves in obtaining subscriptions to the author's edition.

CHARLES MCILVAINE.

PUBLICATIONS

ILLUSTRATIONS TO SUPPLEMENT

INDEX TO SUPPLEMENT

Supplement

Amanita Frostiana pallidipes n. var. (See A. Frostiana, page 16.) Amanita. In his report of the New York State Botanist for 1899, Prof. Charles H. Peck describes a new variety of Amanita Frostiana as follows:

The typical form of this species, which is common in our cool northern woods, has the pileus and annulus, and usually the stem also, of a yellow color, that of the pileus sometimes verging to orange. But in warmer and more open or bushy places forms occur in which the whole plant is whitish, but in other respects has the characters of the species. Sometimes the pileus is pale-yellow and the stem and annulus white. The warts are soft and flocculent, are sometimes numerous and persistent, and again are few or wanting. The form with yellow stem and annulus and yellow or orange pileus may be considered the typical form of the species, but forms having the stem and annulus pale or white may be designated as variety pallidipes. *Peck*, 53d Rep. N. Y. State Bot.

Undoubtedly POISONOUS. *McIlvaine.*

Lepiota Morgani Pk. (See page 37.) The majority of mycopha- Lepiota. gists are immune to the poison of this species. Yet many cases of severe, but not fatal poisoning by it came within the writer's knowledge during the season of 1900-1901.

A valuable report is contained in a letter from George B. Clementson, attorney, Lancaster, Wis.:

" * * * Lepiota Morgani has grown in this locality this season in unusual abundance. While I was absent last week, my father picked a number, mistaking them for L. procera, and my mother, in preparing them for the table, ate a small piece of the cap of one—a piece, she assures me, no larger than a hickory nut. About two hours afterward and shortly after dinner (at which the mushrooms were not served, and

711

Lepiota. at which nothing indigestible was eaten) she experienced a peculiar numbness and nausea, with constriction of the throat. Vomiting set in within half an hour and was excessive, lasting several hours and giving no relief. She was very greatly weakened and thought herself dying, being so reduced at one time that she was unable to see. Purging set in not long after the vomiting. The constriction of the throat did not disappear until after the vomiting stopped.

"Whisky and nitroglycerine (by the stomach) were given to keep up the heart's action.

"It seems probable that the poison itself did not directly affect the heart, but that the alarming weakness was due to the vomiting and purging. That is my mother's own opinion. After being in bed for a day she was able to get around, but suffered considerable pain in the abdomen for forty-eight hours.

"I presume that owing to the fact that my mother is not very strong and has a weak stomach, she was more violently affected than many might be. But a poison that in any person can produce such symptoms, when taken in so small a quantity, ought to be labeled decidedly dangerous.

"There can be no question that the specimens were L. Morgani, as I examined some that were left of those picked, and also gathered others from the same patch where these were obtained.

"As everything relating to mushroom poisoning should be of interest to the mycologist and mycophagist, I take the liberty of reporting this case."

The Lepiota Morgani appears to be spreading. In 1901 I found large specimens of it outside a stable in Lebanon, Pa. Its appearance and luxuriance are so much in its favor, that the toadstool lover will be tempted to try it. Experiments in eating it should be conducted with the greatest caution.

Lepiota clypeolaria (Bull.) Fr. Shield Lepiota. (Plate II.) **Pileus** thin, soft, convex or subcampanulate, becoming nearly plane, obtuse or umbonate, squamose, whitish or yellowish, the center or umbo smooth, yellowish or brownish, the margin often appendiculate with fragments of the veil. **Flesh** white. **Lamellæ** thin, close, free, white. **Stem** slender, equal or slightly tapering upward, hollow, fragile, pallid, adorned with soft, loose, white or yellowish floccose scales or filaments. **Spores** Lepiota. oblong or subfusiform, 12–20μ long, 6–8μ broad.

The cap of the shield lepiota is at first somewhat ovate or bell-shaped, but with advancing age it becomes convex above or nearly flat. It is white or whitish, but spotted with numerous small scales of a yellowish or brownish-yellow color. These scales are the result of the breaking up of the thin cuticle that covers the very young plant, and they have the same color as it. A small space in the center is brown or yellowish-brown, or darker than the rest of the cap, because the cuticle covering it remains unbroken and retains its color. The center in some specimens is more prominent than in others, giving what is called an umbonate cap.

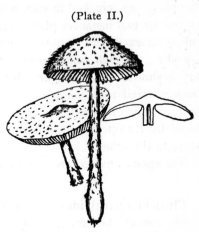

(Plate II.)

LEPIOTA CLYPEOLARIA PK.
About one-half nat. size. After Peck.

The margin of the cap is sometimes shaggy, specially in young plants, by the adhering fragments of the whitish veil.

The gills are thin, closely placed side by side and rounded at the end next the stem, but they are not attached to the stem. They are white. The stem is rather long and slender, fragile and adorned with loose, soft fibrils or flocculent, cottony tufts, which give it a somewhat shaggy appearance, but it becomes smoother as the plant grows older.

The cap is usually from 1–2.5 inches broad, and the stem from 1.5–3 inches long and 1.5–3 lines thick. The plants grow in woods, specially in hilly and mountainous regions, and are generally solitary or few in a place, but in favorable seasons they are of frequent occurrence and may be found from July to October. Though small and thin, the caps are well flavored and make a desirable dish. *L. metulaespora* B. and Br. scarcely differs from this species, except in the striate margin of its cap. *Peck*, 54th Rep. N. Y. State Bot.

Lepiota naucinoides Pk. (See page 45.) This valuable food species is spreading and rapidly increasing in many of the states. Prof.

Agaricaceæ

Charles H. Peck, in 54th Annual Report of the New York State Museum, says of it: "It has shown considerable variability in some of its characters. Usually its pileus is very white and smooth, clean and attractive, but specimens have been found this year having the pileus dingy or smoky brown, others have been seen in which the cuticle of the pileus was cracked in such a way as to form minute squamules, and in one or two instances plants were observed having the surface of the pileus adorned with minute granules, a character attributed to *L. naucina* Fr. In such cases the importance of recognizing the spore characters is shown. By disregarding this character our plant has sometimes been referred to *L. naucina* and sometimes to *Agaricus cretaceus* Fr., both of which it closely resembles, and with which it appears to be confused by European mycologists, some referring it to one species and some to the other."

The species named are equally excellent.

Clitocybe patuloides Pk. (Plate III.) **Pileus** fleshy, firm, rather thick, convex, becoming nearly plane or somewhat centrally depressed,

(Plate III.)

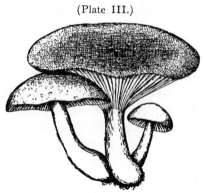

CLITOCYBE PATULOIDES PK.
About two-thirds nat. size. After Peck.

glabrous, even and white when young, with the margin incurved, becoming pale ochraceous with age and often squamose or rimosely areolate. **Flesh** white, taste mild, odor like that of mushrooms. **Lamellæ** thin, close, slightly or strongly decurrent, forked or anastomosing at the base, white. **Stem** usually short, equal or slightly tapering upward, solid, white. **Spores** broadly elliptic, 6–8μ long, 5μ broad.

Pileus 1–4 inches broad. **Stem** 1–3 inches long, 4–12 lines thick.

Gregarious or cespitose. Woods, especially of pine. When growing in tufts the stem is often eccentric and the pileus irregular. The base of the stem is often white tomentose. Its agreeable odor and mild taste led to a trial of its edible qualities, but it developed a bitter taste in cooking. *Peck*, 54th Rep. N. Y. State Bot.

Clitocybe Adirondackensis Pk. Adirondack Clitocybe. **Pileus** thin, convex or nearly plane and umbilicate, or centrally depressed and funnel-form, glabrous, moist, white or pale tan color. **Flesh** white. **Lamellæ** thin, narrow, close, very decurrent, white. **Stem** nearly equal, glabrous, stuffed or hollow, colored like the pileus. **Spores** subglobose or broadly elliptic, 4–5μ long, 3–4μ broad.

The Adirondack clitocybe is common in the northern forests of the state, but is not limited to them. Its cap is thin, and soon becomes nearly flat with a decurved margin and a central depression or umbilicus, or very concave by the elevation of the margin, and then it resembles a wineglass in shape. Its margin is sometimes wavy or irregular. In color it varies from white to a very pale red or tan color. White specimens sometimes have the center slightly darker than the rest.

The gills are very narrow, being scarcely broader than the thickness of the flesh of the cap. They are closely placed, white and decurrent.

The stem is nearly cylindric, smooth and stuffed or hollow. It is colored like the cap. Often there is a white tomentum or cottony substance at its base.

The cap varies in size and is 1–2 inches broad; the stem 1.5–3 inches long and 1–2 lines thick. It may be found from July to October. Its flavor is suggestive of that of the common mushroom. *Peck*, 54th Rep N. Y. State Bot.

Clitocybe maculosa Pk. Spotted Clitocybe. **Pileus** fleshy, convex, often centrally depressed, glabrous, centrally marked with numerous small round spots, yellowish-white, the young margin involute and minutely downy. **Flesh** white, taste mild. **Lamellæ** narrow, close, very decurrent, whitish or slightly yellowish, some of them forked. **Stem** equal or slightly tapering upward, glabrous or sparingly fibrillose, stuffed, sometimes becoming hollow, whitish. **Spores** subglobose or orbicular, 4–5μ broad.

The peculiar mark by which the spotted clitocybe may be distinguished consists in the small round definite spots in the central part of the cap. They have a slightly darker or watery or yellowish color and appear as if depressed below the rest of the surface. The cap is smooth and whitish or yellowish white and is generally depressed in the center and decurved on the margin. The margin is usually adorned with slight, short radiating ridges. The flesh is white and the taste mild.

Clitocybe. The gills are closely placed side by side, narrow and prolonged downward on the stem. They have nearly the same color as the cap. The stem is nearly cylindric, smooth or adorned with a few silky fibrils, whitish and spongy within or sometimes hollow when old.

The cap is from 1–3 inches broad; the stem 2–3 inches long and 2–4 lines thick. This mushroom grows among fallen leaves in woods. It appears in August and September. I have found it in the Adirondack forests only. Its range is probably northward, and its rarity detracts from its importance as an edible species. *Peck*, 54th Rep. N. Y. State Bot.

Hygrophorus. **Hygrophorus lauræ** Morg. **Pileus** fleshy, convex and umbonate, then expanded and depressed, more or less irregular, glutinous, white, clouded with a reddish or brownish tinge especially on the disk. **Stem** solid, more or less curved or crooked, tapering downward, yellowish-white; the apex scabrous with scaly points. **Lamellæ** unequally adnate-decurrent, distant, white. **Spores** pellucid, elliptic, apiculate, .0083x.0055μ.

Growing in rich soil among the leaves in hilly woods. Pileus 2–4 in. broad, stipe 2–4 in. long and ½ an inch thick. This is a much larger plant than *H. eburneus*, has a wash of red or brown upon the disk, and is covered with a thick gluten. It is more like *H. cossus*, but has no odor. Journal Cincinnati Soc. Nat. Hist. Vol. VI, 180, 1883. Edible. *Prof. C. H. Peck.*

Lactarius. **Lactarius subpurpureus** Pk. Purplish Lactarius. (Plate IV. See

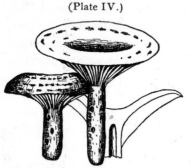

(Plate IV.)

page 172.) "When fresh, their taste is slightly acrid, but, when they are cooked, it is scarcely inferior to that of *L. deliciosus.*" *Peck*, 54th Rep. N. Y. State Bot.

I had not seen this species when the first edition of this work went to press, consequently could not report its edible qualities. The favorable testing by Professor Peck adds weight to the opinion I have frequently expressed, that acrid-

LACTARIUS SUBPURPUREUS PK.
About one-half nat. size. After Peck.

ity of species when raw is no evidence whatever that they are harmful. Acridity usually disappears in cooking.

Lactarius distans Pk. Distant-gilled Lactarius. **Pileus** firm, Lactarius.
broadly convex or nearly plane, umbilicate or slightly depressed in the
center, with a minute velvety pruinosity, yellowish tawny or brownish
orange. **Lamellæ** rather broad, distant, adnate or slightly decurrent,
white or creamy yellow, the interspaces venose, milk white, mild.
Stem short, equal or tapering downward, solid, pruinose, colored like
the pileus. **Spores** subglobose. 9–11μ broad.

The distant-gilled Lactarius is similar to the orange Lactarius in
color, but in other respects it is quite distinct. The short stem, widely
separated gills and pruinose surface of the cap are distinctive features.
The cap is broadly convex and often has a small central depression or
umbilicus. In some cases it becomes nearly plane or even slightly funnel-shape by the spreading or elevation of the margin. The surface,
specially in young and in well-developed specimens, has a soft pruinose
or almost velvety appearance to the naked eye, and when viewed
through a magnifying glass it is seen to be covered with minute persistent granules. The surface is sometimes wrinkled and frequently it
cracks in such a way as to form small angular or irregular areas.
The color is a peculiar one, varying somewhat in shade, but with tawny
hues prevailing. It has been described as yellowish tawny and brownish orange. The flesh is white or whitish and has a mild taste.

The gills are wide apart, somewhat arched in specimens having a convex cap and slightly decurrent in those with fully expanded or centrally depressed caps. Their color is white or creamy yellow and in
old and dried specimens they have a white pruinosity as if frosted by
the spores. The milk is white and mild.

The stem is short, rarely more than an inch long, and is cylindric or
tapering downward. It is solid and colored and clothed like the cap.

The cap is 1 to 4 in. broad; the stem is usually about 1 in. long, 4
to 8 lines thick. It is found in thin woods, bushy places and pastures
from July to September. It is similar to the orange Lactarius, *L. volemus*, in its edible qualities. *Peck*, 52d Rep. N. Y. State Bot.

Russula rugulosa n. sp. Rugulose Russula. **Pileus** rather thin, Russula.
fragile, convex, becoming nearly plane or centrally depressed, viscid
when moist, roughened or uneven with small tubercles and rugæ, even
on the margin when young, becoming tuberculate striate with age, the
viscid pellicle separable on the margin. **Flesh** white, reddish under the

Russula. cuticle, taste tardily acrid.   **Lamellæ** rather close, adnate or slightly rounded behind, white.   **Stem** nearly equal, spongy within, white. **Spores** white, rough, subglobose, 8–10μ broad, shining in transmitted light.

The rugulose russula is closely related to the emetic russula, but differs from it in the uneven or rugulose surface of the cap, in the tardily acrid taste and in its closer adnate gills.   Its cap is red, varying from pale-red to dark-red, viscid when moist, even on the margin when young, but somewhat tuberculate and striate when old.   Its surface is roughened by minute tubercles or pimples, which sometimes appear to run together and form short ridges.   These are sometimes absent from the center of the cap.   The viscid cuticle easily peels from the margin of the cap, but not from the center.   The flesh is white, except just under the cuticle, where it is reddish.   It is soft and fragile, and its taste is slowly and much less sharply acrid than in the emetic russula. Its gills are closely placed, attached to the stem and persistently white. The stem is brittle, soft and spongy within, smooth and white.   The cap is 2–4 inches broad, the stem 2–3 inches long, 4–8 lines thick.

It grows in woods among mosses and fallen leaves or on the bare ground, and appears in August and September.   It is an inhabitant of the Adirondack forests.   Its slightly acrid flavor is destroyed in cooking, and it affords a harmless, tender and agreeable food.   *Peck*, 54th Rep. N. Y. State Bot.

Russula abietina n. sp.   Fir Tree Russula.   **Pileus** thin, fragile, convex, becoming nearly plane or slightly depressed in the center, viscid when moist, the viscid pellicle separable, tuberculate striate on the margin.   **Flesh** white, taste mild.   **Lamellæ** subdistant, ventricose, narrowed toward the stem, rounded behind and nearly free, whitish, becoming pale yellow, the interspaces venose.   **Stem** equal or tapering toward the top, stuffed or hollow, white.   **Spores** bright yellowish ochraceous, subglobose, rough, 8–10μ broad.

The fir tree russula is closely related to the youthful russula, *R. puellaris* Fr., from which it is separated by the viscid cap, the gills rather widely separated from each other and nearly free, the stem never yellowish nor becoming yellow where wounded, and the spores having an ochraceous hue.   They are much brighter and more highly colored in the mass than the mature gills.   The cap varies much in color, but the center is generally darker than the rest. It may be dull purple or greenish purple with a brownish or blackish center, or sometimes with an olive green center, or it may be olive green or smoky green with a brownish center. Olive green and purplish hues of various shades are variously combined, but sometimes the margin is grayish and the center olive green. The flesh is white and its taste mild. The gills are white when young, or barely tinged with yellow, but they become pale yellow with age. They are neither crowded nor widely attached to the stem, and are connected with each other by cross veins, which can be seen at the bottom of the interspaces. The stems are rather slender, soft or spongy, sometimes becoming hollow and occasionally tapering upward. They are very constantly and persistently white. The **cap** is 1–2.5 inches broad, the **stem** 1–2.5 inches long, 3–5 lines thick. This russula grows under or near pine, spruce or balsam fir trees. It occurs from July to October. It is tender and palatable. The stems also are tender and may be cooked with the caps. *Peck*, 54th Rep. N. Y. State Bot.

Cantharellus cinnabarinus Schw. Cinnabar Chantarelle. (Plate V.)
Pileus firm, convex or slightly depressed in the center, often irregular with a wavy or lobed margin, glabrous, cinnabar red. **Flesh** white. **Lamellæ** narrow, distant, branched, decurrent, red. **Stem** equal or tapering downward, glabrous, solid or stuffed, red. **Spores** elliptic, 8–10µ long, 4–5µ broad.

(Plate V.)

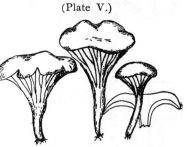

CANTHARELLUS CINNABARINUS Pk.
About one-half nat. size. After Peck.

The cinnabar Chantarelle is readily recognized by its color. It is externally red in all its parts, the interior only being white. It is a small species but often quite irregular in shape. Small specimens are more likely to be regular than large ones. Sometimes the cap is more fully developed on one side than on the other. This makes the stem eccentric or in some cases almost lateral. The color is quite constant, but in some instances it is paler and approaches a pinkish hue. It is apt to fade or even disappear in dried specimens. The gills are blunt on the edge as in other species of this genus. They are forked or branched, narrow and decurrent.

719

Cantharellus. The stem is small, smooth and usually rather short. It is generally solid, but in the original description it is characterized as stuffed. The cap is 8 to 18 lines broad; the stem 6 to 12 lines long and 1 to 3 broad. It grows gregariously in thin woods and open places and may be found from July to September. It sometimes occurs in great abundance, which adds to its importance as an edible species. The fresh plant has a tardily and slightly acrid flavor, but this disappears in cooking. In Epicrisis, Fries referred this species to the genus Hygrophorus, and in Sylloge also it is placed in that genus, but it is a true Cantharellus and belongs in the genus in which Schweinitz placed it. *Peck,* 52d Rep. N. Y. State Bot.

Cortinarius. **Cortinarius corrugatus** Pk. Corrugated Cortinarius. (Plate VI.) **Pileus** fleshy, broadly campanulate or very convex, viscid when moist,

(Plate VI.)

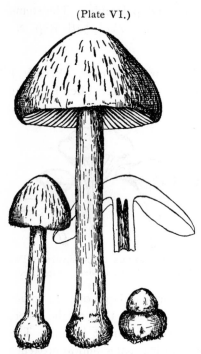

CORTINARIUS CORRUGATUS PK.
About two-thirds nat. size. After Peck.

coarsely corrugated, bright-yellow, reddish-yellow, tawny or ochraceous. **Flesh** white. **Lamellæ** close, pallid when young, becoming tawny with age. **Stem** rather long, equal, hollow, bulbous, pallid or yellowish, the bulb viscid and usually colored like the pileus. **Spores** broadly elliptical, rough, 11–16μ long, 8–10μ broad.

The corrugated Cortinarius is a well-marked and easily-recognized species, quite distinct from its allies. Although the color of the pileus is variable, its viscid, corrugated surface and the viscid bulb of the stem afford distinctive and easily-recognized characters. Sometimes the corrugations or wrinkles anastomose with each other in such a way as to give a reticulated appearance. The color varies from yellow to reddish-tawny or reddish-ochraceous. The margin in young plants is incurved.

There is a variety in which the cap is adorned with darker-colored spots or scales. This bears the name, variety *subsquamosus*. In all Cortinarius. other respects it is like the species.

The gills are closely placed side by side. They are at first of a pale hue, but assume a darker and more definite tawny color with age. They are usually minutely uneven or eroded on the edge and transversely striate on the sides. They are slightly narrowed toward the stem.

The stem is generally a little longer than the width of the cap. It is commonly smooth, but sometimes sprinkled near the top with minute yellowish particles and adorned below with a few fibrils. It is hollow and has a distinct viscid bulbous base, the viscidity of which is a peculiar feature. This bulb in the very young plant is even broader than the young cap, that at this stage of development appears to rest upon it. The color of the bulb is usually like that of the cap, but the stem is commonly paler than either.

The cap is 2 to 4 inches broad, the stem 3 to 5 inches long, 3 to 8 lines thick. The plants are gregarious in woods and bushy places, and may be found from June to September. It sometimes grows in considerable abundance, and as an edible species it is not to be despised. *Peck*, 52d Rep. N. Y. State Bot.

Agaricus hæmorrhoidarius Schulz. Bleeding Mushroom. The Agaricus. bleeding mushroom is easily recognized, when fresh, by the red color assumed by wounds of the flesh either of the cap or stem. This character is also found in the seashore mushroom, *A. maritimus*, a species that has a solid stem and has not yet been found growing far from the sea. The cap is generally some shade of brown, but sometimes when young it is white. It is adorned with darker fibrils or scales, though these sometimes become obscure or disappear with age. When young it is hemispheric or very convex, but it soon becomes broadly convex or nearly flat, with the center either slightly depressed or somewhat prominent. The flesh is generally whitish or grayish white when first exposed to the air. It assumes the red color rather slowly and after a time loses it again.

The gills are pink or rarely whitish when young, but become brown or blackish brown with age. The stem is long or short, cylindric or tapering upward, sometimes slightly thickened or bulbous at the base, sometimes not. It is hollow, but the cavity small, at first fibrillose and more or less adorned with floccose scales toward the base, but these

Agaricus. generally disappear with age, and the primary white color of the stem is apt to become darker with age. The collar is membranaceous and at first conceals the gills. It is persistent, silky and white or whitish, sometimes tinged with brown.

The cap is 2–4 inches broad; the stem 2–4 inches long, 3–5 lines thick. It grows in woods or bushy places and seems to prefer damp soil rich in vegetable mold. It may be found from August to October. It sometimes grows in clusters. It gives to milk in which it is stewed a brownish color. Its flavor is similar to that of the common mushroom. A variety in which the stem is commonly shorter and the pileus of a darker smoky brown color is sometimes abundant in low damp ground on Long Island. It may be called *variety fumosus*. *Peck*, 54th Rep. N. Y. State Bot.

Agaricus abruptus Pk. (A. silvicola Vitt., A. arvensis var. abruptus Pk.)

(Plate I, page 722.) Agaricus abruptus Pk. is described on page 343 as A. silvicola Vitt. It is very common in the woods of West Virginia, New Jersey and Pennsylvania. In the summer of 1901, I found it in Rockingham Co., N. C. The probabilities are that its spread is extensive.

Being the wood cousin of the field mushroom (A. campester) it deserves more than ordinary attention. It is found during months which do not favor the growth of the mushroom. It is equally good, though not so fleshy. It gives the true mushroom flavor to less flavored edible species when cooked with them.

When seen at a distance, growing in the woods, it has the appearance of an Amanita, but the color of the gills, which are never white after the cap opens and become as the spores ripen a blackish brown, distinguishes it at once. Neither has it a volva.

The excellent photograph of the species, taken by the late Dr. J. R. Weist, Richmond, Ind., presents a life-like picture of it.

Boletus. **Boletus granulatus albidipes** n. var. "Under pine trees. Westport. October. This variety differs from the typical form of the species in having the flesh of the pileus white, except next the tubes, where it is faintly yellowish, the stem white externally and internally, and in having a slight membranaceous veil which forms a very thin

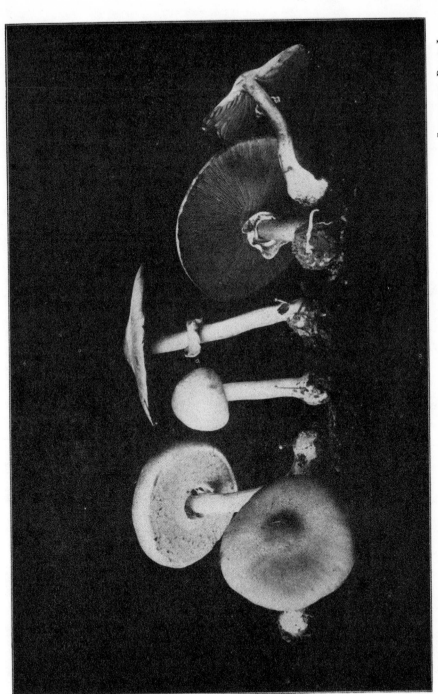

AGARICUS ABRUPTUS PK.

Photograph by Dr. J. R. Weist.

annulus on the stem of the young plant, or forms fragments which adhere to the margin of the pileus.'' *Peck*, 54th Rep. N. Y. State Bot.

For typical form B. granulatus, see page 416.

Boletus chrysenteron albocarneus n. var. White Flesh Boletus.

Pileus fleshy, convex above, dry, subglabrous, varying from brick red to bay red. **Flesh** white, sometimes tinged with red near the surface; tubes rather long, adnate or slightly depressed around the stem, greenish yellow, their mouths small, subrotund. **Stem** equal or nearly so, solid, subglabrous, colored like or a little paler than the pileus, white within.

The white flesh boletus is quite common in the Adirondack forests and quite constant in its characters. I have not seen it with yellow flesh, though in other respects it agrees very well with the description of *B. chrysenteron*. The cap is not often cracked, but, when it is, the cracks are sometimes red, sometimes yellowish, though the flesh is constantly white except just beneath the cuticle, where it is sometimes reddish. The tubes are long and greenish yellow. They are at first nearly plane in the mass, but with the expansion of the cap the mass often becomes ventricose. The mouths of the tubes are small and nearly round. Wounds or bruises of the mass become bluish or greenish blue. The stem is firm, solid and colored like the cap, though it is sometimes a little paler.

The **Cap** is 1–2.5 inches broad; the **Stem** 1–2 inches long, 2–4 lines thick. The trial specimens were fried in butter and found to be harmless, palatable and digestible. *Peck*, 54th Rep. N. Y. State Bot.

For typical species—*B. chrysenteron*—see page 431.

RAISING MUSHROOMS AT HOME

By the courtesy of the publishers of The Woman's Home Companion, the author is permitted to republish his article, " Raising Mushrooms at Home," which appeared in the October, 1901, number of that excellent monthly—encyclopedic in all home matters.

IN October is the time to prepare the manure and beds for house-raising of mushrooms. During the warm months they can not be cultivated without trial of one's temper and test of one's taste. Any one having control of a cellar can raise a fine crop of expectations, and may raise a crop of mushrooms by either accident or experience. They are at all times the most contrary of growths, and require the nicest management and much patience. The first thing to do is to select a well-ventilated spot away from direct drafts, where the temperature can be maintained at from fifty to sixty degrees and a moist atmosphere assured. Thoroughly cleanse the cellar and give it an entire covering of whitewash.

Decide upon the size of bed desired. In width the bed should not exceed reaching distance to its center when there is a pathway on each side of it, say six feet. The length of the bed should reach to its useful stopping-place. If the cellar has a portable heater in it, and is warm, the bed should be ten to twelve inches in depth; if the heater is walled in, or the cellar is cool, the bed should be fifteen inches deep.

Calculate how much fresh horse-manure, with the long straw only removed from it and that has not been rained upon, it will take to make a bed of desired dimensions solidly tramped. Get it, put it in a compact heap, and keep it covered from rain. It will heat rapidly and get smoking-hot, because a fermentation sets in which produces heat. If loam can be procured from a pasture or elsewhere it is well to add one-fifth (in bulk) of it to the manure, mixing it thoroughly. This addition retards the fermentation and absorbs the ammonia—a valuable fertilizer—which would otherwise be driven off by the heat. It also takes up any surplus of moisture.

After the compact pile has been thus prepared it should stand two or three days, then be well forked over and again piled. This forking should be repeated from four to six times, at intervals of from two to four days, depending upon the use or not of loam, which affects the rapidity of heating. If loam is used the forking should be at longer intervals unless the heat becomes excessive. The manure will probably then be in good order to go into beds. It is upon proper, careful preparation of this medium that successful mushroom-raising greatly depends. All work and hopes are thrown away if the greatest care is not exercised. Just as it is folly to buy poor seeds upon which to expend costly labor, so it is folly to make beds of poorly prepared manure.

The manure must neither contain too much nor too little water. By

BRICK CUT FOR PLANTING.

far the largest percentage of failures is due to too much. It rots the spawn vine (mycelium), and thus destroys the starting place of the fruit, or mushroom. The object in forking the manure so frequently is to sweeten it (as the operation is called) and to prevent overheating from fermentation. If it gets too hot it "burns"—gets too dry. Molding, too, is avoided. Moldy manure will not produce. If, in forking over the pile, dry places are found, they should be sprinkled with water; if, when the fermentation grows less active, the manure is too wet, spread it out to air and dry somewhat. It is in good condition and properly moist when tight squeezing will not press water from it. Far better that it should be too dry than too wet. The manure now ready should be moved to the cellar and made into beds while warm.

Good ventilation is a necessity. Two thermometers are needed— one to mark the temperature of the cellar, the other to place well and solidly down in the bed to record what it is doing in the heat way. It is probable that the mercury will rise slowly. It may go as high as one hundred and twenty-five or one hundred and thirty degrees. Do not disturb the bed, however high it goes. When it falls to between ninety and eighty degrees plant the spawn. If possible, keep the temperature up for several days. It should then fall slowly to sixty degrees, but go down no farther. Never plant on a rising temperature.

Mushroom-spawn comes in brick-shaped blocks. They can be purchased, of good quality, from any reliable seedsman. These blocks are made of a mixture of dungs, through which the mycelium, or vine, from which mushrooms grow, has been run. After this mixture is filled with the vine (badly named spawn) it is pressed into blocks and dried. It should be kept dry until used. Spawning a bed is nothing more than placing cuttings of this exceedingly fine vine under the influence of moisture and heat in a soil fitted for its growth (such as the bed should be), then inducing it to run and fruit. Spawn is originally made to grow by planting the seed of mushrooms in specially prepared dungs and germinating them. The mycelium, or vine, coming from this germination is called "virgin spawn," and is perpetuated in its growth by running (training) it through manures, pieces of which form the spawn of commerce.

With a sharp hatchet cut the bricks into twelve pieces of equal size; a fine, clean meat-saw may be used, as it reduces breakage. With the hand make holes in the bed ten inches apart each way. These holes must be so deep that when the lumps of spawn are thrust firmly down into them the top of the lumps will be not less than one inch or more than two inches below the surface of the bed. Cover the lumps firmly. Have the surface of the bed as even as possible. Without having to go very far into the cold region of mathematics, the number of bricks of spawn needed is easily figured. Ascertain the number of holes, ten inches apart, that can be made in the bed. Divide this number by twelve, and lo! you have it.

After the bed is spawned it is well to lay a double thickness of newspapers over it, putting a few plastering-laths or light sticks upon them to keep them in place. This is to keep the heat in the bed, as it is desirable that the temperature should not run down too rapidly. It should be two weeks falling to sixty degrees.

Ten days after spawning, if the heat of the bed has gone down to sixty-five or sixty degrees, cover the bed with two inches of loam and pat it solid with spade or board. The bed should not be covered with loam when the temperature is too high. Removing the papers will allow the heat to escape. At the time of covering with loam the spawn should have begun to spread. It will show plainly in the manure close to the lumps of spawn. Its odor is unmistakable, being musky, spicy, much like mushrooms, but stronger. Care should be taken not to disturb the new mycelium, as all breakage of the fine, web-like threads lessens its product. The mycelium should start and grow quickly up to the time of covering with loam. After that a slow increase is best. To effect this the surrounding outside temperature should be from fifty-seven to sixty-two degrees. Ventilation should be upward and good, but not directly upon the bed.

The mycelium will now run and completely fill the bed. Minute white nodules will appear upon the threads of it; these are the beginnings of the mushrooms to come. In from seven to eight weeks after spawning tiny button mushrooms should appear on top of the bed. If the cellar has been cool it may be a few days longer. Mr. Falconer says, "If the temperature of the bed falls below fifty-seven degrees, and the atmospheric temperature below forty-five degrees, the beds should be covered with matting or other material." Newspapers will do. Upon the appearance of the mushrooms is the time a moist atmosphere is needed. This is obtained by sprinkling the walks and cellar well with warm water. This moisture should be kept up all the while the crop is growing. Unless the fruiting beds show a marked dryness they should not be watered. If watering is required, do it very carefully with a fine rose or syringe. Have the pure water at ninety degrees, and do not more than moisten the loam covering. Never let the water settle in pools or wash the surface.

After the bed is in bearing the addition of strong, liquid manure plentifully applied between the bunches (never on them) will add to their weight and size. It should be done with a long spout without rose. A sprinkling of salt on bare places is beneficial.

GATHERING THE CROP.

The mushrooms will now show in various sizes, from pin-head to large, full-grown specimens, singly and in dense clusters. As fast as they reach the desired size twist them from their sockets. Do not cut or pull them. Keep the gills downward, to prevent dirt getting in them.

727

Take care not to disturb those left in the beds more than is necessary. It is unnecessary to add, cook them, but it is very necessary to tell how, because many excellent cooks commit the outrageous sacrilege of peeling mushrooms. A large amount of the flavor and deliciousness of a mushroom is in the skin—as it is in the apple. One might just as well peel a strawberry. First, always holding the plant gills downward and not over others, cut away the extreme base of the stem and brush off any adhering dirt. If the cap shows much scruff, rub it off with a piece of coarse flannel or cloth. Throw the mushrooms thus cleaned into cold

BERTH BEDS AGAINST CELLAR WALLS.

water; they will float. Run the fingers through them several times, then lift them to a fresh pan of water, wash them and place them, gills downward, on a cloth to drain, or put them in a colander. Then cook them to taste. Here, again, sacrilege is frequent. Many foods are simply mediums for added flavors. Not so the mushroom; it has a decided, exquisite flavor of its own. It should not be made in cooking to taste like something else. Put the mushrooms in a stew-pan with a little water; cover them, and stew slowly for twenty minutes, adding butter, salt and pepper to taste. Cream or milk may be added. Another very good way is to butter well a medium-hot pan; cut the mushrooms into equal-sized pieces, put them in it, cover, and fry. Stir them from time to time, and when quite done season with salt and pepper. A good gravy is made for them by using water, milk or cream. Now if you must have a meat of some sort, put the meat on one dish and the mushrooms on another. By doing this you spoil the taste of neither.

Beds will continue to produce for several weeks if properly cared for. As soon as they cease bearing remove them, clean up, white-wash, coal-oil every inch of wood, salt the floor, and be ready to try again. After the amateur has his or her hand in, the bed area can be largely increased by building rough berths, one above the other, in which beds can be made. An important bit of advice is: Start in a small way. Do not expend any more money than you can afford to lose.